Studies in Computational Intelligence

Volume 576

Series editor

Janusz Kacprzyk, Polish Academy of Sciences, Warsaw, Poland
e-mail: kacprzyk@ibspan.waw.pl

About this Series

The series "Studies in Computational Intelligence" (SCI) publishes new developments and advances in the various areas of computational intelligence—quickly and with a high quality. The intent is to cover the theory, applications, and design methods of computational intelligence, as embedded in the fields of engineering, computer science, physics and life sciences, as well as the methodologies behind them. The series contains monographs, lecture notes and edited volumes in computational intelligence spanning the areas of neural networks, connectionist systems, genetic algorithms, evolutionary computation, artificial intelligence, cellular automata, self-organizing systems, soft computing, fuzzy systems, and hybrid intelligent systems. Of particular value to both the contributors and the readership are the short publication timeframe and the world-wide distribution, which enable both wide and rapid dissemination of research output.

More information about this series at http://www.springer.com/series/7092

Ahmad Taher Azar · Quanmin Zhu
Editors

Advances and Applications in Sliding Mode Control systems

Editors
Ahmad Taher Azar
Faculty of Computers and Information
Benha University
Benha
Egypt

Quanmin Zhu
Department of Engineering Design and Mathematics
University of the West of England
Bristol
UK

ISSN 1860-949X ISSN 1860-9503 (electronic)
ISBN 978-3-319-35469-9 ISBN 978-3-319-11173-5 (eBook)
DOI 10.1007/978-3-319-11173-5

Springer Cham Heidelberg New York Dordrecht London

Softcover reprint of the hardcover 1st edition 2015

Printed on acid-free paper

Springer is part of Springer Science+Business Media (www.springer.com)

Foreword

In control theory of linear and nonlinear dynamical systems, sliding mode control (SMC) is a nonlinear control method. The sliding mode control method alters the dynamics of a given dynamical system (linear or nonlinear) by applying a discontinuous control signal that forces the system to "slide" along a cross-section (manifold) of the system's normal behaviour.

Sliding mode control (SMC) is a special class of variable-structure systems (VSS). In sliding mode control method, the state feedback control law is not a continuous function of time. Instead, the state feedback control law can switch from one continuous structure to another based on the current position in the state space.

Variable-structure systems (VSS) and the associated sliding mode behavior was first investigated in the early 1950s in the USSR and seminal papers on SMC were first published by Profs. S.V. Emelyanov (1967) and V.I. Utkin (1968). The early research on VSS dealt with single-input and single-output (SISO) systems. In recent years, the majority of research in SMC deals with multi-input and multi-output (MIMO) systems.

For over 50 years, the sliding mode control (SMC) has been extensively studied and widely used in many scientific and industrial applications due to its simplicity and robustness against parameter variations and disturbances.

The design procedure of the sliding mode control (SMC) consists of two major steps, (A) Reaching phase and (B) Sliding-mode phase. In the reaching phase, the control system state is driven from any initial state to reach the sliding manifold in finite time. In the sliding-mode phase, the system is confined into the sliding motion on the sliding manifold. The stability results associated with the sliding mode control are established using the direct method of the Lyapunov stability theory.

Hence, the sliding mode control scheme involves (1) the selection of a hypersurface or a manifold (i.e. the sliding manifold) such that the system trajectory exhibits desirable behavior when confined to this sliding manifold and (2) finding feedback gains so that the system trajectory intersects and stays on the sliding manifold.

The merits of sliding mode control (SMC) are robustness against disturbances and parameter variations, reduced-order system design, and simple control structure. Some of the key technical problems associated with sliding mode control (SMC) are chattering, matched and unmatched uncertainties, unmodeled dynamics, etc. Many new approaches have been developed in the last decade to address these problems.

Important types of sliding mode control (SMC) are classical sliding mode control, integral sliding mode control, second-order sliding mode control, and higher order sliding mode control. The new SMC approaches show promising dynamical properties such as finite time convergence and chattering alleviation.

Sliding mode control has applications in several branches of Engineering like Mechanical Engineering, Robotics, Electrical Engineering, Control Systems, Chaos Theory, Network Engineering, etc.

One of the key objectives in the recent research on sliding mode control (SMC) is to make it more intelligent. Soft computing (SC) techniques include neural networks (NN), fuzzy logic (FL), and evolutionary algorithms like genetic algorithms (GA), etc. The integration of sliding mode control and soft computing alleviates the shortcomings associated with the classical SMC techniques.

It has been a long road for the sliding mode control (SMC) from early VSS investigations in the 1950s to the present-day investigations and applications. In this book, Dr. Ahmad Taher Azar and Dr. Quanmin Zhu have collected and edited contributions of well-known researchers and experts in the field of sliding mode control theory in order to provide a comprehensive view of the recent research trends in sliding mode control theory. Their efforts have been very successful. Therefore, it has been a great pleasure for me to write the Foreword for this book.

Sundarapandian Vaidyanathan
Professor and Dean, R & D Centre
Vel Tech University
Chennai
Tamil Nadu, India

Preface

Sliding mode control, also known as variable structure control, is an important robust control approach and has attractive features to keep systems insensitive to uncertainties on the sliding surface. For the class of systems to which it applies, sliding mode controller design provides a systematic approach to the problem of maintaining stability and consistent performance in the face of modeling imprecision. On the other hand, by allowing tradeoffs between modeling and performance to be quantified in a simple fashion, it can illuminate the whole design process. Sliding mode schemes have become one of the most exciting research topics in several fields such as electric drives and actuators, power systems, aerospace vehicles, robotic manipulators, biomedical systems, etc. In its earlier approach, an infinite frequency control switching was required to maintain the trajectories on a prescribed sliding surface and then eventually to enforce the orbit tending to the equilibrium point along the sliding surface. However, in practice the system states do not really locate on the designed sliding surface after reaching it due to numerically discretizing errors, signal noise, as well as structural uncertainties in the dynamical equations. Since the controller was fast switched during operation, the system underwent oscillation crossing the sliding plane. Around the sliding surface is often irritated by high frequency and small amplitude oscillations known as chattering. The phenomenon of chattering is a major drawback of SMC, which makes the control power unnecessarily large. To eliminate chattering, some methods are being developed.

This book consists of 21 contributed chapters by subject experts specialized in the various topics addressed in this book. The special chapters have been brought out in this book after a rigorous review process. Special importance was given to chapters offering practical solutions and novel methods for recent research problems in the main areas of this book. The objective of this book is to present recent theoretical developments in sliding mode control and estimation techniques as well as practical solutions to real-world control engineering problems using sliding mode methods. The contributed chapters provide new ideas and approaches, clearly indicating the advances made in problem statements, methodologies, or applications with respect to the existing results. The book is not only a valuable title on the publishing market, but is also a successful synthesis of sliding mode control in the world literature.

As the editors, we hope that the chapters in this book will stimulate further research in sliding mode control methods for use in real-world applications. We hope that this book, covering so many different aspects, will be of value to all readers.

We would like to thank also the reviewers for their diligence in reviewing the chapters.

Special thanks go to Springer publisher, especially for the tireless work of the series editor "Studies in Computational Intelligence," Dr. Thomas Ditzinger.

Benha, Egypt — Ahmad Taher Azar
Bristol, UK — Quanmin Zhu

Contents

Adaptive Sliding Mode Control of the Furuta Pendulum

Ahmad Taher Azar and Fernando E. Serrano

Abstract In this chapter an adaptive sliding mode controller for the Furuta pendulum is proposed. The Furuta pendulum is a class of underactuated mechanical systems commonly used in many control systems laboratories due to its complex stabilization which allows the analysis and design of different nonlinear and multivariable controllers that are useful in some fields such as aerospace and robotics. Sliding mode control has been extensively used in the control of mechanical systems as an alternative to other nonlinear control strategies such as backstepping, passivity based control etc. The design and implementation of an adaptive sliding mode controller for this kind of system is explained in this chapter, along with other sliding mode controller variations such as second order sliding mode (SOSMC) and PD plus sliding mode control (PD + SMC) in order to compare their performance under different system conditions. These control techniques are developed using the Lyapunov stability theorem and the variable structure design procedure to obtain asymptotically stable system trajectories. In this chapter the adaptive sliding mode consist of a sliding mode control law with an adaptive gain that makes the controller more flexible and reliable than other sliding mode control (SMC) algorithms and nonlinear control strategies. The adaptive sliding mode control (ASMC) of the Furuta pendulum, and the other SMC strategies shown in this chapter, are derived according to the Furuta's pendulum dynamic equations making the sliding variables, position errors and velocity errors reach the zero value in a specified reaching time. The main reason of deriving two well known sliding mode control strategy apart from the proposed control strategy of this chapter (adaptive sliding mode control) is for comparison purposes and to evince the advantages and disadvantages of adaptive sliding mode control over other sliding mode control strategies for the stabilization of the Furuta

A.T. Azar (✉)
Faculty of Computers and Information, Benha University, Benha, Egypt
e-mail: ahmad_t_azar@ieee.org

F.E. Serrano
Department of Electrical Engineering, Florida International University,
10555 West Flagler St, Miami, FL 33174, USA
e-mail: fserr002@fiu.edu

A.T. Azar and Q. Zhu (eds.), *Advances and Applications in Sliding Mode Control systems*,
Studies in Computational Intelligence 576, DOI 10.1007/978-3-319-11173-5_1

pendulum. A chattering analysis of the three SMC variations is done, to examine the response of the system, and to test the performance of the ASMC in comparison with the other control strategies explained in this chapter.

1 Introduction

In this chapter an adaptive sliding mode control of the Furuta pendulum is proposed. The Furuta pendulum is a class of underactuated mechanical system used in laboratories to test different kinds of control strategies that are implemented in aerospace, mechanical and robotics applications. A mechanical system is underactuated when the number of actuators is less than the degrees of freedom of the system, for this reason, the research on the control of this kind of systems is extensively studied.

There are different kinds of control strategies for the Furuta pendulum found in literature, these approaches take in count the complexity of the dynamic model considering that is coupled and nonlinear. In Ramirez-Neria et al. 2013 an active disturbance rejection control (ADRC) is proposed for the tracking of a Furuta pendulum, specially, when there are disturbances on the system; the ADRC cancels the effects of the disturbance on the system by an on line estimation of the controller parameters. In Hera et al. (2009), the stabilization of a Furuta pendulum applying an efficient control law to obtain the desired trajectory tracking is corroborated by the respective phase portraits. Some authors propose the parameter identification of the model (Garcia-Alarcon et al. 2012) implementing a least square algorithm; becoming an important technique that can be used in adaptive control strategies. Another significant control approach is implemented by Fu and Lin (2005) where a backstepping controller is applied for the stabilization of the Furuta pendulum where a linearized model of the pendulum is used to stabilizes this mechanical system around the equilibrium point.

Sliding mode control SMC has been extensively implemented in different kinds of systems, including mechanical, power systems, etc. this is a kind of variable structure controllers VSC that is becoming very popular in the control systems community due to its disturbance rejection properties yielded by external disturbance or unmodeled dynamics (Shtessel et al. 2014). It consist on stabilizing the system by selecting an appropriate sliding manifold until these variables reach the origin in a determined convergence time; during the last decades the SMC control strategy has evolved, from first order SMC to higher order sliding mode control HOSMC (Kunusch et al. 2012), which has been implemented in recent years due to its chattering avoidance properties (Bartolini and Ferrara 1996). Due to the discontinous control action of the SMC, sometimes the chattering effect is found in the system producing unwanted system responses. Chattering basically is a high frequency oscillations in the control input that can yield instability and unwanted system response, due to the chattering avoidance properties of the HOSMC and their disturbance rejection (Utkin 2008), this control technique has replaced the classical sliding mode approach.

In order to solve the chattering problem, some SMC techniques have been proposed to deal with this effect, like the twisting and super twisting algorithms (Fridman 2012), even when they are first generation algorithms, they have some advantages and disadvantages when they are implemented in the control of underactuated mechanical systems. Even when the previous algorithms are commonly implemented, a second order sliding mode control SOSMC for the stabilization of the Furuta pendulum is proposed in Sect. 2 where an specially design control algorithm is implemented in the control and tracking of this mechanism (Moreno 2012). This approach is developed in this section to provide a different point of view on how to deal with this kind of problem and because this is the theoretical background for the development of other control algorithms, including adaptive sliding mode control (Ferrara and Capisani 2012). In Sect. 3 the derivation and application of a proportional derivative plus sliding mode control (PD + SMC) for the stabilization of a Furuta pendulum is explained to show the advantages and disadvantages of this hybrid control strategy over SOSMC and compare it with the adaptive sliding mode control strategy proposed in this article. The reason because these two sliding mode control approaches are explained in this chapter, is because it is necessary to compare these two sliding mode control strategy with the main contribution of this chapter, in which the stabilization of the Furuta pendulum by an adaptive sliding mode control strategy for the Furuta pendulum is proposed to be compared and analyzed with other sliding mode control approaches and to understand the theoretical background of adaptive sliding mode controllers for mechanical underactuated systems. The derivation of the adaptive sliding mode control ASMC strategy for the Furuta pendulum is shown in Sect. 4, where an adaptive gain control strategy is obtained (Fei and Wu 2013; Liu et al. 2013; Chen et al. 2014) exploiting the advantage of a classical sliding mode controller with the on line tuning of a variable parameter controller. ASMC has been demonstrated to be an effective control strategy for similar mechanical systems (Yao and Tomizuka 1994) and other mechanical devices (Jing 2009; Li et al. 2011) where the improved parameter adjustment make this strategy ideal for the control and stabilization of this underactuated mechanical systems. In Sect. 5 a chattering analysis of the three control approaches shown in this chapter is done to find the oscillation period yielded by the discontinuous control law, then some conclusions are obtained from this controller's comparison. In Sect. 6 a discussion about the performance of the three approaches explained in this chapter are analyzed to explain the advantages, disadvantages and characteristic of the proposed control technique; finally, in Sect. 7 the conclusions of this chapter are shown to summarize the results obtained in this chapter.

2 Second Order Sliding Mode Control of the Furuta Pendulum

In this section the derivations of a second order sliding mode control (SOSMC) is shown to stabilizes the Furuta pendulum. The main idea of this control approach is to find a suitable control law which stabilizes the system reducing the chattering effects and making the sliding manifold to reach the origin in finite time (Bartolini et al. 1998). SOSMC has been extensively implemented in the control of different kind of mechanisms, where the dynamic model of the system is considered to develop an appropriate switching control law (Su and Leung 1992; Zhihong et al. 1994; Gracia et al. 2014) therefore it has became in an attractive strategy for the control of the Furuta pendulum.

Apart from the chattering avoidance nature of the SOSMC, another advantage of the SOSMC is the disturbance rejection properties of this approach, making it a suitable choice for the control of mechanical systems, (Punta 2006; Chang 2013; Estrada and Plestan 2013), considering that the Furuta pendulum is an underactuated system (Nersesov et al. 2010), generating exponentially stable sliding manifolds to reach the origin in a prescribed time. The SOSMC strategy allows the design of appropriate sliding manifolds which converge to zero in a defined time, for MIMO and coupled dynamic systems (Bartolini and Ferrara 1996) making this approach ideal for the control of the Furuta pendulum.

Higher order sliding mode control HOSMC (Levant 2005) has demonstrated its effectiveness in the control of different kinds of systems (Rundell et al. 1996; Shkolnikov et al. 2001; Fossas and Ras 2002), for this reason a SOSMC is designed to stabilizes the Furuta pendulum with specific initial conditions implementing a Lyapunov approach to obtain a suitable control law. This control strategy is designed considering the dynamics of the Furuta pendulum (Fridman 2012) instead of implementing well known SOSMC control algorithms such as the twisting and the super twisting algorithms (Moreno 2012). The SOSMC strategy is done by designing a control algorithm for arbitrary order SMC (Levant 2005; Fridman 2012) and test the stability of the SOSMC by the Lyapunov theorem. The stabilization of the Furuta pendulum is not a trivial task, even when different nonlinear control techniques are proposed by some authors (Fu and Lin 2005; Ramirez-Neria et al. 2013) an ideal control law that improves the system performance and reduces the tracking error with smaller oscillations in the system that can be harmful for the mechanical system. It is important to avoid these unwanted effects on the system considering an appropriate second order sliding mode control law which decreases the deterioration of the system performance, then a stabilizing control law that makes the sliding variables and its first derivative to reach the origin in finite time is chosen for the stabilization of this underactuated mechanical system.

The first subsection of this chapter is intended to explain the dynamic equations of the Furuta pendulum that are determined by the respective kinematics equations and the Euler–Lagrange formulation of this mechanical system. This model is implemented in the rest of this chapter to derive the sliding mode controllers explained in the following sections. In Sect. 2.2 the design of a second order sliding mode controller for the Furuta pendulum is explained where this control strategy is designed according to the system dynamics of the model while keeping the tracking error as small as possible and driving the sliding variable to zero in finite time. Finally in Sect. 2.3 an illustrative example of this control approach is done visualizing the system performance and analyzing the controlled variables behavior; such as the angular position, velocity and tracking error of this mechanical system.

The intention in this section is to compare the SOSMC algorithm with the proposed strategy of this chapter, then the discussion and analysis of this control approach are explained in Sects. 5 and 6.

2.1 Dynamic Model of the Furuta Pendulum

The Furuta pendulum is an underactuated mechanism which consists of a rotary base with a pendulum connected to a arm. The angle of the rotary base is denoted as ϕ and the angle of the pendulum is denoted as θ. This mechanism is a perfect example of underactuated nonlinear mechanical system that is implemented in the development and design of different kind of nonlinear architectures for several kinds of applications such as aeronautics, aerospace, robotics and other areas in the control systems field. This mechanism works by rotating the base of the pendulum and then the arm rotates according to the interaction of the pendulum and base of the arm. As it is explained in the introduction of this section the stabilization of this mechanical system is a difficult task, so in this section it is proved that a suitable second order sliding mode control for the stabilization of this system is possible, while considering the system dynamics of the model. The dynamical model shown in this section has two angles, that must be controlled efficiently in order to keep the base and pendulum positions in the desired values. This mathematical model is necessary for the design of efficient sliding mode techniques where in order to design the proposed control strategy the linearization of the model is essential to develop the adaptive gain SMC technique. The design of a second order sliding mode controller for the Furuta pendulum leads the path to the development of other sliding mode controller variations, so an efficient control system design is important in this section to improve the performance of the controlled system, therefore well defined dynamical systems equations lead to an efficient design of the sliding mode control strategies that are developed in this and the following sections.

In Fig. 1 the Furuta pendulum configuration is depicted showing the respective rotational angles; meanwhile, in Fig. 2 a CAD model of the Furuta pendulum is depicted for a clear understanding of the model.

The dynamic equations of the Furuta pendulum are given by (Fu and Lin 2005; Hera et al. 2009; Garcia-Alarcon et al. 2012; Ramirez-Neria et al. 2013):

$$(p_1 + p_2\sin^2(\theta))\ddot{\phi} + p_3\cos(\theta)\ddot{\theta} + 2p_2\sin(\theta)\cos(\theta)\dot{\theta}\dot{\phi} - p_3\sin(\theta)\dot{\theta}^2 = \tau_\phi \quad (1)$$

$$p_3\cos(\theta)\ddot{\phi} + (p_2 + p_5)\ddot{\theta} - p_2\sin(\theta)\cos(\theta)\dot{\phi}^2 - p_4\sin(\theta) = 0 \quad (2)$$

where:

$$p_1 = (M + m_p)\ell_a^2 \quad (3)$$

$$p_2 = (M + (1/4)m_p)\ell_p^2 \quad (4)$$

$$p_3 = (M + (1/2)m_p)\ell_p\ell_a \quad (5)$$

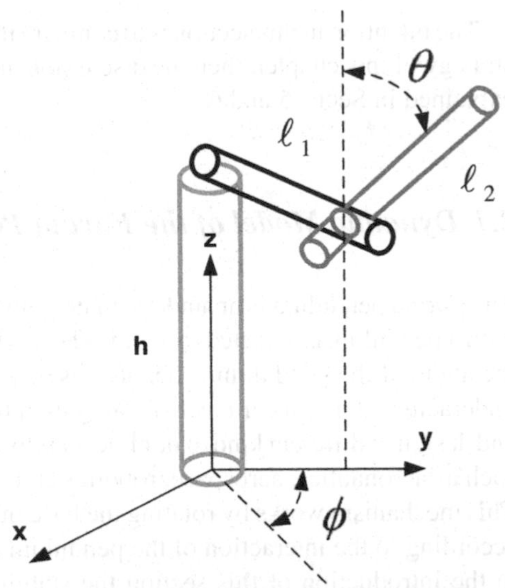

Fig. 1 Furuta pendulum system

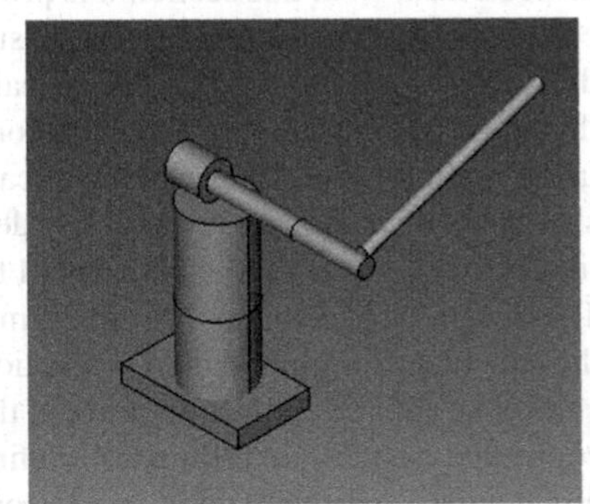

Fig. 2 CAD drawing of the furuta pendulum

$$p_4 = (M + (1/2)m_p)\ell_p g \tag{6}$$
$$p_5 = (1/12)m_p\ell_p^2 \tag{7}$$

where ℓ_a is the length of the arm, ℓ_p is the length of the pendulum, m_p is the pendulum mass, M is the mass of the bob at the end of the pendulum and g is the gravity constant.

Now, to establish the dynamic equations in the standard form it is necessary to define the next vector $q = [\phi, \theta]^T = [q_1, q_2]^T$, then the dynamic equations are represented by:

$$D(q)\ddot{q} + C(q, \dot{q})\dot{q} + g(q) = \begin{bmatrix} \tau_\phi \\ 0 \end{bmatrix} \tag{8}$$

defining the following state variables:

$$x_1 = q$$
$$x_2 = \dot{q}$$

The following state space representation is obtained:

$$\dot{x}_1 = x_2$$
$$\dot{x}_2 = -D^{-1}(x_1)C(x_1, x_2)x_2 - D^{-1}(x_1)g(x_1) + D^{-1}(x_1)\begin{bmatrix} 1\,0 \\ 0\,0 \end{bmatrix}\tau$$

where τ, the inertia matrix, coriolis matrix and gravity vector are defined as:

$$D(q) = \begin{bmatrix} (p_1 + p_2\sin^2(q_2)) & p_3\cos(q_2) \\ p_3\cos(q_2) & p_2 + p_5 \end{bmatrix} \tag{9}$$

$$C(q, \dot{q}) = \begin{bmatrix} 2p_2\sin(q_2)\cos(q_2)\dot{q}_2 & -p_3\sin(q_2)\dot{q}_2 \\ -p_2\sin(q_2)\cos(q_2)\dot{q}_1 & 0 \end{bmatrix} \tag{10}$$

$$g(q) = \begin{bmatrix} 0 \\ -p_4\sin(q_2) \end{bmatrix} \tag{11}$$

$$\tau = \begin{bmatrix} \tau_\phi \\ \tau_\theta \end{bmatrix} \tag{12}$$

where $D(q)$ is the inertia matrix, $C(q, \dot{q})$ is the coriolis matrix and $g(q)$ is the gravity vector. With these equations, the SMC can be derived in this and the following sections, stablishing a theoretical background for the development of the sliding mode controllers because they are settled on the dynamic equations of the Furuta pendulum. In the next subsection a SOSMC is derived for the stabilization of the Furuta pendulum, where its performance is analyzed and compared in the following sections.

2.2 Second Order Sliding Mode Control of the Furuta Pendulum

In this section a SOSMC is designed for the stabilization of the Furuta pendulum. Second order sliding mode control has been proved to be an effective control strategy for different kind of mechanical systems (Punta 2006), therefore an appropriate control algorithm is developed considering the dynamics of the model (Fridman 2012).

The second order sliding mode controller for this mechanism is designed to ensure that the sliding variables and their derivatives reach the origin in finite time $\sigma = \dot{\sigma} = 0$, in order to calculate this convergence time the reader should check Sect. 6. The convergence of the sliding variables of the system ensures that the controlled variables of the model, angular positions and velocities, reach and keep the desired values in steady state. Second order sliding mode control (SOSMC) and higher order sliding mode control (HOSMC) are appropriate control strategies for this kind of mechanical systems, due to the chattering avoidance properties, disturbance rejection and robustness to unmodelled dynamics, therefore an appropriate SOSMC strategy is implemented in the stabilization of this underactuated mechanism to keep the controlled variables in the desired values by moving the joint positions from their initial conditions to the final position of the pendulum and base. Despite of the control of the Furuta pendulum with other nonlinear control techniques such as backstepping or robust control, second order sliding mode control remains acceptable for the control and stabilization of different kind of mechanism due to the performance enhancement properties such as robustness and disturbance rejection properties, for these reasons, a higher order sliding mode controller is proposed in this section instead of well known sliding mode control algorithms such as the twisting and super twisting. The implementation of well defined dynamical equations of the Furuta pendulum by the Euler Lagrange formulation is an important fact that must be considered in the design of an efficient second order sliding mode controller that yields an efficient trajectory tracking by minimizing the system errors. Another important fact shown in this subsection is the design of an appropriate sliding mode control strategy that reduces chattering and avoids the saturation of the system actuator, so this SOSMC strategy suppress these effects on the system.

In order to design the desired SOSMC, the first step is the design of the sliding manifold that in this case is given by:

$$\sigma = \dot{e} + \Phi e \tag{13}$$

where σ is the sliding manifold, q is the position vector, q_d is the desired position vector, Φ is a $\Re^{2\times 2}$ positive definite matrix and:

$$e = q_d(t) - q(t) \tag{14}$$

Then in order to design the required controller the variable ϕ must be defined before deriving the control law (Bartolini et al. 1998; Levant 2005; Fridman 2012):

$$\phi = \dot{\sigma} + \beta_i \left|\sigma\right|^{\frac{1}{2}} \operatorname{sign}(\sigma) \tag{15}$$

Then the established control law is given by (Fridman 2012):

$$u = -\sigma + \alpha \operatorname{sign}(\phi) = \tau \tag{16}$$

where $\alpha > 0$ is a positive constant. Before proving the stability of the system the following property must be explained:

Definition 1 An n-degrees of freedom mechanical system has the following property:

$$\sigma^T \left(\frac{1}{2}\dot{D}(q) - C(q,\dot{q})\right)\sigma = 0 \tag{17}$$

where $\sigma \in \Re^n$.

In order to test the stability of the system with the proposed control law, the following theorem is necessary to assure the convergence of the states and sliding manifold of the system

Theorem 1 *The second order sliding mode controller assures the stability of the system if the defined Lyapunov function indicates that the system is asymptotically stable.*

Proof Define the following Lyapunov function with the established sliding manifold:

$$V(\sigma) = \frac{1}{2}\sigma^T D(q)\sigma \tag{18}$$

where $D(q)$ is the inertia matrix of the system. Then the derivative of the Lyapunov function yields:

$$\dot{V}(\sigma) = \sigma^T D(q)\dot{\sigma} + \frac{1}{2}\sigma^T \dot{D}(q)\sigma \tag{19}$$

The term $D(q)\dot{\sigma}$ can be described as:

$$D(q)\dot{\sigma} = -\tau + \xi - C(q, \dot{q})\sigma \tag{20}$$

where

$$\xi = D(q)(\ddot{q}_d + \Phi\dot{e}) + C(q, \dot{q})(\dot{q}_d + \Phi e) + g(q) \tag{21}$$

Definition 2 The term ξ has the following property (Liu 1999; Xiang and Siow 2004)

$$\xi \leq \alpha_1 + \alpha_2 \|e\| + \alpha_3 \|\dot{e}\| + \alpha_4 \|e\| \|\dot{e}\| \tag{22}$$

where α_1,α_2,α_3, and α_4 are positive constants. Then applying (20) and Definition 1 the Lyapunov function derivative becomes in:

$$\dot{V}(\sigma) = -\sigma^T\tau + \sigma^T\xi \tag{23}$$

Then applying the norm and Definition 2, the Lyapunov function derivative becomes:

$$\dot{V}(\sigma) \leq -\left\|\sigma^T\right\| \|\tau\| + \left\|\sigma^T\right\| \|\xi\| \tag{24}$$

Therefore asymptotically stability is assured due to the upper bound of $\|\xi\|$, explained in Definition 2, implementing the SOSMC.

With these conditions the stability of the SOSMC is assured in the stabilization of the Furuta pendulum. In the next subsection an illustrative example of the stabilization of the Furuta pendulum by a SOSMC is shown, the system is tested under certain initial conditions to analyze the performance of the measured variables and the sliding manifold.

2.3 Example 1

In this section an example of the stabilization of a Furuta pendulum by a SOSMC is shown to test the performance of the system under specified initial conditions $(\pi/2, 0)$. The purposes of the second order sliding mode control is to make the sliding variables and their derivatives to reach the origin in a finite time $\sigma = \dot{\sigma} = 0$ in order to make the controlled variables such as angular position and velocity reach the desired final value in steady state when a disturbance is applied on the model. In this example the idea is to illustrate the theoretical background of the second order sliding mode controller when it is implemented in the control and stabilization of the Furuta pendulum by a mathematical model of the system. In this example the simulation results of the Furuta pendulum controlled by a second order sliding mode controller is shown, depicting the angular position trajectories, the angular velocities, the phase portraits, the tracking errors and the control input. With these simulation results the performance of the Furuta pendulum, represented by a mathematical model, show the system variables performance and evinces important conclusions on the stabilization of this underactuated system with specified initial conditions.

The Furuta pendulum parameters are given in Table 1 The gains of the SOSMC are given by:

$$\alpha = 0.01 \tag{25}$$

$$\Phi = \begin{bmatrix} 1000 & 0 \\ 0 & 1000 \end{bmatrix} \tag{26}$$

$$\beta = \begin{bmatrix} 0.7 & 0 \\ 0 & 0.7 \end{bmatrix} \tag{27}$$

The simulations where done in $MATLAB^{®}$ and $SimMechanics^{®}$ and the results are depicted in Figs. 3 and 4

In Figs. 3 and 4 the angle position for ϕ and its angular velocity respectively, show how these variables reach the final positions in a considerable time. As it is noticed, even when there are some oscillations, these variables reaches the zero position. These results confirm that is possible to stabilize an underactuated mechanical system, in this case the Furuta pendulum, by a second order sliding mode controller. As it is

Table 1 Furuta pendulum parameters

Parameter	Values
ℓ_a	0.15 m
m_a	0.298 Kg
ℓ_p	0.26 m
m_p	0.032 m
J	0.0007688 Kg.m^2
g	9.81 m/s^2

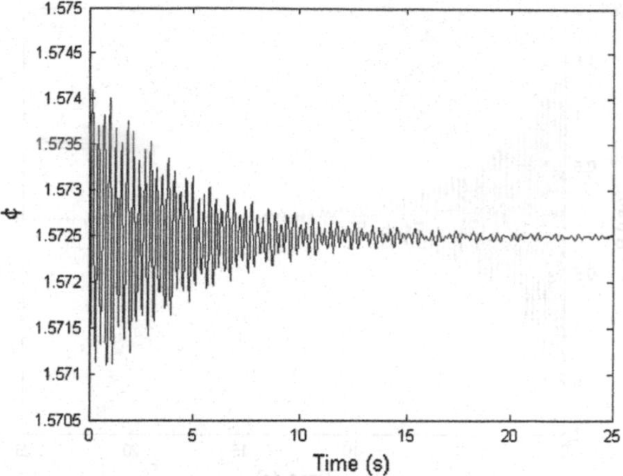

Fig. 3 Angular position for ϕ

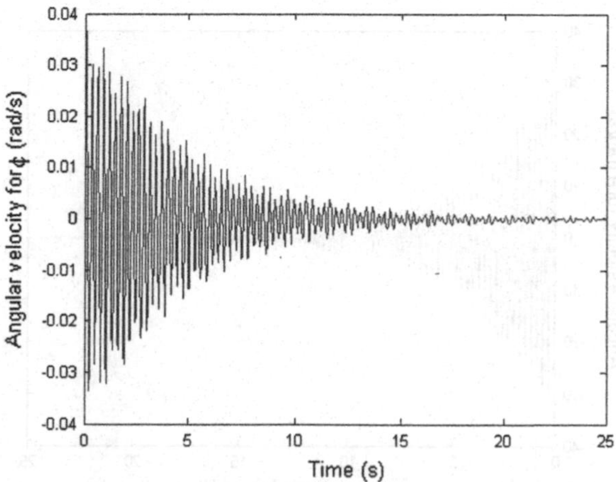

Fig. 4 Angular velocity for ϕ

explained later in this section, these results are obtained due to the performance of the sliding variables and their derivatives. In the following sections an analysis of the oscillation or chattering is done to find the oscillation characteristics and compare it with other control strategies.

In Figs. 5 and 6 the variables for the position and angular velocity of θ are shown. Even when stabilizes the Furuta pendulum variable θ is not an easy task, in this example the angular rotation and velocity of the pendulum are stabilized satisfactorily due to the performance of the sliding mode variables and their derivatives. As can be noticed, these variable reach zero in a specified time, even when there are some

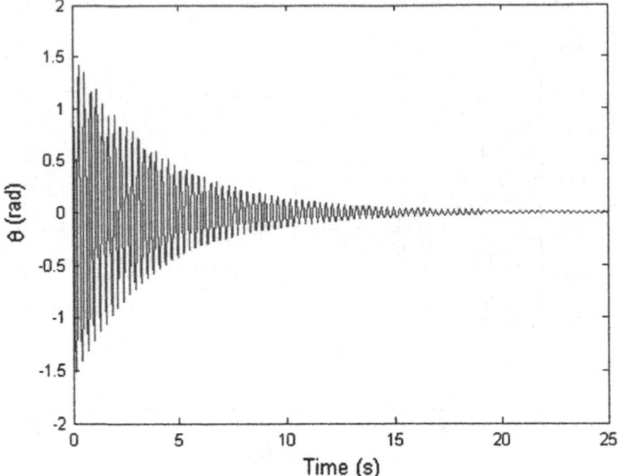

Fig. 5 Angular position for θ

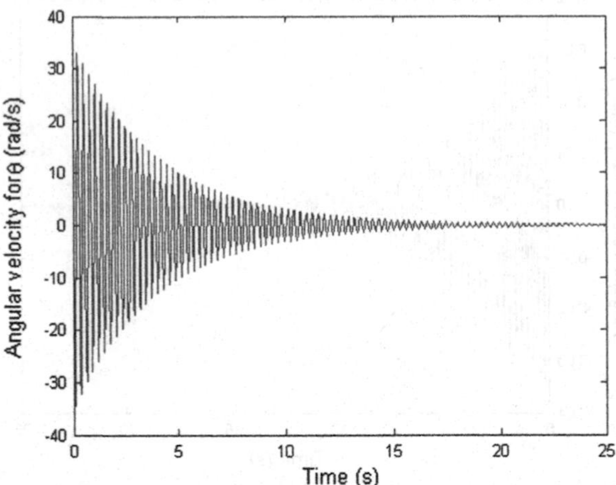

Fig. 6 Angular velocity for θ

oscillations the system reach the origin in steady state, proving that the SOSMC is effective.

In Figs. 7 and 8 the phase portrait of ϕ and θ are shown depicting the phase trajectories of the measured variables. It can be noticed how the trajectories of the system reach the equilibrium points, proving that the system is stable under these conditions. The limit cycles generated by the periodic orbits of the system are stabilized by the second order sliding mode control that avoids instabilities and the control system drives the state trajectories of the system until they reach the desired final values ensuring the asymptotical stability of the system as proved theoretically in the previous section. This fact is very important for the chattering analysis, because the limit

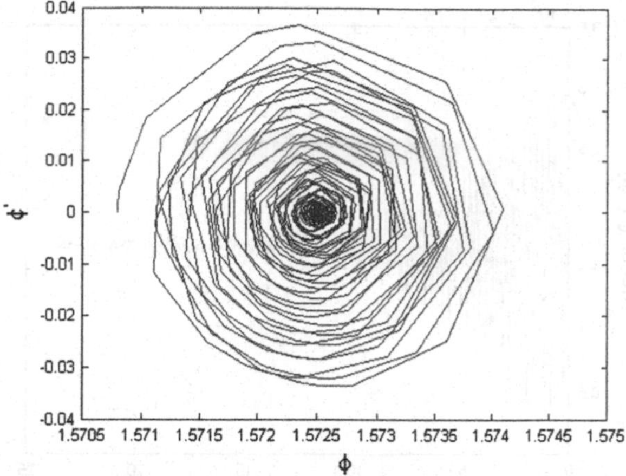

Fig. 7 Phase portrait of ϕ

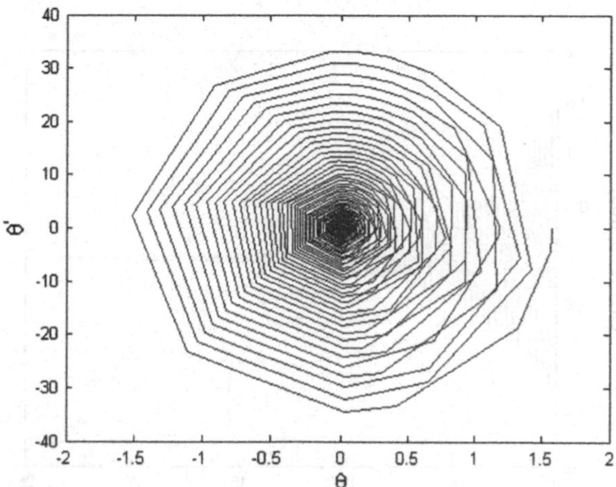

Fig. 8 Phase portrait for θ

cycles yielded by the periodic oscillations provides crucial information that can be analyzed by Poincare maps as explained in Sect. 5 to calculate the oscillation period.

In Figs. 9 and 10 the error signals for ϕ and θ are shown respectively. The results depicted in these figures, shown that a very small tracking error for both variables is obtained and they reach very small values in steady state. While keeping the tracking errors as small as possible, the trajectory tracking of the two controlled variables of the system is done effectively by the second order sliding mode control. As it is explained before, the tracking error is reduced to zero in steady state by the convergence of the sliding variables in finite time, proving that this control strategy is suitable for the trajectory tracking of this underactuated mechanical system.

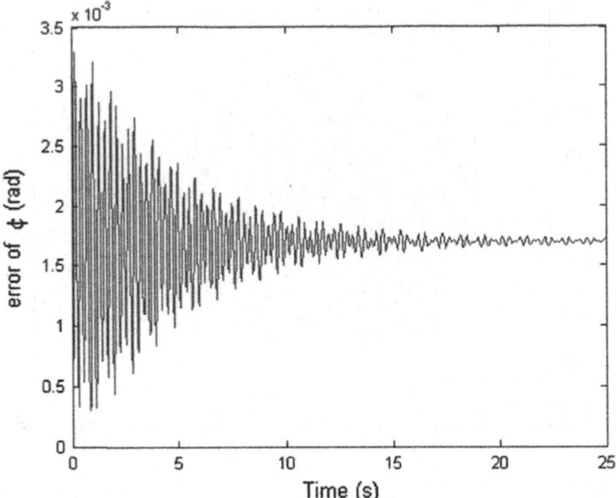

Fig. 9 Error of ϕ

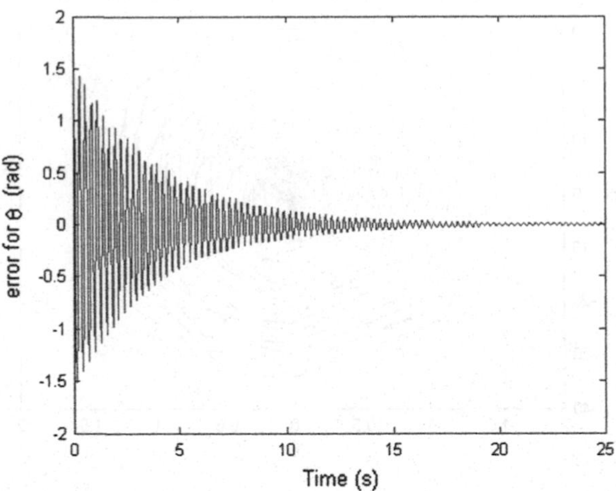

Fig. 10 Error of θ

The input torque that controls joint 1 ϕ is shown in Fig. 11 where a considerable control effort is necessary to stabilizes the measured variables in a considerable time. It is important to notice the oscillations yielded by the control switching function and the necessary control effort applied by actuator 1 ϕ in order to keep the joint in the desired position. As it is expected, even when the sliding mode control law reduces chattering, it is still present, therefore it is necessary to analyze this effect for comparison with the other strategies explained in this chapter.

In Figs. 12 and 13 the sliding variables for σ_1 and σ_2 are shown. The sliding variables reach the origin in a specified time assuring that the state variables achieve

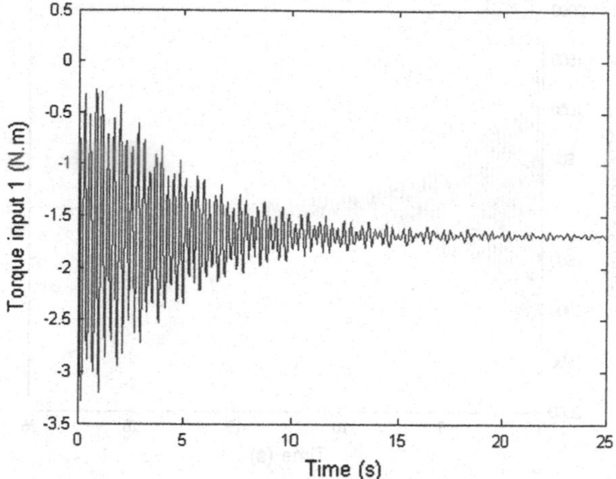

Fig. 11 Torque for input 1

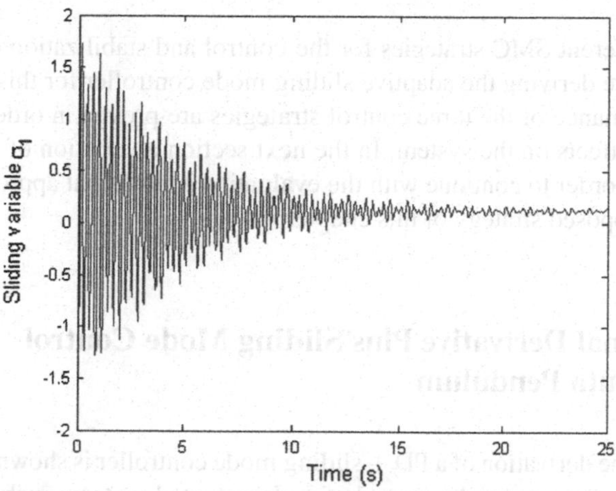

Fig. 12 Sliding variable 1

the zero value in steady state. This fact is very important since the stabilization of the control variables such as the position and velocity depends on the convergence of the sliding mode variable, so as it is shown theoretically the selection of an appropriate control law algorithm is crucial for the efficiency of the SOSMC to stabilize the Furuta pendulum.

In this section the design of a second order sliding mode controller for the Furuta pendulum is shown, a higher order sliding mode control law is implemented to make the state variables to reach the desired steady state value. A convenient control law is proposed instead of applying classical second order sliding mode approaches such as the twisting or super twisting algorithms. The objective of the SOSMC design is

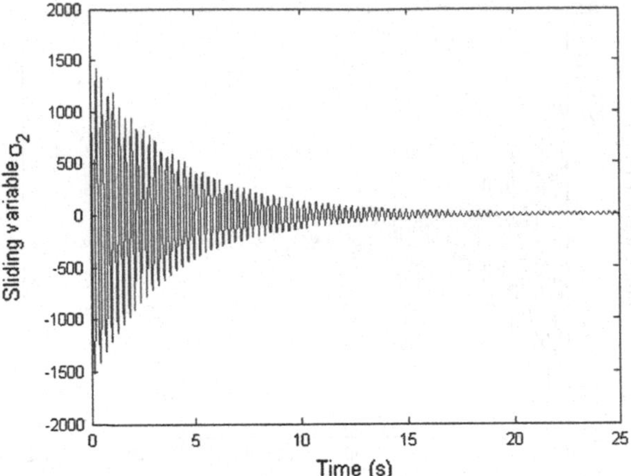

Fig. 13 Sliding variable 2

to elucidate different SMC strategies for the control and stabilization of the Furuta pendulum before deriving the adaptive sliding mode controller for this mechanism, then the performance of the three control strategies are proven in order to evaluate the chattering effects on the system. In the next section a variation of SMC control is explained, in order to continue with the evaluation of different approaches before deriving the proposed strategy of this chapter.

3 Proportional Derivative Plus Sliding Mode Control of the Furuta Pendulum

In this section the derivation of a PD + sliding mode controller is shown to prove that is an efficient alternative for the control of underactuated systems such as the Furuta pendulum. Proportional derivative control (PD) has been proved to be an effective and simple control architecture for mechanical systems, for this reason, a combined control strategy along with a sliding mode controller is shown in this section. The main idea in this section is to show that the Furuta pendulum can be stabilized by this control law, even when this controller is simple. A combined linear control law, given by the PD part of the controller, and a nonlinear part, given by the sliding mode controller (Ouyang et al. 2014) make the system variables to reach the desired values while the sliding surface reach the origin in a defined time interval.

Descentralized PD controllers are very popular in the control of different kind of mechanical system; including robotic arms, parallel robots, "etc", due to the simplicity of their tuning parameters this kind of controllers at least ensure the local stability of the controlled system. Even when this kind of controllers are very popular and simple they have some disadvantages such a poor disturbance rejection and robustness; for this reason in order to improve the properties of the PD controller sometimes it is necessary to combine this control strategy with a nonlinear control law. There are some control strategies found in literature in which the PD controller is combined with nonlinear control to improve the system performance, for example in Xiang and Siow (2004) a combined PD + nonlinear + neural network control is implemented for the stabilization of a robotic arm, the hybrid control law improves the system performance in which the trajectory tracking of a two links robotic arm is done by following a desired trajectory. In Liu (1999) another PD controller variation is implemented in the control of a two links robotic arm, where a nonlinear part is added to the proportional derivative controller for the trajectory tracking of this mechanism, this descentralized control strategy make the system variables to follow the desired trajectories when disturbance are applied to the system. Then finally, a PD + sliding mode controller for the trajectory tracking of a robotic system is explained in Ouyang et al. (2014), where the controller properties are improved by adding a nonlinear discontinuous function to the combined control law. Therefore based on the previous cases a suitable PD + sliding mode controller is suggested in this section for the stabilization of the Furuta pendulum, considering the similarities of the properties of some mechanical systems with the Furuta pendulum, the control approach presented in this section is not only suitable for the control of this underactuated system, it allows the tracking of the mechanical system properties efficiently while keeping the tracking error as small as possible, with small chattering effect and control effort.

In the following sections the design of a PD + sliding mode controller is explained, where a proposed sliding surface is defined to ensure that the system is stable, proved by an appropriate selection of a Lyapunov function (Liu 1999; Xiang and Siow 2004). Then, an example of the stabilization of a Furuta pendulum is shown to illustrate the implementation of this control law in this underactuated system, to analyze its performance under a specified initial condition. The idea of this section is to provide an alternative to the adaptive sliding mode control of the Furuta pendulum, that is analyzed and compared in Sect. 5, then some conclusions are obtained according to the chattering analysis of these controllers.

3.1 Derivation of the PD + Sliding Mode Controller

In this section the derivation of a proportional derivative plus sliding mode controller for the Furuta pendulum is developed. A stabilizing PD + sliding mode controller has been proved to be effective in the control of different kind of mechanical systems, considering that this is an underactuated mechanical model, the control of this system by PD + SMC is appropriate due to the combined advantages and properties of this control strategy. The development of this control technique consists in designing an appropriate sliding manifold considering the dynamical system properties of the model that are common in many mechanical systems. Chattering avoidance is one of the properties of the model that is required in order to avoid the instability and system variables deterioration; this controller is very effective in order to cancel this unwanted effect. Even when this technique is efficient in cancelling the chattering effects in the system, this phenomenon is still present but with smaller negative results than classical sliding mode controllers. Therefore, the analysis of this phenomenon on the system is shown later in this chapter for comparison purposes with the other sliding mode controllers explained in this chapter. The intention of this section is to evince a combined sliding mode controller technique to understand and compare with the main controller derived in this chapter, then some interesting conclusions are obtained from all of these sliding mode control approaches, so all of these control strategies are developed to show different alternatives and as a preview and comparison with the proposed adaptive sliding mode controller explained in the next section. A complete analysis of this controller with the respective simulation is shown in this section in order to clarify the theoretical background of this control approach by deriving the PD + SMC strategy and show an illustrative example in order to verify the performance of this controller.

The first step in the derivation of the PD + SMC for the Furuta pendulum, is to define the following error signal (Liu 1999; Ouyang et al. 2014):

$$e = q_d - q \tag{28}$$

where based on the error signal the sliding surface r is given by:

$$r = \dot{e} + \Phi e \tag{29}$$

where Φ is a positive definite matrix. Then the PD + sliding mode control for the Furuta pendulum is given by (Liu 1999; Ouyang et al. 2014):

$$\tau = k_c r + k_1 \text{sign}(r) \tag{30}$$

where k_c and k_1 are positive definite matrices for the PD and the sliding mode parts of the control law respectively. The stability properties of this control law will be examined later according to the Lyapunov stability theorem.

Substituting r in the dynamic system of the Furuta pendulum yields:

$$D(q)\dot{r} + C(q, \dot{q})r = -\tau + \xi \tag{31}$$

where

$$\xi = D(q)(\ddot{q}_d + \Phi\dot{e}) + C(q, \dot{q})(\dot{q}_d + \Phi e) + g(q) \tag{32}$$

As explained in the previous section, ξ has a property that is very important for the analysis of the stability of the closed loop system as described in Definition 2.

Definition 3 An n-degrees of freedom mechanical system has the following properties according to the dynamical systems characteristics:

$$\mu_{\min} I < D(q) < \mu_{\max} I \tag{33}$$

and

$$\|C(q)\| \le C_H \tag{34}$$

$$\|g(q)\| \le C_g \tag{35}$$

where $\mu_{\max} > \mu_{\min} > 0$ and $C_H, C_g > 0$

Definitions 2 and 3 are very important in order to prove the stability of the systems, according to the dynamical systems of the Furuta pendulum. Now, with these properties and the dynamical system characteristics, the stability of the systems with the specified control law is done as explained in the following theorem.

Theorem 2 *The PD + sliding mode controller ensure the stability of the system if the defined Lyapunov function indicates that the system is asymptotically stable.*

Proof Consider the following Lyapunov function

$$V(r) = \frac{1}{2} r^T D(q) r \tag{36}$$

The derivative of the Lyapunov function is given by:

$$\dot{V}(r) = r^T D(q)\dot{r} + \frac{1}{2} r^T \dot{D}(q) r \tag{37}$$

where

$$D(q)\dot{r} = -\tau + \xi - C(q, \dot{q}) r \tag{38}$$

Then by applying Definition 1 and (38) the derivative of the Lyapunov function becomes in:

$$\dot{V}(r) = -r^T \tau + r^T \xi \tag{39}$$

Then applying the norm on both sides of (39) and substituting the control law τ_ϕ yields:

$$\dot{V}(r) \le -\left\| r^T \right\| \|k_c r + k_1 \mathrm{sign}(r)\| + \left\| r^T \right\| \|\xi\| \tag{40}$$

Converting this inequality in:

$$\dot{V}(r) \le -k_{c\min} \left\| r^T \right\| \|r\| - k_{1\min} \left\| r^T \right\| \|\mathrm{sign}(r)\| + \left\| r^T \right\| \|\xi\| \tag{41}$$

where $k_{c\min} = \min_{i \in n} k_c$ with $k_{c\min} > 0$ and $k_{1\min} = \min_{i \in n} k_1$ with $k_{1\min} > 0$ (Liu 1999; Xiang and Siow 2004).

Using the properties explained in Definition 2 and 3 the Lyapunov function indicates that the system, representing the Furuta pendulum, is asymptotically stable with the specified PD + sliding mode control law.

In the next section an illustrative example of the control of the Furuta pendulum with a PD + sliding mode control is done to prove the validity of the theoretical background demonstrated in this subsection.

3.2 Example 2

In this section an illustrative example of the control of the Furuta pendulum by a proportional derivative plus sliding mode control is shown to clarify the application of this controller to this underactuated mechanical system. Even when the control of underactuated systems is difficult, it is shown theoreticaly and by an illustrative example that is possible to stabilize this mechanism by selecting an appropriate control law algorithm. In this example the angle trajectories and velocities are depicted to prove that these variables are stable and reach the desired values in steady state. The phase portraits shown in this section, verify the asymptotical stability of the system while minimizing the tracking error of the model.

In this example the stabilization of a Furuta pendulum with a PD + sliding mode control is shown with appropriate parameter selection. The parameters of the Furuta pendulum are specified in Table 1 with $(\pi/2, 0)$ as the initial conditions of the system. The gains of the PD + SMC are given as follow:

$$k_c = \begin{bmatrix} 0.7 & 0 \\ 0 & 0.7 \end{bmatrix} \tag{42}$$

$$k_1 = \begin{bmatrix} 0.01 & 0 \\ 0 & 0.01 \end{bmatrix} \tag{43}$$

$$\Phi = \begin{bmatrix} 90 & 0 \\ 0 & 90 \end{bmatrix} \tag{44}$$

The simulations were done in $MATLAB^{®}$ and $SimMechanics^{®}$ where the specified parameters are used in all the simulation process. In Fig. 14 the angle trajectory ϕ is depicted, where as it is noticed the trajectory of this variable reaches the specified value in steady state, proving that PD + sliding mode controller stabilizes the system with the desired performance. In Fig. 15, the angular velocity for the variable ϕ is shown, where this variable reaches the zero value in a specified value as the corresponding variable is stabilized. As it is noticed these variables reaches the expected values in finite time, this result is achieved due to the appropriate sliding mode manifold is selected in order to stabilize the controlled variables.

In Fig. 16 the angular trajectory for the pendulum angle θ is shown, where this angle reach the value of zero in steady state as defined by the controller and system specifications, so the PD + sliding mode controller of the Furuta stabilizes this variable in the required time keeping the two controlled variables in the desired

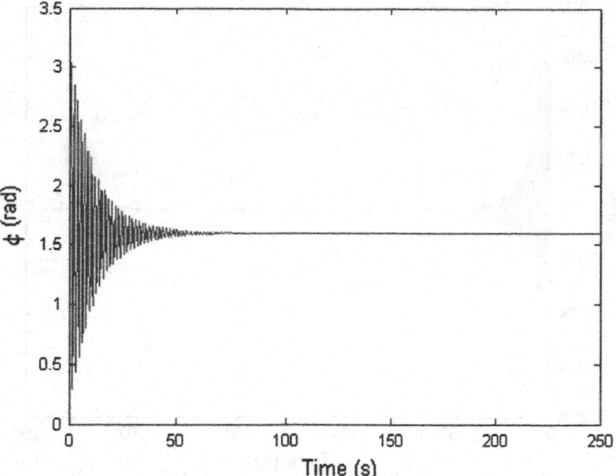

Fig. 14 Angle position for ϕ

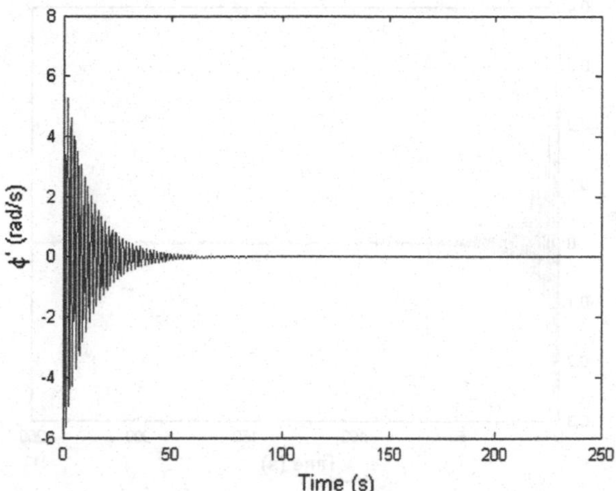

Fig. 15 Angular velocity for ϕ

mechanism positions. In Fig. 17 the angular velocity of the controlled variable θ is shown, where this variable reaches the zero value in steady state as defined by the controller and system specifications. With these results the stability of all the controlled variables is ensured by the implementation of a PD + sliding mode controller, keeping the Furuta pendulum stable when external disturbances are applied in the system. As it is proven theoretically, the appropriated sliding manifold selection is very important in order to stabilize these variables, reaching and keeping the desired values in finite time.

The respective phase portraits for ϕ and θ are shown in Figs. 18 and 19. As it is noticed, the two phase portraits show that these variables are stable, according

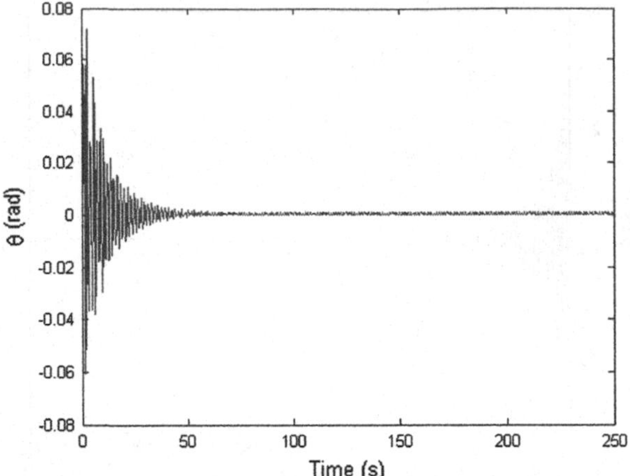

Fig. 16 Angle position for θ

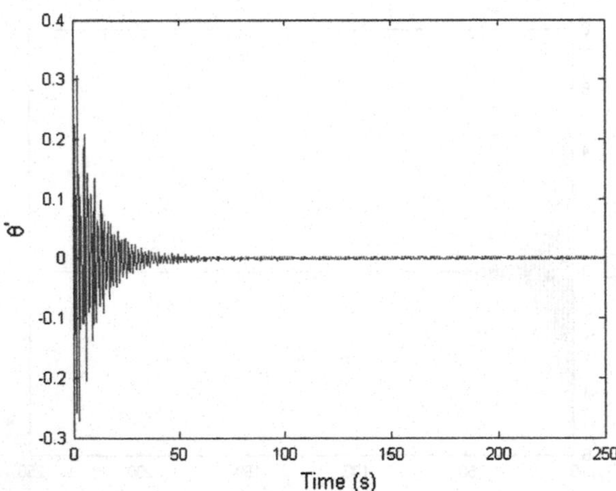

Fig. 17 Angular velocity for θ

to their respective phase trajectories. The two limit cycles depicted in these figures show that the oscillations follow a prescribed trajectory until the variables reach the desired values when a disturbance is applied on the system. The phase portraits show that the limit cycles yielded by the periodic oscillations are stable, proving that the PD + sliding mode control law meets the required specifications according to the stabilization of the state variables of the system.

In Fig. 20 the respective input torque for joint 1 (base) is shown. As it is noticed the control effort for the joint is reasonable so it is not necessary to saturate the actuator. The torque input applied to the base joint behaves in an oscillatory manner as it is

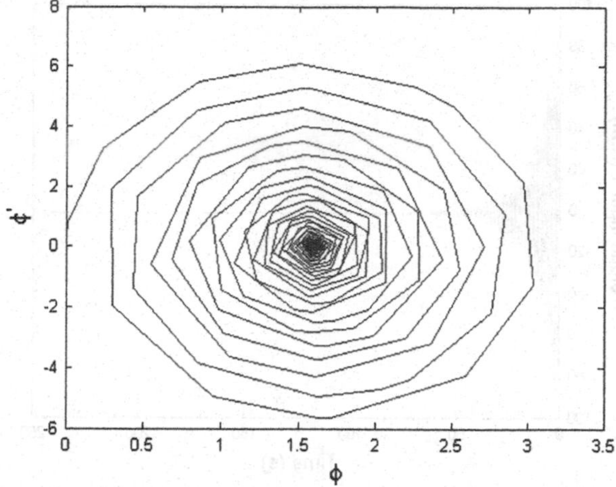

Fig. 18 Phase portrait of ϕ

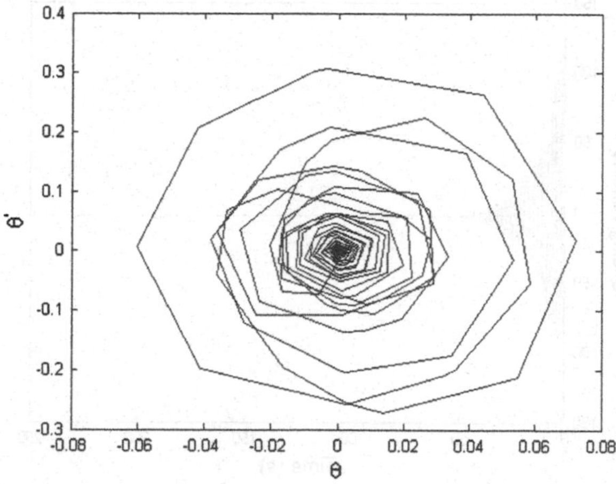

Fig. 19 Phase portrait of θ

expected generating some oscillations until the system variables reach the desired values in steady state.

The corresponding oscillation analysis of the variables and the torque inputs is done in Sect. 5 where the chattering effect is evaluated according to the oscillation frequencies of this and the other SMC strategies explained in this chapter.

In Figs. 21 and 22 the respective sliding variables of the PD + sliding mode controller are shown, where the two variables converge to zero in a determined time. Ensuring that the sliding variables reach the origin in an expected time allowing the system to reach the specified steady state values with a considerable small control effort generated by the switching control law.

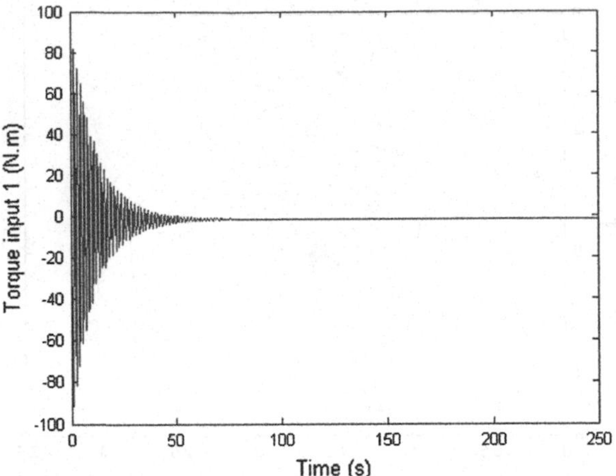

Fig. 20 Input torque 1

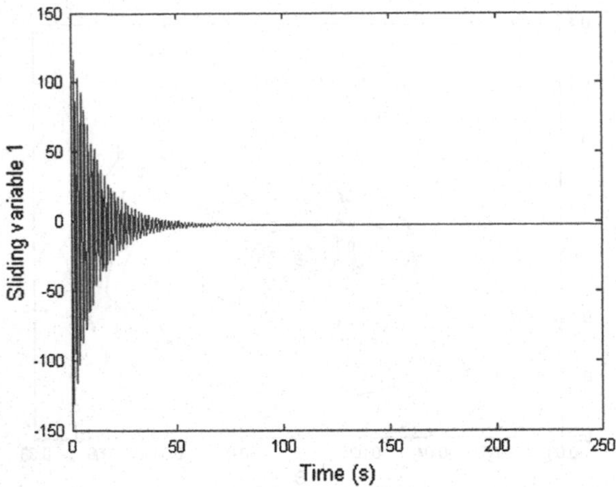

Fig. 21 Sliding variable 1

The error signals for ϕ and θ are shown in Figs. 23 and 24 respectively, where the PD + sliding mode controller makes the error signal to reach the zero value in an expected time, proving the efficiency of this controller to stabilizes mechanical systems of different kind.

In this section a PD + sliding mode controller for the control and stabilization of the Furuta pendulum is explained, to prove their suitability in the control of this kind of underactuated mechanical system. This control approach is advantageous because it combines the simplicity of a proportional derivative controller and the efficiency of a nonlinear sliding mode control making this strategy ideal for the stabilization of this kind of mechanism. The stability of the PD + SMC is corroborated by the selection of

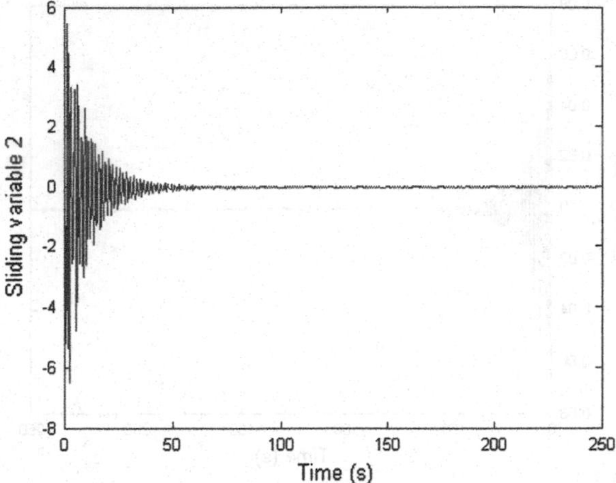

Fig. 22 Sliding variable 2

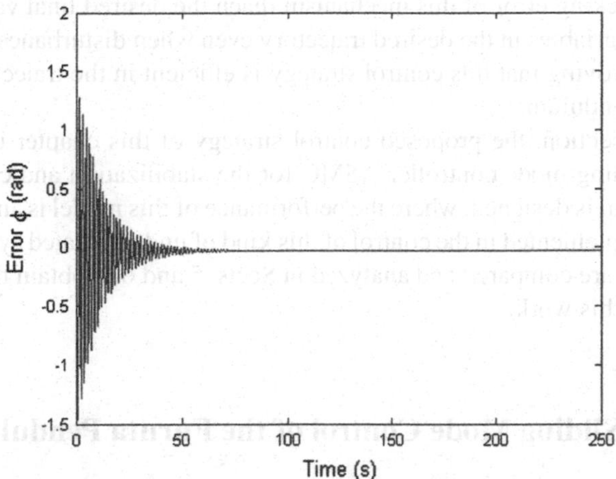

Fig. 23 Error signal for ϕ

an appropriate Lyapunov function and this fact is confirmed by a numerical example and simulation of the Furuta pendulum with this control strategy.

As it is confirmed in this chapter, all the variables are stabilized according with the system design specifications and initial condition of the model; reaching the expected value in steady state. This system behavior is illustrated in the phase plot of each variable, where the state trajectories reach the specified point in these diagrams, so this control strategy yields stable limit cycle oscillations when a disturbance is applied to the system.

As it is noticed in Example 2, the control effort generated by the controller output is significantly small to keep the controlled variables in the equilibrium point of the

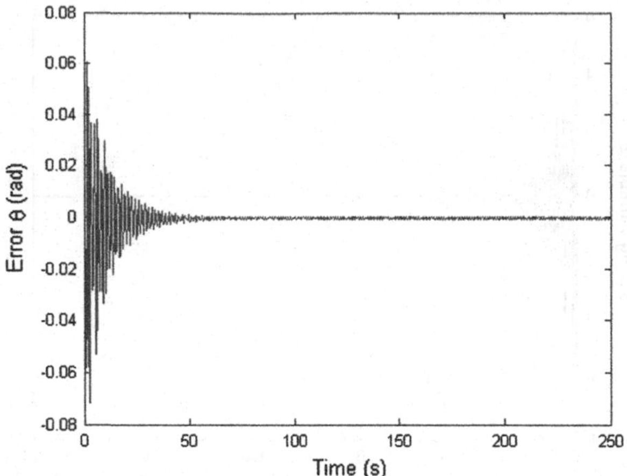

Fig. 24 Error signal for θ

system. The tracking error of this mechanism reach the desired final value, keeping the controlled variables in the desired trajectory even when disturbances are present in the model, proving that this control strategy is efficient in the trajectory tracking of the Furuta pendulum.

In the next section, the proposed control strategy of this chapter is developed, an adaptive sliding mode controller ASMC for the stabilization and control of the Furuta pendulum is designed, where the performance of this model is analyzed when an ASMC is implemented in the control of this kind of underactuated system. All the SMC strategies are compared and analyzed in Sects. 5 and 6 to obtain the respective conclusions of this work.

4 Adaptive Sliding Mode Control of the Furuta Pendulum

In this section the main control technique of this chapter is explained, an adaptive sliding mode control for the stabilization of the Furuta pendulum (ASMC). Adaptive sliding mode control is a control approach that has been implemented extensively in different kinds of applications due to the flexibility of the sliding mode parameters (Chang 2013; Cheng and Guo 2010); this is a very useful control approach used in different kinds of systems such as electrical (Liu et al. 2013; Chen et al. 2014) and mechanical systems (Fei and Wu 2013) yielding the desired performance when disturbances are applied to the system.

An adaptive gain SMC is implemented for the stabilization of the Furuta pendulum considering the system dynamics of the model and its stability properties to keep the mechanism trajectory in the desired position. Another property of this control approach is that the chattering effect is minimized and then, the oscillations of the system are cancelled by the controller characteristics.

Adaptive sliding mode control is a control technique that consists in the implementation of an adaptive gain parameter obtained according to the Lyapunov stability theorem along with a sliding mode controller to ensure the asymptotical stability of the system. The design procedure for this kind of controller consists in designing a feasible controller that reaches the sliding manifold in finite time while stabilizing the state variables for the trajectory tracking. The advantage of this control strategies is that the effects yielded by perturbation and disturbances are suppressed by the adaptive control along with the sliding controller action. For this reason, this control technique is appropriate for the control and stabilization of different kind of mechanical system, obtaining a very robust controller that adjusts its parameters in real time moving and keeping the state variables in the desired trajectory. The control law of this controller is designed by combining an adaptive part with a sliding mode controller that makes the Furuta pendulum variables reach the desired values in steady state while keeping the system in the sliding manifold to obtain an asymptotically stable system. Another advantage of this controller is the chattering suppresion effects yielded by the adaptive sliding mode controller, so the action produced by the discontinuos sliding mode controller algorithm is cancelled by the effect of the adaptive gain of the system. Taking in count that the parameters of the mechanical system such as the gravitational, coriolis and inertia matrices change in time an adaptive sliding mode controller is suitable for this kind of mechanism (Yao and Tomizuka 1994) making a flexible control strategy that vary the controller parameters in real time while ensuring the asymptotical stability of the system. This control strategy has another advantage that is related to the smaller control effort that is necessary in order to stabilizes the system variables while keeping the sliding variables in the origin, this is a desirable property that means that it is not necessary to saturate the actuator due to higher values of the control action. The adaptive sliding mode controller is proved to be a strong control strategy that is applied in the control and stabilization of the Furuta pendulum as demonstrated in this section, the sliding mode control strategies developed in this chapter proved that are effective and they are the fundamental control approaches for the stabilization of the Furuta pendulum. The SMC approaches shown in this chapter are developed to shown the fundamentals of adaptive sliding mode control for the stabilization of the Furuta pendulum and for comparison purposes with the proposed control approach of this chapter, that even when these control strategies are effective in the control of this underactuated system, they lack of important properties that adaptive sliding mode control has for the stabilization of the Furuta pendulum; for this reason the comparison and analysis of these control strategies are shown in Sects. 5 and 6.

In the following sections the development of an ASMC for the Furuta pendulum is derived and explained as the proposed control strategy of this chapter and then the system performance is corroborated with an illustrative example of the ASMC for the Furuta pendulum with specified initial conditions. The proposed control strategy is compared later in the following sections according to the chattering effects on the system and other characteristics of the control system.

4.1 Derivation of the Adaptive Sliding Mode Controller of the Furuta Pendulum

Adaptive sliding mode has successfully proved that is an efficient control technique for different kinds of systems (Yu and Ozguner 2006) therefore this control approach has better disturbance rejection properties than classical SMC, for this reason this control strategy is convenient for the control of underactuated mechanical systems as explained in this section. Adaptive sliding mode control is suitable for the control of underactuated mechanical systems due to its robustness and disturbance rejection properties when unmodelled dynamics and disturbance are present in the system, this controller updates its adaptive gain online improving the performance of the controller and therefore the asymptotical stability of the system is ensured by this control strategy. The derivation of this adaptive gain is done by the Lyapunov stability theorem in order to ensure the stability of the system and the sliding variables convergence for a better trajectory tracking of the system. The objective of this chapter is to proved that a feasible adaptive sliding mode controller can be designed in order to improve the disturbance rejection and chattering avoidance properties of the Furuta pendulum by ensuring the stability of the system with a small control effort and reducing the chattering effects on the system.

Before deriving the ASMC for the Furuta pendulum, an important property for mechanical systems is described as follow: Consider the dynamics equation of the Furuta pendulum as described in (8), then this system is linearly parametrizable as described in the next equation (Liu 1999; Xiang and Siow 2004).

$$\hat{D}(q)\ddot{q}_r + \hat{C}(q,\dot{q})\dot{q}_r + \hat{g}(q) = Y(q,\dot{q},\dot{q}_r,\ddot{q}_r)\psi \tag{45}$$

where $\hat{D}(q)$, $\hat{C}(q,\dot{q})$, $\hat{g}(q)$ are the estimated dynamical systems matrices and vector respectively and ψ is the parameter of the dynamical system model that is adjusted by the adaptive control law. Then $\dot{q}_r$ is defined as:

$$\dot{q}_r = \dot{q}_d + \Phi\tilde{q} \tag{46}$$

where $\tilde{q} = q_d(t) - q(t)$, $q(t)$ is the position vector of the Furuta pendulum, $q_d(t)$ is the desired position vector and Φ is a positive definite matrix.

Based on the previous variables, the sliding surface is defined as:

$$S = \dot{q}_r - \dot{q} = \dot{\tilde{q}} + \Phi\tilde{q} \tag{47}$$

where the derivative of S is given by:

$$\dot{S} = \ddot{q}_r - \ddot{q} = \ddot{\tilde{q}} + \Phi\dot{\tilde{q}} \tag{48}$$

The proposed control law for the stabilization of the Furuta pendulum is (Xiang and Siow 2004; Fei and Wu 2013; Liu et al. 2013; Chen et al. 2014):

$$\tau = \hat{D}(q)\ddot{q}_r + \hat{C}(q,\dot{q})\dot{q}_r + \hat{g}(q) - k_d S - k_1 \text{sign}(S) \tag{49}$$

Then using the linear parametrization property of the Furuta pendulum dynamics, (49) is converted to:

$$\tau = Y(q,\dot{q},\dot{q}_r,\ddot{q}_r)\psi - k_d S - k_1 \text{sign}(S) \tag{50}$$

where k_d and k_1 are constant positive definite matrices. The following theorem ensures asymptotical stability of the system with the proposed adaptive sliding mode control law and it is necessary in order to find the adaptive parameter of the system.

Theorem 3 *The adaptive sliding mode controller ensures the stability of the system if the defined Lyapunov function indicates that the system is asymptotically stable.*

Proof Consider the following Lyapunov function:

$$V(S,\psi) = \frac{1}{2}S^T D(q)S + \frac{1}{2}\psi^T \Gamma^{-1}\psi \tag{51}$$

where Γ is a positive definite adaptive gain matrix. Then the derivative of the Lyapunov function is given by:

$$\dot{V}(S,\psi) = S^T D(q)\dot{S} + \frac{1}{2}S^T \dot{D}(q)S + \dot{\psi}^T \Gamma^{-1}\psi \tag{52}$$

Then with

$$D(q)\dot{S} = -\tau + \xi - C(q,\dot{q})S \tag{53}$$

where

$$\xi = D(q)(\ddot{q}_d + \Phi\dot{\tilde{q}}) + C(q,\dot{q})(\dot{q}_d + \Phi\tilde{q}) + g(q)$$
$$\xi = D(q)\ddot{q}_r + C(q,\dot{q})\dot{q}_r + g(q) \tag{54}$$

Then $\dot{V}(S,\psi)$ becomes in

$$\dot{V}(S,\psi) = -S^T\tau - S^T C(q,\dot{q})S + S^T\xi + \frac{1}{2}S^T\dot{D}(q)S + \dot{\psi}^T\Gamma^{-1}\psi \tag{55}$$

Then considering the estimation of the parameter ξ

$$\xi = Y(q,\dot{q},\dot{q}_r,\ddot{q}_r)\psi \tag{56}$$

and rearrenge to apply Definition 1, $\dot{V}(S,\psi)$ becomes in:

$$\dot{V}(S,\psi) = -S\tau + \dot{\psi}^T\Gamma^{-1}\psi + S^T Y(q,\dot{q},\dot{q}_r,\ddot{q}_r)\psi \tag{57}$$

Then in order to stabilize the system the updating law of the variable parameter must be:

$$\dot{\psi}^T = -S^T Y(q, \dot{q}, \dot{q}_r, \ddot{q}_r)\Gamma \tag{58}$$

Therefore the derivative of the Lyapunov function becomes in:

$$\dot{V}(S, \psi) \leq -|S| \, \|\tau\| \tag{59}$$

So the system is asymptotically stable with the updating law of the adaptive parameter ψ. This completes the proof of the theorem.

In the next section an example of the stabilization of the Furuta pendulum by an ASMC is shown to illustrate the system performance.

4.2 Example 3

In this section an illustrative example of the control and stabilization of the Furuta pendulum by an ASMC is shown. The main idea of this example is to illustrate the system performance by a numerical simulation, where the system is tested with specified initial conditions and trajectories. Then the results obtained for the angular positions, velocities, phase portraits, tracking errors and input torques are analyzed to obtain the respective conclusions of the system performance. In this section is proved that the sliding mode controller with adaptive gain meets the requirement of the stabilization and tracking error reduction by an appropriate adaptive control law with a well defined updating gain algorithm. The theoretical background of the adaptive sliding mode controller for the stabilization of the Furuta pendulum is corroborated in this example by selecting the appropriate controller parameters for the adaptive gain, in order to stabilize the system by the on line adaptation of the adaptive gain.

The controller parameters are:

$$k_1 = \begin{bmatrix} 0.0000001 & 0 \\ 0 & 0.0000001 \end{bmatrix} \tag{60}$$

$$k_d = \begin{bmatrix} 7 & 0 \\ 0 & 7 \end{bmatrix} \tag{61}$$

The adaptive gain evolution is shown later in this section and the initial condition of the model is $(\pi, 0)$. The simulations were done in $MATLAB^{®}$ and $SimMechanics^{®}$ where the specified parameters are used in all the simulation process. In Figs. 25 and 26 the angular position and velocity of the base ϕ are shown respectively, where as it is noticed these variables reach the desired final values in steady state, with no oscillations in comparison with the previous SMC alternatives.

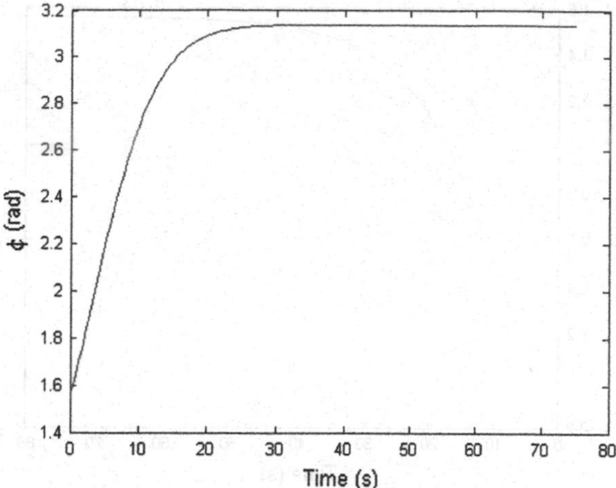

Fig. 25 Angle position for ϕ

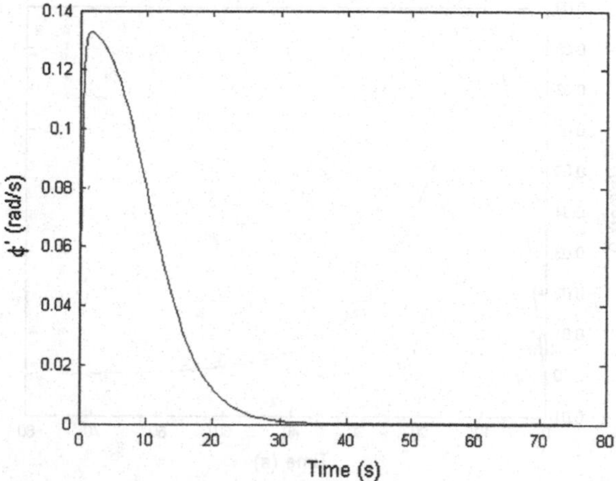

Fig. 26 Angular velocity for ϕ

The positions and angular velocities of the base are stabilized as it is defined by the adaptive control law, this requirement is met due to the convergence of the sliding variables in finite time. In comparison with the previous sliding mode control strategies, the trajectory tracking of these variables evince less oscillations, and a better system performance.

In Figs. 27 and 28 the angular position and velocity of the pendulum θ are shown respectively where the desired final values in steady state of the system are reached in a determined time when a disturbance is applied on the system. Practically there are no oscillations on the pendulum parameters, and then this proves that the proposed ASMC is an effective technique for the stabilization of this underactuated mechanical

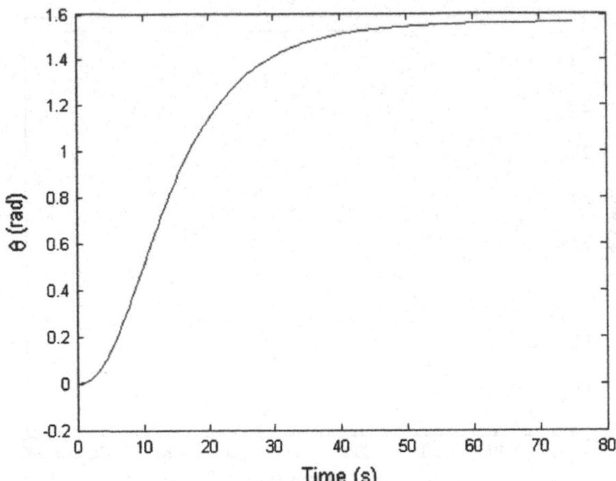

Fig. 27 Angle position for θ

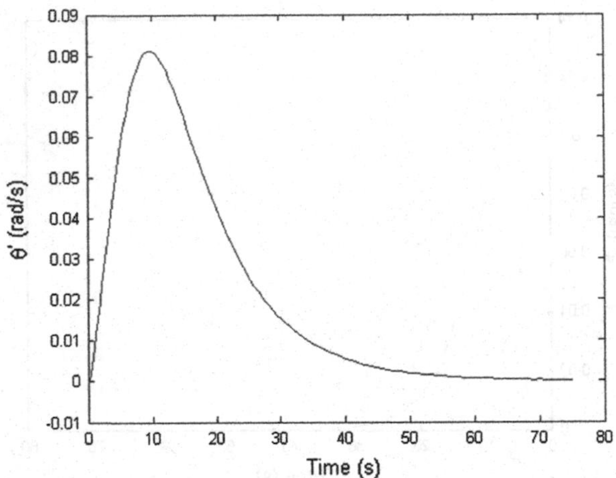

Fig. 28 Angular velocity for θ

system. These variables reach the desired position due to the convergence of the sliding variables, and in comparison with the other control strategies, the adaptive sliding mode controller for the stabilization of the Furuta pendulum shows a better performance in the stabilization of these variables due to a small chattering, that can be considered as oscillations of the system, and less control effort.

The input torque for the joint actuator is shown in Fig. 29, where the necessary control effort is necessary to be applied to stabilizes the controlled variables of the system. As explained before, it can be noticed that practically there are not oscillations on this control input, so this undesirable effect is eliminated by the adaptive characteristic of this adaptive gain system.

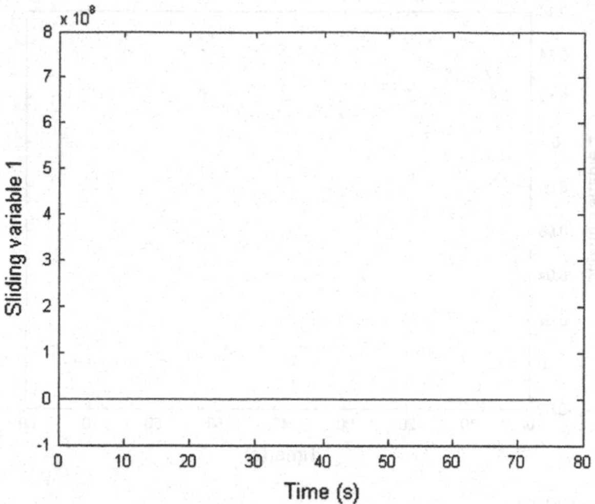

Fig. 29 Input torque 1

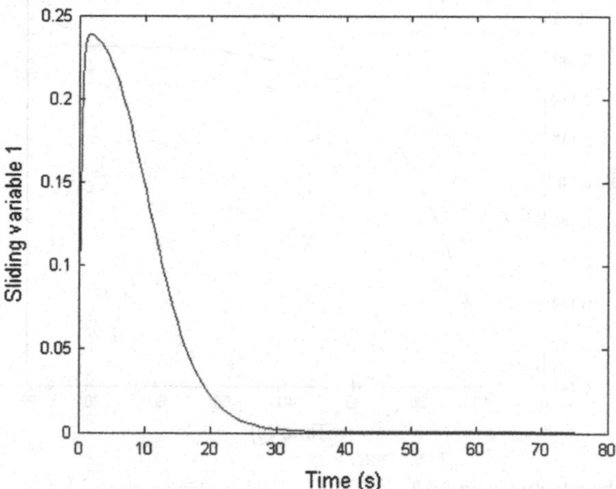

Fig. 30 Sliding variable 1

In Figs. 30 and 31 the respective sliding variables are shown, where as it is noticed these variables reach the origin in a considerable time, yielding the convergence of the controlled variable in finite time. These facts corroborates the theory behind this control strategy implemented in the control of the Furuta pendulum, where the sliding manifold must be reached in finite time to ensure the stability of the system.

In Fig. 32 the norm of the adaptive gain $\|\psi\|$ is depicted in this figure. As can be noticed, the evolution of the adaptive gain goes from the initial value to the final value of this parameter until the system variables and the adaptive gain reach the desired value in finite time.

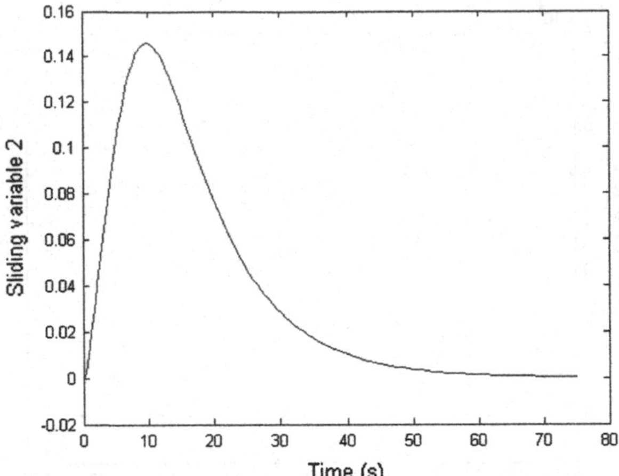

Fig. 31 Sliding variable 2

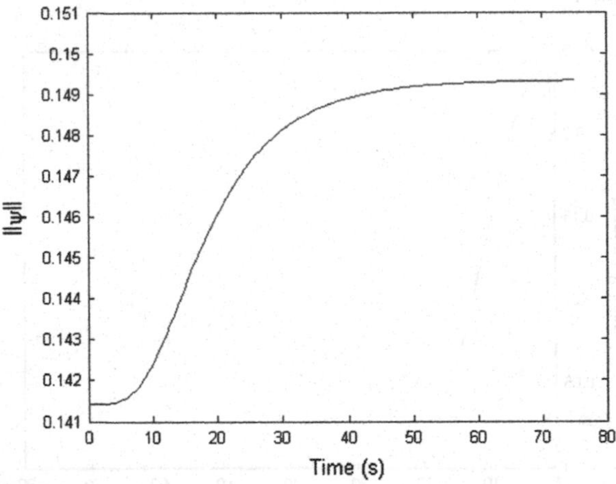

Fig. 32 Norm of the adaptive gain $\|\psi\|$

Finally in Figs. 33 and 34, the error signals of the model reach the zero value in finite time as specified in the ASMC design, so the controlled variables ϕ and θ are stabilized in finite time while keeping the tracking error about the zero.

In this section the proposed control strategy of this chapter is shown, for the stabilization and control of the Furuta pendulum, keeping the tracking error of the controlled variable about zero. This objective was proved theoretically and corroborated later by a simulation example. The adaptive gain of ASMC improves the performance of the system considerably in comparison with the control strategies developed in the previous sections, and as it is confirmed in the following section the

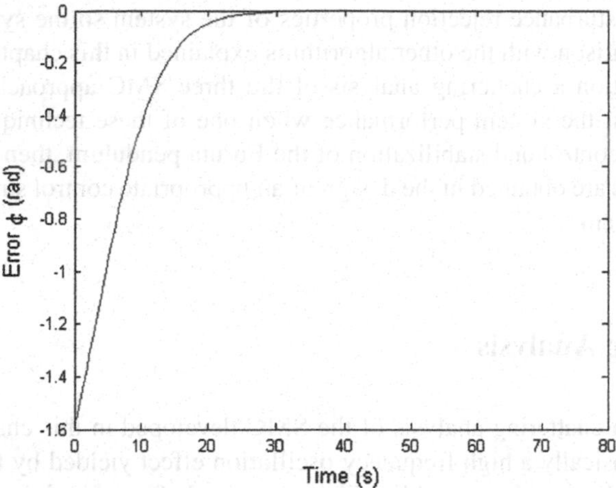

Fig. 33 Error of ϕ

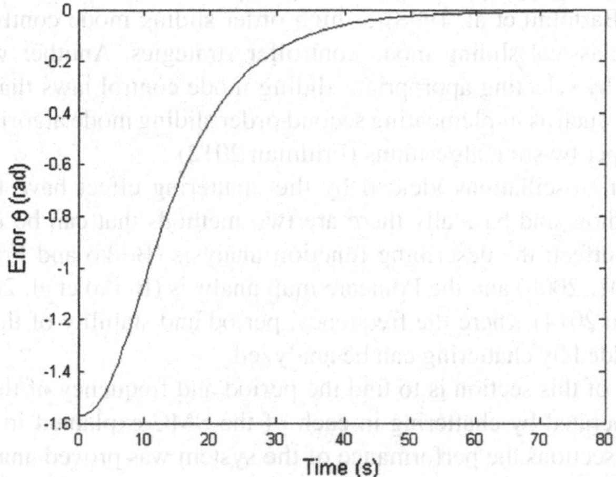

Fig. 34 Error of θ

performance and chattering avoidance properties of the proposed controller is better than the other two control alternatives explained in this chapter.

The main objective of this chapter is to analyze and develop different sliding mode control strategies, from the classical to new kinds of SMC variations to find the theoretical basis of well known sliding mode control algorithms and compare them with novel SMC strategies for the control of underactuated mechanical systems.

In this section the adaptive sliding mode controller is developed exploiting the linear parametrization of the system, that is an important properties of different kinds of mechanical systems, and it is a contrasting characteristic of the ASMC in comparison with the other algorithms explained in this chapter. The adaptive gain

increases the disturbance rejection properties of the system so the system is more robust in comparison with the other algorithms explained in this chapter.
In the next section a chattering analysis of the three SMC approaches is clearly explained to test the system performance when one of these techniques is implemented for the control and stabilization of the Furuta pendulum, then some important conclusions are obtained in the design of an appropriate control strategy for this mechanical system.

5 Chattering Analysis

In this section a chattering analysis of the SMC developed in this chapter is done. Chattering is basically a high frequency oscillation effect yielded by the switching inputs of the sliding mode control law. This unwanted effect deteriorates the system performance and could lead to the instability of the system. One way to avoid this effect is by designing appropriated control strategies such as the second order sliding mode control (Bartolini et al. 1998) or high order sliding mode control instead of implementing classical sliding mode controller strategies. Another way to solve this problem is by selecting appropriate sliding mode control laws that reduce this unwanted effect such as implementing second order sliding mode algorithms like the twisting and super twisting algorithms (Fridman 2012).

The chattering oscillations yielded by the chattering effect have been studied by different authors and basically there are two methods that can be implemented to analyze this effect; the describing function analysis (Boiko and Fridman 2005; Boiko et al. 2007, 2008) and the Poincare map analysis (Boiko et al. 2008; Haddad and Chellaboina 2014) where the frequency, period and stability of the limit cycle oscillations yielded by chattering can be analyzed.

The purpose of this section is to find the period and frequency of the limit cycle oscillations generated by chattering in each of the SMC explained in this section. In the previous sections the performance of the system was proved analytically and corroborated by examples; therefore there is a clear idea of the limit cycle properties yielded by chattering in the input of the mechanical system, so the intention here is to show an analytical procedure to find the oscillation periods by Poincare maps.

In order to define the Poincare maps, the first step is to transform the dynamical system (8) of the Furuta pendulum to state space by linearizing the model

$$\dot{x} = Ax + B\tau \tag{62}$$

$$Y = Cx \tag{63}$$

where

$$A = \begin{bmatrix} 0 & 0 & 1 & 0 \\ 0 & 0 & 0 & 1 \\ 0 & \frac{-p_3 p_4}{p_1 p_2 + p_1 p_5 - p_3^2} & 0 & 0 \\ 0 & \frac{p_1 p_4}{p_1 p_2 + p_1 p_5 - p_3^2} & 0 & 0 \end{bmatrix} \tag{64}$$

$$B = \begin{bmatrix} 0 & 0 \\ 0 & 0 \\ \frac{p_2 + p_5}{p_1 p_2 + p_1 p_5 - p_3^2} & \frac{-p_3}{p_1 p_2 + p_1 p_5 - p_3^2} \\ \frac{p_3}{p_1 p_2 + p_1 p_5 - p_3^2} & \frac{p_1}{p_1 p_2 + p_1 p_5 - p_3^2} \end{bmatrix} \tag{65}$$

$$C = I_{4\times 4} \tag{66}$$

where $I_{4\times 4}$ is an identity matrix. Now defining the solution of the linear system in a specified point by (Haddad and Chellaboina 2014):

$$s(t, p) = x(o)e^{t} + \int_{0}^{t} e^{(t-\lambda)A} B(\lambda)\tau(\lambda)d\lambda \tag{67}$$

Define the function:

$$\zeta(x) \triangleq \left\{ \hat{\zeta} > 0 : S(\hat{\zeta}, x) \in S \text{and } S(t, x) \notin S, 0 < t < \hat{\zeta} \right\} \tag{68}$$

Then the Poincare map is given by:

$$P(x) \triangleq s(\zeta(x), x) \tag{69}$$

Finally, from the Poincare map shown in (70) a dynamic system in discrete time is obtained as shown in (71)

$$z(k+1) = P(z(k)) \tag{70}$$

Therefore proving the stability of the discrete function shown in (71) the stability of the periodic orbit can be determined in a fixed point $x = p$ since the period is $T = \zeta(p)$ and consequently $p = P(p)$ (Haddad and Chellaboina 2014).

Then making the Poincare map

$$P(z(k)) = 0 \tag{71}$$

With $t = T$, then the smaller positive period T obtained from this equation is the resulting period of the chattering oscillations. Making the Poincare map for each

Table 2 Period and frequency of the chattering oscillations

Example	Chattering period (s)	Chattering frequency (rad/s)
1	3.10	2.02
2	5.10	1.23
3	8.10	0.77

example of this chapter the following periods and chattering oscillation frequencies are obtained:

In Table 2 the chattering period and frequency is shown, as can be noticed, the frequency of the oscillations of Example 3 is the smallest of the three examples corroborating that the adaptive sliding mode controller avoids the chattering effects better than the other two approaches as it is seen in the simulation results. It is important to notice, that the PD + sliding mode controller implemented in Example 2 yield a small chattering frequency in comparison with the second order sliding mode controller of Example 1, so this combined control strategy yields better results than the approach shown in Sect. 2. With this analysis the results obtained by simulation were proved by an analytical method, therefore the proposed adaptive sliding mode controller for the Furuta pendulum yields better results than the other control strategies as it is corroborated in Sect. 4 due to the chattering avoidance properties of this controller.

6 Discussion

In this chapter, three control strategies are shown for the control and stabilization of the Furuta pendulum. A SOSMC, a PD + SMC; and the proposed control strategy of this paper, the ASMC for the Furuta pendulum is shown. In order to verify the performance and properties of each controller there are two important properties that must be considered to evaluate the system performance; these parameters are the convergence time, chattering period and frequency.

The converge time can be computed considering the Lyapunov function V as follow (Shtessel et al. 2014):

$$t_r \leq \frac{2V^{\frac{1}{2}}(0)}{\alpha} \tag{72}$$

where α is a positive constant and t_r is the upper bound of the convergence time. All the sliding mode controllers for the stabilization of the Furuta pendulum reach the sliding manifold with this convergence time, so all the alternatives shown in this chapter meet the requirement to control this underactuated mechanical system. This is an important property that must be considered in the design of sliding mode controller for coupled nonlinear systems because it ensures the stability and converge of the controlled variables in finite time and keeping the desired final value in steady state. All the simulation experiments shown in this chapter reach the sliding manifold in a defined convergence time, proving that these control strategies are suitable for the stabilization of the Furuta pendulum and similar mechanical underactuated systems. The main purpose of this chapter is to show the design and analysis of an adaptive sliding mode controller for the Furuta pendulum, but some control alternatives were developed to show the evolution and comparisons of the proposed control strategy with similar control techniques. The second order or higher order sliding mode control was chosen because of its chattering avoidance properties while ensuring the convergence of the sliding mode variables in finite time, even when this alternative has been extensively used in different kind of mechanical systems (Nersesov et al. 2010; Chang 2013) it has some disadvantages in comparison with other sliding mode control variations. Instead of implementing an classical sliding mode algorithm, such as the twisting and super twisting algorithms, in this example a higher order sliding mode control law was implemented in order to make the system variables to reach the sliding manifold in a defined time.

A proportional derivative + sliding mode controller for the Furuta pendulum yields better results than the conventional second order sliding mode controller, due to the best controller characteristics of the linear PD control law and the nonlinear sliding mode control, reaching the sliding manifold in a finite time ensuring the convergence of the controlled variables in steady state and keeping these variables in the final value as it is defined by the designer.

The adaptive sliding mode controller for the Furuta pendulum shown in this chapter is designed taking in count the mechanical system properties of the model, designing the required adaptive gain sliding mode control law to make the sliding surface to reach zero in finite time, this condition is met due to the adaptive gain was designed by the Lyapunov stability conditions making it suitable for the control of this kind of underactuated mechanism. Therefore it was proven that an effective adaptive sliding mode technique can be implemented for this kind of mechanical systems with the desired stability conditions for the model, where the sliding manifold reaches the origin in a determined time as expected.

Chattering is the second effect that is analyzed in this chapter, where as it is explained before this unwanted effect can deteriorate the system performance. Even when high order sliding mode control strategy has improved the chattering avoidance properties of control systems, this unwanted effect is still found, then as the chattering analysis show in the previous section, this effect is found in the system, but it is not so harmful as it is found in other kinds of classical sliding mode control implementations. The proportional derivative plus sliding mode controller for the Furuta pendulum, has better chattering avoidance properties in comparison with the second order sliding mode control due to improvement of a standard SMC with a PD control law, even when it is a simple control strategy, this combined SMC strategy has better chattering avoidance and tracking error properties. The proposed adaptive sliding mode control strategy for the stabilization of the Furuta pendulum is proved to stabilizes this mechanical system decreasing the chattering effects on the system due to the adaptive gain of the controller that makes it more robust and reliable in comparison with the previous control strategies shown in this chapter, for this reason, this strategy is proved to be superior to the previous control approaches, where the stability, small chattering and convergence of the sliding manifold in finite time make it the best alternative for the control of this kind of underactuated system.

7 Conclusions

In this chapter the design of an adaptive sliding mode controller for the stabilization of the Furuta pendulum is shown. Apart from this control approach, the stabilization of this mechanism by a second order sliding mode controller and a proportional derivative plus sliding mode controller are derived for comparison purposes and to explain the fundamentals of the proposed control technique. The second order sliding mode controller for this mechanism is explained due to its disturbance rejection properties making this control strategy ideal for the stabilization of this kind of mechanism; a higher order sliding mode control law design is proposed instead of standard SMC algorithms such as the twisting and super twisting algorithms. This fact allows the design of a chattering avoidance control technique where this effect is decreased by the selection of an appropriate control law that makes the system stable and makes the controlled variables, such as positions and velocities, to reach the desired values in steady state. The second sliding mode approach shown in this chapter deals with the design of a proportional derivative plus sliding mode controller for the Furuta pendulum, where this combined control structure stabilizes the system due to the linear PD action and the nonlinear sliding mode control, decreasing the chattering effects on the system and making the sliding variables to reach the origin in finite time. This control approach is more reliable and convenient for the stabilization of this underactuated mechanism due to its chattering avoidances, small tracking error and because it drives the positions and velocities of the system to reach and keep the desired values in steady state.

The adaptive sliding mode controller for the Furuta pendulum was designed according to the stability properties of the system ensuring that the adaptive gain of the controller meets the Lyapunov stability theorem requirements. As it is proved analytically and confirmed by a simulation example, this strategy is more convenient in comparison with the other control approaches explained in this chapter, due to its disturbance rejection properties, chattering avoidance and smaller tracking error than the other control approaches explained in this chapter.

For future research, other sliding mode control variations suitable for this kind of underactuated mechanical systems, such as integral sliding mode control and disturbance rejection control, will be investigated due to the importance of this control approach in several fields such as aeronautic, aerospace, robotics, mechatronics systems. Another issue to be considered in future research is the robustness analysis of the system considering the chattering avoidance properties and other characteristics of underactuated systems when different control approaches, such as passivity based or backstepping control are implemented.

References

Bartolini, G., Ferrara, A.: Multi-input sliding mode control of a class of uncertain nonlinear systems. IEEE Trans Autom Control **41**(11), 1662–1666 (1996)

Bartolini, G., Ferrara, A., Usai, E.: Chattering avoidance by second-order sliding mode control. IEEE Trans Autom Control **43**(2), 241–246 (1998)

Boiko, I., Fridman, L.: Analysis of chattering in continuous sliding-mode controllers. IEEE Trans Autom Control **50**(9), 1442–1446 (2005)

Boiko, I., Fridman, L., Pisano, A., Usai, E.: Performance analysis of second-order sliding-mode control systems with fast actuators. IEEE Trans Autom Control **52**(6), 1053–1059 (2007)

Boiko I, Fridman L, Pisano A, Usai E (2008) A comprehensive analysis of chattering in second order sliding mode control systems. En: modern sliding mode control theory. Lecture notes in control and information sciences. Springer, Berlin, pp 23–49

Chang, J.L.: Dynamic compensator-based second-order sliding mode controller design for mechanical systems. IET Control Theory Appl **7**(13), 1675–1682 (2013)

Chang, Y.: Adaptive sliding mode control of multi-input nonlinear systems with perturbations to achieve asymptotical stability. IEEE Trans Autom Control **54**(12), 2863–2869 (2009)

Cheng CC, Guo CZ (2010) Design of adaptive sliding mode controllers for systems with mismatched uncertainty to achieve asymptotic stability. In: Proceedings of the IEEE American control conference, Baltimore, MD, USA

Chen, X., Shen, W., Cao, C., Kapoor, A.: A novel approach for state of charge estimation based on adaptive switching gain sliding mode observer in electric vehicles. J Power Sour **246**(15), 667–678 (2014)

Estrada A, Plestan F (2013) Second order sliding mode output feedback control with switching gains application to the control of a pneumatic actuator. J Franklin Inst

Fei, J., Wu, D.: Adaptive sliding mode control using robust feedback compensator for MEMS gyroscope. Math Probl Eng **2013**(2013), 1–10 (2013)

Ferrara A, Capisani LM (2012) Second order sliding modes to control and supervise industrial robot manipulators. Sliding modes after the first decade of the 21st century. Lecture notes in control and information sciences. Springer, Berlin, pp 541–567

Fossas E, Ras A (2002) Second order sliding mode control of a buck converter. In: Proceedings of the 41st IEEE conference on decision and control, Las Vegas, Nevada

Fridman L (2012) Sliding mode enforcement after 1990: main results and some open oroblems. Sliding modes after the first decade of the 21st century. Lecture notes in control and information sciences. Springer, Berlin, pp 3–57

Fu Y-C, Lin J-S (2005) Nonlinear backstepping control design of the furuta pendulum. In: Proceedings of the 2005 IEEE conference on control applications, Toronto, Canada

Garcia-Alarcon, O., Puga-Guzman, S., Moreno-Valenzuela, J.: On parameter identification of the furuta pendulum. Proced Eng **35**(2012), 77–84 (2012)

Gracia, L., Sala, A., Garelli, F.: Robot coordination using task-priority and sliding-mode techniques. Robot Comput Integr Manuf **30**(1), 74–89 (2014)

Haddad WM, Chellaboina V (2014) Nonlinear dynamical systems and control: a lyapunov based approach. Princeton University Press, Princeton

Jing Y et al (2009) Adaptive global sliding mode control strategy for the vehicle antilock braking systems. In: Proceedings of the IEEE American control conference. St. Louis, MO, USA

Kunusch C, Puleston P, Mayosky M (2012) Fundamentals of sliding-mode control design. Sliding-mode control of PEM fuel cells. Springer, New York, pp 35–71

La Hera, P.X., Freidovich, L.B., Shiriaev, A.S., Mettin, U.: New approach for swinging up the furuta pendulum: theory and experiments. Mechatronics **19**(8), 1240–1250 (2009)

Levant, A.: Quasi-continuous high-order sliding-mode controllers. IEEE Trans Autom Control **50**(11), 1812–1816 (2005)

Li H, Bai Y, Su Z (2011) Adaptive sliding mode control of electromechanical actuator with improved parameter estimation. In: Proceedings of the IEEE Asian control conference (ASCC). Kaohsiung, Taiwan

Liu J, Laghrouche S, Harmouche M, Wack M (2013) Adaptive-gain second-order sliding mode observer design for switching power converters. Control Eng Pract 2–8

Liu, M.: Decentralized control of robot manipulators: nonlinear and adaptive approaches. IEEE Trans Autom Control **44**(2), 357–363 (1999)

Moreno JA (2012) Lyapunov approach for analysis and design of Ssecond order sliding mode algorithms. Sliding modes after the first decade of the 21st century. Lecture notes in control and information sciences. Springer, Berlin, pp 113–149

Nersesov SG, Ashrafiuon H, Ghorbanian P (2010) On the stability of sliding mode control for a class of underactuated nonlinear systems.. In: Proceedings of the IEEE American control conference marriott waterfront. Baltimore, MD, USA

Ouyang, P.R., Acob, J., Pano, V.: PD with sliding mode control for trajectory tracking of robotic system. Robot Comput Integr Manuf **30**(2), 189–200 (2014)

Punta E (2006) Multivariable second order sliding mode control of mechanical systems. In: Proceedings of the 45th IEEE conference on decision and control. San Diego, CA, USA

Ramirez-Neria M, Sira-Ramirez H, Garrido-Moctezuma R, Luviano-Jurez A (2013) Linear active disturbance rejection control of underactuated systems: the case of the furuta pendulum. ISA Trans 1–8 (available online)

Rundell, A.E., Drakunov, S.V., DeCarlo, R.A.: A sliding mode observer and controller for stabilization of rotational motion of a vertical shaft magnetic bearing. IEEE Trans Control Syst Technol **4**(5), 598–608 (1996)

Shkolnikov IA, Shtessel YB, Brown MDJ (2001) A second-order smooth sliding mode control. In: Proceedings of the 40th IEEE conference on decision and control. Orlando, Florida, USA

Shtessel Y, Edwards C, Fridman L, Levant A (2014) Introduction: intuitive theory of sliding mode control. Sliding mode control and observation control engineering. Springer, pp 1–42

Su CY, Leung TH, Zhou QJ (1992) Force/motion control of constrained robots using sliding mode. IEEE Trans Autom Control 37(5):668–672

Utkin VI (2008) Sliding mode control: mathematical tools, design and applications. En: nonlinear and optimal control theory. Lecture notes in mathematics. Springer, Berlin, pp 289–347

Xiang C, Siow SY (2004) Decentralized control of robotic manipulators with neural networks. In: Proceedings of the IEEE, 8th international conference on control, automation, robotics and vision kunming. China, Dec 2004

Yao B, Tomizuka M (1994) Smooth robust adaptive sliding mode control of manipulators with guaranteed transient performance. In: Proceedings of the IEEE American control conference. Baltimore, MD, USA

Yu H, Ozguner U (2006) Adaptive seeking sliding mode control. In: Proceedings of the IEEE, 2006 American control conference. Minneapolis, Minnesota, USA

Zhihong, M., Paplinski, A.P., Wu, H.R.: A robust MIMO terminal sliding mode control scheme for rigid robotic manipulators. IEEE Trans Autom Control **39**(12), 2464–2469 (1994)

Optimal Sliding and Decoupled Sliding Mode Tracking Control by Multi-objective Particle Swarm Optimization and Genetic Algorithms

M. Taherkhorsandi, K.K. Castillo-Villar, M.J. Mahmoodabadi, F. Janaghaei and S.M. Mortazavi Yazdi

Abstract The objective of this chapter is to present an optimal robust control approach based upon smart multi-objective optimization algorithms for systems with challenging dynamic equations in order to minimize the control inputs and tracking and position error. To this end, an optimal sliding and decoupled sliding mode control technique based on three multi-objective optimization algorithms, that is, multi-objective periodic CDPSO, modified NSGAII and Sigma method is presented to control two dynamic systems including biped robots and ball and beam systems. The control of biped robots is one of the most challenging topics in the field of robotics because the stability of the biped robots is usually provided laboriously regarding the heavily nonlinear dynamic equations of them. On the other hand, the ball and beam system is one of the most popular laboratory models used widely to challenge the control techniques. Sliding mode control (SMC) is a nonlinear controller with characteristics of robustness and invariance to model parametric uncertainties and nonlinearity in the dynamic equations. Hence, optimal sliding mode tracking control tuned by multi-objective optimization algorithms is utilized in this study to present a controller having exclusive qualities, such as robust performance and optimal control inputs. To design an optimal control approach, multi-objective particle swarm optimization (PSO) called multi-objective periodic CDPSO introduced by authors in their previous research and two notable smart multi-objective optimization algorithms, i.e. modified NSGAII and the Sigma method are employed to ascertain the optimal parameters of the control approach with regard to the design criteria. In comparison, genetic algorithm optimization operates based upon reproduction, crossover and mutation; however particle swarm optimization functions by means of

M. Taherkhorsandi (✉) · K.K. Castillo-Villar
Department of Mechanical Engineering, The University of Texas at San Antonio, San Antonio, TX 78249, USA
e-mail: m.taherkhorsandi@gmail.com; milad.taherkhorsandi@utsa.edu

M.J. Mahmoodabadi · F. Janaghaei · S.M. Mortazavi Yazdi
Department of Mechanical Engineering, Sirjan University of Technology, Sirjan, Iran
e-mail: mahmoodabadi@sirjantech.ac.ir

A.T. Azar and Q. Zhu (eds.), *Advances and Applications in Sliding Mode Control systems*, Studies in Computational Intelligence 576, DOI 10.1007/978-3-319-11173-5_2

a convergence and divergence operator, a periodic leader selection method, and an adaptive elimination technique. When the multi-objective optimization algorithms are applied to the design of the controller, there is a trade-off between the tracking error and control inputs. By means of optimal points of the Pareto front obtained from the multi-objective optimization algorithms, plenty of opportunity is provided to engineers to design the control approach. Contrasting the Pareto front obtained by multi-objective periodic CDPSO with two noteworthy multi-objective optimization algorithms i.e. modified NSGAII and Sigma method dramatizes the excellent performance of multi-objective periodic CDPSO in the design of the control method. Finally, the optimal sliding mode tracking control tuned by CDPSO is applied to the control of a biped robot walking in the lateral plane on slope and the ball and beam system. The results and analysis prove the efficiency of the control approach with regard to providing optimal control inputs and low tracking and position errors.

1 Introduction

Two-legged robots named biped robots are the most similar kind of robots to human. As a unique advantage of this robot in comparison to other robots, they can be used in any situation where working of a human is unsafe or hazardous (Cha et al. 2011; Dehghani et al. 2013; Ding et al. 2013; Feng et al. 2013; Lee et al. 2014). One important challenging characteristic of this robot is having the dynamic equations which are nonlinear and demanding to control. To this end, it is crucial to employ a nonlinear controller which is robust with regard to disturbances and uncertainty in order to provide stable walking for biped robots (Andalib Sahnehsaraei et al. 2013). In this regard, sliding mode control is an effectual control method in terms of providing low tracking error in contrast to PID control (Lu et al. 2011) and linear feedback control (Basin et al. 2012). Owing to its exclusive benefits, a number of researchers have applied it to a variety of problems with challenging dynamic equations and successful application of it has been reported. For instance, Nizar et al. applied the sliding mode controllers for the time delay systems (Nizar et al. 2013). Han et al. proposed the sliding mode control of T-S fuzzy descriptor systems with time-delay (Han et al. 2012). Yakut applied an intelligent sliding mode controller with moving sliding surface for overhead cranes (Yakut 2014). Eker utilized the second-order sliding mode controller with PI sliding surface for an electromechanical plant (Eker 2012). Sira-Ramirez et al. proposed a robust input-output sliding mode control for the buck converter (Sira-Ramirez et al. 2013). Bayramoglu and Komurcugil used time-varying sliding-coefficient-based terminal sliding mode control methods for a class of fourth-order nonlinear systems (Bayramoglu and Komurcugil 2013). Li et al. proposed a sliding mode controller for uncertain chaotic systems with input nonlinearity (Li et al. 2012). Yin et al. designed a sliding mode controller for a class of fractional-order chaotic systems (Yin et al. 2012). Zhang et al. applied the second-order terminal sliding mode controller for a hypersonic vehicle in cruising flight with sliding mode disturbance observer (Zhang et al. 2013). Moreover, it is reliable and efficient when it is applied for the control of robots (Hu et al. 2012; Sun et al. 2011). In illustration, Nikkhah et al. utilized a robust sliding mode tracking control algorithm for a biped robot modeled as a five-link planar robot with four actuators following a human-like gait trajectory in the sagittal plane (Nikkhah et al. 2007). Lin et al. proposed a hybrid control approach based upon the sliding mode method and a recurrent cerebellar model articulation controller to control biped robots (Lin et al. 2007). Moreover, the Taylor linearization approach was used to enhance the learning ability of the recurrent cerebellar model articulation controller. Since chattering is a crucial unresolved issue in designing sliding mode controllers, some researchers have tried to reduce the amount of chattering in actuators (Mondal and Mahanta 2013a; Pourmahmood Aghababa and Akbari 2012; Ramos et al. 2013; Cerman and Husek 2012; Lin et al. 2011; Singla et al. 2014). In particular, Shahriari kahkeshi et al. constructed smooth sliding mode control for a class of high-order nonlinear systems having no prior knowledge about uncertainty (Shahriari kahkeshi et al. 2013). They could eliminate the chattering problem completely via proposing a scheme which involves an adaptive fuzzy wavelet neural controller to construct equivalent control term and an adaptive proportional-integral (A-PI) controller for applying switching term to deliver smooth control input. Adhikary and Mahanta proposed an integral backstepping sliding mode control approach to control underactuated systems (Adhikary and Mahanta 2013). Rejecting matched and mismatched uncertainties, providing a chattering free control law, and using less control effort than the sliding mode controller were reported as the most important advantages of the proposed controller. Liu used Lie-group differential algebraic equation method to design a sliding mode controller through adding a compensated control force which resulted in steering rapidly and enforcing continuously the state trajectory on the sliding surface (Liu 2014). The proposed control methodology is chattering-free for a class of regulator problems and finite-time tracking problems of nonlinear systems. Mondal and Mahanta used the derivative of the control input in the proposed control law to design an adaptive integral higher order sliding mode controller for uncertain systems (Mondal and Mahanta 2013b). The actual control signal gained through integrating the derivative control signal is chattering free and smooth.

Designing the parameters of control approaches is an interesting and challenging issue in industry and academia. Multi-objective optimization algorithms are appropriate approaches to gain these parameters via considering both tracking error and control effort. In particular, the genetic algorithm and particle swarm optimization are two notable effectual optimization algorithms to gain optimal solutions (Castillo-Villar et al. 2012, 2014; Martínez-Soto et al. 2013; Elshazly et al. 2013; Aziz et al. 2013). In the literature, particle swarm optimization has been successfully used to augment the performance of type-1 and type-2 fuzzy control (Martínez-Soto et al. 2013), fractional fuzzy control (Pan et al. 2012), PID control (Jadhav and Vadirajacharya 2012), constrained multivariable predictive controllers (Júnior 2014), and other controllers (Lari et al. 2014). PSO, first introduced by Kennedy and Eberhart, is one of the modern smart heuristic algorithms (Kennedy and Eberhart 1995). It is a robust optimization algorithm developed via simulation of simplified social systems to solve nonlinear optimization problems (Angeline 1998). Successful applications of this robust optimization algorithm have been reported not only in the area of control theory but also in a wide variety field of research. For instance, in Wang and Jun Zheng (2012), a new particle swarm optimization algorithm was applied to optimum design of the armored vehicle scheme. In Yildiz and Solanki (2012), multi-objective optimization of vehicle crashworthiness was performed using a new particle swarm based approach. In Kim and Son (2012), a probability matrix based particle swarm optimization was applied for the capacitated vehicle routing problem. In Belmecheri et al. (2013), the particle swarm optimization algorithm was used for a vehicle routing problem with heterogeneous fleet, mixed backhauls, and time windows. In Nejat et al. (2014), the optimization of the airfoil shape was performed using the improved multi-objective territorial particle swarm algorithm with the objective of improving stall characteristics. In Hart and Vlahopoulos (2010), an integrated multidisciplinary particle swarm optimization approach was introduced for conceptual design of ships. In Aparecida de Pina et al. (2011), the particle swarm optimization algorithm was applied to the design of steel catenary risers in a lazy-wave configuration. In Nwankwor et al. (2013), hybrid differential evolution and particle swarm optimization were utilized for optimal well placement. In Zhang et al. (2013), the sequential quadratic programming particle swarm optimization was applied for wind power system operations considering emissions. In Zheng and Wu (2012), Power optimization of gas pipelines was performed with aid of an improved particle swarm optimization algorithm. In Biswas et al. (2013), constriction factor based particle swarm optimization was introduced for analyzing tuned reactive power dispatch. Short calculation time and more stable convergence are two important characteristics of the PSO technique (Eberhart and Shi 1998; Yoshida et al. 2000). Depending on the case study, PSO can show better performance than genetic algorithm optimization (Lin and Lin 2012). Appropriate performance of PSO in combination with genetic algorithm optimization has been reported in the literature as a hybrid optimization algorithm (Mousa et al. 2012; Kuo et al. 2012). Furthermore, PSO is an effectual algorithm to solve multi-objective problems (Carvalho and Pozo 2012). The techniques, a self-adaptive diversity control strategy (Wang and Tang 2012), chaotic local search (Jia et al. 2011), and migration of some particles from one complex to another have been employed to prevent from premature convergence of PSO (Gang et al. 2012). Lately, several methods have been proposed to develop the PSO algorithm to deal with multi-objective optimization problems. To this end, dynamic neighborhood PSO (Hu and Eberhart 2002), dominated tree (Fieldsend and Singh 2002), Sigma method (Mostaghim and Teich 2003), vector evaluated PSO (Parsopoulos et al. 2004), etc. were introduced to solve the multi-objective optimization problems. The principal difference among these methods is the leader selection technique.

In the present chapter, multi-objective periodic CDPSO introduced in authors' previous works (Mahmoodabadi et al. 2011, 2012a) and two prominent smart evolutionary algorithms (Mostaghim and Teich 2003; Atashkari et al. 2007) are used to eliminate the tedious trial-and-error process and design the optimal nonlinear sliding mode tracking control. This controller is applied to a biped robot modeled and walking in the lateral plane on slope (Mahmoodabadi et al. 2014a, b). The outline of the rest of this chapter is as follows: Sect. 2 discusses the multi-objective particle swarm optimization including the convergence and divergence operators, the periodic leader selection approach, and the adaptive elimination technique. Section 3 involves the control approach where Canonical and Non-canonical forms of both sliding mode control and decoupled sliding mode control are discussed. The dynamic model of the biped robot and the Pareto design of sliding mode control of the biped robot based on multi-objective periodic CDPSO, modified NSGAII and sigma method are presented in Sect. 4. Section 5 includes the dynamic model of the Ball and Beam system and the Pareto design of decoupled sliding mode control for the Ball and Beam system using multi-objective periodic CDPSO, modified NSGAII and sigma method. Section 6 presents the conclusions of this study.

2 Multi-objective Particle Swarm Optimization

In this chapter, multi-objective periodic CDPSO is used to address the problem of the proper selection of parameters of sliding mode control. Indeed, this optimization method was successfully employed to acquire the Pareto frontiers of non-commensurable objective functions in the design of linear state feedback controllers (Mahmoodabadi et al. 2011) and the suspension system for a vehicle vibration model (Mahmoodabadi et al. 2012b). This method is a combination of the particle swarm optimization, convergence and divergence operators. Moreover, a new leader selection method is applied to produce a set of Pareto optimal solutions which has good diversity and distribution. The archive is pruned in this algorithm by implementing an adaptive elimination technique. The algorithm was named multi-objective periodic CDPSO. PSO, convergence divergence operator, periodic leader selection method and adaptive elimination technique are described concisely, as follows:

2.1 Particle Swarm Optimization

PSO is a population-based evolutionary algorithm and is inspired by the simulation of social behavior (Kennedy and Eberhart 1995). Although PSO had been initially employed to balance weights in neural networks (Eberhart et al. 1996), it became a very popular global optimizer, mostly in the problems where the decision variables were real numbers (Engelbrecht 2002, 2005). Each candidate solution in PSO is associated with a velocity (Kennedy and Eberhart 1995; Ratnaweera and Halgamuge 2004), and it is assumed that the particles will move toward better solution areas. Mathematically, the particles are functioning based upon the following equations.

$$\vec{x}_i(t+1) = \vec{x}_i(t) + \vec{v}_i(t+1) \tag{1}$$

$$\vec{v}_i(t+1) = W\vec{v}_i(t) + C_1 r_1(\vec{x}_{pbest_i} - \vec{x}_i(t)) + C_2 r_2(\vec{x}_{gbest} - \vec{x}_i(t)) \tag{2}$$

In which, $\vec{x}_i(t)$ and $\vec{v}_i(t)$ stand for the position and velocity of particle i at the time step (iteration) t. $r_1, r_2 \in [0, 1]$ denote random values. C_1 stands for the cognitive learning feature and represents the attraction that a particle has toward its own success. C_2stands for the social learning feature and represents the attraction that a particle has toward the success of the entire swarm. It was obtained that the best solutions were gained when C_1 is linearly decreased and C_2 is linearly increased over the iterations (Ratnaweera and Halgamuge 2004). W stands for the inertia weight and controls the impact of the previous history of velocities on the current velocity of particle i. Based on experimental results, PSO functioning improves when the inertia weight diminishes linearly over iterations (Kennedy and Eberhart 1995). Moreover, $\vec{x}_{pbest_i}$ is the personal best position of the particle i and $\vec{x}_{gbest}$ represents the position of the best particle of the whole swarm.

2.2 The Convergence Operator

In the present chapter, the convergence formula, which contains four parent particles proposed in Mahmoodabadi et al. (2011, 2012a) is employed. Let $\rho \in [0, 1]$ be a random number. If $\rho \le P_{\text{Convergence}}$ ($P_{\text{Convergence}}$ is convergence probability), then one of the following operators should be operated to generate the new particle position $\vec{x}_i(t+1)$ from the old particle position $\vec{x}_i(t)$:

If fitness $\vec{x}_i(t)$ is smaller than fitness $\vec{x}_j(t)$ and fitness $\vec{x}_k(t)$ then:

$$\vec{x}_i(t+1) = \vec{x}_{gbest} + \sigma_1 \left(\frac{\vec{x}_{gbest}}{\vec{x}_i(t)} \right) (2\vec{x}_i(t) - \vec{x}_j(t) - \vec{x}_k(t)) \tag{3}$$

If fitness $\vec{x}_j(t)$ is smaller than fitness $\vec{x}_i(t)$ and fitness $\vec{x}_k(t)$ then:

$$\vec{x}_i(t+1) = \vec{x}_{gbest} + \sigma_2 \left(\frac{\vec{x}_{gbest}}{\vec{x}_i(t)} \right) (2\vec{x}_j(t) - \vec{x}_i(t) - \vec{x}_k(t)) \tag{4}$$

If fitness $\vec{x}_k(t)$ is smaller than fitness $\vec{x}_j(t)$ and fitness $\vec{x}_i(t)$ then:

$$\vec{x}_i(t+1) = \vec{x}_{gbest} + \sigma_3 \left(\frac{\vec{x}_{gbest}}{\vec{x}_i(t)} \right) (2\vec{x}_k(t) - \vec{x}_j(t) - \vec{x}_i(t)) \tag{5}$$

In which, particles $\vec{x}_j(t)$ and $\vec{x}_k(t)$ are chosen from swarm by a uniform selection approach. σ_1, σ_2, and σ_3 are arbitrary numbers chosen from [0, 1] and $\vec{x}_{gbest}$ stands for the position of the best particle of the entire swarm. After calculating Eqs. (3), (4) or (5), the superior one between $\vec{x}_i(t)$ and $\vec{x}_i(t+1)$ should be selected. If $\rho \ge P_{\text{Convergence}}$, then no convergence operation is operated for $\vec{x}_i(t)$.

2.3 The Divergence Operator

The divergence operator provides a feasible leap on some selected particles. Let $\vartheta \in [0, 1]$ be an arbitrary number. If $\vartheta \leq P_{Divergence}$, ($P_{Divergence}$ is divergence probability) and particle $\vec{x}_i(t)$ was not improved by the convergence operator, the following divergence operator is operated to create a new particle.

$$\vec{x}_i(t+1) = \text{Norm rand}(\vec{x}_i(t), S_D) \tag{6}$$

Norm rand($\vec{x}_i(t)$, S_D) creates arbitrary numbers from the normal distribution with mean parameter $\vec{x}_i(t)$ and standard deviation parameter S_D (S_D is a positive constant). If particle $\vec{x}_i(t)$ was augmented by convergence operator or $\vartheta \geq P_{Divergence}$, then no divergence operation is operated. More features of this operator are discussed in Mahmoodabadi et al. (2011, 2012a).

2.4 The Periodic Leader Selection Approach

This methodology is based on the density measures. A neighborhood radius $R_{neighborhood}$ is defined for leaders and if their Euclidean distance (measured in the objective domain) is less than $R_{neighborhood}$, two leaders are neighbors. In this respect, the number of neighbors of each leader is computed in the objective function area. The particle with fewer neighbors is preferred as the leader; even though, the leader position and its density will change after several iterations. Hence, the leader selection operation should be repeated and a new leader must be ascertained. Thus, the maximum iteration is divided into several equal periods and each period has the same iteration T. The relation among maximum iteration, number of periods and T satisfies Eq. (7):

$$\text{maximum iteration} = \text{number of periods} \times T \tag{7}$$

In each period, the leader selection operation could be performed, and the non-dominated solution which has fewer neighbors is preferred as the leader. Moreover, if a particle dominates the leader in the start of the iteration in a period, then this particle will be regarded as a new leader.

2.5 The Adaptive Elimination Technique

This approach is employed to prune the archive; and in this approach, the archive's members have an elimination radius which equals $\varepsilon_{elimination}$. If the Euclidean distance (in the objective function space) between two particles is less than $\varepsilon_{elimination}$, then one of them will be omitted. The following equation is used to determine the value of $\varepsilon_{\text{elimination}}$ named adaptive $\varepsilon_{\text{elimination}}$:

$$\varepsilon_{\text{elimination}} = \frac{\text{t}}{\zeta \times \text{maximum iteration}} \tag{8}$$

In which, ζ is a positive constant, t is the current iteration number, and maximum iteration presents the maximum number of permissible iterations (Mahmoodabadi et al. 2011, 2012a).

3 The Control Approach

Sliding Mode Control (SMC) is an effective control methodology for nonlinear systems. One of the main advantages of SMC is that the uncertainties and external disturbances of the system can be handled under the invariance characteristics of sliding conditions. Nevertheless, the SMC technique can be applied only to the systems with the canonical form. In this respect, the basic idea of Decoupled Sliding Mode Control (DSMC) is proposed to design a controller for systems with the non-canonical form. However, for the optimum design of DSMC, it is difficult to determine the parameters of the sliding surface. This problem could be solved by using the evolutionary optimization techniques. Fleming and Purshouse (2002) is an appropriate reference to overview the application of the evolutionary algorithms in the field of the design of controllers. In particular, the design of controllers in Fonseca and Fleming (1994) and Sanchez et al. (2007) was formulated as a multi-objective optimization problem and solved using Genetic Algorithms (GAs). Furthermore, in Javadi-Moghaddam and Bagheri (2010), the GA was utilized to select the parameters of SMC for an underwater remotely operated vehicle. In Ker-Wei and Shang-Chang (2006), the sliding mode control configurations were designed for an alternating current servo motor while a Particle Swarm Optimization (PSO) algorithm was used to select the parameters of the controller. Also, PSO was applied to tune the linear control gains in Gaing (2004); Qiao et al. (2006). These works have shown that PSO is a fast and reliable tool to design the optimal controllers, and also can outperform other evolutionary algorithms. In Chen et al. (2009), three parameters associated with the control law of the sliding mode controller for the inverted pendulum system were properly chosen by a modified PSO algorithm. Wai et al. proposed a total sliding-model-based particle swarm optimization approach to design a controller for the linear induction motor (Wai et al. 2007). More recently, in Gosh et al. (2011), an ecologically inspired direct search method was applied to solve the optimal control problems with Bezier parameterization. Moreover, in Tang et al. (2011), a controllable probabilistic particle swarm optimization (CPPSO) algorithm was applied to design a memoryless feedback controller.

3.1 Canonical and Non-canonical Forms

Consider a forth-order nonlinear system, which could be represented by the following canonical form.

$$\begin{aligned} \dot{x}_1 &= x_3 \\ \dot{x}_2 &= x_4 \\ \dot{x}_3 &= f_1(x) + b_1(x) u_1 \\ \dot{x}_4 &= f_2(x) + b_2(x) u_2 \end{aligned} \tag{9}$$

where $x = \left[x_1\ x_2\ x_3\ x_4\right]^T$ is the state vector, $f_1(x)$, $f_2(x)$, $b_1(x)$, and $b_2(x)$ are nonlinear functions. u_1 and u_2 are control inputs.

This type of forth-order systems with the canonical form could be controlled by using many kinds of techniques, such as fuzzy control, proportional–integral–derivative (PID) control, sliding mode control, etc. In fact, the control laws u_1 and u_2 can be easily designed to control the system introduced by Eq. (9). However, for some nonlinear models such as the ball and beam system, the system dynamic equations are not in the canonical form. The state space model of a system with the non-canonical form is presented, as follows.

$$\begin{aligned} \dot{x}_1 &= x_3 \\ \dot{x}_2 &= x_4 \\ \dot{x}_3 &= f_1(x) + b_1(x) u \\ \dot{x}_4 &= f_2(x) + b_2(x) u \end{aligned} \tag{10}$$

where $x = \left[x_1\ x_2\ x_3\ x_4\right]^T$ stands for the state vector, $f_1(x)$, $f_2(x)$, $b_1(x)$, and $b_2(x)$ are nonlinear functions. u is a control input. The control techniques mentioned above could control only one of the subsystems in Eq. (10). In other words, these methodologies cannot simultaneously control both subsystems by only one control input u. Hence, the idea of decoupling is employed to design a control law u to control the whole system. In the following sections, the general concepts of the sliding mode control and decoupled sliding mode control are briefly presented.

3.2 Sliding Mode Control for Canonical Forms

The basic concepts of sliding mode control are presented in Wang and Jun Zheng (2012). By regarding the dynamical system introduced according Eq. (9) and having desired trajectories $x_{1d}(t)$ and $x_{2d}(t)$, the errors are defined as $e_1(t) = x_1(t) - x_{1d}(t)$ and $e_2(t) = x_2(t) - x_{2d}(t)$. After a reaching phase, the sliding mode controller forces the system to track the following sliding surfaces.

$$s_1(x) = \dot{e}_2 + \lambda_1 e_1 = x_3 - x_{3d} + \lambda_1 (x_1 - x_{1d}) = 0$$
$$s_2(x) = \dot{e}_2 + \lambda_2 e_2 = x_4 - x_{4d} + \lambda_2 (x_2 - x_{2d}) = 0 \tag{11}$$

where $x_{3d}(t) = \dot{x}_{1d}(t), x_{4d}(t) = \dot{x}_{2d}(t)$, and sliding constants λ_1 and λ_2 are strictly positive. In the steady state conditions, the system follows the desired trajectories when $S_1(x(t_{1reach})) = 0$ and $S_2(x(t_{2reach})) = 0$. t_{1reach} and t_{2reach} represent reaching times. Hence, suitable control actions based on sliding surfaces introduced by Eq. (11) would be achieved. Lyapunov functions could be chosen as Eq. (12).

$$V_1 = \frac{1}{2} S_1^2(x)$$
$$V_2 = \frac{1}{2} S_2^2(x) \tag{12}$$

with the following controller actions:

$$u_1 = \hat{u}_1 - K_1\, sign\, (S_1(x) b_1(x)) \text{ and } K_1 > 0$$
$$u_2 = \hat{u}_2 - K_2\, sign\, (S_2(x) b_2(x)) \text{ and } K_2 > 0 \tag{13}$$

where K_1 and K_2 are the design parameters or functions of $x(t)$ such that $K_1 = K_1(x)$ and $K_2 = K_2(x)$. $sign$represents sign function. $\hat{u}_1$ and $\hat{u}_2$ would be obtained using Eq. (14).

$$\hat{u}_1 = -b_1^{-1}(x)(f_1(x) - \ddot{x}_{1d} + \lambda_1 \dot{e}_1)$$
$$\hat{u}_2 = -b_2^{-1}(x)(f_2(x) - \ddot{x}_{2d} + \lambda_2 \dot{e}_2) \tag{14}$$

Derivations of Lyapunov functions introduced via Eq. (12) are written in the following forms:

$$\dot{V}_1 \leq -\eta_1 |S_1(x)|$$
$$\dot{V}_2 \leq -\eta_2 |S_2(x)| \tag{15}$$

where $K_1 > \eta_1$ and $K_2 > \eta_2$. Hence, $\dot{V}_1$ and $\dot{V}_2$ are negative definite in the switching surfaces. Moreover, if $x_1(t = 0) \neq x_{1d}(t = 0)$ and $x_2(t = 0) \neq x_{2d}(t = 0)$, Eq. (15) shows that $S_1(x) = 0$ and $S_2(x) = 0$ will be reached in the finite times t_{1reach} and t_{2reach}, respectively.

It is clear that with starting from initial conditions, trajectories reach the manifold $S_1(x) = 0$ and $S_2(x) = 0$ in the finite times and slide toward the origins of the error phase planes according to Eq. (11). But, function $sign$ in Eq. (13) causes the high frequency switching near the sliding surfaces. Thus, in order to reduce this chattering phenomenon, the sign function is replaced with the saturation function as follows.

$$u_1 = \hat{u}_1 - K_1\, sat(S_1(x)b_1(x)\phi_1) \text{ and } K_1 > 0$$
$$u_2 = \hat{u}_2 - K_2\, sat(S_2(x)b_2(x)\phi_2) \text{ and } K_2 > 0 \quad (16)$$

which sat is the saturation function. ϕ_1 represents the inverse of the width of boundary layer for S_1. ϕ_2 represents the inverse of the width of boundary layer for S_2.

3.3 Decoupled Sliding Mode Control for Non-canonical Forms

The sliding mode control technique described in the previous section could not be applied to a system with the form of Eq. (10), which is not in the canonical form and includes the coupled subsystems. The basic idea of the decoupled sliding mode control is the design of a control law such that the single input u simultaneously controls several subsystems to accomplish the desired performance. To achieve this goal, the following sliding surfaces are defined.

$$S_1(x) = \lambda_1(x_2 - x_{2d} - z) + x_4 - x_{4d} = 0 \quad (17)$$
$$S_2(x) = \lambda_2(x_1 - x_{1d}) + x_3 - x_{3d} = 0 \quad (18)$$

where variable Z is used to transfer S_2 to S_1. Furthermore, its value is proportional to S_2 and its range is proper to x_2. Comparing Eq. (17) with (11) shows that the control objectives for the subsystem are $x_2 = x_{2d} + z$ and $x_4 = x_{4d}$. On the other hand, Eq. (18) means that the control objectives are $x_1 = x_{1d}$ and $x_3 = x_{3d}$. Now, let the control law for Eq. (17) be the sliding mode control with a boundary layer which is similar to Eq. (16):

$$u_1 = \hat{u}_1 - G_{f1}\, sat\,(S_1(x)b_1(x)G_{s1}) \text{ and } G_{f1}, G_{s1} > 0 \quad (19)$$
$$\text{with } \hat{u}_1 = -b_1^{-1}(x)\,(f_1(x) - \ddot{x}_{2d} + \lambda_1 x_4 - \lambda_1 \dot{x}_{2d}) \quad (20)$$

Sliding constant λ_1 is strictly positive. Now, let the control law for Eq. (18) be another sliding mode controller with a boundary layer as follows.

$$z = G_{f2}\, sat\,(S_2(x)G_{s2}), \text{ and } 0 < G_{f2} < 1 \quad (21)$$

Note that in (21), z is a decaying oscillation signal due to $0 < G_{f2} < 1$. Moreover, in Eq. (17), if $S_1 = 0$, then $x_2 = x_{2d} + z$ and $x_4 = x_{4d}$.

Therefore, the control sequence is as follows. When $S_2 \to 0$, then $z \to 0$ in Eq. (17), and it forces Eq. (19) to generate a controller action for reducing S_2; as S_2 decreases, z decreases too. Hence, at the limit $S_2 \to 0$ with $x_1 \to x_{1d}$, then $z \to 0$ with $x_2 \to x_{2d}$, so $S_1 \to 0$, and the goal will be achieved.

4 Biped Robot

4.1 The Dynamic Model of the Biped Robot

The robot is modeled in the lateral plane by using a three-link-planar model. Figure 1 shows the model of the robot. The first link is anchored to the ground surface, while the third link moves freely along the lateral plane and the second link represents the head, arms and trunk. Four characteristics of mass, length, inertia and the center of gravity are used to define each link. Anthropometric parameters are obtained from Winter (1990) for a humanoid model which is 171 cm in height and 74 kg in weight and are illustrated in Table 1. The distance between two legs of the model ($2d_2$) is equal to 32.7 cm.

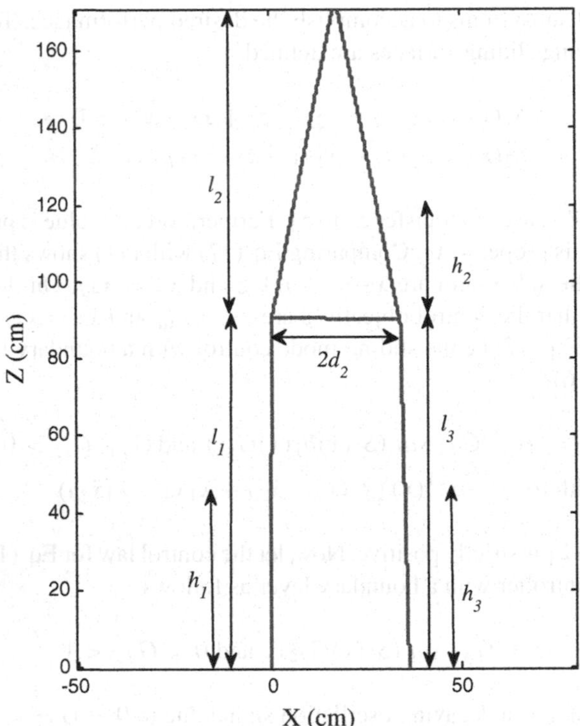

Fig. 1 The parameters of the robot based on the anthropometric table

Table 1 The anthropometric parameters of the model

	First link	Second link	Third link	Unit
Mass	$m_1 = 13.75$	$m_2 = 46.5$	$m_3 = 13.75$	kg
Inertia	$I_1 = 1.4$	$I_2 = 3.25$	$I_3 = 1.4$	kg m^2
Length	$l_1 = 0.91$	$l_2 = 0.8$	$l_3 = 0.91$	m
CG	$h_1 = 0.50$	$h_2 = 0.27$	$h_3 = 0.50$	m

The Newton-Euler method is employed to gain the dynamic equations of the model (Mahmoodabadi et al. 2014a, b). The dynamic equations of the model for θ_1, θ_2 and θ_3 are (Fig. 2):

$$
\begin{aligned}
I_1\ddot{\theta}_1 &= u_1 - u_2 + h_1 m_1 g \sin\theta_1 + l_1 \sin\theta_1\, g\,(m_2+m_3) + h_1 m_1\left(-h_1\ddot{\theta}_1\right) \\
&\quad + l_1 m_2\,[-\ddot{\theta}_1 l_1 + \ddot{\theta}_2\{d_2\sin(\theta_2-\theta_1) - h_2\cos(\theta_2-\theta_1)\} + \dot{\theta}_2^2\{d_2\cos(\theta_2-\theta_1) \\
&\quad + h_2\sin(\theta_2-\theta_1)\}] + l_1 m_3[-\ddot{\theta}_1 l_1 + \ddot{\theta}_2\{2d_2\sin(\theta_2-\theta_1) - \theta_3\,(l_2-h_3)\cos(\theta_3-\theta_1) \\
&\quad + \dot{\theta}_2^2\, 2d_2\cos(\theta_2-\theta_1) + \dot{\theta}_3^2\,(l_3-h_3)\sin(\theta_3-\theta_1)\}]
\end{aligned}
\tag{22}
$$

$$
\begin{aligned}
I_2\ddot{\theta}_2 &= u_2 - u_3 + m_2 d_2 g\cos\theta_2 + m_2 h_2 g\sin\theta_2 + 2m_3 d_2 g\cos\theta_2 \\
&\quad - \ddot{\theta}_1\,[\cos(\theta_2-\theta_1) m_2 h_2 l_1 - \sin(\theta_2-\theta_1) m_3 2 d_2 l_1 - \sin(\theta_2-\theta_1) m_2 d_2 l_1] \\
&\quad + \dot{\theta}_1^2[\sin(\theta_1-\theta_2) m_2 l_1 h_2 - \cos(\theta_2-\theta_1) 2 m_3 d_2 l_1 - \cos(\theta_2-\theta_1) m_2 d_2 l_1] \\
&\quad + \ddot{\theta}_2[-m_2 d_2^2 - 4 m_3 d_2^2 - m_2 h_2^2] + \ddot{\theta}_3[2 m_3 d_2 (l_3-h_3)\sin(\theta_2-\theta_3)] \\
&\quad - \dot{\theta}_3^2[2 m_3 d_2 (l_3-h_3)\cos(\theta_2-\theta_3)]
\end{aligned}
\tag{23}
$$

$$
\begin{aligned}
I_3\ddot{\theta}_3 &= (l_3-h_3)\, m_3 g\sin\theta_3 + u_3 - m_3\,(l_3-h_3)\, l_1\cos(\theta_3-\theta_1)\,\ddot{\theta}_1 \\
&\quad + m_3\,(l_3-h_3)\, l_1\sin(\theta_1-\theta_3)\,\dot{\theta}_1^2 + 2d_2 m_3\,(l_3-h_3)\sin(\theta_2-\theta_3)\,\ddot{\theta}_2 \\
&\quad + 2d_2 m_3\,(l_3-h_3)\cos(\theta_3-\theta_2)\,\ddot{\theta}_2 - m_3(l_3-h_3)^2\ddot{\theta}_3
\end{aligned}
\tag{24}
$$

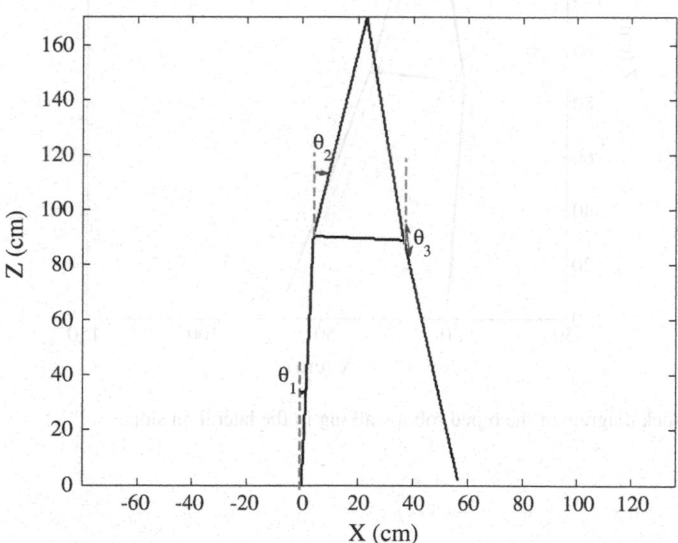

Fig. 2 The angles of the robot

The walking of biped robot in the lateral plane is periodic, and it can be divided into two phases (Mahmoodabadi et al. 2014a, b). These two phases are Double Support Phase (DSP) and Single Support Phase (SSP). DSP term is used for situations where the biped robot has two isolated contact surfaces with the floor. Indeed, this situation happens when the biped robot is supported by both feet. The time of this phase is regarded as 20 percent of the whole time. SSP term is employed for situations where the biped robot has only one contact surface with the floor. This situation occurs when the biped robot is supported with only one foot. According to Fig. 3, the biped robot passes DSP and SSP, respectively. The swing foot trajectory which has the first-order continuity is generated, and it maintains the ZMP on the inside of the support polygon. Then, the inverse kinematic is utilized to obtain the desired trajectories of the joints. The desired trajectories should have first-order and second-order continuity. The first-order derivative continuity guarantees smoothness of the joint velocity, while the second order continuity guarantees smoothness of the acceleration or torque on the joints (Mahmoodabadi et al. 2014a, b).

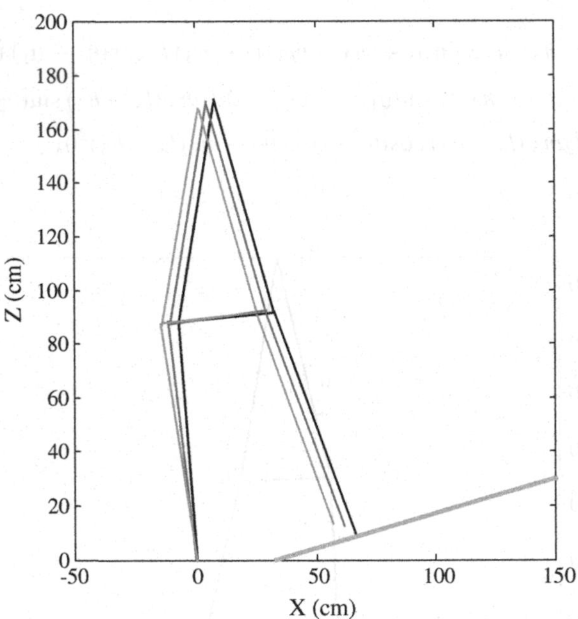

Fig. 3 The stick diagram of the biped robot walking in the lateral on slope

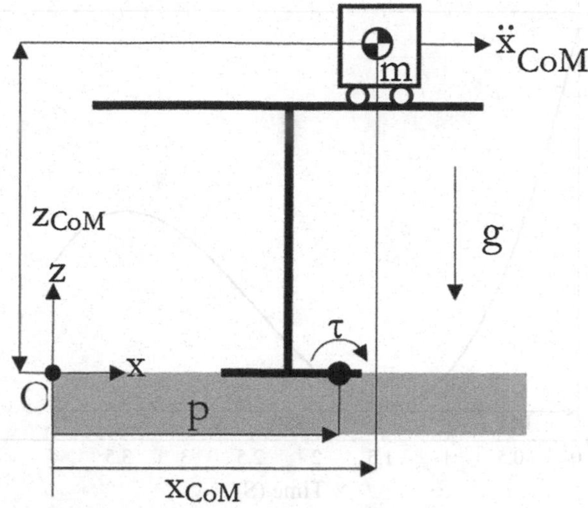

Fig. 4 Cart table model

If the biped robot's joint velocities augment, the dynamic forces will dominate the static forces. The faster the movements, the more dominate these dynamic forces will be. ZMP can be regarded as the dynamical equivalent of the Floor projection of the Center Of Mass (FCOM). The ZMP criterion does take dynamic forces, as well as static forces, into consideration. In order to achieve a dynamically stable gait, ZMP should be within on the inside of the support polygon at every time instance. The support polygon in DSP is the area between both feet. The support polygon i.e. x_{ZMP} in this problem ranges from −11.5 to 44.2 cm in the DSP and ranges from −11.5 to 11.5 cm in the SSP (Mahmoodabadi et al. 2014a, b). The cart-table model is used to compute the ZMP. Figure 4 illustrates the simplified model of the biped robot, which consists of a running cart on a mass-less table. The cart has mass m, and its position (x, z) corresponds to the equivalent center of the mass of the biped robot. The center of reference frame is considered in the middle of the stance foot. Moreover, the table is assumed to have the same support polygon as the biped robot. The torque around point p can be written as (Mahmoodabadi et al. 2014a, b):

$$\tau = -\mathrm{mg}\,(x_{CoM} - p) + m\ddot{x}_{CoM} z_{CoM} \tag{25}$$

g is the gravitational acceleration downwards. Now, using the ZMP definition: torque must be zero and, thus $x_{ZMP} = p$, we have:

$$x_{ZMP} = p = x_{CoM} - \frac{\ddot{x}_{CoM}}{g} z_{CoM} \tag{26}$$

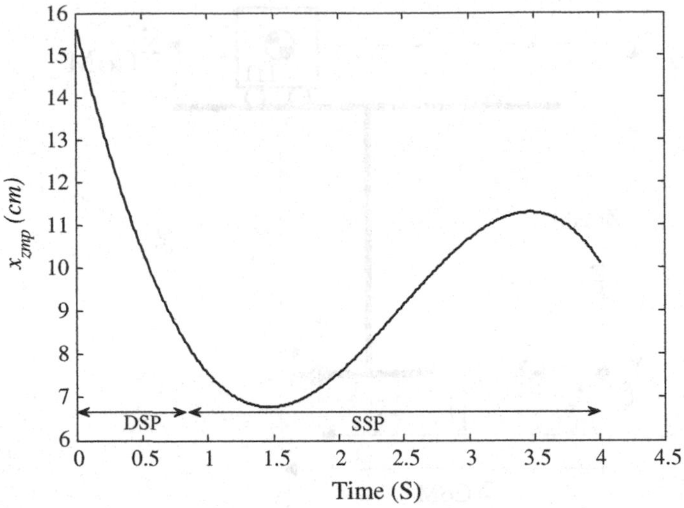

Fig. 5 Zero moment point on the inside of support polygon

The robot tracks the desired trajectories and maintains the ZMP on the inside of the support polygon, simultaneously (Fig. 5).

4.2 Sliding Mode Control for the Biped Robot

The sliding mode controller has been successfully employed to control the robots (Jing and Wuan 2006). Islam et al. applied multiple model/control-based sliding mode control to a 2-DOF robot manipulator (Islam and Liu 2011). Xiuping et al. controlled a biped robot in the double support phase by sliding mode control (Xiuping and Qiong 2004). Moosavian et al. controlled the biped robot in the sagittal plane with aid of sliding mode control and utilized the fuzzy system to regulate major control parameters (Moosavian et al. 2007). In the present chapter, there are six controlling coefficients, three sliding surfaces (s_i), and three equivalent control inputs(u_{eqi}), which must be chosen appropriately. The linear dynamic equations are utilized to obtain the equivalent control inputs u_{eqi} and $i = 1, 2, 3$ (appendix). Since the problem has three system states, the sliding surfaces and control inputs can be written, as follows:

$$s_i\left(e_i, t\right) = \left(\frac{d}{dt} + \lambda_i\right)^{n-1} e_i = 0 \qquad i = 1, 2 \text{ and } 3 \tag{27}$$

$$u_i = u_{eqi} - k_i \operatorname{sat}(\Phi) \qquad i = 1, 2 \text{ and } 3 \tag{28}$$

4.3 The Pareto Design of Sliding Mode Control for the Biped Robot

The objective of the sliding mode control method is to define asymptotically stable surfaces in such a manner that all system trajectories converge to these surfaces and slide along them until reaching the origin at their intersection (Utkin 1978). However, the heuristic sliding parameters are required to be chosen properly. Hence, the multi-objective periodic CDPSO (Mahmoodabadi et al. 2011, 2012a), Sigma method (Mostaghim and Teich 2003), and modified NSGAII (Atashkari et al. 2007) are utilized to ascertain the proper parameters and eliminate the tedious and repetitive trial-and-error process. The performance of a controlled closed loop system is usually evaluated by a variety of goals (Toscana 2005; Wolovich 1994). In this chapter, normalized summation of angles errors and normalized summation of control effort are regarded as the objective functions. These objective functions have to be minimized simultaneously. The vector $[k_1, k_2, k_3, \lambda_1, \lambda_2, \lambda_3]$ is the vector of selective parameters of sliding mode control. k_1, k_2 and, k_3 are positive constants. λ_1, λ_2 and, λ_3 are coefficients of the sliding surfaces. The normalized summation of angles errors and normalized summation of control effort are functions of this vector's components. That is to say, we can make changes in the normalized summation of angles errors and normalized summation of control effort by choosing various values for the selective parameters. This is noticeably an optimization problem with two objective functions (normalized summation of angles errors and normalized summation of control effort) and six decision variables ($k_1, k_2, k_3, \lambda_1, \lambda_2, \lambda_3$). The regions of the selective parameters are:

k_1, k_2, k_3 : Positive constants $0 \leq k_1, k_2, k_3 \leq 10$

$\lambda_1, \lambda_2, \lambda_3$: Coefficients of the sliding surfaces $100 \leq \lambda_1, \lambda_2, \lambda_3 \leq 1000$

The parameters of the multi-objective periodic CDPSO algorithm are selected as follows. In each period, the inertia weight W is linearly decreased from $W_1 = 0.9$ to $W_2 = 0.4$, C_1 is linearly decreased from $C_{1i} = 2.5$ to $C_{1f} = 0.5$, and C_2 is linearly increased from $C_{2i} = 0.5$ to $C_{2f} = 2.5$, over time. The related variables used in the convergence and divergence operators are: $P_{Convergence} = 0.1$, $P_{Divergence} = 0.1$, and $S_D = \frac{x_{max} - x_{min}}{2}$. The term v" is limited to the range of $[-v_{ave}, +v_{ave}]$, in which $v_{ave} = \frac{x_{max} - x_{min}}{2}$. While the velocity violates this range, it will be multiplied by a random number between [0, 1]. Furthermore, the positive constant for $\varepsilon_{elimination}$ is $\zeta = 300$ and the neighborhood radius for the leader selection is $R_{neighborhood} = 0.04$. The number of iterations in a period equals $T = 7$, the swarm size is 50 and the maximum iteration equals 150. The Pareto front of multi-objective periodic CDPSO (Mahmoodabadi et al. 2011, 2012a) for this issue is shown in Fig. 6, and multi-objective periodic CDPSO's feasibility and efficiency is assessed in comparison with Sigma method (Mostaghim and Teich 2003) and modified NSGAII (Atashkari et al. 2007).

Although the functioning of these algorithms is competitively appropriate in the present chapter, the most interesting result is that the multi-objective periodic CDPSO algorithm has more uniformity and diversity. In Fig. 6, points A and C stand for the

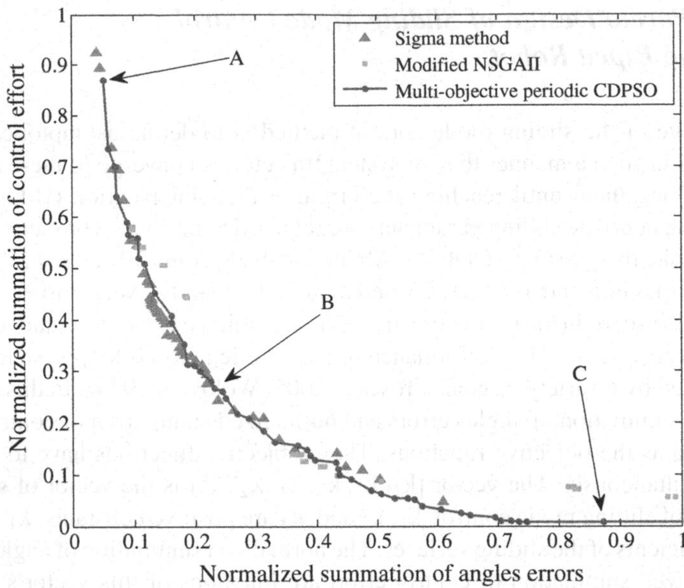

Fig. 6 The obtained Pareto fronts obtained by using Sigma method (Mostaghim and Teich 2003), modified NSGAII (Atashkari et al. 2007), and multi-objective periodic CDPSO (Mahmoodabadi et al. 2011, 2012a) for optimal control design of the biped robot

best normalized summation of angles errors and normalized summation of control effort, respectively. According to this figure, all the optimum design points in the Pareto front are non-dominated and can be selected by the designer as optimal sliding mode tracking controllers. Furthermore, choosing a better value for any objective function in the Pareto front causes a worse value for another objective. The corresponding decision variables (vector of sliding mode tracking controllers) of the Pareto front shown in Fig. 6 are the best possible design points. In this regard, if any other set of decision variables is chosen, the corresponding values of the pair of those objective functions will place an inferior point in Pareto front. Indeed, the inferior area in the space of the two objective functions is top/right side of Fig. 6. Hence, there are some crucial optimal design facts between these two objective functions which have been ascertained by the Pareto optimum design approach. Point B in Fig. 6 demonstrates important optimal design facts; in fact, it can be the trade-off optimum choice when considering minimum values of both of the normalized summation of angles errors and normalized summation of control effort. Design variables and objective functions according to theoptimum design points A, B, and C are illustrated in Table 2. The real tracking trajectories of the optimum design points A, B, and C are shown in Figs. 7, 8 and 9. The tracking error of the optimum design points A, B, and C are shown in Figs. 10, 11 and 12. In addition, Figs. 13, 14 and 15 illustrate the sliding surfaces of the optimum design points A, B, and C.

Table 2 The objective functions and their associated design variables for the optimum points of Fig. 6

Optimum design point	A	B	C
Normalized summation of angles errors	5.18×10^{-2}	2.44×10^{-1}	8.72×10^{-1}
Normalized summation of control effort	8.69×10^{-1}	2.61×10^{-1}	8.30×10^{-3}
Design variable k_1	7.04×10^{-2}	3.71×10^{0}	1.98×10^{0}
Design variable k_2	3.54×10^{-5}	6.69×10^{-5}	7.75×10^{-2}
Design variable k_3	6.53×10^{-3}	1.59×10^{-3}	8.24×10^{-1}
Design variable λ_1	5.76×10^{2}	2.24×10^{2}	1.00×10^{2}
Design variable λ_2	5.98×10^{2}	2.60×10^{2}	1.01×10^{2}
Design variable λ_3	6.02×10^{2}	2.54×10^{2}	1.00×10^{2}

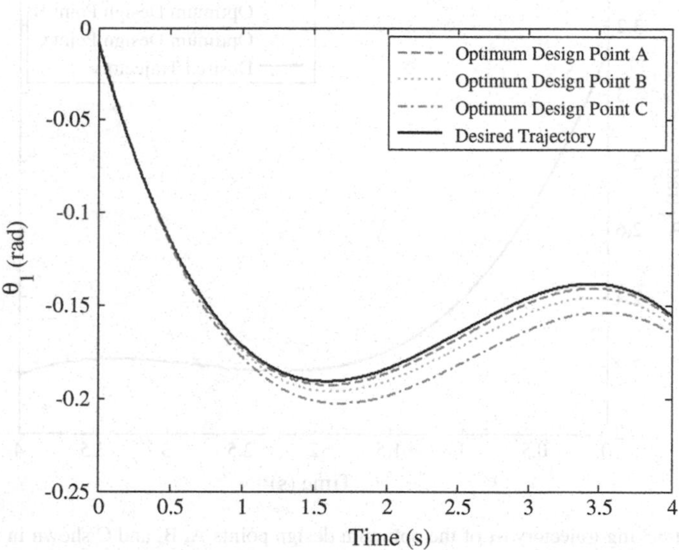

Fig. 7 The tracking trajectory θ_1 of the optimum design points A, B, and C shown in the Pareto front (Fig. 6)

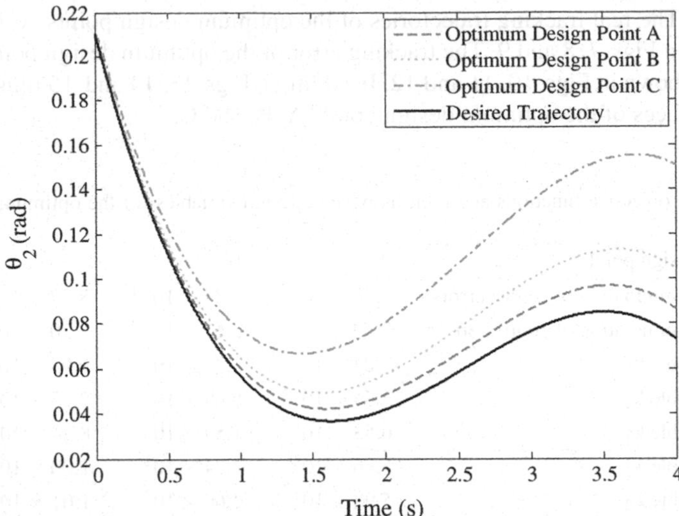

Fig. 8 The tracking trajectory θ_2 of the optimum design points A, B, and C shown in the Pareto front (Fig. 6)

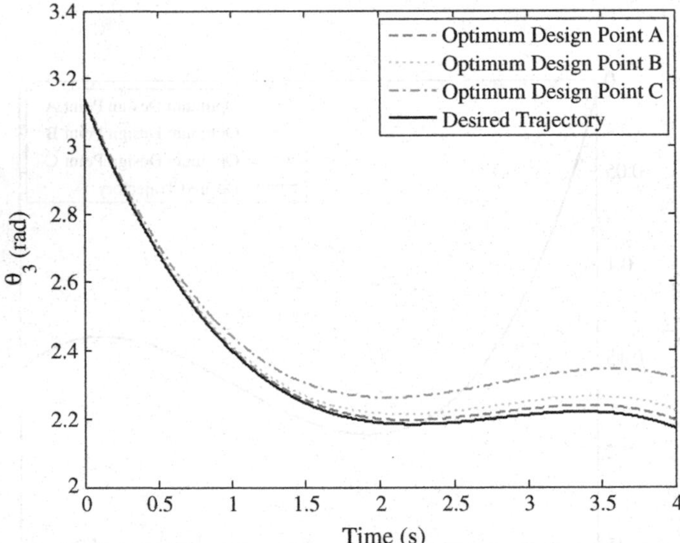

Fig. 9 The tracking trajectory θ_3 of the optimum design points A, B, and C shown in the Pareto front (Fig. 6)

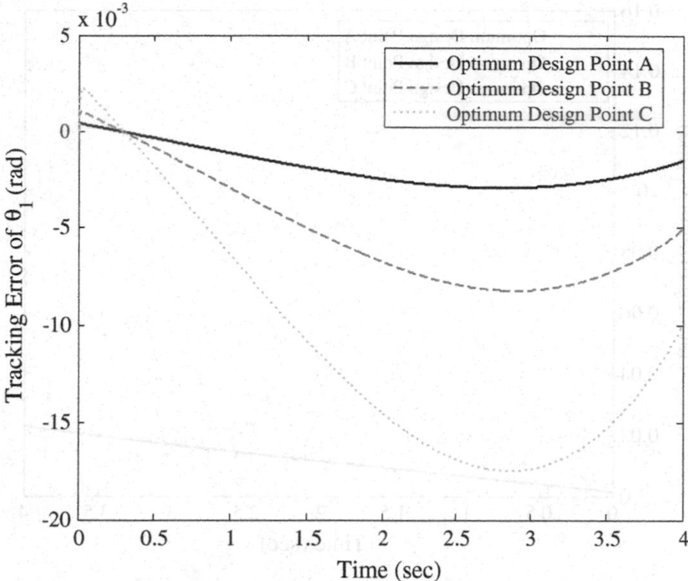

Fig. 10 The tracking error of θ_1 for the optimum design points A, B, and C shown in the Pareto front (Fig. 6)

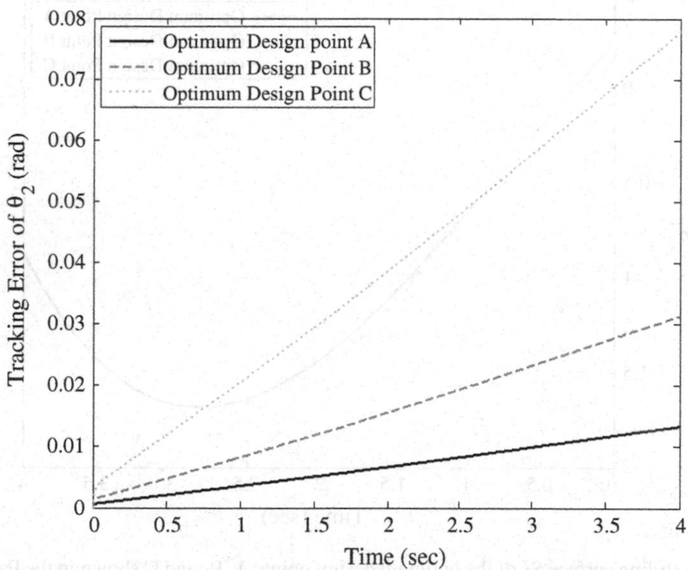

Fig. 11 The tracking error of θ_2 for the optimum design points A, B, and C shown in the Pareto front (Fig. 6)

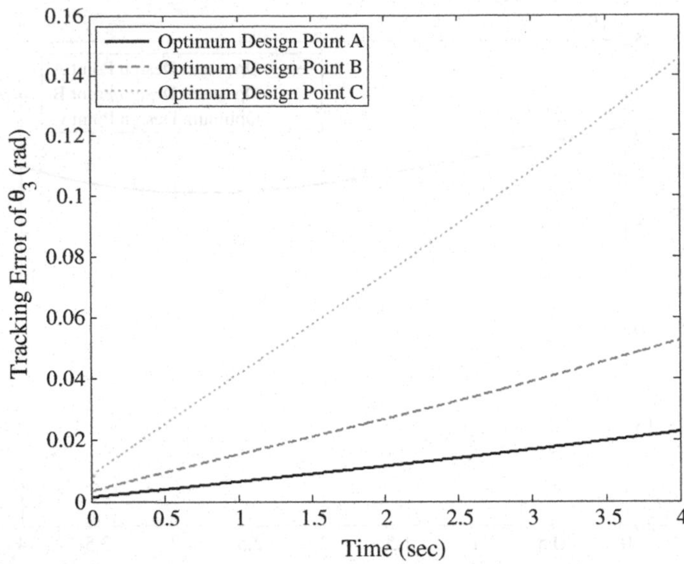

Fig. 12 The tracking error of θ_3 for the optimum design points A, B, and C shown in the Pareto front (Fig. 6)

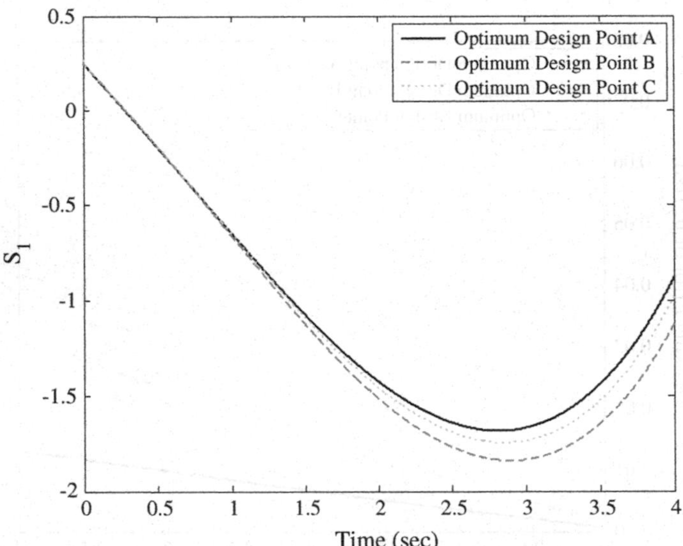

Fig. 13 The sliding surface S_1 of the optimum design points A, B, and C shown in the Pareto front (Fig. 6)

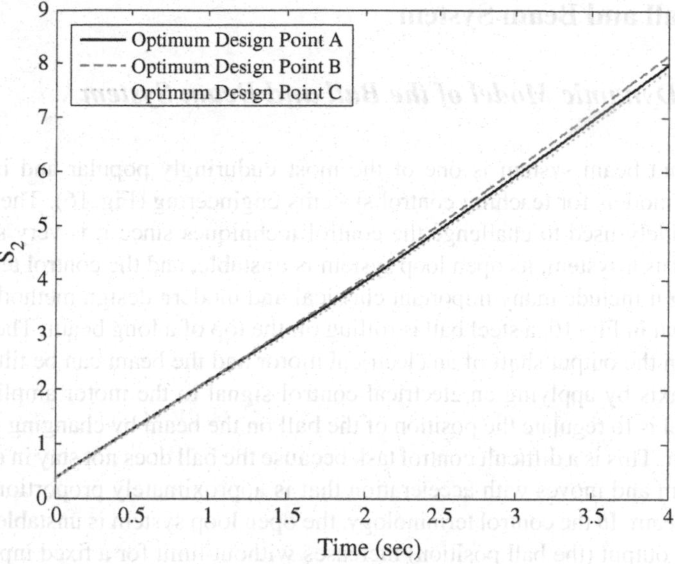

Fig. 14 The sliding surface S_2 of the optimum design points A, B, and C shown in the Pareto front (Fig. 6)

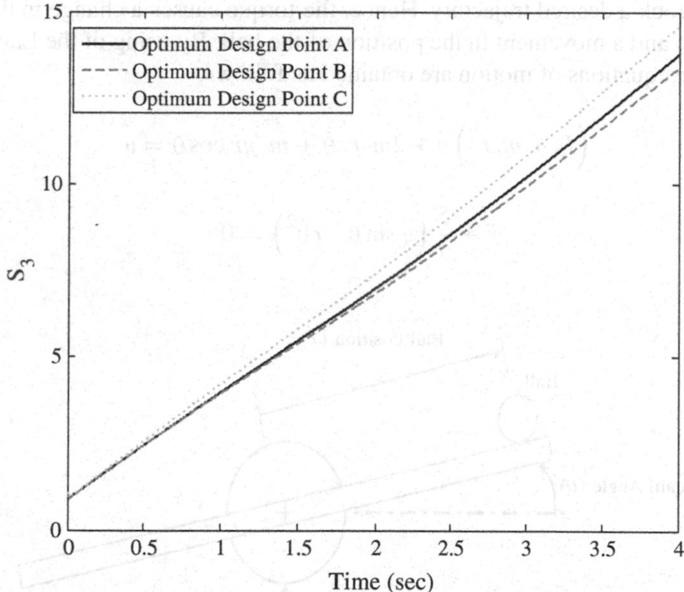

Fig. 15 The sliding surface S_3 of the optimum design points A, B, and C shown in the Pareto front (Fig. 6)

5 The Ball and Beam System

5.1 The Dynamic Model of the Ball and Beam System

The ball and beam system is one of the most enduringly popular and important laboratory models for teaching control systems engineering (Fig. 16). The ball and beam is widely used to challenge the control techniques since it is very simple to understand as a system, its open loop system is unstable, and the control techniques of studying it include many important classical and modern design methods.

As shown in Fig. 16, a steel ball is rolling on the top of a long beam. The beam is mounted on the output shaft of an electrical motor and the beam can be tilted about its center axis by applying an electrical control signal to the motor amplifier. The control goal is to regulate the position of the ball on the beam by changing the angle of the beam. This is a difficult control task because the ball does not stay in one place on the beam and moves with acceleration that is approximately proportional to the tilt of the beam. In the control terminology, the open loop system is unstable because the system output (the ball position) increases without limit for a fixed input (beam angle). Feedback control must be used to stabilize the system and keep the ball in a desired position on the beam. In other words, the goal of the control approach is to control the torque u applied at the pivot of the beam, such that the ball can roll on the beam and track a desired trajectory. Hence, the torque causes a change in the angle of the beam and a movement in the position of the ball. By using of the Lagrangian method, the equations of motion are obtained as follows.

$$\left(I_b + m_s r^2\right)\ddot{\theta} + 2m_s r\dot{r}\dot{\theta} + m_s gr\cos\theta = u \tag{29}$$

$$\ddot{r} + \frac{5}{7}\left(g\sin\theta - r\dot{\theta}^2\right) = 0 \tag{30}$$

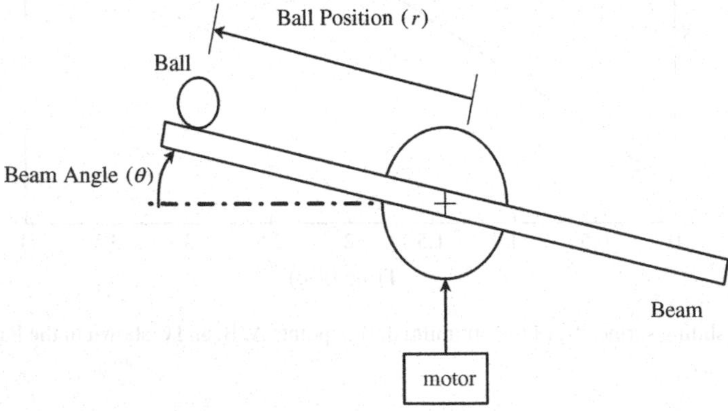

Fig. 16 The schematic model of the ball and beam system

where u is the torque applied to the beam, $I_b = \frac{m_b a^2}{12}$ is beam's moment of inertia, $m_b = 1\,\text{kg}$ is mass of beam, $a = 1\,\text{m}$ is the length of the beam , θ is the angle of the beam, $m_s = 0.05\,\text{kg}$ is the mass of the ball, I_s is the ball's moment of inertia ($I_s = \frac{2}{5}\left(m_s r_s^2\right)$), $r_s = 0.01\,\text{m}$ is the radius of the sphere, r stands for the position of the ball. If the states are defined as follows,

$$x_1 = \theta \quad , \quad x_2 = \dot{\theta} \quad , \quad x_3 = r \quad , \quad x_4 = \dot{r}$$

Then, the state-space equations would be written as Eq. (31).

$$\begin{cases} \dot{x}_1 = x_2 \\ \dot{x}_2 = \frac{-m_s x_3 (2x_4 x_2 - g\cos(x_1))}{m_s x_3^2 + I_b} + \frac{u}{m_s x_3^2 + I_b} \\ \dot{x}_3 = x_4 \\ \dot{x}_4 = -\frac{5}{7}\left(g\sin(x_1) - x_3 x_2^2\right) \end{cases} \tag{31}$$

5.2 *Decoupled Sliding Mode Control for the Ball and Beam System*

The optimal decoupled sliding mode controller has been successfully employed by researchers to control the ball and beam system. Alfaro-Cid et al. used the genetic algorithm to design the decoupled sliding mode controllers (Alfaro-Cid et al. 2005). Chang et al. designed optimal fuzzy sliding-mode control for the ball and beam system using fuzzy ant colony optimization (Chang et al. 2012). In Mahmoodabadi et al. (2012a), a multi-objective genetic algorithm was applied to Pareto design of decoupled sliding-mode controllers for nonlinear systems. In (Andalib Sahnehsaraei et al. 2013), multi-objective particle swarm optimization was utilized to design the decoupled sliding mode controller for an inverted pendulum system. Mahmoodabadi et al. proposed an online optimal decoupled sliding mode controller using the moving least squares and particle swarm optimization (Mahmoodabadi et al. 2014c). Regarding the DSMC of the ball and beam system in this chapter, the sliding surfaces can be written as follows:

$$S_1 = c_1(\theta - z) + \dot{\theta} = c_1(x_1 - z) + x_2 \tag{32}$$

$$S_2 = c_2 r + r = c_2 x_3 + x_4 \tag{33}$$

which

$$z = sat\left(\frac{S_2}{\varphi_2}\right) z_u \quad 0 < z_u < 1 \tag{34}$$

By using the decoupled sliding mode control strategy, we have:

$$\dot{S}_1 = c_1 x_2 - c_1 \dot{z} + f_1 + b_1 u_1 \tag{35}$$

which

$$\begin{cases} \dot{z} = 0 & if \left|\frac{S_2}{\varphi_2}\right| \geq 1 \\ \dot{z} = \frac{z_u}{\varphi_2}\dot{S}_2 & if \left|\frac{S_2}{\varphi_2}\right| < 1 \end{cases} \tag{36}$$

Furthermore

$$\dot{S}_2 = c_2\,\dot{x}_3 + \dot{x}_4 = c_2\,x_4 + f_2 + b_2\,u \tag{37}$$

$\hat{u}$ would be achieved if $\dot{S}_1 = 0$,

$$\begin{cases} \dot{S}_1 = c_1\,x_2 - \frac{c_1\,z_u}{\varphi_2}(c_2 x_4 + f_2 + b_2 u) + f_1 + b_1 u = 0 & if \left(\frac{S_2}{\varphi_2}\right) < 1 \\ \dot{S}_1 = c_1\,x_2 + f_1 + b_1 u = 0 & if \left(\frac{S_2}{\varphi_2}\right) \geq 1 \end{cases} \tag{38}$$

Therefore

$$\begin{cases} \hat{u} = \frac{-1}{b_1 - \frac{c_1 z_u}{\varphi_2} b_2}\left[f_1 + c_1\,x_2 - \frac{c_1 z_u}{\varphi_2}(f_2 + c_2 x_4)\right] & if \left(\frac{S_2}{\varphi_2}\right) < 1 \\ \hat{u} = \frac{-1}{b_1}(f_1 + c_1\,x_2) & if \left(\frac{S_2}{\varphi_2}\right) \geq 1 \end{cases} \tag{39}$$

and

$$u = \hat{u} - k\,sat\left(\frac{S_1\,b_1\,(x)}{\varphi_1}\right) \tag{40}$$

5.3 The Pareto Design of Decoupled Sliding Mode Control for the Ball and Beam System

In this section, the CDPSO approach is employed to select the parameters of DSMC for the ball and beam system with respect to two objective functions. To compare the performance of the optimizer technique, the optimization process is also performed via Sigma method (Mostaghim and Teich 2003) and modified NSGAII (Atashkari et al. 2007). The performance of a controlled closed-loop system is evaluated by various goals. In this chapter, the normalized integral of the absolute value of the ball distance and the normalized integral of the absolute value of the beam are considered as the objective functions. In other words, the objective functions and constrains are as follows:

Constraint: max $(abs\ (u)) \leq 40\, N\, m$
Objective functions:

$$f_1 = \text{Normalized}\ (\int |r\ (t)\ |dt)\ \text{ and }\ f_2 = \text{Normalized}\ (\int |\theta\ (t)\ |dt)$$

These objective functions have to be minimized simultaneously. When solving the optimization problem, the population size and maximum iteration are set at 50 and 150, respectively. Vector $[c_1,\ c_2,\ \varphi_1,\ \varphi_2, k,\ z_u]$ is the vector of the selective parameters (design variables) of the decoupled sliding mode control.

In this chapter, we are concerned with choosing values for the selective parameters to minimize the objective functions. Clearly, this is an optimization problem with two object functions and six decision variables.

Now, it is supposed that the initial values for the states of the ball and beam system are as follows.

$$\left(x_1 = \frac{\pi}{3}\ ,\ x_2 = 0\ ,\ \ x_3 = 0.1\ ,\ \ x_4 = 0\right) \tag{41}$$

The regions of the selective parameters are:

$$1 \leq c_1 \leq 100,\quad 0.3 \leq c_2 \leq 10,\quad 1 \leq \varphi_1 \leq 100,\quad 0.01 \leq \varphi_2 \leq 1,\quad 0 \leq k \leq 10,\quad 0.1 \leq z_u \leq 1$$

The parameters of the multi-objective periodic CDPSO algorithm are chosen as follows. In each period, the inertia weight W is linearly decreased from $W_1 = 0.9$ to $W_2 = 0.4$, C_1 is linearly decreased from $C_{1i} = 2.5$ to $C_{1f} = 0.5$, and C_2 is linearly increased from $C_{2i} = 0.5$ to $C_{2f} = 2.5$, over time. The related variables used in the convergence and divergence operators are: $P_{Convergence} = 0.1$, $P_{Divergence} = 0.1$, and $S_D = \frac{X_{max} - X_{min}}{2}$. The term v'' is limited to the range of $[-v_{ave}, +v_{ave}]$, in which $v_{ave} = \frac{X_{max} - X_{min}}{2}$. While the velocity violates this range, it will be multiplied by a random number between [0, 1]. Furthermore, the positive constant for $\varepsilon_{elimination}$ is $\zeta = 300$ and the neighborhood radius for leader selection is $R_{neighborhood} = 0.04$. The number of iterations in a period is equal to T = 7, the swarm size is 50 and the maximum iteration is equal to 150. The Pareto front of multi-objective periodic CDPSO (Mahmoodabadi et al. 2011, 2012a) for this problem is shown in Fig. 17, and multi-objective periodic CDPSO's feasibility and efficiency are assessed in comparison with Sigma method (Mostaghim and Teich 2003) and modified NSGAII (Atashkari et al. 2007).

In Fig. 17, points A and C stand for the best normalized integral of the absolute value of the ball distance and the normalized integral of the absolute value of the beam angle, respectively. According to this figure, the multi-objective periodic CDPSO algorithm has more uniformity and diversity in comparison to Sigma method (Mostaghim and Teich 2003) and modified NSGAII (Atashkari et al. 2007). Moreover, all the optimum design points in the Pareto front are non-dominated and can be chosen by the designer as optimal decoupled sliding mode controllers. It is noticeable

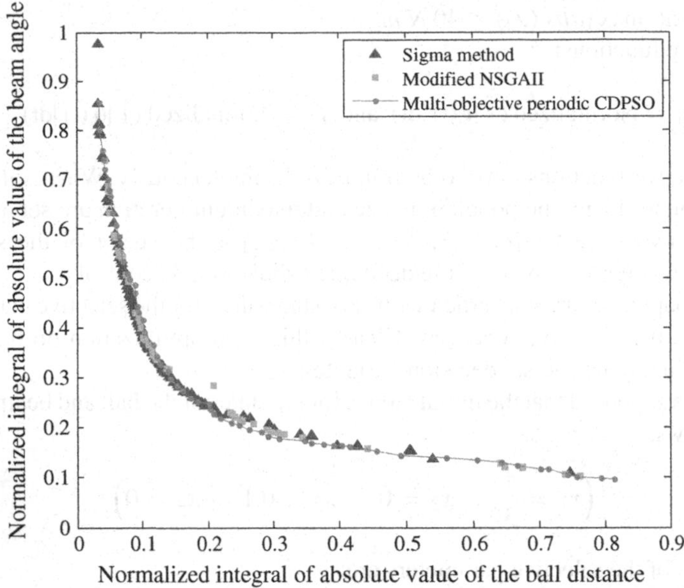

Fig. 17 The obtained Pareto fronts by using Sigma method (Mostaghim and Teich 2003), modified NSGAII (Atashkari et al. 2007), and multi-objective periodic CDPSO (Mahmoodabadi et al. 2011, 2012a) for optimal control design of the ball and beam system

Table 3 The objective functions and their associated design variables for the optimum points of Fig. 17

	c_1	c_2	φ_1	φ_2	k	z_u	f_1	f_2
Point A	5.8655	1.7532	20.2309	0.9514	0.6817	0.8729	0.0353	0.8561
Point B	30.648	1.9889	43.0080	0.2075	0.7272	0.7370	0.2919	0.1832
Point C	4.8042	1.6336	32.0405	0.9434	0.0266	0.4044	0.8143	0.0975

that choosing a better value for any objective function in the Pareto front causes a worse value for another objective function. Point B in Fig. 17 demonstrates important optimal design facts. This point can be the trade-off optimum choice when considering minimum values of both of the normalized integral of the absolute value of the ball distance and the normalized integral of the absolute value of the beam angle. Design variables and objective functions according to the optimum design points A, B, and C are illustrated in Table 3. The response time of the ball and beam system for the optimum design points A, B, and C are shown in Figs. 18, 19 and 20.

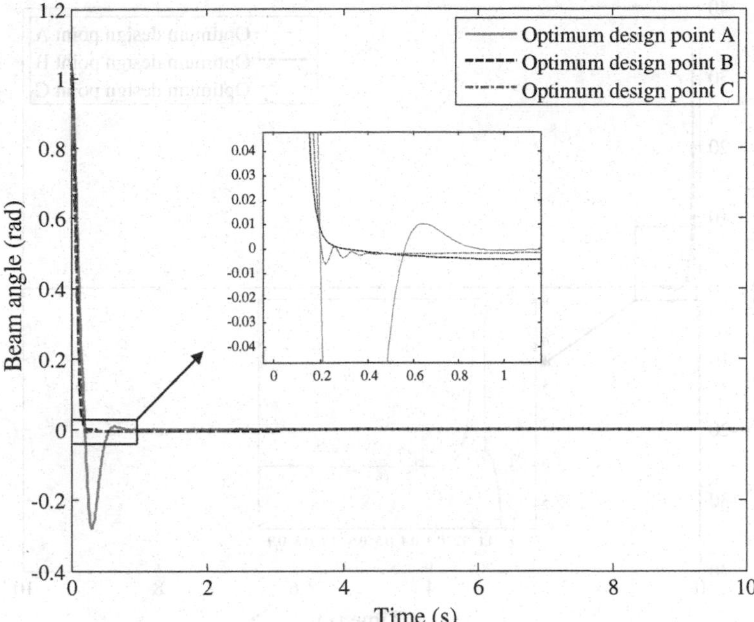

Fig. 18 The time response of the beam angle for the optimum design points A, B, and C shown in the Pareto front (Fig. 17)

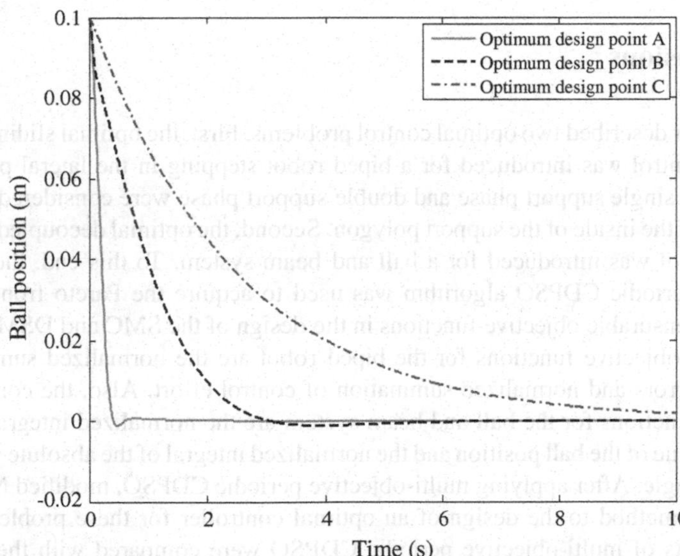

Fig. 19 The time response of the ball distance of the optimum design points A, B, and C shown in the Pareto front (Fig. 17)

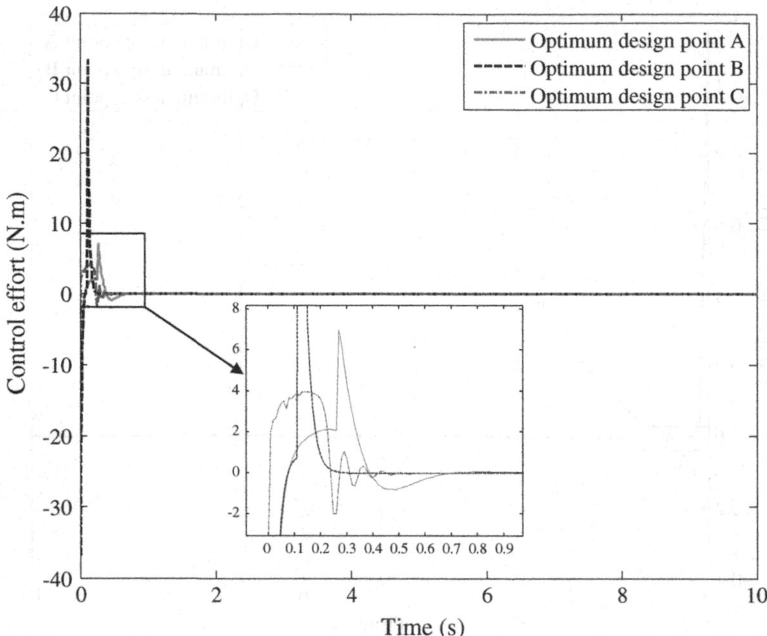

Fig. 20 The time response of the control effort for the ball and beam system of the optimum design points A, B, and C shown in the Pareto front (Fig. 17)

6 Conclusions

This chapter described two optimal control problems. First, the optimal sliding mode tracking control was introduced for a biped robot stepping in the lateral plane on slope. Both single support phase and double support phase were considered to take the ZMP on the inside of the support polygon. Second, the optimal decoupled sliding mode control was introduced for a ball and beam system. To this end, the multi-objective periodic CDPSO algorithm was used to acquire the Pareto front of the non-commensurable objective functions in the design of the SMC and DSMC. Two conflicting objective functions for the biped robot are the normalized summation of angles errors and normalized summation of control effort. Also, the conflicting objective functions for the ball and beam system are the normalized integral of the absolute value of the ball position and the normalized integral of the absolute value of the beam angle. After applying multi-objective periodic CDPSO, modified NSGAII and Sigma method to the design of an optimal controller for these problems, the Pareto fronts of multi-objective periodic CDPSO were compared with the Pareto fronts of modified NSGAII and Sigma method. The Pareto fronts of multi-objective periodic CDPSO were much more scattered than the other two algorithms. Hence, the designer has the ample opportunity to select the finest points. Finally, three points of each Pareto front of multi-objective periodic CDPSO were selected to compute the parameters of the SMC and DSMC. The results demonstrated the efficacy of multi-objective periodic CDPSO in the design of the SMC and DSMC for problems with challenging dynamic equations.

Acknowledgments The authors would like to thank the anonymous reviewers for their valuable suggestions that enhance the technical and scientific quality of this chapter.

Appendix

The equivalent sliding mode control inputs for the biped robot are as follows

$$\begin{aligned}
u_{eq1} = {} & -0.2260855175\,\dot{\theta}_2 - 5.376931543\,\theta_3 + 0.5609845516t^3 - 24.77733929t \\
& + 11.04563973 + 0.3042022263\dot{\theta}_3 - 52.53069026\theta_1 \\
& - 23.33466928\lambda_2\dot{\theta}_2 - 2.116394625\lambda_3\dot{\theta}_3 - 61.09883670\lambda_1\dot{\theta}_1 \\
& + 1.093385806\theta_1 - 2.335441469\lambda_3 - 5.878003190\lambda_2 \\
& - 16.56389463\lambda_1 - 14.61433450\theta_2 - 1.029058915\lambda_2 t^2 \\
& + 5.282969124\ \lambda_2 t + 11.12677748t^2 - 0.3123798467\lambda_3 t^2 \\
& + 1.744755729\lambda_3 t - 3.024392417\lambda_1 t^2 + 15.22583011\lambda_1 t
\end{aligned}$$

$$\begin{aligned}
u_{eq2} = {} & 2.127074613\left(10^{-10}\right)\dot{\theta}_2 + 3.014817654\theta_3 - 0.1832721690t^3 \\
& - 7.435154298t + 0.06684314392\dot{\theta}_3 + 0.6081082555\theta_1 \\
& - 12.97075008\lambda_2\dot{\theta}_2 - 2.116394625\lambda_3\dot{\theta}_3 - 6.759058956\lambda_1\dot{\theta}_1 \\
& + 1.093385810\dot{\theta}_1 - 2.335441469\lambda_3 - 3.267331946\lambda_2 \\
& - 1.832380883\lambda_1 - 14.19340124\theta_2 - 0.5720100787\lambda_2 t^2 \\
& + 2.936577819\lambda_2 t + 3.182845520t^2 + 9.317127676 \\
& - 0.3123798467\lambda_3 t^2 + 1.744755729\lambda_3 t - 0.3345734183\lambda_1 t^2 \\
& + 1.684357492\lambda_1 t
\end{aligned}$$

$$\begin{aligned}
u_{eq3} = {} & -3.400956231\left(10^{-11}\right)\dot{\theta}_2 + 3.029347274\theta_3 - 0.2748341166t^3 \\
& - 3.658549547t + 11.95619521 + 1.750512761\left(10^{-10}\right)\dot{\theta}_3 \\
& + 0.1293520741\theta_1 - 2.497500599\lambda_2\dot{\theta}_2 - 3.630907800\lambda_3\dot{\theta}_3 \\
& + 3.604860241\lambda_1\dot{\theta}_1 + 0.2233967865\theta_1 - 4.006706757\ \lambda_3 \\
& - 0.6291204009\ \lambda_2 + 0.9772776114\lambda_1 + 1.33021167t^2 \\
& - 0.01647731197\theta_2 - 0.1101397764\ \lambda_2 t^2 + 0.5654341356\lambda_2 t \\
& - 0.8983311722\ \lambda_1 t - 0.5359219913\ \lambda_3 t^2 + 2.993320390\ \lambda_3 t \\
& + 0.1784405819\lambda_1 t^2
\end{aligned}$$

References

Adhikary, N., Mahanta, C.: Integral backstepping sliding mode control for underactuated systems: swing-up and stabilization of the cart-pendulum system. ISA Trans. **52**(6), 870–880 (2013)

Alfaro-Cid, E., McGookin, E.W., Murray-Smith, D.J., Fossen, T.I.: http://www.sciencedirect.com/science/article/pii/S0967066104001741Genetic algorithms optimisation of decoupled sliding mode controllers: simulated and real results. Control Eng. Pract. 13(6), 739–748 (2005)

Andalib Sahnehsaraei, A., Mahmoodabadi, M.J., Taherkhorsandi, M.: Optimal robust decoupled sliding mode control based on a multi-objective genetic algorithm. In: Proceedings of International Symposium on Innovations in Intelligent Systems and Application, Albena, 19–21 June 2013, pp. 1–5 doi:10.1109/INISTA.2013.6577641

Angeline, P.J.: Using selection to improve particle swarm optimization. Proceedings of Congress on Evolutionary Computation, Anchorage 4–9 May 1998, pp. 84–89 doi:10.1109/ICEC.1998.699327

de Aparecida, P.A., Horst Albrecht, C., de Souza Leite Pires, L.B., Pinheiro Jacob, B.: Tailoring the particle swarm optimization algorithm for the design of offshore oil production risers. Optim. Eng. **12**(1–2), 215–235 (2011)

Atashkari, K., Nariman-Zadeh, N., Golcu, M., Khalkhali, A., Jamali, A.: Modelling and multi-objective optimization of a variable valve-timing spark-ignition engine using polynomial neural networks and evolutionary algorithms. Energy Convers. Manag. **48**(3), 1029–1041 (2007)

Aziz, A.S.A., Azar, A.T., Salama, M.A., Hassanien, A.E., Hanafy, S.E.O.: Genetic algorithm with different feature selection techniques for anomaly detectors generation. Proceedings of Federated Conference On Computer Science And Information Systems Kraków, 8–11 Sept 2013, pp. 769–774

Basin, M., Rodriguez-Ramirez, P., Ferrara, A., Calderon-Alvarez, D.: Sliding mode optimal control for linear systems. J. Franklin Inst. **349**(4), 1350–1363 (2012)

Bayramoglu, H., Komurcugil, H.: Time-varying sliding-coefficient-based terminal sliding mode control methods for a class of fourth-order nonlinear systems. Nonlinear Dyn. **73**(3), 1645–1657 (2013)

Belmecheri, F., Prins, C., Yalaoui, F., Amodeo, L.: Particle swarm optimization algorithm for a vehicle routing problem with heterogeneous fleet, mixed backhauls, and time windows. J. Intell. Manuf. **24**(4), 775–789 (2013)

Biswas, S., Mandal, K.K., Chakraborty, N.: Constriction factor based particle swarm optimization for analyzing tuned reactive power dispatch. Front. Energy **7**(2), 174–181 (2013)

De Carvalho, A.B., Pozo, A.: Measuring the convergence and diversity of CDAS multi-objective particle swarm optimization algorithms: a study of many-objective problems. Neurocomputing **75**(1), 43–51 (2012)

Castillo-Villar, K.K., Smith, N.R., Herbert-Acero, J.F.: Design and optimization of capacitated supply chain networks including quality measures. Math. Probl. Eng., Article ID 218913, pp. 17 (2014)

Castillo-Villar, K.K., Smith, N.R., Simonton, J.L.: The impact of the cost of quality on serial supply-chain network design. Int. J. Prod. Res. **50**(19), 5544–5566 (2012)

Cerman, O., Husek, P.: Adaptive fuzzy sliding mode control for electro-hydraulic servo mechanism. Expert Syst. Appl. **39**(11), 10269–10277 (2012)

Cha, Y.S., Kim, K.G., Lee, J.Y., Lee, J., Choi, M., Jeong, M.H., Kim, C.H., You, B.J., Oh, S.R.: MAHRU-M: a mobile humanoid robot platform based on a dual-network control system and coordinated task execution. Robot. Auton. Syst. **59**(6), 354–366 (2011)

Chang, Y.H., Chang, C.W., Tao, C.W., Lin, H.W., Taur, J.S.: Fuzzy sliding-mode control for ball and beam system with fuzzy ant colony optimizationhttp://www.sciencedirect.com/science/article/pii/S0957417411013613. Expert Syst. Appl. vol. 39(3), 3624–3633 (2012)

Chen, Z., Meng, W., Zhang, J., Zeng, J.: Scheme of sliding mode control based on modified particle swarm optimization. Syst. Eng.Theory Pract. **29**(5), 137–141 (2009)

Dehghani, R., Fattah, A., Abedi, E.: Cyclic gait planning and control of a five-link biped robot with four actuators during single support and double support phases. Multibody Syst. dyn. (2013). doi:10.1007/s11044-013-9404-5

Ding, C.-T., Yang, S.-X., Gan, C.-B.: Input torque sensitivity to uncertain parameters in biped robot. Acta. Mech. Sin. **29**(3), 452–461 (2013)

Eberhart, R., Simpson, P., Dobbins, R.: Computational Intelligence PC Tools. Academic Press Professional, Inc, San Diego (1996)

Eberhart, R.C., Shi, Y.: Comparison between genetic algorithms and particle swarm optimization. Evolutionary Programming VII, pp. 611–616. Springer, Berlin (1998)

Eker, I.: Second-order sliding mode control with PI sliding surface and experimental application to an electromechanical plant. Arab. J. Sci. Eng. **37**(7), 1969–1986 (2012)

Elshazly, H.I., Azar, A.T., Elkorany, A.M., Hassanien, A.E.: Hybrid system based on rough sets and genetic algorithms for medical data classifications. Int. J. Fuzzy Syst. Appl. **3**(4), 31–46 (2013)

Engelbrecht, A.P.: Computational Intelligence: An Introduction. Wiley, New York (2002)

Engelbrecht, A.P.: Fundamentals of Computational Swarm Intelligence. Wiley, New York (2005)

Feng, S., Yahmadi Amur, S.A., Sun, Z.Q.: Biped walking on level ground with torso using only one actuator. Sci. China Inf. Sci. **56**(11), 1–9 (2013)

Fieldsend, J.E., Singh, S.: A multi-objective algorithm based upon particle swarm optimization and efficient data structure and turbulence, In: Workshop on Computational Intelligence, 2002, pp. 37–44

Fleming, P., Purshouse, R.: Evolutionary algorithms in control systems engineering: a survey. Control Eng. Pract. **10**(11), 1223–1241 (2002)

Fonseca, C., Fleming, P.: Multi-objective optimal controller design with genetic algorithms. Proceedings of the International Conference on Control, Coventry, 21–24 (1994) pp. 745–749 doi:10.1049/cp:19940225

Gaing, Z.L.: A particle swarm optimization approach for optimum design of PID controller in AVR system. IEEE Trans. Energy Convers. **19**(2), 384–391 (2004)

Gang, M., Wei, Z., Xiaolin, C.: A novel particle swarm optimization algorithm based on particle migration. Appl. Math. Comput. **218**(11), 6620–6626 (2012)

Gosh, A., Das, S., Chowdhury, A., Giri, R.: An ecologically inspired direct search method for solving optimal control problems with Bezier parameterization. Eng. Appl. Artif. Intell. **24**(7), 1195–1203 (2011)

Han, C., Zhang, G., Wu, L., Zeng, Q.: Sliding mode control of T-S fuzzy descriptor systems with time-delay. J. Franklin Inst. **349**(4), 1430–1444 (2012)

Hart, C.G., Vlahopoulos, N.: An integrated multidisciplinary particle swarm optimization approach to conceptual ship design. Struct. Multi. Optim. **41**(3), 481–494 (2010)

Hu, X., Eberhart, R.: Multi-objective optimization using dynamic neighborhood particle swarm optimization. In: Proceedings of the 2002 Congress on Evolutionary Computation, Honolulu, 2002, pp. 1677–1681. doi:10.1109/CEC.2002.1004494

Hu, J., Wang, Z., Gao, H., Stergioulas, L.K.: Robust H_∞ sliding mode control for discrete time-delay systems with stochastic nonlinearities. J. Franklin Inst. **349**(4), 1459–1479 (2012)

Islam, S., Liu, X.P.: Robust sliding mode control for robot manipulators. IEEE Trans. Industr. Electron. **58**(6), 2444–2453 (2011)

Jadhav, A.M., Vadirajacharya, K.: Performance verification of PID controller in an interconnected power system using particle swarm optimization. Energy Procedia **14**, 2075–2080 (2012)

Javadi-Moghaddam, J., Bagheri, A.: An adaptive neuro-fuzzy sliding mode based genetic algorithm control system for under water remotely operated vehicle. Expert Syst. Appl. **37**(1), 647–660 (2010)

Jia, D., Zheng, G., Qu, B., Khurram Khan, M.: A hybrid particle swarm optimization algorithm for high-dimensional problems. Comput. Ind. Eng. **61**(4), 1117–1122 (2011)

Jing, J., Wuan, Q.H.: Intelligent sliding mode control algorithm for position tracking servo system. Int. J. Inf. Technol. **12**(7), 57–62 (2006)

Kennedy, J., Eberhart, R.: Particle swarm optimization. In: Proceedings of IEEE International Conference on Neural Networks, Perth, 27 Nov–01 Dec 1995, pp. 1942–1948. doi:10.1109/ICNN.1995.488968

Ker-Wei, Y., Shang-Chang, H.: An application of AC servo motor using particle swarm optimization based sliding mode controller. Proceedings of IEEE International Conference on Systems, Man and Cybernetics, Taipei **8–11**, 4146–4150 (2006). doi:10.1109/ICSMC.2006.384784

Kim, B.-I., Son, S.-J.: A probability matrix based particle swarm optimization for the capacitated vehicle routing problem. J. Intell. Manuf. **23**(4), 1119–1126 (2012)

Kuo, R.J., Syu, Y.J., Chen, Z.-Y., Tien, F.C.: Integration of particle swarm optimization and genetic algorithm for dynamic clustering. Inf. Sci. **195**, 124–140 (2012)

Lari, A., Khosravi, A., Rajabi, F.: Controller design based on μ analysis and PSO algorithm. ISA Trans. **53**(2), 517–523 (2014)

Lee, J.H., Okamoto, S., Koike, H., Tani, K.: Development and motion control of a biped walking robot based on passive walking theory. Artif. Life Rob. **19**(1), 68–75 (2014)

Li, J., Li, W., Li, Q.: Sliding mode control for uncertain chaotic systems with input nonlinearity. Commun. Nonlinear Sci. Simul. **17**(1), 341–348 (2012)

Lin, C., Chen, L., Chen, C.: RCMAC hybrid control for MIMO uncertain nonlinear systems using sliding-mode technology. IEEE Trans. Neural Netw. **18**(3), 708–720 (2007)

Lin, T.C., Lee, T.Y., Balas, V.E.: Adaptive fuzzy sliding mode control for synchronization of uncertain fractional order chaotic systems. Chaos, Solitons Fractals **44**(10), 791–801 (2011)

Lin, C.J., Lin, P.T.: Particle swarm optimization based feedforward controller for a XY PZT positioning stage. Mechatronics **22**(5), 614–628 (2012)

Liu, C.-S., et al.: A new sliding control strategy for nonlinear system solved by the Lie-group differential algebraic equation method. Commun. Nonlinear Sci. Numer. Simul. **19**(6), 2012–2038 (2014)

Lu, C.H., Hwang, Y.R., Shen, Y.T.: Backstepping sliding mode tracking control of a vane-type air motor X-Y table motion system. ISA Trans. **50**(2), 278–286 (2011)

Mahmoodabadi, M.J., Arabani Mostaghim, S., Bagheri, A., Nariman-zadeh, N.:http://www.sciencedirect.com/science/article/pii/S0895717712001641 Pareto optimal design of the decoupled sliding mode controller for an inverted pendulum system and its stability simulation via Java programming. Math. Comput. Model. 57(5–6), 1070–1082 (2013)

Mahmoodabadi, M.J., Bagheri, A., Nariman-zadeh, N., Jamali, A., Abedzadeh Maafi, R.: Pareto design of decoupled sliding-mode controllers for nonlinear systems based on a multiobjective genetic algorithm. J. Appl. Math. Article ID 639014, pp. 22 (2012b)

Mahmoodabadi, M.J., Taherkhorsandi, M., Bagheri, A.: Pareto design of state feedback tracking control of a biped robot via multi-objective PSO in comparison with sigma method and genetic algorithms: modified NSGAII and MATLAB's toolbox. Sci. World J. 2014, p. 8 (2014b)

Mahmoodabadi, M.J., Bagheri, A., Arabani-Mostaghim, S., Bisheban, M.: Simulation of stability using Java application for Pareto design of controllers based on a new multi-objective particle swarm optimization. Math. Comput. Model. **54**(5–6), 1584–1607 (2011)

Mahmoodabadi, M.J., Bagheri, A., Nariman-zadeh, N., Jamali, A.: A new optimization algorithm based on a combination of particle swarm optimization, convergence and divergence operators for single-objective and multi-objective problems. Eng. Optim. **44**(10), 1167–1186 (2012a)

Mahmoodabadi, M.J., Taherkhorsandi, M., Bagheri, A.: Optimal robust sliding mode tracking control of a biped robot based on ingenious multi-objective PSO. Neurocomputing **124**, 194–209 (2014a)

Mahmoodabadi, M.J., Momennejad, S., Bagheri, A.: Online optimal decoupled sliding mode control based on moving least squares and particle swarm optimization. Inf. Sci. **268**, 342–356 (2014c)

Martínez-Soto, R., Castillo, O., Aguilar, L.T., Rodriguez, A.: A hybrid optimization method with PSO and GA to automatically design type-1 and type-2 fuzzy logic controllers. Int. J. Mach. Learn. Cybern. (2013). doi:10.1007/s13042-013-0170-8

Mondal, S., Mahanta, C.: Chattering free adaptive multivariable sliding mode controller for systems with matched and mismatched uncertainty. ISA Trans. **52**(3), 335–341 (2013a)

Mondal, S., Mahanta, C.: Adaptive integral higher order sliding mode controller for uncertain systems. J. Control Theory Appl. **11**(1), 61–68 (2013b)

Moosavian, S.A.A., Alghooneh, M., Takhmar, A.: Stable trajectory planning, dynamics modeling and fuzzy regulated sliding mode control of a biped robot. In: Proceedings of 7th IEEE-RAS International Conference on Humanoid Robots, Pittsburgh, Nov. 29–Dec. 1 2007, pp. 471–476. doi:10.1109/ICHR.2007.4813912

Mostaghim, S., Teich, J.: Strategies for finding good local guides in multi- objective particle swarm optimization (MOPSO). In: Proceedings of the 2003 IEEE Swarm Intelligence Symposium, 24–26 April 2003, pp. 26–33. doi:10.1109/SIS.2003.1202243

Mousa, A.A., El-Shorbagy, M.A., Abd-El-Wahed, W.F.: Local search based hybrid particle swarm optimization algorithm for multi objective optimization. Swarm Evol. Comput. **3**, 1–14 (2012)

Nejat, A., Mirzabeygi, P., Shariat Panahi, M.: Airfoil shape optimization using improved multiobjective territorial particle swarm algorithm with the objective of improving stall characteristics. Struct. Multi. Optim. **49**(6), 953–967 (2014)

Nery Júnior, G.A., Martins, M.A.F., Kalid, R.: A PSO-based optimal tuning strategy for constrained multivariable predictive controllers with model uncertainty. ISA Trans. **53**(2), 560–567 (2014)

Nikkhah, M., Ashrafiuon, H., Fahimi, F.: Robust control of under actuated bipeds using sliding modes. Robatica **25**(3), 367–374 (2007)

Nizar, A., Mansour Houda, B., Ahmed Said, N.: A new sliding function for discrete predictive sliding mode control of time delay systems. Int. J. Autom. Comput. **10**(4), 288–295 (2013)

Nwankwor, E., Nagar, A.K., Reid, D.C.: Hybrid differential evolution and particle swarm optimization for optimal well placement. Comput. Geosci. **17**(2), 249–268 (2013)

Pan, I., Korre, A., Das, S., Durucan, S.: Chaos suppression in a fractional order financial system using intelligent regrouping PSO based fractional fuzzy control policy in the presence of fractional Gaussian noise. Nonlinear Dyn. **70**(4), 2445–2461 (2012)

Parsopoulos, K.E., Tasoulis, D.K., Vrahatis, M.N.: Multi-objective optimization using parallel vector evaluated particle swarm optimization. Proceedings of the IASTED International Conference on Artificial Intelligence and Applications. vol. 2, 823–828 (2004)

Pourmahmood Aghababa, M., Akbari, M.E.: A chattering-free robust adaptive sliding mode controller for synchronization of two different chaotic systems with unknown uncertainties and external disturbances. Appl. Math. Comput. **218**(9), 5757–5768 (2012)

Qiao, W., Venayagamoorthy, G., Harley, R.: Design of optimal PI controllers for doubly fed induction generators driven by wind turbines using particle swarm optimization. In: Proceedings of the International Joint Conference on Neural Networks, Vancouver, 2006, pp. 1982–1987. doi:10.1109/IJCNN.2006.246944

Ramos, R., Biel, D., Fossas, E., Griño, R.: Sliding mode controlled multiphase buck converter with interleaving and current equalization. Control Eng. Pract. **21**(5), 737–746 (2013)

Ratnaweera, A., Halgamuge, S.K.: Self-organizing hierarchical particle swarm optimizer with time-varying acceleration coefficient. IEEE Trans. Evol. Comput. **8**(3), 240–255 (2004)

Sanchez, G., Villasana, M., Strefezza, M.: Multi-objective pole placement with evolutionary algorithms. Lect. Notes Comput. Sci. **4403**, 417–427 (2007)

Shahriari kahkeshi, M., Sheikholeslam, F., Zekri, M.: Design of adaptive fuzzy wavelet neural sliding mode controller for uncertain nonlinear systems. ISA Trans. **52**(3), 342–350 (2013)

Singla, M., Shieh, L.-S., Song, G., Xie, L., Zhang, Y.: A new optimal sliding mode controller design using scalar sign function. ISA Trans. **53**(2), 267–279 (2014)

Sira-Ramirez, H., Luviano-Juarez, A., Cortes-Romero, J.: Robust input-output sliding mode control of the buck converter. Control Eng. Pract. **21**(5), 671–678 (2013)

Sun, T., Pei, H., Pan, Y., Zhou, H., Zhang, C.: Neural network-based sliding mode adaptive control for robot manipulators. Neurocomputing **74**(14–15), 2377–2384 (2011)

Tang, Y., Wang, Z., Fang, J.: Controller design for synchronization of an array of delayed neural networks using a controllable probabilistic PSO. Inf. Sci. **181**(20), 4715–4732 (2011)

Toscana, R.: A simple robust PI/PID controller design via numerical optimization approach. J. Process Control **15**(1), 81–88 (2005)

Utkin, V.I.: Sliding Modes and their Application in Variable Structure Systems. Central Books Ltd, London (1978)

Wai, R.J., Chuang, K.L., Lee, J.D.: Total sliding-model-based particle swarm optimization controller design for linear induction motor, In: Proceedings of IEEE Congress on Evolutionary Computation, 25–28 Sept. 2007, Singapore. pp. 4729–4734

Wang, K., Jun Zheng, Y.: A new particle swarm optimization algorithm for fuzzy optimization of armored vehicle scheme design. Appl. Intell. **37**(4), 520–526 (2012)

Wang, X., Tang, L.: A discrete particle swarm optimization algorithm with self-adaptive diversity control for the permutation flowshop problem with blocking. Appl. Soft Comput. **12**(2), 652–662 (2012)

Winter, D.A.: Biomechanics and Motor Control of Human Movement. Wiley, New York (1990)

Wolovich, W.A.: Automatic Control Systems: Basic Analysis and Design. Saunders College Publishing, USA (1994)

Xiuping, M., Qiong, W.: Dynaimc modeling and sliding mode control of a five-link biped during the double support phase. In the Proceedings of the 2004 American Control Conference, vol. 3, pp. 2609-2614 (2004)

Yakut, O.: Application of intelligent sliding mode control with moving sliding surface for overhead cranes. Neural Comput. Appl. **24**(6), 1369–1379 (2014)

Yildiz, A.R., Solanki, K.N.: Multi-objective optimization of vehicle crashworthiness using a new particle swarm based approach. Int. J. Adv. Manuf. Technol. **59**(1–4), 367–376 (2012)

Yin, C., Zhong, S.M., Chen, W.F.: Design of sliding mode controller for a class of fractional-order chaotic systems. Commun. Nonlinear Sci. Simul. **17**(1), 356–366 (2012)

Yoshida, H., Kawata, K., Fukuyama, Y.: A particle swarm optimization for reactive power and voltage control considering voltage security assessment. IEEE Trans. Power Syst. **15**(4), 1232–1239 (2000)

Zhang, R., Sun, C., Zhang, J., Zhou, Y.: Second-order terminal sliding mode control for hypersonic vehicle in cruising flight with sliding mode disturbance observer. J. Control Theory Appl. **11**(2), 299–305 (2013)

Zhang, Y., Yao, F., Iu Ho-Ching, H., Fernando, T., Po Wong, K.: Sequential quadratic programming particle swarm optimization for wind power system operations considering emissions. J. Mod. Power Syst. Clean Energy **1**(3), 231–240 (2013)

Zheng, Z., Wu, C.: Power optimization of gas pipelines via an improved particle swarm optimization algorithm. Petrol. Sci. **9**(1), 89–92 (2012)

Robust Control of Robot Arms via Quasi Sliding Modes and Neural Networks

Maria Letizia Corradini, Andrea Giantomassi, Gianluca Ippoliti, Sauro Longhi and Giuseppe Orlando

Abstract This chapter presents a control approach for robotic manipulators based on a discrete-time sliding mode control which has received much less coverage in the literature with respect to continuous time sliding-mode strategies. This is due to its major drawback, consisting in the presence of a sector, of width depending on the available bound on system uncertainties, where robustness is lost because the sliding mode condition cannot be exactly imposed. For this reason, only ultimate boundedness of trajectories can be guaranteed, and the larger the uncertainties affecting the system are, the wider is the bound on trajectories which can be guaranteed. As a possible solution to this problem, in this chapter a discontinuous control law has been proposed, employing a controller inside the sector based on an estimation, as accurate as possible, of the overall effect of uncertainties affecting the system. Different solutions for obtaining this estimate have been considered and the achievable performances have be compared using experimental data. The first approach consists in estimating the uncertain terms by a well established method which is an adaptive on-line procedure for autoregressive modeling of non-stationary multivariable time series by means of a Kalman filtering. In the second solution, radial basis neural

M.L. Corradini
Scuola di Scienze e Tecnologie, Università di Camerino, via Madonna delle Carceri, 62032 Camerino, MC, Italy
e-mail: letizia.corradini@unicam.it

A. Giantomassi · G. Ippoliti (✉) · S. Longhi · G. Orlando
Dipartimento di Ingegneria dell'Informazione, Università Politecnica delle Marche, Via Brecce Bianche, 60131 Ancona, Italy
e-mail: gianluca.ippoliti@univpm.it

A. Giantomassi
e-mail: a.giantomassi@univpm.it

S. Longhi
e-mail: sauro.longhi@univpm.it

G. Orlando
e-mail: giuseppe.orlando@univpm.it

A.T. Azar and Q. Zhu (eds.), *Advances and Applications in Sliding Mode Control systems*, Studies in Computational Intelligence 576, DOI 10.1007/978-3-319-11173-5_3

networks are used to perform the estimation of the uncertainties affecting the system. The proposed control system is evaluated on the ERICC robot arm. Experimental evidence shows satisfactory trajectory tracking performances and noticeable robustness in the presence of model inaccuracies and payload perturbations.

1 Introduction

Robotic manipulators are highly nonlinear and uncertain dynamic systems which, being commonly used in industrial tasks, are expected to maintain good dynamic performance in face of unmodeled dynamics and uncertainties (Dixon 2007; Marton and Lantos 2011; Corradini et al. 2012). Indeed, the design of ideal controllers for such systems is a challenge for control engineers because of the nonlinearities and the coupling effects typical of robotic systems. A number of different approaches have been followed in order to cope with this problem, such as, for instance, feedback linearization (Abdallah et al. 1991; Melhem and Wang 2009; Li and Su 2013), model predictive control (Copot et al. 2012; Nikdel et al. 2014), and sliding mode control (Islam and Liu 2011; Capisani and Ferrara 2012; Corradini et al 2009, 2010). In general, control approaches not accounting for neglected dynamics and uncertainties can make the performance of the system, in terms of convergence, quite poor. As discussed in Capisani et al. (2007), global feedback linearization is possible in theory, but is difficult to achieve in practice as a consequence of uncertainties coming from incomplete knowledge of the kinematics and dynamics, from joint and link flexibility, actuator dynamics, friction, sensor noise, and unknown loads. This imposes the coupling of the inverse dynamics approach with robust control methodologies (Abdallah et al. 1991). It is well known that sliding mode methods provide noticeable robustness and invariance properties to matched uncertainties (Utkin 1992; Zinober 1994), and are computational simpler with respect to other robust control approaches. Recent literature contains a number of results about Sliding Mode Control (SMC) of manipulators, in some cases coupled with fuzzy control and/or neuro-fuzzy techniques (Wai and Muthusamy 2013; Han and Lee 2013; Chen 2008; Corradini et al. 2012). The largest part of these papers, however, uses the continuous-time dynamic model of the manipulator for design, leaving not addressed the issue of digitalization of the control law. Digital control systems are currently receiving considerable credit as a consequence of the recent advances in digital microprocessor technology, and relevant interest is currently growing in the design of controllers based on the digital model of the system. Nevertheless, the discrete time counterpart of sliding mode control design has received only a limited attention (Corradini et al. 2013; Ignaciuk and Bartoszewicz 2011; Veselic et al. 2010; Cimini et al. 2013; Lin et al. 2013; Xu 2013; Raspa et al. 2013; Milosavljevic et al. 2013; Furuta 1993; Corradini et al. 2012). Indeed, compared with continuous time sliding-mode strategies, the design problem in discrete-time has received much less coverage in the literature. This is due to its major drawback, consisting in the presence of a sector, of width depending on the available bound on uncertainties, where robustness is lost because the sliding mode condition cannot be exactly imposed. For this reason, only ultimate boundedness of trajectories can be guaranteed, and the larger the uncertainties affecting the system are, the wider is the bound on trajectories which can be guaranteed.

Therefore the main contribution of this chapter is to investigate, on a real industrial manipulator, a possible solution to this problem. In particular a discontinuous control law has been proposed, employing a controller inside the sector based on an estimation, as accurate as possible, of the overall effect of uncertainties affecting the system. In this chapter, different solutions for obtaining this estimate have been considered, and the achievable performances have been compared using experimental data.

The first approach consists in estimating the uncertain terms by a well established method which is an adaptive on-line procedure for autoregressive (AR) modeling of non-stationary multivariable time series by means of a Kalman filtering (KF) (Arnold et al. 1998).

In the second solution, Neural networks (NNs) are used to perform the estimation of the uncertainties affecting the system. It is well known in fact, that the learning ability of neural networks has been widely utilized in robotics to make controllers learn nonlinear characteristics of robots through experimental data, without a prior knowledge of their parameters and structure. Early NN-based control schemes for robotic manipulators produced good simulations or even experimental results (Ozaki et al. 1991; Ishiguro et al. 1992). More recently, stable neural network control schemes have been investigated, such as nonlinearly parameterized NN-based adaptive control (Ge et al. 2013; Chaoui and Sicard 2012) and linearly parameterized NN-based adaptive control (Sun et al. 2001; Sanner and Slotine 1995) for robotic manipulators. All these results proved that the stable NN-based control have the potential to deal with the difficulties for the control of robotic manipulators with unmodeled dynamics and uncertainties. The two books (Ge et al. 1998; Lewis et al. 1999) provide a good review of neural networks for the control of robotic manipulators.

An identification procedure is proposed in this chapter to estimate the uncertainties affecting the system using Radial Basis Function Networks (RBFNs). These networks have been widely used for nonlinear system identification (Yassin et al. 2011; Ko 2012; D'Amico et al. 2001; Ciabattoni et al. 2012, 2014) because of their ability to approximate complex nonlinear mappings from input-output data, of their simple topological structure allowing to avoid lengthy calculations (Giantomassi et al. 2011), and because of the chance they offer to reveal how learning proceeds in an explicit manner (Sundararajan et al. 2002).

The chapter is organized as follows. Algorithms to estimate uncertainties affecting the system are described in Sect. 2. In Sect. 3 details on the considered control are discussed. Results on robot arm experimental tests are reported in Sect. 4. The paper ends with comments on the performance of the proposed controller.

2 Uncertainty Estimators

In this section, different solutions to obtain an estimation of the uncertainties affecting the systems are proposed. In particular in Sect. 2.1 an adaptive AR model is considered while neural networks (NNs) are introduced in Sect. 2.2.

2.1 Adaptive AR Model

An Autoregressive (AR) model is considered in this subsection with the aim to estimate the uncertainties affecting the robotic manipulator. AR models belong to the family of equation error models, do not consider any input and are thus used to model time series (Box et al. 1994; Arnold et al. 1998). A scalar AR process of order d is given by:

$$y(k) = a_1(k)\, y(k-1) + a_2(k)\, y(k-2) + \cdots + a_d(k)\, y(k-d) + e(k), \quad (1)$$

where $e(k)$ is a sequence of independent and normal distributed random variables with zero expectation and variance of σ_q^2 (i.d. $\sim\mathcal{N}(0, \sigma_q^2)$). This variable can be interpreted as the uncertainty of the next signal value prediction by regressing the previous observations with the AR coefficients (prediction error) (Guidorzi 2003; Arnold et al. 1998). If the scalar values $y(k)$, $e(k)$ and $a_i(k)$, $i = 1, \ldots, d$ are replaced by vectors and matrices respectively, the AR process given in (1) results:

$$\boldsymbol{y}(k) = \boldsymbol{A}_1(k)\boldsymbol{y}(k-1) + \boldsymbol{A}_2(k)\boldsymbol{y}(k-2) + \cdots + \boldsymbol{A}_d(k)\boldsymbol{y}(k-d) + \boldsymbol{e}(k), \quad (2)$$

where $\boldsymbol{y}(\cdot)$ and $\boldsymbol{e}(\cdot)$ have dimension p of the modeled output space. Matrices $\boldsymbol{A}_i(k)$, $i = 1, \ldots, d$ are square and with the same dimension of $\boldsymbol{y}(\cdot)$. Typically, models in the forms (1) and (2) allow computing, at any time, a one-step ahead prediction of the output on the basis of the observations performed until that moment. Since the equation error is modeled by a white process and the output measurements are known until time sample $k-1$, the optimal predictor, characterized by whiteness and minimal variance of the prediction error, is obtained substituting the left term of (2) with its estimated expression, obtaining:

$$\hat{\boldsymbol{y}}(k) = \boldsymbol{A}_1(k)\boldsymbol{y}(k-1) + \boldsymbol{A}_2(k)\boldsymbol{y}(k-2) + \cdots + \boldsymbol{A}_d(k)\boldsymbol{y}(k-d) + \boldsymbol{e}(k). \quad (3)$$

It is important to note that predictor (3) does not rely on previous predictions (the prediction is a simple regression of observed output samples) and thus is free from stability constraints (Åström and Eykhoff 1971; Ljung 1999).

The computation of the models parameters, given by scalars elements $a_i(k)$, $i = 1, \ldots, d$ in expression (1) and by the elements of matrices $\boldsymbol{A}_i(k)$, $i = 1, \ldots, d$ in (2) and (3), can be seen as an optimization problem (the selection of the "best" model in the considered class) or, from another point of view, as a way of "tuning" the model on the data. Considering the non-stationarity of the signals which are samples from times-series the AR model is tuned by an on-line recursive and adaptive learning algorithm, as a Kalman Filter, instead of the classical approaches of least-squares, forward-backward or Yule-Walker that use off-line batch computation (Ljung 1999; Giantomassi 2012).

To make use of Kalman filter algorithm for parameter estimation of (3) a state space representation of the AR process with stochastic (time-varying) coefficients is given. The multivariable case is addressed developing a state-space representation of the model (3). This is achieved by rearranging the elements of coefficients matrices $\boldsymbol{A}_i(k), i = 1, \ldots, d$ in vector form using the vex-operator, which stacks the columns of a matrix on top of each other. Then, consider the following notation:

$$\begin{aligned} \boldsymbol{a}(k) &= vec((\boldsymbol{A}_1(k), \ldots, \boldsymbol{A}_d(k))^T) \\ \boldsymbol{v}(k) &= (\boldsymbol{y}^T(k-1), \boldsymbol{y}^T(k-2), \ldots, \boldsymbol{y}^T(k-d))^T \\ C(k) &= I_p \otimes \boldsymbol{v}^T(k) \end{aligned} \tag{4}$$

where the symbol $\otimes$ denotes the Kronecker-product of matrices, d the dimension of the vector process regression and I_p the identity matrix of dimension p. An appropriate state-space representation of the multivariate AR model (3) with stochastic coefficients is given by:

$$\begin{aligned} \boldsymbol{a}(k+1) &= \boldsymbol{a}(k) + \boldsymbol{v}(k) \\ \boldsymbol{v}(k) &= C(k)\boldsymbol{a}(k) + \boldsymbol{e}(k), \end{aligned} \tag{5}$$

where $\boldsymbol{v}(k) \sim \mathcal{N}(0, Q(k))$ and $\boldsymbol{e}(k) \sim \mathcal{N}(0, R(k))$. Thus the Kalman filter equations for the defined parameter estimation problem are introduced substituting the term $\boldsymbol{a}(k+1)$ in (5) with its estimation to be compute i.e. $\hat{\boldsymbol{a}}(k+1)$:

$$\begin{aligned} P(k+1|k) &= P(k) + Q(k) \\ K(k+1) &= P(k+1|k)C(k)[C(k)P(k+1|k)C^T(k) + R(k)] \\ \hat{\boldsymbol{a}}(k+1) &= \boldsymbol{a}(k) + K(k+1)[\boldsymbol{v}(k) - C(k)\boldsymbol{a}(k)] \\ P(k+1) &= [I - K(k+1)C(k)]P(k+1|k). \end{aligned} \tag{6}$$

2.2 Radial Basis Function Network

A RBFN with input pattern $\boldsymbol{\ell} \in \mathbb{R}^m$ and an output $\hat{\psi} \in \mathbb{R}^n$ implements a mapping $f_r : \mathbb{R}^m \to \mathbb{R}^n$ according to

$$\hat{\psi} = f_r(\boldsymbol{\ell}) = \kappa_0 + \sum_{i=1}^{n_r} \kappa_i \phi\left(\|\boldsymbol{\ell} - \boldsymbol{c}_i\|\right), \tag{7}$$

where $\phi(\cdot)$ is a given function from $\mathbb{R}^+$ to $\mathbb{R}^n$, $\kappa_i \in \mathbb{R}^n$, $i = 0, 1, \ldots, n_r$ are the weights or parameters, $\boldsymbol{c}_i \in \mathbb{R}^m$, $i = 1, 2, \ldots, n_r$, are the radial basis functions centers (called also units or neurons) and n_r is the number of centers (Chen et al. 1991). The RBFN is used for the estimation of the uncertainties affecting the robotic manipulator. The uncertainty dynamics can be taken into account through the network input pattern $\boldsymbol{\ell}$, that is composed of a proper set of system input and output samples acquired in a finite set of past time instants (Haykin 1999) as specified in (27).

For the non-linearity $\phi(\cdot)$, a function of the distance d_i between the current input $\boldsymbol{\ell}$ and the centre $\boldsymbol{c}_i$, the following gaussian function is considered:

$$\phi(d_i) = \exp\left(-d_i^2/\beta^2\right), \qquad i = 1, 2, \ldots, n_r \tag{8}$$

where $d_i = \|\boldsymbol{\ell} - \boldsymbol{c}_i\|$ and the real constant β is a scaling or "width" parameter (Chen et al. 1991).

2.2.1 Orthogonal Least Squares Algorithm

By providing a set of network input pattern $\boldsymbol{\ell}(k)$ and the corresponding desired output $\psi(k)$ to be approximated by the net, for $k = 1, 2, \ldots, D$, the centers $\boldsymbol{c}_i$, $i = 1, 2, \ldots, n_r$ are generally chosen from the data set $\{\boldsymbol{\ell}(k)\}_{k=1}^{D}$. The Orthogonal Least Squares (OLS) algorithm (Chen et al. 1991) is an efficient method for selecting centers from the data set obtaining adequate and parsimonious RBFN thus reducing computational complexity and numerical ill-conditioning. To apply this method, the RBFN needs to be expressed by (7) as a linear regression model:

$$\psi(k) = \sum_{i=1}^{M} \gamma_i(k)\delta_i + \epsilon(k) \tag{9}$$

where $M := n_r + 1$, $\psi(k)$ is the desired output to be approximated by the net, $\epsilon(k) := \psi(k) - \hat{\psi}(k)$ is the error signal, $\delta_i := \kappa_{i-1}$ are the parameters to be estimated, $\gamma_i(k) := \phi(\|\boldsymbol{\ell}(k) - \boldsymbol{c}_i\|)$ are given fixed function of $\boldsymbol{\ell}(k)$ where the centers $\boldsymbol{c}_i$ have to be fixed and $\gamma_1(k) = 1$. The error signal $\epsilon(k)$ is assumed to be uncorrelated with $\gamma_i(k)$. The OLS method, is considered for the selection of centers $\boldsymbol{c}_i$ from the data set $\{\boldsymbol{\ell}(k)\}_{k=1}^{D}$, with a reduced number M of $\gamma_i(\cdot)$. These significant regressors $\gamma_i(k)$, $i = 1, 2, \ldots, M$ can be selected using the OLS algorithm operating in a forward regression way (Chen et al. 1991).

Arranging (9) for $k = 1, 2, \ldots, D$ in the following matrix form:

$$\boldsymbol{\Psi} = \boldsymbol{\Gamma}\boldsymbol{\Delta} + \boldsymbol{E} \tag{10}$$

where $\boldsymbol{\Psi} := [\psi(1) \ldots \psi(D)]^T$, $\boldsymbol{\Delta} := [\delta_1 \ldots \delta_M]^T$, $\boldsymbol{E} := [\epsilon(1) \ldots \epsilon(D)]^T$ and $\boldsymbol{\Gamma} := \left[\boldsymbol{\gamma}_1 \ldots \boldsymbol{\gamma}_M\right]$ with $\boldsymbol{\gamma}_i := \left[\gamma_i(1)\ \gamma_i(2) \ldots \gamma_i(D)\right]^T$, $i = 1, 2, \ldots, M$ the OLS method involves the transformation of the set of $\boldsymbol{\gamma}_i$ into a set of orthogonal basis vectors, z_i, $i = 1, 2, \ldots, M$, where the space spanned by the set of orthogonal basis vectors z_i, $i = 1, 2, \ldots, M$, is the same space spanned by the set of $\boldsymbol{\gamma}_i$, $i = 1, 2, \ldots, M$, and (10) can be rewritten as

$$\boldsymbol{\Psi} = \boldsymbol{Z}\boldsymbol{h} + \boldsymbol{E}, \tag{11}$$

with $\boldsymbol{Z} = [z_1,\ z_2, \ldots, z_M]$. This algorithm makes it possible to calculate the individual contribution to the desired output energy from each basis vector. Because z_i and z_j are orthogonal for $i \neq j$, the sum of squares or energy of $\boldsymbol{\Psi}(k)$ is

$$\boldsymbol{\Psi}^T\boldsymbol{\Psi} = \sum_{i=1}^{M} \boldsymbol{h}_i^2 z_i^T z_i + \boldsymbol{E}^T\boldsymbol{E}, \tag{12}$$

and this relation suggests to consider as the error reduction ratio due to z_i the following quantity:

$$[err]_i := \boldsymbol{h}_i^2 z_i^T z_i \Big/ \left(\boldsymbol{\Psi}^T\boldsymbol{\Psi}\right), \quad i = 1, 2, \ldots, M. \tag{13}$$

The regressors selection procedure terminates at the M_sth step when

$$1 - \sum_{j=1}^{M_s} [err]_j < \Xi \tag{14}$$

where $0 < \Xi < 1$ is a chosen tolerance. This gives rise to a model containing only M_s significant regressors. The orthogonal property makes the whole selection procedure simple and efficient and the tolerance Ξ is an important parameter in balancing the accuracy and the complexity of the final network.

2.2.2 *K*-Means Clustering Algorithm

The OLS method can be employed as a forward regression procedure (Chen et al. 1991) to select a suitable set of centers $n_s \leq n_r$, from a large initial set of candidates, for the RBFN $f_r : \mathbb{R}^m \to \mathbb{R}^n$. In the developed solution the initial set of centers is obtained using the k-means unsupervised clustering algorithm that starting from a reasonable high number of centers randomly chosen in the input space, moves them in the most significative regions of the input space (Chen et al. 1991).

The k-means clustering algorithm is given by the sequential execution of the following steps:

1 Initialize the cluster centers $\boldsymbol{c}_j$, $j = 1, 2, \ldots, n_c$. This centers are randomly chosen from the input data set $\{\boldsymbol{\ell}(k)\}_{k=1}^{D}$.

2 Group all the inputs of the data set $\{\boldsymbol{\ell}(k)\}_{k=1}^{D}$ in the sets Ω_j, $j = 1, 2, \ldots, n_c$, where Ω_j is the set of input vectors $\boldsymbol{\ell}(\cdot)$ closest to the cluster center $\boldsymbol{c}_j$, i.e. $\Omega_j := \{\boldsymbol{\ell}(\cdot) \in \{\boldsymbol{\ell}(k)\}_{k=1}^{D} \mid \|\boldsymbol{\ell}(k) - \boldsymbol{c}_j\| = \min_{i=1}^{D} \|\boldsymbol{\ell}(k) - \boldsymbol{c}_i\|\}$.
3 For all $j = 1, 2, \ldots, n_c$ do $\boldsymbol{c}_j = \frac{1}{M_j} \sum_{\boldsymbol{\ell} \in \Omega_j} \boldsymbol{\ell}(k)$, where M_j is the number of input vectors $\boldsymbol{\ell}(k) \in \Omega_j$.
4 If there is no change in the cluster assignments of step 3 from one iteration to the next the algorithm is stopped otherwise go back to step 2.

2.2.3 "Width" of the Radial Basis Functions

Once the centers have been selected, the normalization parameter β^2 of Eq. (8), that represents a measure of the spread of the data associated with each centre, has to be determined. No rigorous method exists to calculate this parameter. In the developed solution an Akaike-type criteria (Chen et al. 1991) is proposed for choosing the normalization parameter obtaining a good compromise between estimate accuracy and network complexity (Haykin 1999). Akaike-type criteria which compromises between the performance and the number of parameters has the following form:

$$AIC(\chi) = N \log(\sigma_\varepsilon^2) + M_s \chi \tag{15}$$

where χ is the critical value of the chi-squared distribution with one degree of freedom and for a given level of significance, N is the number of data set, σ_ε^2 is the variance of the net estimate residuals and M_s is the number of significant regressors. The procedure terminates when $AIC(\chi)$ reaches its minimum.

3 Control Design

3.1 Preliminaries

From the Euler-Lagrangian formulation, the equations of motion of a robot manipulator can be written as (Siciliano et al. 2009)

$$\boldsymbol{B}(\boldsymbol{q})\ddot{\boldsymbol{q}} + \boldsymbol{C}(\boldsymbol{q}, \dot{\boldsymbol{q}})\dot{\boldsymbol{q}} + \boldsymbol{F}_v\dot{\boldsymbol{q}} + \boldsymbol{G}(\boldsymbol{q}) = \boldsymbol{\tau} \tag{16}$$

where $\boldsymbol{q} \in \mathbb{R}^n$ is the vector of generalized coordinates (rotational joint configurations), $\boldsymbol{B}(\boldsymbol{q}) \in \mathbb{R}^{n \times n}$ is the inertia matrix, $\boldsymbol{C}(\boldsymbol{q}, \dot{\boldsymbol{q}})\dot{\boldsymbol{q}} \in \mathbb{R}^n$ represents centrifugal and Coriolis torques, $\boldsymbol{F}_v \in \mathbb{R}^{n \times n}$ is the diagonal matrix of the viscous friction coefficients, $\boldsymbol{G}(\boldsymbol{q}) \in \mathbb{R}^n$ is the vector of gravitational torques and $\boldsymbol{\tau} \in \mathbb{R}^n$ is the vector of torques acting at the joints. As well known, the robot model (16) is characterized by the structural properties given in Siciliano et al. (2009). Introducing the state vector

$\boldsymbol{x} = \left[x_1 \ldots x_n \; x_{n+1} \ldots x_{2n}\right]^T = \left[\boldsymbol{q}^T \dot{\boldsymbol{q}}^T\right]^T$, the control input $\boldsymbol{u} = \boldsymbol{\tau}$, and considering possible uncertainties affecting model (16), this latter can be expressed as:

$$\begin{aligned}\dot{\boldsymbol{x}} &= \boldsymbol{f}_c(\boldsymbol{x}) + \boldsymbol{g}_c(\boldsymbol{x})\boldsymbol{u} \\ &= \boldsymbol{f}_c^0(\boldsymbol{x}) + \boldsymbol{\Delta f}_c(\boldsymbol{x}) + (\boldsymbol{g}_c^0(\boldsymbol{x}) + \boldsymbol{\Delta g}_c(\boldsymbol{x}))\boldsymbol{u}\end{aligned} \tag{17}$$

where $\boldsymbol{\Delta f}_c(\boldsymbol{x})$, $\boldsymbol{\Delta g}_c(\boldsymbol{x})$ depend on the uncertainties, while the nominal model is given by $\boldsymbol{f}_c^0(\boldsymbol{x})$, $\boldsymbol{g}_c^0(\boldsymbol{x})$:

$$\boldsymbol{f}_c^0(\boldsymbol{x}) = \begin{bmatrix} x_{n+1} \\ \vdots \\ x_{2n} \\ f_1(\boldsymbol{x}) \\ \vdots \\ f_n(\boldsymbol{x}) \end{bmatrix}; \; \boldsymbol{g}_c^0(\boldsymbol{x}) = \begin{bmatrix} & \boldsymbol{0}_{n\times n} & \\ g_{1,1}(\boldsymbol{x}) & \cdots & g_{1,n}(\boldsymbol{x}) \\ \cdots & \cdots & \cdots \\ g_{n,1}(\boldsymbol{x}) & \cdots & g_{n,n}(\boldsymbol{x}) \end{bmatrix} \tag{18}$$

Note that all the terms present in (17), (18) can be easily computed from (16).

Assumption 1 In view of the existence of physical bounds on achievable positions and velocities by the robot arm, it is assumed that the uncertain terms $\boldsymbol{\Delta f}_c(\boldsymbol{x})$, $\boldsymbol{\Delta g}_c(\boldsymbol{x})$ are norm bounded.

A planar two-link manipulator with revolution joints (Siciliano et al. 2009) will be considered in this paper, in order to illustrate the feasibility of the proposed control algorithm. Therefore the variable $\boldsymbol{q}$ is $\boldsymbol{q} = [q_2 \; q_3]^T$, where q_2, q_3 denote the joint displacements of the two considered rotational joints 2 and 3 of Fig. 1. The arm dynamics is described by (16) with $n = 2$ and the detailed model can be found in Nicosia and Tomei (1990), Siciliano et al. (2009).

3.2 Sliding Mode Controller Design

In this section, the development of an estimation-based discrete-time sliding mode control law is described, aimed at solving the trajectory tracking problem in the joint space of the considered planar two-link manipulator.

Control design will be carried out in the discrete time framework, and discretization after control design (performed in the continuous-time framework) will be avoided, in accordance to the discussion reported in Young et al. (1999). In this sense, the design approach used belongs to the so called 'classical' sliding mode design techniques in the framework of discrete-time sliding modes (Furuta 1993; Gao et al. 1995). In this context, several approaches are available in literature for the discretization of a linear plant using Zero Order Hold (ZOH) method (Wang et al. 2008, 2009, 2010), showing that inherent properties of SMC are not maintained

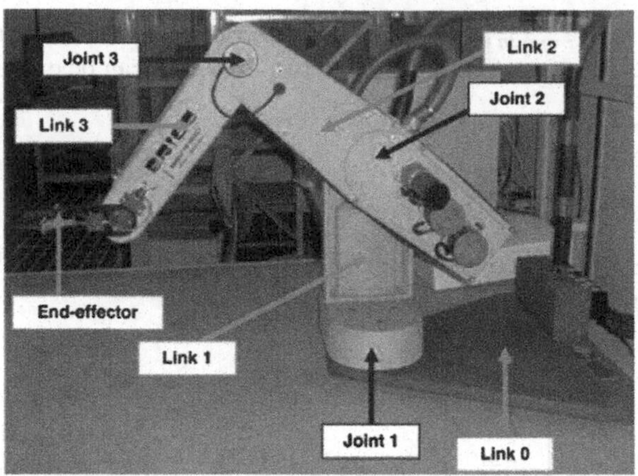

Fig. 1 ERICC manipulator

after discretization. However, due to the presence of strong non linearities in the robot model, a simpler approach has to be preferred for the plant discretization, using the Euler method. Finally, consider that the plant discretization method is likely not to seriously affect closed loop performances, since the control design is performed directly in the discrete time domain (Young et al. 1999).

Considering a sampling time T_c, and discretizing the uncertain model (17) by Euler method, one has:

$$\begin{cases} \boldsymbol{q}_s(k+1) = \boldsymbol{q}_s(k) + T_c\dot{\boldsymbol{q}}_s(k) \\ \dot{\boldsymbol{q}}_s(k+1) = \dot{\boldsymbol{q}}_s(k) + T_c\boldsymbol{f}(k) + T_c\boldsymbol{g}(k)\boldsymbol{u}(k) + \boldsymbol{n}(k) \end{cases} \tag{19}$$

with:

$$\boldsymbol{q}_s(k) = [x_1(kT_c)\ x_2(kT_c)]^T, \quad \dot{\boldsymbol{q}}_s(k) = [x_3(kT_c)\ x_4(kT_c)]^T.$$

Moreover:

$$\boldsymbol{f}(k) = [f_1(k)\ f_2(k)]^T, \quad \boldsymbol{g}(k) = \begin{bmatrix} g_{1,1}(k)\ g_{1,2}(k) \\ g_{2,1}(k)\ g_{2,2}(k) \end{bmatrix}$$

[see (18) for $n = 2$], where with some abuse of notation we have written $f_i(k) = f_i(\boldsymbol{x}(kT_c))$, $i = 1, 2$, $g_{i,j}(k) = g_{i,j}(\boldsymbol{x}(kT_c))$, $i, j = 1, 2$. Finally, $\boldsymbol{n}(k)$ is given by:

$$\boldsymbol{n}(k) = T_c(\boldsymbol{\Delta f}(k) + \boldsymbol{\Delta g}(k)\boldsymbol{u}(k)) \tag{20}$$

with

$$\boldsymbol{\Delta f}(k) = \boldsymbol{\Delta f}_c(\boldsymbol{x}(kT_c)), \quad \boldsymbol{\Delta g}(k) = \boldsymbol{\Delta g}_c(\boldsymbol{x}(kT_c)).$$

Assumption 2 It is assumed that $\boldsymbol{g}(k)$ and $\boldsymbol{g}(k) + \boldsymbol{\Delta g}(k)$ are invertible matrices $\forall\, \boldsymbol{x}(kT_c)$, $\forall k$ for the chosen T_c.

Remark 1 According to Assumption 1, the matrix $\boldsymbol{g}(k)$ and the uncertain terms $\boldsymbol{\Delta f}(k)$ and $\boldsymbol{\Delta g}(k)$ are bounded by known constants:

$$||\boldsymbol{g}(k)|| \leq g_M; \quad ||\boldsymbol{\Delta f}(k)|| \leq \rho_f;$$

$$||\boldsymbol{\Delta g}(k)|| \leq \rho_g; \quad ||\boldsymbol{g}(k) + \boldsymbol{\Delta g}(k)|| \geq g_{min}.$$

The control law ensuring the robust tracking of a reference variable $\boldsymbol{q}_d(k) = [x_{1,d}(kT_c)\ x_{2,d}(kT_c)]^T$ by the sampled position $\boldsymbol{q}_s(k) = [x_1(kT_c)\ x_2(kT_c)]^T$ will be described in the following.

Define the discrete-time tracking error as $\boldsymbol{\pi}(k) = [\pi_1(kT_c)\ \pi_2(kT_c)]^T = \boldsymbol{q}_s(k) - \boldsymbol{q}_d(k)$, and consider the following discrete-time variable:

$$\boldsymbol{s}(k) = \boldsymbol{\pi}(k+1) - \boldsymbol{\Lambda}\boldsymbol{\pi}(k) \tag{21}$$

with $eig(\boldsymbol{\Lambda}) = \lambda_i$, $i = 1, 2$ such that $|\lambda_i| < 1$. Using (19), it can be shown that:

$$\boldsymbol{s}(k) = (\boldsymbol{I} - \boldsymbol{\Lambda})\boldsymbol{q}_s(k) + T_c\dot{\boldsymbol{q}}_s(k) - \boldsymbol{q}_d(k+1) + \boldsymbol{\Lambda}\boldsymbol{q}_d(k) \tag{22}$$

Remark 2 Note that $\boldsymbol{s}(k)$ is always computable at the time instant k. In fact, from (19) the term $\boldsymbol{q}_s(k+1)$ present in (21) can be replaced by $\boldsymbol{q}_s(k+1) = \boldsymbol{q}_s(k) + T_c\dot{\boldsymbol{q}}_s(k)$, producing (22).

Moreover, consider the following sliding surface:

$$\boldsymbol{\sigma}(k) = \boldsymbol{s}(k) - \alpha\boldsymbol{s}(k-1) = \boldsymbol{0}; \quad 0 < |\alpha| < 1. \tag{23}$$

Theorem 1 *Consider the arm model* (19) *and Remark 1. A quasi-sliding motion on the surface* (23) *is enforced by the control law* $\boldsymbol{u}(k) = \boldsymbol{u}^{eq}(k) + \boldsymbol{u}^n(k)$, *with*

$$\begin{aligned} T_c^2\boldsymbol{g}(k)\boldsymbol{u}^{eq}(k) = &-(\boldsymbol{I} - \boldsymbol{\Lambda})\boldsymbol{q}_s(k) - T_c(2\boldsymbol{I} - \boldsymbol{\Lambda})\dot{\boldsymbol{q}}_s(k) \\ &- T_c^2\boldsymbol{f}(k) - \boldsymbol{\Lambda}\boldsymbol{q}_d(k+1) + \boldsymbol{q}_d(k+2) + \alpha\boldsymbol{s}(k) \end{aligned} \tag{24}$$

and

$$\boldsymbol{u}^n(k) = \begin{cases} \theta\,(||\tilde{\boldsymbol{\sigma}}(k)|| - \rho_s)\cdot\begin{bmatrix}\frac{1}{\sqrt{2}}\\ \frac{1}{\sqrt{2}}\end{bmatrix} & if\ ||\tilde{\boldsymbol{\sigma}}(k)|| - \rho_s > 0 \\ -[\boldsymbol{g}(k)T_c^2]^{-1}\hat{\boldsymbol{n}}(k) & otherwise \end{cases} \tag{25}$$

with $0 < |\theta| < 1$, $0 < |\alpha| < 1$ *and:*

$$\tilde{\boldsymbol{\sigma}}(k) = \frac{\boldsymbol{\sigma}(k)}{T_C^2(g_M + \rho_g)}; \quad \rho_s = \frac{\rho_f + \rho_g U_M}{g_{min}} \tag{26}$$

where U_M is the maximum input torque supplied by joint actuators, i.e. $||\boldsymbol{u}|| \leq U_M$. The approximation $\hat{\boldsymbol{n}}(k)$ of $\boldsymbol{n}(k)$, can be performed or by the AR model (3) *with $\hat{\boldsymbol{y}}(k) := \hat{\boldsymbol{n}}(k) \in \mathbb{R}^2$ or by a neural network $f_r : \mathbb{R}^6 \to \mathbb{R}^2$ of the form* (7) *with the output $\hat{\psi}(k) := \hat{\boldsymbol{n}}(k) \in \mathbb{R}^2$ and the input pattern $\boldsymbol{\ell} \in \mathbb{R}^6$ defined as*

$$\boldsymbol{\ell}(k) := [\boldsymbol{q}_s(k)\, \boldsymbol{u}^n(k-1)\, \boldsymbol{n}(k-1)]. \tag{27}$$

The desired output of the AR model or the NN has the form $\boldsymbol{n}(k) \in \mathbb{R}^2$ given by Eq. (20).

Proof Inserting (24) in $\boldsymbol{\sigma}(k+1)$ gives:

$$\boldsymbol{\sigma}(k+1) = T_C^2 \boldsymbol{g}(k)\boldsymbol{u}^n(k) + T_C \boldsymbol{n}(k). \tag{28}$$

The imposition of the condition $||\boldsymbol{\sigma}(k+1)|| < ||\boldsymbol{\sigma}(k)||$ produces, considering (20) and (28):

$$\left|\left|\left[\boldsymbol{g}(k) + \Delta \boldsymbol{g}(k)\right] \cdot \left[\boldsymbol{u}^n(k) + \boldsymbol{d}(k)\right]\right|\right| < \frac{||\boldsymbol{\sigma}(k)||}{T_c^2} \tag{29}$$

with $\boldsymbol{d}(k)$ given by:

$$\boldsymbol{d}(k) = \left[\boldsymbol{g}(k) + \Delta \boldsymbol{g}(k)\right]^{-1} \left[\Delta \boldsymbol{f}(k) + \boldsymbol{\Delta g}(k)\boldsymbol{u}^{eq}(k)\right]. \tag{30}$$

Condition (29) is fulfilled if:

$$||\boldsymbol{u}^n(k) + \boldsymbol{d}(k)|| < ||\tilde{\boldsymbol{\sigma}}(k)||. \tag{31}$$

Unlikely continuous-time sliding modes, for discrete-time systems the plant cannot be permanently restricted to the designed surface. What can be ensured is the following decreasing condition $||\boldsymbol{\sigma}(k+1)|| < ||\boldsymbol{\sigma}(k)||$, which unfortunately cannot be ensured $\forall k$ but can be guaranteed outside a given region. In fact, it is easy to verify that condition (31) is guaranteed by $\boldsymbol{u}^n(k)$ given in (25) when $||\tilde{\boldsymbol{\sigma}}(k)|| > \rho_s$. On the contrary, when $||\tilde{\boldsymbol{\sigma}}(k)|| \leq \rho_s$, i.e. inside the sector, the sliding mode condition cannot be imposed exactly and an estimation $\hat{\boldsymbol{n}}(k)$ of $\boldsymbol{n}(k)$ is used, given by the AR model of Eq. (3) or a neural network of the form (7). Replacing $\boldsymbol{n}(k)$ by $\hat{\boldsymbol{n}}(k)$ in (28), and setting $\boldsymbol{\sigma}(k+1) = \boldsymbol{0}$, control law (25) is obtained, for $||\tilde{\boldsymbol{\sigma}}(k)|| \leq \rho_s$. The previous developments can be summarized as follows: the variable $\boldsymbol{\sigma}(k)$ tends to the region $||\tilde{\boldsymbol{\sigma}}(k)|| \leq \rho_s$ because of the choice of $\boldsymbol{u}^n(k)$ given in (25). Once such region is entered, it approximately holds $\boldsymbol{\sigma}(k+1) \simeq \boldsymbol{0}$ in view of the approximation capability of AR models and NNs.

4 Experimental Implementation

The proposed controller has been implemented on an ERICC robot arm (Fig. 1), built by Barras Provence (France). The robot is installed in the Robotics Laboratory at the Dipartimento di Ingegneria dell'Informazione, of the Università Politecnica delle Marche. In Fig. 1 is shown the robot with labels indicating the three base joints. Other two joints (not indicated in Fig. 1) are for wrist movements. In this section, the experimental setup and results are discussed.

4.1 Experimental Setup

The considered robot has five degrees of freedom but for the sake of simplicity only links 2 and 3 have been utilized in the experiments. Anyway, the developed experimental validation over a real planar robot can give the feasibility of this industrial application. The two considered rotational joints 2 and 3 are actuated by two dc motors with reduction gears. Position measurements are obtained by means of potentiometers and velocity measurements by tachometers. The ERICC command module consists of a power supply module, which provides the servo power for the system; a joint interface module, which contains the hardware to drive the motors and provides sensor feedback from each joint; and a processor module to run user developed software. In order to implement complex control algorithms, a new controller is used in this setup in place of the original ERICC processor module. This system, including hardware and software, combines an experimental apparatus with an easy-to-use software platform based on a dSPACE controller board (http://www.dspaceinc.com 2011). In particular the control law is implemented on a dSPACE DS1102 real-time controller board. A sampling time of 0.01 s has been used.

4.2 Structure and Validation of Implemented Estimators

Training and testing phases of considered AR models and NNs have been performed off-line with data acquired on a set of planned trajectories chosen with different shapes and considering different payload configurations for the robot. In particular an AR model and a RBFN are designed to estimate uncertainties $\boldsymbol{n}(k) = [n_2(k)\, n_3(k)]^T$ of controlled links 2 and 3. As measure of the performance of the proposed estimation algorithms residuals have been calculated, $\boldsymbol{e}(\cdot) = \boldsymbol{n}(\cdot) - \hat{\boldsymbol{n}}(\cdot) = [e_2(\cdot)\, e_3(\cdot)]^T$, and whiteness test on the estimation errors $e_i(\cdot)$, $i = 2, 3$ (residuals) has been used for validation (Ljung 1999). The whiteness of residuals is evaluated by computing the sample covariances

$$\hat{R}^K_{e_i}(\tau) = \frac{1}{K}\sum_{k=1}^{K} e_i(k)e_i(k+\tau),\, i = 2,\, 3 \tag{32}$$

with $\tau = 1, \ldots, S$. If $e_i(\cdot)$, $i = 2, 3$ are white-noise sequences, then the quantities

$$\zeta_i^{K,S} = \frac{K}{(\hat{R}^K_{e_i}(0))^2}\sum_{\tau=1}^{S}(\hat{R}^K_{e_i}(\tau))^2,\, i = 2,\, 3 \tag{33}$$

will have, asymptotically, a chi-square distribution $\chi^2(S)$ (Ljung 1999). The independence between residuals can be verified by testing whether $\zeta_i^{K,S} < \chi^2_{\varrho}(S)$, $i = 2, 3$, the ϱ level of the $\chi^2(S)$-distribution, for a significant choice of ϱ. Typical choices of ϱ range from 0.05 to 0.005.

The order of the AR model is chosen by the Minimum Description Length (MDL) test (Ljung 1999). In Fig. 2 the AR model order for the uncertain terms $n_2(k)$ and $n_3(k)$ is chosen as $d = 15$.

For the AR model tests are made to tune the covariance matrices of the process and measurements noise. In the considered experimental tests the numeric values of these parameters are $P(0) = 0.4, \sigma^2_q(0) = 0.1$ and $\sigma^2_{r,i}(0) = 0.0001, i = 1, \ldots, pd$.

The set of experimental data used to train the NNs is given by the pairs $(\boldsymbol{\ell}(k), \boldsymbol{n}(k))$, $k = 1, 2, \ldots$, where $\boldsymbol{n}(k)$ and $\boldsymbol{\ell}(k)$have the form specified in Eqs. (20) and (27), respectively. These data sets are used to train the nets offline by the OLS-based algorithm of Sect. 2.2 (Chen et al. 1991; Antonini et al. 2006). Data have been also normalized in order to have the same range. The complexity of RBFNs (i.e. the number of centers), has been chosen to match the approximation capability of nets to a low nets complexity which is necessary for the implementation of the control scheme. The number of centers has been chosen to obtain a good trade off among the time complexity of learning, computation efforts of the resulting NN control schemes and the accuracy of predictions for obtaining satisfactory control performance. Therefore, hidden layers of these nets are chosen with a number of 19 centers.

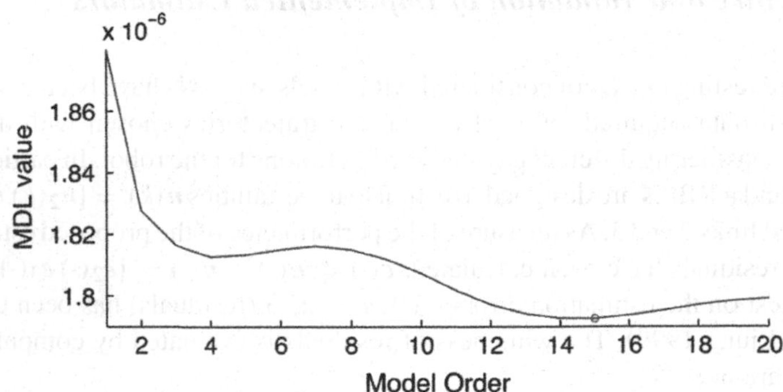

Fig. 2 MDL test for the AR model of the uncertain terms $n_i(k)$, $i = 2, 3$. The minimum is highlighted by the *circle*

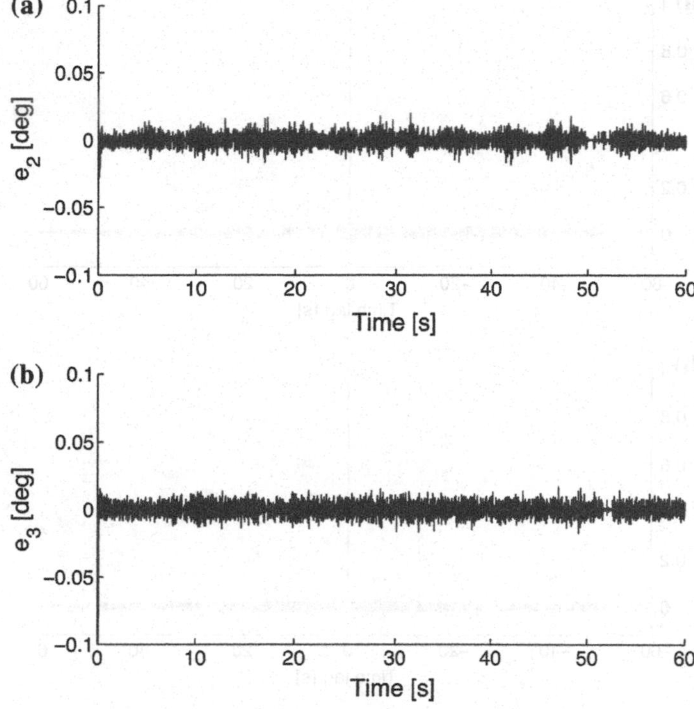

Fig. 3 Residuals obtained by the estimation performed by the network **a** $e_2(\cdot)$; **b** $e_3(\cdot)$

A sample of the performed estimation tests is given in Figs. 3 and 4 for the estimation of the uncertainty $\boldsymbol{n}(k)$. Residuals of the performed estimation shown in Fig. 3a, b confirm that the implemented NN is accurate, in particular the Mean Square of the Error (MSE) is 1.27×10^{-4} and 1.52×10^{-4} for joint 2 and 3, respectively.

In Fig. 4a, b the sample covariances of residuals $e_2(\cdot)$ and $e_3(\cdot)$ are reported; the whiteness test passes with $\varrho = 0.005$.

4.3 Experimental Results

Experimental results have been collected for trajectory tracking tasks performed in the robot joint space. The parameters of the discrete-time SMC law for both joints, are reported in Table 1.

A set of experimental results is reported in Figs. 6 through 15, obtained for the robot following the reference trajectories depicted in Fig. 5. In these figures, the performance produced by the proposed solutions for the uncertainties compensation in the SMC are illustrated for the robot following the reference trajectories of Fig. 5 with and without a payload.

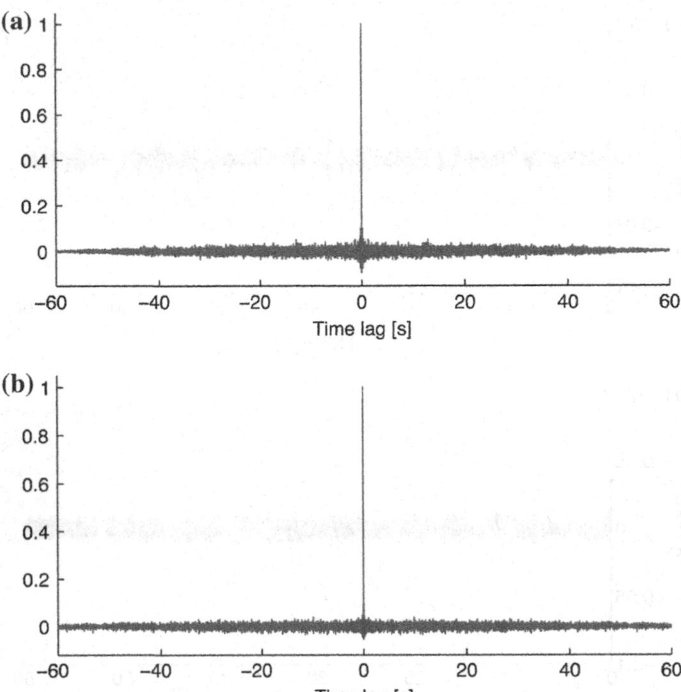

Fig. 4 Sample covariance of the residuals obtained by the estimation performed by the network **a** $e_2(\cdot)$; **b** $e_3(\cdot)$. The whiteness test passes with $\varrho = 0.005$

Table 1 SMC parameters for both joints

Param.	Value
ρ_s	0.0859
$\min_q \|\|\mathbf{g}(k) + \Delta\mathbf{g}(k)\|\|$	10.3385
θ_{q2}	0.1624
θ_{q3}	0.5019
$eig(\mathbf{\Lambda})$	$[0.92\ 0.91]^T$
α	0.95

Figures 6, 7, 8 and 9 show the performance when the robot is without a payload. In particular in Figs. 6 and 7 performance of the RBFN based SMC are shown. The tracking error of the RBFN based SMC is shown in Fig. 6. The voltage control inputs from the dSPACE controller board are depicted in Fig. 7.

The performance using the adaptive AR based SMC are shown in Figs. 8 and 9. The tracking errors are shown in Fig. 8 and voltage control inputs are shown in Fig. 9.

Figures 10 and 11 show the performance of the proposed control solutions when the reference robot motion trajectories are the same as before (see Fig. 5) and the robot moves a payload of 2 Kg. For the RBFN based SMC the tracking errors are

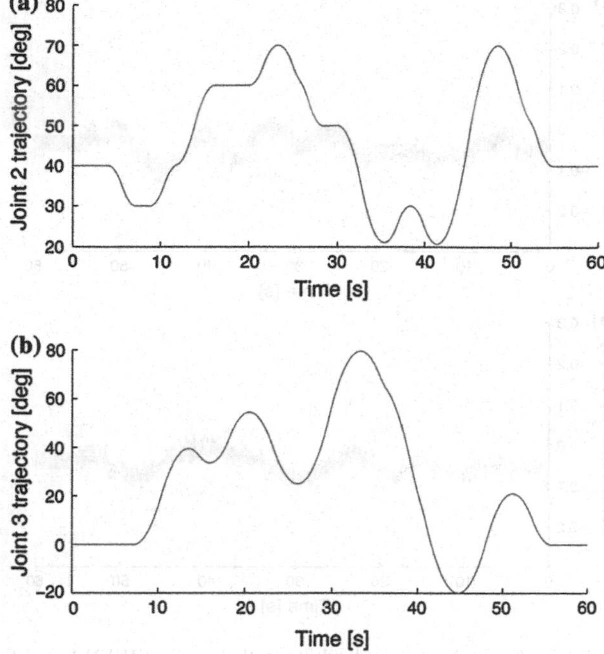

Fig. 5 Reference trajectories used for experiments: **a** joint 2; **b** joint 3

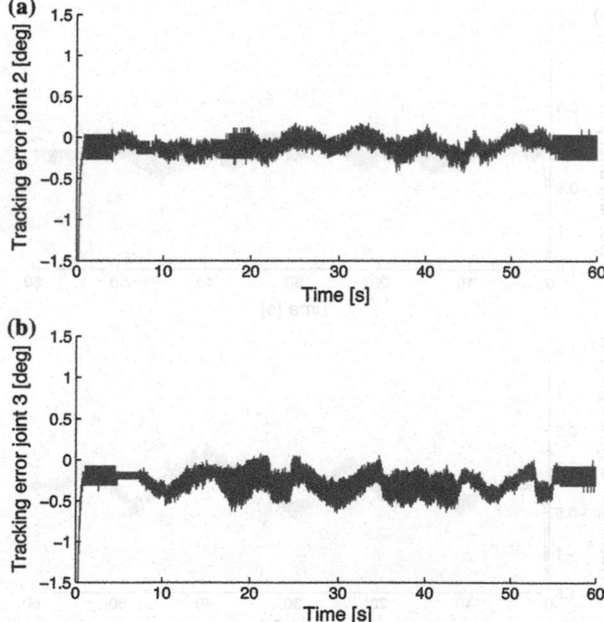

Fig. 6 Results for the robot without a payload controlled by the RBFN based SMC—Tracking errors: **a** joint 2; **b** joint 3

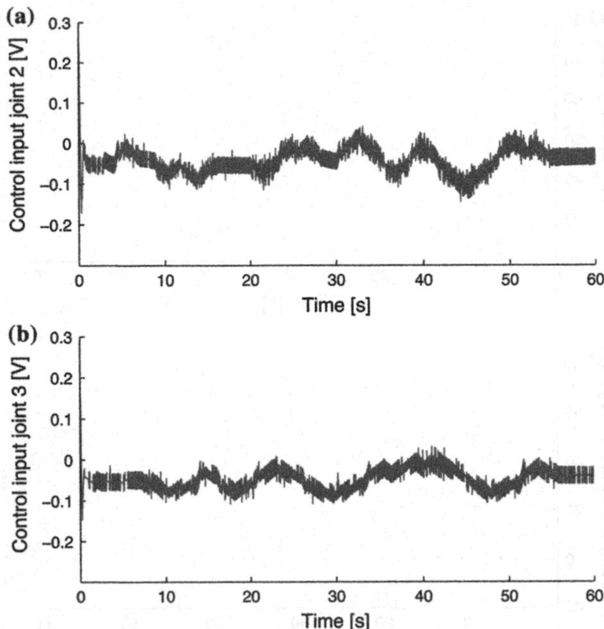

Fig. 7 Results for the robot without a payload controlled by the RBFN based SMC—Control inputs: **a** joint 2; **b** joint 3

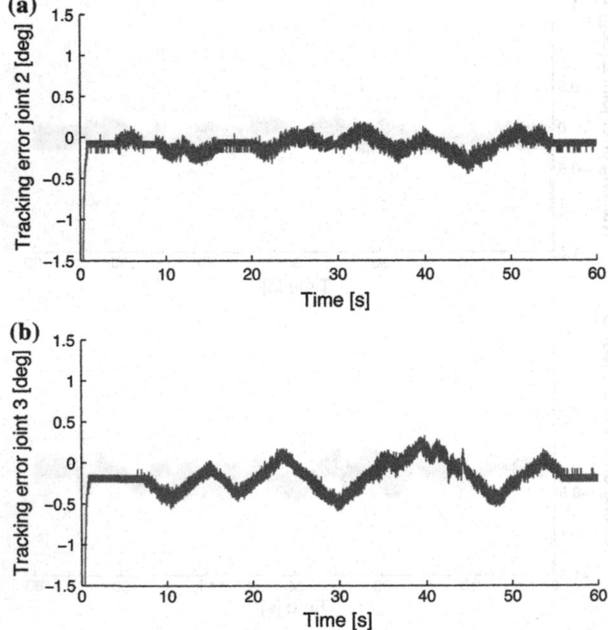

Fig. 8 Results for the robot without a payload controlled by the adaptive AR based SMC—Tracking errors: **a** joint 2; **b** joint 3

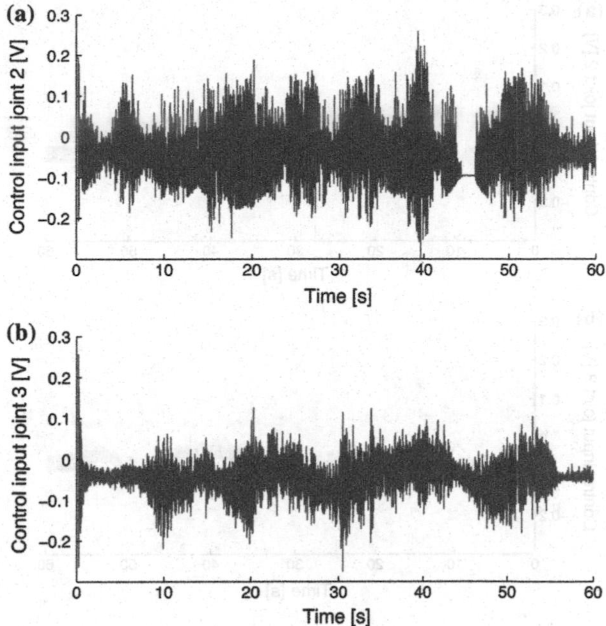

Fig. 9 Results for the robot without a payload controlled by the adaptive AR based SMC—Control inputs: **a** joint 2; **b** joint 3

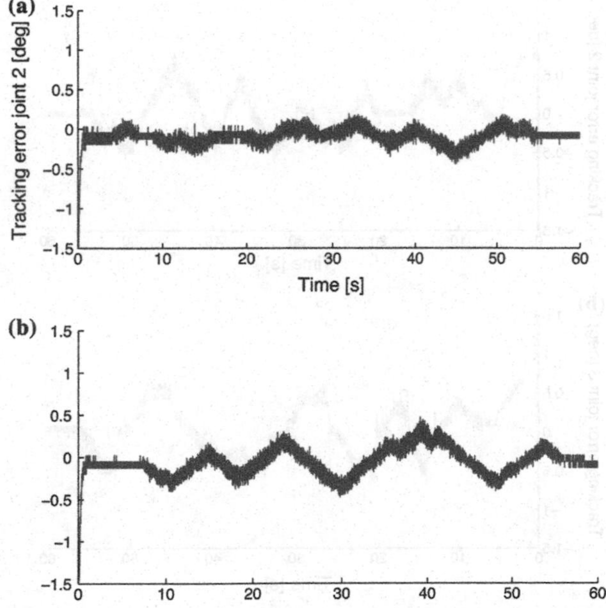

Fig. 10 Results for the robot with a payload of 2 Kg controlled by the RBFN based SMC—Tracking errors: **a** joint 2; **b** joint 3

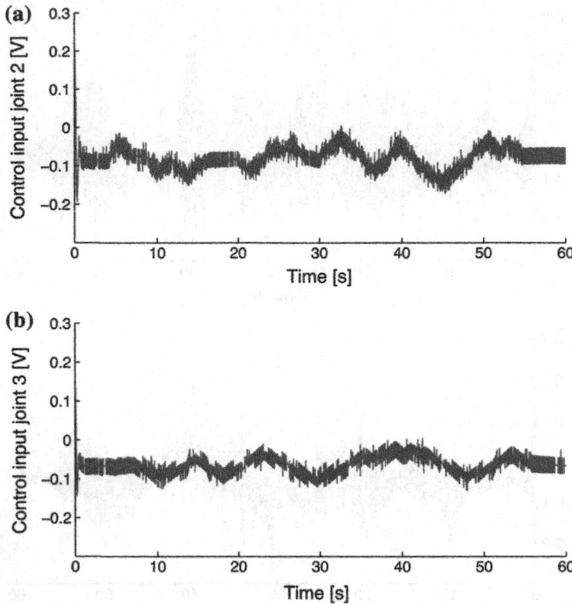

Fig. 11 Results for the robot with a payload of 2 Kg controlled by the RBFN based SMC—Control inputs: **a** joint 2; **b** joint 3

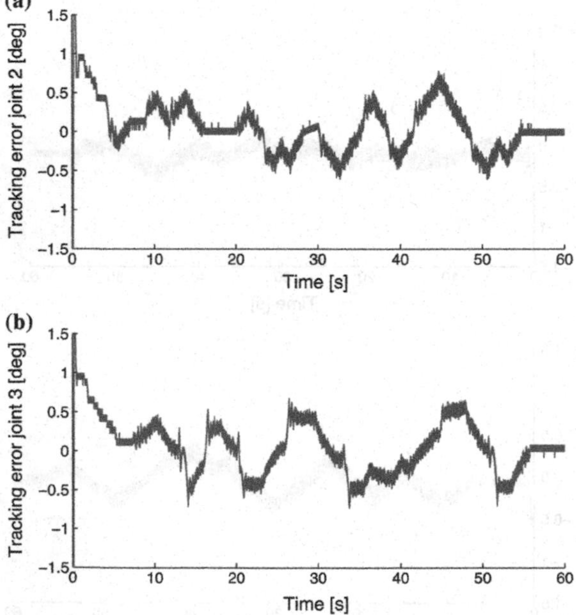

Fig. 12 Results for the robot with a payload of 2 Kg controlled by the adaptive AR based SMC—Tracking errors: **a** joint 2; **b** joint 3

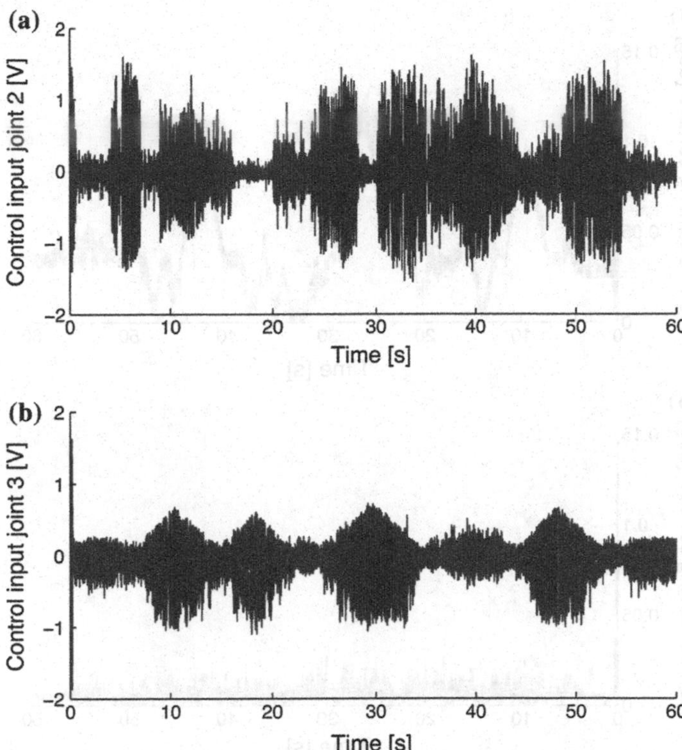

Fig. 13 Results for the robot with a payload of 2 Kg controlled by the adaptive AR based SMC—Control inputs: **a** joint 2; **b** joint 3

Table 2 Performance comparison—Figs. 6, 7, 8, 9, 10, 11, 12 and 13

Joint	Controllers	No payload deg	Payload deg
q_2	RBFN-based SMC	5.73	6.60
	adaptive AR-based SMC	6.86	7.16
	standard SMC	14.15	16.69
q_3	RBFN-based SMC	7.47	9.89
	adaptive AR-based SMC	12.65	13.69
	standard SMC	15.60	19.76

displayed in Fig. 10 and the voltage control inputs from the dSPACE controller board are depicted in Fig. 11.

Figures 12 and 13 show the performance of the adaptive AR based SMC; the tracking errors are displayed in Fig. 12 and the voltage control inputs from the dSPACE controller board are depicted in Fig. 13.

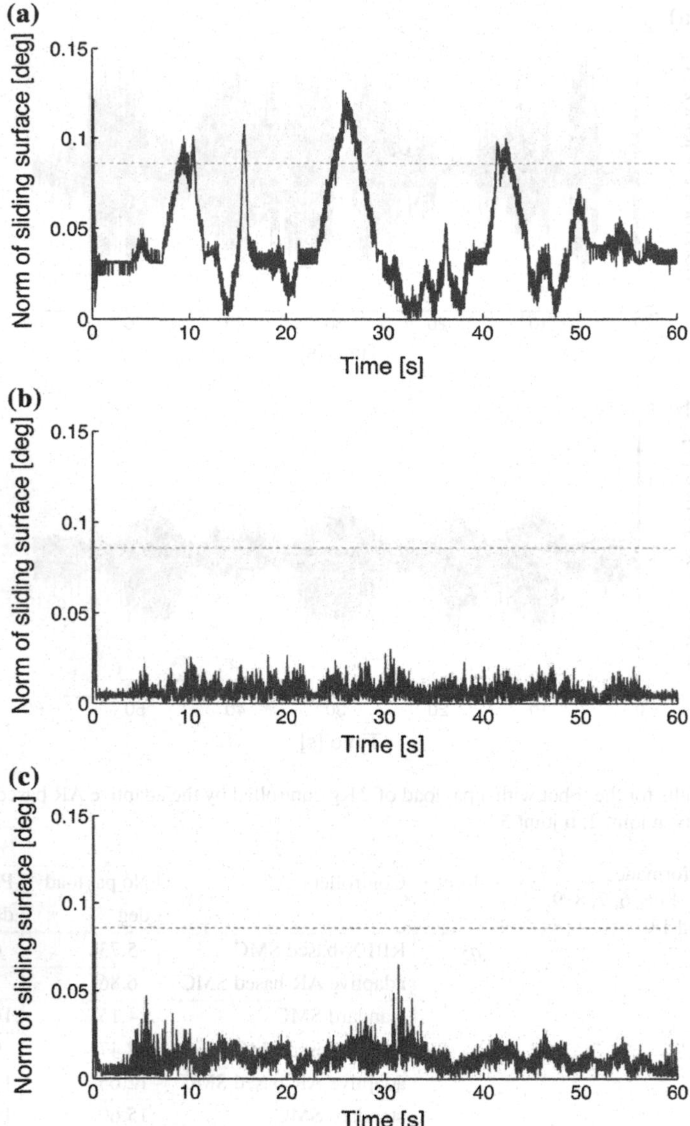

Fig. 14 Results for the robot without a payload—Norm of the sliding surface: *dashed line* denotes the threshold ρ_s, *continuous line* denotes the norm of the sliding surface $||\tilde{\sigma}(k)||$ [see Eq. (26); **a** standard SMC; **b** RBFN based SMC. **c** adaptive AR based SMC

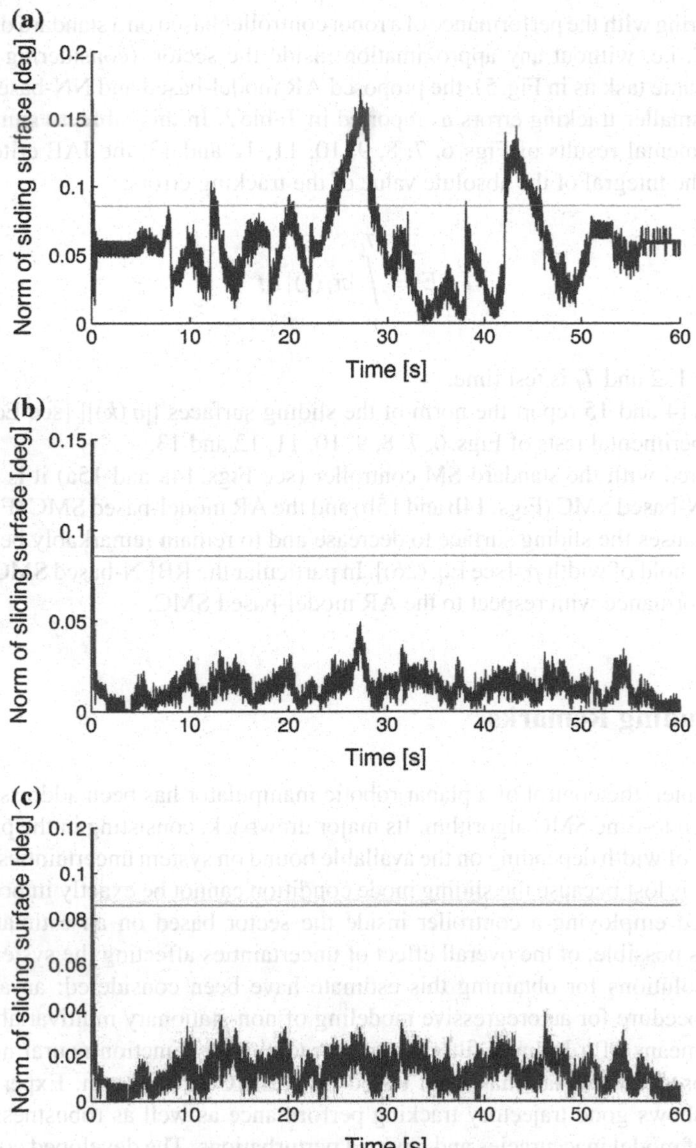

Fig. 15 Results for the robot with a payload—Norm of the sliding surface: *dashed line* denotes the threshold ρ_s, *continuous line* denotes the norm of the sliding surface $||\tilde{\sigma}(k)||$ [see Eq. (26)]; **a** standard SMC; **b** RBFN based SMC. **c** adaptive AR based SMC

Comparing with the performance of a robot controller based on a standard discrete-time SMC, i.e. without any approximation inside the sector, (considering for the robot the same task as in Fig. 5), the proposed AR model-based and NN-based SMC produces smaller tracking errors as reported in Table 2. In this table, to summarize the experimental results of Figs. 6, 7, 8, 9, 10, 11, 12 and 13, the IAE criterion is used, i.e. the integral of the absolute value of the tracking errors:

$$IAE = \int_0^{T_t} |\pi_i(t)|\, dt \tag{34}$$

where $i = 1, 2$ and T_t is test time.

Figures 14 and 15 report the norm of the sliding surfaces $||\tilde{\sigma}(k)||$ [see Eq. (26)], for the experimental tests of Figs. 6, 7, 8, 9, 10, 11, 12 and 13.

Compared with the standard SM controller (see Figs. 14a and 15a) it is evident that the NN-based SMC (Figs. 14b and 15b) and the AR model-based SMC (Figs. 14c and 15c) causes the sliding surface to decrease and to remain remarkably below the sector threshold of width ρ_s [see Eq. (26)]. In particular the RBFN-based SMC shows better performance with respect to the AR model-based SMC.

5 Concluding Remarks

In this chapter, the control of a planar robotic manipulator has been addressed by a robust discrete-time SMC algorithm. Its major drawback, consisting in the presence of a sector, of width depending on the available bound on system uncertainties, where robustness is lost because the sliding mode condition cannot be exactly imposed has been solved employing a controller inside the sector based on an estimation, as accurate as possible, of the overall effect of uncertainties affecting the system. Two different solutions for obtaining this estimate have been considered: an adaptive on-line procedure for autoregressive modeling of non-stationary multivariable time series by means of a Kalman filtering and a radial basis function neural network. The proposed control law has been tested on a ERICC robot arm. Experimental evidence shows good trajectory tracking performance as well as robustness in the presence of model inaccuracies and payload perturbations. The developed controller based on neural networks provided improved tracking performance with respect to the standard discrete-time SMC law and to the adaptive AR model based SMC. As future research activity an online learning algorithm for the neural network is under investigation as well as a method to improve the run-time performance for the real-time implementation of the learning algorithm.

References

Abdallah, C., Dawson, D., Dorato, P., Jamshidi, M.: Survey of robust control for rigid robots. IEEE Control Syst. Mag. **11**(2), 24–30 (1991)

Antonini, P., Ippoliti, G., Longhi, S.: Learning control of mobile robots using a multiprocessor system. Control Eng. Pract. **14**(11), 1279–1295 (2006)

Arnold, M., Milner, X., Witte, H., Bauer, R., Braun, C.: Adaptive AR modeling of nonstationary time series by means of kalman filtering. IEEE Trans. Biomed. Eng. **45**(5), 553–562 (1998)

Åström, K.J., Eykhoff, P.: System identification—a survey. Automatica **7**, 123–162 (1971)

Box, G., Jenkins, G.M., Reinsel, G.: Time Series Analysis: Forecasting and Control. Holden-Day, San Francisco (1994)

Capisani, L., Ferrara, A.: Trajectory planning and second-order sliding mode motion/interaction control for robot manipulators in unknown environments. IEEE Trans. Ind. Electron. **59**(8), 3189–3198 (2012)

Capisani, L., Ferrara, A., Magnani, L.: Second order sliding mode motion control of rigid robot manipulators. In Decision and Control, 2007 46th IEEE Conference on, pp. 3691–3696 (2007)

Chaoui, H., Sicard, P.: Adaptive neural network control of flexible-joint robotic manipulators with friction and disturbance. In: IECON 2012–38th Annual Conference on IEEE Industrial Electronics Society, pp. 2644–2649 (2012)

Chen, C.-S.: Dynamic structure neural-fuzzy networks for robust adaptive control of robot manipulators. IEEE Trans. Ind. Electron. **55**(9), 3402–3414 (2008)

Chen, S., Cowan, C., Grant, P.: Orthogonal least squares learning algorithm for radial basis function networks. IEEE Trans. Neural Net. **2**(2), 302–309 (1991)

Ciabattoni, L., Corradini, M., Grisostomi, M., Ippoliti, G., Longhi, S., Orlando, G.: A discrete-time vs controller based on rbf neural networks for pmsm drives. Asian J. Control **16**(2), 396–408 (2014)

Ciabattoni, L., Ippoliti, G., Longhi, S., Cavalletti, M., Rocchetti, M.: Solar irradiation forecasting using rbf networks for pv systems with storage. In: 2012 IEEE International Conference on Industrial Technology, ICIT 2012, Proceedings, pp. 699–704 (2012)

Cimini, G., Corradini, M., Ippoliti, G., Malerba, N., Orlando, G.: Control of variable speed wind energy conversion systems by a discrete-time sliding mode approach. In: 2013 IEEE International Conference on Mechatronics, ICM 2013, pp. 736–741 (2013)

Copot, C., Lazar, C., Burlacu, A.: Predictive control of nonlinear visual servoing systems using image moments. IET Control Theory Appl. **6**(10), 1486–1496 (2012)

Corradini, M., Fossi, V., Giantomassi, A., Ippoliti, G., Longhi, S., Orlando, G.: Minimal resource allocating networks for discrete time sliding mode control of robotic manipulators. IEEE Trans. Ind. Inform. **8**(4), 733–745 (2012a)

Corradini, M., Giantomassi, A., Ippoliti, G., Longhi, S., Orlando, G.: Discrete time variable structure control of robotic manipulators based on fully tuned rbf neural networks. In: IEEE International Symposium on Industrial Electronics, pp. 1840–1845 (2010)

Corradini, M., Ippoliti, G., Longhi, S., Marchei, D., Orlando, G.: A quasi-sliding mode observer-based controller for pmsm drives. Asian J. Control **15**(2), 380–390 (2013)

Corradini, M., Ippoliti, G., Longhi, S., Orlando, G.: A quasi-sliding mode approach for robust control and speed estimation of pm synchronous motors. IEEE Trans. Ind. Electron. **59**(2), 1096–1104 (2012b)

Corradini, M., Ippoliti, G., Longhi, S., Orlando, G., Signorini, R.: Neural-network-based discrete-time variable structure control of robotic manipulators. In: International Conference on Advanced Robotics, 2009. ICAR 2009, pp. 1–6, (2009)

Corradini, M.L., Fossi, V., Giantomassi, A., Ippoliti, G., Longhi, S., Orlando, G.: Discrete time sliding mode control of robotic manipulators: development and experimental validation. Control Eng. Pract. **20**(8), 816–822 (2012c)

D'Amico, A., Ippoliti, G., Longhi, S.: A radial basis function networks approach for the tracking problem of mobile robots. IEEE/ASME Int. Conf. Adv. Intell. Mechatron. AIM **1**, 498–503 (2001)

Dixon, W.: Adaptive regulation of amplitude limited robot manipulators with uncertain kinematics and dynamics. IEEE Trans. Autom. Control **52**(3), 488–493 (2007)

Furuta, K.: Vss type self-tuning control. IEEE Trans. Ind. Electron. **40**(1), 37–44 (1993)

Gao, W., Wang, Y., Homaifa, A.: Discrete-time variable structure control systems. IEEE Trans. Ind. Electron. **42**, 117–122 (1995)

Ge, S., Lee, T., Harris, C.: Adaptive Neural Network Control of Robotic Manipulators. World Scientific, Singapore (1998)

Ge, S. S., He, W., Xiao, S.: Adaptive neural network control for a robotic manipulator with unknown deadzone. In: 2013 32nd Chinese Control Conference (CCC), pp. 2997–3002, (2013)

Giantomassi, A.: Modeling Estimation and Identification of Complex System Dynamics. LAP Lambert Academic Publishing, Germany (2012)

Giantomassi, A., Ippoliti, G., Longhi, S., Bertini, I., Pizzuti, S.: On-line steam production prediction for a municipal solid waste incinerator by fully tuned minimal RBF neural networks. J. Process Control **21**(1), 164–172 (2011)

Guidorzi, R.: Multivariable System Identification. Bononia University Press, Italy (2003)

Han, S., Lee, J.: Precise positioning of nonsmooth dynamic systems using fuzzy wavelet echo state networks and dynamic surface sliding mode control. IEEE Trans. Ind. Electron. **60**(11), 5124–5136 (2013)

Haykin, S.: Neural Networks: A Comprehensive Foundation, 2nd edn. Prentice Hall, London (1999)

http://www.dspaceinc.com (2011)

Ignaciuk, P., Bartoszewicz, A.: Discrete-time sliding-mode congestion control in multisource communication networks with time-varying delay. IEEE Trans. Control Syst. Technol. **19**(4), 852–867 (2011)

Ishiguro, A., Furuhashi, T., Okuma, S., Uchikawa, Y.: A neural network compensator for uncertainties of robotics manipulators. IEEE Trans. Ind. Electron. **39**(6), 565–570 (1992)

Islam, S., Liu, P.: Robust sliding mode control for robot manipulators. IEEE Trans. Ind. Electron. **58**(6), 2444–2453 (2011)

Ko, C.-N.: Identification of non-linear systems using radial basis function neural networks with time-varying learning algorithm. IET Signal Process. **6**(2), 91–98 (2012)

Lewis, F., Jagannathan, S., Yeşildirek, A.: Neural Network Control of Robot Manipulators and Nonlinear Systems. Taylor & Francis, Abingdon (1999)

Li, Z., Su, C.-Y.: Neural-adaptive control of single-mastermultiple-slaves teleoperation for coordinated multiple mobile manipulators with time-varying communication delays and input uncertainties. IEEE Trans. Neural Netw. Learn. Syst. **24**(9), 1400–1413 (2013)

Lin, Y., Shi, Y., Burton, R.: Modeling and robust discrete-time sliding-mode control design for a fluid power electrohydraulic actuator (EHA) system. IEEE/ASME Trans. Mechatron. **18**(1), 1–10 (2013)

Ljung, L.: System Identification, Theory for the User. Prentice Hall PTR, New Jersey (1999)

Marton, L., Lantos, B.: Control of robotic systems with unknown friction and payload. IEEE Trans. Control Syst. Technol. **19**(6), 1534–1539 (2011)

Melhem, K., Wang, W.: Global output tracking control of flexible joint robots via factorization of the manipulator mass matrix. IEEE Trans. Robot. **25**(2), 428–437 (2009)

Milosavljevic, C., Perunicic-Drazenovic, B., Veselic, B.: Discrete-time velocity servo system design using sliding mode control approach with disturbance compensation. IEEE Trans. Ind. Inform. **9**(2), 920–927 (2013)

Nicosia, S., Tomei, P.: Robot control by using only joint position measurements. IEEE Trans. Autom. Control **35**(9), 1058–1061 (1990)

Nikdel, N., Nikdel, P., Badamchizadeh, M., Hassanzadeh, I.: Using neural network model predictive control for controlling shape memory alloy-based manipulator. IEEE Trans. Ind. Electron. **61**(3), 1394–1401 (2014)

Ozaki, T., Suzuki, T., Furuhashi, T., Okuma, S., Uchikawa, Y.: Trajectory control of robotic manipulators using neural networks. IEEE Trans. Ind. Electron. **38**(3), 195–202 (1991)

Raspa, P., Benetazzo, F., Ippoliti, G., Longhi, S., Srensen, A.: Experimental results of discrete time variable structure control for dynamic positioning of marine surface vessels. In: 9th IFAC Conference on Control Applications in Marine Systems, pp. 55–60 (2013)

Sanner, R.M., Slotine, J.-J.E.: Stable adaptive control of robot manipulator using neural networks. Neural Comput. **7**(3), 753–790 (1995)

Siciliano, B., Sciavicco, L., Villani, L., Oriolo, G.: Robotics Modelling, Planning and Control. Advanced Textbooks in Control and Signal Processing. Springer, Heidelberg (2009)

Sun, F., Sun, Z., Woo, P.: Neural network-based adaptive controller design of robotic manipulators with an observer. IEEE Trans. Neural Netw. **12**(1), 54–67 (2001)

Sundararajan, N., Saratchandraw, P., Li, Y.: Fully Tuned Radial Basis Function Neural Networks for Flight Control. Kluwer Academic Publishers, Norwell (2002)

Utkin, V.: Sliding Modes in Control and Optimization. Springer, Berlin (1992)

Veselic, B., Perunicic-Drazenovic, B., Milosavljevic, C.: Improved discrete-time sliding-mode position control using euler velocity estimation. IEEE Trans. Ind. Electron. **57**(11), 3840–3847 (2010)

Wai, R.-J., Muthusamy, R.: Fuzzy-neural-network inherited sliding-mode control for robot manipulator including actuator dynamics. IEEE Trans. Neural Netw. Learn. Syst. **24**(2), 274–287 (2013)

Wang, B., Yu, X., Chen, G.: ZOH discretization effect on single-input sliding mode control systems with matched uncertainties. Automatica **45**, 118–125 (2009)

Wang, B., Yu, X., Li, X.: Zoh discretization effect on higher-order sliding-mode control systems. IEEE Trans. Ind. Electron. **55**(11), 4055–4064 (2008)

Wang, B., Yu, X., Wang, L.: Convergence accuracy analysis of discretized sliding mode control systems. In: Proceedings of the 11th International Conference on Control, Autom., Robotics and Vision, pp. 1370–1374 (2010)

Xu, Q.: Enhanced discrete-time sliding mode strategy with application to piezoelectric actuator control. IET Control Theory Appl. **7**(18), 2153–2163 (2013)

Yassin, I., Taib, M., Abdul Aziz, M., Rahim, N., Tahir, N., Johari, A.: Identification of DC motor drive system model using radial basis function (RBF) neural network. In: Industrial Electronics and Applications (ISIEA), 2011 IEEE Symposium on, pp. 13–18 (2011)

Young, K., Utkin, V., Ozguner, U.: A control engineer's guide to sliding mode control. IEEE Trans. Contr. Syst. Technol. **7**, 328–342 (1999)

Zinober, A.: Variable Structure and Lyapunov Control. Springer-Verlag New York Inc, Secaucus, NJ, USA (1994)

A Robust Adaptive Self-tuning Sliding Mode Control for a Hybrid Actuator in Camless Internal Combustion Engines

Benedikt Haus, Paolo Mercorelli and Nils Werner

Abstract This contribution deals with an adaptive sliding mode control for a hybrid actuator consisting of a piezo, a mechanicacal and a hydraulic part that can be used for camless engine motor applications. The control structure comprises a feedforward controller and a sliding mode controller. The general approach of this actuator is to use the advantages of both systems, the high precision of the piezoelectric actuator and the force of the hydraulic part. In fact, piezoelectric actuators (PEAs) are commonly used for precise positioning, despite PEAs present nonlinearities, such as hysteresis, saturations, and creep. A sliding mode control is proposed and for deriving the structure of such a controller a Lyapunov approach is used. An adaptive self-tuning algorithm is realised. The conceived sliding mode control takes the hydraulic actuator in a resonance operating point which corresponds to the rotational speed of the engine. When the engine speed changes, the sliding mode controller adapts its parameter in a way that the resonance frequency of the controlled hydraulic part of the actuator changes and corresponds to the working frequency of the engine. The resulting controller is therefor totally self-tuning and robust with respect to the model parameter variation. Asymptotic tracking is shown using Lyapunov approach. Moreover, the proposed technique avoids a switching function for the calculation of the equivalent signal of the sliding mode controller. In this way the chattering problem is completely avoided. Simulations with real data of a camless engine are presented.

B. Haus (✉) · P. Mercorelli
Institute of Product and Process Innovation, Leuphana University of Lueneburg,
Volgershall 1, 21339 Lueneburg, Germany
e-mail: benedikt.haus@stud.leuphana.de

P. Mercorelli
e-mail: mercorelli@uni.leuphana.de

N. Werner
Faculty of Automotive Engineering, Ostfalia University of Applied Sciences,
Kleiststr. 14-16, 38440 Wolfsburg, Germany
e-mail: n.werner@ostfalia.de

A.T. Azar and Q. Zhu (eds.), *Advances and Applications in Sliding Mode Control systems*,
Studies in Computational Intelligence 576, DOI 10.1007/978-3-319-11173-5_4

1 Introduction

Recently, a lot of attention has been given to variable engine valve control because it can lessen the pumping losses and improve torque performance over a wider range of working conditions than a conventional spark-ignition engine. Thus, it improves the whole efficiency of the engine. The main reason is that variable valve timing permits the control of internal exhaust gas recirculation, so, as already mentioned, making fuel economy better and lower NOx emissions. Combined with microprocessor control, very important functions of the motor management can be controlled well by hybrid actuators. For moving distances between 5 and 8 mm there are many types of actuators with different benefits. Besides mechanical and hydraulic variable valve train options, electromagnetic valve actuators have been proposed in the past. Recent works mark technical progress in this area, in particular, (Tai and Tsao 2003; Hoffmann and Stefanopoulou 2001; Peterson and Stefanopoulou 2003). Theoretically, electromagnetic valve actuators offer the highest potential to improve fuel economy due to their control flexibility. In real applications, however, the electromagnetic valve actuators developed so far mostly suffer from high power consumption. A new general orientation is to use electromagnetic actuators such as very recently described in Mercorelli (2012a, b). This kind of actuators presents a high value of inductance which can generate some problems of the electromagnetic compatibility with the environment. Moreover, a high value of inductance represents an inertial aspect in the control system, in particular for some kinds of control strategies such as sliding mode control. Sliding mode control is one of the most used control strategy thanks to its robustness against model uncertainties and signal noise. For instance, in Ran et al. (2012) it was shown that for a sewage treatment system, which normally is a system with large internal parameter perturbation and strong external disturbance, the controlled system using sliding mode control with the help of linear matrix inequalities toolbox has good robustness properties. In automotive context sliding mode is often used, too. This choice is motivated by the well-known robustness features of the sliding mode control approach, which are particularly appropriate dealing with the automotive context. For instance in Jie et al. (2008) an application of sliding mode control combining a fuzzy controller for longitudinal brake control of hybrid electric vehicles is proposed. This kind of approach, despite the presence of external disturbance and model uncertainties, represents a compensation control to achieve good tracking performance. In Ferrara and Vecchio (2008) the proposed approach produces a considerable reduction of the chattering phenomenon, which can determine undesired mechanical wear in the actuators. In Loukianov et al. (2008) sliding mode control techniques form a stabilising controller for an internal combustion engine with a throttle driven by a DC motor. This approach enables the inherent non-linearities of the engine to be compensated and high-level external disturbances to be rejected. Innovative and alternative concepts in the conception of actuators are required to reduce the losses and drawbacks while keeping high actuator dynamics which is characterized by high values of velocity and generated force. An original approach to control an electromagnetic actuator has recently been proposed in

Jou et al. (2012): a non-conventional electromagnetic actuator together with a fuzzy controller to overcome the problem of the unknown parameters of the model and achieve precision positioning by friction compensation. Somewhat earlier (Gan and Cheung 2003), a precision manufacturing system was developed based on the variable reluctance principle. This work included an effective nonlinear control method based on a cascade structure. The aim of this paper is showing

- A new hybrid actuator for camless engines
- A model of this hybrid actuator
- Sliding mode based controller for a holding valve to realise soft landing
- Self-tuning of the proposed sliding mode parameters based on the resonance principle

Typically, in the technical literature, soft landing for moving masses in general is defined in relation to its landing velocity which should be no more than a value which depends on the specific application. In the considered application a landing velocity not greater than 1 m/s is considered. The theoretical literature in the field of sliding mode control points to possible applications in mechanical systems and actuators for trajectory tracking (e.g., Corradini et al. 2004). The applications in Betin et al. (2006) show very interesting results for position control in induction machines in term of robustness, across a wide range of mechanical configurations. More recently, there has been a notable interest in applications of sliding mode control for actuators. For example, in Jian-Xin and Abidi (2008), Xinkai and Hisayama (2008) and Pan (2008) position controls using a sliding mode technique are proposed for various different actuator structures. The robustness of this approach against parameter uncertainties is demonstrated. In more recent publications, intelligent control designs have been proposed for electromagnetic systems—for example, the development of a robust adaptive sliding mode controller (She et al. 2011) and the proposal of a cascade controller which could be used in a maglev train (Lee and Duan 2011). In sliding mode control one of the most important issues is the adaptation of the parameters of the controller. This topic is the most recent one and the most important in the application field. In Alanisa et al. (2014) real-time discrete adaptive output trajectory tracking for induction motors in the presence of bounded disturbances is proposed. A controller is designed which combines discrete-time block control and sliding modes techniques. Good results in terms of tracking and robustness for an adaptive sliding mode strategy with the help of neural networks are shown. In Yang et al. (2013), the authors proposed a multiobjective optimal design and energy compensation control to achieve soft landing. The landing velocity can be greatly reduced by adjusting the duty cycle of the landing current. It is notable that the trend in controlling electromagnetic actuators with fast dynamics is to keep the system from using switching modalities which can be attributed to two main factors. The first is electromagnetic compatibility. In fact, switching signals can generate dangerous interferences. In particular, to achieve a soft landing, a high switching frequency is required near the landing point. For these reasons, approaches such as those in Nguyen et al. (2007) or Betin et al. (2006) are not suitable for the target application. The second is that if there is high inductance in the electrical circuit in which the switching signals are involved, it is known to be difficult to switch the current quickly. This kind of problem has been considered in recent literature. In particular, in sliding mode control the phenomena associated with a high switching frequency are referred to as chattering. In Levant (2010) and Levant and Fridman (2010) a detailed analysis of the chattering phenomena and how to avoid them is done. Chattering phenomena are dangerous but it was proven in Levant (2010) that they can be eliminated by proper use of high-order sliding modes. Moreover, in Levant and Fridman (2010) it is shown that, for some sliding magnitudes in higher orders of the homogeneous sliding modes, the chattering effect is not amplified in the presence of actuators. In general, a high switching frequency should be used to achieve a soft landing. A detailed analysis of this issue was presented in Mercorelli (2012a), along with a control strategy for the proposed design. Notably, in that paper, though another kind of actuator was considered, an alternative approach to dealing with the switching mode was suggested. In particular, a pre-action current was injected to prepare the actuator and to "slide" in the final part of the trajectory tracking phase. In investigations related to that study, it became clear that an important prerequisite for obtaining good control performance and soft landing would be more robustness against noise and uncertainties in general. Because of the structural complexity of the hybrid actuator the proposed controller consists of a feedforward structure that consists of a quasi discrete inversion of the piezoelectric and mechanical part of the actuator. The second part of the controller consists of a sliding mode structure which guarantees asymptotical tracking convergence, robustness, efficiency in term of energy consumption and soft landing. The proposed technique avoids a switching function for the calculation of the equivalent signal of the sliding mode controller. In this way the chattering problem is completely avoided. To derive the sliding mode structure a Lyapunov approach is used. The conceived sliding mode control takes the hydraulic actuator in a resonance working point which corresponds to the running frequency of the engine. When the engine frequency changes, the sliding controller adapts its parameter in a way that the resonance frequency of the controlled actuator changes and corresponds to the working frequency of the engine. In this sense, the sliding mode based controller is totally self-tuning. The control strategy consists of a sliding structure to take the hydraulic part of the actuator around a unique resonant operating frequency. This condition is an optimal condition in terms of energy consumption of the whole actuator. In fact, the hydraulic part of the the actuator is the part dedicated to power the valve in order to actually move it. The sliding mode controller is realised by a holding valve collocated on the back hydraulic line of the hydraulic part related to the upper compartment of the servo valve in order to realize the break phase which is needed for a soft landing. The paper is divided into the following sections. Section 2 is devoted to the general specification of the considered problem. In Sect. 3 a description of the proposed hybrid actuator is given together with the piezoelectric, mechanic and hydraulic part of the actuator. In particular this section is dedicated to the description of the model in which the Preisach dynamic model of the piezo actuator with the above mentioned nonlinearities is considered. Moreover the description of the mechanical part of the actuator, which basically consists of the hydraulic transmission ratio and the hydraulic part, which is the final part of the actuator is given. In Sect. 4 the control law is presented which consists of a combination of a feedforward and a sliding based controller. Section 5 is devoted to the description of an innovative adaptive self-tuning parameter of the obtained controller. In Sect. 6 tracking and energy simulation results using real data of an actuator are presented. Moreover, energy considerations based on the simulation results are drawn. The conclusions close the paper.

The main nomenclature

$V_{in}(t)$: input voltage
$V_z(t)$: internal piezo voltage
$i(t)$: piezo input current
R_0: input resistance in the piezo model
R_a: parasite resistance in the piezo model
C_a: parasite capacitance in the piezo model
C_z: internal capacitance in the piezo model
$x_p(t)$: internal position of the piezo part
$x_1(t)$: position of the piezo mass
$x_2(t)$: position of the servo piston
$x_V(t)$: position of the valve
$v_V(t)$: velocity of the valve
$H(x_p(t), V_{in}(t))$: hysteresis characteristic of the piezo
$M_p/3$: moving piezo mass
K_x: internal spring constant of the piezo
K: spring constant acting on the piezo
D: damping constant acting on the piezo
D_{oil}: damping constant of the oil chamber acting on the piezo
M_{SK}: mass of the servo piston
K_{SK}: spring constant acting on the servo piston
D_{SK}: damping constant acting on the servo piston
$Q_{th}(t)$: volumetric flow of the hydraulic part
V_H: steady-state factor of the model of the hydraulic part of the actuator
A_1: surface of the conic hydraulic transmission from the piezo side
A_2: surface of the conic hydraulic transmission ratio from the servo piston side
A_{VP}: surface of the valve piston
A_L: effective surface which characterises internal hydraulic leakage
T_s: sampling time

2 General Specifications and Tracking Problem in Camless Engines

Figure 1 shows the principle of the engine valves to be controlled. The intake valve allows air and fuel to rush into the cylinder so combustion can take place. The exhaust valve releases the spent fuel and air mixture from the cylinder. Clearly, the timing of the valve opening and closing strongly influences the engine efficiency

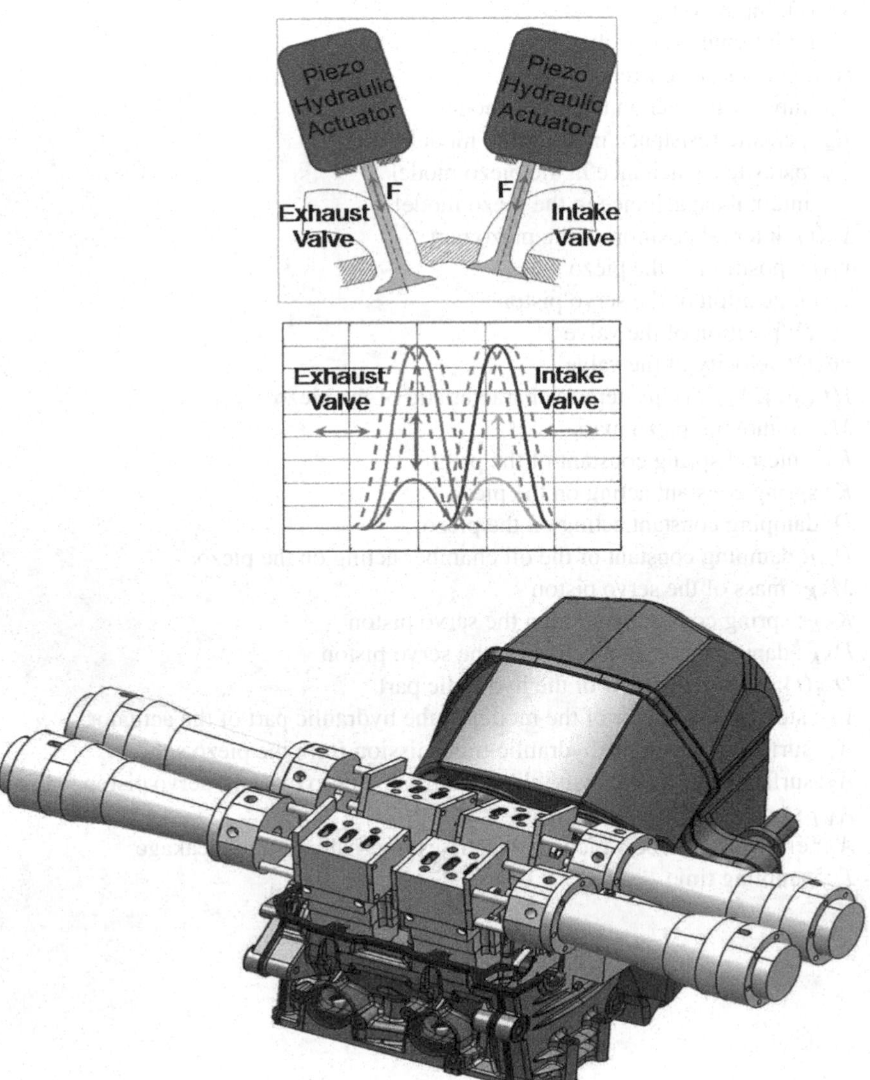

Fig. 1 New structure of the engine

and fuel economy. The optimal choice of the opening and closing timing depends on the simultaneous operation conditions of the engine. In conventional spark-ignition engines the valves are driven by the camshaft and their timing is fixed by the engine speed. In general the use of electromagnetic as well as of a hybrid valve actuators decouples the valve timing from the engine speed and ensures fully timing variability. If, for instancef a frequency of around 6,000 rpm is considered and a distance of 10 mm must be covered, then a time interval of about 4 ms for opening and closing of the valve is required. This is one of the worst practical cases in which high accelerations up to 4,500 m/s^2 have to be achieved, even in case of large disturbances due to the strong cylinder gas force acting against the exhaust valve opening.

Figure 1 shows the phase diagram of the positions of an engine intake and exhaust valves. In this figure the intake and the exhaust valve position profiles are indicated. Figure 1 demonstrates the new engine structure with, evidently, four piezo actuators.

3 General Structure of the Hybrid Actuator and Its Model

Figure 2 shows the whole hybrid structure of the actuator. In particular, on the left side of the diagram the piezo part which is in contact with a piston of surface equal to A_1 is visible. Through a conic structure filled with oil, the piston of surface A_1 transfers the movement to the mechanical servo piston with surface equal to A_2. It should be noted that $A_1 \gg A_2$. This ratio allows to realize a favorable position transfer from the piezo part to the mechanical part. Thus, the short strokes of the

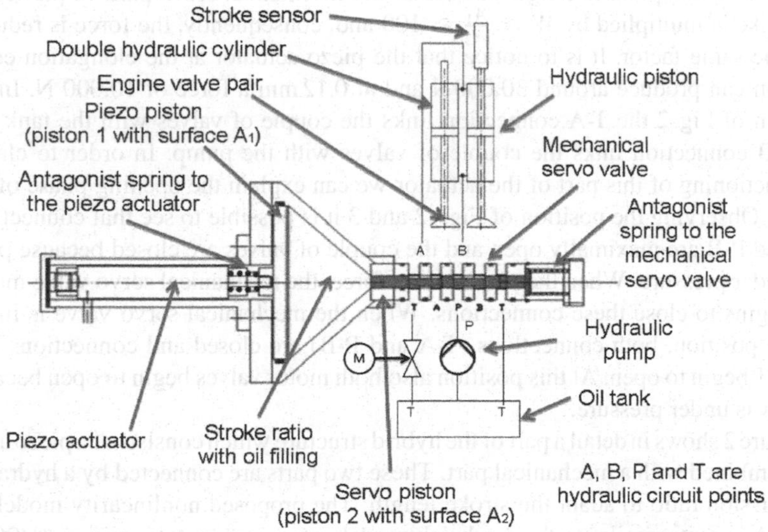

Fig. 2 Scheme of the whole hybrid piezo hydraulic actuator

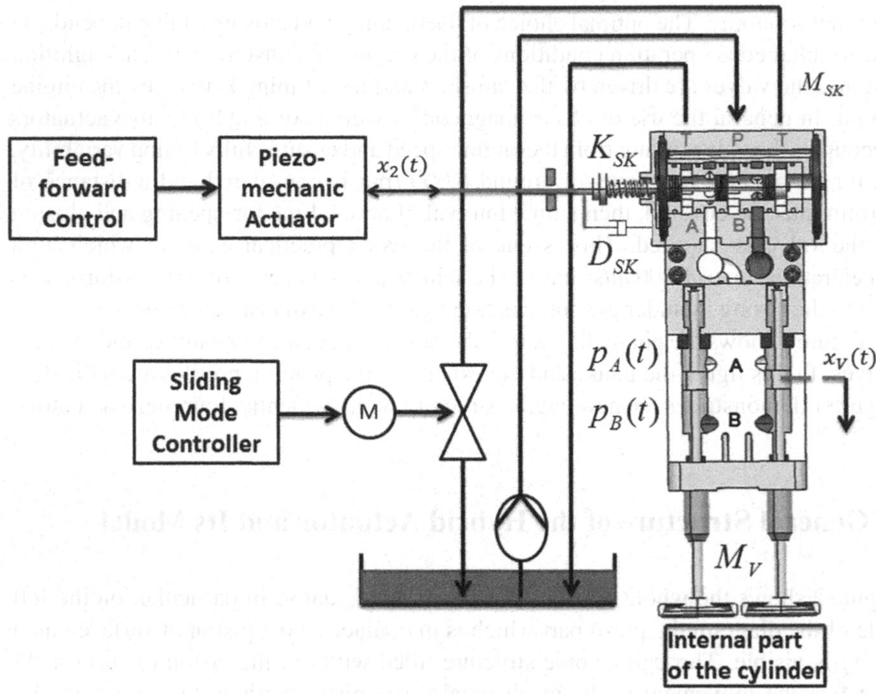

Fig. 3 Control scheme of the whole hybrid piezo hydraulic actuator

piezo part correspond to longer strokes of the mechanical servo part. In particular, the stroke is multiplied by $W = \frac{A_1}{A_2} \approx 100$ and, consequently, the force is reduced with the same factor. It is to notice that the piezo actuator at the elongation equal to 0 mm can produce around 30,000 N and at 0.12 mm a force of 10,000 N. In the diagram of Fig. 2 the T-A connection links the couple of valves with the tank and the P-B connection links the couple of valves with the pump. In order to clarify the functioning of this part of the actuator we can explain the opening phase of the valves. Observing the position of Figs. 2 and 3 it is possible to see that connections T-A and P-B are maximally open and the couple of valves are closed because point B is under pressure. When the piezo exerts force, the mechanical servo valve moves and begins to close these connections. When the mechanical servo valve is in the middle position, both connections (T-A and P-B) are closed and connections A-P and B-T begin to open. At this position also both motor valves begin to open because point A is under pressure.

Figure 2 shows in detail a part of the hybrid structure which consists of a piezo actuator combined with a mechanical part. These two parts are connected by a hydraulic transmission ratio to adapt the stroke length. The proposed nonlinearity model for the PEA is quite similar to the sandwich models presented in Adriaens et al. (2000) and Yu and Lee (2005).

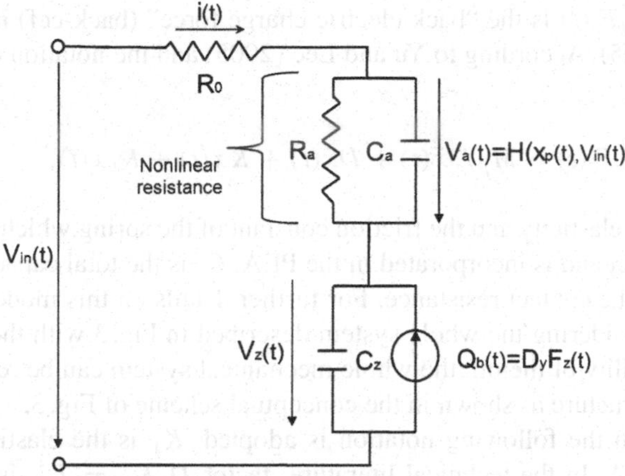

Fig. 4 Scheme of the electrical part of the piezo actuator

3.1 Mathematical Model of the Actuator

Figure 4 shows the equivalent circuit for a PEA with the I-layer nonlinearities of hysteresis and creep, in which two I-layers are combined together as C_a and R_a. The I-layer capacitor, C_a, is an ordinary one, which might be varied slightly with some factors, but here it would be assumed constant first for simplicity. The I-layer resistor, R_a , however, is really an extraordinary one with a significant nonlinearity. The resistance is either fairly large, say $R_a > 10^6\ \Omega$, when the voltage $\| V_a \| < V_h$, or is fairly small, say $R_a < 1{,}000$, when $\| V_a \| > V_h$. In Yu and Lee (2005), the threshold voltage, V_h, is defined as the hysteresis voltage of a PEA. Yu and Lee (2005) gave this definition due to the observation that there is a significant difference and an abrupt change in resistance across this threshold voltage and it is this resistance difference and change across V_h that introduces the nonlinearities of hysteresis and creep in a PEA. The hysteresis effect could be seen as a function of input $V_{in}(t)$ and output $y(t)$ as follows: $H(y(t), V_{in}(t))$. According to this model, if $V_h = 0$, then the hysteresis will disappear, and if $R_a = \infty$ when $\| V_a \| < V_h$, then the creep will also disappear. In Mercorelli and Werner (2014) a detailed analysis on the hysteresis effect and its identification with measurement validation was presented. Based on this proposed sandwich model and the equivalent circuit as shown in Fig. 4, we can further derive the state model as follows:

$$\dot{V}_a(t) = -\Big(\frac{1}{R_a} + \frac{1}{R_o}\Big)\frac{V_a(t)}{C_a} - \frac{V_z(t)}{C_a R_o} + \frac{V_{in}(t)}{C_a R_o} \tag{1}$$

$$\dot{V}_z(t) = \frac{\dot{Q}_b}{C_z} + \frac{1}{C_z}\Big(-\frac{V_a(t)}{R_o} - \frac{V_z(t)}{R_o} + \frac{V_{in}(t)}{R_o}\Big), \tag{2}$$

where $Q_b = D_y F_z(t)$ is the "back electric charge force" (back-ecf) in a PEA, see Yu and Lee (2005). According to Yu and Lee (2005) and the notation of Fig. 3, it is possible to write:

$$F_z(t) = M_p/3\ddot{x}(t) + D\dot{x}(t) + Kx(t) + K_x x(t). \tag{3}$$

K and D are the elasticity and the friction constant of the spring which is antagonist to the piezo effect and is incorporated in the PEA. C_z is the total capacitance of the PEA and R_o is the contact resistance. For further details on this model see Yu and Lee (2005). Considering the whole system described in Fig. 3 with the assumption of incompressibility of the oil, the whole mechanical system can be represented by a spring mass structure as shown in the conceptual scheme of Fig. 3.

In this system the following notation is adopted. K_x is the elasticity constant factor of the PEA. In the technical literature, factor $D_x K_x = T_{em}$ is known with the name "transformer ratio" and states the most important characteristic of the electromechanical transducer. M_{SK} is the sum of the mass of the piston with the oil and the moving actuator and M_v is the mass of the valve. $M_p/3$ is, in our case, the moving mass of the piezo structure which is a fraction of whole piezo mass. The value of this fraction is given by the constructor of the piezo device and it is determined by experimental measurements. K_{SK} and D_{SK} are the characteristics of the antagonist spring to the mechanical servo valve, see Fig. 3. D_{oil} is the friction constant of the oil. Moreover, according to Yu and Lee (2005), motion $x_p(t)$ of the piezo is:

$$x_p(t) = D_x V_z(t). \tag{4}$$

According to diagram of Fig. 4, it is possible to state:

$$V_z = V_{in}(t) - R_0 i(t) - H(x_p(t), V_{in}(t)), \tag{5}$$

where R_0 is the connection resistance and $i(t)$ is the input current as shown in Fig. 4.

For a piezo actuator, in the technical literature, factor $D_x K_x = T_{em}$ is known as the "transformer ratio" and states the the most important characteristic of the electromechanical transducer in which K_x is the elasticity constant factor of the PEA and D_x is the parameter which is responsible to transform voltage into movement. In fact, another well-known physical relation is $= F_1 = D_x K_x V_z$ which represents the piezo force in which V_z is the internal voltage. Ideally, $V_z = V_{in}$ where V_{in} is the input voltage.

Considering the whole system with the assumption of compressibility of the oil, the whole mechanical system can be represented by a spring mass structure as shown in the conceptual scheme of Fig. 5. If this model is considered with the assumption

$$A_1 = A_2, \tag{6}$$

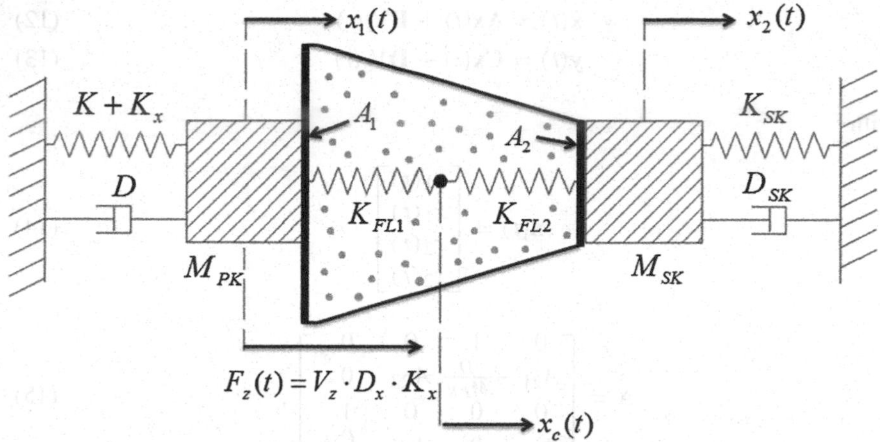

Fig. 5 Model of the piezo mechanical part of the actuator

then:

$$M_{PK} \cdot \ddot{x}_1(t) = V_z(t) D_x K_x - K_{FL1} \cdot (x_1(t) - x_c(t)) \\ - (K_x + K) \cdot x_1(t) - D\dot{x}_1(t), \tag{7}$$

together with

$$0 = 0 - K_{FL1} \cdot (x_c(t) - x_1(t)) - K_{FL2} \cdot (x_c(t) - x_2(t)), \tag{8}$$

in which $m_{Oil} = 0$ is considered, and

$$M_{SK} \cdot \ddot{x}_2(t) = K_{FL2} \cdot (x_2(t) - x_c(t)) - K_{SK} \cdot x_2(t) - D_{SK} \cdot \dot{x}_2(t). \tag{9}$$

Considering Eq. (8), the following expression is obtained:

$$x_c(t) = \frac{K_{FL1} \cdot x_1(t) + K_{FL2} \cdot x_2(t)}{K_{FL1} + K_{FL2}}, \tag{10}$$

and substituting Eq. (10) in Eq. (9), the following expression is obtained:

$$M_{SK} \cdot \ddot{x}_2(t) = -K_{FL2} \cdot x_2(t) - K_{SK} \cdot x_2(t) - D_{SK} \cdot \dot{x}_2(t) \\ + K_{FL2} \cdot \frac{K_{FL1} \cdot x_1(t) + K_{FL2} \cdot x_2(t)}{K_{FL1} + K_{FL2}}. \tag{11}$$

If a matrix representation of these two differential equations is considered, the system can be transformed into a linear state space representation

$$\dot{\mathbf{x}}(t) = \mathbf{A}\mathbf{x}(t) + \mathbf{B}V_z(t) \tag{12}$$

$$\mathbf{y}(t) = \mathbf{C}\mathbf{x}(t) + \mathbf{D}V_z(t) \tag{13}$$

with

$$\mathbf{x}(t) = \begin{bmatrix} x_1(t) \\ \dot{x}_1(t) \\ x_2(t) \\ \dot{x}_2(t) \end{bmatrix} \tag{14}$$

$$\mathbf{A} = \begin{bmatrix} 0 & 1 & 0 & 0 \\ A_{21} & -\frac{D}{M_{PK}} & A_{23} & 0 \\ 0 & 0 & 0 & 1 \\ A_{41} & 0 & A_{43} & -\frac{D_{SK}}{M_{SK}} \end{bmatrix}, \tag{15}$$

$$A_{21} = \frac{\frac{K_{FL1}^2}{K_{FL1}+K_{FL2}} - K - K_x - K_{FL1}}{M_{PK}} \tag{16}$$

$$A_{23} = \frac{K_{FL1}K_{FL2}}{(K_{FL1} + K_{FL2})M_{PK}} \tag{17}$$

$$A_{41} = \frac{K_{FL1}K_{FL2}}{(K_{FL1} + K_{FL2})M_{SK}} \tag{18}$$

$$A_{43} = \frac{\frac{K_{FL2}^2}{K_{FL1}+K_{FL2}} - K_{SK} - K_{FL2}}{M_{SK}} \tag{19}$$

$$\mathbf{B} = \frac{A_1}{A_2} \begin{bmatrix} 0 \\ \frac{D_x K_x}{M_{PK}} \\ 0 \\ 0 \end{bmatrix}, \tag{20}$$

$$\mathbf{C} = \begin{bmatrix} 0 \; 0 \; 1 \; 0 \end{bmatrix}, \tag{21}$$

$$\mathbf{D} = 0, \tag{22}$$

in which in (20) the effect of the two different surfaces is considered.

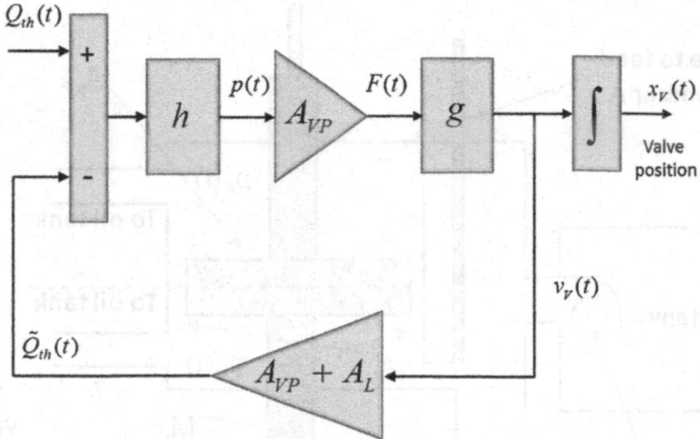

Fig. 6 Block diagram structure of the hydraulic part of the actuator

3.2 Hydraulic Part of the Actuator

In Fig. 6 a nonlinear possible model often utilised in practical applications is presented. The model was presented in Murrenhoff (2002) but in a linear approximation form which is very often used in industrial applications.

In Fig. 6 a more general representation of the hydraulic part of the actuator is visible. In Fig. 6 the following notation is adopted:

$$h \rightarrow \dot{p}(t) = V_H N_p\big(p(t), Q_{th}(t) - \tilde{Q}_{th}(t)\big) \tag{23}$$

and

$$g \rightarrow \dot{v}_V(t) = -\frac{k_v}{M_V} N_v(v_V(t)) + \frac{1}{M_V} F(t) \tag{24}$$

Moreover, in Fig. 6 some parameters are visible. V_H represents the steady state factors of the hydraulic part of the model and establishes the static connection between the pressure and the volumetric flow. Parameter A_{VP} is the surface of the moving part (valve piston). The other parameter which characterises the hydraulic-mechanical model is A_L. In fact, parameter A_L is a characteristic value of the velocity-dependent internal leakage. A_L is *not* a real surface, but an equivalent surface that quantifies the flow losses through the valve piston.

Observing Fig. 6 and considering that Q_{th} is the volumetric flow involved in the hydraulic actuator, the following mathematical models are derived:

$$\begin{cases} \dot{x}_V(t) = v_v(t) \\ \dot{v}_V(t) = -\frac{k_v}{M_V} N_v(v_V(t)) + \frac{1}{M_V} F(t) \\ \dot{p}(t) \;\; = V_H N_p\big(p(t), Q_{th}(t) - \tilde{Q}_{th}(t)\big) \end{cases} \tag{25}$$

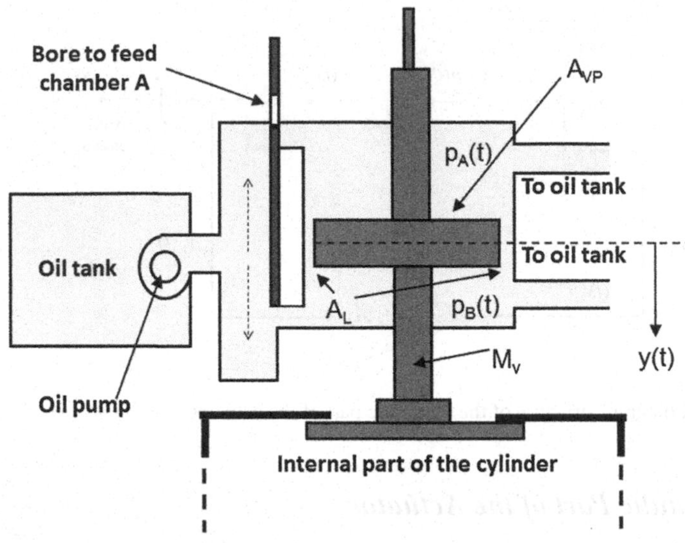

Fig. 7 Schematic of the hydraulic part with A_L

and

$$\begin{cases} F(t) &= A_{VP}\,p(t) \\ \tilde{Q}_{th}(t) &= (A_{VP} + A_L)v_V(t), \end{cases} \tag{26}$$

where parameter M_V represents the mass of the moving part (valve piston+valve), k_v is the friction factor which states the mechanical losses due to the movement, $N_p\big(p(t), Q_{th}(t) - (A_{VP} + A_L)N_l(v_V(t))\big)$ is a nonlinear function which states the mathematical connection between state $v_V(t)$ (velocity of the valve piston) and input Q_{th}. $N_v(v_V(t))$ and $N_l(v_V(t))$ are nonlinear functions of the velocity valve piston which state friction and volumetric flow leakage losses respectively. V_H is the steady-state parameter between $p(t)$ and $\tilde{Q}_{th}(t) - Q_{th}(t)$ where $\tilde{Q}_{th}(t)$ represents the leakage volumetric flow.

4 Combining Feedforward and Sliding Mode Control

The mathematical description of the system in the above sections indicates that the considered system consist of nine states variables in a cascade form. This consideration shows that the system can present a relevant phase delay between the controlling piezo input voltage and the final position of the valve to be controlled. It is known that typically in such kind of hybrid systems, in order to minimize the phase delay a feedforward controller is used. The strategy adopted in the proposed control system consist of a feedforward controller combined with a feedback one. The feedback

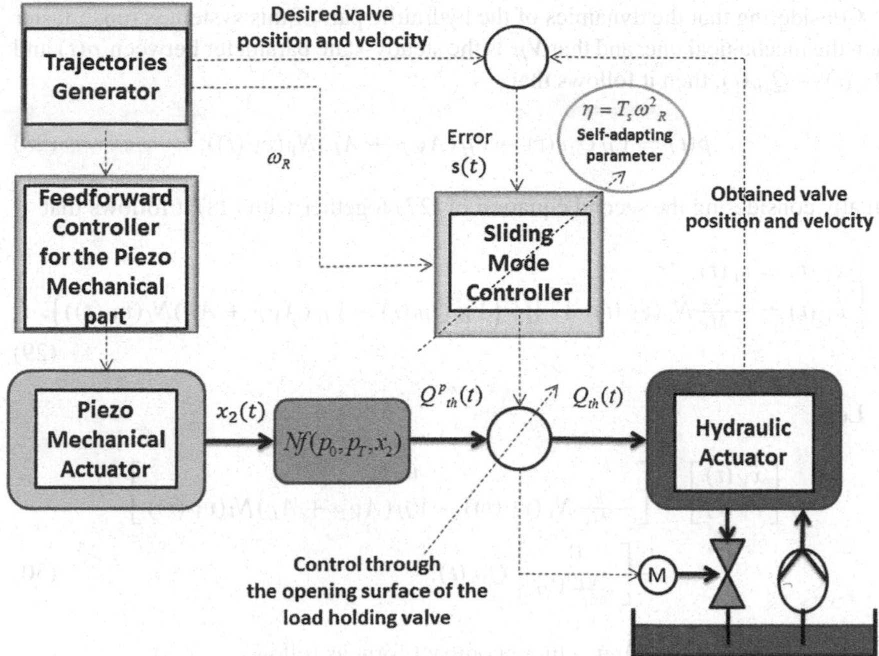

Fig. 8 Control scheme structure

controller is realized using basically a sliding mode control strategy. The feedforward controller works to compensate the phase delay of the piezo and the mechanical part of the actuator which consists of the hydraulic transmission ratio together with the servo piston and its antagonist spring. The sliding mode controller is applied to a hydraulic valve which is located at the return hydraulic on the upper part of the valve chamber to realise a soft landing of the valve. Figure 8 shows the conceptual scheme of the adopted control system.

4.1 Sliding Mode Control Using a Hydraulic Valve

The sliding structure presented in this section is a typical one and is similar to that presented in Lee et al. (2010). Considering Fig. 6 and Eq. (25), the following representation can be derived:

$$\begin{cases} \dot{x}_V(t) = v_v(t) \\ \dot{v}_V(t) = -\frac{k_v}{M_V} N_v(v_V(t)) + \frac{1}{M_V} A_{VP}\, p(t) \\ \dot{p}(t) \;\; = V_H N_p\big(p(t), Q_{th}(t) - (A_{VP} + A_L) N_l(v_V(t))\big), \end{cases} \quad (27)$$

Considering that the dynamics of the hydraulic part of this system is much faster than the mechanical one, and that V_H is the steady-state parameter between $p(t)$ and $\tilde{Q}_{th}(t) - Q_{th}(t)$, then it follows that:

$$p(t) = V_H Q_{th}(t) - V_H(A_{VP} + A_L)N_v(v_V(t)). \tag{28}$$

Finally considering the second equation of (27) together with (28) it follows that

$$\begin{cases} \dot{x}_V(t) = v_v(t), \\ \dot{v}_V(t) = -\frac{k_v}{M_V}N_v(v_V(t)) + \frac{A_{VP}}{M_V}\Big(V_H Q_{th}(t) - V_H(A_{VP} + A_L)N_l(v_V(t))\Big). \end{cases} \tag{29}$$

Let

$$\begin{bmatrix} \dot{x}_V(t) \\ \dot{v}_V(t) \end{bmatrix} = \begin{bmatrix} v_V(t) \\ -\frac{k_v}{M_V}N_v(v_V(t)) - V_H(A_{VP} + A_L)N_l(v_V(t)) \end{bmatrix} + \begin{bmatrix} 0 \\ \frac{A_{VP}}{M_V}V_H \end{bmatrix} Q_{th}(t), \tag{30}$$

then we can write Eq. (30) in a more compact form as follows

$$\dot{\mathbf{x}}_H(t) = \mathbf{f}(\mathbf{x}_H(t)) + \mathbf{B}_H Q_{th}(t), \tag{31}$$

where field

$$\mathbf{f}(\mathbf{x}_H(t)) = \begin{bmatrix} v_V(t) \\ -\frac{k_v}{M_V}N_v(v_V(t)) - V_H(A_{VP} + A_L)N_l(v_V(t)) \end{bmatrix}, \tag{32}$$

and

$$\mathbf{B}_H = \begin{bmatrix} 0 \\ \frac{A_{VP}}{M_V}V_H \end{bmatrix}, \quad \mathbf{x}_H(t) = \begin{bmatrix} x_V(t) \\ v_V(t) \end{bmatrix}. \tag{33}$$

Proposition 1 *Let's consider the system $\dot{\mathbf{x}}_H(t) = \mathbf{f}(\mathbf{x}_H(t)) + \mathbf{B}_H Q_{th}(t)$, a possible stabilizing control law with weighting matrix* $\mathbf{G}$ *is*

$$Q_{th}(kT_s) = Q_{th}((k-1)T_s) + \big(\mathbf{G}\mathbf{B}_H T_s\big)^{-1}\big(\eta T_s s(kT_s)\big). \tag{34}$$

Proof The following sliding function is defined:

$$s(t) = \mathbf{G}\big(\mathbf{x}_{H_d}(t) - \mathbf{x}(t)\big), \tag{35}$$

where $\mathbf{G} = \begin{bmatrix} \lambda_1 & \lambda_2 \end{bmatrix}$, and $\mathbf{x}_{H_d}(t)$ represents the vector of the desired trajectories (valve position and velocity).

$$s(t) = \begin{bmatrix} \lambda_1 & \lambda_2 \end{bmatrix} \begin{bmatrix} xv_d(t) - xv(t) \\ vv_d(t) - vv(t) \end{bmatrix}, \tag{36}$$

thus

$$s(t) = \lambda_1 \Big(xv_d(t) - xv(t) \Big) + \lambda_2 \Big(vv_d(t) - vv(t) \Big). \tag{37}$$

It is reasonable, but not necessary to weight both errors equally: $\lambda_1 = \lambda_2 = 1$. If the following Lyapunov function is defined:

$$V(s) = \frac{s^2(t)}{2}, \tag{38}$$

then it follows that:

$$\dot{V}(s) = s(t)\dot{s}(t). \tag{39}$$

In order to find the stability of the solution $s(t) = 0$, it is possible to choose the following function:

$$\dot{V}(s) = -\eta s^2(t), \tag{40}$$

with $\eta > 0$. Comparing (39) with (40), the following relationship is obtained:

$$s(t)\dot{s}(t) = -\eta s^2(t), \tag{41}$$

and finally

$$s(t)\big(\dot{s}(t) + \eta s(t)\big) = 0. \tag{42}$$

The non-trivial solution follows from the condition

$$\dot{s}(t) + \eta s(t) = 0. \tag{43}$$

From (35) it follows:

$$\dot{s}(t) = \mathbf{G}\big(\dot{\mathbf{x}}_{H_d}(t) - \dot{\mathbf{x}}_H(t)\big) = \mathbf{G}\dot{\mathbf{x}}_{H_d}(t) - \mathbf{G}\dot{\mathbf{x}}_H(t). \tag{44}$$

The main idea is to find a $u_{eq}(t)$, an equivalent input, and after that a $Q_{th}(t)$, such that $\dot{\mathbf{x}}_H(t) = \dot{\mathbf{x}}_{H_d}(t)$.

For that, from (31) it follows that:

$$\dot{\mathbf{x}}_H(t) = \dot{\mathbf{x}}_{H_d}(t) = \mathbf{f}(\mathbf{x}_{H_d}(t)) + \mathbf{B}_H Q_{th}(t), \tag{45}$$

and from (44) the following relationship is obtained:

$$\dot{s}(t) = \mathbf{G}\dot{\mathbf{x}}_{H_d}(t) - \mathbf{G}\mathbf{f}(\mathbf{x}_{H_d}(t)) - \mathbf{G}\mathbf{B}_H Q_{th}(t) = \mathbf{G}\mathbf{B}_H\big(u_{eq}(t) - Q_{th}(t)\big), \tag{46}$$

where $u_{eq}(t)$ is the equivalent input which, in our case, assumes the following expression:

$$u_{eq}(t) = \big(\mathbf{G}\mathbf{B}_H\big)^{-1}\mathbf{G}\Big(\dot{\mathbf{x}}_{H_d}(t) - \mathbf{f}(\mathbf{x}_{H_d}(t))\Big). \tag{47}$$

After inserting (46) in (43) the following relationship is obtained:

$$\mathbf{G}\mathbf{B}_H\big(u_{eq}(t) - Q_{th}(t)\big) + \eta s(t) = 0, \tag{48}$$

and in particular

$$Q_{th}(t) = u_{eq}(t) + \big(\mathbf{G}\mathbf{B}_H\big)^{-1}\eta s(t). \tag{49}$$

From the description of the hydraulic model it is a difficult job to calculate $u_{eq}(t)$. If Eq. (46) is rewritten in a discrete form using explicit Euler approximation, then it follows:

$$\frac{s((k+1)T_s) - s(kT_s)}{T_s} = \mathbf{G}\mathbf{B}_H\big(u_{eq}(kT_s) - Q_{th}(kT_s)\big). \tag{50}$$

If Eq. (49) is also rewritten in a discrete form, then

$$Q_{th}(kT_s) = u_{eq}(kT_s) + \big(\mathbf{G}\mathbf{B}_H\big)^{-1}\eta s(kT_s). \tag{51}$$

Equation (50) can be also rewritten as

$$u_{eq}(kT_s) = Q_{in}(kT_s) + \big(\mathbf{G}\mathbf{B}_H\big)^{-1}\frac{s((k+1)T_s) - s(kT_s)}{T_s}. \tag{52}$$

Equation (52) can be estimated to one-step backward in the following way:

$$u_{eq}((k-1)T_s) = Q_{in}((k-1)T_s) + \big(\mathbf{G}\mathbf{B}_H\big)^{-1}\frac{s(kT_s) - s((k-1)T_s)}{T_s}. \tag{53}$$

Because of function $u_{eq}(t)$ is a continuous one, we can write

$$u_{eq}(kT_s) \approx u_{eq}((k-1)T_s). \tag{54}$$

Considering Eq. (54), then Eq. (53) becomes

$$u_{eq}(kT_s) = Q_{in}((k-1)T_s) + \left(\mathbf{GB}_H\right)^{-1} \frac{s(kT_s) - s((k-1)T_s)}{T_s}. \tag{55}$$

Inserting (55) into (51) gives

$$Q_{th}(kT_s) = Q_{th}((k-1)T_s) + \left(\mathbf{GB}_H\right)^{-1} \left(\eta s(kT_s) + \frac{s(kT_s) - s((k-1)T_s)}{T_s}\right), \tag{56}$$

and finally

$$Q_{th}(kT_s) = Q_{th}((k-1)T_s) + \left(\mathbf{GB}_H T_s\right)^{-1} \left(\eta T_s s(kT_s) + s(kT_s) - s((k-1)T_s)\right). \tag{57}$$

The controller stated by Eq. (57) can be seen basically as a integral action on the error which is represented by the sliding function $s(t)$. In fact, according to the Lyapunov analysis it is obtained that

$$\lim_{k \to \infty} s(kT_s) = 0 \tag{58}$$

and thus, it exists a $\overline{k}$ such that for $k > \overline{k}$

$$s(kT_s) \approx s((k-1)T_s). \tag{59}$$

So after a transient period Eq. (57) can be rewritten as follows:

$$Q_{th}(kT_s) = Q_{th}((k-1)T_s) + \left(\mathbf{GB}_H T_s\right)^{-1} \left(\eta T_s s(kT_s)\right). \tag{60}$$

Equation (60) states an integral action with a gain η. □

5 Resonance Condition for a Robust Self-tuning of the Sliding Mode Control

One of the most important issues in sliding mode control approach is the adaptive self-tuning of its parameters. As explained in the introduction, there are many approaches to realise this specification. In this paper a resonance condition of the hydraulic part of the actuator is proposed. The idea is, as already explained, to maintain this part of the actuator in a resonance condition at the frequency of the engine changing adaptively parameter η of the obtained sliding control law of Eq. (57). The hydraulic part is the most important part in terms of energy of the hybrid actuator. Therefore the efficiency of the whole controlled actuator can obtain a benefit from this resonance condition. To analyse the resonance condition of the hydraulic part of the actuator let us take into consideration the following fundamental background. It is known that if the following transfer function is considered

$$F(s) = \frac{1}{\frac{1}{\omega_n^2}s^2 + \frac{2\zeta}{\omega_n}s + 1}, \tag{61}$$

with

$$s = j\omega, \tag{62}$$

then

$$F(j\omega) = \frac{1}{\frac{-\omega^2}{\omega_n^2} - \frac{2j\zeta\omega}{\omega_n} + 1}. \tag{63}$$

Considering

$$|F(j\omega)| = \frac{1}{\sqrt{(1 - \frac{\omega^2}{\omega_n^2})^2 + \frac{4\zeta^2\omega^2}{\omega_n^2}}}, \tag{64}$$

in which

$$u = \frac{\omega}{\omega_n}, \tag{65}$$

then

$$|F(j\omega)| = \frac{1}{\sqrt{(1-u^2)^2 + 4\zeta^2 u^2}}. \tag{66}$$

If

$$D = (1-u^2)^2 + 4\zeta^2 u^2, \tag{67}$$

then

$$\max_{\omega} |F(j\omega)| = \min_{u} D \tag{68}$$

and to find this maximum it follows that

$$0 = -4(1-u^2)u + 8\zeta^2 u. \tag{69}$$

From (69) it follows that

$$u_R = \pm\sqrt{1 - 2\zeta^2} \tag{70}$$

and

$$\omega_R = \omega_n\sqrt{1 - 2\zeta^2}. \tag{71}$$

It is know that ω_R exists for $0 \leq \zeta \leq \frac{1}{\sqrt{2}}$.

In general, if a linear system, characterized by transfer function $G(s)$, is controlled in closed loop, the following transfer function is obtained:

$$W(s) = \frac{G_R(s)G(s)}{1 + G_R(s)G(s)}, \tag{72}$$

in which $G_R(s)$ represents the transfer function of a possible linear controller. Considering that in the frequency range of the engine, the valve velocity remains between 0 and 5 m/s and the nonlinearity effects, which are basically represented by the losses, are very small with respect to the input effect. Moreover, this controller plays a decisive role in the region of the soft landing in which the velocity is very small and it is possible to consider the following linear system as an approximated model of the hydraulic part of the actuator:

$$G(s) = \frac{X_V(s)}{Q_{th}(s)} = \frac{\frac{V_H}{M_V A_{VP}}}{s}, \tag{73}$$

in which $X_V(s)$ and Q_{th} represent the valve position and the volumetric flow. The following proposition states the adaptive law of the proposed Sliding Mode Control.

Proposition 2 *Let's consider the following transfer function*

$$G(s) = \frac{X_V(s)}{Q_{th}(s)} = \frac{K}{s}, \tag{74}$$

where

$$K = \frac{V_H}{M_V A_{VP}} \tag{75}$$

is a constant originating from Eq. (30), *$G(s)$ represents the approximated model of the hydraulic part of the model. If the following I controller is considered*

$$G_R(s) = K_I \frac{1}{s}, \tag{76}$$

then η *of* Eq. (40) *results to be*

$$\eta = \lambda_2 T_s \omega_R^2. \tag{77}$$

Proof The closed-loop transfer function (72) can now be written as

$$\frac{X_V(s)}{X_{Vd}(s)} = \frac{K_I \frac{1}{s} K \frac{1}{s}}{1 + K_I \frac{1}{s} K \frac{1}{s}} \tag{78}$$

or

$$\frac{X_V(s)}{X_{Vd}(s)} = \frac{1}{\frac{1}{K_I K} s^2 + 1}. \tag{79}$$

Equating the coefficients of that with Eq. (61) gives

$$\omega_n = \sqrt{K_I K} \tag{80}$$

and, ideally,

$$\zeta = 0 \tag{81}$$

and thus

$$\omega_R = \omega_n \sqrt{1 - 2\zeta^2} = \omega_n \tag{82}$$

where ω_R is the dominant frequency of the speed of the combustion engine. This is the resonance condition.

$$\omega_R = \sqrt{K_I K}, \tag{83}$$

so

$$K_I = \frac{\omega_R^2}{K}. \tag{84}$$

Considering Eq. (75) the following equation is obtained:

$$K_I = \omega_R^2 \frac{M_V A_{VP}}{V_H}. \tag{85}$$

If Eq. (60) is rewritten as follows:

$$\frac{Q_{th}(kT_s) - Q_{th}((k-1)T_s)}{T_s} = \left(\mathbf{GB}_H T_s\right)^{-1} \left(\eta s(kT_s)\right), \tag{86}$$

then, substituting $\mathbf{GB}_H = \lambda_2 K$ from Eq. (75) in (86) it results:

$$K_I = \left(\lambda_2 \frac{V_H}{M_V A_{VP}} T_s\right)^{-1} \eta \tag{87}$$

To conclude, if Eq. (85) is combined with (87), then the following robust adaptive law is derived:

$$\eta = \lambda_2 T_s \omega_R^2 \tag{88}$$

□

Remark 1 Equation (88) states an interesting and effective robust adaptive self-tuning controller. The robustness is stated by the independence of the parameters of the model. The factor λ_2 in η is cancelled out in the controller equation (57). However, together with λ_1, it remains as a weighting factor for the position error and the velocity error (see Fig. 15). It is possible to justify and to interpret Eq. (88) just by thinking that the proportionality of parameter η with respect to the engine revolution cylces are due to the necessity to adapt the controller to the velocity. The proportionality with respect to parameter T_s is due to the necessity to adapt the controller to the delay of the sampling rate. The wider the sampling rate is, the stronger the action of the controller is.

6 Simulation Results

The designed sliding mode controller (Fig. 15) was tested using the complete model including both the plant and the feedforward controller (see 3).

The disturbance force caused by exhaust gas after the combustion is implemented like a step followed by an exponential drop right at the beginning of the valve opening period. Technically speaking, as soon as the valve has opened even a very small slit the pressure drops fast.

The desired valve trajectory (Fig. 10) is generated as a gaussian curve and thus as an exponential function. This makes it possible to calculate derivates symbolically instead of numerically.

$$xV_{\text{desired}}(t) = He^{-(\frac{mt+a}{apt})^2} \tag{89}$$

where H ist the valve lift height, mt is a periodical ramp from 0° to 720° and a is a constant phase delay (−360°). apt is the aperture of the gaussian curve and is proportional to its full width at half maximum (with a factor of $\frac{1}{2\sqrt{\ln 2}}$).

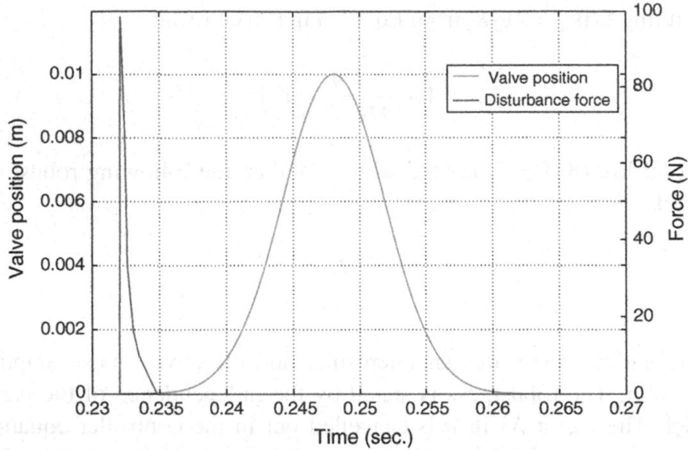

Fig. 9 Timing of the modeled disturbance relative to the valve trajectory

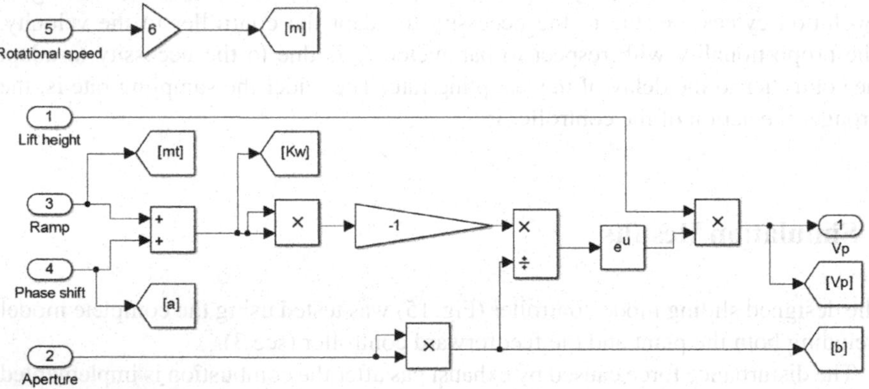

Fig. 10 Trajectory generation

Due to the nature of the exponential function the derivatives of the trajectory can be expressed as a polynomial multiplied by the original trajectory:

$$\dot{x}_{V_{\text{desired}}}(t) = -2m\frac{mt + a}{apt^2} \cdot He^{(\frac{mt+a}{apt})^2} \tag{90}$$

This helps to avoid noisy numerical derivatives, at least for the desired trajectory, and improves several aspects of the feedforward control. Further derivatives can be calculated likewise.

Without any feedback controller, the valve trajectory is almost acceptable for low engine speeds (Fig. 12). This indicates that the implemented feedforward control is a somewhat successful inversion of the plant. There is some high-frequency oscillation that is caused by the hysteresis of the piezo actuator (setting $V_h = 0$ makes it

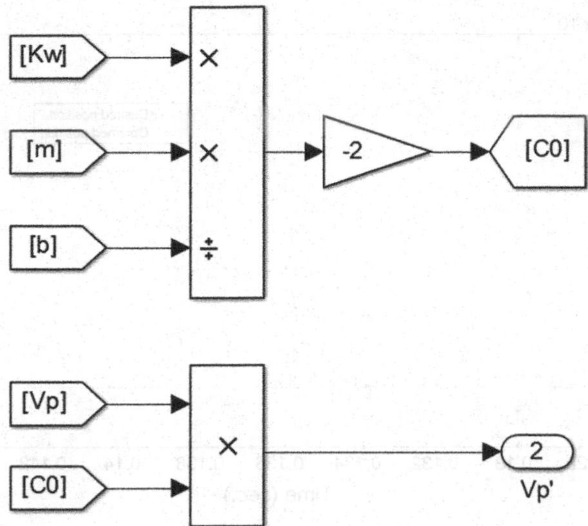

Fig. 11 First derivative of the valve position: Valve velocity. Note that $b = apt^2$

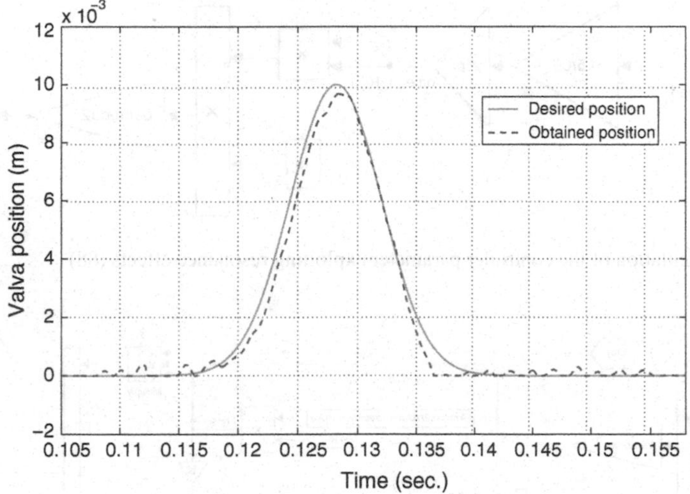

Fig. 12 Obtained valve position without a feedback controller at 2,000 rpm

disappear, see 3.1). It does not bother us too much at 2,000 rpm, but at 8,000 rpm (Fig. 13) undesired resonance effects cause the valve to open between cycles, which is completely unacceptable.

Futhermore, the closing velocity is always too high (about 1.5 m/s). One goal of this contribution is to lower it to achieve soft landing in order to protect the valves and reduce noise.

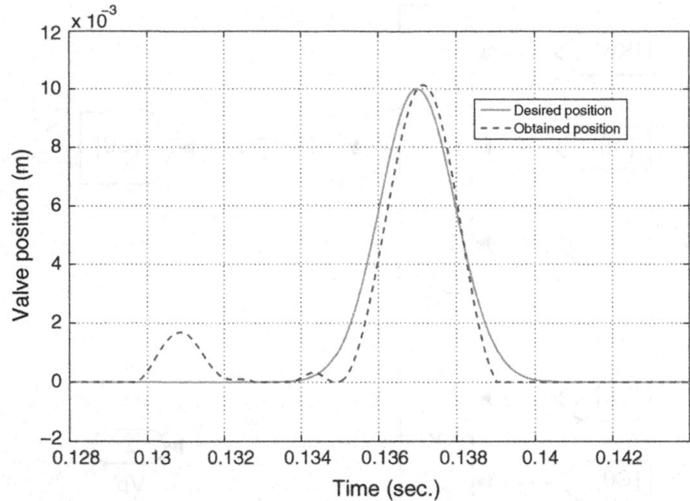

Fig. 13 Obtained valve position without a feedback controller at 8,000 rpm

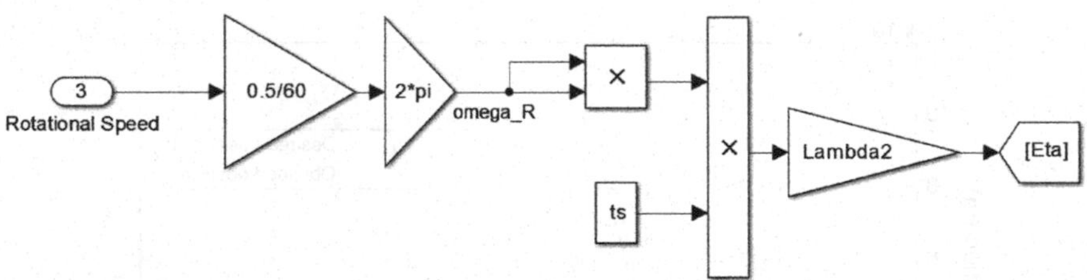

Fig. 14 Calculation of the controller parameter exploiting resonance effects (88)

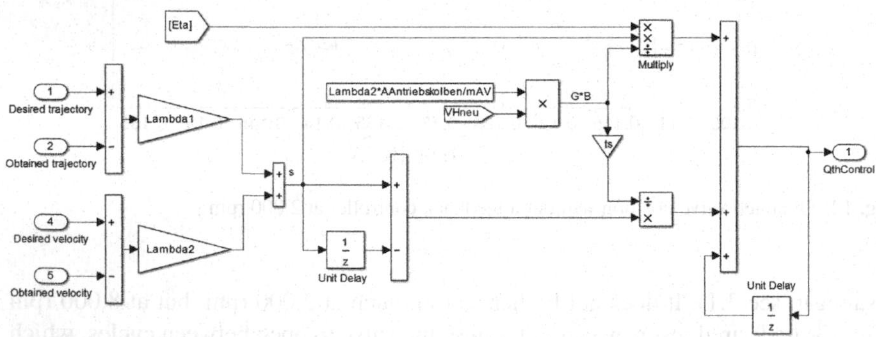

Fig. 15 Sliding mode controller

It will be shown hereinafter that such a soft landing can be achieved by controlling the volumetric flow as described in 4.

The main controller parameter Eta is calculated using the rotational speed of the engine and T_s (88). The goal is to choose an η that alters the resonance frequency of the system *online* matching the current rotational speed to obtain amplification.

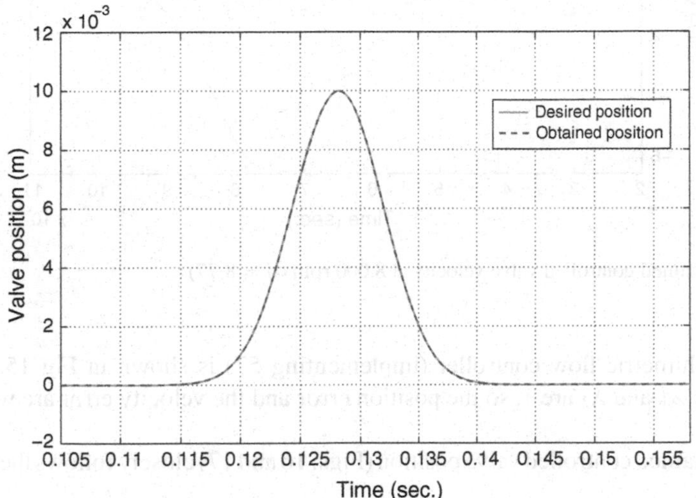

Fig. 16 Obtained controlled valve position at 2,000 rpm ($\eta = 0.55$)

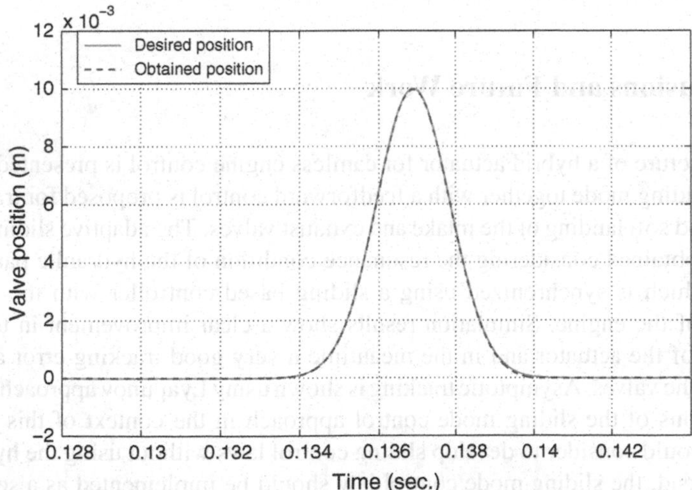

Fig. 17 Obtained controlled valve position at 8,000 rpm ($\eta = 8.77$)

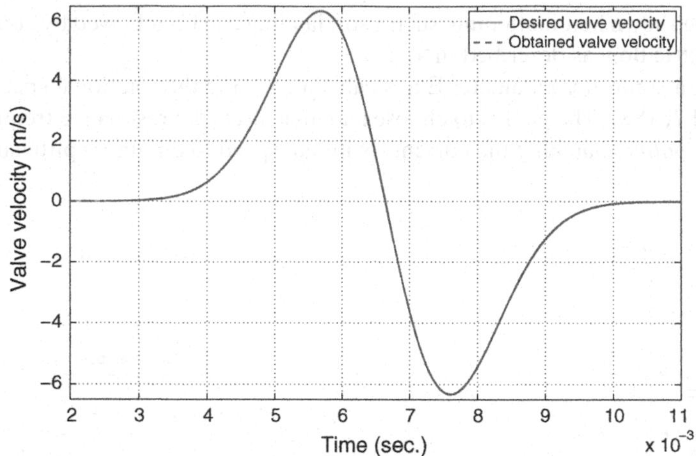

Fig. 18 Obtained controlled valve velocity at 8,000 rpm ($\eta = 8.77$)

The volumetric flow controller (implementing 57) is shown in Fig. 15. In this simulation λ_1 and λ_2 are 1, so the position error and the velocity error are weighted the same.

The obtained controlled valve position (Figs. 16 and 17) closely follows the desired trajectory.

The valve velocity is looking fine, too. The valve is landing softly in the valve seat.

7 Conclusions and Future Work

A new structure of a hybrid actuator for camless engine control is presented and an adaptive sliding mode together with a feedforward control is proposed for trajectory tracking and soft landing of the intake and exhaust valves. The adaptive sliding mode control is obtained considering the resonance condition of the hydraulic part of the actuator which is synchronized using a sliding based controller with the number of cycles of the engine. Simulation results show a clear improvement in terms of efficiency of the actuator and in the meantime a very good tracking error and soft landing of the valves. Asymptotic tracking is shown using Lyapunov approach. Future investigations of the sliding mode control approach in the context of this kind of actuator should consider to develop sliding control laws without using the hydraulic valve. Instead, the sliding mode control law should be implemented as a software-only trajectory governor without the help of supplementary hardware structures.

Acknowledgments This work was financially supported by Bundesministerium für Bildung und Forschung of Germany. BMBF-Projekt-Nr.: 17N2111.

References

Adriaens, H.J.M.T.A., de Koning, W.L., Banning, R.: Modeling piezoelectric actuators. IEEE/ASME Trans. Mechatron. **5**(4), 331–341 (2000)

Alanisa, A.Y., Sanchezb, E.N., Loukianov, A.G.: Real-time output trajectory tracking neural sliding mode controller for induction motors. Journal of The Franklin Institute **351**(4), 2315–2334 (2014)

Corradini, M.L., Jetto, L., Parlangeli, G.: Robust stabilization of multivariable uncertain plants via switching control. IEEE Trans. Autom. Control **49**(1), 107–114 (2004)

Ferrara, A., Vecchio, C.: Second-order sliding mode control of a platoon of vehicles. Int. J. Model. Ident. Control **3**(3), 277–285 (2008)

Gan, W.C., Cheung, N.C.: Development and control of a low-cost linear variable-reluctance motor for precision manufactoring for automation. IEEE/ASME Trans. Mechatron. **8**(3), 326–333 (2003)

Hoffmann, W., Stefanopoulou, A.G.: Iterative learning control of electromechanical camless valve actuator. In: Proceedings of American Control Conference, vol. 4, pp. 2860–2866, Arlington, June 2001

Jian-Xin, X., Abidi, K.: Discrete-time output integral sliding-mode control for a piezomotor-driven linear motion stage. IEEE Trans. Industr. Electron. **55**(11), 3917–3926 (2008)

Jie, S., Yong, Z., Chengliang, Y.: Longitudinal brake control of hybrid electric bus using adaptive fuzzy sliding mode control. Int. J. Model. Ident. Control **3**(3), 270–276 (2008)

Jou, C.-H., Lu, J.-S., Chen, M.-Y.: Adaptive fuzzy controller for a precision positioner using electromagnetic actuator. Int. J. Fuzzy Syst. **4**(1), 110–116 (2012)

Jou, C.-H., Lu, J.-S., Chen, M.-Y.: Adaptive fuzzy controller for a precision positioner using electromagnetic actuator. Int. J. Fuzzy Syst. **4**(1), 110–116 (2012)

Lee, V., Lee, D., Won, S.: Precise tracking control of piezo actuator using sliding mode control with feedforward compensation. In: Proceedings of the SICE Annual Conference 2010 (2010)

Lee, J.-D., Duan, R.-Y.: Cascade modeling and intelligent control design for an electromagnetic guiding system. IEEE/ASME Trans. Mechatron. **16**(3), 470–479 (2011)

Levant, A.: Chattering analysis. IEEE Trans. Autom. Control **55**(6), 1380–1389 (2010)

Levant, A., Fridman, L.M.: Accuracy of homogeneous sliding modes in the presence of fast actuators. IEEE Trans. Autom. Control **55**(3), 810–814 (2010)

Loukianov, A.G., Vittek, J., Castillo-Toledo, B.: A robust automotive controller design. Int. J. Model. Ident. Control **3**(3), 270–276 (2008)

Mercorelli, P., Werner, N.: A hybrid actuator modelling and hysteresis effect identification in camless internal combustion engines control. Int. J. Model. Ident. Control **21**(3), 264–269 (2014)

Mercorelli, P.: An anti-saturating adaptive pre-action and a slide surface to achieve soft landing control for electromagnetic actuators. IEEE/ASME Trans. Mechatron. **17**(1), 76–85 (2012)

Mercorelli, P.: A two-stage augmented extended kalman filter as an observer for sensorless valve control in camless internal combustion engines. IEEE Trans. Industr. Electron. **59**(11), 4236–4247 (2012)

Mercorelli, P.: A hysteresis hybrid extended kalman filter as an observer for sensorless valve control in camless internal combustion engines. IEEE Trans. Ind. Appl. **48**(6), 1940–1949 (2012)

Murrenhoff, H.: Servohydraulik. Shaker, Aachen (2002)

Nguyen, T., Leavitt, J., Jabbari, F., Bobrow, J.E.: Accurate sliding-mode control of pneumatic systems using low-cost solenoid valves. IEEE/ASME Trans. Mechatron. **12**(2), 216–219 (2007)

Pan, Y., Ozgiiner, 0., Dagci, O.H.: Variable-structure control of electronic throttle valve. IEEE Trans. Industr. Electron. **55**(11), 3899–3907 (2008)

Peterson, K.S., Stefanopoulou, A.G.: Rendering the elecreomechanical valve actuator globally asymptotically stable. In: Proceedings of the 42nd IEEE Conference on Decision and Control, vol. 2, pp. 1753–1758, Maui, Dec 2003

Ran, Z., Xueli, W., Zhang, Q., Quanmin, Z., Nouri, H.: The modelling and sliding mode control study of sewage treatment system. Proceedings of International Conference on Modelling, Identification and Control (ICMIC) 2012, pp. 908–913. Wuhan, Hubei, June 2012

She, J.H., Xin, X., Pan, Y.: Equivalent-input-disturbance approach: analysis and application to disturbance rejection in dual-stage feed drive control system. IEEE/ASME Trans. Mechatron. **16**(2), 330–340 (2011)

Tai, C., Tsao, T.: Control of an electromechanical actuator for camless engines. In: Proceedings of American Control Conference, Denver, June 2003

Xinkai, C., Hisayama, T.: Adaptive sliding-mode position control for piezo-actuated stage. IEEE Trans. Industr. Electron. **55**(11), 3927–3934 (2008)

Yang, Y.-P., Liu, J.-J., Ye, D.-H., Chen, Y.-R., Lu, P.-H.: Multiobjective optimal design and soft landing control od an electromagnetic valve actuator fo a camless engine. IEEE/ASME Trans. Mechatron. **18**(3), 963–972 (2013)

Yu, Y.-C., Lee, M.-K.: A dynamic nonlinearity model for a piezo-actuated positioning system. In: Proceedings of the 2005 IEEE International Conference on Mechatronics, ICM 10th–12th July, Taipei (2005)

Sliding Mode Control of Class of Linear Uncertain Saturated Systems

Bourhen Torchani, Anis Sellami and Germain Garcia

Abstract This chapter proposes a new design approach of continuous sliding mode control of linear systems in presence of uncertainty and saturation. The saturation constraint is reported on inputs vector and it is subject to constant limitations in amplitude. The uncertainty is being norm bounded reported on both dynamic and control matrices. In general, sliding mode control strategy consists on two essential phases. The design of the sliding surface is the first phase which is formulated as a pole assignment of linear uncertain and saturated system in a specific region through convex optimization. The solution to this problem is therefore numerically tractable via linear matrix inequalities (LMI) optimization. The controller design is the second phase of the sliding mode control design, which leads to the development of a continuous and non-linear control law. This nonlinear control law is build by choosing switched feedback gain capable of forcing the plant state trajectory to the sliding surface and maintaining a sliding mode condition. To give provider of robustness of the proposed nonlinear control, an approximation on the trajectory deviation of the uncertain saturated system compared to the ideal behavior is proposed. Finally, the validity and the applicability of this approach are illustrated by a multivariable numerical example of a robot pick and place.

1 Introduction

The control of dynamical systems in presence of uncertainties and disturbances is a common problem to deal with when considering real plants. The effect of these uncertainties on the system dynamics should be carefully taken into account in the

---

B. Torchani (✉)
School of Rural Equipment Engineers, ESIER, Medjez Elbeb, Tunisia
e-mail: bourhen@gmail.com

A. Sellami
National Higher School of Engineering, ENSIT, Tunis, Tunisia
e-mail: anis.sellami@esstt.rnu.tn

G. Garcia
University of Toulouse, LAAS-CNRS, Toulouse, France
e-mail: garcia@laas.fr

A.T. Azar and Q. Zhu (eds.), *Advances and Applications in Sliding Mode Control systems*,
Studies in Computational Intelligence 576, DOI 10.1007/978-3-319-11173-5_5

controller design phase since they can worsen the performance or even cause system instability. For this reason, during recent years, the problem of controlling dynamical systems in presence of heavy uncertainty conditions has become an important subject of research. Many considerable progresses have been attained in robust control techniques. Let's quote the sliding mode control, the nonlinear adaptive control, predictive control model and many others. All these techniques are capable of guaranteeing the attainment of the control objectives in spite of modeling errors and uncertainties affecting the controlled plant.

Among the existing methodologies, the sliding mode control technique turns out to be characterized by high simplicity and robustness. This controller utilizes discontinuous control laws to drive the system state trajectory onto a specified surface in the state space, the so-called sliding or switching surface, and to keep the system state on this manifold for all the subsequent times. The main advantages of this approach are two: first, while the system is on the sliding manifold it behaves as a reduced order system with respect to the original plant and, second, the dynamic of the system while in sliding mode is insensitive to model uncertainties and disturbances. The variable structure control VSC (Abiri and Rashidi 2009; Afshari et al. 2009; Dorling and Zinober 1986; Emel'yanov 1967; Itkis 1976; Utkin 1977; Zinober 1994), is a nonlinear control strategy and is characterized by a sliding mode. The VSC was first proposed and elaborated by several researchers from the former Russia, starting from the sixties (Emel'yanov 1970; Emel'yanov and Taran 1962; Utkin 1974). The ideas did not appear outside of Russia until the seventies when a book by (Itkis 1976) and a survey paper by (Utkin 1977) were published in English. Since then, sliding mode control has developed into a general design control method applicable to a wide range of system types including nonlinear systems, MIMO systems, discrete time models and infinite dimensional systems.

The uncertainties are one of the most important problems in robust control design. This requires us to look for a control strategy able to surmount this problem. The purpose of this chapter is to describe the performance of sliding mode control when applied to uncertain systems. The motivation for exploring uncertain systems is the fact that model identification of real world systems introduces parameter errors. Hence, models contain uncertain parameters which are often know to lie within upper and lower bounds. A whole body of literature has arisen in recent years concerned with the stabilization of systems having uncertain parameters lying within know bounds (Bartolini and Zolezzi 1996; Edwards and Spurgeon 1998; Isidori 1999; Krstic et al. 1995; Young et al. 1999). Indeed, the VSC with sliding mode has exceptional invariance proprieties with respect to the classes of matched uncertainties (Arzelier et al. 1993; Ryan and Corless 1984; Sarcheshmah and Seifi 2009; Sellami et al. 2007). The plant uncertainties are required to lie in the image of the initial function for all values of t and x. this requirement is the so-called matching condition (Edwards and Spurgeon 1998; Perruquetti and Barbot 2002; Slotine and Li 1991; Utkin 1977). Assuming that the matching conditions are satisfied, it is possible to lump the total plant uncertainty into a single vector. However, the so-called matched uncertainties, has no effect on the dynamics because they lie within the range space of the input distribution matrix. The sliding mode control is not completely insensitive to some classes of uncertainties like unmatched types, but it has a significant amount of robustness.

Since a long time, the attention of the control engineers is attracted towards the problem of saturation. Though often ignored, as happens in classical control theory, it cannot be avoided in practice. From practical perspective, one of the key problems in feedback control systems is that the signal $u(t)$, generated by the control law, cannot be implemented due to physical constraints. A common example of such constraint is input saturation, which imposes limitations on the amplitude of the control input. The phenomenon of amplitude saturation in actuators is due to inherent physical limitations of devices. In some applications this problem is crucial, especially in combination with nonlinear control, which tends to be aggressive in seeking the desired tracking performance and can cause the instability of the system. These problems lead to an inevitable physical limitations and severe deterioration of closed loop system performance. Towards dealing with the problem of saturation, a sufficient condition guaranteeing the satisfaction of a bound constraint on the magnitude of the control signal is derived. In this way, the saturation is prevented and guarantees that the control signal will not exceed its maximum allowable value. Many rigorous design of control put into consideration this type of problem and are available to provide guarantee properties on systems stability. Let us quote, the anti-windup design (Hippe 2006; Gomes et al. 2004), and many other methods which consider saturation function conditions (Corradini and Orlando 2007; Haijun et al. 2004; Klai 1994; Torchani et al. 2009; Reinelt 2001).

The present work introduces the problem of sliding mode control design of linear systems affected by two major constraints: Presence of unmatched uncertainties and saturation on the entries. The considered class of unmatched uncertainties is norm bounded type and reported both on the dynamic and control matrices. The saturation constraint is a constant limitation in the magnitude of the control vector. The sliding surface design is treated as a problem of root clustering in LMI region (Woodham and Zinober 1986) and gives a chattering free control law. The development and the robustness analysis take into account the constraint of saturation in the different steps. This chapter is organised as follows: In the beginning, we present the saturation structure reported on the control vector and its implementation in the system. Then we give a short introduction to the sliding mode control and the uncertainty. We will then present a design of robust saturated sliding mode control. To show the robustness of this control, an approximation of the trajectory deviation of the uncertain system compared to the ideal behavior is presented. Finally, we consider a multivariable example of robot “pick and place” to validate the theoretical concepts of this work.

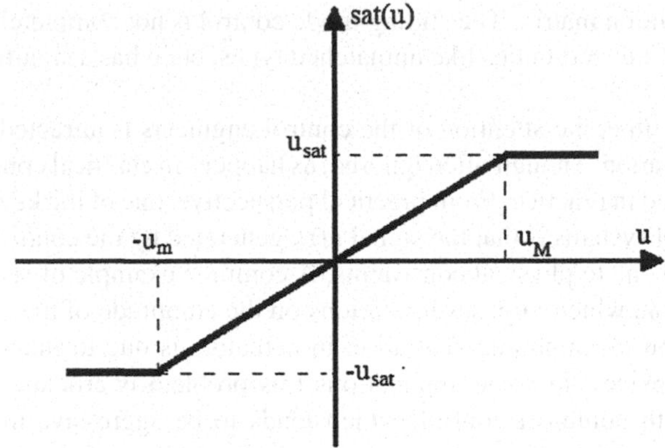

Fig. 1 Structure of the saturation constraint

2 Problem Statement

2.1 Saturation Structure

The control vector is subject to constant limitations in amplitude. It is defined by

$$u(t) \in \Omega \subset \Re^m = \left\{u(t) \in \Re^m / -u_m^i \le u^i(t) \le u_M^i;\, u_m^i,\, u_M^i > 0\, \forall\, i = 1 \ldots m\right\} \quad (1)$$

Assumption 1 The form of the linear control of static feedback state is $u(t) = kx(t)$ where $k \in \Re^{m \times n}$.

In what follows, we give the structure of the saturation.

Then, the control can saturate in the following form

$$u(t) = sat(kx(t)) \quad (2)$$

where

$$sat(kx(t)) = \left[sat(kx(t))^1, \ldots, sat(kx(t))^m\right]^T \quad (3)$$

and the saturation function form is given by

$$sat(kx(t))^i = \begin{cases} u_M^i & \text{if } (kx(t))^i > u_M^i \\ (kx(t))^i & \text{if } -u_m^i \le (kx(t))^i \le u_M^i \\ -u_m^i & \text{if } (kx(t))^i < -u_m^i \end{cases}, \quad \forall \mathrm{i} = 1, \ldots, \mathrm{m} \quad (4)$$

The term of saturation $sat(kx(t))$ can be written as follows

$$sat(kx(t)) = \Gamma(\varsigma(x(t)))kx(t) \tag{5}$$

where the elements $\varsigma^i(x(t))$ of the diagonal matrix $\Gamma(\varsigma(x(t)))$ are defined by

$$\varsigma^i(x(t)) = \begin{cases} \frac{u_M^i}{(kx(t))^i} & \text{if } (kx(t))^i > u_M^i \\ 1 & \text{if } -u_m^i \le (kx(t))^i \le u_M^i, \\ -\frac{u_m^i}{(kx(t))^i} & \text{if } (kx(t))^i < -u_m^i \end{cases} \quad \forall i = 1, \ldots, m \tag{6}$$

and

$$0 < \varsigma^i(x(t)) \le 1 \tag{7}$$

2.2 Presentation of the Uncertain Saturated System

The uncertain saturated system is given by the following expression

$$\begin{cases} \dot{x}(t) = (A + \Delta A)x(t) + (B + \Delta B)\Gamma(\varsigma(x(t)))u(t) \\ y(t) = Cx(t) \end{cases} \tag{8}$$

where $A \in \Re^{n\times n}$, $B \in \Re^{n\times m}$, $x(t) \in \Re^n$ and $u(t) \in \Re^m$. and

$$\lfloor \Delta A \; \Delta B \rfloor = D_1 \nabla \lfloor E_1 \; E_2 \rfloor \tag{9}$$

$D \in \Re^{n\times d}$, $E_1 \in \Re^{e\times n}$ and $E_2 \in \Re^{e\times m}$ are constant matrices characterizing the structure of uncertainty, and $\nabla \in \Re^{d\times e}$ is an uncertain matrix such as

$$\nabla \in F_{bn} = \left\{ \nabla \in \Re^{d\times e} / \nabla^T \nabla \le I_{ee} \right\} \tag{10}$$

where ∇^T and I_{ee} are, respectively, the transposed of ∇ and the identity matrix $(e \times e)$. We take

$$f(x(t)) = \Delta A x(t) \tag{11}$$

and

$$g(x(t), u(t)) = \Delta B u(t) \tag{12}$$

Knowing that

$$\|f(x(t))\| \le k_f \|x(t)\| + k_d \tag{13}$$

$$\|g(x(t), u(t))\| \le k_g \|u(t)\| + \alpha(x(t)) \tag{14}$$

Assumption 2 The pair (A, B) is controllable, B has full rank m and $n > m$.

To simplify, we consider $\Gamma(\varsigma(x(t))) = \Gamma(x)$ and we can write

$$\begin{cases} \dot{x} = (A + \Delta A)x + (B + \Delta B)\Gamma(x)u \\ y = Cx \end{cases} \tag{15}$$

3 Design of the Sliding Surface

3.1 Hyper-Plan Design Procedure

In this section we present the existence of the sliding mode. The canonical form used for VSC design can be extended to uncertain saturated systems to select the matrix F. Existence of a sliding (Edwards and Spurgeon 1998; Itkis 1976; Utkin 1992, 1977) requires stability f the state trajectory to the sliding surface $S(x) = 0$ at least in a neighborhood of $\{x|\, S(x) = 0\}$, the system state must approach the surface at least asymptotically. The largest such neighborhood is called the region of attraction. The method given by Filippov is one possible technique for determining the system motion in the sliding mode as outlined in the previous section. A more straightforward technique easily applicable to multi-input systems is the equivalent control method, as proposed in (Drazenovic 1969; Utkin 1992, 1977). This method of equivalent control can be used to determine the system motion restricted to the switching surface S and the analytical nature of this ethos makes it a powerful tool for both analysis and design purposes.

We consider the system (15). The condition of the sliding mode is verified, if the state variables achieves and remains on the sliding surface or switching surfaces S. these surfaces are defined by the intersection of the hyper-plans passing by the origin of the space of state defined by:

$$S = \bigcap_{j=1}^{m} S_j = \left\{x \in \Re^n : F^j x = 0,\, j = 1\; to\; m\right\} \tag{16}$$

where S is the null space (or Kernel) of F and t_R is the time for which the sliding mode is reached.

Then, for any $t \leq t_R$, the trajectory in sliding mode is on the surface

$$S = Fx = 0 \tag{17}$$

To determine the dynamics of the uncertain system we shall adopt the method of the equivalent control vector u_{eq}, presented in what follows.

While differentiating (17), we obtain

$$\dot{S} = F\dot{x} = 0 \quad (18)$$

Using (15), we obtain

$$\dot{S} = F(A + \Delta A)x + F(B + \Delta B)\Gamma(x)u = 0 \quad (19)$$

We pose $u = u_{eq}$ and if $(FB)^{-1}$ exists. Then, the control law u_{eq} to achieve $\dot{S} = 0$ is

$$u_{eq} = -(F(B + \Delta B)\Gamma(x))^{-1}F(A + \Delta A)x \quad (20)$$

The form of the equivalent control is given by

$$u_{eq} = -Kx \quad (21)$$

with

$$K = (F(B + \Delta B)\Gamma(x))^{-1}F(A + \Delta A) \quad (22)$$

While introducing u_{eq} in (15), we obtain

$$\dot{x} = (A + \Delta A)x - (B + \Delta B)\Gamma(x)((C(B + \Delta B)\Gamma(x))^{-1}C)(A + \Delta A)x \quad (23)$$

$$\dot{x} = (I_n - (B + \Delta B)\Gamma(x))((C(B + \Delta B)\Gamma(x))^{-1}C)(A + \Delta A)x \quad (24)$$

The dynamics of the system in the sliding mode can be written by

$$\dot{x} = (I_n - (B + \Delta B)\Gamma(x))((F(B + \Delta B)\Gamma(x))^{-1}F)(A + \Delta A)x \quad (25)$$

The new dynamic matrix is given by

$$A_{eq} = (I_n - (B + \Delta B)\Gamma(x))((C(B + \Delta B)\Gamma(x))^{-1}C)(A + \Delta A) \quad (26)$$

with A_{eq} describes the motion on the sliding surface which is independent of the actual value of the control and depends only on the choice of the matrix F, the structure of the saturation and the uncertainties ΔA and ΔB.

The basic form used for the design of the variable structure control VSC can be expanded to the systems with saturation constraint for the choosing of the gain matrix C, which gives a good stable dynamics in the sliding mode.

The following part shows the existence of the sliding mode.

Remark 1 There exists an $n \times n$ orthogonal transformation matrix T knowing that $TB = \begin{bmatrix} 0 & B_2 \end{bmatrix}'$, where $B_2 \in \Re^{m \times m}$ is non-singular.

Note that the choice of an orthogonal matrix T avoids inverting T when transforming back to the original system. The transformed state variable vector is defined as

$$y = Tx \tag{27}$$

The transformed variable is shared according to:

$$y^T = \left[y_1^T \; y_2^T \right] \tag{28}$$

with

$$\begin{cases} y_1 \in \Re^{n-m} \\ y_2 \in \Re^{n} \end{cases} \tag{29}$$

Remark 2 The sliding mode is invariant in such transformation that is u_{eq} and A_{eq} are always unchanged.

The state equation becomes

$$\dot{y} = T\left(A + \Delta A\right)T^T y + T\left(B + \Delta B\right)\Gamma(y)u \tag{30}$$

and

$$\dot{y} = TAT^T y + T\Delta AT^T y + TB\Gamma(y)u + T\Delta B\Gamma(y)u \tag{31}$$

with

$$TAT^T = \begin{bmatrix} A_{11} & A_{12} \\ A_{21} & A_{22} \end{bmatrix}, T\Delta AT^T = TD\nabla E_1 T^T = \begin{bmatrix} D_1\nabla E_{11} & D_1\nabla E_{12} \\ D_2\nabla E_{11} & D_2\nabla E_{12} \end{bmatrix} \tag{32}$$

and

$$TB = \begin{bmatrix} 0 \\ B_2 \end{bmatrix}, T\Delta B = TD\nabla E_2 = \begin{bmatrix} 0 \\ D_2\nabla E_2 \end{bmatrix} \tag{33}$$

The transformed system is shared and can be given by the following expression

$$\begin{cases} \dot{y}_1 = A_{11}y_1 + A_{12}y_2 + D_1\nabla E_{11}y_1 D_1\nabla E_{12}y_2 \\ \dot{y}_2 = A_{21}y_1 + A_{22}y_2 + B_2\Gamma(y)u + D_2\nabla E_{11}y_1 + D_2\nabla E_{12}y_2 + D_2\nabla E_2\Gamma(y)u \end{cases} \tag{34}$$

The structures of the uncertainties are replaced by $f(y)$ and $g(y, \Gamma(y)u)$.

we obtain

$$\begin{cases} \dot{y}_1 = A_{11}y_1 + A_{12}y_2 + f(y_1, y_2) \\ \dot{y}_2 = A_{21}y_1 + A_{22}y_2 + B_2\Gamma(y)u + g(y_1, y_2, \Gamma(y)u) \end{cases} \tag{35}$$

with

$$\begin{cases} f\left(y_1, y_2\right) = D_1\nabla E_{11}y_1 + D_1\nabla E_{12}y_2 \\ g\left(y_1, y_2, \Gamma(y)u\right) = D_2\nabla E_{11}y_1 + D_2\nabla E_{12}y_2 + D_2\nabla E_2\Gamma(y)u \end{cases} \tag{36}$$

f and g represent the uncertainty in the system and satisfy the following conditions.

Condition 1 The norm of the uncertainty on the dynamic matrix is bounded and given by

$$\|f(y_1, y_2)\| \leq k_f \sqrt{\|y_1\|^2 + \|y_2\|^2} \tag{37}$$

where

$$k_f = \|D_1\| \left\| \begin{bmatrix} E_{11} \ E_{12} \end{bmatrix} \right\| = \overline{\sigma}(D_1) \overline{\sigma}\left(\begin{bmatrix} E_{11} \ E_{12} \end{bmatrix} \right) \tag{38}$$

Condition 2 The norm of the uncertainty on the control matrix is bounded, given by

$$\|g(y_1, y_2, \Gamma(y)u)\| \leq \alpha(y_1, y_2) + k_g \|\Gamma(y)u\| \tag{39}$$

with

$$k_g = \|D_2\| \|E_2\| \tag{40}$$

and

$$\alpha(y_1, y_2) = k_\alpha \sqrt{\|y_1\|^2 + \|y_2\|^2} \tag{41}$$

where

$$k_\alpha = \|D_2\| \left\| \begin{bmatrix} E_{11} \ E_{12} \end{bmatrix} \right\| = \overline{\sigma}(D_2) \overline{\sigma}\left(\begin{bmatrix} E_{11} \ E_{12} \end{bmatrix} \right) \tag{42}$$

$\|\cdot\|$ and $\bar{\sigma}(.)$ denote respectively the 2-norm, and the largest singular value. The defining sliding condition is

$$F_1 y_1 + F_2 y_2 = 0 \tag{43}$$

with

$$FT^T = \begin{bmatrix} F_1 \ F_2 \end{bmatrix} \tag{44}$$

Remark 3 FB is non-singular implies that F_2 must be non-singular too.

The sliding condition is

$$y_2 = -F_2^{-1} F_1 y_1 = -G y_1 \tag{45}$$

and

$$G = F_2^{-1} F_1 \tag{46}$$

G being an $m \times (n-m)$ matrix and the order of the uncertain system is $(n-m)$. Then, in the sliding mode, the equivalent system must satisfy not only the n-dimentional state dynamics, but also the m algebraic equations given by $S(x) = 0$. The use of both constraints reduces the system dynamics from an nth order model to an $(n-m)$th order model. The sliding mode is governed by

$$\begin{cases} \dot{y}_1 = A_{11} y_1 + A_{21} y_2 + \tilde{f}(y_1, y_2) \\ y_2 = -G y_1 \end{cases} \tag{47}$$

y_2 becomes a state feedback control.

The closed loop system will then have the dynamics $\dot{y}_1 = [A_{11} - A_{12}G + D_1\nabla (E_1 - E_2G)]\, y_1$ and the design of a stable sliding mode requires the selection of a matrix G knowing that $A_{11} - A_{12}G + D_1\nabla(E_1 - E_2G)$ has $(n - m)$ left-half plan eigenvalues. If the gain G has been determined, F is given by

$$F = \left[G \; I_m \right] T \tag{48}$$

3.2 Pole Assignment of a Reduced System

To determine the gain G and the matrix F, LMI method seems to us very effective. Moreover, in order to improve the performances of the system response, we can use the root clustering approach in LMI region, which enables us to obtain a good result. For that, we propose to choose all the eigenvalues of the matrix $A_{11} - A_{12}G + D_1\nabla(E_1 - E_2G)$ in the region defined by a disc of center q and radius r of the complex plan.

Let $(A_{11} - A_{12}G) = \Sigma_1$ and $D_1\nabla(E_1 - E_2G) = \Delta\tilde{A}$, then (42) is stable and its eigenvalues are localised in the disc, if the two following inequalities are verified:

$$(\Sigma_1 + \Delta\tilde{A} + qI)^T P(\Sigma_1 + \Delta\tilde{A} + qI) - r^2 P < 0 \tag{49}$$

$$P = P^T > 0 \tag{50}$$

The Schur complement is given by the following lemma

Lemma 1 *Let $Q(x) = Q(x)^T$, $R(x) = R(x)^T$ and $S(x)$ denote for $n \times n$ matrices and suppose that P and R commute.*

The following LMI are equivalent

$$\begin{pmatrix} Q(x) & S(x) \\ S(x)^T & R(x) \end{pmatrix} > 0 \tag{51}$$

$$R(x) > 0,\ Q(x) - S(x)R(x)^{-1}S(x) > 0 \tag{52}$$

Let's pose $S = P^{-1}$ and into pre and post multiplying by $\begin{pmatrix} S & 0 \\ 0 & I \end{pmatrix}$, we obtain:

$$\begin{pmatrix} -r^2 S & S\Sigma_1^T + qS + S\Delta\tilde{A}^T \\ \Sigma_1 S + qS + \Delta\tilde{A}S & -S \end{pmatrix} < 0 \tag{53}$$

We take $FS = R$, (52) can be written as

$$\begin{pmatrix} -r^2S & SA_{11}^T - R^TA_{12}^T + qS \\ A_{11}S - A_{12}R + qS & -S \end{pmatrix} + \begin{pmatrix} 0 \\ D_1 \end{pmatrix} \nabla \begin{pmatrix} E_1S - E_2R & 0 \end{pmatrix} + \begin{pmatrix} 0 & D_1^T \end{pmatrix} \nabla^T \begin{pmatrix} (E_1S - E_2R)^T \\ 0 \end{pmatrix} < 0 \tag{54}$$

Let Y, H, E_1 and E_2 are appropriate matrices and Y is symmetric. The inequality

$$Y + H\nabla(E_1 - E_2G) + (E_1 - E_2G)^T\nabla^T H^T < 0 \tag{55}$$

is verified for ∇ satisfying $\nabla\nabla^T \leq I_{ee}$ and there is a scalar $\varepsilon > 0$, with

$$\varepsilon^{-1}HH^T + Y + \varepsilon(E_1 - E_2G)^T(E_1 - E_2G) < 0 \tag{56}$$

Then we can rewrite the LMI (49) as (56)

$$\begin{pmatrix} -r^2S - (E_1S^T - E_2R^T)^T(-\varepsilon I)(E_1S - E_2R) & SA_{11}^T - R^TA_{12}^T + qS \\ A_{11}S - A_{12}R + qS & -S + \frac{1}{\varepsilon}D_1D_1^T \end{pmatrix} < 0 \tag{57}$$

This carries us to write the following result.

Theorem 1 *The system is stable and its eigenvalues are localised in the disc of center $-q$ and radius r, if there is a constant $\alpha = \frac{1}{\varepsilon} > 0$, a matrix R and a symmetric matrix positive S knowing that*

$$\begin{pmatrix} -r^2S & SA_{11}^T - R^TA_{12}^T + qS & (E_1S^T - E_2R^T)^T \\ A_{11}S - A_{12}R + qS & -S + \alpha D_1D_1^T & 0 \\ E_1S - E_2R & 0 & -\alpha I \end{pmatrix} < 0 \tag{58}$$

The stabilizing gain is given by $G = RS^{-1}$.

Once the matrix F is determined, the existent problem has been solved. Attention must be turned to solving the reaching problem.

4 Control Law Design

The controller design is the second phase of the sliding control design procedure mentioned earlier. The problem is to choose switched feedback gains capable of forcing the plant state trajectory to the switching surface and maintaining a sliding mode condition. The assumption is that the sliding surface has already been designed. In the considered case, the control is an m-vector u.

In this part we are interested in the reaching problem. In fact, the controller design procedure consists of two steps. First, to reach the sliding surface and ensure that trajectories are directed towards the switching surface, this involves the selection of a nonlinear feedback control function u which ensures that trajectories are directed towards the switching surface from any point in the state space. However, in order to account for the presence of modeling imprecision and disturbances, the control law has to be discontinuous across S. since the implementation of the associated control switching is imperfect, this leads to chattering, which is undesirable in practice, since it involves high control activity and may excite high frequency dynamics neglected in course of modeling.

Thus, in a second step, the discontinuous control law u is suitably smoothed to achieve an optimal trade-off between control bandwidth and tracking precision. The first step achieves robustness for parametric uncertainty; the second step achieves robustness to high-frequency unmodeled dynamics. An ideal sliding mode exists only when the state trajectory x of the controlled plant agrees with the desired trajectory at every $t \geq t_1$ for some t_1. This may require infinitely fast switching. In real systems, a switched controller has imperfections which limit switching to a finite frequency. The representative point then oscillates within a neighborhood of the switching surface. This oscillation, called chattering.

The general form of the used strategy control is the following

$$u = u^L + u^N \tag{59}$$

where u^L and u^N are the linear and non linear control law parts. The first part of the control is to make the derivative of the sliding surface equal zero to stay on the sliding surface, same the equivalent control. The second part is to compensate the deviations from the sliding surface to reach the sliding surface.

Also, it can be given by

$$u = Lx + \rho(x, u)\frac{Nx}{\|Mx\| + \delta} \tag{60}$$

with

$$u^L = Lx \tag{61}$$

$$u^N = \rho(x, u)\frac{Nx}{\|Mx\| + \delta} \tag{62}$$

where L, N and M are appropriate matrices, with $Ker(N) = Ker(M) = Ker(C)$. $\rho(x, u)$ is a design function expressed from the structure of the saturation and the structure of the bounded norm uncertainties, affecting simultaneously the dynamic and the control matrices. δ is a smoothing parameter to reduce the effect of chattering phenomenon.

The necessity of using smoothing parameter, is that in real life applications, it is not reasonable to assume that the control signal time evolution can switch at infinite frequency, while it is more realistic, due to the inertias of the actuators and sensors, and the presence of noise and exogenous disturbances, to assume that it commute at a very high finite frequency.

The control oscillation frequency turns out to be almost unpredictable. An interesting class of smoothing functions, characterized by a time-varying parameters, was proposed in (Slotine and Li 1991), the most recent and interesting approach for the elimination of chattering is represented by the second order sliding mode methodology (Bartolini et al. 1998b; Levant 1993).

The saturated control law is given by the following form

$$u = \Gamma(x)Lx + \rho(x, \Gamma(x)u)\frac{Nx}{\|Mx\| + \delta} \tag{63}$$

4.1 Design of the Linear Control Low U^L

Proposing a second linear transformation $T_2 : \Re^n \to \Re^n$, given by

$$z = T_2 y = T_2 T x \tag{64}$$

knowing that $z = \begin{bmatrix} z_1^T & z_2^T \end{bmatrix}$, with $z_1 \in \Re^{n-m}$ and $z_2 \in \Re^m$.

The transformation matrix T_2 is non-singular and given by

$$T_2 = \begin{bmatrix} I_{n-m} & 0 \\ G & I_m \end{bmatrix} \tag{65}$$

with

$$T_2^{-1} = \begin{bmatrix} I_{n-m} & 0 \\ -G & I_m \end{bmatrix} \tag{66}$$

The new state variables are then

$$\begin{cases} z_1 = y_1 \\ z_2 = Gy_1 + y_2 \end{cases} \tag{67}$$

After transformation the uncertain saturated system is given by the following expression

$$\begin{cases} \dot{z}_1 = \sum_1 z_1 + \sum_2 z_2 + \tilde{f}(z_1, z_2) \\ \dot{z}_2 = \sum_3 z_1 + \sum_4 z_2 + B_2\Gamma(z)u + \tilde{g}(z_1, z_2, \Gamma(z)u) \end{cases} \tag{68}$$

where

$$\begin{aligned}\Sigma_1 &= A_{11} - A_{12}G \\ \Sigma_2 &= A_{12} \\ \Sigma_3 &= G\Sigma_1 + A_{21} - A_{22}G \\ \Sigma_4 &= GA_{12} + A_{22}\end{aligned} \tag{69}$$

$$\tilde{f}(z_1, z_2) = D_1\nabla(E_{11} - E_{12}G)z_1 + D_1\nabla E_{12}z_2 \tag{70}$$

$$\tilde{g}(z_1, z_2, \Gamma(z)u) = (D_2\nabla(E_{11} - E_{12}G))z_1 + D_2\nabla E_{12}z_2 + G\tilde{f}(z_1, z_2) + D_2\nabla E_2\Gamma(z)u \tag{71}$$

$\tilde{f}$ and $\tilde{g}$ satisfy Conditions 1 and 2, and we have the following conditions.

Condition 3 The norm of the uncertainty on the dynamic matrix is bounded and given by

$$\left\|\tilde{f}(z_1, z_2)\right\| \le k_{\tilde{f}}\sqrt{\|z_1\|^2 + \|z_2 - Gz_1\|^2} \tag{72}$$

Condition 4 The norm of the uncertainty on the control matrix is bounded and given by

$$\|\tilde{g}(z_1, z_2, \Gamma(z)u)\| \le \tilde{\alpha}(z_1, z_2) + k_{\tilde{g}}\|\Gamma(z)u\| \tag{73}$$

where $k_g = k_g$ and

$$\tilde{\alpha}(z_1, z_2) = \left(k_\alpha + k_f\overline{\sigma}(G)\right)\sqrt{\|z_1\|^2 + \|z_2 - Gz_1\|^2} \tag{74}$$

Remark 4 We can write $\tilde{\alpha}(z_1, z_2)$ as follows

$$\tilde{\alpha}(z_1, z_2) = k_\alpha\sqrt{\|z_1\|^2 + \|z_2 - Gz_1\|^2} + k_f\overline{\sigma}(G)\sqrt{\|z_1\|^2 + \|z_2 - Gz_1\|^2} \tag{75}$$

The constants k_α and k_f are previously given according to D_1, D_2, E_{11} and E_{12}.

The new sliding condition is

$$z_2 = 0 \tag{76}$$

The linear part of control law $u = u^L$ impose the below condition

$$\dot{z}_2 = 0 \tag{77}$$

what gives

$$\Sigma_3 z_1 + \Sigma_4 z_2 + B_2\Gamma(z)u^L = 0 \tag{78}$$

and

$$u^L(z) = -(B_2\Gamma(z))^{-1}(\Sigma_3 z_1 + \Sigma_4 z_2) \tag{79}$$

Furthermore

$$u^L(z) = -(B_2\Gamma(z))^{-1}\left(\Sigma_3 z_1 + \left(\Sigma_4 - \Sigma_4^*\right) z_2\right) \tag{80}$$

where $\Sigma_4^* = diag\{\mu_i\} \in \Re^{m\times m}$ is any design matrix with stable eigenvalues, with $Re(\mu_i) < 0$ for $i = 1$ to m.

Transforming back into x-space we obtain

$$u^L = -(B_2\Gamma(x))^{-1}\left[\Sigma_3\left(\Sigma_4 - \Sigma_4^*\right)\right] T_2 T x \tag{81}$$

with

$$L = -(B_2\Gamma(x))^{-1}\left[\Sigma_3\left(\Sigma_4 - \Sigma_4^*\right)\right] T_2 T \tag{82}$$

Once the linear part of the control is obtained, the nonlinear part is determined in what follows.

4.2 Design of the Nonlinear Control Low U^N

The general nonlinear form of the control law is given by

$$u^N\left(z_1, z_2, \Gamma(z)u^L\right) = -\rho\left(z_1, z_2, \Gamma(z)u^L\right)\frac{B_2^{-1}\Gamma(z)P_2 z_2}{\|P_2 z_2\|} \tag{83}$$

Consider the Lyapunov equation with

$$V_2(z_2) = \frac{1}{2} z_2^T P_2 z_2 \tag{84}$$

P_2 denotes the positive defined unique solution of the following equation

$$P_2\Sigma_4^* + \Sigma_4^* P_2 + I_m = 0 \tag{85}$$

Then $P_2 z_2 = 0$ if and only if $z_2 = 0$ we have

$$u^N\left(z_1, z_2, \Gamma(z)u^L\right) = -\rho\left(z_1, z_2, \Gamma(z)u^L\right)\frac{B_2^{-1}\Gamma(z)P_2 z_2}{\|P_2 z_2\| + \delta} \tag{86}$$

Differentiating Lyapunov equation

$$\dot{V}_2(z_2) = z_2^T P_2 B_2 \Gamma(z) u^N + z_2^T P_2 \Sigma_4^* z_2 \tag{87}$$

while using

$$P_2\Sigma_4^* = -\left(\frac{I_m}{2}\right) \tag{88}$$

we obtain

$$\dot{V}_2(z_2) = -\frac{1}{2}\,\|z_2\|^2 + z_2^T P_2 B_2 \Gamma(z) u^N \tag{89}$$

After some intermediate calculations

$$z_2^T P_2 B_2 \Gamma(\varsigma(x)) u^N = -\rho(z_1, z_2)\frac{\|P_2 z_2\|^2}{\|P_2 z_2\| + \delta} \tag{90}$$

we obtain

$$\dot{V}_2(z_2) < 0 \tag{91}$$

The structure of the uncertainty $\rho(z_1, z_2, \Gamma(z)u^L)$, is given by

$$\rho(z_1, z_2, \Gamma(z)u^L) = (1 - k_g)^{-1}\gamma_1(((z_1, z_2, \Gamma(z)u^L) + k_g \left\|\Gamma(z)u^L\right\|) + \gamma_2) \tag{92}$$

Transforming back into x-space, we obtain

$$\rho(x, \Gamma(x)u^L) = \gamma_4(((\gamma\;\|T_2 T x\|) + \gamma_3 \left\|\Gamma(x)u^L\right\|) + \gamma_2) \tag{93}$$

with

$$\begin{aligned} \gamma_4 &= \left(1 - k_g\right)^{-1} \gamma_1 \\ \gamma_3 &= k_g \\ \gamma_2 &> 0 \\ \gamma_1 &> 1 \\ \gamma &= k_\alpha + k_f \overline{\sigma}(G) \end{aligned} \tag{94}$$

From (86), matrices M and N according to the variable of state are given by

$$N = -\Gamma(x) B_2^{-1} \begin{bmatrix} 0 & P_2 \end{bmatrix} T_2 T \tag{95}$$

and

$$M = \begin{bmatrix} 0 & P_2 \end{bmatrix} T_2 T \tag{96}$$

5 Robustness Analysis

- If $\delta = 0$: The control law will be discontinuous and guaranteed to reach the sliding surface S in a finished time. However, it results a phenomenon of very severe chattering, especially during the practical applications.
- If $\delta > 0$: To get solution for reaching phenomenon, the continuous control ensures that the trajectory is in the specified region defined by ellipsoids, where the system is sliding, and therefore $z_2 = 0$.

We give then the following results:

Theorem 2 *The continuous nonlinear control structure* $u = u^L + u^{cN}$ *defined in* (60)*:*

(i) *Ensure that the sliding subspace S is bounded by* Ψ_1 *given by:*

$$S \subset \Psi_1 = \left\{ \left[z_1 \in \Re^{n-m} z_2 \in \Re^m \right] / V_2 (z_2) = \frac{1}{2} z_2^T P_2 z_2 \leq \varepsilon_1 \right\} \tag{97}$$

where

$$\varepsilon_1 = \left(\frac{\delta}{\gamma_4 - 1} \right)^2 \times \frac{1}{2\lambda_{min} (P_2)} \tag{98}$$

(ii) *Drive an arbitrary initial state* $(z_1 (t_0), z_2 (t_0)) = (z_1^0, z_2^0) \in \Psi_1$ *to the sliding subspace* Ψ_1 *in a time* T_R *satisfying the following condition*

$$T_R \leq \frac{1}{\gamma_2} \frac{\sqrt{2 z_2^{0T} P_2 z_2^0} - \sqrt{2\varepsilon_1}}{\sqrt{\lambda_{min} (P_2)}} \tag{99}$$

Theorem 3 *If*

$$k_f \leq \left(2\sqrt{1 + \|G\|^2} \, \|P_1\| \right)^{-1} \tag{100}$$

with

$$(z_1(t_0), z_2(t_0)) = (z_1^0, z_2^0) \in \Psi_1 \tag{101}$$

Then

(i) *Every trajectory* z_1 *must ultimately enter and remain within the ellipsoid* $\Psi_2 (\varepsilon_2)$*:*

$$\Psi_2 = \left\{ z_1 \in \Re^{n-m} / V_1 (z_1) \leq \varepsilon_2 \right\} \tag{102}$$

with

$$\varepsilon_2 = \varepsilon + \frac{1}{2} \left(\frac{1}{2} - k_f \sqrt{1 + \|G\|^2} \, \|P_1\| \right)^2 \|P_1\|^3 (k_d + k_n)^2 \tag{103}$$

and

$$k_n = \sqrt{2\varepsilon_1}\left(\left\|\varepsilon_2 P^{-\frac{1}{2}}\right\| + k_f \left\|P^{-\frac{1}{2}}\right\|\right) \tag{104}$$

where $\varepsilon > 0$ *is arbitrary small.*

(ii) *The uncertain system* (8) *will approximate the prescribed dynamic behavior* $z_{1m} = e^{\Sigma_1(t-t_0)z_1^0}$ *in the sense that every deviation* $\Delta z_1 = z_1 - z_{1m}$ *of the motion* z_1 *from the ideal sliding motion* z_{1m} *will be bounded with respect to the ellipsoid:*

$$\Psi_3 = \left\{\Delta z_1 \in \Re^{n-m} / V_1(\Delta z_1) \leq \varepsilon_3\right\} \tag{105}$$

where

$$\varepsilon_3 = \begin{cases} 2\|P_1\|^3\left(k_f\sqrt{1+\|G\|^2}\left\|P_1^{-\frac{1}{2}}\right\|\left\|P_1^{-\frac{1}{2}}z_1^0\right\| + k_n\right)^2 if z_1^0 \notin \Psi_2 \\ 2\|P_1\|^3\left(k_f\sqrt{1+\|G\|^2}\left\|P_1^{-\frac{1}{2}}\right\|\sqrt{2\varepsilon_1} + k_n\right)^2 if z_1^0 \in \Psi_2 \end{cases} \tag{106}$$

Proof of Theorem 2

Knowing that $u = u^L + u^{cN}$, then

$$\begin{cases} \dot{z}_1 = \Sigma_1 z_1 + \Sigma_2 z_2 + \tilde{f}(z_1, z_2) \\ \dot{z}_2 = \Sigma_4^* z_2 + B_2\Gamma(z)u^{cN} + \tilde{g}\left(z_1, z_2, \Gamma(z)u^L\right) \end{cases} \tag{107}$$

The most important task is to design a switched control that will drive the plant state to the switching surface and maintain it on the surface upon interception. A Lyapunov approach is used to characterize this task. The Lyapunov method is usually used to determine the stability properties of an equilibrium point without solving the state equation.

Let *V(x)* be a continuously differentiable scalar function defined in a domain that contains the origin. This method is to assure that the function is positive definite when it is negative and function is negative definite if it is positive. In that way the stability is assured.

Differentiating the Lyapunov function and substituting (107), we obtain

$$\dot{V}_2(z_2) = \frac{1}{2}\dot{z}_2^T P_2 z_2 + \frac{1}{2} z_2^T P_2 \dot{z}_2 \tag{108}$$

We replace (107) in (108)

$$\begin{aligned} \dot{V}_2(z_2) =& \frac{1}{2}\left(\Sigma_4^* z_2 + B_2\Gamma(z)u^{cN} + \tilde{g}\left(z_1, z_2, \Gamma(z)u^L\right)\right)^T P_2 z_2 + \frac{1}{2} z_2^T P_2\left(\Sigma_4^* z_2 + B_2\Gamma(z)u^{cN}\right. \\ &\left.+\tilde{g}\left(z_1, z_2, \Gamma(z)u^L\right)\right) \end{aligned} \tag{109}$$

After some calculations

$$\dot{V}_2(z_2) = z_2^T P_2 B_2 \Gamma(z) u^{cN} + z_2^T P_2 \tilde{g}\left(z_1, z_2, \Gamma(z) u^L\right) + z_2^T P_2 \Sigma_4^* z_2 \tag{110}$$

$$\dot{V}_2(z_2) = z_2^T P_2 B_2 \Gamma(z) u^{cN} + z_2^T P_2 \tilde{g}\left(z_1, z_2, \Gamma(z) u^L\right) - z_2^T \left(\frac{I_m}{2}\right) z_2 \tag{111}$$

$$\dot{V}_2(z_2) = -\frac{1}{2}\|z_2\|^2 + z_2^T P_2 B_2 \Gamma(z) u^{cN} + z_2^T P_2 \tilde{g}\left(z_1, z_2, \Gamma(z) u^L\right) \tag{112}$$

We have

$$z_2^T P_2 B_2 \Gamma(z) u^{cN} = -\rho\left(z_1, z_2, \Gamma(z) u^L\right) \frac{\|P_2 z_2\|^2}{\|P_2 z_2\| + \delta} \tag{113}$$

Considering

$$\left\|\tilde{f}(z_1, z_2)\right\| \le k_{\tilde{f}} \sqrt{\|z_1\|^2 + \|z_2 - F z_1\|^2} \tag{114}$$

$$\left\|\tilde{g}\left(z_1, z_2, \Gamma(z) u^L\right)\right\| \le \tilde{\alpha}(z_1, z_2) + k_{\tilde{g}} \left\|\Gamma(z) u^L\right\| \tag{115}$$

$\tilde{g}$ Can be expressed in the following way

$$\left\|\tilde{g}\left(z_1, z_2, \Gamma(z) u^L\right)\right\| \le \tilde{\alpha}(z_1, z_2) \tag{116}$$

Using

$$\rho\left(x, \Gamma(x) u^L\right) = \gamma_4\left(\tilde{\alpha} + \gamma_3 \left\|\Gamma(x) u^L\right\| + \gamma_2\right) \tag{117}$$

we obtain the follow expression

$$\begin{aligned} z_2^T P_2 \|\tilde{g}\| \le \tilde{\alpha}(z_1, z_2) \|P_2 z_2\| = \frac{1}{\gamma_4} \rho\left(z_1, z_2, \Gamma(x) u^L\right) \|P_2 z_2\| \\ - \left(\gamma_3 \left\|\Gamma(X) u^L\right\| + \gamma_2\right) \|P_2 z_2\| \end{aligned} \tag{118}$$

Afterwards

$$\begin{aligned} \dot{V}_2(z_2) \le -\frac{1}{2}\|z_2\|^2 + z_2^T P_2 B_2 \Gamma(x) u^{cN} + \frac{1}{\gamma_4} \rho\left(z_1, z_2, \Gamma(x) u^L\right) \|P_2 z_2\| \\ - \left(\gamma_3 \left\|\Gamma(x) u^L\right\| + \gamma_2\right) \|P_2 z_2\| \end{aligned} \tag{119}$$

$$\dot{V}_2(z_2) \leq -\frac{1}{2}\|z_2\|^2 - \rho\left(z_1, z_2, \Gamma(z)u^L\right)\frac{\|P_2z_2\|^2}{\|P_2z_2\|+\delta} + \frac{1}{\gamma_4}\rho\left(z_1, z_2, \Gamma(z)u^L\right)\|P_2z_2\| - \left(\gamma_3\left\|\Gamma(z)u^L\right\| + \gamma_2\right)\|P_2z_2\| \tag{120}$$

and we obtain the relation

$$\dot{V}_2(z_2) \leq -\frac{1}{2}\|z_2\|^2 - \left(\gamma_3\left\|\Gamma(z)u^L\right\| + \gamma_2\right)\|P_2z_2\| - \rho\left(z_1, z_2, \Gamma(z)u^L\right)\|P_2z_2\|\left(\frac{\|P_2z_2\|}{\|P_2z_2\|+\delta} - \frac{1}{\gamma_4}\right) \tag{121}$$

from where

$$\gamma_4\|P_2z_2\| - \|P_2z_2\| - \delta > 0 \tag{122}$$

$$\|P_2z_2\|(\gamma_4 - 1) - \delta > 0 \tag{123}$$

$$\|P_2z_2\| > \frac{\delta}{\gamma_4 - 1} \tag{124}$$

with $\gamma_4 = \left(1 - k_g\right)^{-1}\gamma_1$, then

$$\dot{V}_2(z_2) < 0 \tag{125}$$

This condition will be satisfied if

$$V_2(z_2) > \left(\frac{\delta}{\gamma_4 - 1}\right)^2 \times \frac{1}{2\lambda_{min}(P_2)} = \varepsilon_1 \tag{126}$$

We conclude that the sliding subspace S is included in the Ψ_1 subspace defined previously.

If $(z_1(t_0), z_2(t_0)) = (z_1^0, z_2^0) \notin \Psi_1$, then the relation can be expressed

$$\dot{V}_2(z_2) \leq -\|P_2\|^{-1}V_2(z_2) - \gamma_2\sqrt{2\lambda_{\min}(P_2)V_2(z_2)} \tag{127}$$

$\lambda_{min}(.)$ represents the minimal eigenvalue of (.). By using the following mathematical propriety.

Propriety 1: Let $\dot{X} \leq -aX - b\sqrt{X}$. The time T_{01}, is necessary so that X changes from X_0 to X_1, satisfies the condition $\left\|\tilde{f}(z_1, z_2)\right\| \leq k_{\tilde{f}}\sqrt{\|z_1\|^2 + \|z_2 - Gz_1\|^2}$, with the form

$$T_{01} \leq \frac{2}{a}ln\left(\frac{a\sqrt{X_0} + b}{a\sqrt{X_1} + b}\right) \leq \frac{2}{b}\left(\sqrt{X_0} - \sqrt{X_1}\right) \tag{128}$$

The variable of state $z_2 \in \Re^m$ satisfies the condition

$$V_2(z_2) = \frac{1}{2} z_2^T P_2 z_2 \leq \varepsilon_1 \tag{129}$$

in the subspace Ψ_1, where ε_1 given previously. We can obtain the necessary time to reach space Ψ_1

$$T_R \leq \frac{1}{\gamma_2} \frac{\sqrt{2V_2(z_2(t_0))} - \sqrt{2V_2(z_2(t_0 + T_R))}}{\sqrt{\lambda_{min}(P_2)}} \tag{130}$$

$$T_R \leq \frac{1}{\gamma_2} \frac{\sqrt{2 z_2^0 P_2 z_2^0} - \sqrt{2\varepsilon_1}}{\sqrt{\lambda_{min}(P_2)}} \tag{131}$$

So we conclude that the continuous law control $u = u^L + u^{cN}$ does not allow reaching the sliding mode exactly, the switching function is not cancelled ($z_2 \neq 0$) but it is bounded.

with

$$V_2(z_2) \leq \varepsilon_1 = \left(\frac{\delta}{\gamma_4 - 1}\right)^2 \frac{1}{2\lambda_{min}(P_2)} \tag{132}$$

Geometrically the sliding surface S is included in Ψ_1 called ellipsoid limited by ε_1 Therefore attainability is done in a finished time, whose maximum value given by T_{01} is ensured in subspace Ψ_1.

Proof of Theorem 3

In presence of uncertainties, the dynamics of the system in the subspace Ψ_1 is given by

$$\begin{cases} \dot{z}_1 = \Sigma_1 z_1 + \Sigma_2 z_2 + \tilde{f}(z_1, z_2) \\ z_1(t_0) = z_1^0 \end{cases} \tag{133}$$

Differentiating Lyapunov equation $V_1(z_1)$

$$\dot{V}_1(z_1) = \frac{1}{2} \dot{z}_1^T P_1 z_1 + \frac{1}{2} z_1^T P_1 \dot{z}_1 \tag{134}$$

and using (133)

$$\dot{V}_1(z_1) = \frac{1}{2}\left((\Sigma_1 z_1)^T + (\Sigma_2 z_2)^T + \tilde{f}(z_1 z_2)\right) P_1 z_1 + \frac{1}{2} z_1^T P_1 \left(\Sigma_1 z_1 + \Sigma_2 z_2 + \tilde{f}(z_1 z_2)\right) \tag{135}$$

$$\dot{V}_1(z_1) = \frac{1}{2} z_1^T \left(\Sigma_1^T P_1 + P_1 \Sigma_1 \right) z_1 + z_1^T P_1 \Sigma_2 z_2 + z_1^T P_1 \tilde{f}(z_1 z_2) \tag{136}$$

$$\dot{V}_1(z_1) = \frac{1}{2} z_1^T (-I_{n-m}) z_1 + z_1^T P_1 \Sigma_2 z_2 + z_1^T P_1 \tilde{f}(z_1 z_2) \tag{137}$$

$$\dot{V}_1(z_1) = -\frac{1}{2} \|z_1\|^2 + z_1^T P_1 \Sigma_2 z_2 + z_1^T P_1 \tilde{f}(z_1 z_2) \tag{138}$$

Then the Condition 3 on the function $\tilde{f}(z_1 z_2)$, leads us to write the following transformations

$$\left\| \tilde{f}(z_1 z_2) \right\| \le k_{\tilde{f}} \sqrt{\|z_1\|^2 + \|z_2 - G z_1\|^2} \le k_{\tilde{f}} \left(\|z_1\| \sqrt{1 + \|G\|^2} + \|z_2\| \right) \tag{139}$$

$$\|z_2\| = \left\| P_2^{\frac{1}{2}} P_2^{\frac{1}{2}} \right\| = \left\| P_2^{\frac{1}{2}} \right\| \left\| P_2^{\frac{1}{2}} z_2 \right\| = \left\| P_2^{\frac{1}{2}} \right\| \sqrt{2 V_2(z_2)} \le \left\| P_2^{\frac{1}{2}} \right\| \sqrt{2 \varepsilon_1} \tag{140}$$

So we can express partly

$$z_1^T P_1 \tilde{f}(z_1 z_2) \le k_{\tilde{f}} \|P_1\| \|z_1\|^2 \sqrt{1 + \|G\|^2} + k_{\tilde{f}} \|P_1\| \|z_1\| \left\| P_2^{\frac{1}{2}} \right\| \sqrt{2 \varepsilon_1} \tag{141}$$

$$z_1^T P_1 \Sigma_2 z_2 \le \|P_1\| \|z_1\| \left\| \Sigma_2 P_2^{\frac{1}{2}} \right\| \sqrt{2 \varepsilon_1} \tag{142}$$

Substituting (141) and (142) in (139), we obtain

$$V_1(z_1) \le -\left(\frac{1}{2} \|z_1\| - \|P_1\| \left(k_{\tilde{f}} \|z_1\| \sqrt{1 + \|G\|^2} + k_n \right) \right) \|z_1\| \tag{143}$$

So if

$$V_1(z_1) > \varepsilon_1 - \varepsilon = \frac{1}{2} \left(\frac{1}{2} - k_{\tilde{f}} \|P_1\| \sqrt{1 + \|G\|^2} \right)^{-2} \|P_1\|^3 k_n^2 \tag{144}$$

then

$$\dot{V}_1(z_1) < 0 \tag{145}$$

k_f is verified in the Condition 3 and (i), which signifies that the trajectory z_1 must definitely enter and remain in ellipsoid $\Psi_2(\varepsilon_2)$ independent from t_0

(ii) $\Delta z_1 = z_1 - z_{1m}$ is the deviation of the system behavior compared to ideal dynamic movement in sliding mode, $z_{1m} = e^{\Sigma_1 (t - t_0)} z_1^0$ for $\tilde{f} = 0$

$$\Delta z_1 = z_1 - z_{1m} \tag{146}$$

$$\begin{aligned} \Delta \dot{z}_1 &= \dot{z}_1 - \dot{z}_{1m} \\ &= \Sigma_1 z_1 + \tilde{f}(z_1 z_2) - \Sigma_1 z_{1m} \\ &= \Sigma_1 \Delta z_1 \tilde{f}(z_1 z_2) \end{aligned} \tag{147}$$

Let the Lyapunov function

$$V_1(\Delta z_1) = \frac{1}{2}(\Delta z_1, P_1, \Delta z_1) = \frac{1}{2}\Delta z_1^T P_1 \Delta z_1 \tag{148}$$

and the derivative

$$\dot{V}_1(\Delta z_1) = \frac{1}{2}\Delta \dot{z}_1^T P_1 \Delta z_1 + \frac{1}{2}\Delta z_1^T P_1 \Delta \dot{z}_1 \tag{149}$$

$$\begin{aligned} \dot{V}_1(\Delta z_1) = &\frac{1}{2}\left((\Sigma_1 \Delta z_1)^T + (\Sigma_2 z_2)^T + \tilde{f}(z_1 z_2)\right) P_1 \Delta z_1 \\ &+ \frac{1}{2}\Delta z_1^T P_1 \left(\Sigma_1 \Delta z_1 + \Sigma_2 z_2 + \tilde{f}(z_1 z_2)\right) \end{aligned} \tag{150}$$

$$\begin{aligned} \dot{V}_1(\Delta z_1) = &\frac{1}{2}\Delta z_1^T \left(\Sigma_1^T P_1 + P_1 \Sigma_1\right) \Delta z_1 + \Delta z_1^T P_1 \Sigma_2 z_2 \\ &+ \Delta z_1^T P_1 \tilde{f}(z_1 z_2) \end{aligned} \tag{151}$$

$$\dot{V}_1(\Delta z_1) = \frac{1}{2}\Delta z_1^T(-I_{n-m})\Delta z_1 + \Delta z_1^T P_1 \Sigma_2 z_2 + \Delta z_1^T P_1 \tilde{f}(z_1 z_2) \tag{152}$$

$$\dot{V}_1(\Delta z_1) = -\frac{1}{2}\|\Delta z_1\|^2 + \Delta z_1^T P_1 \Sigma_2 z_2 + \Delta z_1^T P_1 \tilde{f}(z_1 z_2) \tag{153}$$

According to (143) and (153), we obtain

$$\Delta z_1^T P_1 \tilde{f}(z_1 z_2) \le k_{\tilde{f}} \|P_1\| \, \|\Delta z_1\|^2 \left(\left\|P_1^{\frac{1}{2}}\right\| \left\|P_1^{\frac{1}{2}} \Delta z_1\right\| \sqrt{1 + \|G\|^2} + \left\|P_2^{-\frac{1}{2}}\right\| \sqrt{2\varepsilon_1} \right) \tag{154}$$

Also we can express

$$\Delta z_1^T P_1 \Sigma_2 z_2 \le \|P_1\| \, \|\Delta z_1\| \left\|\Sigma_2 P_2^{\frac{1}{2}}\right\| \sqrt{2\varepsilon_1} \tag{155}$$

Replacing in (154), we obtain

$$V_1(\Delta z_1) \leq \frac{1}{2}\|\Delta z_1\|^2 + \|P_1\| \left(k_{\tilde{f}} \left\|P_1^{\frac{1}{2}}\right\| \left\|P_1^{\frac{1}{2}} \Delta z_1\right\| \sqrt{1+\|G\|^2} + k_n\right) \|\Delta z_1\| \tag{156}$$

If $k_{\tilde{f}}$ verifies the Condition 3 so the condition (i) is verified too. Consequently, every trajectory z_1 definitely enters and remains in the ellipsoid $\Psi_2(\varepsilon_2)$ independent from t_0

$$\Psi_2 = \left\{z_1 \in \Re^{n-m} / V_1(z_1) \leq \varepsilon_2\right\} \tag{157}$$

We have

$$V_1(z_1) = \frac{1}{2} z_1^T P_1 z_1 \leq \varepsilon_2 \tag{158}$$

$$z_1^T P_1 z_1 = z_1^T P_1^{\frac{1}{2}} P_1^{\frac{1}{2}} z_1 = \left\|P_1^{\frac{1}{2}} z_1\right\|^2 \leq 2\varepsilon_2 \tag{159}$$

$$\left\|P_1^{\frac{1}{2}}(z_1)\right\| \leq 2\varepsilon_2 \tag{160}$$

From where

$$V_1(z_1) = \begin{cases} V_1^0(z_1) & \text{if } z_1^0 \notin \Psi_2 \\ \varepsilon_2 & \text{if } z_1^0 \in \Psi_2 \end{cases} \tag{161}$$

and

$$\left\|P_1^{\frac{1}{2}} z_1\right\| = \begin{cases} \left\|P_1^{\frac{1}{2}} z_1^0\right\| & \text{if } z_1^0 \notin \Psi_2 \\ \sqrt{2\varepsilon_2} & \text{if } z_1^0 \in \Psi_2 \end{cases} \tag{162}$$

Using (161) and (162), we obtain (ii).

6 Numerical Application

We study the system of the robot "pick and place robot". We realize simulations in presence of the norm bounded uncertainties, which allows us to judge the robustness control towards the parametric variations. The real system contains some uncertainties reported on the weight of the arm of the robot m_2, thus we can add a weight of the load m_3, which can be expressed by:

$$0Kg \leq \Delta m_2 \leq m_{3max} = 0.1Kg \tag{163}$$

The norm bounded uncertainty, can be written as the following

$$m_2 = m_2 + \Delta m_2 \tag{164}$$

The state matrices in presence of uncertainties, are described by

$$A_{inc} = \begin{bmatrix} 0\,1 & 0 & 0\,0 & 0 \\ 0\,0 & 1 & 0\,0 & 0 \\ 0\,0 & -k_1(\frac{1}{m_1+m_2} + \Delta_1) & 0\,0 & 0 \\ 0\,0 & 0 & 0\,1 & 0 \\ 0\,0 & 0 & 0\,0 & 1 \\ 0\,0 & 0 & 0\,0 & -k_2(\frac{1}{m_2} + \Delta_2) \end{bmatrix} \tag{165}$$

$$B_{inc} = \begin{bmatrix} 0\,0 & 0 & 0\,0 & \frac{1}{m_2} + \Delta_4 \\ 0\,0 & \frac{1}{m_1+m_2} + \Delta_3 & 0\,0 & 0 \end{bmatrix}^T \tag{166}$$

Also, we can write

$$A_{inc} = A + D_1 \nabla E_1 \tag{167}$$

and

$$B_{inc} = B + D_2 \nabla E_2 \tag{168}$$

with

$$E_1 = \begin{bmatrix} 0\,0 & -k_1 & 0\,0 & 0 \\ 0\,0 & 0 & 0\,0 & -k_2 \end{bmatrix}, \quad E_2 = \begin{bmatrix} 1\,0 \\ 0\,1 \end{bmatrix} \tag{169}$$

$$D_1 = D_2 = \begin{bmatrix} 0\,0 & 0 & 0\,0 & \frac{-m_{3max}}{m_2(m_2+m_{3max})} \\ 0\,0 & \frac{-m_{3max}}{(m_1+m_2)(m_1+m_2+m_{3max})} & 0\,0 & 0 \end{bmatrix}^T \tag{170}$$

and

$$0 \le \nabla \le 1 \tag{171}$$

The simulations are done with the numerical values of the system parameters, as follows: $m_1 = m_2 = 1\text{Kg}$, $0\text{Kg} \le m_3 \le 0.1\text{Kg}$, $k_1 = k_2 = 0.1\text{Kgs}^{-1}$ and we have

$$x_0 = \begin{bmatrix} 0\; 0.4\; 0\; 0\; 0.7\; 0 \end{bmatrix}^T \tag{172}$$

$$-5 \le u(t) \le 5 \tag{173}$$

Different matrices are presented

$$A = \begin{bmatrix} 0 & 1 & 0 & 0 & 0 & 0 \\ 0 & 0 & 1 & 0 & 0 & 0 \\ 0 & 0 & -0.05 & 0 & 0 & 0 \\ 0 & 0 & 0 & 0 & 1 & 0 \\ 0 & 0 & 0 & 0 & 0 & 1 \\ 0 & 0 & 0 & 0 & 0 & -0.1 \end{bmatrix}, B = \begin{bmatrix} 0 & 0 \\ 0 & 0 \\ 0.5 & 0 \\ 0 & 0 \\ 0 & 0 \\ 0 & 1 \end{bmatrix} \quad (174)$$

$$D_1 = D_2 = \begin{bmatrix} 0 & 0 & 0 & 0 & 0 & -0.0909 \\ 0 & 0 & -0.0238 & 0 & 0 & 0 \end{bmatrix}^T \quad (175)$$

$$E_1 = \begin{bmatrix} 0 & 0 & -0.1 & 0 & 0 & 0 \\ 0 & 0 & 0 & 0 & 0 & -0.1 \end{bmatrix}, \quad E_2 = \begin{bmatrix} 1 & 0 \\ 0 & 1 \end{bmatrix} \quad (176)$$

The simulations given by Figs. 2, 3 and 4 are carried out to show the behaviour of the multivariable system with uncertainties and saturation constraint. Figures 2 and 3 present respectively the evolution of state variables and control law from different initial conditions. Figure 4 show the evolution of the switching surface. These simulations show that the control effectively corrects the deviation between the certain and uncertain system behaviour in the presence of uncertainty. Also, it is clear that the response of the state trajectories and control are almost confused and the deviation is very small, which judges the robustness of the control towards these parametric perturbations as shown in Fig. 2. Also, these simulations show a typical stable sliding mode convergence of the system in the two cases. The control law is very smooth and not showing a chattering phenomenon as illustrated in Fig. 3. This gives an idea onto the performance of the approach proposed to resolve the problem of chattering.

The control input is saturated and always inferior to its maximal value. The evolution of the sliding surface norm is displayed in Fig. 4, which proves that the control law enables to reach the sliding surface, in both certain and uncertain cases. In addition, we can notice that the convergence is done in a relatively short time.

For the class of systems to which it applies, sliding controller design provides a systematic approach to the problem of maintaining stability and consistent performance in the face of modelling imprecision.

7 Conclusion

In this chapter, a sliding mode control design approach for linear saturated systems affected by norm bounded uncertainty has been proposed. The saturation constraint is reported on inputs vector and uncertainty being norm bounded reported on both dynamic and control matrices. The two design parts of the variable structure control

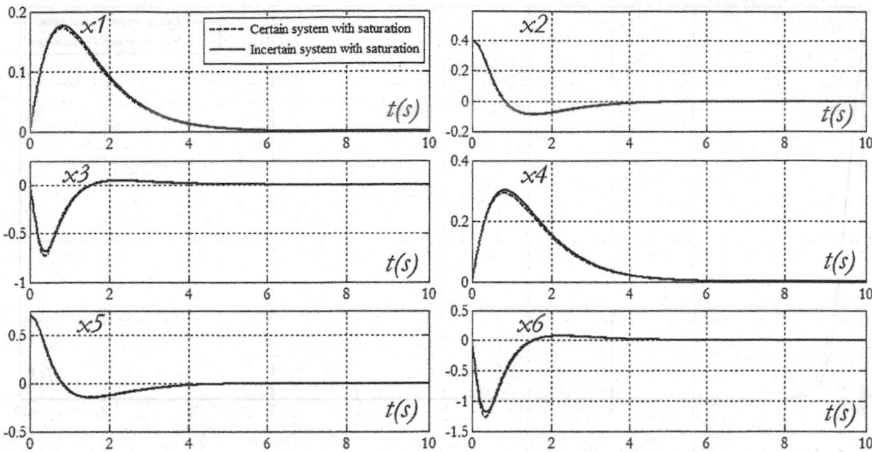

Fig. 2 Evolution of state variables

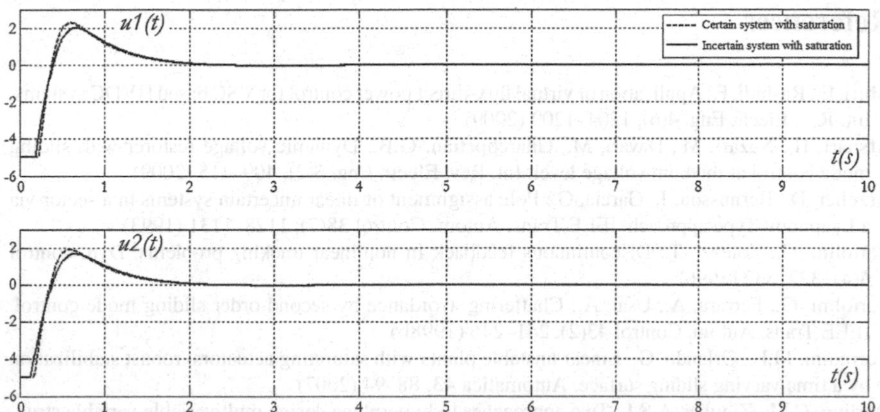

Fig. 3 Evolution of control law of the closed loop saturated system

methodology have been exposed. In the first step, the design of the sliding surface is formulated as a pole assignment of a reduced system. In the second step, a nonlinear saturated control scheme is introduced. It totally eliminates the undesirable chattering phenomenon and ensures a stable sliding mode motion. Numerical application is presented to show the applicability, the efficiency, and the robustness of the proposed control.

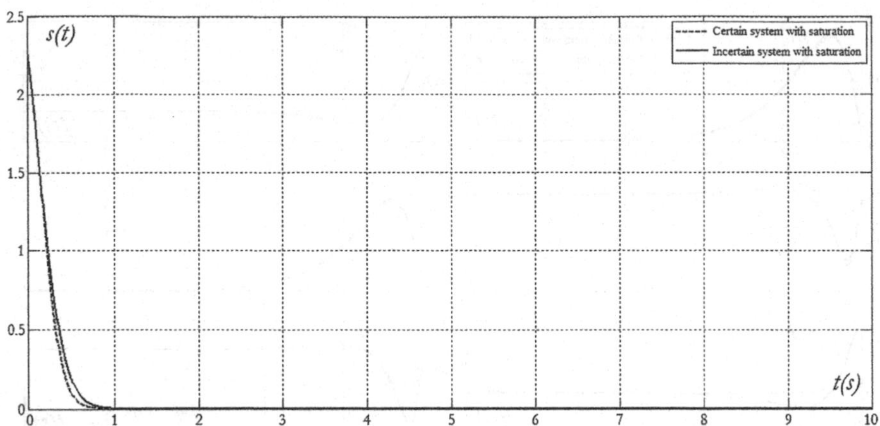

Fig. 4 Evolution of the switching surface

References

Abiri, E., Rashidi, F.: Application of virtual flux-direct power control for VSC based HVDC systems. Int. Rev. Electr. Eng. **4**(6), 1204–1209 (2009)

Afshari, H., Nazari, M., Davari, M., Gharehpetian, G.B.: Dynamic voltage restorer with sliding mode control at medium voltage level. Int. Rev. Electr. Eng. **5**(2), 409–415 (2009)

Arzelier, D., Bernussou, J., Garcia, G.: Pole assignment of linear uncertain systems in a sector via a Lyapunov-Type approach. IEEE Trans. Autom. Control **38**(7), 1128–1131 (1993)

Bartolini, G., Zolezzi, T.: Discontinuous feedback in nonlinear tracking problems. Dyn. control **6**(4), 323–332 (1996)

Bartolini, G., Ferrara, A., Usai, A.: Chattering avoidance by second-order sliding mode control. IEEE Trans. Autom. Control **43**(2), 241–246 (1998b)

Corradini, M.L., Orlando, G.: Linear unstable plants with saturating actuators: robust stabilization by a time varying sliding surface. Automatica **43**, 88–94 (2007)

Dorling, C.M., Zinober, A.S.I.: Two approaches to hyperplane design multivariable variable structure control systems. I. J. Control **44**(1), 65–82 (1986)

Drazenovic, B.: The invariance conditions in variable structure systems. Automatica **5**(3), 287–295 (1969)

Edwards, C., Spurgeon, K.S.: Sliding Mode Control: Theory and Applications. Taylor & Francis, London (1998)

Emel'yanov, S.V., Taran, V.A.: On a class of variable structure control systems. In: USSR Academy of Sciences, Energy and Automation. Moscow, Russia. (In Russia), (1962)

Emel'yanov, S.V.: Automatic Control Systems of Variable Structure. Nauka, Moscow (1967)

Emel'yanov, S.V.: Theory of variable structure systems. Moscow, Russia (1970). (In Russian)

Galeani, S., Tarbouriech, S., Turner, M.C., Zaccarian, L.: A tutorial on modern anti-windup design. Eur. J. Control **15**(3–4), 418–440 (2009)

da Gomes Jr, Silva, Reginatto, R., Tarbouriech, S.: Anti-windup design guaranteed regions of stability for discrete-time linear systems with saturating controls. Revista Controle & automaçao **14**(1), 3–9 (2004)

Haijun, F., Zongli, L., Tingshu, H.: Analysis of linear systems in the presence of actuator saturation and l2-disturbances. Automatica **40**, 1229–1238 (2004)

Hippe, P.: Windup prevention for stable and unstable MIMO systems. Int. J. Syst. Sci. **37**(2), 67–78 (2006)

Isidori, A.: Nonlinear Control Systems II. Springer, Berlin (1999)
Itkis, U.: Control Systems of Variable Structures. Wiley, New York (1976)
Kapila, V., Grigoriadis, K.: Actuator Saturation Control. Marcel Dekker Inc, New York (2002)
Klai, M.: Stabilisation des systèmes linèaires continus contraints sur la commande par retour d'état et de sortie saturés. Ph.D.dissertation, Paul sabatier Univ., Toulouse (1994)
Krstic, M., Kokotovic, P.V., Kanellakopoulos, I.: Nonlinear and Adaptive Control Design. Wiley, New york (1995)
Levant, A.: Sliding order and sliding accuracy in sliding mode control. Int. J. Control **58**(6), 1247–1263 (1993)
Perruquetti, W., Barbot, J.P.: Sliding Mode Control in Engineering. Marcel Dekker Inc, New York (2002)
Reinelt, W.: Control of Multivariable Systems with Hard Contraints. Available: http://www.control.isy.liu.se/publications/#divaReports. Last accessed 07th Feb 2011 (2001)
Ryan, E.P., Corless, M.: Ultimate boundedness and asymptotic stability of a class of uncertain dynamicall systems via continuous and discontinuous feedback control. IMA J. Math. Control Inform. **1**, 223–242 (1984)
Sarcheshmah, M.S., Seifi, A.R.: A new fuzzy power flow analysis based on uncertain inputs. Int. Rev. Electr. Eng. **4**(1), 122–128 (2009)
Sellami, A., Arzelier, D., M'Hiri, R., Zrida, J.: A sliding mode control approach for systems subjected to a norm-bounded uncertainty. Int. J. Robust Nonlinear Control **17**(4), 327–346 (2007)
Slotine, J.J.E., Li, W.: Applied Nonlinear Control. Prentice Hall, Englewood Cliffs (1991)
Tarbouriech, S., Turner, M.C.: Anti-windup synthesis: an overview of some recent advances and open problems. IET Control Theory Appl. **3**(1), 1–19 (2009)
Torchani, B., Sellami, A., M'hiri, R., Garcia, G.: Control of saturated systems with sliding mode. In: International Conference on Electronics, Circuits, and Systems ~ ICECS, IEEE, Hammamet, Tunisia, pp. 567–570 (2009)
Utkin, V.I.: Sliding Modes and Their Application in Variable Structure Systems. Nauka, Moscow (1974). (In Russia)
Utkin, V.I.: Sliding modes and their applications in variable structure systems. IEEE Trans. Autom. Control **22**(22), 279–281 (1977)
Utkin, V.I.: Sliding Modes in Control Optimization. Springer, Berlin (1992)
Woodham, C.A., Zinober, A.S.I.: Eigenvalue placement in a specified sector for variable structure control systems. I. J. Control **57**(5), 1021–1037 (1986)
Young, K., Utkin, V.I., Ozguner, U.: A control engineers guide to sliding mode control. IEEE Trans. Control Syst. Technol. **7**(3), 328–342 (1999)
Zinober, A.S.I.: An Introduction to Sliding Mode Variable Structure Control, Lecture Notes in Control and Information Sciences N193, Springer (1994)

Sliding Mode Control Scheme of Variable Speed Wind Energy Conversion System Based on the PMSG for Utility Network Connection

Youssef Errami, Mohammed Ouassaid, Mohamed Cherkaoui and Mohamed Maaroufi

Abstract The study of a Variable Speed Wind Energy Conversion System (VS-WECS) based on Permanent Magnet Synchronous Generator (PMSG) and interconnected to the electric network is presented. The system includes a wind turbine, a PMSG, two converters and an intermediate DC link capacitor. The effectiveness of the WECS can be greatly improved by using an appropriate control. Furthermore, the system has strong nonlinear multivariable with many uncertain factors and disturbances. Accordingly, the proposed control law combines Sliding Mode Variable Structure Control (SM-VSC) and Maximum Power Point Tracking (MPPT) control strategy to maximize the generated power from Wind Turbine Generator (WTG). Considering the variation of wind speed, the grid-side converter injects the generated power into the AC network, regulates DC-link voltage and it is used to achieve unity power factor, whereas the PMSG side converter is used to achieve Maximum Power Point Tracking (MPPT). Both converters used the sliding mode control scheme considering the variation of wind speed. The employed control strategy can regulate both the reactive and active power independently by quadrature and direct current components, respectively. With fluctuating wind, the controller is capable to maximize wind energy capturing. This work explores a sliding mode control approach to achieve power efficiency maximization of a WECS and to enhance system robustness to parameter variations. The performance of the system has been demonstrated under

Y. Errami (✉)
Department of Physical, Faculty of Science, University Chouaib Doukkali,
Eljadida, Morocco
e-mail: errami.emi@gmail.com

M. Ouassaid
Department of Industrial Engineering, Ecole Nationale des Sciences Appliquées-Safi,
Cadi Ayyad University, Safi, Morocco
e-mail: ouassaid@emi.ac.ma

M. Cherkaoui · M. Maaroufi
Department of Electrical Engineering, Ecole Mohammadia D'Ingénieur,
Mohammed V-Agdal University, Rabat, Morocco
e-mail: cherkaoui@emi.ac.ma

M. Maaroufi
e-mail: maaroufi@emi.ac.ma

A.T. Azar and Q. Zhu (eds.), *Advances and Applications in Sliding Mode Control systems*,
Studies in Computational Intelligence 576, DOI 10.1007/978-3-319-11173-5_6

varying wind conditions. A comparison of simulation results based on SMC and PI controller is provided. The system is built using Matlab/Simulink environment. Simulation results show the effectiveness of the proposed control scheme.

1 Introduction

During the last few decades, the progress in the use of renewable green energy resources is becoming the key solution to the environment contamination caused by the traditional energy sources and to the serious energy crisis (Tseng et al. 2014; Chou et al. 2014). Thus, power generation systems based on renewable energy are making more and more contributions to the total energy production all over the world (He et al. 2014; Guo et al. 2014). On the other hand, among available renewable energy technologies, wind energy source is the most promising options, as it is omnipresent, environmentally friendly, and freely available (Chen et al. 2013a). Compared to other types, wind energy system is regarded as an important renewable green energy resource, mainly as a consequence of its high reliability and cost effectiveness. So, wind energy conversion, has become a fast increasing energy source in the global market (Ma et al. 2014; She et al. 2013). In addition, it is predicted that the wind power system could be supplying 29.1 % of the world energy by 2030 and higher later on (Meng et al. 2013). Consequently, this increasing trend must be accompanied by continuous technological advance and optimization, leading to better options concerning integration to the electric network, reductions in expenses, and improvements concerning turbine performance and dependability in the electricity deliverance (Che et al. 2014; Wang et al. 2014; Nguyen et al. 2014).

In addition, wind energy source could be utilized by mechanically converting it to electrical energy using wind turbine (WT). During the last two decades, various WT concepts have been developed into wind power technologies and led to significant augmentation of wind power capacity. Wind turbine systems can be classified into two main types: fixed speed and variable speed. The fixed velocity system operates almost at constant speed even in variable wind speed which allows direct connection of the generator to the electric network. Recently, fixed speed wind energy conversion systems, due to poor power quality, poor energy capture and stress in mechanical parts have given way to variable velocity systems (Meng et al. 2013). Furthermore, variable speed wind generation system has distinct advantages over fixed-speed generation system, such as lower mechanical stress, operation at maximum power point, less power fluctuation and increased energy capture (Chen et al. 2013a; Li et al. 2012). So, to design reliable and effective systems to utilize this energy, variable speed wind generation systems are better then fixed velocity systems. This is due to the fact that variable velocity systems can accomplish reliability at all wind speeds and the maximum efficiency, improved electric network disturbance rejection characteristics, and the reduction of the flicker problem (Patil and Bhosle 2013; Melo et al. 2014; Li et al. 2013).

In the field of wind energy generation technology, Permanent Magnet Synchronous Generator (PMSG) and Doubly Fed Induction Generators (DFIG) are emerging as the preferred equipment which is used to transform the wind power into electrical energy (Nian et al. 2014; Zhang et al. 2013; She et al. 2013; Tong et al. 2013; Chen et al. 2013b). At present, one of the troubles associated with VS-WECS is the existence of gearbox coupling the generator to the wind turbine and which causes problems. So, the gearbox suffers from faults and requires regular maintenance (Cheng et al. 2009; Najafi et al. 2013). In contrast, PMSG with higher numbers of poles has been used to eliminate the need for gearbox which can be translated into higher generation efficiency (Orlando et al. 2013). Besides, wind power generation based on the PMSG has gained increasing popularity due to several advantages, including its higher power density and better controllability, the elimination of a dc excitation system, low maintenance requirements, higher efficiency compared to other kinds of generators and low energy loss (Alizadeh et al. 2013; Xia et al. 2013; Alshibani et al. 2014; Zhang et al. 2014). Besides, the performance of PMSG equipment has been improving and the price has been decreasing recently (Yaramasu et al. 2013). Therefore, it has been considered a promising candidate for new designs in Wind Energy Conversion Systems (WECS). With those advantages, PMSGs are attracting great attention and interests all over the world. So, some of them have become commercially accessible, for example, Enercon E70 (2.5 MW), Vestas V112 (3.0 MW) and Goldwind 1.5 MW series products (Cárdenas et al. 2013; Yaramasu et al. 2013).

To control the PMSG based WECS, power electronic converter systems are commonly adopted as the interface between the WECS and the power grid (Blaabjerg et al. 2013; Ma et al. 2013). They give the ease for integrating the WECS units to achieve high performance and efficiency when connected to the electric network (Cespedes et al. 2014). Thus, the wind power converters have various power rating coverages of the WECS (Blaabjerg et al. 2013), as shown in Fig. 1. Then, under variable speed operation, the power converters are used to transfer the PMSG output power in the form of variable frequency and variable voltage to the fixed frequency also fixed voltage electric network (Vazquez et al. 2014; Ma et al. 2014). Several

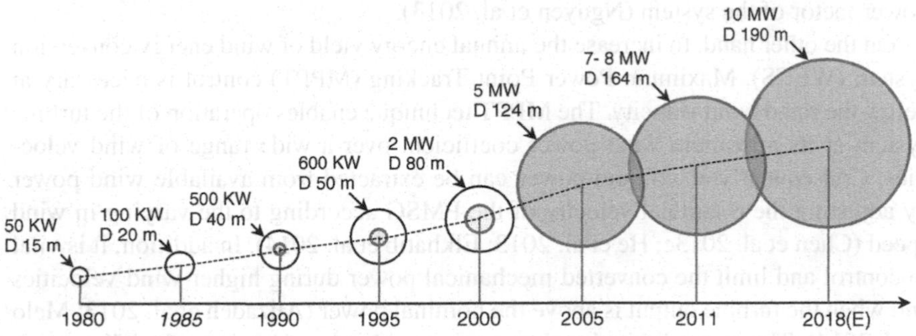

Fig. 1 Evolutions of wind turbine dimension and the corresponding capacity coverage by power electronics converters seen from 1980 to 2018 (estimated)

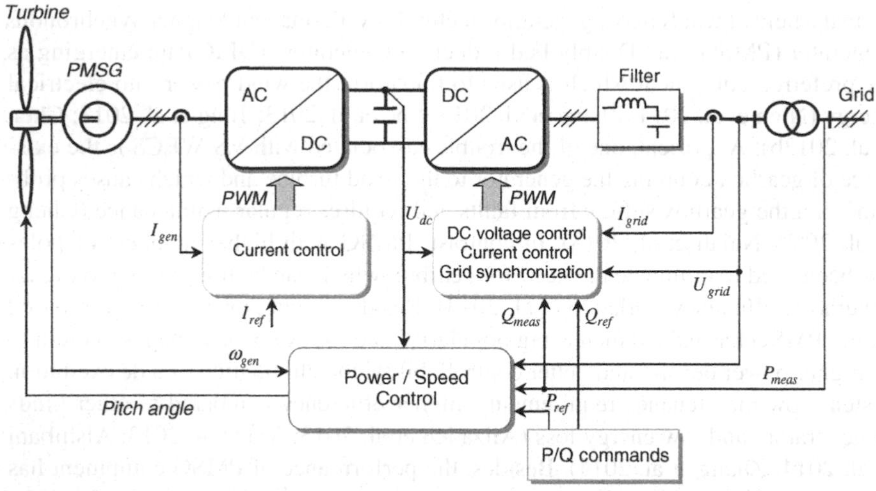

Fig. 2 Control of active and reactive power in a WECS based PMSG

power converter configurations were presented in the literature for PMSG based WECS (Nuno et al. 2014; Li et al. 2013; Xia et al. 2013).

Figure 2 shows the schematic diagram of a typical WECS connected to an electric network. The power electronic conversion system consists of a back-to-back PWM power converter which is composed of a PMSG side converter, a grid-side converter and a dc link. The capacitor decoupling, offers the opportunity of separate control for each power converter. The generator side converter works as a rectifier and it is used to control the torque, the speed or power for PMSG (Chen et al. 2013b; Giraldo et al. 2013; Xin et al. 2013). The grid side converter works as an inverter. The main role of the inverter is to remain the dc-link voltage constant and to synchronize the ac power generated by the WECS with the electric network (Alizadeh et al. 2013; Zhang et al. 2014). Besides, the inverter should have the capability of adjusting active and reactive power that the WECS exchange with the power grid and achieve unity power factor of the system (Nguyen et al. 2013).

On the other hand, to increase the annual energy yield of wind energy conversion system (WECS), Maximum Power Point Tracking (MPPT) control is necessary at below the rated wind velocity. The MPPT technique enables operation of the turbine system at its maximum wind power coefficient over a wide range of wind velocities. Consequently, maximum power can be extracted from available wind power by adjusting the rotational velocity of the PMSG according to the varying in wind speed (Chen et al. 2013c; He et al. 2013; Elkhatib et al. 2014). In addition, it is vital to control and limit the converted mechanical power during higher wind velocities and when the turbine output is above the nominal power (Alizadeh et al. 2013; Melo et al. 2014). The power limitation may be done either by pitch control, stall control, or active stall (Polinder et al. 2007; Spruce et al. 2013). Figure 2 shows the general control structure for modern WECS.

With high penetration of wind power resources in the modern electric network, the power quality from WECS is attracting great attention and interests all over the world (Nguyen et al. 2013). The recent development has been focused mainly on control methodologies for maximum electrical power production (Karthikeya et al. 2014). Accordingly, various control methods have been proposed. Conventional design of WECS control systems is based on Vector Control with d–q decoupling (Xin et al. 2013; Alepuz et al. 2013; Shariatpanah et al. 2013). The control strategy involves relatively complex transformation of currents, voltages and control outputs. Also, the standard design methods consist of properly tuned proportional integral (PI) controllers. Thus, the performance highly relies on the modification of the PI parameters (Chen et al. 2013a; Giraldo et al. 2013; Corradini et al. 2013). So, this technique requires accurate information of WECS parameters. Consequently, the performance is degraded when the actual system parameters differ from those values used in the control system. In addition, VC requires complex reference transformation. Due to the advantages of simple structure and low dependency on the parameters, direct control techniques such as Direct Power Control (DPC) and Direct Torque Control (DTC) were widely used into the WECS (Rajaei et al. 2013; Zhang et al. 2013; Harrouz et al. 2013). They are an alternative to the VC control for WECS because they reduce the complexity of the VC strategy and minimize the employ of generator parameters. The voltage vectors are selected directly according to the differences between the reference and actual value of torque and stator flux or between active and reactive power. Consequently, the converter switching states were selected from an optimal switching table. Besides, DPC and DTC do not necessitate coordinate transformations, specific modulations and current regulators. But, there are high ripples in flux/torque or reactive/active powers at stable state and the switching frequency is variable with operating point due to the employ of predefined switching table and hysteresis regulator (Rajaei et al. 2013). Also, its performance deteriorates during very low speed operation.

For WECS integration into power network and, because the VC and direct control techniques show a limited performances, especially against uncertainties and cannot follow the changes in WECS parameters (Leonhard 1990), Sliding Mode Control approach (SMC) can be used. It has low sensitivity with respect to uncertainty, dynamic performance and good robustness (Slotine and Li 1991; Utkin 1993; Utkin et al. 1999; Sabanovic et al. 2002; Evangelista et al. 2013a). It is one of the powerful control approaches for systems with unknown trouble and uncertainties (Li et al. 2013a, b). Besides, SMC is insensitive to parameter variations of systems. Thus, SMC is suitable for wind power applications (Evangelista et al. 2013a, b; Huang et al. 2013; Susperregui et al. 2013) propose sliding mode control to maximize the energy production of a WECS. Subudhi et al. (2012), Xiao et al. (2013) propose a pitch control based siding mode approach to control the extracted power above the rated wind speed. Xiao et al. (2011), Martinez et al. (2013) introduce sliding mode regulator to control the WECS for fault conditions. Chen et al. (2013), Bouaziz and Bacha (2013), Guzman et al. (2013) present a sliding mode control methodology of power converters.

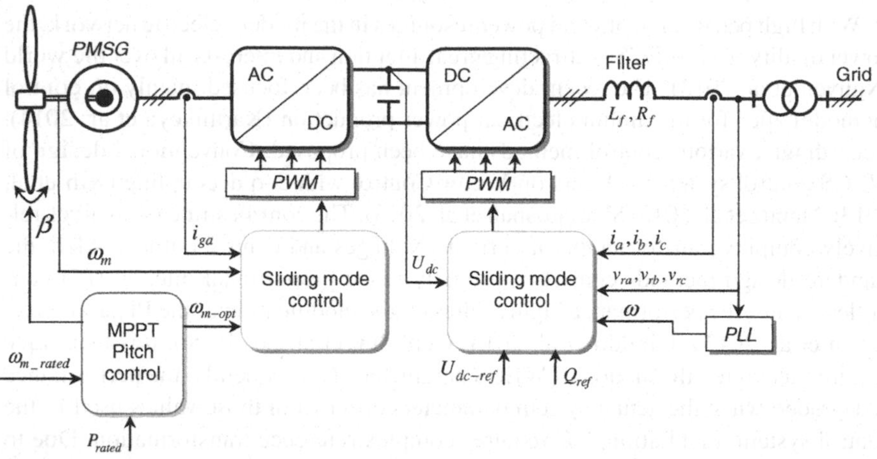

Fig. 3 Schematic of control strategy for WECS

Among the main research subjects in the WECS field, there is the study of novel control methodologies which can maintain MPPT despite the effects of the parameter variations of system, uncertainties in both the electrical and the aerodynamic models and variations of wind speed. In this context, this chapter presents proposes a nonlinear power control strategy for a grid connected VS-WECS topology based on Permânent Magnet Synchronous Generator (PMSG). The schematic diagram of proposed system is shown in Fig. 3. The system under consideration employs WECS based PMSG with a back-to-back voltage source converter (VSC). The generator side converter is employed to control the speed of the PMSG with MPPT. The grid-side converter is used in order to control the DC link voltage and to regulate the power factor during wind variations. This work explores a sliding mode control approach to achieve power efficiency maximization of a WECS and to enhance system robustness to parameter variations. Also, a pitch control scheme for WECS is proposed so as to prevent wind turbine damage from excessive wind velocity.

The rest part of the study is organized as follows. In Sect. 2, the models of the wind turbine system and the PMSG are developed. SMC strategy for WECS is proposed, designed, and analyzed in Sect. 3. Section 4 presents the simulation results to demonstrate the performance of the proposed SMC strategy. Finally, the conclusion is made in Sect. 5.

2 Modelling Description of WECS

The block diagram of proposed WECS is shown in Fig. 3. So, it is seen that the system consists of different components including: wind turbine, PMSG, voltage source converter, and controllers. The wind turbine is used to capture the wind energy that is converted to the electricity by PMSG with variable frequency. Consequently, the generated voltages are rectified by a rectifier and an inverter. The extracted power will be transferred to the grid through a filter.

2.1 Wind Turbine

Wind turbine is used to convert the wind power to a mechanical power. The power generated by a wind turbine can be written as Chen et al. (2013a):

$$P_{Turbine} = \frac{1}{2}\rho A C_P(\lambda, \beta) v^3 \tag{1}$$

where, $P_{Turbine}$ is the mechanical power of the turbine in watts, $\rho$is the air density (typically 1.225 kg/m^3), A is the area swept by the rotor blades (in m^2), C_P is the power performance coefficient of the turbine, v is the wind velocity (in m/s), β is the turbine blade pitch angle, and λ is the Tip Speed Ratio (TSR). Thus, if the air density, swept area and wind speed are constants, the output aerodynamic power is determined by the power performance coefficient of wind turbine system.

The wind turbine mechanical torque output T_m given as:

$$T_m = \frac{1}{2}\rho A C_P(\lambda, \beta) v^3 \frac{1}{\omega_m} \tag{2}$$

In addition, C_P is influenced by the tip-speed ratio λ which is defined as the ratio between the rotor blade tip and the speed of the wind, and is given by Errami et al. (2013):

$$\lambda = \frac{\omega_m R}{v} \tag{3}$$

where ω_m and R are the rotor angular speed (in rad/sec) and the radius of the swept area by turbine blades (in m), respectively. The computation of the power performance coefficient C_P requires the use of the information of blade geometry and blade element theory. Consequently, these complex issues are usually empirical considered and a generic equation is used so as to model the power performance coefficient $C_P(\lambda, \beta)$ based on the modeling turbine system characteristics described in Errami et al. (2013) as:

$$C_P = \frac{1}{2}(\frac{116}{\lambda_i} - 0.4\beta - 5)e^{-(\frac{21}{\lambda_i})}$$
$$\frac{1}{\lambda_i} = \frac{1}{\lambda + 0.08\beta} - \frac{0.035}{\beta^3 + 1} \tag{4}$$

where β is the blade pitch angle (in degrees). C_P is a nonlinear function of both blade pitch angle (β) and the tip speed ratio (λ).

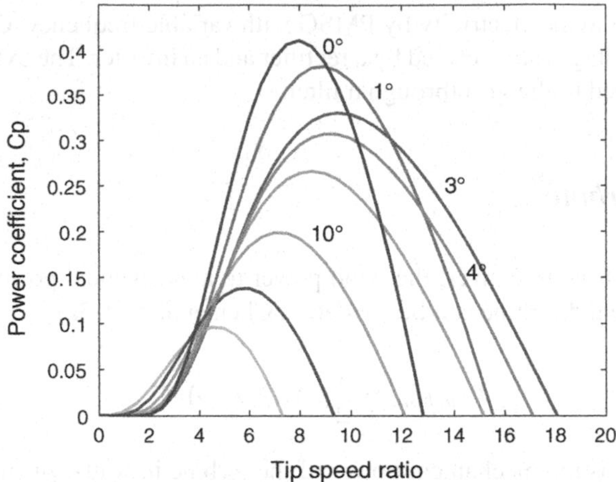

Fig. 4 Characteristics C_p versus λ; for various values of the pitch angle β

The $C_P(\lambda, \beta)$ characteristics, for various values of the pitch angle β , are illustrated in Fig. 4. The maximum value of C_P, that is $C_{P\,\max} = 0.41$, is achieved for $\lambda_{opt} = 8.1$ and for $\beta = 0$. According to Fig. 4, there is one specific λ at which the turbine is most efficient. This optimal value of C_P occurs in different values of λ. Consequently, if β is fixed, there is an optimal value λ_{opt} at which the turbine system follows the $C_{P\,\max}$ to capture the maximum power up to the rated velocity by adjusting rotor speed. Besides, if wind speed is supposed constant, C_P value will be dependent on rotor velocity of the wind turbine. Accordingly, for a given wind velocity, there is an optimal value for rotor velocity which maximizes the power supplied by the wind. If the PMSG velocity can always be controlled to make the turbine operate under optimum tip-speed-ratio λ_{opt} during wind velocity variations, then the power coefficient reaches its maximum value $C_{P\,\max}$. That is equally saying, the turbine system realizes Maximum Power Point Tracking (MPPT) function (Chen et al. 2013c).

Then, for a given wind velocity the system can operate at the peak of the $P(\omega_m)$ curve and the maximum power is extracted continuously from the wind. Consequently, the curve connecting the peaks of these curves will generate the maximum output power and will follow the path for maximum power operation. That is illustrated in Fig. 5.

When the rotor velocity is adjusted to maintain its optimal value, the maximum power can be gained as:

$$P_{Turbine} = \frac{1}{2}\rho A C_{P\max} \nu^3 \tag{5}$$

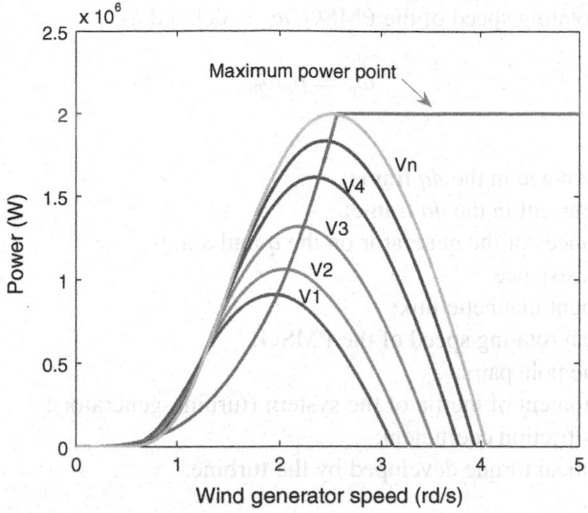

Fig. 5 Wind generator power curves at various wind speed

2.2 Mathematical Model of the PMSG

The PMSG dynamic model is given in a rotative frame (*dq*) where the d axis is aligned with the rotor flux. So, the generator model in the *d-q* frame can be described by the following equations. The electrical equations of the PMSG are shown in (6), (7)and (8), the torque equation in (9) and the mechanical equation in (10) (Shariatpanah et al. 2013).

$$v_{gq} = R_g i_q + \omega_e \phi_d + \frac{d}{dt}\phi_q \tag{6}$$

$$v_{gd} = R_g i_d - \omega_e \phi_q + \frac{d}{dt}\phi_d \tag{7}$$

The quadratic and direct magnetic flux are given by:

$$\begin{aligned} \phi_q &= L_q i_q \\ \phi_d &= L_d i_d + \psi_f \end{aligned} \tag{8}$$

$$T_e = \frac{3}{2} p_n \left[\psi_f i_q + \left(L_d - L_q \right) i_d i_q \right] \tag{9}$$

$$J \frac{d\omega_m}{dt} = T_e - T_m - F\omega_m \tag{10}$$

The electrical rotating speed of the PMSG, ω_e is defined as:

$$\omega_e = p_n \omega_m \tag{11}$$

where

v_{gq}, v_{gd} stator voltage in the dq frame;
i_q, i_d stator current in the dq frame;
L_q, L_d inductances of the generator on the q and d axis;
R_g stator resistance;
ψ_f permanent magnetic flux;
ω_e electrical rotating speed of the PMSG;
p_n machine pole pairs;
J total moment of inertia of the system (turbine-generator);
F viscous friction coefficient;
T_m mechanical torque developed by the turbine.

Thus,

$$\frac{di_q}{dt} = \frac{1}{L_q}(v_{gq} - R_g i_q - \omega_e L_d i_d - \omega_e \psi_f) \tag{12}$$

$$\frac{di_d}{dt} = \frac{1}{L_d}(v_{gd} - R_g i_d + \omega_e L_q i_q) \tag{13}$$

If the PMSG is assumed to have equal d-axis, q-axis in inductances ($L_q = L_d = L_s$), the expression for the electromagnetic torque can be described as:

$$T_e = \frac{3}{2} p_n \left[\psi_f i_q\right] \tag{14}$$

3 Control Strategy of the WECS

3.1 Adopted MPPT Control Algorithm

The reference velocity of the PMSG corresponding to the maximum power extractable from the wind turbine system at a given wind speed is retrieved by the MPPT technique. This algorithm is operated when the wind speed is below the threshold. Then, for each instantaneous wind velocity, the PMSG optimal rotational speed ω_{m-opt} can be computed on the basis of the following expression (Zhang et al. 2014):

$$\omega_{m-opt} = \frac{v\lambda_{opt}}{R} \tag{15}$$

Each wind turbine can produce maximum power by (5). Therefore, the maximum mechanical output power of the turbine system is given as follows:

$$P_{Turbine\_\max} = \frac{1}{2}\rho A C_{P\max}\left(\frac{R\omega_{m-opt}}{\lambda_{opt}}\right)^3 \tag{16}$$

Accordingly, we can get the maximum power $P_{Turbine\_\max}$ by regulating the turbine velocity in different wind speed under rated power of the WECS. In addition, if the speed of generator can always be controlled in order to make turbine system work under optimum tip speed ratio λ_{opt}, regardless of the wind velocity, then the power coefficient reaches its maximum value $C_{P\max}$ (Kuschke et al. 2014). The P_{MPPT} curve is defined as function of ω_{m-opt}, the speed referred to the generator side:

$$P_{MPPT} = K\omega_{m-opt}^3 \tag{17}$$

$$K = \frac{1}{2}\rho A C_{P\max}\left(\frac{R}{\lambda_{opt}}\right)^3 \tag{18}$$

K depends on the blade aerodynamics and wind turbine parameters. The MPPT controller system computes this optimal speed ω_{m-opt} and an optimum value of tip speed ratio λ_{opt} can be maintained. Thus, maximum wind power of the turbine can be captured. Depending on the wind velocity, the MPPT algorithm regulates the electric output power, bringing the turbine system operating points onto the "maximum power point," like in Fig. 5 (Errami et al. 2013).

3.2 Pitch Angle Control System

At wind velocities below the rated power area, the wind turbine system regulator maintains the power performance coefficient C_P of the turbine at its maximum. But, at higher wind velocities, the power coefficient decreases to limit the turbine speed. So, most high power wind turbine systems are equipped with pitch control to achieve power limitation, and where wind speed is low or medium, the pitch angle is controlled to allow turbine system to operate at its optimum condition. On the contrary for high wind speeds, the pitch control is active and it is designed to prevent wind turbine system damage from excessive wind speed (Polinder et al. 2007). This means that, when the wind speed reaches the rated value, the pitch angle controller enters in operation to decrease the performance coefficient of power. The angle of blades β, will increase until the wind turbine system is at the rated velocity. Figure 6 illustrates the schematic diagram of the implemented turbine blade pitch angle controller. P_g is the generated power.

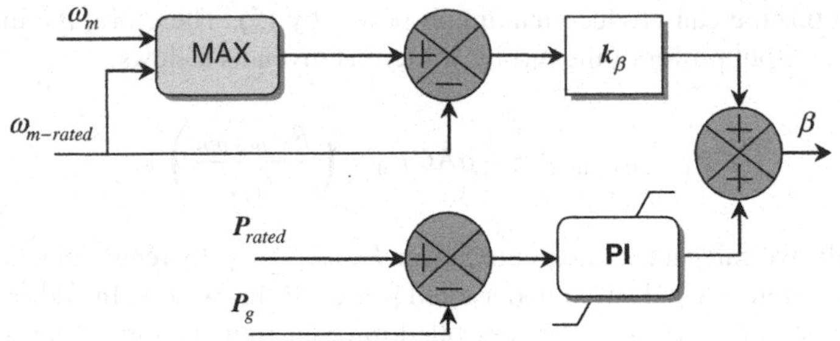

Fig. 6 WECS Pitch angle controller

3.3 Sliding Mode Control (SMC) and MPPT Algorithm for Generator Side converter

The generator side converter controls the PMSG rotational speed to produce the maximum power extractable from the wind turbine system. Thus, the generator side three phase converter is used as a rectifier and it is used to keep the PMSG velocity at an optimal value obtained from the MPPT algorithm. This controller makes the WECS working at highest efficiency. The proposed control strategy for the generator side converter is based on SMC methodology. The adopted MPPT algorithm generates ω_{m_opt}, the reference speed. On the other hand, it is deduced from equations (10) and (14) that the generator velocity can be controlled by regulating the q-axis stator current component (i_{qr}).

According to the theory of SMC, the error of PMSG speed is selected as sliding surface:

$$S_\omega = \omega_{m\_opt} - \omega_m \tag{19}$$

$\omega_{m_opt}$is generated by a MPPT controller.

Consequently:

$$\frac{dS_\omega}{dt} = \frac{d\omega_{m\_opt}}{dt} - \frac{d\omega_m}{dt} \tag{20}$$

Using (10), the time derivative of S_ω can be calculated as:

$$\frac{dS_\omega}{dt} = \frac{d\omega_{m\_opt}}{dt} - \frac{1}{J}(T_e - T_m - F\omega_m) \tag{21}$$

When the trajectory of PMSG speed coincides with the sliding surface (Evangelista et al. 2013a),

$$S_\omega = \frac{dS_\omega}{dt} = 0 \tag{22}$$

In order to obtain commutation around the sliding surface, each component of the control algorithm is proposed to be calculated as the addition of two terms (Evangelista et al. 2013b):

$$u_c = u_{eq} + u_n \tag{23}$$

where u_{eq} is the equivalent control concept of a sliding surface. It is the continuous control that allows the continuance of the state trajectory on the sliding surface. The expression for the equivalent control term is obtained from the equation formed by equalling to zero the first time derivative of S_ω. As a result, during the sliding mode and in permanent regime, u_{eq} is calculated from the expression:

$$\frac{dS_\omega}{dt} = 0 \tag{24}$$

Although, u_n is used so as to guarantee the attractiveness of the variable to be controlled towards the commutation sliding mode surface. So, it maintains the state on the sliding surface in the presence of the parametric variations and external disturbances for all subsequent time. Also, the system state slides on the sliding surface until it reaches the equilibrium point. Then it is restricted to the surface

$$u_n = k_\omega \mathrm{sgn}(S_\omega) \tag{25}$$

where k_ω is a positive constant, which is the gain of the sliding mode regulator. u_n keeps the system dynamic on the sliding surface $S_\omega = 0$ for all the time.

Moreover, SMC is a discontinuous control. So as to reduce the chattering, the continuous function as exposed in (26) where $\mathrm{sgn}(S_\omega)$ is a sign function defined as (Xiao et al. 2011):

$$\mathrm{sgn}(S_\omega) = \begin{cases} 1 & S_\omega \succ \varepsilon \\ \frac{S_\omega}{\varepsilon} & \varepsilon \geq |S_\omega| \\ -1 & -\varepsilon \succ S_\omega \end{cases} \tag{26}$$

where ε the width of the boundary layer. It is a small positive number and it should be chosen attentively, otherwise the dynamic quality of the system will be reduced.

On the other hand, to ensure the PMSG speed convergence to the optimal velocity and to reduce the copper loss by setting the d axis current to be zero, current references are derived. Based on equations (10), (14), (19), (22), (23), (24) and (25), the following equation for the system of speed can be obtained:

$$i_{qr} = \frac{2}{3p_n\psi_f}\left(T_m + J\frac{d\omega_{m\_opt}}{dt} + F\omega_m + Jk_\omega \mathrm{sgn}(S_\omega)\right) \tag{27}$$

where $k_\omega \succ 0$.

$$i_{dr} = 0 \tag{28}$$

In addition, from (21) the following equation can be deduced:

$$S_\omega \frac{dS_\omega}{dt} = S_\omega \frac{d\omega_{m\_opt}}{dt} - \frac{S_\omega}{J}(T_e - T_m - F\omega_m)$$
$$= -k_\omega S_\omega \text{sgn}(S_\omega) + \frac{S_\omega}{J}(T_m - \frac{3}{2} p_n \psi_f i_q + F\omega_m + Jk_\omega \text{sgn}(S_\omega) + J\frac{d\omega_{m\_opt}}{dt}) \tag{29}$$

Then, to regulate the currents components i_d and i_q to their references, it is necessary to define the sliding surfaces, as follows:

$$S_d = i_{dr} - i_d \tag{30}$$

$$S_q = i_{qr} - i_q \tag{31}$$

Substituting (12) and (13) into above equations gives:

$$\frac{dS_d}{dt} = \frac{di_{dr}}{dt} - \frac{di_d}{dt} = -\frac{1}{L_s}(v_{gd} - R_g i_d + L_s \omega_e i_q) \tag{32}$$

$$\frac{dS_q}{dt} = \frac{di_{qr}}{dt} - \frac{di_q}{dt}$$
$$= \frac{di_{qr}}{dt} - \frac{1}{L_s}(v_{gq} - R_g i_q - L_s \omega_e i_d - \omega_e \psi_f) \tag{33}$$

when the sliding mode occurs on the sliding mode surfaces:

$$S_q = \frac{dS_q}{dt} = 0 \tag{34}$$

$$S_d = \frac{dS_d}{dt} = 0 \tag{35}$$

Consequently, the control voltages of q axis and d axis are defined by:

$$v_{qr} = R_g i_q + L_s \omega_e i_d + \omega_e \psi_f + L_s \frac{di_{qr}}{dt} + L_s k_q \text{sgn}(S_q) \tag{36}$$

$$v_{dr} = R_g i_d - L_s \omega_e i_q + L_s k_d \text{sgn}(S_d) \tag{37}$$

where $k_q \succ 0$ and $k_d \succ 0$.

In addition:

$$S_d \frac{dS_d}{dt} = S_d \left[-\frac{1}{L_s}(v_{gd} - R_g i_d + L_s \omega_e i_q) \right]$$
$$= -k_d S_d \text{sgn}(S_d) + \frac{S_d}{L_s} \left[-v_{gd} + R_g i_d - L_s \omega_e i_q + L_s k_d \text{sgn}(S_d) \right] \tag{38}$$

$$S_q\frac{dS_q}{dt}=S_q\left[\frac{di_{qr}}{dt}-\frac{di_q}{dt}\right]$$
$$=-k_qS_q\text{sgn}(S_q)+\frac{S_q}{L_s}\left[L_s\frac{di_{qr}}{dt}-v_{gq}+R_gi_q+L_s\omega_ei_d+\omega_e\psi_f+L_sk_q\text{sgn}(S_q)\right] \tag{39}$$

Theorem 1 *If the dynamic sliding mode control laws are designed as* (27), (28), (36) *and* (37), *therefore the global asymptotical stability is ensured.*

Proof The proof of the theorem 1 will be carried out using the Lyapunov theory of stability. To determine the required condition for the existence of the sliding mode, it is fundamental to design the Lyapunov function. So, the Lyapunov function can be chosen as (Evangelista et al. 2013b):

$$\Upsilon_1=\frac{1}{2}S_\omega^2+\frac{1}{2}S_q^2+\frac{1}{2}S_d^2 \tag{40}$$

From Lyapunov theory of stability, to ensure controller stability and convergence of the state trajectory to the sliding mode, Υ_1 can be derived that (Huang et al. 2013),

$$\frac{d\Upsilon_1}{dt}\prec 0 \tag{41}$$

According to the definition of Υ_1, the time derivative of Υ_1 can be calculated as:

$$\frac{d\Upsilon_1}{dt}=S_\omega\frac{dS_\omega}{dt}+S_d\frac{dS_d}{dt}+S_q\frac{dS_q}{dt} \tag{42}$$

Using Equations (29), (38) and (39), we can rewrite (42) as:

$$\frac{d\Upsilon_1}{dt}=-k_\omega S_\omega\text{sgn}(S_\omega)+\frac{S_\omega}{J}(T_m-\frac{3}{2}p_n\psi_fi_q+F\omega_m+Jk_\omega\text{sgn}(S_\omega)+J\frac{d\omega_{m\_opt}}{dt})$$
$$-k_dS_d\text{sgn}(S_d)+\frac{S_d}{L_s}\left[-v_{gd}+R_gi_d-L_s\omega_ei_q+L_sk_d\text{sgn}(S_d)\right]$$
$$-k_qS_q\text{sgn}(S_q)+\frac{S_q}{L_s}\left[L_s\frac{di_{qr}}{dt}-v_{gq}+R_gi_q+L_s\omega_ei_d+\omega_e\psi_f+L_sk_q\text{sgn}(S_q)\right] \tag{43}$$

Substituting (27), (36) and (37) into above equation gives:

$$\frac{d\Upsilon_1}{dt}=-k_\omega S_\omega\text{sgn}(S_\omega)-k_dS_d\text{sgn}(S_d)-k_qS_q\text{sgn}(S_q) \tag{44}$$

As a result:

$$\frac{d\Upsilon_1}{dt}=-k_\omega\left|S_\omega\right|-k_d\left|S_d\right|-k_q\left|S_q\right|\prec 0 \tag{45}$$

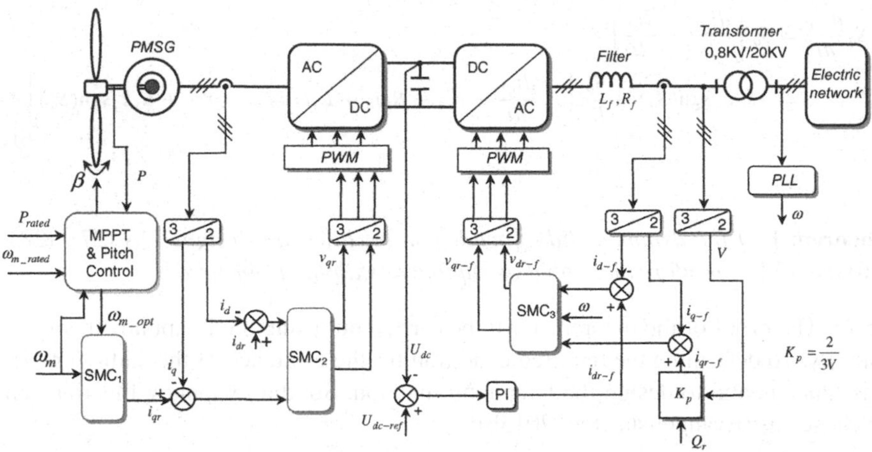

Fig. 7 Schematic of SMC strategy for WECS

Accordingly, the global asymptotical stability is ensured and the velocity control tracking is achieved.

Finally, PWM is used to generate the control signal to implement the SMC for the PMSG. The double closed-loop control diagram for generator side converter is shown as Fig. 7.

3.4 Grid Side Controller Methodology with SMC

The grid side converter (GSC) works as an inverter. The main function of the GSC is to keep constant dc bus voltage, regulates the reactive and active power flowing into the grid and to provide grid synchronization. So, it can regulate the grid side power factor during wind variation. Besides, there are many strategies used to control GSC (Yaramasu et al. 2013; Blaabjerg et al. 2013; Ma et al. 2013). In this study, Pulse Width Modulation (PWM) associated with SMC is used in order to control the converter. Double-loop structure is used: the inner control loops regulates q-axis current and d-axis current, but outer voltage loop regulates the dc-link voltage via controlling the output power. The schematic diagram of the GSC based on the proposed control strategy is shown by Fig. 7.

The voltage balance across the inductor L_f and R_f is given by:

$$\begin{bmatrix} e_a \\ e_b \\ e_c \end{bmatrix} = R_f \begin{bmatrix} i_a \\ i_b \\ i_c \end{bmatrix} + L_f \frac{d}{dt} \begin{bmatrix} i_a \\ i_b \\ i_c \end{bmatrix} + \begin{bmatrix} v_a \\ v_b \\ v_c \end{bmatrix} \quad (46)$$

where

e_a, e_b, e_c	voltages at the inverter system output;
v_a, v_b, v_c	grid voltage components;
i_a, i_b, i_c	line currents;
L_f	filter inductance;
R_f	filter resistance.

Transferring equation (46) in the rotating dq reference frame gives:

$$\frac{di_{d-f}}{dt} = \frac{1}{L_f}(e_d - R_f i_{d-f} + \omega L_f i_{q-f} - v_d) \tag{47}$$

$$\frac{di_{q-f}}{dt} = \frac{1}{L_f}(e_q - R_f i_{q-f} - \omega L_f i_{d-f} - v_q) \tag{48}$$

where

e_d, e_q	inverter d-axis and q-axis voltage components;
v_d, v_q	grid voltage components in the d-axis and q-axis;
i_{d-f}, i_{q-f}	d-axis current and q-axis current of grid.
ω	network angular frequency

The network angular frequency is computed by a Phase Locked Loop (PLL). The instantaneous powers are given by:

$$P = \frac{3}{2}(v_d i_{d-f} + v_q i_{q-f}) \tag{49}$$

$$Q = \frac{3}{2}(v_d i_{q-f} - v_q i_{d-f}) \tag{50}$$

Thus, the DC-link system equation can be given by:

$$C\frac{dU_{dc}}{dt} = \frac{3}{2}\left(\frac{v_d}{U_{dc}} i_{d-f} + \frac{v_q}{U_{dc}} i_{q-f}\right) - i_{dc} \tag{51}$$

where

U_{dc}	dc-link voltage;
i_{dc}	grid side transmission line current;
C	dc-link capacitor.

If the grid voltage space vector $\vec{u}$ is oriented on d-axis, then:

$$v_d = V \text{ and } v_q = 0 \tag{52}$$

Therefore, using Eq. (52), we can rewrite Eqs. (47–48) as:

$$L_f \frac{di_{d-f}}{dt} = e_d - R_f i_{d-f} + \omega L_f i_{q-f} - V \quad (53)$$

$$L_f \frac{di_{q-f}}{dt} = e_q - R_f i_{q-f} - \omega L_f i_{d-f} \quad (54)$$

Also, the active power and reactive power can be expressed as:

$$P = \frac{3}{2} V i_{d-f} \quad (55)$$

$$Q = \frac{3}{2} V i_{q-f} \quad (56)$$

As a result, reactive and active power control can be achieved by controlling quadrature and direct grid current components, respectively. So, the q-axis current reference is set to zero for unity power factor, but the d-axis current is determined by dc-bus voltage controller to control the converter output active power. The GSC controller is implemented based on the electric network current *d-q* components, as it is depicted in Fig. 7. The control method consists of a two closed loop controls to regulate the reactive power and the dc link voltage independently. Then, the fast dynamic is associated with the line current control, in the inner loop, where the SMC is adopted to track the line current control. Moreover, in the outer loop, slow dynamic is associated with the dc-bus control. The outer control loop uses the Proportional Integral (PI) controller to generate the reference source current i_{dr-f} and regulate the DC voltage, although the reference signal of the q-axis current i_{qr-f} is produced by the reactive power $Q_r$according to (56).

We adopt the following surfaces for i_{d-f} and i_{q-f}:

$$S_{d-f} = i_{dr-f} - i_{d-f} \quad (57)$$

$$S_{q-f} = i_{qr-f} - i_{q-f} \quad (58)$$

where i_{dr-f} and i_{qr-f} are the desired value of d-axis current and q-axis current, respectively. Also, i_{dr-f} is produced by the loop of DC-bus control and the reference signal of the q-axis current i_{qr-f}, is directly given from the second loop outside of the controller and it sets to zero to reach unity power factor control.

Using equations (53) and (54), the time derivatives of $S_{d-f}$and S_{q-f}can be calculated as:

$$\frac{dS_{d-f}}{dt} = \frac{di_{dr-f}}{dt} - \frac{di_{d-f}}{dt} = \frac{di_{dr-f}}{dt} - \frac{1}{L_f}(e_d - R_f i_{d-f} + \omega L_f i_{q-f} - V) \quad (59)$$

$$\frac{dS_{q-f}}{dt} = \frac{di_{qr-f}}{dt} - \frac{di_{q-f}}{dt} = \frac{di_{qr-f}}{dt} - \frac{1}{L_f}(e_q - R_f i_{q-f} - \omega L_f i_{d-f}) \quad (60)$$

when the sliding mode takes place on the sliding mode surface, then:

$$S_{d-f} = \frac{dS_{d-f}}{dt} = 0 \tag{61}$$

$$S_{q-f} = \frac{dS_{q-f}}{dt} = 0 \tag{62}$$

Combining (53), (54) and (57)–(62) the controls voltage of d axis and q axis are defined by:

$$v_{dr-f} = L_f \frac{di_{dr-f}}{dt} + R_f i_{d-f} - L_f \omega i_{q-f} + V + L_f k_{d-f} \operatorname{sgn}(S_{d-f}) \tag{63}$$

$$v_{qr-f} = R_f i_{q-f} + L_f \omega i_{d-f} + L_f k_{q-f} \operatorname{sgn}(S_{q-f}) \tag{64}$$

where $k_{d-f} \succ 0$ and $k_{q-f} \succ 0$.

Besides, from (59) and (60), the following equations can be deduced:

$$\begin{aligned} S_{d-f}\frac{dS_{d-f}}{dt} &= S_{d-f}\left[\frac{di_{dr-f}}{dt} - \frac{1}{L_f}(e_d - R_f i_{d-f} + L_f \omega i_{q-f} - V)\right] \\ &= -k_{d-f} S_{d-f} \operatorname{sgn}(S_{d-f}) \\ &\quad + \frac{S_{d-f}}{L_f}\left[L_f \frac{di_{dr-f}}{dt} - e_d + R_f i_{d-f} - L_f \omega i_{q-f} + V + k_{d-f} L_f \operatorname{sgn}(S_{d-f})\right] \end{aligned} \tag{65}$$

$$\begin{aligned} S_{q-f}\frac{dS_{q-f}}{dt} &= S_{q-f}\left[-\frac{di_{q-f}}{dt}\right] \\ &= -k_{q-f} S_{q-f} \operatorname{sgn}(S_{q-f}) \\ &\quad + \frac{S_{q-f}}{L_f}\left[-e_q + R_f i_{q-f} + L_f \omega i_{d-f} + k_{q-f} L_f \operatorname{sgn}(S_{q-f})\right] \end{aligned} \tag{66}$$

Theorem 2 *If the Dynamic sliding mode control laws are designed as* (63) *and* (64) *therefore the global asymptotical stability is ensured.*

Proof The proof of the Theorem 2 will be carried out using the Lyapunov theory of stability. To determine the required condition for the existence of the sliding mode, it is fundamental to design the Lyapunov function. So, the Lyapunov function can be chosen as:

$$\Upsilon_2 = \frac{1}{2} S_{d-f}^2 + \frac{1}{2} S_{q-f}^2 \tag{67}$$

From Lyapunov theory of stability, to ensure controller stability and convergence of the state trajectory to the sliding mode, Υ_2 can be derived that,

$$\frac{d\Upsilon_2}{dt} \prec 0 \tag{68}$$

By differentiating the Lyapunov function (67), we obtain:

$$\frac{d\Upsilon_2}{dt} = S_{d-f}\frac{dS_{d-f}}{dt} + S_{q-f}\frac{dS_{q-f}}{dt} \tag{69}$$

Based on equations (65) and (66), it can be obtained:

$$\begin{aligned}\frac{d\Upsilon_2}{dt} &= -k_{d-f}S_{d-f}\mathrm{sgn}(S_{d-f}) \\ &+ \frac{S_{d-f}}{L_f}\left[L_f\frac{di_{dr-f}}{dt} - e_d + R_f i_{d-f} - L_f\omega i_{q-f} + V + k_{d-f}L_f\mathrm{sgn}(S_{d-f})\right] \\ &- k_{q-f}S_{q-f}\mathrm{sgn}(S_{q-f}) + \frac{S_{q-f}}{L_f}\left[-e_q + R_f i_{q-f} + L_f\omega i_{d-f} + k_{q-f}L_f\mathrm{sgn}(S_{q-f})\right]\end{aligned} \tag{70}$$

Substituting (63) and (64) into above equation gives:

$$\frac{d\Upsilon_2}{dt} = -k_{d-f}S_{d-f}\mathrm{sgn}(S_{d-f}) - k_{q-f}S_{q-f}\mathrm{sgn}(S_{q-f}) \tag{71}$$

Therefore:

$$\frac{d\Upsilon_2}{dt} = -k_{d-f}\left|S_{d-f}\right| - k_{q-f}\left|S_{q-f}\right| \prec 0 \tag{72}$$

As a result, the asymptotic stability in the current loop is guaranteed and the dc-bus voltage control tracking is achieved. Finally, PWM is used to produce the control signal. The structure of the dc-link voltage and current controllers for grid-side converter, for the WECS, is illustrated in Fig. 7.

4 Simulation Result Analysis

This paragraph presents the simulated responses of the WECS under varying wind conditions. In this example simulation, Matlab/Simulink simulations were carried out for a 2 MW PMSG variable speed wind energy conversion system to verify the feasibility of the proposed method. The parameters of the system are given in the Tables 1 and 2. Besides, during the simulation, for the PMSG side converter control, the d axis command current component, i_{dr}, is set to zero; while, for the grid side inverter system, Q_{ref}, is set to zero. On the other hand, the DC link voltage reference and the grid frequency value are $U_{dc-r} = 1500$ V and 50 Hz, respectively. The topology of the studied WECS based on PMSG connected electric network is depicted in Fig. 7. The grid voltage phase lock loop (PLL) system is implemented to track the fundamental phase and frequency. On the other hand, this paragraph is divided into two parts, that is, Sect. 4.1 demonstrates the satisfactory performance of the WECS under varying wind conditions, while Sect. 4.2 reflects the robustness of both the rectifier and inverter control systems against electrical and mechanical parameter deviations.

4.1 WECS Characteristics with SMC Approach

The WECS response under SMC strategy is illustrated by Figs. 8, 9, 10 and 11. Figures 8 and 9 show, respectively, the wind speed profile, the simulation results of pitch angle, coefficients of power conversion C_p, tip speed ratio, rotor angular velocity of the PMSG and total power generated of WECS. The rated wind speed considered in the simulation is v_n =12.4 m/s. Then, it can be seen, that when the wind velocity increases, the rotor angular speed increases proportionally too with a limitation, the power coefficient will drop to maintain the rated output power. Then, at the wind velocities less than the rated rotor velocity, the pitch angle is fixed at 0° and the power performance coefficient of the turbine is fixed at its maximum value, around 0.41. The speed of PMSG is controlled in order to make the turbine system operates under its optimum tip speed ratio $\lambda_{opt} = 8.1$, regardless of the wind speed. So, the PMSG velocity is regulated at an optimal value obtained from the MPPT algorithm. Thus, this control makes the wind turbine working at highest efficiency. On the contrary, for high wind speeds and if v is greater than the rated velocity v_n, the operation of the pitch angle control is actuated and the pitch angle β is increased. Then, the pitch control is used to maintain the PMSG power at rated power and it is designed to prevent damage from excessive wind velocity. So, the power performance coefficient of the turbine decreases to limit the rotor speed. Consequently, extracted power is optimized with MPPT algorithm and keeps at his nominal value when the wind speed exceeds the nominal value. Figure 9c shows the power extracted. As can be seen, if the wind velocity is up the rated wind speed, the power extracted reaches its maximum level. Figure 9b illustrates the waveforms of the mechanical velocity of the PMSG tracks the optimum velocity obtained from MPPT algorithm so as to guarantee the maximum power conversion at the optimal tip speed ratio. Then, it is clearly shown that the PMSG speed tracks the reference velocity closely. As the WECS successfully operates with MPPT ($\lambda = \lambda_{opt}$), the primary control objective is adequately attained, and the PMSG power finely follows the maximum value. Figure 10a depicts the simulation result of reactive power. As can be seen, the WECS supplies grid system with a purely active power. The fulfilment of the second control objective can be appreciated in Fig. 10b, where the WECS dc link voltage and the external reference U_{dc-r} are depicted together. The dc link voltage is regulated toward its reference of 1,500 V. This proves the effectiveness of the established controller systems. Figure 11 illustrates the variation and a closer observation of three phase current and voltage of grid. Besides, the frequency is controlled and maintained at 50 Hz through a Phase Lock Loop (PLL) process. It is obvious that the grid voltage is in phase with the current since the reference of reactive current is set to zero. So,

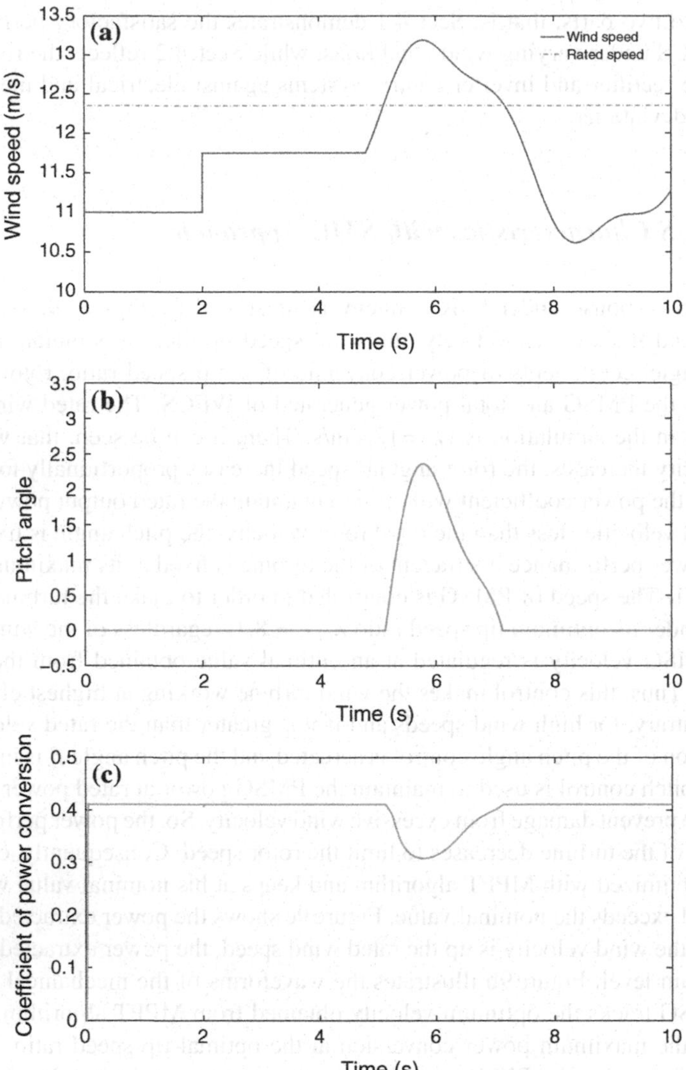

Fig. 8 Waveforms of WECS characteristics with SMC(Part 1). **a** Instantaneous wind speeds (m/s). **b** Pitch angles β (in degree). **c** Coefficients of power conversion C_p

unity power factor of wind energy conversion system is achieved approximatively and is independent of the variation of the wind velocity but only on the reactive power reference (Q_{ref}). Consequently, the simulation results demonstrate that the SMC strategy shows very good dynamic and steady state performance and works very well.

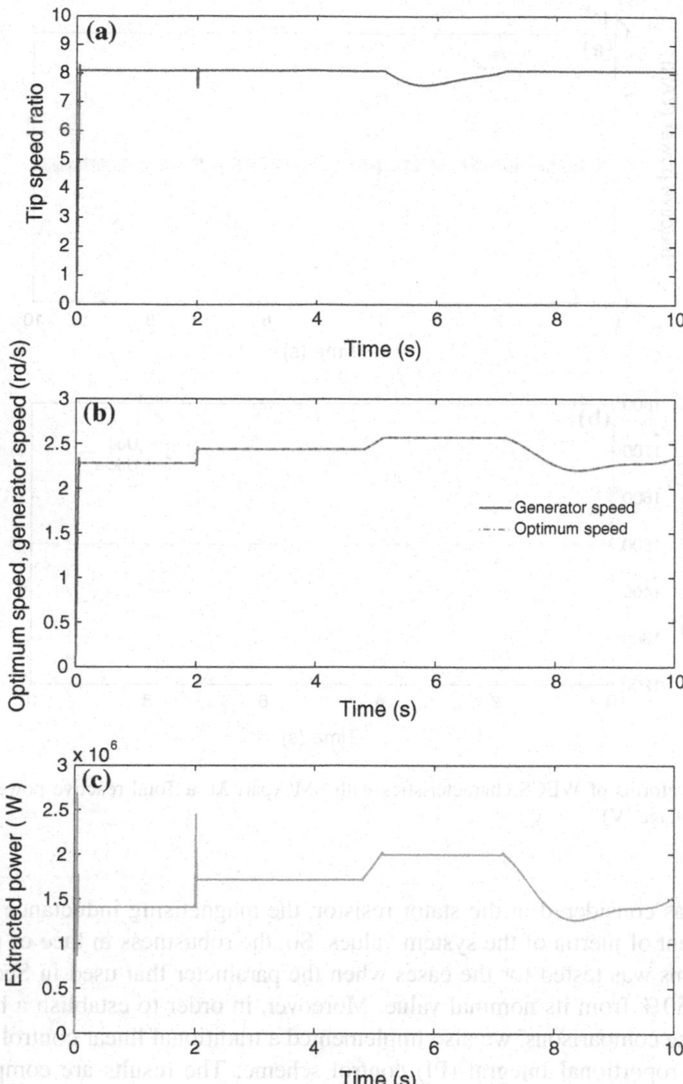

Fig. 9 Waveforms of WECS characteristics with SMC(Part 2). **a** Tip speed ratio. **b** Speed of PMSG (rd/s). **c** Power generated (W)

4.2 Robustness of the SMC Controller Under Electric and Mechanic Parameter Variations of WECS

In order to prove the robustness of the proposed controllers, model uncertainties were included considering the parametric uncertainties. Besides, these variations were also used in a percentage scale that takes the respective nominal values as references. An

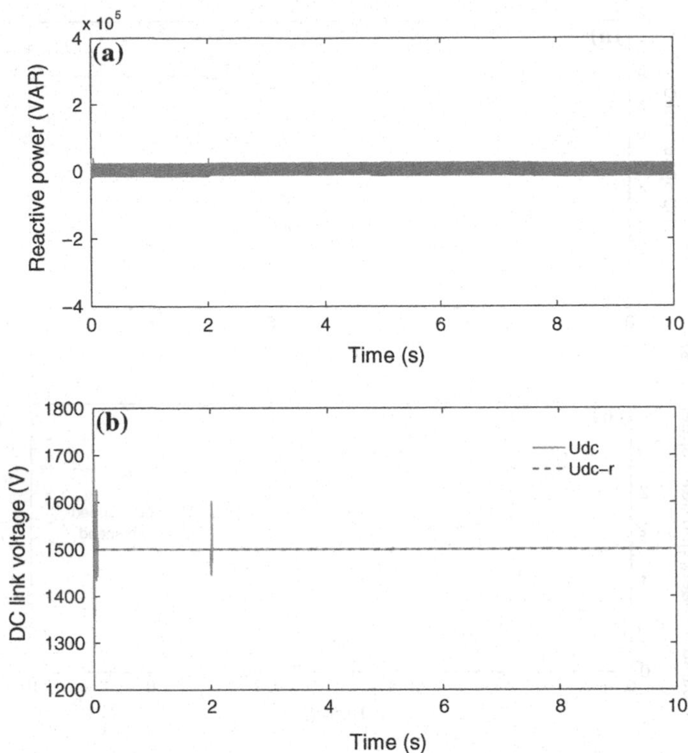

Fig. 10 Waveforms of WECS characteristics with SMC(part 3). **a** Total reactive power (VAR). **b** DC link voltage (V)

increase was considered in the stator resistor, the magnetising inductance and the total moment of inertia of the system values. So, the robustness in face of parameter variations was tested for the cases when the parameter that used in Sect. 4.1 is perturbed 50 % from its nominal value. Moreover, in order to establish a basis for performance comparisons, we also implemented a traditional linear control scheme based on Proportional Integral (PI) control scheme. The results are compared in Figs. 12, 13, 14 and 15. With the similar control parameter values and grid voltage condition as used in Sect. 4.1, superscripts 'A' and 'B' in Figs. 12 and 13 refer to present section (affected by parameter variations) and to waveforms corresponding to Sect. 4.1 (with nominal parameters), respectively. Figure 12 shows the comparison of the generator speed between the SMC method and PI strategy. In Fig. 13, the curves describe the test of sensitivity in face of parameter deviations for the coefficient of power. The responses of both control strategies shown in Figs. 14 and 15 verify the parametric robustness of the proposed scheme for the dc link voltage. As it is shown in the simulation result, the SMC strategy gives lower overshoot and faster response. The result indicates that the SMC control has faster velocity response and shorter settling time. Furthermore, it can be concluded that the proposed nonlinear

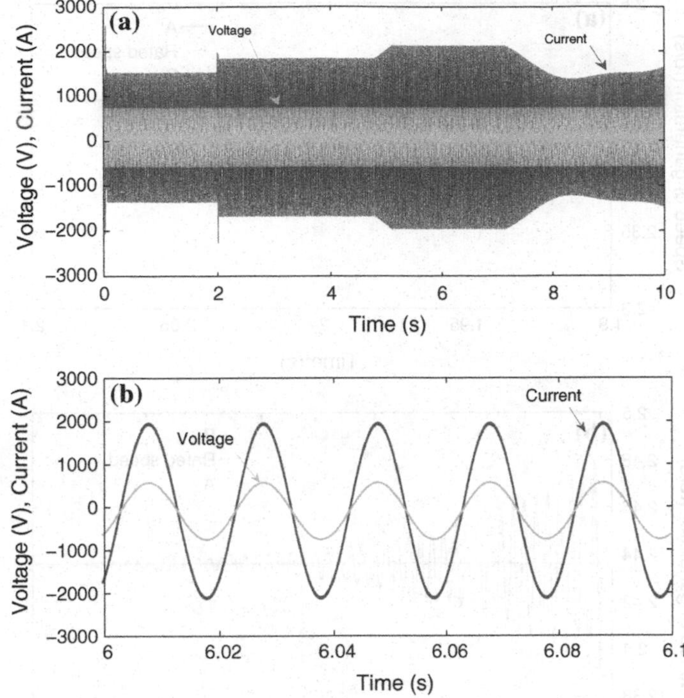

Fig. 11 The waveforms of three phase current and voltage of GRID

sliding mode control is rather robust against parameter variations than its PI counterpart. Consequently, the robustness of proposed SMC to the parameter deviations is convincingly verified.

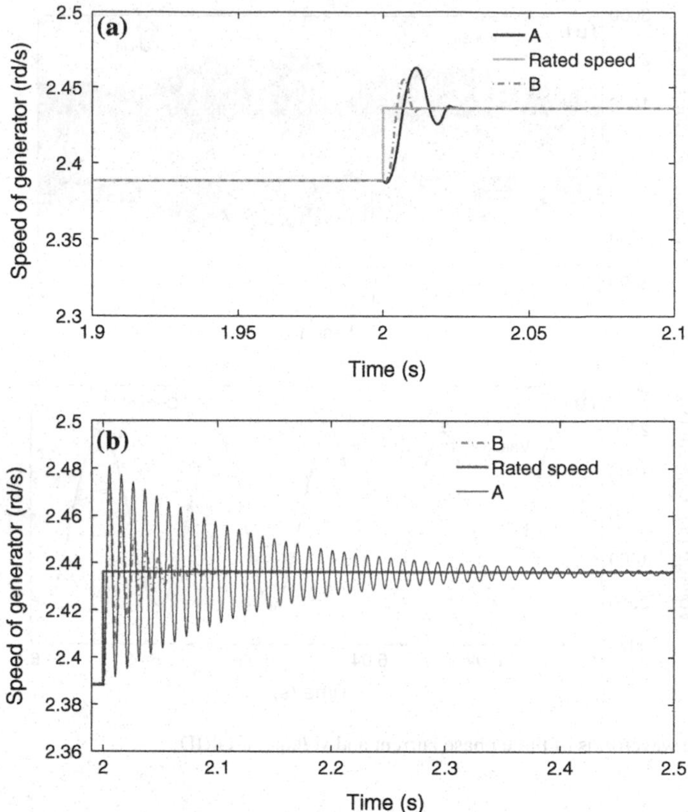

Fig. 12 Generator speed (rd/s). **a** Sliding mode. **b** PI controller

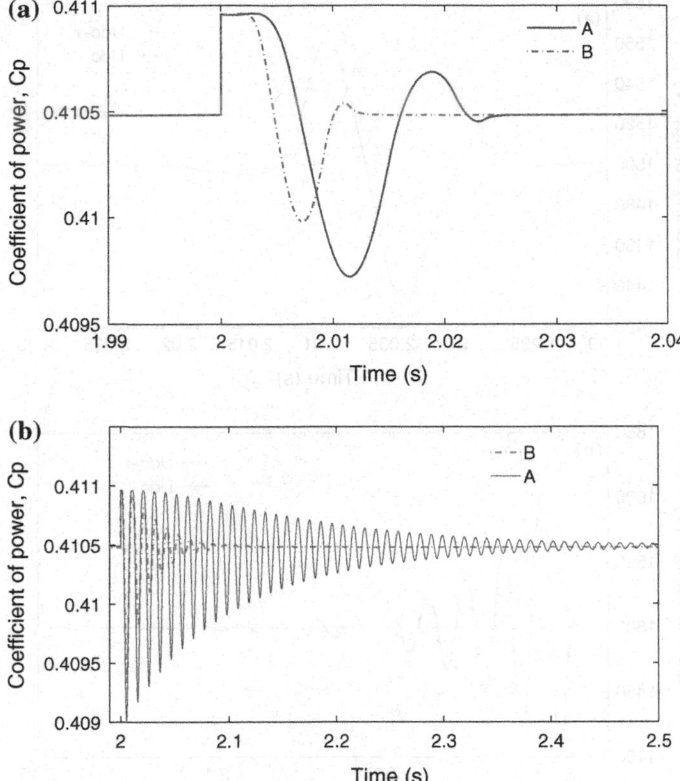

Fig. 13 Coefficient of power. **a** Sliding mode. **b** PI controller

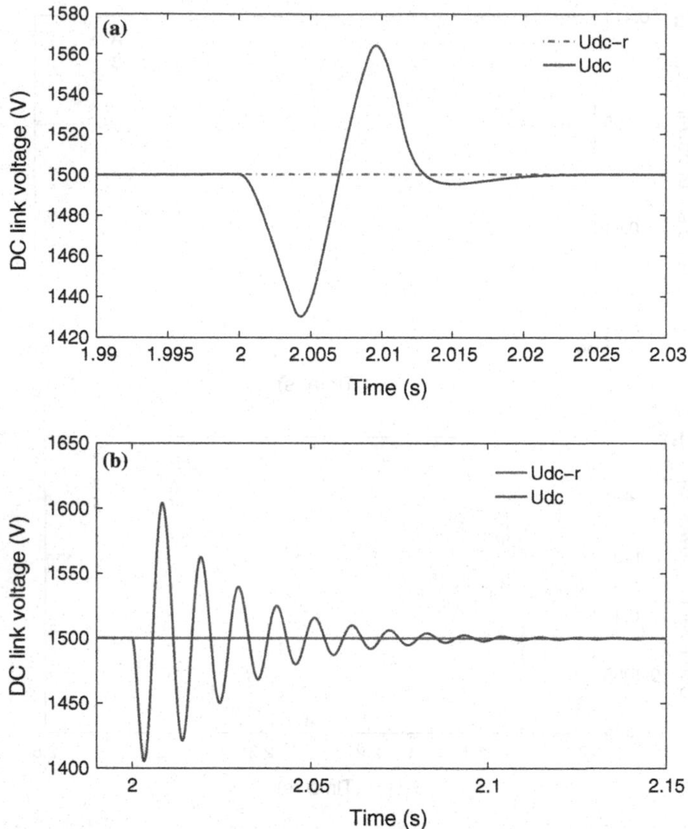

Fig. 14 DC link voltage (V) (with nominal values). **a** Sliding mode. **b** PI controller

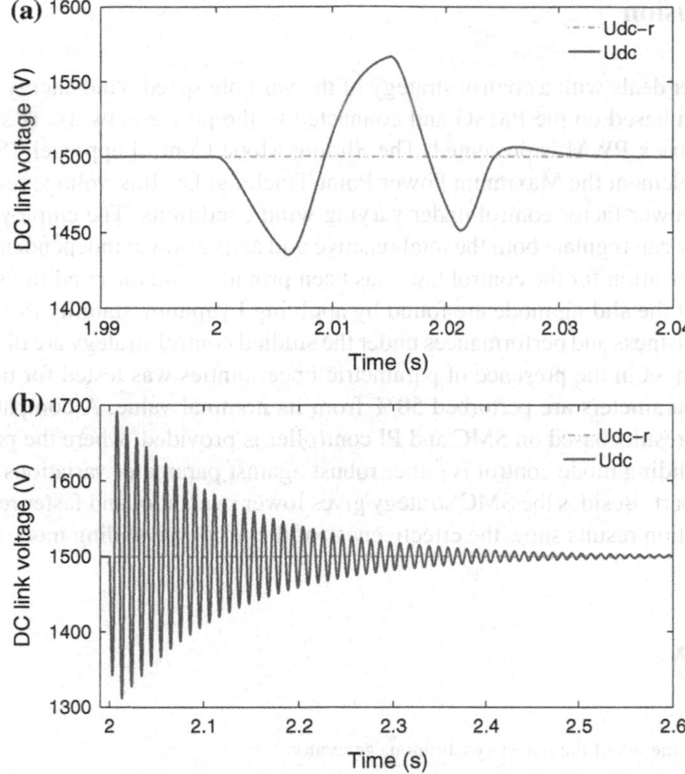

Fig. 15 DC link voltage (V) (with parameter variations). **a** Sliding mode. **b** PI controller

5 Conclusion

This chapter deals with a control strategy of the variable speed wind energy conversion system based on the PMSG and connected to the power network. A structure of back to back PWM is presented. The Sliding Mode Control approach (SMC) is used to implement the Maximum Power Point Tracking, DC link voltage regulation and unity power factor control under varying wind conditions. The employed control strategy can regulate both the total reactive and active power independently. The detailed derivation for the control laws has been provided and the conditions for the existence of the sliding mode are found by applying Lyapunov stability theory. The WECS robustness and performances under the studied control strategy are discussed. The robustness in the presence of parametric uncertainties was tested for the cases when the parameters are perturbed 50 % from its nominal value. A comparison of simulation results based on SMC and PI controller is provided, where the proposed nonlinear sliding mode control is rather robust against parameter variations than its PI counterpart. Besides the SMC strategy gives lower overshoot and faster response. The simulation results show the effectiveness of the proposed sliding mode control.

Appendix

Table 1 Parameters of the power synchronous generator

Parameter	Value
P_r rated power	2 (MW)
ω_m rated mechanical speed	2.57 (rd/s)
Rstator resistance	0.008 (Ω)
L_s stator d-axis inductance	0.0003 (H)
ψ_f permanent magnet flux	3.86 (wb)
p_n pole pairs	60

Table 2 Parameters of the turbine

Parameter	Value
ρ the air density	1.08 kg/m^3
A area swept by blades	4775.94 m^2
v_n base wind speed	12.4 m/s

References

Alepuz, S., Calle, A., Busquets-Monge, S., Kouro, S.: Use of stored energy in PMSG rotor inertia for low-voltage ride-through in back-to-back npc converter-based wind power systems. IEEE Trans. Indus. Electr. **60**(5), 1787–1796 (2013)

Alizadeh, O., Yazdani, A.: A strategy for real power control in a direct-drive PMSG-based wind energy conversion system. IEEE Trans. Power Delivery **28**(3), 1297–1305 (2013)

Alshibani, S., Agelidis, V.G., Dutta, R.: Lifetime cost assessment of permanent magnet synchronous generators for MW level wind turbines. IEEE Trans. Sustain. Energy **5**(1), 10–17 (2014)

Blaabjerg, F., Ma, K.: Future on power electronics for wind turbine systems. IEEE J. Emerg. Sel. Topics Power Electr. **1**(3), 139–152 (2013)

Bouaziz, B., Bacha, F.: Direct power control of grid-connected converters using sliding mode controller. In: IEEE International Conference on Electrical Engineering and Software Applications (ICEESA), pp. 1–6 (2013)

Cárdenas, R., Peña, R., Alepuz, S., Asher, G.: Overview of control systems for the operation of DFIGs in wind energy applications. IEEE Trans. Ind. Electr. **60**(7), 2776–2798 (2013)

Cespedes, M., Sun, J.: Impedance modeling and analysis of grid-connected voltage-source converters. IEEE Trans. Power Electr. **29**(3), 1254–1261 (2014)

Che, H., Levi, Emil, Jones, Martin, Duran, Mario J., Hew, Wooi-Ping, Abd, Nasrudin, Rahim, : Operation of a Six-Phase Induction Machine Using Series-Connected Machine-Side Converters. IEEE Transactions On Industrial Electronics **61**(1), 164–176 (2014)

Chen, B., Ong, F., Minghao, Z.: Terminal sliding-mode control scheme for grid-side PWM converter of DFIG-based wind power system. In: IEEE Conference of the Industrial Electronics Society (IECON), pp. 8014–8018 (2013)

Chen, H., David, N., Aliprantis, D.C.: Analysis of permanent-magnet synchronous generator with vienna rectifier for wind energy conversion system. IEEE Trans. Sustain. Energy **4**(1), 154–163 (2013)

Chen, J., Jie, C., Chunying, G.: New overall power control strategy for variable-speed fixed-pitch wind turbines within the whole wind velocity range. In: IEEE Transactions On Industrial Electronics, Vol. 60, No. 7, pp. 2652–2660 (2013a)

Chen, J., Jie, C., Chunying, G.: On optimizing the aerodynamic load acting on the turbine shaft of PMSG-based direct-drive wind energy conversion system. In: IEEE Transactions on Industrial Electronics, vol 99 (2013b)

Chen, J., Jie, C.M., Chunying, G.: On optimizing the transient load of variable-speed wind energy conversion system during the MPP tracking process. In: IEEE Transactions On Industrial Electronics, Vol. 99, pp. 1–9 (2013c)

Cheng, K.W.E., Lin J.K., Bao, Y.J., Xue, X.D.: ReView of the wind energy generating system. In: International Conference on Advances in Power System Control, Operation and Management (APSCOM 2009), pp. 1–7, (2009)

Chou, S., Chia-Tse, L., Hsin-Cheng, K., Po-Tai, C.: A low-voltage ride-through method with transformer flux compensation capability of renewable power grid-side converters. In: IEEE Transactions On Power Electronics, Vol. 29, No. 4, pp. 1710–1719 (2014)

Corradini, M.L., Ippoliti, G., Orlando, G.: Robust control of variable-speed wind turbines based on an aerodynamic torque observer. IEEE Trans. Control Syst. Technol. **21**(4), 1199–1206 (2013)

Elkhatib, K., Aitouche, A., Ghorbani, R., Bayart, M.: Fuzzy Scheduler fault-tolerant control for wind energy conversion systems. IEEE Trans. Control Syst. Technol. **22**(1), 119–131 (2014)

Errami, Y., Ouassaid, M., Maaroufi, M.: A MPPT vector control of electric network connected wind energy conversion system employing PM synchronous generator. In: IEEE International Renewable and Sustainable Energy Conference (IRSEC), pp. 228–233 (2013)

Evangelista, C., Fernando, V., Paul, P.: Active and reactive power control for wind turbine based on a MIMO 2-sliding mode algorithm with variable gains. In: IEEE Transactions On Energy Conversion, Vol. 28, No. 3, pp. 682–689 (2013b)

Evangelista, C., Puleston, P., Valenciaga, F., Fridman, L.M.: Lyapunov-designed super-twisting sliding mode control for wind energy conversion optimization. IEEE Trans. Indust. Electr. **60**(2), 538–545 (2013)

Giraldo, E., Garces, A.: An Adaptive control strategy for a wind energy conversion system based on PWM-CSC and PMSG. IEEE Trans. Power Syst. textbf99, 1–8 (2013)

Guo, X., Zhang, X., Wang, B., Guerrero, J.M.: Asymmetrical grid fault ride-through strategy of three-phase grid-connected inverter considering network impedance impact in low-voltage grid. IEEE Trans. Power Electr. **29**(3), 1064–1068 (2014)

Guzman, R., de Luís G., Vicuña, Antonio, C., José, M., Miguel, C., Jaume, M.: Active damping control for a three phase grid- connected inverter using sliding mode control. In: IEEE Conference of the Industrial Electronics Society (IECON), pp. 382 (2013)–387.

Harrouz, A., Benatiallah, A., Moulay Ali, A., Harrouz, O.: Control of machine PMSG dedicated to the conversion of wind power off-grid. In: IEEE International Conference on Power Engineering, Energy and Electrical Drives Istanbul, pp. 1729–1733 (2013)

He, J., Li, Y.W., Blaabjerg, F., Wang, X.: Active harmonic filtering using current-controlled, grid-connected dg units with closed-loop power contro. IEEE Trans. Power Electr. **29**(2), 642–653 (2014)

He, L., Li, Y., Harley, R.G.: Adaptive multi-mode power control of a direct-drive PM wind generation system in a microgrid. IEEE J. Emerg. Sel. Topics Power Electr. **1**(4), 217–225 (2013)

Huang, N., He, J., Nabeel, A., Demerdash, O.: Sliding Mode observer based position self-sensing control of a direct-drive PMSG wind turbine system fed by NPC converters. In: IEEE International Electric Machines Drives Conference (IEMDC), pp. 919–925 (2013)

Karthikeya, B.R., Schütt, R.J.: Overview of wind park control strategies. IEEE Trans. Sustain. Energy **99**, 1–7 (2014)

Kuschke, M., Strunz, K.: Energy-efficient dynamic drive control for wind power conversion with PMSG: modeling and application of transfer function analysis. IEEE J. Emerg. Sel. Top. Power Electron. **2**(1), 35–46 (2014)

Leonhard, W.: Control of Electric Drives. Springer, London (1990)

Li, R., Dianguo, X.: Parallel operation of full power converters in permanent-magnet direct-drive wind power generation system. IEEE Trans. Industr. Electron. **60**(4), 1619–1629 (2013)

Li, S., Du, H., Yu, X.: Discrete-time terminal sliding mode control systems based on Euler's discretization. IEEE Trans. Autom. Control, **99** (2013a)

Li, S., Zhou, M., Yu, X.: Design and implementation of terminal sliding mode control method for PMSM speed regulation system. IEEE Trans. Indust. Inform. **9**(4) 1879–1891 (2013b)

Li, S., Haskew, T.A., Swatloski, R.P., Gathings, W.: Optimal and direct-current vector control of direct-driven PMSG wind turbines. IEEE Trans. Power Electr. **27**(5), 2325–2337 (2012)

Ma, K., Blaabjerg, F.: Modulation methods for neutral-point-clamped wind power converter achieving loss and thermal redistribution under low-voltage ride-through. IEEE Trans. Indus. Electr. **61**(2), 835–845 (2014)

Ma, K., Marco, L., Frede, B.: Comparison of multi-MW converters considering the determining factors in wind power application. IEEE Energy Conversion Congress and Exposition (ECCE), pp. 4754–4761 (2013)

Martinez, M.I., Susperregui, A., Tapia, G.: Sliding-mode control of a wind turbine-driven double-fed induction generator under non-ideal grid voltages. IET Renew. Power Gener. **7**(4), 370–379 (2013)

Melo, D.F.R., Chang-Chien, L.-R.: Synergistic control between hydrogen storage system and offshore wind farm for grid operation. IEEE Trans. Sustain. Energy **5**(1), 18–27 (2014)

Meng, W., Yang, Q., Ying, Y., Sun, Y., Yang, Z., Sun, Y.: Adaptive power capture control of variable-speed wind energy conversion systems with guaranteed transient and steady-state performance. IEEE Trans. Energy Convers, **28**(3), 716–725 (2013)

Najafi, P., Rajaei, A., Mohamadian, M., Varjani, A.Y.: Vienna rectifier and B4 inverter as PM WECS grid interface. In: IEEE Conference on Electrical Engineering (ICEE), pp. 1–5 (2013)

Nguyen, T.H., Lee, D.-C., Kim, C.-K.: A series-connected topology of a diode rectifier and a voltage-source converter for an HVDC transmission system. IEEE Trans. Power Electron. **29**(4), 1579–1584 (2014)

Nguyen, T., Lee, D.-C.: Advanced fault ride-through technique for PMSG wind turbine systems using line-side converter as STATCOM. IEEE Trans. Indust. Electr. **60**(7), 2842–2850 (2013)

Nian, H., Song, Y.: Direct power control of doubly fed induction generator under distorted grid voltage. IEEE Trans. Power Electr. **29**(2), 894–905 (2014)

Nuno, M.A.F., Marques, António J.C.: A Fault-tolerant direct controlled PMSG drive for wind energy conversion systems. IEEE Trans. Indust. Electr. **61**(2), 821–834 (2014)

Orlando, N.A., Liserre, M., Mastromauro, R.A., Dell'Aquila, A.: A survey of control issues in PMSG-based small wind-turbine systems. IEEE Trans. Indust. Inf. **9**(3), 1211–1221 (2013)

Patil, N.S., Bhosle, Y.N.: A review on wind turbine generator topologies. In: IEEE International Conference on Power, Energy and Control (ICPEC), pp. 625–629 (2013)

Polinder, H., Bang, D., R.P.J.O.M., van Rooij, McDonald, A.S., Mueller, M.A.: 10 MW wind turbine direct-drive generator design with pitch or active speed stall control. In: IEEE International Conference On Electric Machines & Drives(IEMDC'07), Vol. 2, pp. 1390–1395 (2007)

Rajaei, A.H., Mohamadian, M., Varjani, A.Y.: Vienna-rectifier-based direct torque control of PMSG for wind energy application. IEEE Trans. Indust. Electr. **60**(7), 2919–2929 (2013)

Sabanovic, K.J., Sabanovic, N.: Sliding modes applications in power electronics and electrical drives in Variable Structure Systems. Towards the 21st Century, vol. 274. Springer, New York, pp. 223–251. (2002)

Shariatpanah, H., Fadaeinedjad, R., Rashidinejad, M.: A new model for PMSG-based wind turbine with yaw control. IEEE Trans. Energy Convers. **28**(4), 929–937 (2013)

She, X., Huang, A.Q., Wang, F., Burgos, R.: Wind energy system with integrated functions of active power transfer, reactive power compensation, and voltage conversion. IEEE Trans. Indust. Electr. **60**(10), 4512–4524 (2013)

Slotine, J.E., Li, W.: Applied Nonlinear Control. Prentice Hall, New Jersey (1991)

Spruce, C.J., Judith, K.T.: Tower vibration control of active stall wind turbines. IEEE Trans. Control Syst. Technol. **21**(4), 1049–1066 (2013)

Subudhi, B., Pedda, S.O.: Sliding Mode Approach to Torque and Pitch Control for a Wind Energy System. IEEE India Conference (INDICON), pp. 244–250 (2012)

Susperregui, A., Martinez, M.I., Tapia, G., Vechiu, I.: Second-order sliding-mode controller design and tuning for grid synchronisation and power control of a wind turbine-driven doubly fed induction generator. IET Renew. Power Gener. **7**(5), 540–551 (2013)

Li, T., Zou, X., Shushuai F., Yu., C., Yong, K., Huang, Q., Huang, Y.: SRF-PLL-Based Sensorless Vector Control Using Predictive Dead-beat Algorithm for Direct Driven Permanent Magnet Synchronous Generator (PMSG), p. 99. IEEE Trans. Power Electron. (2013)

Tseng, K., Huang, C.-C.: High step-up high-efficiency interleaved converter with voltage multiplier module for renewable energy system. IEEE Trans. Indust. Electr. **61**(3), 1311–1319 (2014)

Utkin, V.I., Guldner, J., Shi, J.: Sliding Mode Control in Electromechanical Systems. CRC Press, Boca Raton, FL, USA (1999)

Utkin, V.I.: Sliding mode control design principles and applications to electrical drives. IEEE Trans. Indust. Electr. **40**(1), 23–36 (1993)

Vazquez, S., Sanchez, J.A., Reyes, M.R., Leon, J.I., Carrasco, J.M.: Adaptive vectorial filter for grid synchronization of power converters under unbalanced and/or distorted grid conditions. IEEE Trans. Indust. Electr. **61**(3), 1355–1367 (2014)

Wang, L., Thi, M.S.-N.: Stability enhancement of large-scale integration of wind, solar, and marine-current power generation fed to an SG-based power system through an LCC-HVDC link. IEEE Trans. Sustain. Energy **5**(1), 160–170 (2014)

Xia, C., Wang, Z., Shi, T., Song, Z.: A novel cascaded boost chopper for the wind energy conversion system based on the permanent magnet synchronous generator. IEEE Trans. Energy Convers. **28**(3), 512–522 (2013)

Xiao, L., Shoudao, H., Lei, Z., Xu, Q., Huang, K.: Sliding mode SVM-DPC for grid-side converter of D-PMSG under asymmetrical faults. In: IEEE International Conference on Electrical Machines and Systems (ICEMS), pp. 1–6 (2011)
Xiao, S., Geng, Y., Hua, G.:Individual pitch control design of wind turbines for load reduction using sliding mode method. In: IEEE International Energy Conversion Congress and Exhibition ECCE Asia Downunder (ECCE Asia), pp. 227–232 (2013)
Xin, W., Cao, M., Li, Q., Chai, L., Qin, B.: Control of direct-drive permanent-magnet wind power system grid-connected using back-to-back PWM converter. In: IEEE International Conference on Intelligent System Design and Engineering Applications, pp. 478–481 (2013)
Yaramasu, V., Bin, W.: Predictive Control of Three-Level Boost Converter and NPC Inverter for High Power PMSG-Based Medium Voltage Wind Energy Conversion Systems. In: IEEE Transactions on Power Electronics, p. 99 (2013)
Yaramasu, V., Wu, B., Rivera, M., Rodriguez, J.: A new power conversion system for megawatt PMSG wind turbines using four-level converters and a simple control scheme based on two-step model predictive strategy—Part II: simulation and experimental analysis. IEEE J. Emerg. Select. Topics Power Electr. pp. 99 (2013)
Zhang, Y., Hu, J., Zhu, J.: Three vectors based predictive direct power control of doubly fed induction generator for wind energy applications. In: IEEE Transactions on Power Electronics pp. 99 (2013)
Zhang, Z., Zhao, V., Wei, Q., Qu, L.: A Discrete-Time direct-torque and flux control for direct-drive PMSG wind turbines. In: IEEE Industry Applications Society Annual Meeting, pp. 1–8 (2013)
Zhang, Z., Zhao, Y., Qiao, W., Qu, L.: A space-vector modulated sensorless direct-torque control for direct-drive PMSG wind turbines. In: IEEE Transactions on Industry Applications, p. 99 (2014)

Super-Twisting Air/Fuel Ratio Control for Spark Ignition Engines

Jorge Rivera, Javier Espinoza-Jurado and Alexander Loukianov

Abstract In this work, a model-based controller for the air to fuel ratio (represented by λ) is designed for spark ignition (SI) engines in order to rise the fuel consumption efficiency and to reduce the emission of pollutant gases to the atmosphere. The proposed control method is based on an isothermal mean value engine model (MVEM) developed by Elbert Hendricks and in the super-twisting sliding mode control algorithm that results to be robust to matched perturbations and alleviates the chattering problem. The dynamics for λ depends on the time derivative of the control input, i.e., the injected fuel mass flow ($\dot{m}_{fi}$). This term is estimated by means of the well-known robust sliding mode differentiator which is feedback to the control algorithm. To solve the time-delay measurement problem (due to combustion process and the transportation of gases) at the Universal Exhaust Gas Oxygen (UEGO) sensor, the delay represented with an exponential function in the frequency domain is approximated by means of a Padé method which yields to a transfer function. Then, this transfer function is taken to a state space representation in order to design an observer based on the super-twisting sliding mode algorithm, where the real λ factor is finally determined by the equivalent control method and used for feedback. Digital simulations were carried on, where the proposed control scheme is simulated with two observers based on a second and third order Padé approximations. Also, the proposed controller is simulated without an observer, where λ is directly taken from the UEGO sensor. Simulations predict a better output behavior in the case of a controller based observer design, and in particular, the observer based on the third order approximation provides the best results. Therefore, the controller based on the third order observer is chosen for parametric uncertainties and noise measurement simulation, where the air to fuel ratio still performs well.

J. Rivera (✉) · J. Espinoza-Jurado
University of Guadalajara, 44430 Guadalajara, México
e-mail: jorge.rivera@cucei.udg.mx
J. Espinoza-Jurado
e-mail: javier_9512@hotmail.com

A. Loukianov
CINVESTAV Guadalajara, 45019 Guadalajara, México
e-mail: louk@gdl.cinvestav.mx

A.T. Azar and Q. Zhu (eds.), *Advances and Applications in Sliding Mode Control systems*, Studies in Computational Intelligence 576, DOI 10.1007/978-3-319-11173-5_7

1 Introduction

One of the main control challenges for SI engines is the preservation of the stoichiometric value (14.67 for gasoline fuel) in the air to fuel (AFR) mixture, in order to keep an efficient fuel management, good output torque, and a more complete combustion (Pulkrabek 2004). Modern SI engines are equipped with electronic fuel injection (EFI) systems, making them more efficient than mechanical options (carburetors and mechanical fuel injection systems) that facilitates the implementation of modern control algorithms (Guzzella and Onder 2010). The measurement of the AFR in the EFI systems is made by an UEGO sensor that measures the λ factor (that is equal to 1 when the engine is running into a stoichiometric value) in the cylinder by measuring the present oxygen at the exhaust gases.

Currently, there exist many works for the AFR control, in which different types of mathematical models for the SI engine are implemented. The models ranging from simple, as those presented in Yildiz et al. (2008) and Muske (2006) where linearized models are used, reducing the complexity of the system by neglecting some dynamics in the process; to more complex models like the one used in Benvenuti and Benedetto (2003) where all the cylinders are modeled in an individual fashion, adding difficulty when comes to design a controller. There are other works like (Bastian 1994) where look-up tables containing the amount of air for a given engine speed and inlet manifold pressure in order to inject the exact quantity of fuel. The problem is that the look-up tables must be updated when engine is modified.

In this work, an isothermal mean value engine model (MVEM) developed by Hendricks (1990, 1992, 1996, 2000, 2001) is considered for the AFR control. This model is an intermediate option to the above mentioned models. It is well know that MVEM is a control oriented model, that neglects discrete cycles of the engine and assumes that all processes and effects are spread out over the engine cycle.

The main control problems to solve for the AFR are the rejection to internal and external perturbations due to environmental circumstances, sensor failures, the wall wetting fuel dynamics, among others. An important issue to take into account is the delay in the λ measurement, which is basically the time between the fuel injection and the burned gases reaching the UEGO sensor. There are several well established control techniques that have been applied for the AFR control problem in the SI engine. In Guzzella and Onder (2010) based in a linearized model of the engine, a H-infinity control to ensure robustness to parametric uncertainties of the engine and to the time delay from the UEGO sensor is designed. The drawback is that results a high order controller that is only valid around an operating point. Soft computing techniques has also been adopted as in Zhai et al. (2011), where an artificial neural network is used and adapted on line in order to deal with nonlinearities and parameter uncertainties. One disadvantage of this strategy is that the engine remains open-loop for about 2 s that correspond to the time that the neural network takes for initial adaptation, generating in that way a large transient peak in the rate of injected fuel at the beginning of the process. In the work presented in Tang et al. (2010), a global linearized control strategy is compared with a classical sliding mode design without considering time delays measurements from λ sensor. In this research it is shown the advantage of using classical sliding mode controller against the global linearized controller due to its robustness property in presence of matched uncertainties and disturbances. However, the main disadvantage with the use of classical sliding modes is the chattering problem that adversely affects the performance of any dynamical system.

On the other hand, the sliding mode control is a popular technique among control engineer practitioners due to the fact that introduces robustness to unknown bounded perturbations that belong to the control sub-space; moreover, the residual dynamic under the sliding regime, i.e., the sliding mode dynamics, can easily be stabilized with a proper choice of the sliding surface. A proof of their good performance in motion control systems can be found in the book by Utkin et al. (1999). One drawback of this technique are small oscillations of finite frequency at the output tracking signal that is known as chattering. The control signal is characterized by a discontinuous control action with an ideal infinite frequency that leads to the chattering problem. This problem is harmful because it leads to low control accuracy; high wear of moving mechanical parts and high heat losses in power circuits (Levant 2010). The chattering phenomenon can be caused by the deliberate use of classical sliding mode control technique. When fast dynamics are neglected in the mathematical model such phenomenon can appear. Another situation responsible for chattering is due to implementation issues of the sliding mode control signal in digital devices operating with a finite sampling frequency, where the switching frequency of the control signal cannot be fully implemented (Rivera et al. 2011).

In order to overcome the chattering phenomenon, the higher-order sliding mode (HOSM) concept was introduced by Levant (2003). Let us consider a smooth dynamic system with an output function S of class $\mathscr{C}^{r-1}$ closed by some static or dynamic discontinuous feedback as in Levant (2007). Then, the calculated time derivatives $S, \dot{S}, \ldots, S^{r-1}$, are continuous functions of the system state, where the set $S = \cdots S = \cdots = S^{r-1} = 0$ is non-empty and consists locally of Filippov trajectories (Filippov 1988). The motion on the set above mentioned is said to exist in r-sliding mode or rth order sliding mode. The rth derivative S^r is considered to be discontinuous or non-existent. Therefore the high-order sliding mode removes the relative-degree restriction and can practically eliminate the chattering problem. There are several algorithms to realize HOSM. In particular, the 2nd order sliding mode controllers are used to zero outputs with relative degree two or to avoid chattering while zeroing outputs with relative degree one. Among 2nd order algorithms one can find the sub-optimal controller, the terminal sliding mode controllers, the twisting controller and the super-twisting controller. In particular, the twisting algorithm forces the sliding variable S of relative degree two in to the 2-sliding set, requiring knowledge of $\dot{S}$. The super-twisting algorithm does not require $\dot{S}$, but the sliding variable has relative degree one. Hence the super-twisting algorithm is nowadays preferable over the classical siding mode, since it eliminates the chattering phenomenon.

In this work we are interested to contribute to the control problem of the AFR in SI engines by exploiting sliding mode control techniques for improving the robust performance of SI engines. Based on a MVEM (Hendricks et al. 1996) a HOSM controller is designed for maintaining the stoichiometric value equal to one in a SI engine. The HOSM technique is based on a super-twisting algorithm (Levant 2003; Utkin 2013; Fridman and Iriarte 2005). To estimate the intrinsic delays (due to exhaust gas transportation and combustion process) in the λ measurements provided by the UEGO sensor, it is necessary to model them properly. To do this, a Padé approximation (Kosiba et al. 2006; Probst et al. 2009; Liu et al. 2009) is used for representing the time delay in the frequency domain with a transfer function, where a state space representation is finally obtained. Then based on second and third order approximations, HOSM observers are designed for λ.

The remaining of this work is organized as follows. Section 2 reviews the mean value engine model and the time delay of the UEGO sensor. Section 3 deals with the control law and observer designs for controlling the AFR ratio. A simulation study is carried on in Sect. 4, and finally some comments conclude the work in Sect. 5.

2 Mean Value Engine Models

The mean value engine model (MVEM) proposed by Hendricks, is an intermediate between large cyclic simulation models and the simplistic phenomenological transfer function models, making it a compact model, easy to adapt for EFI, turbocharged, diesel, and emission control systems. The MVEMs for the SI engine primarily consist of 3 subsystems explained in Hendricks and Sorenson (1990), Hendricks et al. (1996, 2000):

- The intake manifold filling dynamics
- The fuel mass flow rate
- The crank shaft speed

2.1 The Intake Manifold Filling Dynamics

The intake manifold filling dynamics are based on an isothermal one, where the temperature exchange between the ambient temperature and the intake manifold temperature occurs slowly, therefore both temperatures are assumed to be the same. The intake manifold filling dynamics are segmented in three equations: (1) the intake manifold pressure, (2) the throttle air mass flow and (3) the intake port air mass flow.

2.1.1 Intake Manifold Pressure

The intake manifold is the volume between the throttle valve and the intake valve of the cylinder. The state equation for the intake manifold is obtained by applying conservation mass to the intake manifold volume m_{man}

$$\dot{m}_{man} = \dot{m}_{at} - \dot{m}_{ap} \tag{1}$$

where $\dot{m}_{at}$ is the air mass flow to the throttle valve and $\dot{m}_{ap}$ is the air mass flow to the intake valve. The pressure in the intake manifold p_{man} can be related to m_{man} using the ideal gas equation

$$p_{man} V_m = m_{man} R T_m \tag{2}$$

with R as the ideal gas constant, T_m is the air temperature in the manifold, V_m is the intake manifold volume. Taking the time derivate of (2) and using (1), the intake manifold pressure equation is obtained as

$$\dot{p}_{man} = \frac{RT_m}{V_m}(-\dot{m}_{ap} + \dot{m}_{at}) \tag{3}$$

2.1.2 Throttle Air Mass Flow

This part of the model is based on the isentropic flow equation for a converging-diverging nozzle, this equation its detailed in Hendricks et al. (2000)

$$\dot{m}_{at} = \dot{m}_{at1}\sqrt{\frac{P_a}{T_m}}\beta_1(\alpha)\beta_2(P_r) + \dot{m}_{at0} \tag{4}$$

were $\dot{m}_{at1}$ and $\dot{m}_{at0}$ are constant parameters, P_a is the ambient pressure, α is the angle of the throttle plate and $\beta_1(\alpha)$ is the ratio of the throttle throat diameter to the throttle plate shaft diameter

$$\beta_1(\alpha) = 1 - cos(\alpha) - \frac{\alpha_0^2}{2} \tag{5}$$

where α_0 is the close angle throttle plate. Function $\beta_1(\alpha)$ is useful only when the throttle plate has a circular shape, in other case it must be found another equation that can describe it in an appropriated fashion. Expression $\beta_2(P_r)$ is the isentropic flow

$$\beta_2(P_r) = \begin{cases} 1 & Pr < Pc \\ \sqrt{1 - (\frac{Pr-Pc}{1-Pc})^2} & Pr \geq Pc \end{cases} \tag{6}$$

where $P_r = p_{man}/P_a$ and P_c is the critical pressure (turbulent flow).

2.1.3 Intake Port Air Mass Flow

The air mass flow at the intake port can be obtained from a speed density equation:

$$\dot{m}_{ap} = \sqrt{\frac{T_m}{T_a}} \frac{V_d}{120RT_m} e_v p_{man} n_e \tag{7}$$
$$e_v p_{man} = s_i P_{man} - y_i$$

where T_a is the ambient temperature, V_d is the engine displacement, e_v is the volumetric efficiency, s_i is the intake manifold slope and y_i is the manifold intercept pressure.

2.2 The Fuel Mass Flow Rate

According to the experiments reported in Hendricks and Sorenson (1990) and Hendricks and Vesterholm (1992). The equations that describe the fuel mass flow rate $\dot{m}_f$ into the cylinder are as follows:

$$\dot{m}_{fv} = (1 - X_f)\dot{m}_{fi}$$
$$\ddot{m}_{ff} = \frac{1}{\tau_f}(-\dot{m}_{ff} + X_f \dot{m}_{fi})$$
$$\dot{m}_f = \dot{m}_{fv} + \dot{m}_{ff} \tag{8}$$

where m_{ff} is the mass of the fuel film adhered to the manifold wall, $\dot{m}_{fi}$ is the fuel flow rate from the injector, X_f is the fraction of injected fuel that remains as fuel film, τ_f is the fuel evaporation time constant, $\dot{m}_{fv}$ is the portion of fuel that enters to the cylinder valve. The fraction in the fuel film X_f is approximated in Hendricks et al. (1996)

$$X_f = X_1 - X_2 \frac{\dot{m}_{ap}}{\dot{m}_{ap,max}} \tag{9}$$

where $\dot{m}_{ap,max}$ its the maximum air mass flow for the engine.

2.3 The Crankshaft Speed

The crank shaft state equation is derived using straight forward energy conservation considerations. Energy is inserted into the crank shaft via the fuel flow. Losses in pumping and friction dissipate rotational energy while some of the energy available goes into the load.

$$\dot{n} = -\frac{1}{In}(P_f + P_p + P_b) + \frac{1}{In} H_u \eta_i \dot{m}_f(t - \Delta\tau_d) \qquad (10)$$

where n is the crankshaft speed, I its the inertial moment of the crankshaft; P_f, P_p, and P_b are the power losses by friction, pumping and load respectively, H_u is the fuel burn value, η_i is the thermal efficiency, and $\dot{m}_f$ is the mass fuel rate into the cylinder with a torque time delay $\Delta\tau_d$. The fiction and pumping losses in the engine can be expressed as polynomials of the crankshaft speed and the intake manifold pressure:

$$P_f(n) + P_p(n, p_{man}) = n(a_0 + a_1 n + a_2 n^2) + n(a_3 + a_4 n) \qquad (11)$$

and $P_b = k_b n^3$ where k_b its the load factor. The thermal efficiency η_i can be expressed as:

$$\eta_i(\theta, \lambda, n, p_{man}) = \eta_1(\theta, n, p_{man})\eta_i(\lambda, n)\eta_1(n)\eta_i(p_{man}) \qquad (12)$$

where

$$\begin{aligned} \eta_i(\theta) &= \Theta_0 + \Theta_1(\theta - \theta_{mbt}) \\ &\quad - \Theta_2(\theta - \theta_{mbt})^2 \\ \eta_i(n) &= \eta_{i0} - \eta_{i1} n^{\eta_{i3}} \\ \eta_i(p_{man}) &= \rho_0 + \rho_1 p_{man} + \rho_2 p_{man}^2 \\ \eta_i(\lambda) &= \begin{cases} \Lambda_0 + \Lambda_1\lambda + \Lambda_2\lambda^2 & \text{if } \lambda \le 1 \\ \Lambda_3 + \Lambda_4\lambda + \Lambda_5\lambda^2 & \text{if } \lambda > 1 \end{cases} \end{aligned} \qquad (13)$$

with θ_{mbt} as the maximum brake torque and Λ_i $(i = 0, \ldots, 5)$ as constant parameters. Figure 1 show the interconnection between MVEMs state equations.

2.4 λ Sensor Model

The λ factor (normalized AFR) is defined by the equation

$$\lambda = \frac{\dot{m}_{ap}}{L_{th}\dot{m}_f} \qquad (14)$$

where Lth is the desire stoichimetric value. The UEGO λ sensor has a linear response in a range of values which represent a lean, rich or stoichiometric mixture of the engine (Vigild et al. 1999). The sensor is approximated by a first order system:

$$\frac{\Lambda_m(s)}{\Lambda_{exh}(s)} = \frac{1}{s\tau_\lambda + 1} \qquad (15)$$

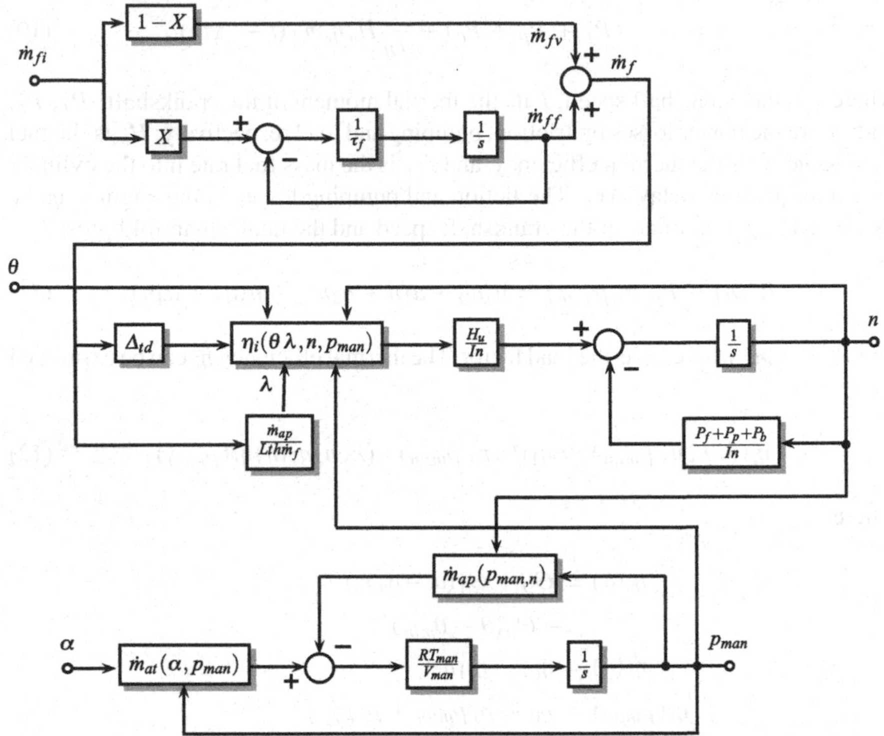

Fig. 1 MVEM block diagram

where $\Lambda_m(s) = \mathscr{L}\{\lambda_m(t)\}$ and $\Lambda_{exh}(s) = \mathscr{L}\{\lambda_{exh}(t)\}$, with $\mathscr{L}$ as the Laplace operator. λ_m is the λ measurement given by the sensor, λ_{exh} represents the λ value available at the sensor and τ_λ as the time constant of the sensor that can depend on the temperature in the exhaust pipe (Vigild et al. 1999). Meanwhile the relation between λ_{exh} and λ is of the following form:

$$\Lambda_{exh}(s) = e^{-\tau_d s}\Lambda(s) \tag{16}$$

where $\Lambda(s) = \mathscr{L}\{\lambda(t)\}$ and τ_d is a time delay that is due by three factors:

(1) τ_{d1} is the time delay due to the fuel injection and the time valve opening

$$\tau_{d1} = \frac{60\Delta\theta_1}{360°n} \tag{17}$$

where $\Delta\theta_1$ is the crank angle between injection and intake valve opening.

(2) τ_{d2} is the time delay due to the combustion to the intake valve opening and to the exhaust valve opening

$$\tau_{d2} = \frac{60\Delta\theta_2}{360°n} \tag{18}$$

where $\Delta\theta_2$ is the crank angle between the intake valve opening and the exhaust valve opening.

(3) τ_{d3} is the time delay due to the transportation of the exhaust matter from the exhaust valve to the sensor

$$\tau_{d3} = \frac{\rho_{exh} l_{exh} A_{exh}}{\dot{m}_{ap}} = \frac{\frac{p_{exh}}{RT_{exh}} l_{exh} A_{exh}}{\dot{m}_{ap}} \tag{19}$$

where ρ_{exh} and p_{exh} are the air density and pressure in the exhaust manifold respectively, A_{ehx} is the cross section of the exhaust pipe, l_{exh} is the distance between the exhaust valve and the λ sensor and T_{exh} is the exhaust gas temperature.

3 Control Design

The control problem consists in forcing the output λ to track a desired lambda factor ($\lambda_r = 1$) with time delay output measurements of λ, where the input controlled variable is $\dot{m}_{fi}$. To tackle this problem, a high order sliding mode control technique based in the super-twisting algorithm is used (Levant 2003). The designed control law depends on the time derivative of control input $\dot{m}_{fi}$. This term is estimated by using a robust exact differentiation via sliding mode technique. On the other hand, to solve the time delay problem from the measurement of the UEGO sensor, an observer based solution is implemented by means of a high order sliding mode methodology. For that, the delay is approximated by means of Padé method in the frequency domain with transfer functions, then it is transformed to a state space representation without delay.

3.1 Super-Twisting Sliding Mode Control of Normalized AFR

Let us define the output error as

$$z = \lambda - \lambda_r \tag{20}$$

where λ_r is the reference signal for λ. The dynamic error equation for (20) can be represented in the general form

$$\dot{z} = f(x, u, \dot{u}) \tag{21}$$

with $x = (n, \lambda, \dot{m}_{at}, \dot{m}_{ff})^T$, $u = \dot{m}_{fi}$ and $f(x, u, \dot{u})$ as

$$f(x,u,\dot{u}) = -\left(\frac{e_v V_d n}{120 V_i} + \frac{(1-X)\dot{u} + \frac{1}{\tau_f}(-\dot{m}_{ff} + Xu)}{(1-X)u + \dot{m}_{ff}}\right)\lambda$$

$$+ \frac{e_v V_d n}{120 V_i}\left(\frac{\dot{m}_{at}}{Lth((1-X)u + \dot{m}_{ff})}\right) - \dot{\lambda}_r. \tag{22}$$

Now a new control input v is introduced as $v = f(x, u, \dot{u})$, that simplifies (21) as follows:

$$\dot{z} = v \tag{23}$$

by choosing the sliding function as z, then v is selected as a super-twisting algorithm (Levant 2003)

$$\begin{aligned} v &= -k_1|z|^{1/2} sign(z) + v_1 \\ \dot{v}_1 &= -k_2 sign(z) \end{aligned} \tag{24}$$

with properly chosen constants $k_1 > 0$ and $k_2 > 0$ (Perruquetti and Barbot 2002), z will decay to zero in finite-time. From the relation $v = f(x, u, \dot{u})$ and by making use of the implicit function theorem (Khalil 2002) one determines the following control law

$$u = \frac{\left(\frac{1}{\tau_f}\lambda - v - \frac{e_v V_d n}{120 V_i}\lambda\right)\dot{m}_{ff} + \frac{e_v V_d n}{120 V_i L_{th}}\dot{m}_{at} - (1-X)\dot{u}\lambda}{v(1-X) + \frac{e_v V_d n}{120 V_i}\lambda(1-X) + \frac{1}{\tau_f}X} \tag{25}$$

It is worth noting that control (25) depends on the time derivative of the control itself. By differentiating (25) one retrieves $\dot{u}$. Then this signal is fedback to reconstruct control law (25) as in Castillo-Toledo and Lopez Cuevas (2009). The time derivative $\dot{u}$ is determined with the following robust sliding mode differentiator:

$$\begin{aligned} \dot{\xi}_0 &= y_0 \\ \dot{\xi}_1 &= -\kappa_2 \gamma_a sign(\xi_0 - u) \\ y_0 &= \xi_1 - \kappa_1 \gamma_a^{1/2} |\xi_0 - u|^{1/2} sign(\xi_0 - u) \end{aligned} \tag{26}$$

where $\hat{u} = \xi_0$ and $\hat{\dot{u}} = \xi_1$ and κ_1, κ_2 and γ_a are positive constant design parameters (Levant 1998). With a bounded and free noise signal u, this differentiator ensures finite–time convergence of the following equalities

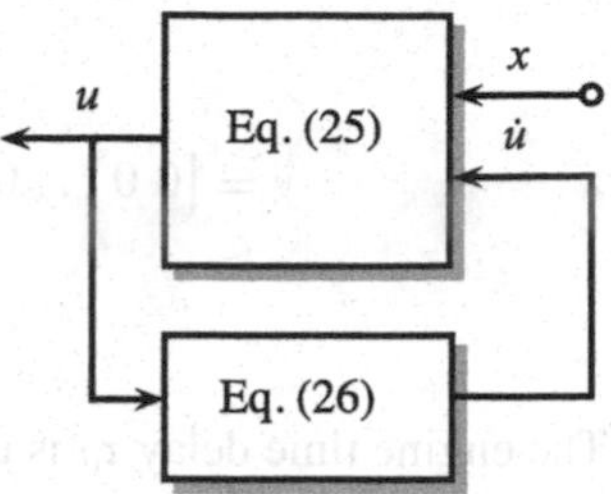

Fig. 2 Block diagram of the proposed control algorithm

$$\xi_0 = u, \quad \xi_1 = \dot{u}, \tag{27}$$

with initial accuracies

$$\begin{aligned} |\xi_0(t_0) - u(t_0)| &\leq \mu_0(t_0) \\ |\xi_1(t_0) - \dot{u}(t_0)| &\leq \mu_1(t_0) \end{aligned} \tag{28}$$

where $\mu_0 > 0$ and $\mu_1 > 0$. It means that the solution of system (26) is Lyapunov stable satisfying (27) for $t \geq t_0 + t_s$. Finally Fig. 2 illustrates a block diagram of the proposed control algorithm.

3.2 Observer Design for Normalized AFR (λ)

The observer design is based on an j order Padé approximation of the time-delay in the frequency domain (Fournodavlos and Nestoridis 2013; Vigild et al. 1999). The Padé approximation of the exponential function that represents the time delay in the frequency domain is now represented as a transfer function with real coefficients as in the following general form:

$$\frac{y(s)}{u(s)} = e^{-s\tau_d} \approx \frac{\sum_{i=0}^{j} N_i(-s)^i}{\sum_{i=0}^{j} D_i(s)^i} = \frac{\sum_{i=0}^{j} \frac{(2j-i)!}{i!(j-i)!}(-s\tau_d)^i}{\sum_{i=0}^{j} \frac{(2j-i)!}{i!(j-i)!}(s\tau_d)^i}. \tag{29}$$

The general state space representation of the Padé approximation is given by the following equations:

$$\begin{bmatrix} \dot{x}_1 \\ \dot{x}_2 \\ \vdots \\ \dot{x}_{j-1} \\ \dot{x}_j \end{bmatrix} = \begin{bmatrix} 0\,0 \ldots 0 & -D_0 \\ 1\,0 \ldots 0 & -D_2 \\ \vdots\,\vdots \quad \vdots & \vdots \\ 0\,0 \ldots 1 & -D_{j-2} \\ 0\,0 \ldots 0 & -D_{j-1} \end{bmatrix} \begin{bmatrix} x_1 \\ x_2 \\ \vdots \\ x_{j-1} \\ x_j \end{bmatrix} + \begin{bmatrix} N_j - D_j N_0 \\ N_{j-1} - D_{j-1} N_1 \\ \vdots \\ N_2 - D_2 N_{j-1} \\ N_1 - D_1 N_j \end{bmatrix} u$$

$$y = \begin{bmatrix} 0 \; 0 \ldots 0 \; 1 \end{bmatrix} \begin{bmatrix} x_1 \\ x_2 \\ \vdots \\ x_{j-1} \\ x_j \end{bmatrix} + \begin{bmatrix} -1^j \end{bmatrix} u. \tag{30}$$

The engine time delay τ_d is modeled as in Vigild et al. (1999)

$$\tau_d = \frac{\xi}{n_o} = \frac{0.187krpm.s}{n_o} \tag{31}$$

where n_o is an operational average value of the engine speed.

In the following, two HOSM observers are designed for $j = 2, 3$, with the purpose of comparing the accuracy of the approximation and the effects when closing the loop.

3.2.1 Second-Order Observer Design for Normalized AFR (λ)

The second-order Padé approximation is given by the transfer function

$$e^{-\tau_d s} = \frac{\Lambda_{exh}(s)}{\Lambda(s)} \approx \frac{\frac{12}{\tau_d^2} - \frac{6}{\tau_d}s + s^2}{\frac{12}{\tau_d^2} + \frac{6}{\tau_d}s + s^2}. \tag{32}$$

The corresponding state space representation results to be

$$\begin{aligned} \dot{\lambda}_1 &= -\frac{12}{\tau_d^2}\lambda_2 \\ \dot{\lambda}_2 &= \lambda_1 - \frac{6}{\tau_d}\lambda_2 - \frac{12}{\tau_d}\lambda \\ \lambda_{exh} &= \lambda_2 + \lambda. \end{aligned} \tag{33}$$

In order to represent the time response provided by the UEGO sensor (denoted as λ_m), the following first order system is proposed (Vigild et al. 1999):

$$\begin{aligned} \dot{\lambda}_m &= -\frac{1}{\tau_\lambda}\lambda_m + \frac{1}{\tau_\lambda}\lambda_2 + \frac{1}{\tau_\lambda}\lambda \\ y_\lambda &= \lambda_m. \end{aligned} \tag{34}$$

The observer is proposed of the following form

$$\dot{\hat{\lambda}}_1 = -\frac{12}{\tau_d^2}\hat{\lambda}_2$$

$$\dot{\hat{\lambda}}_2 = \hat{\lambda}_1 - \frac{6}{\tau_d}\hat{\lambda}_2 + k_\lambda \nu$$
$$\dot{\hat{\lambda}}_m = -\frac{1}{\tau_\lambda}\hat{\lambda}_m + \frac{1}{\tau_\lambda}\hat{\lambda}_2 - \nu$$
$$\hat{y}_\lambda = \hat{\lambda}_m. \tag{35}$$

where ν is the injected signal to the observer that will be defined in the following lines. Now the estimation errors are introduced as $\tilde{\lambda}_1 = \lambda_1 - \hat{\lambda}_1$, $\tilde{\lambda}_2 = \lambda_2 - \hat{\lambda}_2$ and $\tilde{\lambda}_m = \lambda_m - \hat{\lambda}_m$. The dynamics of the estimation errors result as follows:

$$\dot{\tilde{\lambda}}_1 = -\frac{12}{\tau_d^2}\tilde{\lambda}_2$$
$$\dot{\tilde{\lambda}}_2 = \tilde{\lambda}_1 - \frac{6}{\tau_d}\tilde{\lambda}_2 - \frac{12}{\tau_d}\lambda - k_\lambda \nu$$
$$\dot{\tilde{\lambda}}_m = -\frac{1}{\tau_\lambda}\tilde{\lambda}_m + \frac{1}{\tau_\lambda}\tilde{\lambda}_2 + \frac{1}{\tau_\lambda}\lambda + \nu$$
$$\tilde{y}_\lambda = \tilde{\lambda}_m. \tag{36}$$

One can choose the sliding function as $\tilde{\lambda}_m$ and the observer injected signal according to a super-twisting sliding mode algorithm (Levant 2003):

$$\nu = -\sigma_1|\tilde{\lambda}_m|^{1/2} sign(\tilde{\lambda}_m) + \nu_1$$
$$\dot{\nu}_1 = -\sigma_2 sign(\tilde{\lambda}_m). \tag{37}$$

With a proper choice of positive observer gains σ_1 and σ_2, the finite-time convergence of $\tilde{\lambda}_m$ to 0 is feasible. Then by applying the equivalent control method (Utkin et al. 1999) one can determine the equivalent injected signal from $\dot{\tilde{\lambda}}_m = 0$ as follows:

$$\nu_{eq} = -\frac{1}{\tau_\lambda}(\tilde{\lambda}_2 + \lambda). \tag{38}$$

If k_λ is chosen equal to $-\tau_\lambda/\tau_d$, then the sliding mode dynamics for the estimation errors result as follows:

$$\begin{bmatrix} \dot{\tilde{\lambda}}_1 \\ \dot{\tilde{\lambda}}_2 \end{bmatrix} = \begin{bmatrix} 0 - \frac{12}{\tau_d^2} \\ 1 - \frac{7}{\tau_d} \end{bmatrix} \begin{bmatrix} \tilde{\lambda}_1 \\ \tilde{\lambda}_2 \end{bmatrix} + \begin{bmatrix} 0 \\ -\frac{13}{\tau_d} \end{bmatrix} \lambda. \tag{39}$$

By assuming that λ is constant, the steady-state solution for (39) is given by

$$\tilde{\lambda}_{1,ss} = \frac{13}{\tau_d}\lambda$$
$$\tilde{\lambda}_{2,ss} = 0 \tag{40}$$

thus, according to (38), $\lim_{t\to\infty} \nu_{eq}(t) = -\lambda/\tau_\lambda$. Therefore λ is estimated as

$$\hat{\lambda} = -\tau_\lambda \nu_1. \tag{41}$$

3.2.2 Third-Order Observer Design for Normalized AFR (λ)

The third-order Padé approximation is given by the transfer function

$$e^{-\tau_d s} = \frac{\Lambda_{exh}(s)}{\Lambda(s)} \approx \frac{\frac{120}{\tau_d^3} - \frac{60}{\tau_d^2}s + \frac{12}{\tau_d}s^2 - s^3}{\frac{120}{\tau_d^3} + \frac{60}{\tau_d^2}s + \frac{12}{\tau_d}s^2 + s^3} \tag{42}$$

where the corresponding state space equations are:

$$\dot{\lambda}_1 = -\frac{120}{\tau_d^3}\lambda_3 + \frac{240}{\tau_d^3}\lambda$$
$$\dot{\lambda}_2 = \lambda_1 - \frac{60}{\tau_d^2}\lambda_3$$
$$\dot{\lambda}_3 = \lambda_2 - \frac{12}{\tau_d}\lambda_3 + \frac{24}{\tau_d}\lambda$$
$$\lambda_{exh} = \lambda_3 - \lambda. \tag{43}$$

The model of the time response of the UEGO sensor is given by the following equations:

$$\dot{\lambda}_m = -\frac{1}{\tau_\lambda}\lambda_m + \frac{1}{\tau_\lambda}\lambda_3 - \frac{1}{\tau_\lambda}\lambda$$
$$y_\lambda = \lambda_m. \tag{44}$$

The third-order observer is proposed of the following form:

$$\dot{\hat{\lambda}}_1 = -\frac{120}{\tau_d^3}\hat{\lambda}_3 + k_{\lambda 1}$$
$$\dot{\hat{\lambda}}_2 = \hat{\lambda}_1 - \frac{60}{\tau_d^2}\hat{\lambda}_3$$

$$
\begin{aligned}
\dot{\hat{\lambda}}_3 &= \hat{\lambda}_2 - \frac{12}{\tau_d}\hat{\lambda}_3 + k_{\lambda 2} \\
\dot{\hat{\lambda}}_m &= -\frac{1}{\tau_\lambda}\hat{\lambda}_m + \frac{1}{\tau_\lambda}\hat{\lambda}_3 - \nu \\
\hat{y}_\lambda &= \hat{\lambda}_m.
\end{aligned} \tag{45}
$$

The estimation errors are introduced as $\tilde{\lambda}_1 = \lambda_1 - \hat{\lambda}_1$, $\tilde{\lambda}_2 = \lambda_2 - \hat{\lambda}_2$, $\tilde{\lambda}_3 = \lambda_3 - \hat{\lambda}_3$ and $\tilde{\lambda}_m = \lambda_m - \hat{\lambda}_m$. The dynamics of the estimation errors result as follows:

$$
\begin{aligned}
\dot{\tilde{\lambda}}_1 &= -\frac{120}{\tau_d^3}\tilde{\lambda}_3 + \frac{240}{\tau_d^3}\lambda - k_{\lambda 1}\nu \\
\dot{\tilde{\lambda}}_2 &= \tilde{\lambda}_1 - \frac{60}{\tau_d^2}\tilde{\lambda}_3 \\
\dot{\tilde{\lambda}}_3 &= \tilde{\lambda}_2 - \frac{12}{\tau_d}\tilde{\lambda}_3 + \frac{24}{\tau_d}\lambda - k_{\lambda 2}\nu \\
\dot{\tilde{\lambda}}_m &= -\frac{1}{\tau_\lambda}\tilde{\lambda}_m + \frac{1}{\tau_\lambda}\tilde{\lambda}_3 - \frac{1}{\tau_\lambda}\lambda + \nu \\
\tilde{y}_\lambda &= \tilde{\lambda}_m.
\end{aligned} \tag{46}
$$

The sliding function is chosen as $\tilde{\lambda}_m$ and the observer injected signal according to a super-twisting sliding mode algorithm (Levant 2003) as follows:

$$
\begin{aligned}
\nu &= -\sigma_3|\tilde{\lambda}_m|^{1/2} sign(\tilde{\lambda}_m) + \nu_1 \\
\dot{\nu}_1 &= -\sigma_4 sign(\tilde{\lambda}_m).
\end{aligned} \tag{47}
$$

With a proper choice of the positive observer gains σ_3 and σ_4, the finite-time convergence of $\tilde{\lambda}_m$ to 0 is again feasible. Then by applying the equivalent control method (Utkin et al. 1999) one determines the equivalent injected signal from $\dot{\tilde{\lambda}}_m = 0$ of the following form:

$$
\nu_{eq} = -\frac{1}{\tau_\lambda}(\tilde{\lambda}_3 - \lambda). \tag{48}
$$

If $k_{\lambda 1}$ is chosen equal to $-120\tau_\lambda/\tau_d^3$ and $k_{\lambda 2}$ as $-12\tau_\lambda/\tau_d$ then the sliding mode dynamics for the estimation errors result as follows:

$$
\begin{bmatrix} \dot{\tilde{\lambda}}_1 \\ \dot{\tilde{\lambda}}_2 \\ \dot{\tilde{\lambda}}_3 \end{bmatrix} = \begin{bmatrix} 0 & 0 & -\frac{240}{\tau_d^3} \\ 1 & 0 & -\frac{60}{\tau_d^2} \\ 0 & 1 & -\frac{24}{\tau_d} \end{bmatrix} \begin{bmatrix} \tilde{\lambda}_1 \\ \tilde{\lambda}_2 \\ \tilde{\lambda}_3 \end{bmatrix} + \begin{bmatrix} \frac{360}{\tau_d^3} \\ 0 \\ \frac{36}{\tau_d} \end{bmatrix} \lambda. \tag{49}
$$

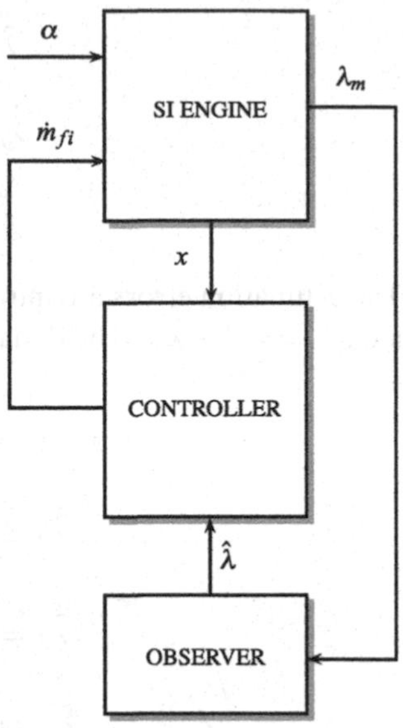

Fig. 3 Block diagram of the proposed observer-based control scheme

By assuming that λ is constant, the steady–state solution for (39) is given by

$$\begin{aligned}\tilde{\lambda}_{1,SS} &= \frac{90}{\tau_d^2}\lambda \\ \tilde{\lambda}_{2,SS} &= 0 \\ \tilde{\lambda}_{3,SS} &= \frac{3}{2}\lambda\end{aligned} \tag{50}$$

thus, according to (48), $\lim_{t\to\infty} \nu_{eq}(t) = -\lambda/\tau_\lambda$. Therefore λ is estimated as

$$\hat{\lambda} = -2\tau_\lambda \nu_1. \tag{51}$$

Finally Fig. 3 shows the proposed observer-based control scheme for the AFR ratio control.

4 Simulations

Simulations are carried out considering the parameters reported in Hendricks et al. (1996) for a 1.27 l British Leyland engine where it is assumed an optimal spark timing. The nominal values for the engine are shown in Table 1. Moreover, the inertia I is described as $I = I_{ac}(\pi/30)^2 1000$ where $I_{ac} = 0.49\,\text{kg/m}^2$

Table 1 Engine nominal values

Parameter	Value	Parameter	Value
R	287.09×10^{-5} bar m^3/kgK	V_m	0.0017 m^3
T_m	293 K	P_a	1.013 bar
$\dot{m}_{at1}$	5.9403	$\dot{m}_{at0}$	0
α_0	10°	P_c	0.4125
V_d	0.001275m^3	H_u	4.3×10^4
s_i	0.961	y_i	0.07
X_1	0.65	X_2	0.27
$\dot{m}_{ap,max}$	0.0597	X_1	0.65
X_2	0.27	a_0	1.673
a_1	0.272	a_2	0.0135
a_3	−0.969	a_4	0.206
k_b	0.22 kW/krpm^3	Θ_0	0.7
Θ_1	0.0240	Θ_2	0.00048
$(\theta - \theta_{mbt})$	27.5°	η_{i0}	η_{i0}
η_{i1}	−0.2187	η_{i2}	−0.360
ρ_0	0.9301	ρ_1	0.2154
ρ_2	1657	Λ_0	−1.299
Λ_1	3.599	Λ_2	−1.332
Λ_3	−0.0205	Λ_4	1.741
Λ_5	−0.745	Lth	14.67

Table 2 Design parameters

	Parameter	Value
Controller	k_1	−0.8
	k_2	−0.0001
Observer	σ_1	−12
	σ_2	−0.8
	σ_3	−10
	σ_4	−0.8
Differentiator	κ_2	−0.4
	κ_1	0.00001

is the load moment inertia, the time delays for $\dot{m}_f$ (τ_f) and λ (τ_λ) are $60/8n$ and $0.187/n$ respectively. The proposed design parameter values are shown in Table 2.

In order to reproduce more accurately the drive of the throttle valve α, the acceleration step commands were passed by a first order filter with a time constant of 0.3 s before entering the engine.

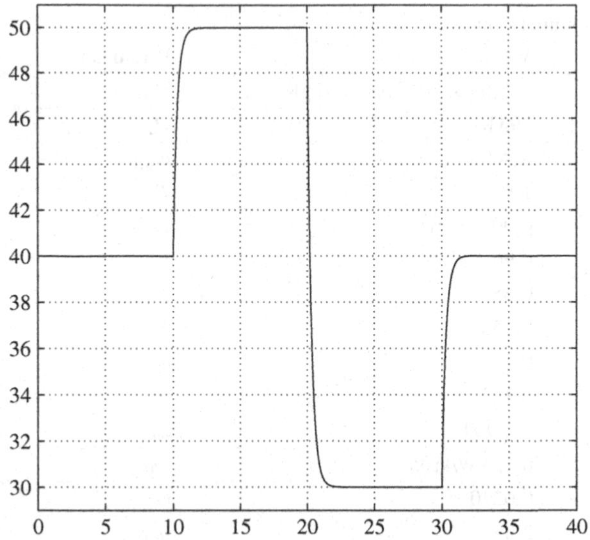

Fig. 4 Throttle valve steps [degree vs s]

4.1 Comparison of Three Control Schemes

A comparison among three controls schemes: (1) control with observer based on a second order Padé approximation; (2) control with observer based on a third order Padé approximation; (3) control without observer. The throttle valve angle is initiated at a value of 40° and remains constant for $t \in [0, 10)$ s, then increases to 50° at time 10s and remains constant until 20 s, then at this time value decreases to 30° and keeps constant for $t \in (20, 30)$ s, finally increases from 30° to 40° remaining constant for subsequent time, Fig. 4 shows the profile of throttle valve angle signal. Figure 5 illustrates the λ response without observer, meanwhile Fig. 6 shows a comparison between the response of λ with the second order and third-order observers.

It can be appreciated a similar behavior in all cases, but in the case of without observer a noise is present in steady-state, this noise is filtered out when using observers for the estimation of λ, where the observer based on third order Padé approximation yields to more accurate results as shown in Fig. 6b.

Figures 7 and 8 show the responses of the second and third order observers respectively. Meanwhile in Fig. 9a, b illustrate the magnitude of the estimation errors for the second and third order observers respectively. In both pair of graphics can be appreciated similar results.

Finally Table 3 shows a quantitative analysis of simulations by presenting the precision error P_e (relative error of the output variable, calculated as the difference between the average steady-state control output and the reference value, divided by the reference value) in steady state among the three control schemes. With

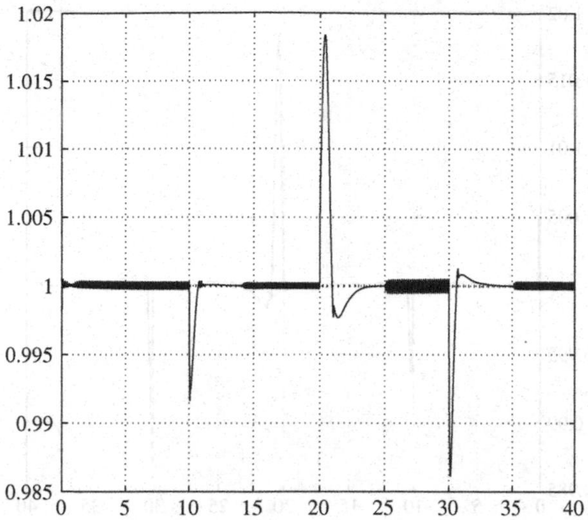

Fig. 5 λ factor response without observer, λ(*solid*) and λ_r(*pointed*) [λ vs s]

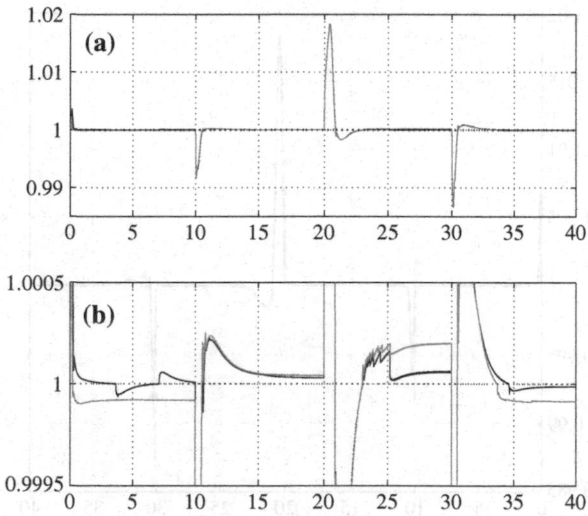

Fig. 6 **a** λ factor comparison between control schemes with second order observer (*grey*) and third order observer (*black*) [λ vs s]. **b** Zoom of the above graphic

$$P_e = 100|S_r - V_m|/S_r \tag{52}$$

where S_r the imposed reference and V_m the average of the output controlled variable in steady state.

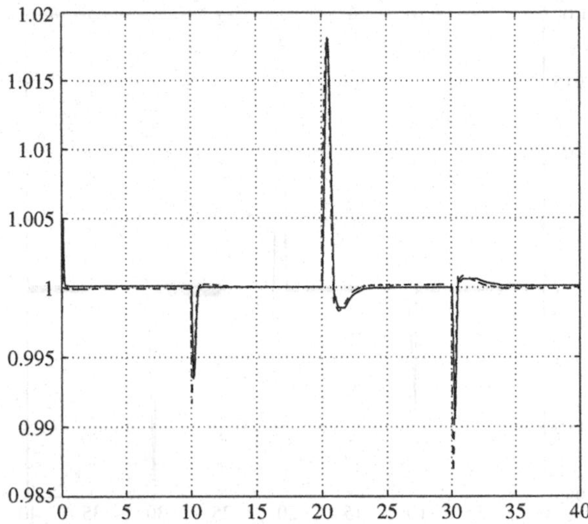

Fig. 7 λ estimation by the 2nd order observer, $\hat{\lambda}$ (*solid*) and λ (*dashed*) [λ vs s]

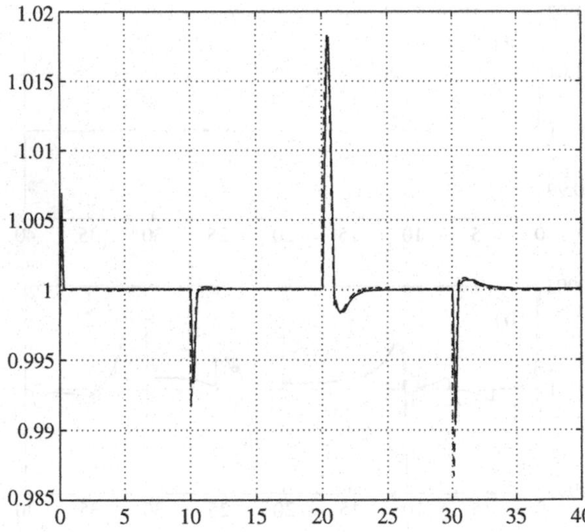

Fig. 8 λ estimation by the 3rd order observer, $\hat{\lambda}$ (*solid*) and λ (*dashed*) [λ vs s]

4.2 Proposed Control Scheme

Based on previous results, the control scheme that performs better is the one with the third-order observer, due to the fact that presents a better estimation of λ and a better tracking of λ_r than the second order observer based controller and the one without observer. Hence the third order observer based controller is simulated with

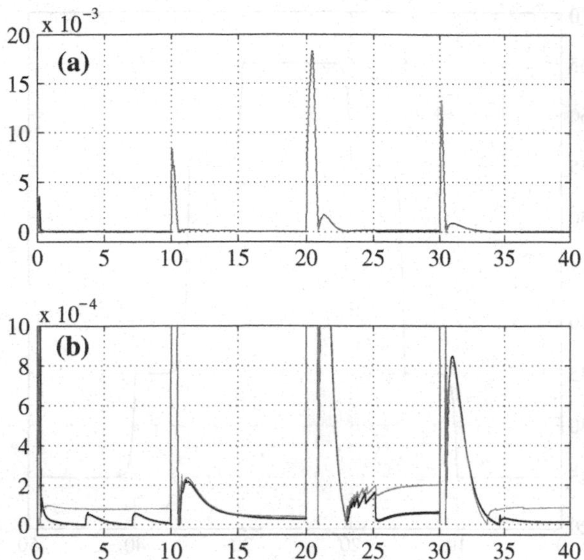

Fig. 9 **a** Comparison of the magnitude of the estimation error for the second and third order observers, $|\tilde{\lambda}|$ for the second order observer (*grey*) and $|\tilde{\lambda}|$ for the third order observer (*black*) [$|\tilde{\lambda}|$ vs s]. **b** Zoom of the above graphic

Table 3 Control schemes comparison

P_e			
Obs.	Without	2nd order	3rd order
α			
30°	0.045 %	0.02 %	0.006 %
40°	0.026 %	0.008 %	0.002 %
50°	0.02 %	0.0042 %	0.003 %

parametric uncertainties by assigning a nominal value of 0.6 to e_v and 0.0014 m^3 to V_d. Moreover, a high frequency noise of ± 0.05 was added to the measurement output of the UEGO sensor in order to simulate an old sensor. The throttle valve angle α starts at 25° and remains constant for $t \in [0, 10)$s, then the angle is increased to 35°, to 55° and to 65° remaining constant in each interval at the time instants of 10, 15 and 20 s respectively. Then the angle is decreased to 55°, 35° and to 25° remaining constant in each interval at the time instants of 30, 35 and 40 s respectively. The throttle valve angle signal is shown in Fig. 10. The output signal, i.e., λ is shown in Fig. 11 where can appreciated that the proposed controller still performs well under parametric uncertainties and measurement noise. The engine velocity signal n and the volumetric efficiency are shown in Figs. 12 and 13 respectively where both signal have similar profiles with respect to the throttle valve angle. The control input signal, i.e., the injected fuel $\dot{m}_{fi}$ is shown in Fig. 14.

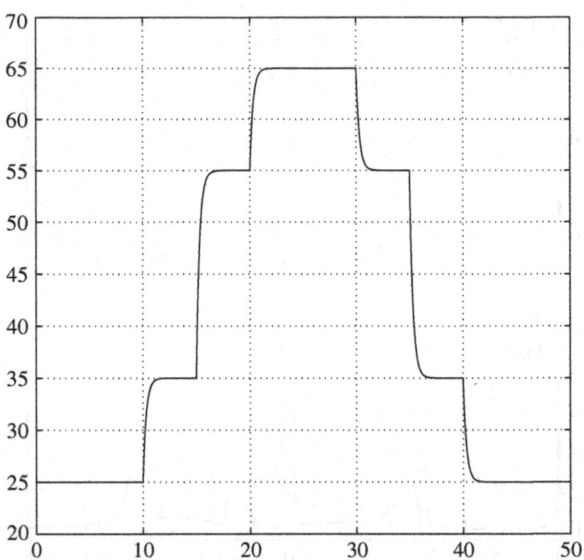

Fig. 10 Throttle valve angle [degree vs s]

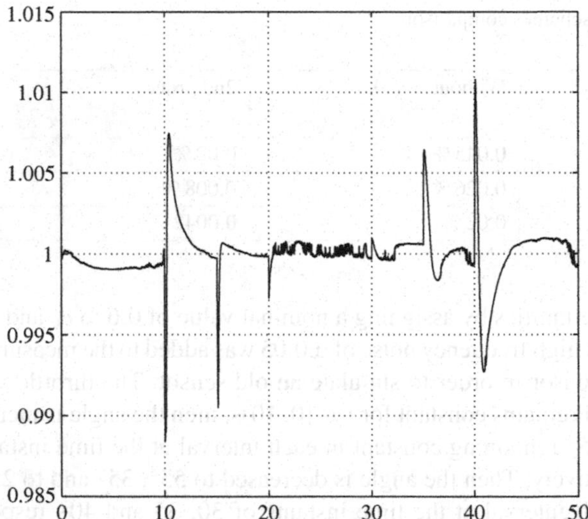

Fig. 11 λ factor in presence of perturbations [λ vs s]

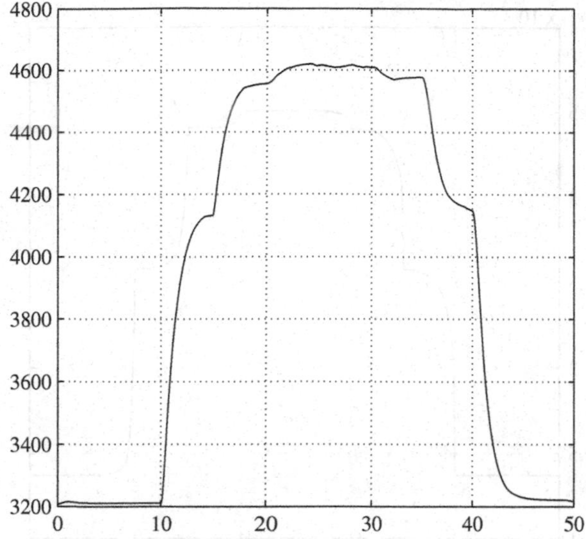

Fig. 12 Engine rotational speed [rpm vs s]

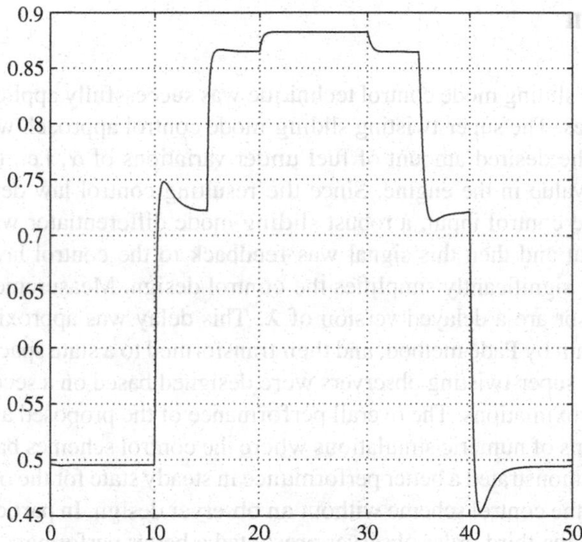

Fig. 13 Volumetric efficiency [*ev* vs s]

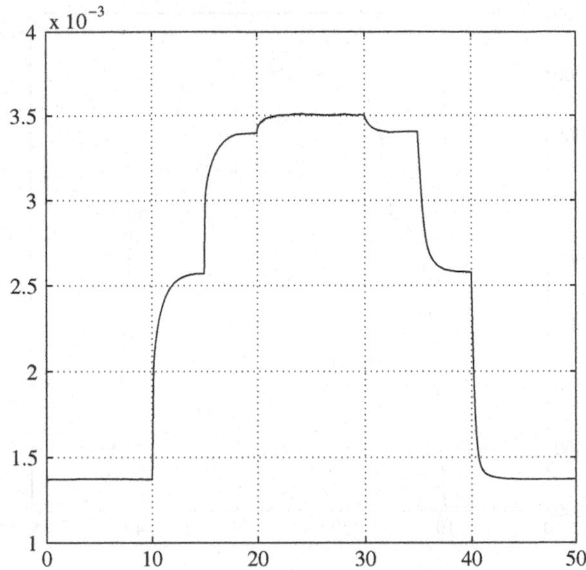

Fig. 14 Control input signal ($\dot{m}_{fi}$) [kg/s vs s]

5 Conclusion

In this work the sliding mode control technique was successfully applied to the control of SI engines. The super-twisting sliding mode control approach was applied in order to track the desired amount of fuel under variations of α, i.e., to maintain a stoichiometric value in the engine. Since the resulting control law depends on the derivative of the control input, a robust sliding mode differentiator was applied to the control input and then this signal was feedback to the control law itself. This control strategy significantly simplifies the control design. Measurements given by the UEGO sensor are a delayed version of λ. This delay was approximated in the frequency domain by Padé method, and then transformed to a state space representation, where two super-twisting observers were designed based on a second and third order Padé approximations. The overall performance of the proposed algorithm was verified by means of numeric simulations where the control schemes based observer designs have demonstrated a better performance in steady state for the output λ when compared with the control scheme without an observer design. In particular the controller based on the third order observer presented a better performance in transient responses and in fuel consumption. Therefore, the latter controller was simulated under parametric uncertainties in the volumetric efficiency and the displacement from the engine as well, moreover a high frequency noise was added to the output measurements in order to simulate an old UEGO sensor. The simulation predicts a good performance for the proposed control scheme, obtaining a smooth control input signal response ($\dot{m}_{fi}$) under adverse conditions. Some interesting issues remain to be investigated, such as the adaptation of plant parameters and/or the adaptation of controller gains.

References

Bastian, A.: Modeling fuel injection control maps using fuzzy logic. In: IEEE World Congress on Computational Intelligence (WCCI), 26–29 June 1994, Orlando, pp. 740–743 (1994). doi:10.1109/FUZZY.1994.343828

Benvenuti, L., Di Benedetto, M.D., Di Gennaro, S., Sangiovanni-Vincentelli, A.: Individual cylinder characteristic estimation for a spark injection engine. Automatica **39**(7), 1157–1169 (2003)

Castillo-Toledo, B., Lopez Cuevas, A.: Tracking through singularities using a robust differentiator. In: 6th International Conference on Electrical Engineering, Computing Science and Automatic Control (CCE), 10–13 January, Toluca, pp. 1–5 (2009). doi:10.1109/ICEEE.2009.5393480

Filippov, A.F.: Differential Equations with Discontinuous Righthand Sides. Kluwer, Boston (1988)

Fournodavlos, G., Nestoridis, V.: Generic approximation of functions by their Padé approximants. J. Math. Anal. Appl. **408**(2), 744–750 (2013)

Fridman, L., Iriarte, R.: Analysis of chattering in continuous sliding mode control. In: American Control Conference (ACC), 8–10 June, Portland, pp. 1442–1446 (2005). doi:10.1109/ACC.2005.1470332

Guzzella, L., Onder, C.H.: Introduction to Modeling and Control of Internal Combustion Engine Systems. Springer, Berlin (2010)

Hendricks, E., Sorenson, S.: Mean value modeling of spark ignition engines. SAE Technical Paper. 900616 (1990)

Hendricks, E., Vesterholm, T.: The analysis of the mean value SI engine models. SAE Technical Paper. 920682 (1992)

Hendricks, E., Chevalier, A., Jensen, M., Sorenson, C.S., Trumpy, D., Asik, J.: Modelling of the intake manifold filling dynamics. SAE Technical Paper. 960037 (1996)

Hendricks, E., Engler, E., Famm, M.: A generic mean value engine model for spark ignition engines. In: SIMS Simulation Conference, 18–19 September, Lyngby, pp. 97–108 (2000)

Hendricks, E., Luther, J.B.: Model and observer based control of internal combustion engines. In: Proceedings of the 1st Workshop on Modeling Emissions and Control in Automotive Engines (MECA), 9–20 September, Fisciano, pp. 1–12 (2001)

Khalil, H.K.: Nonlinear Systems. Macmillan, New York (2002)

Kosiba, E.A., Liu, G., Shtessel, Y.B., Zinober, A.S.I.: Output tracking via sliding modes in causal systems with time delay modeled by higher order Padé approximation. In: Proceedings of the International Workshop on Variable Structure Systems (VSS), 5–7 June, Alghero, pp. 250–255 (2006)

Levant, A.: Robust exact differentiation via sliding mode technique. Automatica **34**(3), 379–384 (1998)

Levant, A.: Higher-order sliding modes differentiation and output-feedback control. Int. J. Control **76**(9–10), 924–941 (2003)

Levant, A.: Principles of 2-sliding mode design. Automatica **43**(4), 576–586 (2007)

Levant, A.: Chattering analysis. IEEE Trans. Automat. Control **55**(6), 1380–1389 (2010)

Liu, G., Zinober, A., Shtessel, Y.B.: Second-order SM approach to SISO time-delay system output tracking. IEEE Trans. Ind. Electron. **56**(9), 3638–3645 (2009)

Muske, K.R.: A Model-based SI engine air fuel ratio controller. In: American Control Conference (ACC), 14–16 June, Minneapolis, pp. 1–6 (2006).doi:10.1109/ACC.2006.1657224

Perruquetti, W., Barbot, J.P.: Sliding Mode Control in Engineering. Marcel Dekker, New York (2002)

Probst, A., Magaña, M.E., Sawodny, O.: Using a Kalman filter and a Padé approximation to estimate random time delays in a networked feedback control system. IET Control Theory Appl. **4**(11), 2263–2272 (2009)

Pulkrabek, W.W.: Engineering Fundamentals of the Internal Combustion Engine. Pearson Prentice Hall, New Jersey (2004)

Rivera, J., Garcia, L., Mora, C., Raygoza, J.J., Ortega, S.: Super-twisting sliding mode in motion control systems. In: Bartoszewicz, A. (ed.) Sliding Mode Control, pp. 237–254. InTech, Rijeka (2011)

Tang, H., Weng, L., Dong, Z.Y., Yan, R.: Engine control design using globally linearizing control and sliding mode. T. I. Meas. Control **32**(2), 225–247 (2010)

Utkin, V.: On convergence time and disturbance rejection of super-twisting control. IEEE Trans. Automat. Control **58**(8), 2013–2017 (2013)

Utkin, V., Guldner, J., Shijun, M.: Sliding Mode Control in Electro-mechanical Systems. CRC Press, Philadelphia (1999)

Vigild, C., Andersen, K., Hendricks, E., Struwe, M.: Towards robust H-infinity control of an SI engine's air/fuel ratio. SAE Technical Paper. 1999–01-0854 (1999)

Yildiz, Y., Annaswamy, A., Yanakiev, D., Kolmanovsky, I.: Adaptive air fuel ratio control for internal combustion engines. In: American Control Conference (ACC), 11–13 June, Seattle, pp. 2058–2063 (2008). doi:10.1109/ACC.2008.4586796

Zhai, Y.J., Yu, D.L., Tafreshi, R., Al-Hamidi, Y.: Fast predictive control for air-fuel ratio of SI engines using a nonlinear internal model. Int. J. Eng. Sci. Technol. **3**(6), 1–17 (2011)

Robust Output Feedback Stabilization of a Magnetic Levitation System Using Higher Order Sliding Mode Control Strategy

Muhammad Ahsan and Attaullah Y. Memon

Abstract This work studies the problem of robust output feedback stabilization of a Magnetic Levitation System using Higher Order Sliding Mode Control (HOSMC) strategy. The traditional (first order) sliding mode control (SMC) design tool provides for a systematic approach to solving the problem of stabilization and maintaining a predefined (user specified) consistent performance of a minimum-phase nonlinear system in the face of modeling imprecision and parametric uncertainties. Recently reported variants of SMC commonly known as Higher Order Sliding Mode Control schemes have gained substantial attention since these provide for a better transient performance together with robustness properties. In this work, we focus on design of an output feedback controller that robustly stabilizes a Magnetic Levitation System with an added objective of achieving an improvement in the transient performance. The proposed control scheme incorporates a higher-order sliding mode controller (HOSMC) to solve the robust semi-global stabilization problem in presence of a class of somewhat unknown disturbances and parametric uncertainties. The state feedback control design is extended to output feedback by including a high gain observer that estimates the unmeasured states. It is shown that by suitable choice of observer gains, the output feedback controller recovers the performance of state feedback and achieves semi-global stabilization over a domain of interest. A detailed analysis of the closed-loop system is given highlighting the various factors that lead to improvement in transient performance, robustness properties and elimination of chattering. Simulation results are included and a performance comparison is given for the traditional SMC and HOSMC designs employing the first and second order sliding modes in the controller structure.

M. Ahsan (✉) · A.Y. Memon
Department of Electronics and Power Engineering, PN Engineering College,
National University of Sciences and Technology, Karachi, Pakistan
e-mail: ahsan_kh05@pnec.nust.edu.pk

A.Y. Memon
e-mail: attaullah@pnec.nust.edu.pk

A.T. Azar and Q. Zhu (eds.), *Advances and Applications in Sliding Mode Control systems*,
Studies in Computational Intelligence 576, DOI 10.1007/978-3-319-11173-5_8

1 Introduction

Output feedback control schemes have long been considered as the preferred and useful design tools for stabilization of control systems. This work focuses on design of an output feedback controller that robustly stabilizes a minimum-phase nonlinear system with an added objective of achieving an improvement in the transient performance. The proposed control scheme incorporates a Higher-Order Sliding Mode Controller (HOSMC) together with a High-Gain Observer (HGO) (Hassan 2008; Atassi and Hassan 1999) to solve the robust output feedback stabilization problem. It is usually required that the controller be able to stabilize the system over a large set of initial conditions, and assure robustness and asymptotic error convergence in presence of somewhat unknown disturbances and parametric uncertainties. Sliding Model Control (SMC) scheme is regarded as one of the most significant control design tools that addresses these requirements effectively (Guldner and Utkin 1999; Edwards and Spurgeon 1998). The variants of SMC, known as Higher Order Sliding Mode Controllers (Pukdeboon 2012; Levant 2001) provide for an improved error convergence, better robustness properties and elimination of chattering in control designs for minimum phase nonlinear systems (Rhif and Zohra 2012; Rhif 2012; Pridor Gitizadeh et al. 2000; Levant 2010).

We consider the problem of robust feedback stabilization of a Magnetic Levitation System, which is widely regarded as a benchmark system for testing various control techniques (Milica Naumovic and Boban 2008; Levine and Ponsart 1996). The system's mathematical model results in a set of coupled nonlinear differential equations which require special treatment (Woodson and Melcher 1968). Furthermore, such systems usually require use of a high gain feedback for achieving the task of stabilization and tracking of the system's output to some desired references, making the control synthesis relatively difficult.

The novelty of this work lies in the application of an HOSMC based Output Feedback Controller which uses an HGO for estimation of system states. This provides us with the leverage of the robust control and control of the convergence speed of the system states. The rest of the chapter is organized as follows: we start with a mathematical description of the magnetic levitation system and formulate the stabilization problem for this system. The following section summarizes some previous work related to the same problem. In Sect. 3, we present control design, first utilizing a first-order SMC, and then incorporating a second-order and third-order SMC structures. In the later part of this section, we extend the state feedback design to output feedback using an HGO. We present performance analysis and simulation results of the proposed control designs in Sect. 4. Finally, Sect. 5 draws the conclusions.

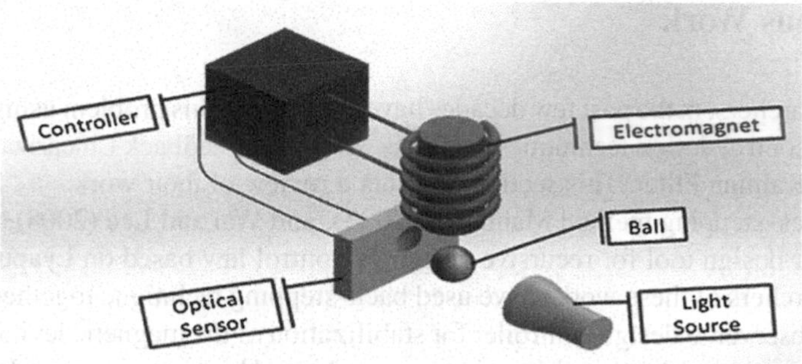

Fig. 1 A 3D view of magnetic levitation system

1.1 The Magnetic Levitation System

We undertake the problem of robust feedback stabilization of a benchmark nonlinear Magnetic Levitation System. A schematic of the system is shown in Fig. 1, where a ferromagnetic ball is required to be precisely levitated using a current controlled electromagnet with position feedback from an optical sensor.

The system is described by the following nonlinear differential equations:

$$\dot{x} = f(x) + g(x)u \tag{1}$$

where

$$x = \begin{bmatrix} x_1 \\ x_2 \\ x_3 \end{bmatrix}, f(x) = \begin{bmatrix} x_2 \\ g - \frac{k}{m}x_2 - \frac{L_o a x_3^2}{2m(a+x_1)^2} \\ \frac{1}{L(x_1)}\left[-Rx_3 + \frac{L_o a x_2 x_3}{(a+x_1)^2}\right] \end{bmatrix}, g(x) = \begin{bmatrix} 0 \\ 0 \\ \frac{1}{L(x_1)} \end{bmatrix}$$

where the states are $x_1 = y$ (position), $x_2 = \dot{y}$ (velocity), $x_3 = i$ (current) and $u = v$ (control input). Other parameters include m as the ball mass, y the measured position, g being the gravitational acceleration coefficient, k as the viscous friction coefficient, L_1, L_0, a are positive constants referred to as the inductance parameters of electromagnet and R is the overall equivalent resistance of the current path. The term $L(x_1)$ and the steady state current value, with r as the desired reference (height) are given as:

$$L(x_1) = L_1 + \frac{L_o}{1 + \frac{x_1}{a}}; \quad I_{ss} = \frac{2mg(a+r)^2}{L_o a} \tag{2}$$

The control objective is to regulate the system output to the desired height while also stabilizing the closed-loop system in the presence of parametric uncertainties. The complex dynamical model of the system along with the requirement of robustness under physical uncertainties make the control design task even more challenging.

2 Previous Work

Many researchers, in the past few decades have considered this problem using various nonlinear control design techniques e.g. Back-stepping, Feedback Linearization, and Extended Kalman Filter. This section presents a review of their work.

The Back-stepping method Mahmoud (2003), and Wai and Lee (2008), provides a nonlinear design tool for recursive design of control law based on Lyapunov theory. Researchers of these works have used back-stepping technique together with an adaptive observer to design controller for stabilization of the magnetic levitation system. Stabilization of the closed-loop system is achieved by incorporating a Lyapunov function whose derivative is rendered negative definite by the control law to achieve stability. In the proposed adaptive control method, a filter mechanism is incorporated with the back-stepping controller to cope with the problem of the finite escape time terms occurring due to repeated differentiations in back-stepping design procedure. Moreover, the observer is designed in such a way to cater for system uncertainties, to solve the trouble of chattering phenomena caused by the sign function in back-stepping and adaptive controller law. The results show that the parameter estimation error converges only locally using Lyapunov methods and to ensure stability of the overall closed-loop system the Lyapunov function is extended with a term penalizing the estimation error. This work shows that the stability was not global because the parameter estimation for control coefficients show to be only locally convergent.

Trumper et al. (1997) used feedback linearization technique to design a suitable controller that stabilizes the system at a desired operating point. The researcher suggests that for applications where large excursions or disturbance forces are not anticipated, a simple linear controller based on a linearized plant model may suffice. This model is derived by writing the states and inputs in terms of operating point, the operating points of the state variables are chosen and evaluating Jacobians at the operating point to get the linearized second-order magnetic suspension system. A major setback of this method is that the model is valid only for small perturbations about the operating point and as the system moves away, quality of this approximation decreases and the performance degrades. The proposed method shows remarkable performance for the single DOF system described described in this work, however only locally since it uses a linearized (i.e. requires accurate) plant model and any modeling errors in actuator input lead to sustained oscillations.

Another way to approach the problem is described by Levis (2003), by designing an ideal LQR controller and then extending the design towards robust control using the Lyapunov redesign method. The system is first represented in a simpler form using a transform and the feedback loop is completed by a standard LQR controller, designed by solving the Riccati equation, without taking uncertainties into account. Using Lyapunov analysis it is shown that the controller is able to stabilize the system but the result is only local as certain limits have to be put on the current input and no variations from the nominal model are allowed. To cater this, a robust controller is designed based on lyapunov redesign by adding an extra term to the linear controller to overcome matched disturbances. An upper bound on the disturbance term is taken and using the augmented controller a Lyapunov analysis is performed. The control input is taken to render the Lyapunov function negative definite using on a smooth switching controller based on Lyapunov redesign. The controller gain is taken greater than the bound on overall system subject to disturbances, to ensure the convergence of steady state error to zero. A comparison with the linear controller depicts that the later strategy is able to handle small parametric variations while resulting in semi-global asymptotic stability of the closed-loop system.

Henley John (2007) proposed a stabilizing controller structure based on feedback linearization in an Extended Kalman Filter (EKF) framework for a single-axis magnetic levitation device. For implementation of the controller, a discrete Extended Kalman Filter provides the system states' estimates. The EKF is based on the standard predict-correct format where the current state estimate and covariance are propagated forward until the next measurement occurs. Then, the Kalman Gain is computed and the state estimate and covariance are updated using appropriate initial conditions on object velocity and input current. The process noise and the sensor noise is taken as a zero mean Gaussian white-noise. The key feature of this method is that the Kalman Filter gain is chosen such that it minimizes the state estimation error. Then using the standard feedback linearizing method a state feedback controller, based on estimated states form the EKF, is used to stabilize the system using the pole placement method. Although this controller formulation is near optimal, it is robust enough that parameter changes and un-modeled plant dynamics do not effect the results.

3 Control Design

This section presents the development of robust stabilizing control for the problem under consideration, by utilizing a first-order SMC initially, and then incorporating a second-order and third-order SMC structures. In the later part of this section, we extend the SMC and HOSMC based state feedback design to output feedback using an HGO.

In order to proceed with systematic control design, we first transform the system into strict feedback normal form by using a suitable state transformation of the form

$$z = T(x) \tag{3}$$

in which T is such that T is invertible; i.e. it must have an inverse map $T^{-1}(.)$ such that $x = T^{-1}(z)$ for all $z \in T(D)$, where D is the domain of T. From (Hassan 2002), the system (1) can be represented in feedback linearizable form if and only if there is a domain $D_o \subset D$ such that:

1. For the system (1), the matrix $G(x) = [g(x) \;\; ad_f g(x) \;\; ad_f^2 g(x)]$ is full rank for all $x \in D_o$
2. The distribution $D = \text{span} g(x), ad_f g(x)$ is involutive in D_o

It can be verified that $G(x) = [g(x) \quad ad_f g(x) \quad ad_f^2 g(x)]$ has rank 3, and the vector represented as $D = \text{span}\{g(x),\ ad_f g(x)\}$ is involutive, because $[g,\ ad_f g]$ becomes a null vector and distribution D has rank 2. The foregoing calculation is valid in the domain $\{D = a + x_1 > 0 \text{ and } x_3 > 0\}$. The system has relative degree 3 (equivalent to the rank of $G(x)$), hence it is full-state linearizable. Therefore, $T(x)$ can be written as follows:

$$T(x) = \begin{bmatrix} h(x) \\ L_f h(x) \\ L_f^2 h(x) \end{bmatrix} = \begin{bmatrix} x_1 \\ x_2 \\ g - \frac{k}{m}x_2 - \frac{L_o a x_3^2}{2m(a+x_1)^2} \end{bmatrix} \tag{4}$$

Using (4) above, system (1) can now be re-written as:

$$\begin{aligned} \dot{z}_1 &= z_2 \\ \dot{z}_2 &= z_3 \\ \dot{z}_3 &= -\frac{k}{m} z_3 + \frac{L_o L_1 a x_2 x_3^2}{mL(x_1)(a+x_1)^3} + \frac{L_o a R x_3^2}{mL(x_1)(a+x_1)^2} - \frac{L_o a x_3}{mL(x_1)(a+x_1)^2} u \end{aligned} \tag{5}$$

The nominal system parameter values are given in Table 1.

3.1 First Order Sliding Mode Control

We start with development of first-order sliding mode controller for System (5). The task is to design a feedback control law to stabilize the system at a desired reference.

The controller is designed such that firstly the system trajectories reach a boundary/manifold (surface) near origin in finite time to ensure a semi-global bounded solution and once the trajectory reaches the manifold, it cannot leave it. This phase is called "reaching phase" as shown in Fig. 2.

Consider the system represented as

$$\dot{z} = f(z) + g(z)u \tag{6}$$

Table 1 Nominal system parameters

m	0.1kg
k	0.01 N/m/s
g	9.81 m/s^2
a	0.05 m
L_o	0.01 H
L_1	0.02 H
R	1 Ω

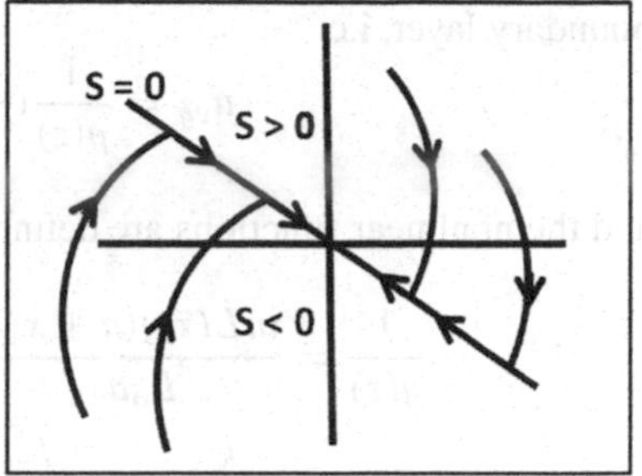

Fig. 2 A typical trajectory of sliding mode control

To start with, we define a Sliding manifold (s) in terms of the system dynamics i.e.

$$s = z_3 + a_2 z_2 + a_1(z_1 - r) = 0 \tag{7}$$

where r is the desired reference (height). The Control Law for the SMC is based on the constraint

$$s \equiv 0 \tag{8}$$

The first task is to design the controller in such a way to bring the trajectory to this manifold $s \equiv 0$ in finite time. The variable s satisfies the equation

$$\dot{s} = a_1 z_2 + a_2 z_3 + f(z) + g(z)u \tag{9}$$

let f and g satisfy the inequality

$$\left| \frac{(a_1 z_2 + a_2 z_3) + f(z)}{g(z)} \right| \leq \rho(z) \quad \forall z \in \Re^n \tag{10}$$

for some known function $\rho(z)$.

To guarantee that the trajectory reaches the manifold we take an energy Lyapunov function

$$V = \frac{1}{2}s^2 \Rightarrow \dot{V} = s\dot{s} < 0 \tag{11}$$

as the Lyapunov function candidate for the system. We get

$$\dot{V} = s\dot{s} = s(a_2 z_3 + a_1 z_2 + f(z)) + g(z)su \tag{12}$$

We take control input to be composed of two parts, i.e.

$$u = u_{eq} + \nu \tag{13}$$

where u_{eq} is taken to cancel all nonlinear part from the above equation and ν is based on the switching controller to make the controller negative definite inside the boundary layer, i.e.

$$u_{eq} = \frac{1}{g(z)}(-f(z) - a_2 z_3 - a_1 z_2) \tag{14}$$

and the nonlinear functions are defined as

$$\frac{1}{g(z)} = \frac{mL(x_1)(a + x_1)^2}{L_o a x_3} \tag{15}$$

$$f(z) = -\frac{k}{m} z_3 + \frac{L_o L_1 a x_2 x_3^2}{mL(x_1)(a + x_1)^3} + \frac{L_o a R x_3^2}{mL(x_1)(a + x_1)^2} \tag{16}$$

We take

$$v = -\gamma sat\left(\frac{s}{\epsilon}\right) \tag{17}$$

where γ is a positive class K function such that $\gamma \geq \rho(z) + \beta_o, \beta_o > 0$, and *sat* is the nonlinear *saturation* function. Under nominal system parameters the gain γ is chosen by using

$$\left|\frac{(a_1 z_2 + a_2 z_3) + f(z)}{g(z)}\right| \leq \frac{((z_2 + z_3)m - kz_3)L(x_1)(a + x_1)^3 + L_o a x_3^2 (x_2 L_1 + R(a + x_1))}{mL_o a x_3 L(x_1)(a + x_1)} \tag{18}$$

$$sat\left(\frac{s}{\epsilon}\right) = \begin{cases} sign(s) & \text{if } |s| > 1 \\ \left(\frac{s}{\epsilon}\right) & \text{if } |s| \leq 1 \end{cases} \tag{19}$$

$$sign(s) = \begin{cases} 1 & \text{if } |s| > 0 \\ 0 & \text{if } |s| = 0 \\ -1 & \text{if } |s| < 0 \end{cases} \tag{20}$$

Using the parametric values given in Table. 1 results in $\rho(z) \leq 6.27$. With this controller, the Lyapunov function becomes

$$\dot{V} \leq g(z)|s|\rho(z) - g(z)s(\rho + \beta_o)sat\left(\frac{s}{\epsilon}\right) \leq -g(z)\beta_o|s| \leq -g_o\beta_o|s| \quad ; |g(z)| \leq g_o$$

where $g_o > 0$. Therefore, under the influence of the controller the trajectory reaches the sliding manifold ($s = 0$) in finite time and once on the manifold it cannot leave it as $\dot{V}$ is negative definite. This motion is called the reaching phase followed by a sliding phase during which the motion is confined to the manifold. This control law $\nu = -\gamma sat\left(\frac{s}{\epsilon}\right)$ is called the continuous sliding mode control where ϵ is the maximum bound of the sliding manifold on either side of origin in the sliding phase.

To check the controller robustness in presence of parametric uncertainties, outside and inside the boundary layer, we define the system parameters within a range i.e. $0.009 < L_o \leq 0.01, 0.01 < L_1 \leq 0.02, 0.1 < m \leq 0.11, 0.9 < R \leq 1.1$. Under these conditions, let $\hat{f}(z)$ and $\hat{g}(z)$ be the nominal models of $f(z)$ and $g(z)$, respectively. Taking

$$u = -\frac{[(a_1 z_2 + a_2 z_3) + \hat{f}(z)]}{\hat{g}(z)} + \nu$$

results in

$$\dot{s} = a_1(z_2 + z_3) + \left[1 - \frac{g(z)}{\hat{g}(z)}\right] + f(z) - \frac{g(z)}{\hat{g}(z)}\hat{f}(z) + g(z)\nu = \delta(z) + g(z)\nu$$

where δ is the perturbation term which satisfies the inequality

$$\left|\frac{\delta(z)}{g(z)}\right| \leq \rho(z) \tag{21}$$

we can take

$$\nu = -\gamma sat\left(\frac{s}{\epsilon}\right) \tag{22}$$

where $\gamma \geq \rho(z) + \beta_o, \beta_o > 0$. Since ρ is an upper bound on the perturbation term, it is likely to be smaller than an upper bound on the whole function.

$$\left|\frac{(a_1 z_2 + a_2 z_3) + \hat{f}(z)}{\hat{g}(z)}\right| \leq \frac{((z_2 + z_3)\hat{m} - k z_3)\hat{L}(x_1)(a + x_1)^3 + \hat{L}_o a x_3^2 (x_2 \hat{L}_1 + \hat{R}(a + x_1))}{\hat{m}\hat{L}_o a x_3 \hat{L}(x_1)(a + x_1)} \tag{23}$$

Taking the parametric values as the upper bound on limits, we get $\rho(z) \leq 7.27$. To analyze the performance of this continuous sliding mode controller in the reaching phase, we take a Lyapunov function $V = \frac{1}{2}s^2$ whose derivative satisfies the inequality

$$\dot{V} \leq -g_o \beta_o |s|$$

when $|s| \geq \epsilon$ outside the boundary layer $\{|s| \leq \epsilon\}$. So until reaching the boundary layer in finite time, $|s(t)|$ will be strictly decreasing and remains inside this set afterwards. Inside the boundary layer, we have

$$z_2 = -a_1(z_1 - r) - z_3 + s$$

where $|s| \leq \epsilon$. The derivative of $V_1 = \frac{1}{2}z_1^2$ satisfies

$$\dot{V}_1 = -a_1(z_1^2 - z_1 r) - z_1 z_3 + z_1 s \leq -a_1 z_1^2 - z_1 z_3 + z_1 \epsilon \leq -(1-\theta) a_1 z_1^2 \quad \forall z_1 \geq \frac{\epsilon}{a_1 \theta}$$

where $0 < \theta < 1$. Thus the trajectory reaches the set $\Omega_\epsilon = \{|z_1| \leq \frac{\epsilon}{a_1\theta}, |s| \leq \epsilon\}$ in finite time. So we get ultimate boundedness with an ultimate bound that can be reduced by decreasing ϵ. Inside the boundary layer $|s| \leq \epsilon$ the control reduces to the linear feedback law $u = -\gamma\left(\frac{s}{\epsilon}\right)$ and the closed loop system can be stabilized by suitable choice of gain γ, to be large enough to overcome the bound ρ. Inside the boundary layer, the closed loop system given as

$$\begin{aligned}
\dot{z_1} &= z_2 \\
\dot{z_2} &= z_3 \\
\dot{z_3} &= f(z) - g(z)\left(\gamma \frac{s}{\epsilon}\right)
\end{aligned}$$

has a unique equilibrium point at $(\bar{x}_1, 0, I_{ss})$, where $\bar{x}_1$ satisfies the equation

$$\dot{z_3} = -kmz_3 L_o a\epsilon + RL_o a\epsilon x_3^2 + L_o a \gamma s I_{ss} - kmz_3 \epsilon L_1(a + \bar{x}_1)$$

and for small ϵ can be approximated by

$$\bar{x}_1 \approx \frac{L_o a}{kL_1}\left(I_{ss} + \frac{\gamma s}{\epsilon}\right)$$

introducing a change of variables to shift to origin results in,

$$\begin{aligned}
y_1 &= z_1 - \bar{z}_1 \qquad \dot{y}_1 = y_2 \\
y_2 &= z_2 \qquad\quad\;\; \dot{y}_2 = y_3 \\
y_3 &= z_3 \\
\dot{y}_3 &= -\frac{k}{m} y_3 + \frac{L_o a R x_3^2}{mL(\bar{x}_1)(a + \bar{x}_1)^2} + \frac{L_o a \gamma x_3 (z_3 + a_2 z_2 + a_1(z_1 - r))}{m\epsilon L(\bar{x}_1)(a + \bar{x}_1)^2} \\
&\approx -\left(\frac{k}{m} y_3 - \frac{L_o a I_{ss} \gamma}{m\epsilon L_1(a + y_1 + \bar{z}_1)^2}\right) y_3^2 - \sigma(y_1)
\end{aligned}$$

where

$$\sigma(y_1) = -\frac{I_{ss} L_o a \gamma y_1 x_3}{m\epsilon L_1(a + y_1 + \bar{z}_1)^2}$$

Consider the Lyapunov function

$$\tilde{V} = \int_o^{y_1} \sigma(s) d(s) - \frac{1}{2} y_3^2$$

where $\tilde{V}$ is positive definite ($|\tilde{V}|$ is radially unbounded) for $y_3 > \frac{L_o a I_{ss} \gamma z_3}{m L_1 \epsilon}$ and its derivative satisfies

$$\dot{\tilde{V}} = \sigma(y_1) + y_3 \dot{y}_3 \leq \frac{L_o a I_{ss} \gamma}{m L_1 \epsilon} y_3^2 < 0$$

Using LaSalle's Invariance Principle we can show that the equilibrium point $(\bar{x}_1, 0, I_{ss})$ is asymptotically stable and attracts every trajectory in Ω_ϵ. For better accuracy, we choose ϵ as small as possible, however we should keep in mind that choosing too small a value may result in chattering. With a suitable choice of ϵ close to zero, the controller achieves ultimate boundedness as all trajectories starting off the manifold $|s| \leq \epsilon$ reach it in finite time and stay there onwards. By suitable choice of $\epsilon \to 0$ and a high enough controller gain γ the proposed controller yields semi-global asymptotic stabilization.

3.2 Higher Order Sliding Control

We now focus our attention to development of stabilizing controllers for the system under consideration that use Higher Order Sliding Modes. This approach has gained substantial attention recently due to its ability to yield in better transient performance, superior robustness properties and removal of chattering when compared to a first-order SMC. The formulation of controller (Korovin and Emeryanov 1996) is as follows:

Consider an uncertain single-input nonlinear system

$$\dot{x} = f(x, t, u), \quad s = s(t, x) \quad t \geq 0 \tag{24}$$

with $x \in X \subseteq \Re^n$ as the state vector, $u \in U \subset \Re$ being the control input and the time varying non-linear function $f(x, t, u) : [0, +\infty) \times \Re^n \times U \to \Re^n$ is a sufficiently smooth uncertain vector field and $s(x, t) : [0, +\infty) \to \Re$ is the function as defined in (7). The relative degree r of the system is defined such that u explicitly appears in only the rth derivative of s and $\frac{d}{du} s^r \neq 0$ at the given point. The task is to achieve the constraint $s \equiv 0$ in finite time and stay there using a discontinuous feedback control. Since $s, \dot{s}, \ddot{s}, \ldots, s^{r-1}$ are continuous functions, the corresponding motion corresponds to an *r-sliding* mode (Levant 2001).

The term Higher Order Sliding Mode specifies a movement on the discontinuity set of the dynamic system in Filippov's sense (i.e. it consists of Filippov's trajectories of the discontinuous dynamic system) (Levant 1999). The controller Sliding Order indicates the dynamic smoothness degree in the vicinity of the mode i.e. it is a number of total continuous derivatives of the manifold (s) (including s^0) in the vicinity of sliding mode. Therefore the rth order sliding mode is determined by the following equalities:

$$s = \dot{s} = \ddot{s} = \cdots s^{r-1} = 0 \quad ; \quad 0 < K_m \leq \frac{\partial}{\partial u} s^r \leq K_M \tag{25}$$

for some positive constants K_m and K_M. This forms an r-dimensional condition on the state of the dynamic system and any motion satisfying (25) is called an *r-sliding* mode with respect to the essential constraint $s \equiv 0$ (Fridman and Levant 2002).

3.2.1 Two Sliding Controller

We start with defining the sliding variable s as the regulated output of the system (5). The second order sliding mode approach provides for the finite time stabilization of the output s and its time derivative $\dot{s}$ by characterizing a discontinuous control input (u) for the system (Perruquetti 2010).

Considering $y_1 = s$, it can been shown that, the second order sliding mode problem is equivalent to the finite time stabilization problem for the following uncertain second order system:

$$\begin{cases} \dot{y_1} &= y_2 \\ \dot{y_2} &= \varphi(t, y) + \gamma(t, y)u \end{cases} \tag{26}$$

where it is considered that only the information about sign of y_2 is available (Fridman and Levant 2002). The nominal functions $\varphi(t, y)$ and $\gamma(t, y)$ are defined as:

$$\begin{cases} |\varphi(t, y)| < \Phi \quad ; \Phi > 0 \\ 0 < \Gamma_m < \gamma < \Gamma_M < 1 \end{cases} \tag{27}$$

$\forall y \in Y \subseteq \Re^2$, such that the system (26) is bounded and stable.

3.2.2 Twisting Algorithm

The Twisting Algorithm is the basic 2-sliding controller (Punta 2006). This algorithm features the twisting of sliding trajectory infinite times around the origin of the 2-sliding plane $y_1 O y_2$. The method is called 'Twisting Controller' because the trajectories perform an infinite number of rotations while converging to the origin along with the vibration magnitudes decays along the axes and the rotation times decreasing in geometric progression (Levant 1999).

The controller, based on the constraint ($s = \dot{s} = 0$), is able to stabilize the dynamic system while achieving semi-global asymptotic output regulation. The control algorithm is defined by the following control law (Floquet and Barbot 2007) in which the condition on $|u|$ provides for $|u| \leq 1$:

$$\dot{u}(t) = \begin{cases} -u & ; |u| > 1 \\ -V_m sign(y_1) & ; (y_1)(y_2) < 0 \text{ and } |u| \leq 1 \\ -V_M sign(y_1) & ; (y_1)(y_2) > 0 \text{ and } |u| \leq 1 \end{cases} \tag{28}$$

The corresponding sufficient conditions to ensure the elimination of reaching phase and finite time convergence to the sliding manifold are:

$$\begin{cases} V_M > V_m & ; V_m > 4\frac{\Gamma_m}{s_o} \\ V_M > \frac{2\Phi + V_M}{\Gamma_M} & ; V_m > \frac{\Phi}{\Gamma_M} \end{cases} \tag{29}$$

where s_o is the max allowed value for manifold s. The results for the controller are discussed in Analysis and Results section and it is shown that this controller results in much better transient phase response as compared to the First Order SMC but due to large relative degree of the system chattering is not completely removed when used to control the nonlinear system.

3.2.3 Three Sliding Controller

For the Magnetic Levitation System with relative degree $\rho = 3$, the 2-sliding controller described above does not completely eliminate chattering. As mentioned in Levant (2010), Korovin and Emeryanov (1996), the main drawbacks of the previously described methods are that when the relative degree ρ of the control variable s is higher than one, the control methods, to completely remove chattering, generally require the knowledge of up to $(\rho - 1)$ derivatives of s. For systems with $\rho = 3$, the usually unavailable quantities $\dot{s}$ and $\ddot{s}$ need to be measured or estimated using an observer (e.g. High-Gain Observer, sliding differentiator) for controller design that completely removes chattering. The 2-sliding controller when applied to a higher relative degree system does not eliminate chattering.

For systems with relative degree higher than 2, the recommended practice is to use a 3-sliding controller (3rd order SMC) to completely eliminate chattering under the constraints described in (29) (Levant 2010). The 3-sliding controller is designed as follows:

Let p be a positive number. Denoting

$$\begin{aligned}
J_{1,r} &= |s|^{(r-1)/r} \\
J_{i,r} &= \left(|s|^{p/r} + |\dot{s}|^{p/(r-1)} + \cdots + |s^{(i-1)}|^{p/(r-i+1)}\right)^{(r-i)/p}, \quad i = 1, \ldots, r-1 \\
J_{r-1,r} &= \left(|s|^{p/r} + |\dot{s}|^{p/(r-1)} + \cdots + |s^{(r-2)}|^{p/2}\right)^{1/p} \\
\psi_{0,r} &= s \\
\psi_{1,r} &= \dot{s} + \beta_1 J_{1,r} sign(s) \\
\psi_{i,r} &= s^{(i)} + \beta_i J_{i,r} sign(\psi_{i-1,r}), \quad i = 1, \ldots, r-1
\end{aligned}$$

where $\beta_1, \ldots, \beta_r - 1$ are positive numbers.

3.2.4 Theorem 1

If the system (26) has relative degree r with respect to the output function s and the condition (25) on $\frac{\partial}{\partial u}s^r$ is satisfied, then with properly choosing the parameters $\beta_1, \ldots, \beta_r - 1$ the controller defined by

$$u = -\alpha sign\left[\psi_{r-1,r}(s, \dot{s}, \ldots, s^{(r-1)})\right] \quad (30)$$

assures the appearance of *r-sliding* mode $s \equiv 0$ while attracting all trajectories in finite time.

The parameters $\beta_1, \ldots, \beta_r - 1$ are chosen to be sufficiently large in the index ordering. Each choice specifies a family of controller applicable to all systems expressed as (26) with relative degree r. The parameter $\alpha > 0$ depends on the choice of positive constants K_m and K_M. Coefficients of $J_{i,r}$ can be chosen as any positive numbers and α needs to be negative when $\frac{\partial}{\partial u}s^r < 0$.

There can be infinite many choices for β_i. A tested example for β_i for $r = 3$ is provided in Fridman and Levant (2002). The 3-sliding controller is given as

$$v = -\alpha sign\left(\ddot{s} + 2(|\dot{s}|^3 + s^2)^{\frac{1}{6}} sign(\dot{s} + |s|^{\frac{2}{3}} sign(s))\right) \quad (31)$$

The idea is that a *1-sliding* mode is established on the smooth parts of the discontinuity set Λ of (31) described by the differential equation $\psi_{r-1,r} = 0$. The resulting movement takes place in some close boundary of the Λ satisfying $\psi_{r-2,r} = 0$, transfers in finite time into some vicinity of the subset satisfying $\psi_{r-3,r} = 0$ and so on. While the trajectory reaches the *r-sliding* set, set Λ shrinks to origin in the coordinates $s, \dot{s}, \ldots, s^{(r-1)}$ (Levant 2012).

This controller placed in (13) makes the overall control input for the system. The parameter α is a positive constant i.e. $\alpha > 0$. For our system we take $\alpha = 20$ with tolerance $\tau = 10^{-3}$ and Euler?s method for integration. The overall 3-sliding controller for the system becomes:

$$u = -\frac{mL(x_1)(a + x_1)^2}{L_o a x_3}\left(\frac{k}{m}z_3 - \frac{L_o L_1 a x_2 x_3^2}{mL(x_1)(a + x_1)^3} - \frac{L_o a R x_3^2}{mL(x_1)(a + x_1)^2} - a_2 z_3 \right.$$
$$\left. -a_1 z_2 - \alpha sign\left(\ddot{s} + 2(|\dot{s}|^3 + s^2)^{\frac{1}{6}} sign(\dot{s} + |s|^{\frac{2}{3}} sign(s))\right)\right) \quad (32)$$

A maximum of rth order accuracy is attainable with the above mentioned 3-sliding controller and with proper choice of parameters $\beta_1, \ldots, \beta_r - 1$ the convergence time is reduced approximately $\kappa(\alpha)$ times where $0 < \kappa \leq 1$.

To analyze the controller performance for reaching phase (to guarantee that the trajectory reaches the manifold in finite time) we follow a similar procedure as for First Order Sliding Controller. Considering a Lyapunov function candidate:

$$V = \frac{1}{2}s^2 \Rightarrow \dot{V} = s\dot{s}$$

Under the influence of the above mentioned 3-sliding controller, we get

$$\dot{V} \leq -\alpha|s||\nu| \leq -\alpha\lambda|s| \qquad ;\alpha > 0$$

where λ is a positive class K function such that $\lambda = \mu+\omega$, $\omega > 0$ and $|\nu| \leq \mu$, $\mu > 0$ is calculated as follows: replacing the $sign$ function by its approximation i.e.

$$sign(s) = \frac{|s|}{s}$$

results in

$$|\nu| = \frac{\left|\ddot{s} + 2\left(|\dot{s}|^3 + s^2\right)^{\frac{1}{6}} \frac{\left|\dot{s}+|s|^{\frac{2}{3}}\frac{|s|}{s}\right|}{\left[\dot{s}+|s|^{\frac{2}{3}}\frac{|s|}{s}\right]}\right|}{\ddot{s} + 2\left(|\dot{s}|^3 + s^2\right)^{\frac{1}{6}} \frac{\left|\dot{s}+|s|^{\frac{2}{3}}\frac{|s|}{s}\right|}{\left[\dot{s}+|s|^{\frac{2}{3}}\frac{|s|}{s}\right]}}$$

using the Triangle Inequity, Preservation of division and Idempotence properties of absolute numbers we get

$$|\nu| \leq \frac{\left|\ddot{s} + 2\left(|\dot{s}|^3 + s^2\right)^{\frac{1}{6}} \frac{\left|\dot{s}+|s|^{\frac{2}{3}}\right|}{\left|\dot{s}+|s|^{\frac{2}{3}}\right|}\right|}{\ddot{s} + 2\left(|\dot{s}|^3 + s^2\right)^{\frac{1}{6}} \frac{\left|\dot{s}+|s|^{\frac{2}{3}}\right|}{\dot{s}+|s|^{\frac{2}{3}}}} \tag{33}$$

we know that $\left(|\dot{s}|^3 + s^2\right)^{\frac{1}{6}} > 0$, using Triangle Inequity and further solving the inequity, we get

$$|\nu| \leq |\ddot{s}| + \left|\left(\dot{s} + s^2\right)^{\frac{1}{6}}\right| \leq |\ddot{s}| \leq W \qquad ; W > 0$$

which makes the Lyapunov function derivative negative definite, i.e.

$$\dot{V} < -\alpha W|s| \qquad ; \alpha > 0 \text{ and } W > 0$$

Therefore, under the influence of the controller the trajectory reaches the sliding manifold ($s = 0$) in finite time and once on the manifold it cannot leave it as $\dot{V}$ is negative definite. So until reaching the boundary layer in finite time, $|s(t)|$ is strictly decreasing and remains inside this set afterwards.

Inside the boundary layer (inside the set Ω_ϵ) a similar analysis can be carried out as in first order sliding mode controller (applying a change of variables and using Invariance Principle) to show that the application of the controller results in an asymptotically stable origin. The results for the controller are discussed in Analysis Section where it is shown that the 3-sliding controller results in much improved performance as compared to the First and Second Order SMC.

3.3 *Output Feedback*

In this section, we extend the state feedback design to output feedback by using a High Gain Observer (HGO) (Esfandiari and Hassan 1992; Atassi and Hassan 1999). Towards that end, we consider the observer as given by the following set of equations:

$$\begin{cases} \dot{\hat{\xi}}_1 &= \hat{\xi}_2 + h_1(y - \hat{\xi}_1) \\ \dot{\hat{\xi}}_2 &= h_2(y - \hat{\xi}_1) \\ \dot{\hat{\xi}}_3 &= h_3(y - \hat{\xi}_1) \end{cases} \tag{34}$$

in which the observer gains are chosen as follows:

$$\begin{bmatrix} h_1 \\ h_2 \\ h_3 \end{bmatrix} = \begin{bmatrix} 2/\epsilon \\ 1/\epsilon^2 \\ 1/\epsilon^3 \end{bmatrix} \tag{35}$$

where ϵ is a design parameter. It is well established that by incorporating an HGO, one can recover the performance of the state feedback controller by a suitable choice of observer gains. This is achieved by choosing the design parameter ϵ sufficiently small which renders the estimation error $(\hat{\xi} - \xi)$ to zero as ϵ approaches zero. However, this process results in a large overshoot for a very limited time in the initial transient phase before the estimation error sharply decays to zero. This overshooting phenomenon is called *peaking* and is usually overcome by saturating the observer for a very brief initial interval during operation.

The output feedback controller incorporating the HGO (34) for 3-sliding feedback controller is given as:

$$\begin{aligned} u = -\frac{mL(x_1)(a+x_1)^2}{L_o a x_3}\Bigg(\frac{k}{m}\hat{z}_3 - \frac{L_o L_1 a x_2 x_3^2}{mL(x_1)(a+x_1)^3} - \frac{L_o a R x_3^2}{mL(x_1)(a+x_1)^2} - a_2\hat{z}_3 \\ -a_1\hat{z}_2 - \alpha sign\Big(\ddot{\hat{s}} + 2\big(|\dot{\hat{s}}|^3 + \hat{s}^2\big)^{\frac{1}{6}} sign\big(\dot{\hat{s}} + |\hat{s}|^{\frac{2}{3}} sign(\hat{s})\big)\Big)\Bigg) \end{aligned} \tag{36}$$

This output feedback controller is applied to the original nonlinear system represented in strict feedback normal form. The inclusion of HGO recovers the performance of the full state feedback controller and, by suitable choice of gains, allows the output feedback to achieve semi-global asymptotic stabilization over a domain of interest.

3.3.1 Theorem 2

Consider the closed loop system comprising of the plant (5) and the output feedback controller (36). Suppose the origin of the closed loop system under state feedback control (32) is asymptotically stable and $\mathcal{R}$ is its region of attraction. Let $\mathcal{S}$ be any compact subset in the interior of $\mathcal{R}$ and $\mathcal{Q}$ be any compact subset of $\mathcal{R}^{\rho}$. Then

- There exists $\epsilon_1^* > 0$ such that for every $0 < \varepsilon \leq \epsilon_1^*$, the solutions of the closed loop system (under state feedback $X(t)$ and under output feedback $(\hat{x}(t))$), starting in $\mathcal{S} \times \mathcal{Q}$, are bounded for all $t > 0$.
- Given any $\mu > 0$, there exists $\epsilon_2^* > 0$ and $T_2 > 0$, both dependent on μ, such that, for every for every $0 < \varepsilon \leq \epsilon_2^*$, the solutions of the closed loop system, starting in $\mathcal{S} \times \mathcal{Q}$, satisfy

$$||X(t)|| \leq \mu \quad ||\hat{x}(t)|| \leq \mu \quad \forall\, t \geq T_2$$

- Given any $\mu > 0$, there exists $\epsilon_3^* > 0$, dependent on μ, such that, for every for every $0 < \varepsilon \leq \epsilon_3^*$, the solutions of the closed loop system, starting in $\mathcal{S} \times \mathcal{Q}$, satisfy

$$||X(t) - X_r(t)|| \leq \mu \quad \forall\, t \geq 0$$

  where X_r is the solution of system under (32) starting at $X(0)$.
- If the origin of system under (32) is exponentially stable and that $f(z)$ is continuously differentiable in some neighborhood of $X = 0$, then there exists $\epsilon_4^* > 0$ such that, for every $0 < \varepsilon \leq \epsilon_4^*$, the origin of the closed loop system is exponentially stable and $\mathcal{S} \times \mathcal{Q}$ is a subset of its region of attraction.

3.3.2 Proof

The proof follows the general outline as given in [Hassan (2002), Sect. 14.5.2] with appropriate modifications as per the problem under consideration. In particular, proof of the theorem establishes that the output feedback controller recovers the performance of the state feedback controller for sufficiently small ϵ. The performance recovery is evident in itself in three points. Firstly, recovery of exponential stability. Second, recovery of region of attraction in the sense that we can recover any compact set in its interior. Third, the solution $X(t)$ under output feedback reaches the solution under state feedback as ϵ tends to zero.

Remark 1 It is well known that if the state feedback controller achieves the state feedback controller achieves global or semi-global asymptotic stabilization with local exponential stability, then for sufficiently small ϵ, the output feedback controller achieves semi-global stabilization with local exponential stability.

4 Performance Analysis

In this section a detailed analysis of the First Order Sliding Mode Controller and the Higher Order Sliding Controllers is carried out and simulation results are shown to demonstrate the performance of different controllers.

4.1 First Order Sliding Mode Controller

The controller $\nu = -\gamma sat\left(\frac{s}{\epsilon}\right)$ is called the First Order Sliding Mode Control.

- *Transient Performance/Reaching Phase:* The First Order SMC behaves poorly in the reaching phase and the system trajectory exhibits large overshoots before reaching the sliding manifold as shown in Fig. 3. But the controller (Fig. 4) guarantees that the trajectory reaches the sliding manifold ($s = 0$) in finite time and once on the manifold it cannot leave it as $\dot{V}$ is negative definite.
- *Sliding Phase:* Inside the boundary layer $|s| \leq \epsilon$ the control reduces to the linear feedback law $\nu = -\gamma sat\left(\frac{s}{\epsilon}\right)$ and the closed loop system can be stabilized by suitable choice of the gain γ, to be large enough to overcome the max bound ρ of the perturbation term u_{eq} i.e. $\gamma > \rho + \beta_o, \beta_o > 0$. The parameter ϵ is a small constant i.e. $0 < \epsilon \leq 1$ defined as the maximum bound of the sliding manifold on either side of origin. So we get ultimate boundedness with an ultimate bound that

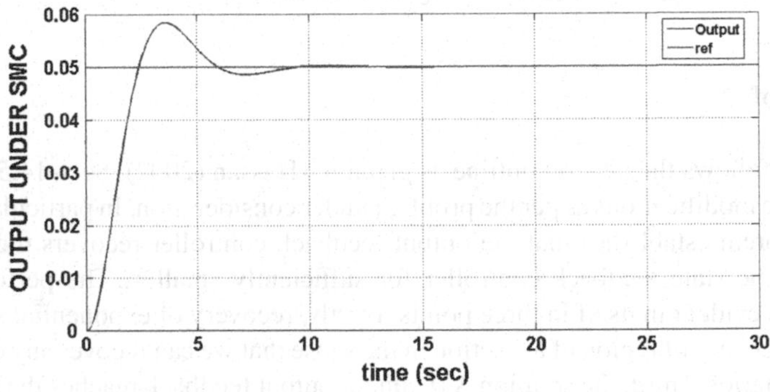

Fig. 3 First order sliding mode control results

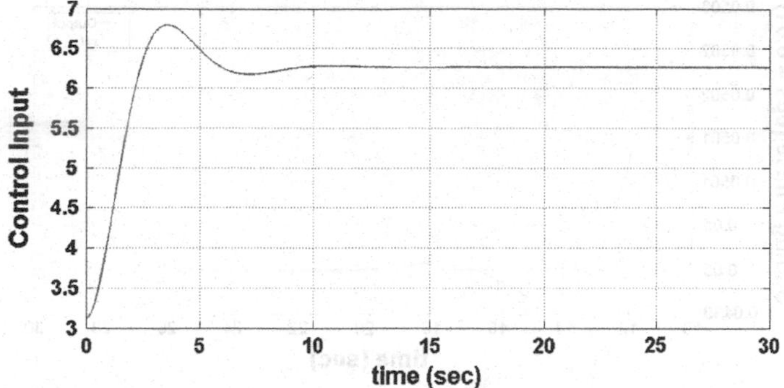

Fig. 4 Control input for sliding mode control

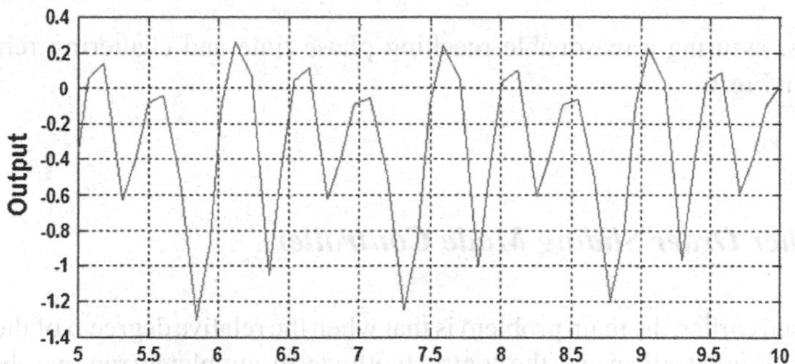

Fig. 5 Sliding phase results For $\epsilon = 0.001$(chattering)

can be reduced by decreasing ϵ i.e. $\epsilon \to 0$.

- *Chattering Analysis and Stability:* The controller is a Continuous Sliding Mode Controller using the sat approximation of the discontinuous $sign$ function i.e. $\nu = -\gamma sat\left(\frac{s}{\epsilon}\right)$ to cater the "chattering" introduced in the system due to switching delay between the sign of s, which causes unwanted oscillations in the system as shown in Fig. 5. The value of ϵ needs to be carefully selected because when ϵ is reduced to zero, the continuous sliding controller approaches a discontinuous sliding controller, i.e. as $\epsilon \to 0 \Rightarrow sat(s) \to sign(s)$ and chattering starts to appear in the system and for the system with input defined as:

$$u = \nu = Ri = i$$

it causes a continuous drawl of current from the source. So, with a proper choice of ϵ close to zero, the controller achieves ultimate boundedness as all trajectories starting off the manifold $|s| \leq \epsilon$ reach it in finite time and stay there onwards. Then by choice of a high enough controller gain γ the controller achieves semi-global asymptotic stability for the system. The simulation values for the controller $\epsilon = 0.01$ and $\gamma = 10$ are based on the same constraints discussed above for

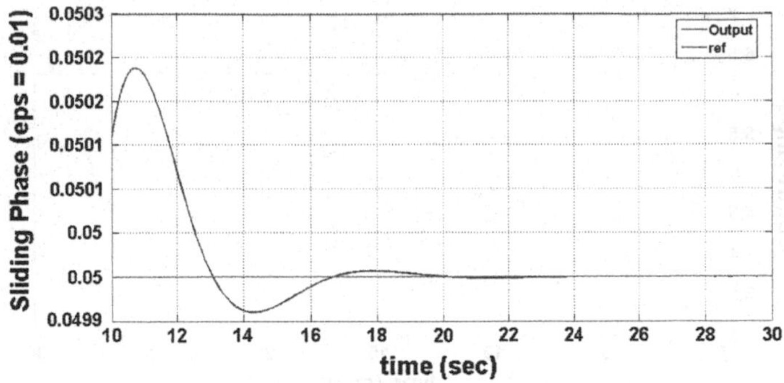

Fig. 6 Sliding phase results For $\epsilon = 0.01$(chattering removed)

γ and ϵ, assuring a reasonable reaching phase time and chattering removal as shown in Fig. 6.

4.2 Higher Order Sliding Mode Controller

As discussed earlier, the main problem is that when the relative degree ρ of the control variable (s) is greater than one, the control methods, to completely remove chattering, generally require the knowledge of up to $(\rho - 1)$ derivatives of s. For the current system with relative degree $\rho = 3$, the usually unavailable quantities $\dot{s}$ and $\ddot{s}$ need to be incorporated for a controller design that completely removes chattering. Using the 2-sliding controller (Fig. 7) for the system also does not completely eliminate chattering as shown in Fig. 8, and the recommended practice is to use a 3-sliding controller.

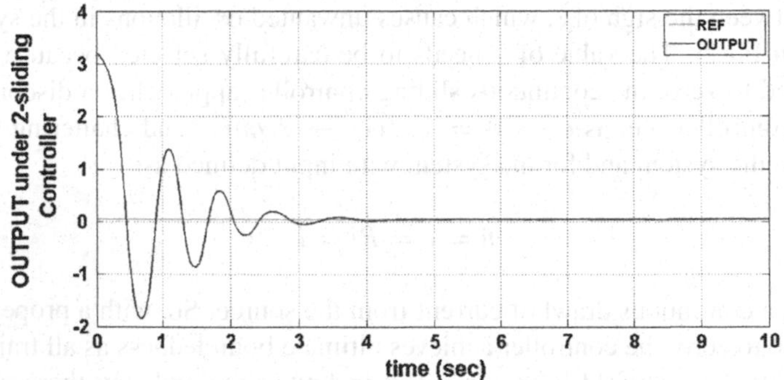

Fig. 7 Output under 2-sliding controller

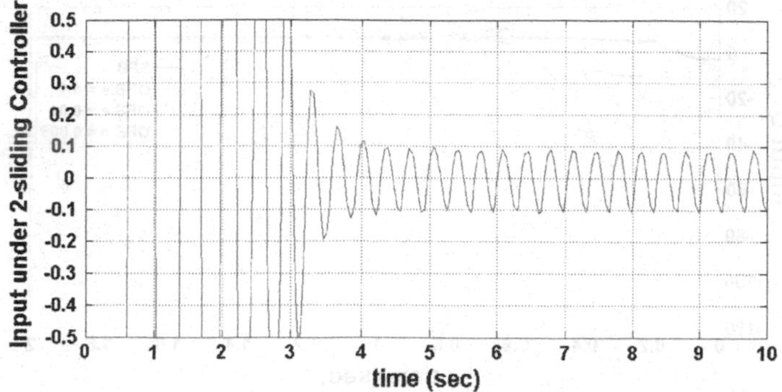

Fig. 8 Control input for 2-sliding controller

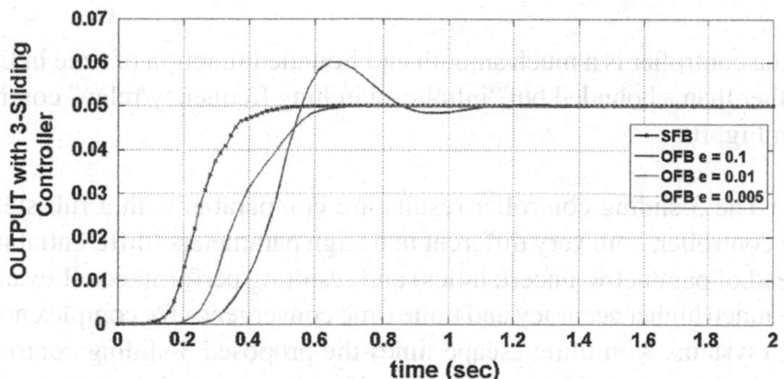

Fig. 9 Output under 3-sliding controller with varying ϵ

- *Reaching Phase Elimination Time:* Using the 2-sliding Twisting Controller and 3-sliding Controller the time required for the system trajectory to reach the sliding surface (s) is considerably reduced as compared to first order sliding mode control results (Levant 2012). With proper choice of parameters $\beta_1, \ldots, \beta_r - 1$ the convergence time is reduced approximately $\kappa(\alpha)$ times where $0 < \kappa \leq 1$ under the constraints defined in (29–30). This is evident from the simulation results that the 3-sliding controller results in much faster global finite time convergence to the origin and the overall control is bounded as shown in Fig. 9.

- *Implementation of the 3-sliding Controller:* The 3-sliding controller implementation requires the availability of the sliding manifold and the knowledge of up to $(\rho - 1)$ derivatives i.e. s, $\dot{s}$ and $\ddot{s}$ at all times. The usually unavailable quantities s, $\dot{s}$ and $\ddot{s}$ need to be incorporated for the controller design that completely removes chattering. With the introduction of these variables as auxiliary variables in the control design procedure, the controller effectively takes care of the discontinuities in the sliding variable (s) and removes the vibrations (harmonics) that may arise due to its higher derivatives, as in the case of first order sliding mode controller. As

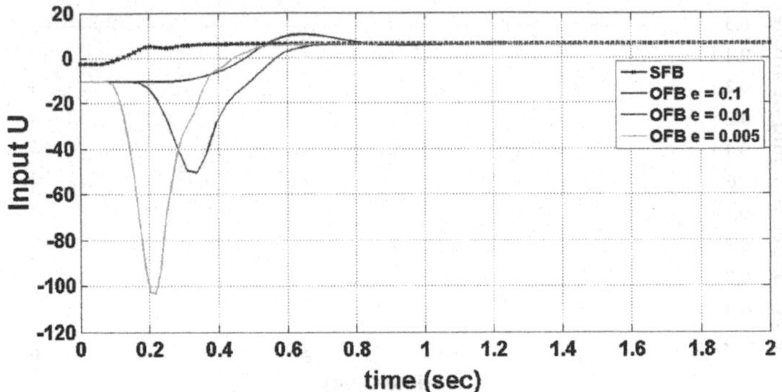

Fig. 10 Control input for 3-sliding controller

a result the controller is a much smooth and bounded function of time in Lipschitz sense rather than a bounded but "infinite switching frequency/relay" controller, as shown in Fig. 10.

- *Stability:* The 3-sliding controller results are comparable with a full-state backstepping controller, both very different in design parameters (differential inequalities instead of parametric uncertainties) and resulting performance (Levant 1999). Offering much higher accuracy and finite time convergence for complex non-linear systems, (systems with finite escape time) the proposed 3-sliding controller features a globally asymptotically stable closed loop system and in some cases locally exponentially stable systems. It is evident from Fig. 9 that the closed loop nonlinear system (5) is stabilized and output is successfully regulated to the desired reference.
- *Chattering Removal:* The above mentioned technique of including the higher derivatives of the sliding manifold in the control design procedure also removes the chattering effect from the system even under very small HGO gain ϵ values (see for $\epsilon = 0.005$ in Fig. 9). When we design the controller based on knowledge of higher derivatives of the sliding variable (s) and cater for the higher derivative terms, the unwanted oscillations (chattering) introduced in the system are considerably reduced. With the proper handling of HOSMC design constraints, chattering is completely removed and we get a local 3-sliding controller rather than the relay controller $u = -\gamma sign(s)$ while achieving a $3rd$ order sliding precision with respect to τ i.e. $O(\tau^3)$ (Levant 2010). The inclusion of HGO for estimation of unmeasured system states in the design of output feedback controller does not degrade the controller performance or stability. The sliding manifold (s) and its derivatives vanish in finite time as shown in Fig. 11.
  To show the controller efficiency and the performance recovery of the State Feedback Controller using the Output Feedback controller based on High Gain Observer, the observer convergence speed control parameter ϵ was varied from

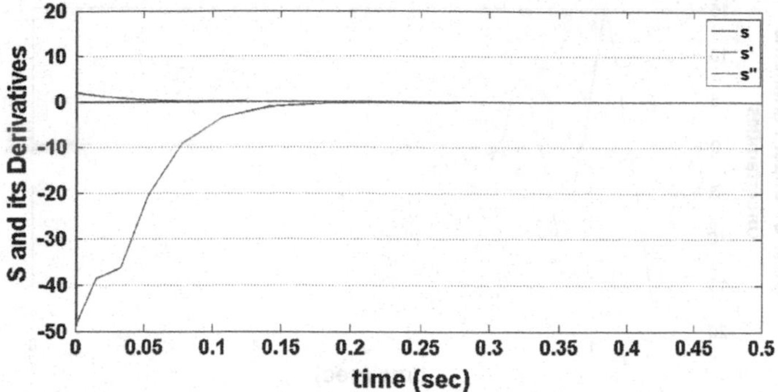

Fig. 11 Sliding manifold and higher derivatives

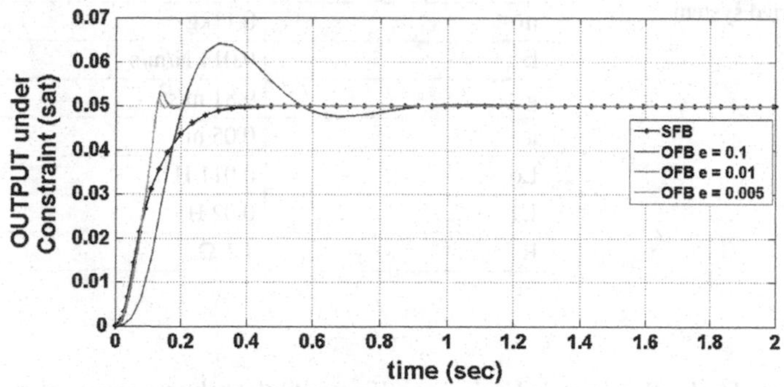

Fig. 12 Output under constraint control input using 3-sliding controller

0.1 to 0.005 to show the difference in the performance recovery. Other simulation parameters taken are: $\mu = 0.01, V_M = 50, V_m = 15, \Gamma_m = 0.5, \Gamma_M = 1, a_1 = a_2 = 1 \;\; \Phi = 2, s_o = I_{ss}$.

- *Discontinuity Regularization/Constraint Control:* The Transient phase overshoot called *peaking* occurring due to the inclusion of HGO is reduced by putting some constraint on control input (regularizing the discontinuity), as per limits, using *sat* function (Fridman and Levant 2002), with the limits $[-2.5 \;\; I_{ss}]$. Due to this, the overshoot magnitudes are considerably reduced without degrading the controller performance as shown in Fig. 12. The control input becomes:

$$u = sat\Bigg[-\frac{mL(x_1)(a+x_1)^2}{L_o a x_3}\Bigg(\frac{k}{m}\hat{z}_3 - \frac{L_o L_1 a x_2 x_3^2}{mL(x_1)(a+x_1)^3} - \frac{L_o a R x_3^2}{mL(x_1)(a+x_1)^2} \\ -a_2\hat{z}_3 - a_1\hat{z}_2 - \alpha sign\Big(\ddot{\hat{s}} + 2\big(|\dot{\hat{s}}|^3 + \hat{s}^2\big)^{\frac{1}{6}} sign\big(\dot{\hat{s}} + |\hat{s}|^{\frac{2}{3}} sign(\hat{s})\big)\Big)\Bigg)\Bigg] \quad (37)$$

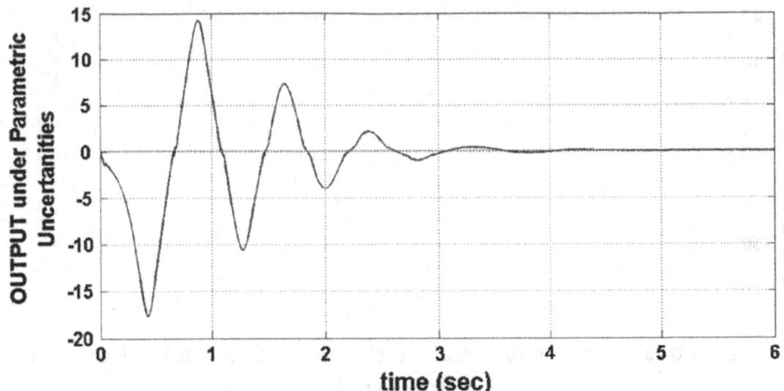

Fig. 13 Output using 3-sliding controller under parametric uncertainties

Table 2 Varied system parameters

m	0.11kg
k	0.011 N/m/s
g	9.81 m/s^2
a	0.05 m
Lo	0.011 H
L1	0.02 H
R	1.1 Ω

- *Robustness Under Parametric Variations:* To verify the robustness properties of the proposed 3-sliding controller, the system nominal parameters were perturbed by 10–20% while keeping the parameters of the controller unchanged. The controller, to stabilize the system at desired reference, has to exert some extra effort but the desired reference is achieved as shown in Fig. 13. The new control input, using nominal system parameters becomes:

$$u = sat\left(\frac{1}{\hat{g}(z)}\left(- \hat{f}(z) - a_2\hat{z}_3 - a_1\hat{z}_2 - \nu\right)\right) \tag{38}$$

The new parameters for system are given in Table 2.

5 Conclusion

We focused on the problem of robust output feedback stabilization of a Magnetic Levitation System using Higher Order Sliding Mode Control (HOSMC) strategy. The traditional (first order) sliding mode control (SMC) design tool provides for a systematic approach to solving the problem of stabilization and maintaining a predefined (user specified) consistent performance of a minimum-phase nonlinear system in the face of modeling imprecision and parametric uncertainties. Recently reported variants of SMC commonly known as Higher Order Sliding Mode Control schemes have gained substantial attention since these provide for a better transient performance together with robustness properties.

We proposed an output feedback controller that robustly stabilizes the closed-loop system with an added objective of achieving an improvement in the transient performance. The proposed control scheme incorporates a higher-order sliding mode controller (HOSMC) to solve the robust semi-global stabilization problem in presence of a class of somewhat unknown disturbances and parametric uncertainties. The state feedback control design is extended to output feedback by including a high gain observer that estimates the unmeasured states. It is shown that by suitable choice of observer gains, the output feedback controller recovers the performance of state feedback and achieves semi-global stabilization over a domain of interest. A detailed analysis of the closed-loop system was given highlighting the various factors that lead to improvement in transient performance, robustness properties and elimination of chattering. Simulation results were included and a performance comparison was given for the traditional SMC and HOSMC designs employing the first, second and third order sliding modes in the controller structure.

A detailed performance analysis showed that the first order SMC was able to stabilize the system at the desired reference point. However, the transient performance of the same was degraded and showed large overshoot, and a slower reaching phase when compared to that of the second-order and third-order SMC, which showed superior transient performance, along with better robustness properties and removal of chattering.

5.1 Future Work

For future work the authors recommend the inclusion of some other observer design technique e.g. an Exact Differentiator or the Internal Model based approach to handle the output feedback control problem for the system. The concept can be extended to Output Regulation of the nonlinear system using the robust HOSMC algorithm based conventional/conditional compensator which may result in further improvement of transient performance and ability to asymptotically track unknown references while rejecting disturbance signals, both produced by some autonomous external system. A natural extension of the HOSMC framework is the control of non-minimum phase systems directly using high gain feedback or incorporate an extended high gain observer and design output feedback control. The incorporation of higher order sliding strategy in controller design opens new dimensions towards robust control design and performance enhancement.

References

Atassi, A.N., Khalil, H.K.: A separation pinciple for the stabilization of a class of nonlinear systems. IEEE Trans. Autom. Control (1999)

BenHadj Braiek, N., Rhif, A., Zohra, K.: A high-order sliding mode observer torpedo guidance application. J. Eng. Technol. **2**, 7 (2012)

Edwards, C., Spurgeon, K.S.: Sliding mode control theory and applications. Taylor & Francis, (1998)

Esfandiari, F., Khalil, H.K.: Output feedback linearization of fully linearizable sytems. Int. J. Control **56**(31), 1007–1037 (1992)

Floquet, T., Barbot, J.P.: Super twisting algorithm based step-by-step sliding mode observers for nonlinear systems with unknown inputs. Int. J. Syst. Sci. **38**, 22 (2007)

Fridman, L., Levant, A.: Higher Order Sliding Modes, chapter 3, page 49. Marcel Dekker Inc, CRC Press (2002)

Guldner, J., Utkin, V.: Sliding Mode Control in Electromechanical Systems. Taylor & Francis, London (1999)

Henley John, A.: Design and implementation of a feedback linearizing controller and kalman filter for a magnetic levitation system. Master's thesis, Department of mechanical engineering, University Of Texas At Arlington, Texas, USA (2007)

Mahmoud N.I.: A backstepping design of a control system for a magnetic levitation system. Master's thesis, Depatmentt of electrical engineering, Linkoping University, Sweden, (2003)

Khalil, H.K.: High-gain observers in nonlinear feedback control. In: Proceedings of the international conference on control, automation and systems, IEEE, 2008. pp. 1527–1528 (2008)

Khalil, H.K.: Nonlinear Systems. Prentice Hall, Upper Saddle River (2002)

Korovin, K.S., Emeryanov, S.V., Levant, A.: High-order sliding modes in control systems. Comput. Math. Model. **3**, 25 (1996)

Levant, A.: Finite-time stability and high relative degrees in sliding-mode control. Tel-Aviv University, Technical report (2012)

Levant, A.: Sliding order and sliding accuracy in sliding mode control. Int. J. Control **58**, 17 (1999)

Levant, Arie: Universal single-input-single-output (siso) sliding-mode controllers with finite-time convergence. IEEE Trans. Autom. Control **46**, 5 (2001)

Levant, A.: Chattering analysis. IEEE Trans. Autom. Control **55**(6), 1380–1389 (2010)

Levis, M.: Nonlinear control of a planar magnetic levitation system. Master's thesis, Electrical and computer engineering, University of Toronto, Canada, (2003)

Lootin, J., Levine, J, Christophe Jean P.: A nonlinear approach to the control of magnetic bearings. In IEEE Trans. Control Syst. Technol., USA (1996)

Milica Naumovic, B., Veselic, B.R.: Magnetic levitation system in control engineering education. Automatic control and robotics, 7:10 (2008)

Olson, S.M., Trumper, D.L., Pradeep, K.: Subrahmanyan student member IEEE. linearizing control of magnetic suspension systems. IEEE Trans. Control Syst. Tech. (1997)

Perruquetti, W.: From 1rst order to higher order sliding modes. Technical report, Ecole Centrale de Lille, Cite Scientilque, BP 48, F-59651 Villeneuve d Ascq Cedex, FRANCE, (2010)

Pridor, A., Gitizadeh, R., Asher-Ben, J.Z., Levant, A., Yaesh, I.: Aircraft pitch control via second order sliding technique. AIAA J. Guid. Control Dyn., 23:31 (2000)

Pukdeboon, Chutiphon: Second-order sliding mode controllers for spacecraft relative translation. Appl. Math. Sci. **6**, 15 (2012)

Punta, E.: Second order sliding mode control of nonlinear multivariable systems. Technical report, Institute of Intelligent Systems for Automation, National Research Council of Italy, ISSIA-CNR, Genoa Italy (2006)

Rhif, A.: High order sliding mode control with pid sliding surface simulation on a torpedo. Int. J. Inf. Technol. Control Autom. **2**, 13 (2012)

Wai, J.R., Lee, J.: Backstepping based levitation control design for magnetic levitation rail system. Control theory and applications series, Department of electrical enggineeing, Institution of engineering and technology, Yuan Ze University, Taiwan (2008)
Woodson, H.H., Melcher, J.R.: Electromechanical Dynamics, Part 1 Discrete Systems. Wiley, New York (1968)

Design and Application of Discrete Sliding Mode Controller for TITO Process Control Systems

A.A. Khandekar and B.M. Patre

Abstract Selection of the proper control system for the multi-variable systems with time delay is a challenging task because of the interacting dynamic behaviour of system variables. Till date most of the multi-variable processes are controlled using proportional-integral-derivative (PID) controllers. The PID controllers for multivariable systems are either having centralized (full structured) or decentralized (diagonal) structure. The design procedure for centralized controllers is very complicated as the loop controllers cannot be designed independently. The decentralized controller design procedure either requires detuning or decoupling of the interactions. The controllers designed with detuning do not perform well for larger interactions. Thus decentralized PID controller with decoupler is the better choice with simple design procedure. In the design procedure for decentralized controllers, initially the decoupler is designed and decoupled subsystems are obtained. Then for each subsystem, the single loop controller is designed and the control signal is applied through decoupler to track the system variables. From the available literature, it can be seen that most of the PID design methods are based on linearized reduced order models. Due to model order reduction, the parametric uncertainty (plant-model mismatch) is introduced, which is not taken into consideration in the design process. Hence the designed PID controller is less robust and even may lead to instability especially in presence of time delay in the system model. Sliding mode control (SMC) is one of the robust control strategy with inherent property of invariance to parametric uncertainty. The continuous time SMC can produce the best response only for very small sampling time in implementation since the implementation sampling time is not taken into account in its design procedure. The discrete time SMC uses the discrete time model of the system and hence considers the sampling time in the design steps. However, it produces chattering in the control signal because of big sampling steps.

A.A. Khandekar (✉)
Department of Electronics and Telecommunication Engineering,
Dnyanganga College of Engineering and Research, Pune, India
e-mail: aniket.khandekar.vit@gmail.com

B.M. Patre
Department of Instrumentation Engineering, S.G.G.S. Institute
of Engineering and Technology, Nanded, India
e-mail: bmpatre@yahoo.com

A.T. Azar and Q. Zhu (eds.), *Advances and Applications in Sliding Mode Control systems*,
Studies in Computational Intelligence 576, DOI 10.1007/978-3-319-11173-5_9

This limitation can be overcome by designing DSMC with convergent quasi-sliding mode. In this chapter, the discrete convergent quasi-sliding mode is presented for interacting two input two output (TITO) systems with time delay. The ideal decoupler is designed to determine the non-interacting subsystem models for each loop. Then each subsystem is reduced to all pole third order plus delay time (TOPDT) model using four point fitting of frequency response. The separate DSMCs are designed for each loop using discrete time state model of the corresponding reduced subsystem. The control signals generated by the DSMCs are applied to the system through the decoupler. The stability condition for the presented controller is derived using Lyapunov stability approach. To validate the performance of the presented controller two well studied systems are simulated. To show the effectiveness of the prosed strategy, its performance is compared with the existing decentralized PID controllers.

1 Introduction

Many industrial chemical processes have multiple input multiple output (MIMO) configuration with interactions among the variables. Also these systems have non-linear dynamics with time delays. Generally the models obtained for these systems are of the linearized form. The controller design methods reported in the literature can be broadly classified as centralized (full structure) and decentralized (diagonal) controllers (Maghade and Patre 2012). In case of centralized multi-variable controllers the loop controllers interact with each other and hence the tuning for individual loop controller can not be done independently which complicates the design procedure. The decentralized controller has independent loop controllers and hence they can be designed and/or tuned separately. The decentralized controllers can be designed by detuning method (Luyben 1986), effective open loop process method (Xiong and Cai 2006) or decoupler methods (Wang et al. 2000; Tavakoli et al. 2006; Nordfeldt and Hagglund 2006).

In detuning methods, the interactions are ignored and the diagonal controllers are designed for the diagonal elements in process transfer function matrix. The diagonal controllers are then detuned by the detuning factor obtained from the interaction measure like relative gain array (RGA). The only advantage of this method is its simplicity. But it produces the loop performance highly affected by the interactions. In effective open loop process method, the effective transfer functions of the individual loops are obtained considering the interactions among the loops and the controllers are independently designed for every loop. Even in this method, the performance may be poor due to the effect of interactions since the interaction is considered in controller design but they are not decoupled in the implementation. In decoupler methods, the decoupler is designed to decouple the interactions and the decoupled subsystems are formed. For every decoupled subsystem, the controller is designed independently and is applied to the process through the decoupler. Hence the control signal is outcome of controller plus decoupler system. This highly reduces the interactions in the final output response. Several dynamic decoupler design methods are reported in the literature including that of (Wang et al. 2000; Tavakoli et al. 2006; Nordfeldt and Hagglund 2006). The static decoupler also can be designed considering only steady state interaction (Chen and Peng 2005) however it removes the interactions only at steady state degrading the dynamic performance. The main advantage of the decoupler method is that allows the use of the single input single output (SISO) controller designed methods which adds great simplicity in the design procedure.

In most of the cases proportional-integral-derivative (PID) controllers are employed due to their simplicity (Wang et al. 2000; Tavakoli et al. 2006; Nordfeldt and Hagglund 2006). Generally PID controller design methods are based on reduced ordered model (Astrom and Hagglund 1995; Wang et al. 1999; Malwatkar et al. 2009). However the performance of PID controllers is less effective for higher order processes since the parametric uncertainty introduced due to plant-model mismatch caused by modelling errors and linearization, model order reduction degrades the overall system performance as these uncertainties are hardly considered in the design of linear controllers like PID. During the past few decades, the robust control system design for plant-model mismatch processes have received considerable attention in control community. Among the established design approaches for robust process control, sliding mode control (SMC) plays an important role because it not only can stabilize certain and uncertain systems but also provide the capability of disturbance rejection and insensitivity to parameter variations (Utkin 1992; Camacho and Smith 2000). The continuous time sliding mode control (CSMC) has already received notable attention within the control community because of the flexibility of implementation, a large class of continuous systems are controlled by digital signal processors and high end micro-controllers. The continuous-time controller produces a good performance only for very small sampling period of actual implementation since it does not take into account the sampling period in the design procedure (Garcia et al. 2005). To analyze the effect of sampling time, discrete-time sliding mode control (DSMC) is studied in literature (Milosavljevic 1985; Gao et al. 1995; Golo and Milosavljevi 2000). The DSMC considers sampling period in the design phase and therefore can give better performance, even if the sampling period is considerably large which is a common case in case very slow systems like chemical processes.

Almost all process control systems have time delay in their dynamics and very few SMC design methods have considered time delay in the design methodology. In the literature, the linear matrix inequality technique was adopted for sliding mode control method to handle a class of uncertain time-delay systems (Hu et al. 2000). The approach has potential to deal with uncertainties and state delay, but the issue of input delay was not considered as a whole. The feasibility of the sliding surface combined with a predictor to compensate for the input delay of the system was investigated in the work reported in (Chen and Peng 2005; Roh and Oh 1999, 2000). Camacho et al. developed an internal model sliding mode controller with smith predictor for chemical processes described by FOPDT dynamics (Camacho et al. 2007). Khandekar et al. (2013) proposed DSMC for the tracking of a general class of higher order time delay systems. In this work, the delay was considered in the system output.

Thus, from the existing results in the literature it can be seen that designing a controller for interacting TITO systems with time delay to have very less effect of interaction and robust performance against parametric uncertainty is a difficult task. Particularly for the systems with considerably large time delay, the linear controllers like PID controller may not be capable of handling the parametric uncertainty. This motivate the authors to design the robust discrete sliding mode controller to tackle with this difficulty.

In this chapter a decentralized DSMC is proposed for two input two output (TITO) systems, an ideal decoupler (Nordfeldt and Hagglund 2006) is used to decouple the interactions. Then an optimal sliding surface combined with a delay ahead predictor is used to design DSMC for the decoupled subsystems. A quadratic performance index is minimized to design the optimal sliding surface. Stability condition of the closed-loop system is derived using Lyapunov approach. Two well studied simulation examples are considered to show the effectiveness of the proposed controller.

2 System Structure and Ideal Decoupler

The structure of the decentralized MIMO control system is as shown in Fig. 1.

Consider a MIMO system with a transfer function matrix

$$G(s) = \begin{bmatrix} G_{11}(s) & G_{12}(s) & \cdots & G_{1n}(s) \\ G_{21}(s) & G_{22}(s) & \cdots & G_{2n}(s) \\ \vdots & \vdots & \vdots & \vdots \\ G_{n1}(s) & G_{n2}(s) & \cdots & G_{nn}(s) \end{bmatrix} \quad (1)$$

and the decoupler of the form

$$D(s) = \begin{bmatrix} D_{11}(s) & D_{12}(s) & \cdots & D_{1n}(s) \\ D_{21}(s) & D_{22}(s) & \cdots & D_{2n}(s) \\ \vdots & \vdots & \vdots & \vdots \\ D_{n1}(s) & D_{n2}(s) & \cdots & D_{nn}(s) \end{bmatrix} \quad (2)$$

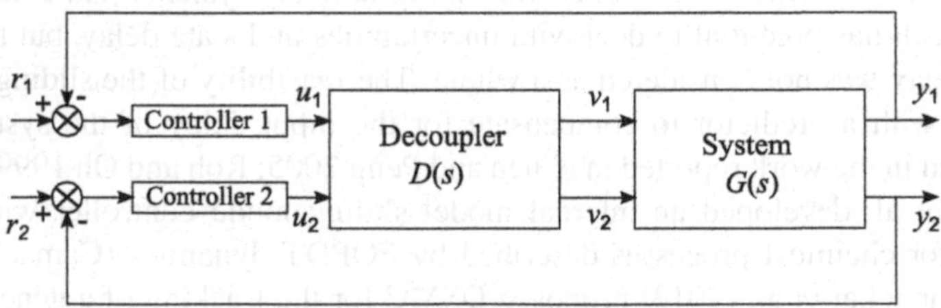

Fig. 1 Structure of the decentralized MIMO control system

The decoupled multi-loop SISO structure is given by

$$G_d(s) = G(S)D(S), \tag{3}$$

where,

$$G_d(s) = \begin{bmatrix} G_{d11}(s) & 0 & \cdots & 0 \\ 0 & G_{d22}(s) & \cdots & 0 \\ \vdots & \vdots & \vdots & \vdots \\ 0 & 0 & \cdots & G_{dnn}(s) \end{bmatrix}$$

Decoupler in Eq. (3) can be represented with little modification in (Nordfeldt and Hagglund 2006) as,

$$D(s) = Adj[G(s)]K(s) \tag{4}$$

where $K(s)$ is a diagonal matrix. The elements $k_{ii}(s)$ are obtained such that common pole-zero, common dead time and smallest gain from ith column of $Adj[G(s)]$ are removed and their inverse is included in $k_{ii}(s)$.

Each decoupled subsystem $G_{dii}(s)$ is reduced to third order plus delay time (TOPDT) all pole structure using four point least square fitting of frequency response of the decoupled subsystem to get (Malwatkar et al. 2009)

$$G_{iir}(s) = \frac{b_{0ii}}{s^3 + a_{1ii}s^2 + a_{2ii}s + a_{3ii}} e^{-t_{dii}s} \tag{5}$$

The frequency response of the higher order subsystem and delay free part of reduced TOPDT subsystem in Eq. (5) are equated at four different frequencies. The frequencies where phase of the higher order system is $-\pi/4$, $-\pi/2$, $-3\pi/4$ and $-\pi$ are considered for frequency response fitting and the delay of the reduced TOPDT model is obtained from the phase difference between higher order model and delay free TOPDT model at the frequency where the phase of the higher order model is $-\pi/2$.

For each subsystem with the transfer function in Eq. (5) the state model is obtained in the form

$$\begin{aligned} \dot{x}(t) &= Ax(t) + Bu(t - t_d), \\ y(t) &= x_1(t) \end{aligned} \tag{6}$$

where $A \in \Re^{3\times3}$, $B \in \Re^{3\times1}$, $C \in \Re^{1\times3}$ are the continuous time state space matrices, $x(t)$ is 3×1 state vector and $u(t)$ and $y(t)$ represent the control input and system output respectively, while t_d is the delay time.

The continuous time state model in Eq. (6) is discretized to get discrete time state model as

$$x(k+1) = Gx(k) + Hu(k-d),$$
$$y(k) = Cx(k) = x_1(k) \tag{7}$$

$G \in \Re^{3\times3}$, $H \in \Re^{3\times1}$, $C \in \Re^{1\times3}$ represent discrete time state space matrices, $x(k)$ is state vector. The term d is the number of delay samples. The matrices G and H in Eq. (7) are computed as,

$$G = e^{AT},$$
$$H = \left[\int_0^T e^{At} dt\right] B \tag{8}$$

where T is the sampling period.

3 DSMC Design and Implementation

A continuous time sliding mode is a first order sliding mode if and only if the sliding surface $s(t)$ obeys the two conditions: $s(t) = 0$ and $s(t)\dot{s}(t) < 0$ when $s(t) \neq 0$, where $s(t)$ is sliding surface. In CSMC, the controller output is updated continuously and therefore the sliding function and its first time derivative have opposite sign, which is the fundamental condition for the existence of sliding mode. The aim of the SMC is to force the control system to move on the sliding surface $s(t) = 0$ with the help of equivalent control and to maintain it on the sliding surface by discontinuous switching control, till error converges to zero.

A first order discrete time approximation of the above mentioned fundamental condition of CSMC is

$$s(k)[s(k+1) - s(k)] < 0. \tag{9}$$

The condition in Eq. (9) results in chattering due to the enlargement of sampling step and hence this condition is necessary but not sufficient. The following condition guaranties a convergent quasi sliding mode (Mihoub et al. 1991, 2009),

$$|s(k+1)| < |s(k)|. \tag{10}$$

3.1 Delay Ahead Predictor

To handle with uncertainty due to plant-model mismatch, the delay ahead predictor is constructed by removing delay from the state model in Eq. (7) as (Chen and Peng 2005; Khandekar et al. 2012),

$$x^*(k+1) = Gx^*(k) + Hu(k),$$
$$y^*(k) = x_1^*(k). \tag{11}$$

To improve the accuracy of state prediction, especially in face with modelling errors and unmeasured disturbances, the following corrections are made for actual implementation

$$\hat{x}_1(k+d|k) = x_1^*(k) + y_p(k) - x_1(k)$$
$$\hat{x}_j(k+d|k) = x_j^*(k), \quad j = 2, 3. \tag{12}$$

where $y_p(k)$ is the output of actual plant. The delay ahead predictor-corrector combination in Eqs. (11) and (12) is analogous to the Smith predictor used for dead time compensation for the transfer function model.

3.2 Equivalent and Switching Control Law of DSMC

The sliding surface is chosen as

$$s(k) = K\left[x^*(k) - x_d\right], \tag{13}$$

where $K = [k_1 \quad k_2 \quad k_3]$ is the tuning parameter matrix and x_d is the desired state vector. The sliding surface at $(k+1)^{\text{th}}$ sample is

$$s(k+1) = K\left[x^*(k+1) - x_d\right]$$
$$s(k+1) = KGx^*(k) + KHu(k) - Kx_d. \tag{14}$$

The equivalent control law that forces the system to reach the sliding surface is obtained by equating $s(k+1)$ in Eq. (14) to zero and is given by,

$$u_{eq}(k) = -(KH)^{-1}[KGx^*(k) - Kx_d]. \tag{15}$$

The robustness is ensured by the addition of a discontinuous term (sign of sliding function $s(k)$) in the control law as

$$u_{sw}(k) = -k_{sw}s(k)\text{sign}(s(k)). \tag{16}$$

to get total control law as

$$u(k) = u_{eq}(k) + u_{sw}(k) \tag{17}$$

3.3 Stability Condition

To evaluate the stability condition of DSMC, direct Lyapunov stability analysis is used. The positive definite Lyapunov function can be chosen as

$$V(k) = |s(k)| \tag{18}$$

The Lyapunov function at instant $k + 1$ is

$$V(k+1) = |s(k+1)| \tag{19}$$

which gives

$$\Delta V(k) = |s(k+1)| - |s(k)| \tag{20}$$

For the system to be stable, right hand term in Eq. (20) must be a negative definite function. This results into $|s(k+1)| < |s(k)|$ which is the reaching condition of DSMC. From Eqs. (13)–(17), (20) we get,

$$\Delta V(k) = |KHu_{sw}(k)| - |s(k)| \tag{21}$$

which gives the stability condition as

$$\begin{aligned} |KHu_{sw}(k)| &< |s(k)|, \\ |KHk_{sw}| &< 1 \end{aligned} \tag{22}$$

However, the discontinuous signum function produces chattering which can be reduced by replacing term sign of $s(k)$ by tan$h$ function with boundary layer β. Also, the delay ahead predicted state $x^*(k)$ is replaced by corrected state $\hat{x}(k)$ to get the final control law

$$u(k) = -(KH)^{-1}[KG\hat{x}(k)] - k_{sw}s(k)\tanh(s(k)/\beta). \tag{23}$$

3.4 Optimization of Sliding Surface

To optimize the sliding surface, the tuning parameter matrix K in the sliding surface is computed by minimizing the steady state quadratic performance index given by,

$$J = \sum_{k=0}^{\infty} x^T k Q x(k) + u^T(k) R u(k) \tag{24}$$

where Q is any 3×3 positive definite real symmetric matrix and R is any positive constant of designer's choice. Minimization of Eq. (24) results into (Ogata 2003)

$$K = 2R^{-1}H^T(G^T)^{-1}[P - Q] \tag{25}$$

where, P is another positive definite real symmetric matrix obtained by solving the equation

$$P = Q + G^T P[I + HR^{-1}H^T P]^{-1}G \tag{26}$$

where I is an identity matrix of dimension 3×3. The sliding surface obtained thus is optimal with respect to performance index in Eq. (24).

4 Simulation Examples

The performance of proposed controller is validated and compared with other controllers for two well studied examples in the literature to show the effectiveness of the proposed controller. The controller is designed and compared with the PID controllers in the literature. *Mathworks*™ MATLAB 7.0.1 is used for simulation.

4.1 Example: Wood Berry Distillation Column

Wood and Berry introduced the transfer function model of a pilot-scale distillation column, which consists of an eight-tray plus re-boiler separating methanol and water (Wood and Berry 1973). The Wood-Berry binary distillation column process is a multi-variable system that has been studied extensively. The process has the transfer function matrix as

$$G(s) = \begin{bmatrix} \frac{12.8}{16.7s+1}e^{-s} & \frac{-18.9}{21s+1}e^{-3s} \\ \frac{6.6}{10.9s+1}e^{-7s} & \frac{-19.4}{14.4s+1}e^{-3s} \end{bmatrix}.$$

The term $K(s)$ in Eq. (4) is

$$K(s) = \begin{bmatrix} \frac{-1}{6.6}e^{3s} & 0 \\ 0 & \frac{1}{12.8}e^{s} \end{bmatrix}.$$

The decoupler determined using Eq. (4) is

$$D(s) = \begin{bmatrix} \frac{2.94}{14.4s+1} & \frac{1.477}{21s+1}e^{-2s} \\ \frac{1}{10.9s+1}e^{-4s} & \frac{1}{16.7s+1} \end{bmatrix}.$$

The Decoupled subsystems are

$$G_{d11}(s) = \frac{37.63}{(16.7s+1)(14.4s+1)}e^{-s} - \frac{18.9}{(21s+1)(10.9s+1)}e^{-7s},$$

$$G_{d12}(s) = 0,$$
$$G_{d21}(s) = 0,$$

and

$$G_{d22}(s) = \frac{9.75}{(10.9s+1)(21s+1)}e^{-9s} - \frac{19.4}{(14.4s+1)(16.7s+1)}e^{-3s}.$$

The reduced TOPDT models of $G_{d11}(s)$ and $G_{d22}(s)$ are

$$G_{11r}(s) = \frac{0.2196}{s^3+1.527s^2+0.3214s+0.01108}e^{-0.3s},$$
$$G_{22r}(s) = \frac{-0.0472}{s^3+0.8915s^2+0.1375s+0.004767}e^{-0.4s}.$$

The continuous time state model matrices for $G_{11r}(s)$ are

$$A_{11} = \begin{bmatrix} 0 & 1 & 0 \\ 0 & 0 & 1 \\ -0.01108 & -0.3214 & -1.5270 \end{bmatrix}, \quad B_{11} = \begin{bmatrix} 0 \\ 0 \\ 0.2196 \end{bmatrix}.$$

The discrete time state model matrices obtained using zero order hold discretization for sampling period of 0.1 s are

$$G_{11} = \begin{bmatrix} 1.0000 & 0.0999 & 0.0048 \\ -0.0001 & 0.9985 & 0.0927 \\ -0.0010 & -0.0298 & 0.8569 \end{bmatrix}, \quad H_{11} = \begin{bmatrix} 0.0000 \\ 0.0010 \\ 0.0204 \end{bmatrix}.$$

The matrix Q and R in Eq. (24) are chosen as,

$$Q = \begin{bmatrix} 0.1 & 0 & 0 \\ 0 & 1 & 0 \\ 0 & 0 & 1 \end{bmatrix}, \quad R = 1$$

The sliding surface parameter matrix K determined using Eq. (25), the switching gain k_{sw} in Eq. (23) and boundary layer constant β in tan$h$ function are

$$K = \begin{bmatrix} 0.2668 & 1.2516 & 0.8381 \end{bmatrix}, \quad k_{sw} = 0.5, \quad \beta = 0.1$$

respectively. Similarly for for $G_{22r}(s)$ the continuous time state model matrices are

$$A_{22} = \begin{bmatrix} 0 & 1 & 0 \\ 0 & 0 & 1 \\ -0.0048 & -0.1375 & -0.8915 \end{bmatrix}, \quad B_{22} = \begin{bmatrix} 0 \\ 0 \\ -0.0472 \end{bmatrix}.$$

The discrete time state model matrices for sampling period of 0.1 s are

$$G_{22} = \begin{bmatrix} 1.0000 & 0.1000 & 0.0049 \\ -0.0000 & 0.9993 & 0.0957 \\ -0.0005 & -0.0132 & 0.9141 \end{bmatrix}, \quad H_{22} = \begin{bmatrix} 0.0000 \\ -0.0002 \\ -0.0045 \end{bmatrix}.$$

The matrix Q and R in Eq. (24) are chosen as,

$$Q = \begin{bmatrix} 0.1 & 0 & 0 \\ 0 & 50 & 0 \\ 0 & 0 & 10 \end{bmatrix}, \quad R = 1$$

The sliding surface parameter matrix K determined using Eq. (25), the switching gain k_{sw} in Eq. (23) and boundary layer constant β in tan$h$ function are

$$K = \begin{bmatrix} -0.2266 & -5.4099 & -5.5606 \end{bmatrix}, \quad k_{sw} = 0.5, \quad \beta = 0.1$$

respectively.

The proposed controller is compared with the Tawakoli et al. (2006) decentralized PI controller, PI and PID controllers proposed by Maghade and Patre (2012). In both control strategies the decoupler given in Tavakoli et al. (2006) is used to get the decoupled subsystems. Then the decoupled subsystems are reduced to first order plus delay time (FOPDT) model and controllers are designed using frequency domain approach. The desired gain and phase margins are specified to obtain the controller parameters. In this example gain and phase margin specification for designing these controllers are chosen $3dB$ and 60° respectively.

The decoupler determined by Tavakoli et al. (2006) and Maghade and Patre (2012) for the Wood Berry distillation column using methodology in Tavakoli et al. (2006) is,

$$D_T(s) = \begin{bmatrix} 1 & \frac{1.477(16s+1)}{21s+1}e^{-2s} \\ \frac{0.3402(14.4s+1)}{10.9s+1}e^{-4s} & 1 \end{bmatrix}.$$

The resulting decoupled subsystems are

$$G_{d11T}(s) = \frac{12.8}{(16.7s+1)}e^{-s} - \frac{6.426}{(21s+1)(10.9s+1)}e^{-7s},$$

$$G_{d12T}(s) = 0,$$
$$G_{d21T}(s) = 0,$$

and

$$G_{d22T}(s) = \frac{9.75(16.7s+1)}{(10.9s+1)(21s+1)}e^{-9s} - \frac{19.4}{(14.4s+1)}e^{-3s}.$$

The reduced FOPDT models of the decoupled subsystems are

$$G_{11FOPDT}(s) = \frac{6.37}{5.411s+1}e^{-1.065s},$$
$$G_{22FOPDT}(s) = \frac{-9.655}{4.684s+1}e^{-2.157s}.$$

Tawakoli et al. (2006) decentralized PI controller is

$$G_{CT}(s) = \begin{bmatrix} 0.41 + \frac{0.074}{s} & 0 \\ 0 & -0.12 - \frac{0.024}{s} \end{bmatrix}.$$

Maghade, Patre's PI and PID controllers are,

$$G_{CMPI}(s) = \begin{bmatrix} 0.4867 + \frac{0.0881}{s} & 0 \\ 0 & -0.1567 - \frac{0.0304}{s} \end{bmatrix}.$$

and

$$G_{CMPID}(s) = \begin{bmatrix} \frac{0.9733+\frac{0.0881}{s}+2.6887s}{5.5252s+1} & 0 \\ 0 & \frac{-0.3134-\frac{0.0304}{s}-0.8070s}{5.1499s+1} \end{bmatrix}.$$

respectively.

To validate the performance proposed controller, unit step change is applied in the set point of output y_1 at time $t = 0$ and unit step change is applied in the set point of output y_1 at time 150. The system outputs y_1 and y_2 and the control inputs applied to process v_1 and v_2 and sliding surface are shown in Figs. 2 and 3 respectively. From Fig. 2 it can be seen that the proposed controller produces output responses with less interaction and overshoot while the responses produced by the other controller are highly oscillatory with large interactions. Figure 3 shows that the control signals produced by the proposed controller are very smooth whereas the other controllers are producing control efforts with sudden jerks which are harmful to the actuating devices.

To evaluate the performance under parametric uncertainty, 20 % uncertainty is introduced in gains, time constants and delays of all the transfer functions. The output responses produced by proposed and other controllers are shown in Fig. 4 and

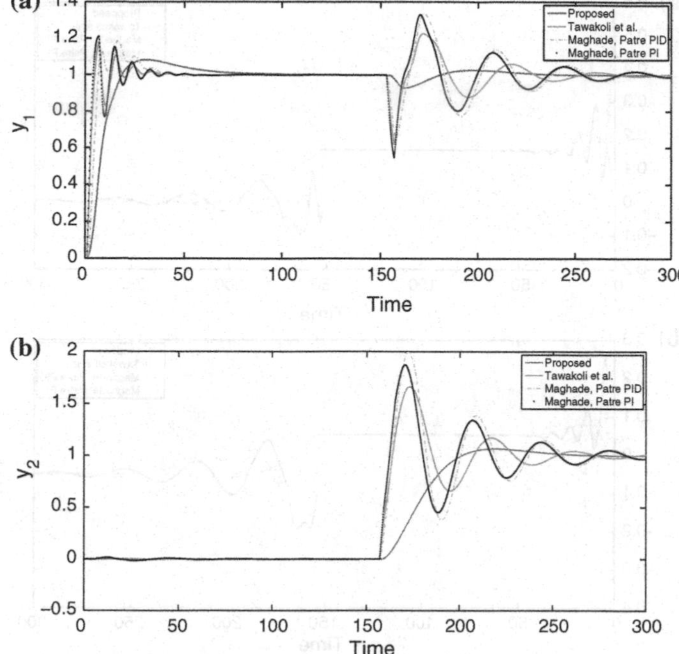

Fig. 2 Output responses. **a** System output y_1 **b** System output y_2

the corresponding control signals are shown in Fig. 5. From Figs. 4 and 5, it can be seen that under the effect of parametric uncertainty, the performance of the proposed controller remains almost same whereas the other controllers produce highly oscillatory output responses with considerable interactions. The control efforts are also highly oscillatory. This shows the robustness of the proposed controller.

4.2 Example: ISP Reactor

ISP reactor is another well studied example of TITO systems with considerable interaction whose transfer function matrix is

$$G(s) = \begin{bmatrix} \frac{22.89}{4.572s+1}e^{-0.2s} & \frac{-11.4}{1.807s+1}e^{-0.4s} \\ \frac{4.689}{2.174s+1}e^{-0.2s} & \frac{5.8}{1.801s+1}e^{-0.4s} \end{bmatrix}.$$

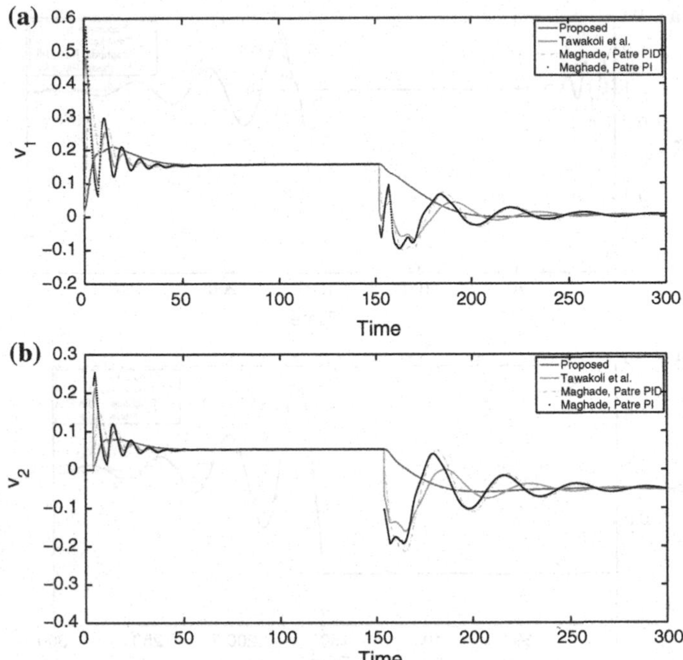

Fig. 3 Control signals. **a** Control signal v_1 **b** Control signal v_2

The term $K(s)$ in Eq. (4) is

$$K(s) = \begin{bmatrix} e^{0.2s} & 0 \\ 0 & e^{0.2s} \end{bmatrix}.$$

The decoupler determined using Eq. (4) is

$$D(s) = \begin{bmatrix} \frac{5.8}{1.801s+1} e^{-0.2s} & \frac{11.64}{1.807s+1} e^{-0.2s} \\ \frac{-4.689}{2.174s+1} & \frac{22.89}{4.572s+1} \end{bmatrix}.$$

The Decoupled subsystems are

$$G_{d11}(s) = \frac{132.762}{(4.572s+1)(1.801s+1)} e^{-0.4s} + \frac{54.58}{(1.807s+1)(2.174s+1)} e^{-0.4s},$$

$$G_{d12}(s) = 0,$$
$$G_{d21}(s) = 0,$$

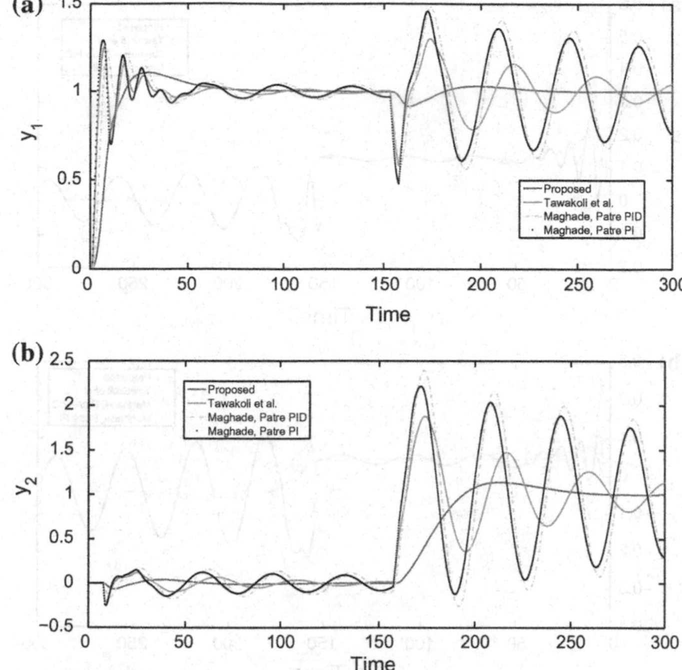

Fig. 4 Output responses under the effect of +20 % parametric uncertainty. **a** System output y_1 **b** System output y_2

and

$$G_{d22}(s) = \frac{132.762}{(4.572s+1)(1.801s+1)}e^{-0.4s} + \frac{54.58}{(1.807s+1)(2.174s+1)}e^{-0.4s},$$

Thus the decoupled subsystems $G_{d11}(s)$ and $G_{d22}(s)$ have same transfer function. The resulting reduced TOPDT model is

$$G_{11r}(s) = \frac{142.1}{s^3+5.33s^2+4.378s+0.7639}e^{-0.5s},$$

The continuous time state model matrices for $G_{11r}(s)$

$$A_{11} = \begin{bmatrix} 0 & 1 & 0 \\ 0 & 0 & 1 \\ -0.7639 & -4.378 & -5.33 \end{bmatrix}, \quad B_{11} = \begin{bmatrix} 0 \\ 0 \\ 142.1 \end{bmatrix}.$$

The discrete time state model matrices obtained using zero order hold discretization for sampling period of 0.1 s are

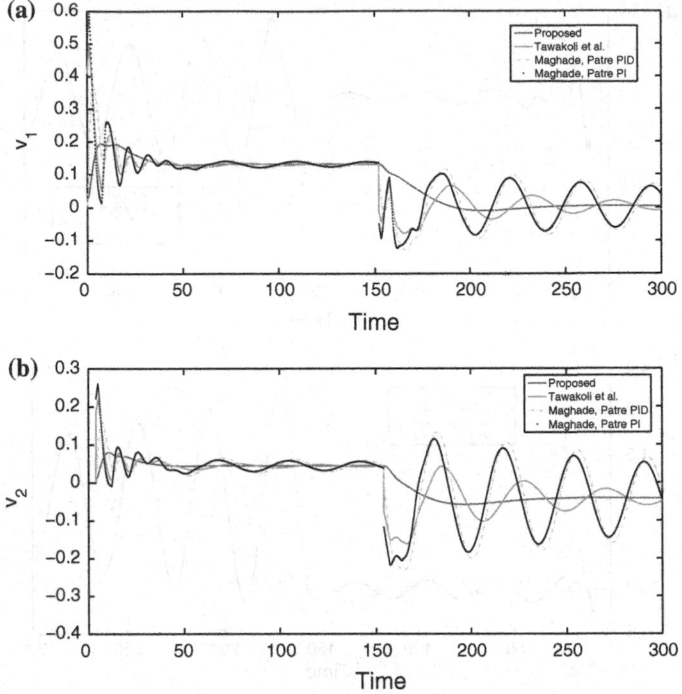

Fig. 5 Control signals under the effect of +20 v_1 **b** Control signal v_2

$$G_{11} = \begin{bmatrix} 0.9999 & 0.0994 & 0.0042 \\ -0.0032 & 0.9815 & 0.0770 \\ -0.0588 & -0.3401 & 0.5713 \end{bmatrix}, \quad H_{11} = \begin{bmatrix} 0.0208 \\ 0.5973 \\ 10.9348 \end{bmatrix}.$$

The matrix Q and R in Eq. (24) are chosen as,

$$Q = \begin{bmatrix} 1 & 0 & 0 \\ 0 & 1 & 0 \\ 0 & 0 & 1 \end{bmatrix}, \quad R = 1$$

The sliding surface parameter matrix K determined using Eq. (25), the switching gain k_{sw} in Eq. (23) and boundary layer constant β in tan$h$ function are

$$K = \begin{bmatrix} 0.0781 & 0.1215 & 0.0587 \end{bmatrix}, \quad k_{sw} = 0.5, \quad \beta = 0.1$$

respectively. Since $G_{11r}(s)$ and $G_{22r}(s)$ are equal, the controller for $G_{22r}(s)$ is same as that for $G_{11r}(s)$

The proposed controller is compared with the Maghade and Patre (2014) PID controllers obtained using FOPDT and SOPDT reduced order models of decoupled subsystems. In both control strategies the decoupler given in Tavakoli et al. (2006) is used to get the decoupled subsystems. Then the decoupled subsystems are reduced into FOPDT and SOPDT models to design the decentralized PID controllers. Both controllers are designed using dominant pole placement approach.

The decoupler determined by Maghade and Patre (2014) for the ISP reactor using methodology in Tavakoli et al. (2006) is,

$$D_M(s) = \begin{bmatrix} e^{-0.2s} & \frac{0.5086(4.572s+1)}{1.807s+1}e^{-0.2s} \\ \frac{-0.8085(1.801s+1)}{2.174s+1} & 1 \end{bmatrix}.$$

The resulting decoupled subsystems are

$$G_{d11M}(s) = \frac{22.89}{(4.572s+1)}e^{-0.4s} + \frac{9.4110(1.801s+1)}{(1.807s+1)(2.174s+1)}e^{-0.4s},$$

$$G_{d12M}(s) = 0,$$
$$G_{d21M}(s) = 0,$$

and

$$G_{d22M}(s) = \frac{5.8}{(1.801s+1)}e^{-0.4s} + \frac{2.3849(4.572s+1)}{(2.174s+1)(1.807s+1)}e^{-0.4s}.$$

The reduced FOPDT models of the decoupled subsystems are

$$G_{11FOPDT}(s) = \frac{32.3003}{3.4712s+1}e^{-0.4107s},$$
$$G_{22FOPDT}(s) = \frac{8.1844}{1.3595s+1}e^{-0.4241s}.$$

The reduced SOPDT models of the decoupled subsystems are

$$G_{11SOPDT}(s) = \frac{e^{-0.1060s}}{0.1820s^2+0.4794s+0.0396},$$
$$G_{22SOPDT}(s) = \frac{e^{-0.1690s}}{0.0156s^2+0.1333s+0.1222}.$$

The decentralized PID controllers designed using FOPDT and SOPDT models are

$$G_{CMFOPDT}(s) = \begin{bmatrix} 0.0006 + \frac{0.0060}{s} - 0.0372s & 0 \\ 0 & 0.0068 + \frac{0.0449}{s} - 0.0383s \end{bmatrix}.$$

and

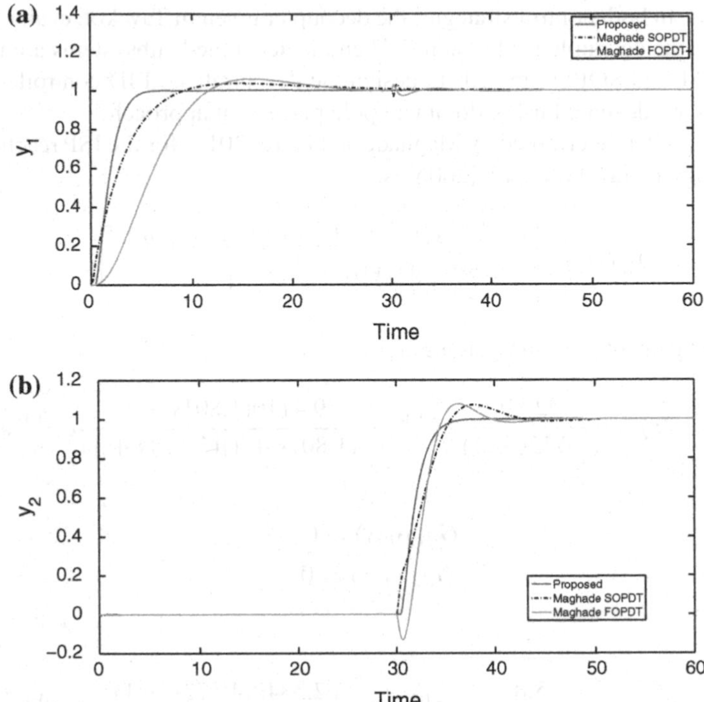

Fig. 6 Output responses. **a** System output y_1 **b** System output y_2

$$G_{CMSOPDT}(s) = \begin{bmatrix} 0.1092 + \frac{0.0138}{s} + 0.2426s & 0 \\ 0 & 0.0443 + \frac{0.0588}{s} + 0.0853s \end{bmatrix}.$$

respectively. To validate the performance proposed controller, unit step change is applied in the set point of output y_1 at time $t = 0$ and unit step change is applied in the set point of output y_2 at time 30. The system outputs y_1 and y_2 and the control inputs applied to process v_1 and v_2 and sliding surface are shown in Figs. 6 and 7 respectively. From Fig. 6 it can be seen that the proposed controller produces output responses with less interaction and overshoot while the responses produced by the other controllers have more overshoot and interactions. Figure 7 shows that the control signals produced by the proposed controller very smooth whereas the other controllers produce the control signals that have sudden jerks.

To evaluate the performance under parametric uncertainty, 20 % uncertainty is introduced in gains, time constants and delays of all the transfer functions. The output responses produced by proposed and other controllers are shown in Fig. 8 and the corresponding control signals are shown in Fig. 9. From Figs. 8 and 9, it can be seen that under the effect of parametric uncertainty, the performance of the proposed controller remains almost same whereas the performance of other controllers are degraded. This shows the robustness of the proposed controller.

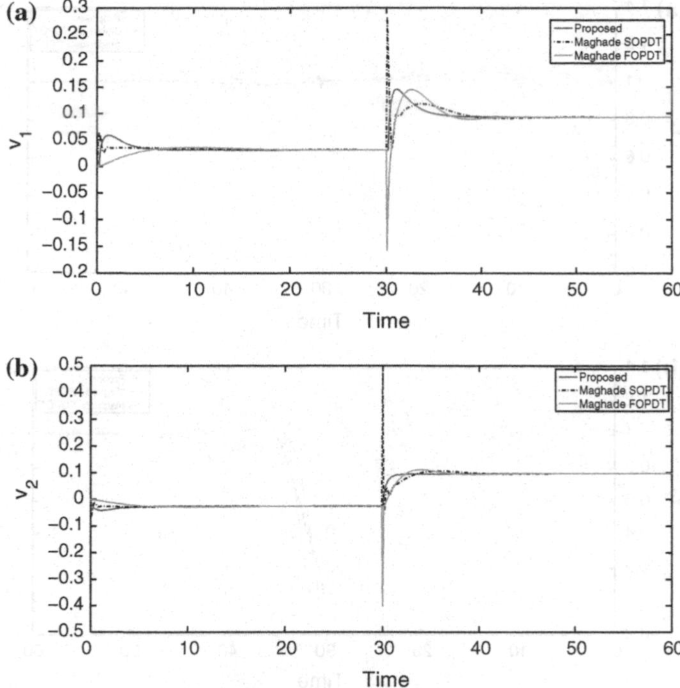

Fig. 7 Control signals **a** Control signal v_1 **b** Control signal v_2

5 Discussions

In this section, the performance comparison of the proposed controller with the representative PI/PID controllers on the basis of three important parameters; time delay, parametric uncertainty and interaction is discussed. The time delay is a process parameter which makes closed loop responses oscillatory and leads to instability and the parametric uncertainty deviates the closed loop system dynamics away from the desired and again may lead to instability if the parametric uncertainty is significant. The interaction among the system variables causes abrupt changes in control signals and deviation of one variable from its set-point due to change in the other variable. Thus for the interacting multi-variable control systems with time delay, a good controller should compensate the time delay, should handle with the parametric uncertainty and should avoid the abrupt changes in control signal in interacting behaviour. Hence it is important the discuss the controller performances with reference to these three effects.

In this chapter, two benchmark process models are considered for simulation study to elaborate the effectiveness and advantages of the proposed controller over the classical linear PI/PID controllers in presence of time delay and under the influence of parametric uncertainty, interacting dynamic behaviour. The time delay in ISP reactor

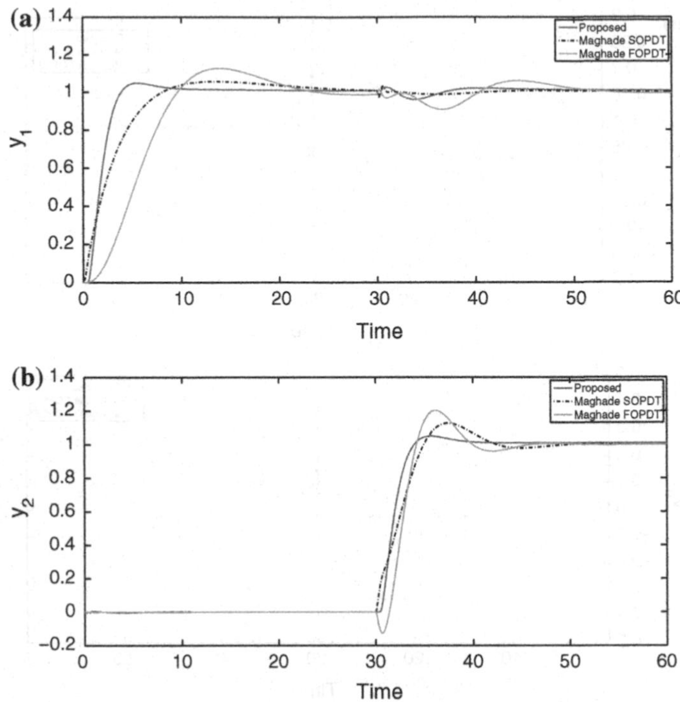

Fig. 8 Output responses under the effect of +20 % parametric uncertainty. **a** System output y_1 **b** System output y_2

is very less as compared to Wood and Berry binary distillation column and hence it can be seen that the PI/PID controllers are producing less oscillatory responses for ISP reactor as compared to Wood and Berry binary distillation column. However, the proposed DSMC is producing smooth responses for both the cases. This indicate that the classical PI/PID controllers can not handle long delays and the DSMC with delay ahead predictor-corrector can cope-up with the long delays.

If we compare the performances of proposed DSMC and PI/PID controllers with reference to effect of parametric uncertainty, from simulation results it can be seen that for both the examples, the output responses resulting from PI/PID controllers are changing a lot and leading to instability. However, the performance of the proposed DSMC is almost the same for nominal and uncertain conditions. This indicates that the DSMC is more capable than PI/PID controllers in handling the parametric uncertainty. This happens because of inherent property of sliding mode controllers that they are insensitive to parametric uncertainties.

Again, if we compare the performances with respect to interactions, from the control signals generated by all controllers in both simulation examples, it can be seen that the DSMC produces very smooth changes in control signal during the step

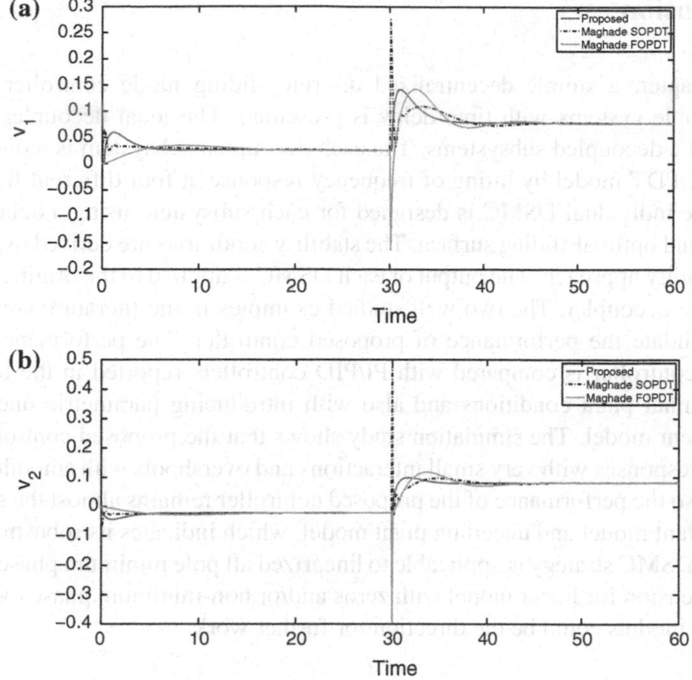

Fig. 9 Control signals under the effect of +20 % parametric uncertainty. **a** Control signal v_1 **b** Control signal v_2

changes in set-points while PI/PID controllers produce abrupt changes which are harmful to actuating devices.

In summary, from the simulation results presented in Sect. 4, it can be seen that presented controller produces far better performance as compared to PI/PID controllers in terms of output responses and control signals. The output responses contain less oscillations and interactions with very smooth control efforts. The comments from the simulation results can be summarized as,

1. The ideal decoupler is the better choice of decoupling the variables as it reduces the interaction to greater extent.
2. The DSMC designed using delay ahead predictor–corrector tackles the time delay in very systematic manner and produces good responses.
3. The optimal sliding surface considered in design generates very smooth control signal which put less jerks on actuating device.
4. The presented controller produces almost same responses for with and without parametric uncertainty. This proves the robustness of the sliding mode control strategy.

6 Conclusion

In this chapter, a simple decentralized discrete sliding mode controller for the multi-variable systems with time delay is presented. The ideal decoupler is used to obtain the decoupled subsystems. The each decoupled subsystem is reduced into all pole TOPDT model by fitting of frequency response at four different frequency points. The individual DSMC is designed for each subsystem using a delay ahead predictor and optimal sliding surface. The stability conditions are derived using Lyapunov stability approach. The output of each DSMC is applied to the original system through the decoupler. The two well studied examples in the literature are considered to validate the performance of proposed controller. The performance of the proposed controller is compared with PI/PID controllers reported in the literature under nominal plant conditions and also with introducing parametric uncertainty in the system model. The simulation study shows that the proposed controller produces the responses with very small interactions and overshoots with smooth control efforts. Also the performance of the proposed controller remains almost the same for nominal plant model and uncertain plant model, which indicates its robustness. The presented DSMC strategy is applicable to linearized all pole minimum phase models and its extension for linear model with zeros and/or non-minimum phase dynamics, non-linear models could be the direction for further work.

References

Astrom, K., Hagglund, T.: PID controllers: theory design and tuning. ISA, NC (1995)
Camacho, O., Rojas, R., Gabin, V.G.: Some long time delay sliding mode control approaches. ISA Trans **46**(1), 95–101 (2007)
Camacho, O., Smith, C.A.: Sliding mode control: an approach to regulate chemical processes. ISA Trans **39**(2), 205–218 (2000)
Chen, C.-T., Peng, S.-T.: Design of a sliding mode control system for chemical processes. J Process Control **15**(5), 515–530 (2005)
Gao, W., Wang, Y., Homaifa, A.: Discrete-time variable structure control systems. IEEE Trans Ind Electron **42**(2), 117–122 (1995)
Garcia, J.P.F., Silva, J.J.F., Martins, E.S.: Continuous-time and discrete-time sliding mode control accomplished using a computer. IEE Proc Control Theory Appl **152**(2), 220–228 (2005)
Golo, G., Milosavljevi, C.: Robust discrete-time chattering free sliding mode control. Sys Control Lett **41**(1), 19–28 (2000)
Hu, J., Chu, J., Su, H.: Smvsc for a class of time-delay uncertain systems with mismatching uncertainties. IEE Proc Control Theory Appl **147**(6), 687–693 (2000)
Khandekar AA, Malwatkar GM, Kumbhar SA, Patre BM (2012) Continuous and discrete sliding mode control for systems with parametric uncertainty using delay-ahead prediction. In: Proceedings of 12th IEEE workshop on variable structure systems, Mumbai, pp 202–207
Khandekar, A.A., Malwatkar, G.M., Patre, B.M.: Discrete sliding mode control for robust tracking of higher order delay time systems with experimental application. ISA Trans **52**(1), 36–44 (2013)
Luyben, W.L.: Simple method for tuning SISO controllers in multivariable systems. Ind Eng Chem Process Des Dev **25**(3), 654–660 (1986)

Maghade, D.K., Patre, B.M.: Decentralized PI/PID controllers based on gain and phase margin specifications for TITO processes. ISA Trans **51**(4), 550–558 (2012)
Maghade, D.K., Patre, B.M.: Pole placement by PID controllers to achieve time domain specifications for TITO systems. Trans Inst Meas Control **36**(4), 506–522 (2014)
Malwatkar, G.M., Sonawane, S.H., Waghmare, L.M.: Tuning PID controllers for higher-order oscillatory systems with improved performance. ISA Trans **48**(3), 347–353 (2009)
Mihoub M, Nouri AS, Abdennour R (2009) Real-time application of discrete second order sliding mode control to a chemical reactor. Control Eng Pract 17(9):1089–1095
Milosavljevic, C.: General conditions for the existence of a quasi-sliding mode on the switching hyperplane in discrete variable structure systems. Autom Remote Control **3**(1), 36–44 (1985)
Nordfeldt, P., Hagglund, T.: Decoupler and PID controller design of TITO systems. J Process Control **16**(9), 923–936 (2006)
Ogata, K.: Discrete time control systems, 2nd edn. Prentice Hall, NJ (2003)
Roh, Y.H., Oh, J.H.: Robust stabilization of uncertain input-delay systems by sliding mode control with delay compensation. Automatica **35**(11), 1861–1865 (1999)
Roh, Y.H., Oh, J.H.: Sliding mode control with uncertainty adaptation for uncertain input-delay systems. Int J Control **73**(13), 1255–1260 (2000)
Sira-Ramirez, H.: Non-linear discrete variable structure systems in quasi-sliding mode. Int J Control **54**(5), 1171–1187 (1991)
Tavakoli, S., Griffin, I., Fleming, P.: Tuning of decentralised PI (PID) controllers for TITO processes. Control Eng Pract **14**(9), 1069–1080 (2006)
Utkin VI (1992) Sliding modes in control and optimization. Springer, Berlin
Wang, Q.G., Huang, B., Guo, X.: Auto-tuning of TITO decoupling controllers from step tests. ISA Trans **39**(4), 407–418 (2000)
Wang, Q.-G., Lee, T.-H., Fung, H.-W., Qiang, B., Zhang, Y.: PID tuning for improved performance. IEEE Trans Control Sys Tech **7**(4), 457–465 (1999)
Wood, R.K., Berry, M.W.: Terminal composition control of a binary distillation column. Chem Eng Sci **28**(9), 1707–1717 (1973)
Xiong, Q., Cai, W.-J.: Effective transfer function method for decentralized control system design of multi-input multi-output processes. J Process Control **16**(8), 773–784 (2006)

Dynamic Fuzzy Sliding Mode Control of Underwater Vehicles

G.V. Lakhekar and L.M. Waghmare

Abstract A novel dynamic fuzzy sliding mode control (DFSMC) algorithm is developed for heading angle control of autonomous underwater vehicles (AUV's) in horizontal plane. At first, we design single input fuzzy sliding mode control (SIFSMC) based on mamdani type fuzzy inference system. The SIFSMC offers significant reduction in rule inferences and simplify the tuning of control parameters. Practically, it can be easily implemented by a look up table using a low cost advanced processor. The control structure provides robustness under the influence of parameter uncertainties and environmental disturbances. Next, we proposed fuzzy adaptation techniques in SIFSMC algorithm to vary the base of input–output membership functions of fuzzy inference engine. This adaptation law provides minimum reaching time to track desired trajectory path and also eliminate chattering effects. So far, the dynamics of AUV's are highly nonlinear, time varying and hydrodynamic coefficients of vehicle are difficult to be accurately estimated a prior, because of the variations of these coefficients with different operating conditions. These types of difficulties cause modeling inaccuracies of AUV's dynamics. Therefore, Traditional control techniques may not be able to handle these difficulties promptly and can't guarantee the desired tracking performance. On the other hand, sliding mode control (SMC) is the suitable choice for control of AUV's, because of its appreciable features such as design simplicity with robustness to parameter uncertainty and external disturbances. But, it has the inherent problem of chattering phenomenon which is the high frequency oscillations of the controller output and another difficulty in the calculation of equivalent control. Therefore, overall knowledge of the plant dynamics is required for this purpose. These problems are suitably circumvented by combining basic principles of sliding mode and fuzzy logic controllers (FLC's). With this scheme, the stability and robustness of the FLC algorithm is ensured by the SMC

G.V. Lakhekar (✉)
Department of Electrical Engineering,
G.H. Raisoni Institute of Engineering and Technology, Pune, India
e-mail: gv_lakhekar@rediffmail.com

L.M. Waghmare
Department of Instrumentation Engineering,
S. G. G. S. Institute of Engineering and Technology, Nanded, India
e-mail: lmwaghmare_@yahoo.com

A.T. Azar and Q. Zhu (eds.), *Advances and Applications in Sliding Mode Control systems*,
Studies in Computational Intelligence 576, DOI 10.1007/978-3-319-11173-5_10

law. By incorporating SMC in to fuzzy logic provides a possible solution to alleviate the chattering phenomena and to achieve zero steady state error. However, the parameters of membership function can't be adjusted to afford optimal control efforts under the occurrence of uncertainties. Therefore, DFSMC is designed for regulating heading angle in horizontal plane, under the influence of parametric uncertainties (as added mass, hydrodynamic coefficients, lift and drag forces), highly coupled nonlinearities and environmental disturbances (like ocean currents and wave effects). This chapter focuses on design of two supervisory fuzzy systems for tuning of boundary layer and hitting gain which are the basic parameters of fuzzy sliding mode control (FSMC) algorithm. The proposed control algorithm is developed from fuzzy inference module, which has single input as a sliding surface and single output as control signal. The input–output membership functions are depends on base values such as boundary layer, equivalent control and hitting gain. The idea behind this control scheme is to update width of boundary layer and hitting gain, due to which the supports of input–output fuzzy membership functions are varied with the help of two fuzzy approximators. Simulation results shows that, the output tracking response has minimum reaching time and tracking error in the approaching phase along with chattering problem can also reduced. The performance of proposed control strategy has been evaluated by comparison with conventional SMC and FSMC. A summary of fuzzy adaptation schemes in FSMC algorithm are given for enhancing tracking performance of AUV's. Finally, research directions for adopting optimal fuzzy supervisory techniques in sliding mode based fuzzy algorithm are suggested.

1 Introduction

In recent years, underwater vehicles have been widely used for scientific inspection of deep sea, long range survey, oceanographic mapping, underwater pipeline tracking, exploitation of underwater resources and so on (Bessa et al. 2008; Guo et al. 2003). While operating an unmanned underwater vehicle, correct positioning is important so that the vehicle can move along the desired path as expected. Thus, equipped with good measuring instruments, tracking sonar, acoustic telemetry modem and automatic control systems, operator can then concentrate on their work without having to worry about the position control. In addition that, underwater vehicles are difficult to control, due to nonlinearity, time variance, unpredictable external disturbances such as the environmental force generated by the sea current fluctuation and the difficulty in accurately modeling the hydrodynamic effect. The well developed linear controllers may fail in satisfying performance requirements, especially when changes in the system and environment occur during the AUV operation. Therefore, it is highly desirable to have a robust control system that has the capacities of learning and adopting to the unknown nonlinear hydrodynamic effects, parameter uncertainties, internal and external perturbations such as water current or sideslip effect. So that, an adaptive PD controller for the dynamic positioning of undersea vehicles working in close proximity of off-sure structures is introduced by Hoang and Kreuzer (2007).

In order to deal with parametric uncertainty and highly nonlinearity in the AUV's dynamics, many researchers concentrated their interests on the applications of robust control for underwater vehicles (Sebastian and Sotelo 2007).

SMC is commonly favored as a powerful robust control method for its independence from parametric uncertainties and external disturbances under matching conditions. It has been successfully applied for dynamic positioning and motion control of underwater vehicles, because of its performance insensitivity to model mismatches and disturbances. Yoerger and Slotine (1985) introduced the basic methodology of using sliding mode control for AUV application, and later Yoerger and Slotine (1991) developed an adaptive sliding mode control scheme in which a nonlinear system model is used. They have investigated the effects of uncertainty of the hydrodynamic coefficients and negligence of cross coupling terms. Goheen and Jefferys (1990) have proposed multivariable self tuning controllers as an autopilot for underwater vehicles to overcome model uncertainties while performing auto positioning and station-keeping. Cristi et al. (1990) proposed an adaptive sliding mode controller for AUV's based on the dominant linear model and the bounds of the nonlinear dynamic perturbations. Fossen and Sagatun (1991) designed a hybrid controller combining an adaptive scheme and a sliding mode term for the motion control of a remotely operated vehicle (ROV). Healey and Lienard (1993) suggested multivariable sliding mode autopilot based on state feedback for the control of decoupled model of underwater vehicles. Da Cunha et al. (1995) developed an adaptive control scheme for dynamic positioning of a ROV, which is based on a sliding mode controller that only used position measurements. Lam and Ura (1996) proposed nonlinear controller along with switched control law for noncruising AUV in path following. Lee et al. (1999) applied a discrete time quasi-sliding mode controller for an AUV with uncertainties of system parameters and with a long sample interval. Choi and Yuh (1996) have designed a multivariable adaptive control scheme based on bound estimation for AUV. Walchko and Nechyba (2003) applied sliding mode control with extended kalman filter estimation for Subjugator as a remotely operated underwater vehicle. Hoang and Kreuzer (2008) proposed a robust adaptive sliding mode control for dynamic of an ROV in which prior knowledge of bounds for uncertainties in parameters was not required.

The main disadvantage of the SMC method is its dependence on system model. On the other hand, even if the system model is known then implementation of SMC is possible. In addition, if all the states to be stabilized and controlled then transform the model into the canonical form. However, these conditions are not met for most AUV models. In order to overcome this problem, researchers use fuzzy logic for AUV's control applications to form a smooth approximation of nonlinear mapping from system input to output space. FLC is therefore well suited and mainly applied to nonlinear control problems. (Kato et al., 1993) applied the fuzzy algorithm to manage the guidance and control of AUV in both attitude control and cable tracking. Smith et al. (1994) proposed a fuzzy logic based autopilots for controlling and guiding a low speed torpedo shaped vehicle. FLC is not depend on a dynamic model, thus allowing for rapid development of a working design and less sensitivity to the plant variations. (DeBitetto et al., 1995) applied fuzzy logic to the low speed ballast control problem for depth control of unmanned underwater vehicles (UUV's). Kanakakis et al. (2004) developed three levels of fuzzy modular control architecture for underwater vehicles, which comprises of the sensor fusion module, the collision avoidance module and the motion control module. Ishaque et al. (2010) proposed single input fuzzy logic controller (SIFLC) for depth and pitch angle regulation of AUV's, reduces the conventional two-input FLC (CFLC) to a single input single output (SISO) controller. The SIFLC offers significant reduction in rule inferences and simplify the tuning of control parameters.

A merit of using fuzzy logic in control methodology is that the dynamics of controlled system need not be fully known. But, rule base of fuzzy controller could not give the guarantee for the stability and robustness of the control system (Azar 2010). The fusion of fuzzy logic and sliding mode control gives the benefit from the both side in nonlinear control technique. Kim and Lee (1995) proposed a fuzzy controller with fuzzy sliding surface for reducing tracking error and eliminating chattering problem due to that stability and robustness is improved. Song and Smith (2000) introduced a sliding mode fuzzy controller that uses pontryagins maximum principle for time optimal switching surface design and uses fuzzy logic to this surface. Guo et al. (2003) applied a sliding mode fuzzy controller to motion control and line of sight guidance of an AUV. Shi et al. (2008) designed to control the AUV's pitch motion under the disturbance of ocean current. Xin and Zaojian (2010) introduced a new type of fuzzy sliding mode control with adaptive disturbance approximation was proposed to deal with the trajectory regulation of underwater robot.

The parameters of FSMC algorithm such as sliding surface slope, boundary layer width and hitting gain are adaptively tune by fuzzy supervisory systems for obtaining better tracking response. A moving sliding surface was designed by Choi et al. (1994) for fast convergence speed with rotating or shifting sliding surface is adaptable to arbitrary initial condition. Ha (1996) introduced a novel sliding mode control with fuzzy logic tuning for accelerating the reaching phase and overcome from the influence of unmodeled uncertainties, due to that robust tracking response is enhanced. Temeltas (1998) employed fuzzy adapted sliding mode controller in which slope of sliding surface and discontinuous gain are tuned by fuzzy logic. Lakhekar (2012, 2013) presented fuzzy tuning technique used in SMC for rotating and shifting sliding surface as well as varying approaching angle towards sliding surface, so that reaching time and tracking error in approaching phase were significantly reduced.

Although combined SMC and FLC techniques have been widely used in various control fields. But, this techniques seems to be much sparse and the studies of this topic is sporadic. In that study, the improvement of their output response is possible by using adaptive technique in control module. Balasuriya and Cong (2003) proposed adaptive fuzzy controller can approximate the unknown system and sliding mode approach provide strong robustness against model uncertainties and external disturbances. Its parameters will be adapted online to utilize control energy more efficiently. Kim and Shin, (2006) developed autopilot for depth control of an underwater flight vehicle (UFV) based on adaptive fuzzy sliding mode control

(AFSMC) with a fuzzy basis function expansion (FBFE) is employed. Sebastion et al. (2007) address the kinematic variables controller based on pioneering algorithm, is utilized in control of underactuated snorkel vehicle. In proposed methodology, adaptive capabilities are provided by several fuzzy estimators, while robustness is provided by the SMC law. Bessa et al. (2008) presented an adaptive fuzzy control algorithm based on sliding mode for depth control of an ROV, which is employed for uncertainty/disturbance compensation with completely eliminating chattering effect. Later, Bessa et al. (2010) applied AFSMC for identification of external disturbances to control the dynamic positioning of underwater vehicles with four controllable degrees of freedom. Marzbanrad (2011) designed a robust adaptive fuzzy sliding mode control (RAFSMC) algorithm for tracking control of ROV, in which sliding mode is a powerful approach to compensate structured and unstructured uncertainties. With fuzzy algorithm is used for on-line estimation of external disturbances as well as unknown nonlinear terms of dynamic model of the ROV. Guo et al. (2012) presented AFSMC to deal with the depth and heading regulation of spherical underwater robots. Furthermore, the designed controller can't only tolerate actuator stuck faults, but also compensate the disturbances with constant components.

Rapid progress in underwater robotics is steadily affording scientist advanced tools for ocean explorations and exploitation. However, much work remains to be done before marine robots can roam the oceans freely, acquiring scientific data on the temporal and spatial scales that are naturally imposed by the phenomena under study. To meet these goals, robots must be equipped with systems to steer them accurately and reliably in the harsh marine environment. For this reason, there has been considerable interest over the last few years in the development of advanced methods for marine vehicle motion control such as, point stabilization, trajectory tracking, and path following control. In typical search or survey scenarios covering large areas required route stability and good turning performance in the horizontal plane of motion. Directional control is thus fundamental problem for the AUV's motion.

In path following control, linear control system design for AUV in horizontal plane is often impossible and difficult to achieve desired path trajectory, due to the dynamics of AUV's are highly nonlinear and the hydrodynamic coefficients of vehicle are difficult to estimate accurately, because variations of these coefficients with different operating conditions. This motivates to break traditional restricting conditions which are usually added to the AUV's motion behavior, while in maneuvering. In order to deal with the unstructured uncertainties in the AUV's dynamics, so that development of the adaptive algorithm is required. Therefore, we proposed DFSMC for directional control in long range survey. Here, two fuzzy approximators are employed to vary the base of input/output membership functions of FSMC algorithm, so that reaching time and chattering effects are minimized. The proposed adaptation algorithm is capable to handle different operating conditions in sea environment.

The outline of this chapter can be summarized as follows: Sect. 2 describes the dynamic model of AUV in horizontal plane. The design procedure of conventional SMC included in Sect. 3. Then, Sect. 4 presented traditional FSMC algorithm for steering control of AUV. While, In Sect. 5, DFSMC algorithm is proposed for regulating heading angle with updating indirectly width of boundary layer and hitting gain. MATLAB/Simulink based numerical simulations for stabilizing horizontal position of AUV presented in Sect. 6. Finally, conclusions are given in Sect. 7.

2 Dynamic Model of AUV in Horizontal Plane

Dynamical behavior of an AUV can be described in a common way through six degree of freedom (DOF) nonlinear equations in the two co-ordinate frames.

$$M(\nu)\dot{\nu} + C_D(\nu)\nu + g(\eta) + d = \tau, \qquad \dot{\eta} = J(\eta)\nu, \tag{1}$$

where, $\eta = [x, y, z, \phi, \theta, \psi]^T$ is the position and orientation vector in earth fixed frame, $\nu = [u, v, w, p, q, r]^T$ is the velocity and angular rate vector in body-fixed frame. $M(\nu) \in \Re^{6\times 6}$ the inertia matrix (including added mass), $C_D(\nu) \in \Re^{6\times 6}$ is the matrix of Coriolis, centripetal and damping term, $g(\eta) \in \Re^6$ the gravitational forces and moments vector, d denotes the disturbances, τ is the input torque vector and $J(\eta)$ is the transformation matrix defined as

$$\mathbf{J}(\eta) = \begin{pmatrix} c\psi c\theta & -s\psi c\phi + c\psi s\theta s\phi & s\psi s\phi + c\psi c\phi s\theta & & & \\ s\psi c\theta & c\psi c\phi + s\phi s\theta s\psi & -c\psi s\phi + s\theta s\psi c\phi & & 0 & \\ -s\theta & c\theta s\phi & c\theta c\phi & & & \\ & & & 1 & s\phi t\theta & c\phi t\theta \\ & 0 & & 0 & c\phi & -s\phi \\ & & & 0 & s\phi/c\theta & c\phi/c\theta \end{pmatrix} \tag{2}$$

where, $s. = \sin(.)$, $c. = \cos(.)$ and $t. = \tan(.)$. Underwater vehicles are generally designed to have symmetric structure; therefore, it is reasonable to assume that the body-fixed co-ordinate is located at the center of gravity with neutral buoyancy. Furthermore, for AUVs, whose shape could be depicted as in Fig. 1 that having one propeller and two stern planes and two rudders to control the vehicle. In horizontal plane, we assume only yaw motion equation for AUV. For control system design purposes, the vehicle was assumed to be commanded directly in thrust. In this case, the simplified horizontal dynamics can be written in dimensional form as,
Yaw motion Equation:

$$I\dot{r} = C_{N\nu}u\nu + C_{N\nu|\nu|}\nu|\nu| + C_{Nr}ur + C_{Nr|r|}r|r| + C_{N\dot{\nu}}\dot{\nu} + C_{N\dot{r}}\dot{r} + \delta_r \tag{3}$$

For small roll and pitch angles, we have that

$$\dot{\psi} = (\sin(\phi)q + \cos(\phi)r)/\cos(\theta) \approx r \tag{4}$$

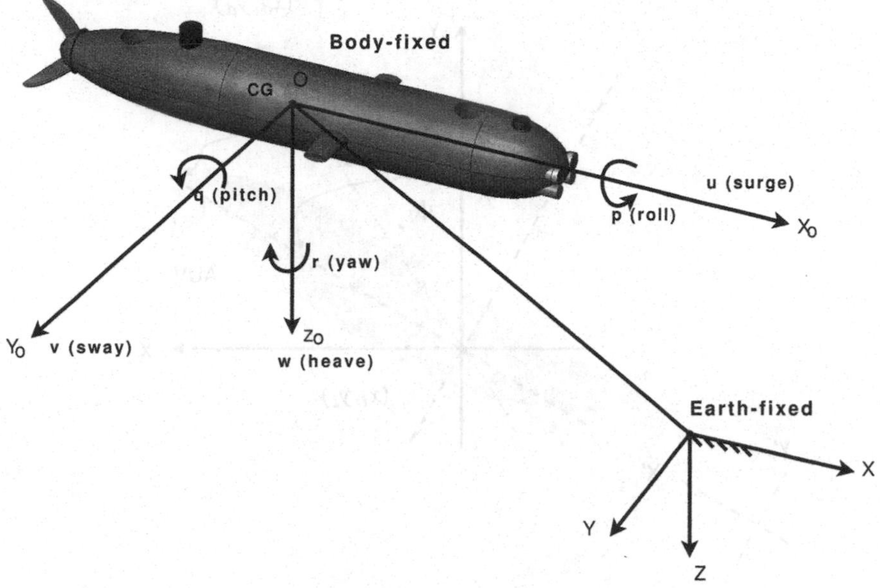

Fig. 1 Body-fixed frame and earth-fixed frame for AUV

Quadratic damping coefficients can be neglected, because of limited magnitude of ν and r. In order to determine a horizontal plane dynamics equation of motion, all unrelated terms (u, ν) will be set to zero, then simplified nonlinear equation of AUV. A one-degree-of-freedom vehicle model is used herein to describe the horizontal turning behavior of the AUV. The model includes drag, added mass, and thrust moment for yaw motion,

$$I\dot{r} + br|r| = u + d \tag{5}$$

where, I denotes the vehicles mass moment of inertia plus the added inertia of the body about the body-fixed z-axis, $r = \dot{\psi}$ represents the body-fixed rate for heading direction, b denotes the square-law damping coefficient, u is the moment generated by commanding differential thrust force on the left and right thrusters, ψ is the heading angle required during horizontal turn and d represents the disturbance caused by ocean currents, modeling errors and unmodeled dynamics. The line-of-sight guidance procedure is illustrated in Fig. 2. With respect to the dynamic model, the following physically motivated assumptions can be made:

Assumptions (1): The vehicle's mass moment of inertia plus the added inertia of the body about the body-fixed z-axis i.e. I is time varying and unknown, but it is positive and bounded in between, i.e. $0 \le I_{\min} \le I_{(t)} \le I_{\max}$.

Assumptions (2): The square-law damping coefficient b is time varying and unknown but it is bounded in between, i.e. $b_{\min} \le b_{(t)} \le b_{\max}$.

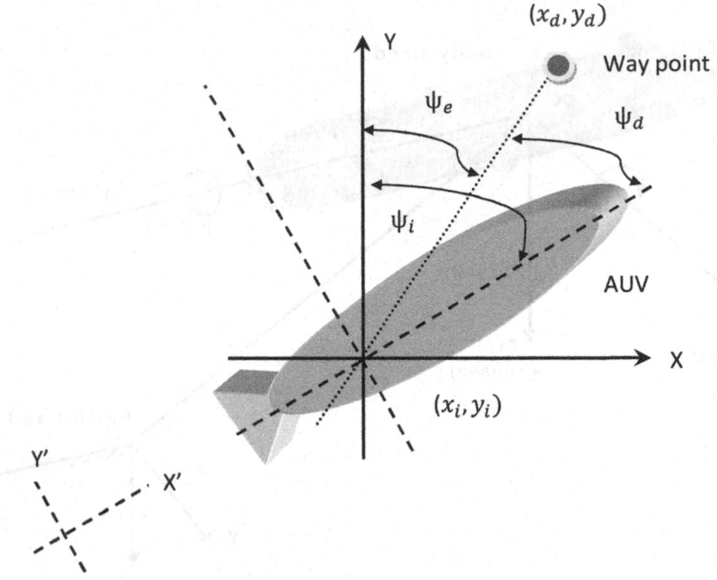

Fig. 2 Rotation of AUV over heading angle

Assumptions (3): The disturbance effect $d(t)$ is time varying and unknown but it is bounded by a known function of ψ, r and t, i.e. $|d_{(t)}| \leq \delta(t, \psi, r)$.

Here, the dynamics of an AUV are highly nonlinear and the hydrodynamic coefficients of vehicle are difficult to be accurately identified, because of variations of these coefficients with different operating environment. For the purpose of simplifications, in this paper, all unstructured uncertainties are assumed to be bounded by known constant.

3 Sliding Mode Control

In this section, we state the general concepts of a SMC for steering control of underwater vehicle in horizontal plane. SMC is well known for its robustness to modeling errors and insensitivity to parameter variations and nonlinearity. Due to this property of SMC that made it find many successful practical applications. The first step of SMC design is to select a sliding surface that models the desired closed loop performance in state variable space. In most of cases, the sliding surfaces were selected as linear hyperplane that resulted in a PD type sliding surface. The second step is to design a hitting control law such that the system state trajectories are forced towards the sliding surface and stay on it.

The control problem is to synthesize a control law such that the state ψ traces the desired trajectory ψ_d within the tolerance error bound defined by

$$\| \psi - \psi_d \| \leq \gamma_1, \| \dot{\psi} - \dot{\psi}_d \| \leq \gamma_2, \quad \gamma_1 > 0, \gamma_2 > 0 \tag{6}$$

It is assumed that $\psi_d(t)$, $\dot{\psi}_d(t)$ and $\ddot{\psi}_d(t)$ are well defined and bounded for all time t. The error signal as $e = \psi - \psi_d$ and let $s(t)$ be a sliding surface defined in the state space by the equation $s(\psi; t) = 0$.

$$s(\psi; t) = \left(\frac{d}{dt} + \lambda\right) e(t) = \dot{e}(t) + \lambda e(t) \tag{7}$$

Since, λ is a positive constant that determines the slope of the sliding surface. The process of sliding mode control can be divided in to two phases, namely the sliding phase with $s(t) = 0$ and $\dot{s}(t) = 0$ and the reaching phase with $s(t) \neq 0$. Corresponding to the two phases, two types of control law, that is, the continuous control and discontinuous control can be derived separately. Based on the Lyapunov theorem, the sliding surface reaching condition is $s.\dot{s} < 0$. If a control input u can be chosen to satisfy this reaching condition, the control system will converge to the origin of the phase plane. Generally, the $sgn(s/\phi)$ is well known and it is a constant or a slow time varying function for practical physical system. It can also be found that $\dot{s}$ increases as u decreases and vice versa. If situation is $s > 0$, then the increasing of u will result in $s.\dot{s}$ decreasing. When the condition is $s < 0$, $s.\dot{s}$ will decrease with the decreasing of u. Based on this qualitative analysis, the control input u can be designed in an attempt to satisfy the inequality $s.\dot{s} < 0$. Now, let the problem of controlling the heading angle of AUV is governed by Eq. (5), be treated in Filippov's way.

The control input to get the state ψ to track a specific time varying desired state ψ_d in the presence of model uncertainty on $(-b/I)\dot{\psi}|\dot{\psi}|$ is made to satisfy the following sliding condition

$$\frac{1}{2}\frac{d}{dt}s^2 \leq \eta|s|, \qquad \eta \geq 0 \tag{8}$$

Let us define a control law composed by an equivalent control and a discontinuous term.

$$u = b\dot{\psi}|\dot{\psi}| - d + I\lambda(\dot{\psi}_d - \dot{\psi}) - Ksgn(s/\phi) \tag{9}$$

where, K is the hitting gain, its value should be selected as a positive real number and $sgn(.)$ is the signum function defined as

$$\mathrm{sgn}(x) = \begin{cases} -1 & \text{if } x < 0 \\ 0 & \text{if } x = 0 \\ 1 & \text{if } x > 0 \end{cases} \tag{10}$$

The discontinuous control term is included to account for the presence of modeling errors and disturbances. It is discontinuous across the sliding surface $s(t)$, which leads to a serious and undesirable phenomenon, namely chattering. To avoid this phenomenon, a boundary layer is introduced with width. Hence, signum function can be easily replaced by a saturation function sat(s/ϕ) that is expressed as follows

$$\operatorname{sat}(s/\phi) = \begin{cases} s/\phi & \text{if } |s/\phi| \leq 1 \\ \operatorname{sgn}(s/\phi) & \text{otherwise} \end{cases} \tag{11}$$

The control law is designed in such manner that, the output trajectory reaches to the sliding surface and slide on it, under that condition it will move towards equilibrium point. The controller is developed by combining the variable structure systems theory and Lyapunov design methods. It possesses the desirable properties of the sliding mode systems while avoiding unnecessary discontinuity of the control and thus, eliminates chattering effect by incorporating fuzzy logic in SMC algorithm, such applications are termed as being indirect. They have the main objective of alleviating practical problems encountered in the implementation of SMC's. Therefore, use of combined fuzzy logic with SMCs is getting more and more popular. In the FSMC approach, a special attention is paid to chattering elimination without system performance degradation. Furthermore, the prior knowledge necessary about the system dynamics for controller design is kept to a minimum.

4 Fuzzy Sliding Mode Control

The dynamic behavior of FLC is characterized by a set of linguistic rules based on expert knowledge. From this set of rules, the inference mechanism of FLC will able to provide appropriate fuzzy control action. Suppose the rules of fuzzy controller are based on SMC, and then it is called the FSMC. In this section, we follow the development established in Kim and Lee (1995) and show that a particular fuzzy controller is an extension of an SMC with a boundary layer. The fuzzy control rules can be represented as the mapping of the input linguistic variable s to output linguistic variable u_f. Let the traditional FSMC algorithm designed in this article is constructed from the following IF-THEN rules,

R^1 : If s is **NL** then u_f is **BB**

R^2 : If s is **NM** then u_f is **B**

R^3 : If s is **ZE** then u_f is **M**

R^4 : If s is **PM** then u_f is **S**

R^5 : If s is **PL** then u_f is **SS**

Equivalently, R^i: If s is $F_s^{\prime i}$ then u_f is $F_{uf}^{\prime i}$, i = 1, 2,...5.

where, NL is *Negative Large*, NM is *Negative Medium*, ZE is *Zero*, PM is *Positive Medium*, PL is *Positive Large*, BB is *Bigger*, B is *Big*, M is *Medium*, S is *Small* and SS is *Smaller*. NL, NM,..., S, SS are labels of fuzzy sets and their corresponding

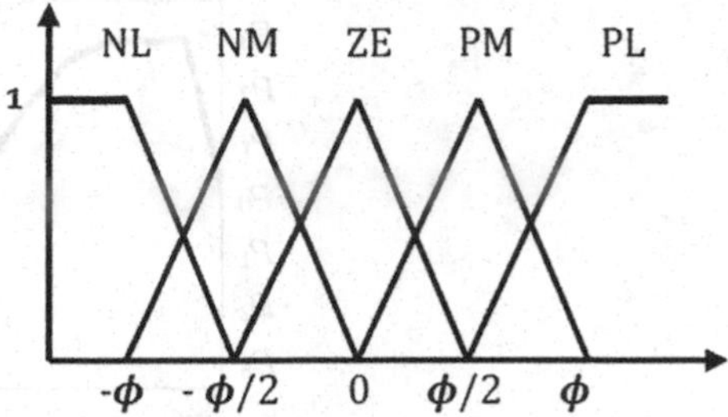

Fig. 3 Membership functions for input s

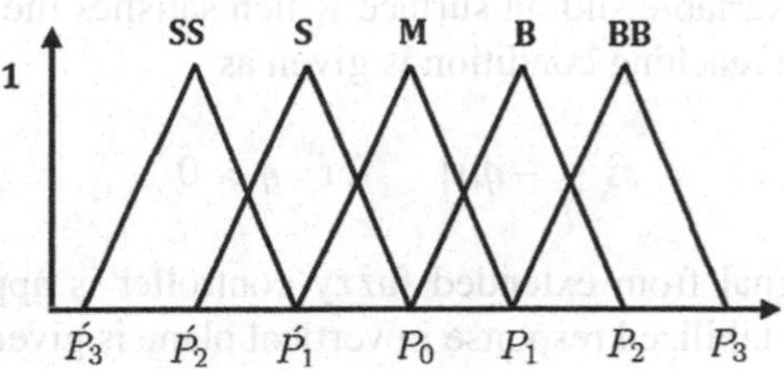

Fig. 4 Membership functions for output u with base values $P_0 : u_{eq}, P_1 : u_{eq} + K_f/2, P_2 : u_{eq} + K_f, P_3 : u_{eq} + 2K_f, \acute{P}_1 : u_{eq} - K_f/2, \acute{P}_2 : u_{eq} - K_f, \acute{P}_3 : u_{eq} - 2K_f$

membership functions are depicted in Figs. 3 and 4, respectively. Let X and Y are the input and output space of the fuzzy rules, respectively. For any arbitrary fuzzy F_x in X, each rule R_i can determine a fuzzy set $F_x * R_i$ in Y.

Use the sup-min compositional rule of inference and suppose F_x be a fuzzy singleton, then

$$\mu_{F_x \circ R^i}(u_f) = \min[\mu_{F^i_s}(\alpha), \mu_{F^i_{uf}}(u_f)] \tag{12}$$

the deduced membership function $F_u^{\prime d}$ of the consequence of all rules is,

$$\mu^d_{\widetilde{F}^d_{uf}}(u_f) = \max[\mu_{\widetilde{F}_x \circ R^i}(u_f), \ldots \mu_{\widetilde{F}_x \circ R^s}(u_f)] \tag{13}$$

where, the output variable in Eq. (13) is fuzzified output. For the defuzzifier, the centroid defuzzification method is used to find the crisp output is given in Eq. (14). Figure 5 is the result of defuzzified output for a fuzzy input as sliding surface and overall control equation of fuzzy sliding mode controller is given as

$$\hat{u} = \frac{\int u_f \ \mu^d_{\widetilde{F}^d_{uf}}(u_f) du_f}{\int \mu^d_{\widetilde{F}^d_{uf}}(u_f) du_f} \tag{14}$$

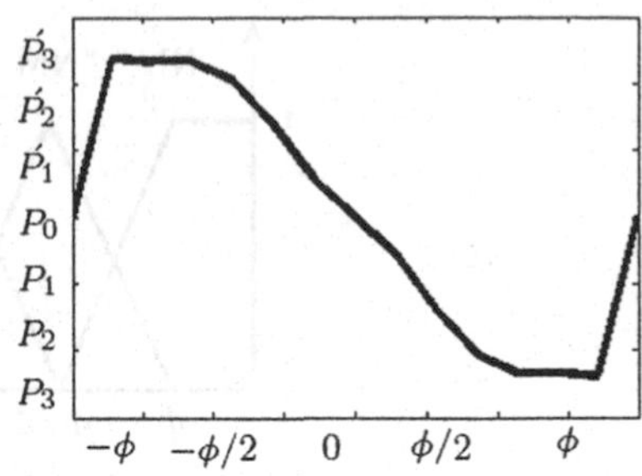

Fig. 5 Result of defuzzification of a fuzzy controller

Then, a FSMC with variable sliding surface which satisfies the reaching condition will be designed. The reaching condition is given as

$$s\dot{s} \leq -\eta|s| \quad \text{for} \quad \eta > 0 \tag{15}$$

The crisp control signal from extended fuzzy controller is applied to the system model for achieving stabilized response in vertical plane is given as follows

$$\hat{u} = u_{eq} - K_f \text{sgn}(s/\phi) \tag{16}$$

Here, fuzzy control is employed as low pass filter for smoothing the control input in SMC due to that chattering problem is prevented. In this technique, minimum rules are designed to satisfy the sliding condition and also capable of adopting uncertainty in the model parameters. In the design of FSMC, FLC scheme have been used as a direct controller, in which, the FLC is non-adaptive in nature. This type of FLC is called non-adaptive if all of its parameters, i.e. scaling factors, membership functions and rules are kept fixed during the operation of the controller. Here, an adaptive FLC is used to fine tune scaling factors and varying support of input–output membership function for improving the output trajectory performance. Therefore, FLC employed as a supervisory control with the FSMC in proposed control algorithm. Here, an AFSMC is designed for improving the output trajectory response in steering control of an AUV model. The AFSMC algorithm is developed in two stages, which employed fuzzy approximators for adaptation of input and output variables of fuzzy inference system.

5 Dynamic Fuzzy Sliding Mode Control

In this section, we propose dynamic tuning methods for input–output linguistic variables of FSMC, through the information of error dynamics, that we call it DFSMC. The motivation is that, we consider the values of boundary layer thickness and hitting gain for adjusting the supports of input–output membership functions in fuzzy inference engine of FSMC algorithm. The proposed control method based on two

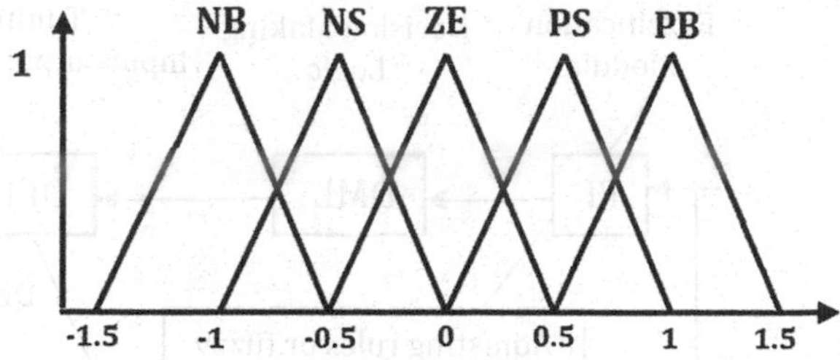

Fig. 6 Membership functions of input variable as error signal and output variable as width of boundary layer

fuzzy approximators for varying boundary layer width and hitting gain, due to which set point tracking response is enhanced with minimum chattering effect. Chattering is a high frequency oscillation around the desired equilibrium point. It is undesirable in practice, because it involves high control activity and can excite high frequency dynamics ignored in the modeling of the system, which can be reduced somewhat by introducing a bound region containing the switching surface to smooth the control behavior. So that, in first stage of design, fuzzy algorithm is developed to choose boundary layer thickness depends on error dynamics for eliminating chattering effect. Due to this first fuzzy approximator, supports of input fuzzy membership function in FSMC algorithm are varying continuously.

The formulation of fuzzy rule for dynamic tuning of boundary layer thickness is based on concepts that, width of the boundary layer indicates the ultimate boundedness of system trajectories, we can arbitrarily adjust the steady state error by proper selection of ϕ. However, a small ϕ might produce a boundary layer so thin that it risks exciting high frequency dynamics. In this fuzzy adaptation, input error signal and output boundary layer thickness are decomposed in to five fuzzy partitions as shown in Fig. 6, which is expressed as *Negative Big* (NB), *Negative Small* (NS), *Zero* (ZE), *Positive Small* (PS) and *Positive Big* (PB). The fuzzy logic rule base is designed as follows

$$\text{Rule}(i) : \text{If} e \text{is } F_1^i \text{ then } \phi_i \text{ is} \gamma_i$$

where, F_1^i, i = 1, 2,...,*m*. are the labels of single input fuzzy set characterized by membership functions and γ_i, i = 1, 2,..., *m* are the triangular membership functions. The sliding inference rules are composed as in Table 1. The defuzzification of the output is accomplished by the method of centroid.

$$\phi^* = \frac{\int \mu_c(\gamma).\gamma \ d\gamma}{\int \mu_c(\gamma) \ d\gamma} \tag{17}$$

Table 1 Rule base for width of boundary layer

Error (e)	NB	NS	ZE	PS	PB
Width (ϕ)	PS	PB	ZE	NB	NS

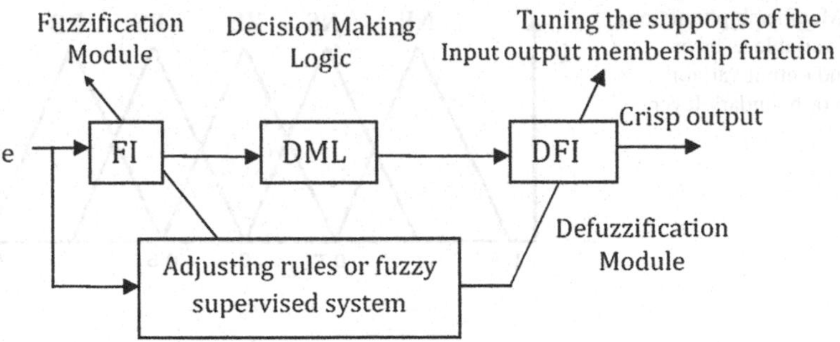

Fig. 7 Block diagram of fuzzy inference module in FSMC algorithm

The crisp output variable ϕ^* of fuzzy logic tuning scheme is used for varying width of boundary layer, due to which chattering problem is eliminated. Here, a single-input fuzzy adaptation is used to continuously compute the width of boundary layer, with the result that boundary layer thickness is time varying and tracking performance of AUV under the heading angle control is enhanced. In the second stage of design, a dynamic fuzzy logic tuning method is developed for estimating hitting gain. Due to that, supports of the output membership function is adjusted through the information of error dynamics. The output membership function of fuzzy inference system composed of two factor such as equivalent control u_{eq} and hitting gain K_f.

A supervisory fuzzy inference system is used to adaptively tune the hitting gain, in order to improve the approaching angle towards sliding surface. The principal of operation can be easily understood from the block diagram of fuzzy inference system as shown in Fig. 7, in which dynamic tunings are used to update the supports of input–output membership functions. In path tracking application, however, the system invariance properties are observed only during the sliding phase, but in reaching phase, tracking may be hindered by disturbances or parameter variations. The straightforward way to reduce tracking error and reaching time by increasing hitting gain, which may causes chattering effect. The chattering can also be reduced by using small boundary layer thickness. The selection of hitting gain value is based on minimization of tracking error and reaching time, whenever the tracking error is negative then choose the small gain value for desired performance of system and vice versa. The sliding hyperplane highly depend upon dynamics of error and change in error, so that consider an error as input variable to the fuzzy logic module for updating hitting gain.

The fuzzy rules are designed such that, as the value of e is in large level then required control effort is bigger, due to which speed of convergence is increased. Therefore, shift the supports of output membership functions towards the right, for providing large control forces. As the value of e is near to zero then the output membership functions return back to the original type to prevent the happening of overshoots and keep the tracking accuracy. In this stage, hitting gain is adapted from the second fuzzy approximator due to which supports of output membership functions are indirectly tuned. This output membership function is the part of fuzzy inference

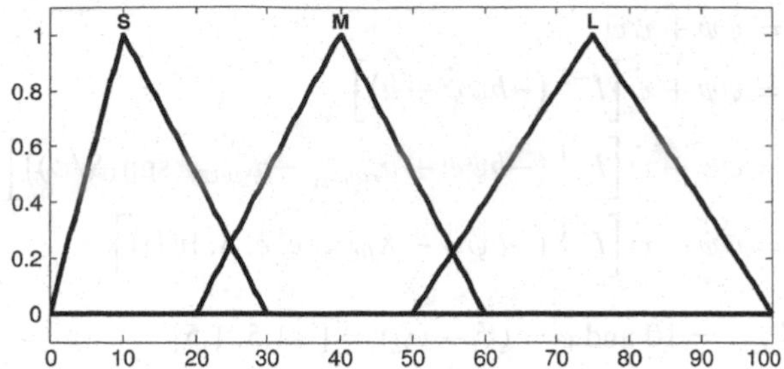

Fig. 8 Membership functions of hitting gain

engine, which included in FSMC algorithm. The second fuzzy approximator plays an important role in an indirect type of adaptation, which has error as input variable and hitting gain is output variable.

By using above consideration, the general rule is composed as, if the error signal in negative region then select small value of hitting gains and vice versa. Let e is the error signal as the input linguistic variable of fuzzy logic is as shown in Fig. 6 and the hitting control gain K_f be the output linguistic variable is as shown in Fig. 8, the associated fuzzy sets for e and K_f are expressed as follows:

The error signal (e) as antecedent proposition can be expressed in to five fuzzy partitions such as *Negative Big* (NB), *Negative Small* (NS), *Zero* (ZE), *Positive Small* (PS) and *Positive Big* (PB). The hitting control gain K_f as consequent proposition can be expressed in to three fuzzy partitions such as *Small* (S), *Medium* (M) and *Large* (L). Then, fuzzy linguistic rule base can be design as follows:

Rule 1 : If e is **NB** then K_f is **S**

Rule 2 : If e is **NS** then K_f is **S**

Rule 3 : If e is **ZE** then K_f is **M**

Rule 4 : If e is **PS** then K_f is **L**

Rule 5 : If e is **PB** then K_f is **L**

In this study, centriod defuzzification method is adopted for estimation of hitting control gain through fuzzy logic inference mechanism. Moreover, the stability of the underwater vehicle in vertical plane can be analyzed by direct lyapunov function approach, which uses DFSMC algorithm. In this investigation, each rule is applied to common lyapunov function.

Select a lyapunov function as follows

$$V = \frac{1}{2}\left(\psi^2 + \dot{\psi}^2\right) \tag{18}$$

which is obviously positive definite and differentiable. Then, it's derivative can represented as

$$\begin{aligned}\dot{V} &= \psi\dot{\psi} + \dot{\psi}\ddot{\psi} \\ &= \psi\dot{\psi} + \dot{\psi}\left[I^{-1}\left(-b\psi\dot{\psi} + \hat{u}\right)\right] \\ &= \psi\dot{\psi} + \dot{\psi}\left[I^{-1}\left(-b\psi\dot{\psi} + u_{eq_{fuzzy}} - K_{fuzzy}\ \mathrm{sgn}(S/\phi)\right)\right] \\ &= \psi\dot{\psi} + \dot{\psi}\left[I^{-1}\left(-b\psi\dot{\psi} - K_{fuzzy}(e, \dot{e}, \lambda, |\psi|)\right)\right]\end{aligned}$$

For Rule (1): $K_{fuzzy} = 10$ and $e = (\psi - \psi_d) = [-1.5, 1.5]$

$$\dot{V} = \psi\dot{\psi} + \dot{\psi}\left[I^{-1}\left(-b\psi\dot{\psi} - 10|\psi|\right)\right] \leq 0$$

For Rule (2): $K_{fuzzy} = 10$ and $e = (\psi - \psi_d) = [-1, 0]$

$$\dot{V} = \psi\dot{\psi} + \dot{\psi}\left[I^{-1}\left(-b\psi\dot{\psi} - 10|\psi|\right)\right] \leq 0$$

For Rule (3): $K_{fuzzy} = 40$ and $e = (\psi - \psi_d) = [-0.5, 0.5]$

$$\dot{V} = \psi\dot{\psi} + \dot{\psi}\left[I^{-1}\left(-b\psi\dot{\psi} - 40|\psi|\right)\right] \leq 0$$

For Rule (4): $K_{fuzzy} = 80$ and $e = (\psi - \psi_d) = [0, 1]$

$$\dot{V} = \psi\dot{\psi} + \dot{\psi}\left[I^{-1}\left(-b\psi\dot{\psi} - 80|\psi|\right)\right] \leq 0$$

For Rule (5): $K_{fuzzy} = 80$ and $e = (\psi - \psi_d) = [0.5, 1.5]$

$$\dot{V} = \psi\dot{\psi} + \dot{\psi}\left[I^{-1}\left(-b\psi\dot{\psi} - 80|\psi|\right)\right] \leq 0$$

Hence, all of the five rules in the FLC can lead to stabilize underwater vehicle and completes the proof.

5.1 Summary of the Proposed Algorithm and Design Procedure

The control algorithm is summarized as follows

1. Determine a stable sliding mode surface from Eq. (7).
2. Calculate the equivalent control from given condition such as $\dot{s} = 0$.
3. Define the membership function for the input variable as sliding surface s and output variable as FSMC output $\hat{u}$.
4. Define rule base for FSMC algorithm.
5. $\hat{u}$ is the output of FLC which calculated via the defuzzification method.
6. Find the boundary layer thickness variation ϕ from fuzzy supervisory system based on error dynamics.

7. Find the K_f using another fuzzy supervisory system which also depend on AUV's error dynamics.
8. Base values of Input–Output membership functions in FSMC algorithm are adaptively tuned by two fuzzy approximators.
9. Calculate the overall control signal applied to AUV in horizontal plane.

As our interests are focused mainly on the application of a fuzzy controller for adopting base values of input–output membership functions of FSMC architecture. The basic parameters of FSMC algorithm are tuned by using fuzzy logic approximators, due to which system performance is enhanced. Fuzzy self tuning of sliding surface slope, boundary layer width and hitting gain are summarized in brief manner.

5.1.1 Fuzzy Self Tuning of Sliding Surface

A conventional time invariant (fixed) sliding surface has the fundamental disadvantage that when the system states are in the reaching mode, the tracking error cannot be controlled directly and hence the system becomes sensitive to the parameter variations. This sensitivity can be minimized or eliminated if the reaching mode duration is shortened. Moreover, finding the optimum value of the slope requires tedious work and usually, it is a complicated task. Thus, how to tune the slope of a sliding surface is an important topic in the sliding mode controlled nonlinear systems. Several methods exist (Liu et al. 2005; Yagiz and Haciogluy 2005; Hung et al. 2007; Yorgancioglu and Komurcugil 2008; Amer et al. 2011) in literature aiming at to eliminate the sensitivity during the reaching mode. The control performance of system using SMC is highly depends on the slope of the sliding mode function with following conditions are considered. When the value of λ becomes larger, the rise time will become smaller, but at the same time, both overshoot and settling time will become larger and vice versa. If the slopes are fixed, the control system may perform differently for different control situations such a control system is difficult to cover all the control situations in good performance. To solve the problem, it is desirable to design a control law to adjust the slope of sliding mode function in real time. In mechanical systems, the value of sliding surface slope is typically limited by three factors such as the frequency of the lowest unmodeled structural mode, the largest unmodeled time delay and the sampling rate.

The movement of sliding surface adapted to arbitrary initial conditions, which was first introduced by Choi et al. (1994). Afterwards, Ha (1996) applied fuzzy tuning to moving sliding surfaces for fast and robust tracking control for a class of nonlinear systems. The fuzzy tuning approach is utilized for accelerating the reaching phase and reducing the influence of unmodeled uncertainties, thus improving system robustness. The fuzzy rule for tuning sliding surface slopes λ_i can be formulated with the help following concepts. If large values of λ_i are available the system will be more stable but the tracking accuracy may be degraded, because of a longer reaching time of the representative point to the surface. Conversely, if small values of λ_i are chosen, the convergence speed on the sliding surface itself will be slow, leading to longer tracking times. In this way fuzzy supervisory system is designed for obtaining optimal value of sliding surface slope, due to which tracking performance is enhanced.

5.1.2 Fuzzy Self Tuning of Boundary Layer Thickness

The chattering describes the phenomenon of finite frequency, finite amplitude oscillations appearing in many sliding mode implementations. These oscillations are caused by the high frequency switching of a SMC, which excites unmodeled dynamics in the closed loop. As one way to alleviate this problem, a boundary layer around sliding surface is typically used. In this case the selection of boundary layer thickness is a crucial problem for trade off between tracking error and chattering. The parameter tuning is usually done by trial-and-error method in practice causing significant effort and time.

In order to attenuate the chattering problem, various methods (Hwang and Tomizuka 1994; Erbatur et al. 1996; Choi et al. 1996; Lee et al. 2001) are describes adaptive tuned boundary layer thickness by using fuzzy approximator. The value of boundary layer thickness should be varying according to the chattering level in the control signal in order to achieve the best performance possible. A variety of chattering measures can be formulated for adjusting boundary layer thickness, due to which smooth tracking performance is obtained. The main idea can be summarized as below. When chattering occurs then width of boundary layer should be increased to force the control input to be smoother. The boundary layer thickness should be decreased if control activity is low. Low control activity can be identified by small values of chattering variable.

The on line tuning of boundary layer thickness with the help of fuzzy approximator, based on $|e|$ and $|\dot{e}|$. The boundary layer thickness is not fixed by an arbitrary value but self tuned by some fuzzy rules, which are formulated by using following conditions. if both $|e|$ and $|\dot{e}|$ have small values, namely, the states approach nearby steady states, then rule decrease the thickness. Similarly, if $|e|$ or $|\dot{e}|$ has a large value, namely, the states are far away from steady states, then rule increases the thickness for alleviating the chattering problem. The determination of a suitable boundary layer thickness which can achieve best performance still eliminating chattering effect, is possible by fuzzy tuning approach.

5.1.3 Fuzzy Self Tuning of Hitting Gain

In SMC, auxiliary control effort should be designed to eliminate the effect of the unpredictable perturbations. The auxiliary control effort is referred to as hitting control effort. The hitting control gain concerned with upper bound of uncertainties and sign function. However, the upper bound of uncertainties, which is required in the control law, is difficult to obtain precisely in advance for practical applications. Several methods exist (Ryu and Park 2001; Liang and Su 2003; Wai and Su 2006; Amer et al. 2011) in literature, which has significant advantage that, convergence speed increased and reaching time is reduced. In path tracking systems, however, the system invariance properties are observed only during the sliding phase. In reaching phase, tracking may be hindered by disturbances or parameter variations.

The straightforward way to reduce tracking error and reaching time by increasing hitting gain, which may causes chattering effect. The chattering can also be reduced by using small boundary layer thickness. The selection of hitting gain value is based on minimization of tracking error and reaching time, whenever the tracking error is negative then we have to choose small gain value for desired performance of system and vice versa. The sliding hyperplane highly depend upon dynamics of error and change in error so that we have to consider this variable as input to the fuzzy logic module for updating hitting gain. By using above consideration, the general rule is composed as, if sliding surface in negative region then select small value of hitting gain and vice versa. In this way, hitting gain can be determined by using fuzzy logic tuning approach.

6 Simulation Results

In this section, some simulation results are provided to demonstrate the effectiveness and robustness of the proposed control technique. Heading angle control of an AUV is chosen as an example for simulation purpose, which can be represented by simplified nonlinear equation having one degree of freedom and described by Guo et al. (2003). Here, the main objective is to control steering of underwater vehicle by using DFSMC method. This control technique is applicable to nonlinear AUV model, because conventional linear control can't handles nonlinearity, modeling error, parametric variation and disturbances.

In order to evaluate the control system performance, three different numerical simulations were performed. The obtained results were presented from Figs. 9, 10, 11, 12 and 13. In the first case, it was considered that the model parameters, I and b, were perfectly known. Regarding controller and model parameters, the following values were chosen $I = 24.13$ kgf.m.s^2, $b = 32.50$ kgf.m.s^2 and $d(t) = 0.25*\sin(t)$. With other control parameters were considered as $\lambda = 0.5$, $\phi = 0.01$ and $k_f = 5$. Figure 9 gives the corresponding results for the tracking of ψ_d, considering that the initial state coincides with the initial desired state. In a first test a piece-wise constant reference position was used, which reports the actual and desired trajectory obtained using the different approaches such as SMC, FSMC and DFSMC. With evidence that, the output heading angle response of AUV, due to DFSMC is most desirable, because it's minimum reaching time, no overshoot and smooth tracking performance.

A tracking test using a sinusoidal reference profile has been also carried out. Figure 10 shows that the actual trajectory response of DFSMC converges to the desired one after a very short transient compared to other approaches.

As observed in Fig. 11, even in the presence of external disturbances, a novel DFSMC and FSMC are able to provide trajectory tracking with a small associated error and no chattering at all. It can be also verified that the proposed control law provides a smaller tracking error when compared with the conventional SMC method.

The improved performance of DFSMC over SMC is due to its ability to recognize and compensate the external disturbances, with better tracking response. In the second simulation study, the parameters for the controller were chosen based on the assumption that exact values are not known but with a maximal uncertainty of

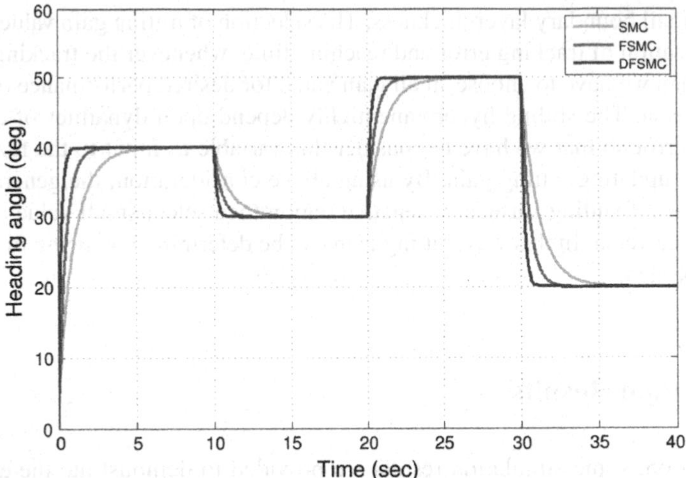

Fig. 9 Set point tracking response of AUV in *horizontal* plane

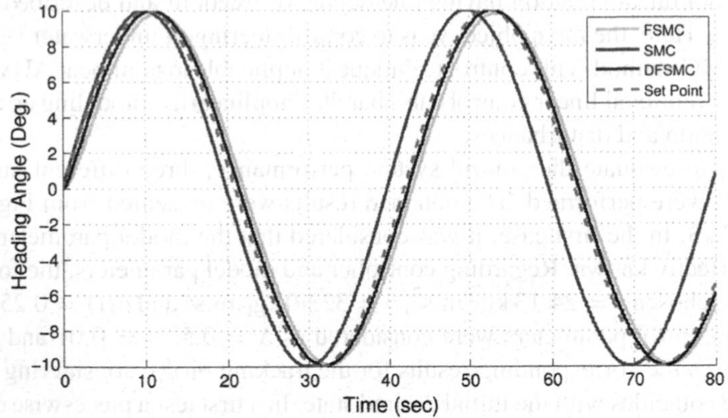

Fig. 10 Sinusoidal tracking response of AUV in *horizontal* plane

$\pm 20\,\%$ over previous adopted values of AUV model parameters. The other control parameters, as well as the disturbance force and the desired trajectory, were defined as before. Figure 12 shows the obtained results, in which proposed controller was able to handle nonlinearity and parametric uncertainty, while output response under the influence of parametric uncertainty due to SMC and FSMC depicts oscillations and small variations respectively.

The phase portrait of AUV model under steering control is shown in Fig. 13, in which reaching time of DFSMC method is better than FSMC and SMC, with no chattering effect. From simulation results, it is clear that proposed DFSMC provide desired tracking response with smooth control signal and minimum reaching time during model uncertainties and disturbances in operating condition.

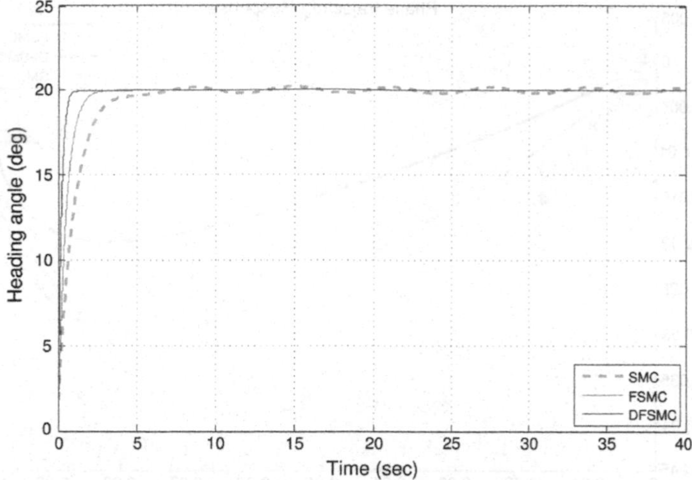

Fig. 11 Response of AUV under the influence of disturbance

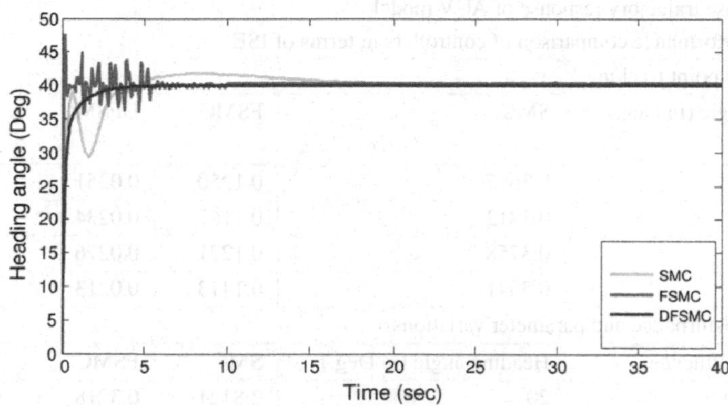

Fig. 12 Response of AUV under the influence of parametric uncertainty

As performance measure for a quantitative comparison, we use integral square of error (ISE) which is defined as

$$ISE = \int_{0}^{t} e^2.dt \tag{19}$$

In performance comparison, three conditions are considered as set point tracking, disturbance rejection and parameter variations. It is observed that, ISE values for above mentioned conditions are considerably reduced in magnitude than other

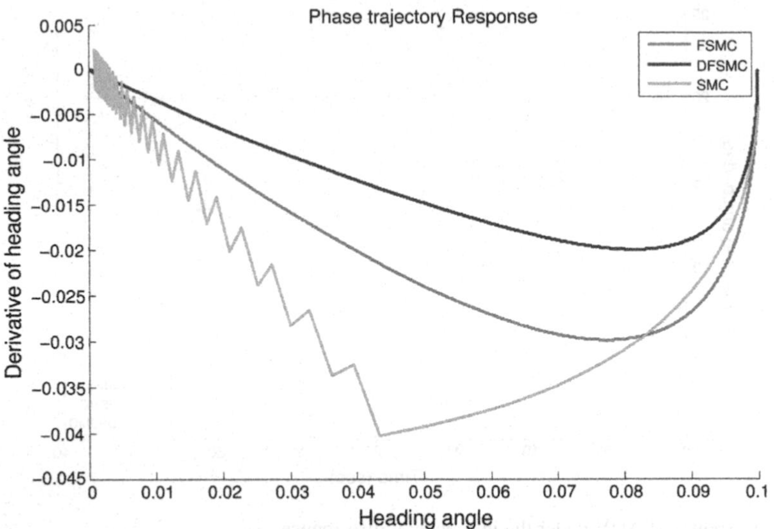

Fig. 13 Phase trajectory response of AUV model

Table 2 Performance comparison of controllers in terms of ISE

Case(I): Set point tracking				
Heading angle (in Deg.)	SMC	FSMC	DFSMC	
40	0.3695	0.1250	0.0251	
30	0.3812	0.1167	0.0234	
50	0.3758	0.1271	0.0276	
20	0.3541	0.1413	0.0213	
Case(II): Disturbance and parameter variations				
Under the influence	Heading angle (in Deg.)	SMC	FSMC	DFSMC
Disturbance	20	2.8121	0.3218	0.1589
Parameter variation	40	3.4813	1.2764	0.1242

techniques dealt within this chapter. The values of different errors for various control strategies and under the influences of different conditions are tabulated in Table 2.

Therefore, from the response curve and Table 2, it is clear that DFSMC control algorithm gives better performance in terms of transient as well as steady state responses.

7 Conclusion and Future Work

7.1 Conclusion

This study has demonstrated the effectiveness and robustness of proposed control algorithm for regulating heading angle of underwater vehicles in horizontal plane. The fuzzy logic approximators were used for enhance the tracking performance of AUV by adapting an input and output parameter as width of boundary layer and hitting control gain of fuzzy inference engine. Due to first fuzzy approximator, width of boundary layer is updated continuously for eliminating chattering effect. The hitting control gain is adapted for tuning the supports of output membership functions, due to which an optimum approaching angle towards sliding surface was determined. By comparing the simulation results, we found that the performance of the proposed DFSMC is superior to that of conventional SMC and FSMC. The attractive features of the controller follow.

- The exact mathematical model and the estimation of upper bounds on uncertainties of the AUV are not required in the controller design. The only necessary information to design the controller is the qualitative knowledge of the AUV, such as it's operation ranges and the form of its nominal model.
- The fuzzy controller is design to learn and compensate nonlinearities and uncertainties, thus allowing a reduction of the SMC's switching gains. The problem of chatter inherent in conventional SMC is therefore managed effectively while ensuring sliding behavior, which implies that chatter is greatly alleviated without sacrificing robustness.
- Two fuzzy supervisory systems are employed in proposed algorithm for updating boundary layer width and hitting gain, due to which chattering problem, reaching time and approaching angle towards sliding surface is effectively reduced.
- The stability of the fuzzy systems is guaranteed by means of the Lyapunov stability criterion, which also gives guidelines in the design of the proposed DFSMC scheme.

Moreover, the fuzzy adaptation is archived by simplified mamdani type single input single output fuzzy inference module with minimum rule base. The single input FLC offers significant reduction in rule inferences and simplify the tuning of control parameters. Practically it can be easily implemented by a look-up table using a low cost microprocessor due its piecewise linear control surface.

7.2 Future Work

Further research will be carried out in the field of nonlinear path following of AUV's for scientific inspection in under sea environment. The following methodologies have scope to develop new control strategies.

- An adaptive Type-2 FSMC can de design to tolerate actuator faults of AUV with external disturbances and measurement noise.
- An alternative approach for fuzzy adaptation scheme can be developed with the help of single layer neural network, which gives minimum computation time for determine optimal values of control parameters.
- Adaptive neuro-fuzzy inference system (ANFIS) can be incorporated in FSMC algorithm for collision avoidance and nonlinear mapping in under sea applications.
- This work can be expanded by applying adaptive fractional FSMC, robust fuzzy terminal SMC and an adaptive fuzzy quasi continuous higher order SMC for regulating various parameters of AUV's.
- The guidance control can be design and tested with the help of navigation algorithms based on forward looking sonar (FLS) images that permit us to obtain the values of the trajectory tracking errors.

Future work will address the problems of reducing controller complexity and evaluating its robustness against parameter uncertainty. The problem of precise tracking of desired trajectory in presence of unknown sea currents also warrants further consideration.

References

Amer, A.F., Sallam, E.A., Elawady, W.M.: Adaptive fuzzy sliding mode control using supervisory fuzzy control for 3 DOF planar robot manipulators. Appl. Soft Comput. **11**(8), 4943–4953 (2011)

Azar, A.T.: Fuzzy Systems. IN-TECH, Vienna, Austria (2010). ISBN: 978-953-7619-92-3

Balasuriya A., Cong L.: Adaptive fuzzy sliding mode controller for underwater vehicles. In: Proceeding of the international conference on control and automation, 12 June 2003, Montreal, Que., Canada, pp. 917–921 (2003). doi:10.1109/ICCA.2003.1595156

Bessa, W.M., Dutra, M.S., Kreuzer, E.: Depth control of remotely operated underwater vehicles using an adaptive fuzzy sliding mode controller. J. Robot. Auton. Syst. **56**(8), 670–677 (2008)

Bessa, W.M., Dutra, M.S., Kreuzer, E.: An adaptive fuzzy sliding mode controller for remotely operated underwater vehicles. Robot. Auton. Syst. **58**(1), 16–26 (2010)

Choi, S.B., Park, D.W., Jayasuriya, S.: A time-varying sliding surface for fast and tracking control of second-order dynamic systems. Automatica **30**(2), 899–904 (1994)

Choi, B.J., Kwak, S.W., Kim, B.K.: Design of a sliding mode controller with self-tuning boundary layer. J. Intell. Robot. Syst. **6**(2), 3–12 (1996)

Choi, S.K., Yuh, J.: Experimental study on a learning control system with bound estimation for underwater vehicles. Int. J. Auton. Robots **3**(2), 187–194 (1996)

Cristi, R., Papoulias, F.A., Healey, A.J.: Adaptive sliding mode control of autonomous underwater vehicles in the dive plane. IEEE J. Ocean. Eng. **15**(3), 152–160 (1990)

Da Cunha, J.P.V.S., Costa, R.R., Hsu, L.: Design of high performance variable structure control of ROV's. IEEE J. Ocean. Eng. **20**(1), 42–55 (1995)

DeBitetto, P.A.: Fuzzy logic for depth control of unmanned undersea vehicles, In Proceedings of IEE of AUV Symposium, 19–20 July 1994, Cambridge, MA, pp. 233–241 (1994). doi:10.1109/AUV.1994.518630

Erbatur, K., Kaynak, O., Sabanovic, A., Rudas, I.: Fuzzy adaptive sliding mode control of a direct drive robot. Robot. Auton. Syst. **19**(2), 215–227 (1996)

Fossen, T.I., Sagatun, S.: Adaptive control of nonlinear systems: a case study of underwater robotic systems. J. Robot. Syst. **8**(3), 393–412 (1991)

Goheen, K.R., Jefferys, E.R.: Multivariable self tuning autopilots for autonomous underwater vehicles. IEEE J. Ocean. Engg. **15**(3), 144–151 (1990)

Guo S., Du J., Xichuan L., Chunfeng Y.: Adaptive fuzzy sliding mode control for spherical underwater robots. In: Proceeding of IEEE international conference on mechatronics and automation, 5–8 Aug. 2012, Chengdu, pp. 1681–1685 (2012). doi:10.1109/ICMA.2012.6284389

Guo, J., Chiu, F.C., Huang, C.C.: Design of a sliding mode fuzzy controller for the guidance and control of an autonomous underwater vehicle. J. Ocean Eng. **30**(16), 2137–2155 (2003)

Ha, Q.P.: Robust sliding mode controller with fuzzy tuning. IEE Electron. Lett. **32**(17), 1626–1628 (1996)

Healey, A.J., Lienard, D.: Multivariable sliding mode control for autonomous diving and steering of unmanned underwater vehicles. IEEE J. Ocean. Eng. **18**(3), 327–339 (1993)

Hoang, N.Q., Kreuzer, E.: Adaptive PD-controller for positioning of a remotely operated vehicle close to an underwater structure: theory and experiments. J. Ocean Eng. **15**(4), 411–419 (2007)

Hoang, N.Q., Kreuzer, E.: A robust adaptive sliding mode controller for remotely operated vehicles. Technische Mechanik **28**(3), 185–193 (2008)

Hung, L.C., Lin, H.P., Chung, H.Y.: Design of self tuning fuzzy sliding mode control for TORA system. Expert Syst. Appl. **32**(1), 201–212 (2007)

Hwang, Y.R., Tomizuka, M.: Fuzzy smoothing algorithms for variable structure systems. IEEE Trans. Fuzzy Syst. **2**(4), 277–284 (1994)

Ishaque, K., Abdullah, S.S., Ayob, S.M., Salam, Z.: Single input fuzzy logic controller for unmanned underwater vehicle. J. Intell. Robot. Syst. **59**(1), 87–100 (2010)

Kanakakis, V., Valavanis, K.P., Tsourveloudis, N.C.: Fuzzy logic based navigation of underwater vehicles. J. Intell. Robot. Syst. **40**(1), 45–88 (2004)

Kato N., Ito Y., Kojjma J., Asakawa K., Shirasaki Y.: Guidance and control of autonomous underwater vehicle AQUA EXPLORER 1000 for inspection of underwater cables, International symposium on unmanned untethered submersible technology, 13–16 Sep. 1994, Brest, pp. 195–211 (1994). doi:10.1109/OCEANS.1994.363845

Kim H.S., Shin Y.K.: Design of adaptive fuzzy sliding mode controller using FBFE for UFV depth control. In: Proceeding of SICE-ICASE iInternational joint conference, 18–21 Oct. 2006, Busan, pp. 3100–3103 (2006). doi:10.1109/SICE.2006.314744

Kim, S.W., Lee, J.J.: Design of a fuzzy controller with fuzzy sliding surface. J. Fuzzy Sets Syst. **71**(3), 359–367 (1995)

Lakhekar, G.V.: Tuning and analysis of sliding mode controller based on fuzzy logic. Int. J. Control Autom. **5**(3), 93–110 (2012)

Lakhekar, G.V.: A new approach to the design of an adaptive fuzzy sliding mode controller. Int. J. Ocean Syst. Eng. **3**(2), 50–60 (2013)

Lam W.C., Ura T.: Nonlinear controller with switched control law for tracking control of noncruising AUV. In: Proceeding of IEEE AUV'96, 2–6 June 1996, Monterey, CA, pp. 78–85 (1996). doi:10.1109/AUV.1996.532403

Lee, P.M., Hong, S.W., Lim, Y.K., Lee, C.M., Jeon, B.H., Park, J.W.: Discrete-time quasi-sliding mode control of an autonomous underwater vehicle. IEEE J. Ocean Eng. **24**(3), 88–395 (1999)

Lee, H., Kim, E., Kang, H.J., Park, M.: A new sliding-mode control with fuzzy boundary layer. Fuzzy Sets Syst. **120**(1), 135–143 (2001)

Liang, C.Y., Su, J.P.: A new approach to the design of a fuzzy sliding-mode controller. Fuzzy Sets and Systems **139**(1), 111–124 (2003)

Liu, D., Yi, J., Zhao, D., Wang, W.: Adaptive sliding mode fuzzy control for a two dimensional overhead crane. Mechatronics **15**(5), 505–522 (2005)

Marzbanrad A. R., Eghtesad M., Kamali R.: A robust adaptive fuzzy sliding mode controller for trajectory tracking of ROVs. In: Proceeding of IEEE conference on decision and control and european control conference, 12–15 Dec. 2011, Orlando FL, pp. 2863–2870 (2011). doi:10.1109/CDC.2011.6160980

Ryu S.H., Park J.H.: Autotuning of sliding mode control parameters using fuzzy logic. In: Proceeding of the American control conference, 25–27 June 2001, Arlington, VA, pp. 618–623 (2001). doi:10.1109/ACC.2001.945615

Sebastian, E., Sotelo, M.A.: Adaptive fuzzy sliding mode controller for the kinematic variables of an underwater vehicle. J. Intell. Robot Syst. **49**(2), 189–215 (2007)

Shi X., Zhou J., Bian X., and Juan L., (2008). Fuzzy Sliding-Mode Controller for the Motion of Autonomous Underwater Vehicle, In Proceeding of IEEE International Conference on Mechatronics and Automation, 5–8 Aug. 2008, Takamatsu, pp. 466–470. doi:10.1109/ICMA.2008.4798800

Smith, S.M., Rae, G.J.S., Anderson, D.T., Shien, A.M.: Fuzzy logic control of an autonomous underwater vehicle. Control Eng. Pract. **2**(2), 321–331 (1994)

Song F., Smith S.M.: Design of sliding mode fuzzy controllers for an autonomous underwater vehicle without system model. In: Proceeding of MTS/IEEE ocean conference, 14 Sep. 2000, Providence, RI, pp. 835–840 (2000). doi:10.1109/OCEANS.2000.881362

Temeltas, H.: A fuzzy adaptation technique for sliding mode controllers, IEEE International symposium of control applications, 7–10 July 1998, Pretoria, pp. 110–115 (1998). doi:10.1109/ISIE.1998.707758

Wai, R.J., Su, K.H.: Adaptive enhanced fuzzy sliding mode control for electrical servo drive. IEEE Trans. Ind. Electron. **53**(2), 569–580 (2006)

Walchko K. J., and Nechyba M.C., (2003). Development of a sliding mode control system with extended Kalman filter estimation for Subjugator, In Proceding of Florida Conference on Recent Advances in Robotics, 18–20 June 2003, Florida, pp. 185–191

Xin S., Zaojian Z.: A fuzzy sliding mode controller with adaptive disturbance approximation for underwater robot. In: Proceeding of international asia conference on informatics in control, automation and robotics, 6–7 March 2010, Wuhan, pp. 50–53 (2010). doi:10.1109/CAR.2010.5456607

Yagiz, Y., Haciogluy, Y.: Fuzzy sliding modes with moving surface for the robust control of a planar robot. J. Vib. Control **11**(7), 903–922 (2005)

Yoerger, D., Slotine, J.: Adaptive sliding control of an experimental underwater vehicle. In: Proceedings of IEEE conference on robotics and aAutomation, 9–11 April 1991, Sacramento, CA, pp. 2746–2751 (1991). doi:10.1109/ROBOT.1991.132047

Yoerger, D., Slotine, J.: Robust trajectory control of underwater vehicles. IEEE J. Ocean. Engg. **10**(4), 462–470 (1985)

Yorgancioglu, F., Komurcugil, H.: Single-input fuzzy-like moving sliding surface approach to the sliding mode control. Electr. Eng. **90**(3), 199–207 (2008)

An Indirect Adaptive Fuzzy Sliding Mode Power System Stabilizer for Single and Multi-machine Power Systems

Saoudi Kamel, Bouchama Ziyad and Harmas Mohamed Naguib

Abstract This chapter presents an indirect adaptive fuzzy sliding mode power system stabilizer (AFSMPSS) that is used to damp out the low frequency oscillations in a single machine infinite bus, local and inter-area oscillations in multi-machine power systems. An adaptive fuzzy control integrates the sliding mode control (SMC) in the design of the proposed controller. The fuzzy logic system is used to approximate the unknown system function and by introducing proportional integral (PI) control term in the design of sliding mode controller in order to eliminate the chattering phenomenon. In addition, the parameters of the controller are optimized using particle swarm optimization (PSO) approach. Based on the Lyapunov theory, the adaptation laws are developed to make the controller adaptive take care of the changes due to the different operating conditions occurring in the power system and guarantees stability converge. The performance of the newly designed controller is evaluated in a single machine infinite bus and two-area four machine power system under the different types of disturbances in comparison with the indirect adaptive fuzzy PSS. Simulation results show the effectiveness and robustness of the proposed stabilizer in damping power system oscillations under various disturbances. Moreover, it is superior in the comparison with other types of PSSs.

1 Introduction

Currently, the power systems wide-area is obliged to function with full power and often in extreme cases of stability. The appearance of low frequency oscillations due to various disturbances is able to induce with a rupture of synchronism of the

S. Kamel (✉)
Department of Electrical Engineering, University of Bouira, 10000 Bouira, Algeria
e-mail: saoudi_k@yahoo.fr

B. Ziyad
Department of Sciences and Technology, University of Borbj Bou Arreridj, 34000 Borbj Bou Arreridj, Algeria

H.M. Naguib
Department of Electrical Engineering, University of Sétif1, 19000 Sétif, Algeria

A.T. Azar and Q. Zhu (eds.), *Advances and Applications in Sliding Mode Control systems*, Studies in Computational Intelligence 576, DOI 10.1007/978-3-319-11173-5_11

generators coupled with the power system and can easily lead to a total collapse of the power system. Also, the improvement of stability in damping of the inter-area mode oscillations becomes more and more very important if an adequate answer is not taken in the seconds or sometimes some cycles which follow.

Power system oscillations are damped by the introduction of a supplementary signal to the excitation system called power system stabilizer (PSS). These stabilizers improve the stability of power systems by creating electrical torques to the rotor, in phase with speed variation to the synchronous machine, that damp out power oscillations (Anderson and Fouad 1977; Kundur 1994). Conventional power system stabilizers (CPSS) are one of the premiere PSSs composed by the use of some fixed lag-lead compensators which are tuned using a linearized model of power system in the specific operating point, shows a good control performance in the specific operating point (Klein et al. 1991; Kundur et al. 1989; Larsen and Swann 1981). But fixed-parameters of conventional stabilizer are difficult to obtain a good control performance in case of changes in operating conditions such as change of load or major disturbances.

Recently PSS design has undergone the advent of artificial intelligence such as fuzzy logic controller (Bhati and Gupta 2013; El-Metwally et al. 1996; El-Metwally and Malik 1995; Hassan et al. 1991; Hiyama 1994; Hussein et al. 2007; Lin 2013) and artificial neural network (Abido and Abdel-Magid 1999; Changaroon et al. 2000; Demirören 2003; Zeynelgil et al. 2002; Zhang et al. 1995) does not require a mathematical model of the system to be controlled, but fixed-parameters are difficult to obtain a good control performance in case of changes in operating conditions such as change of load or major disturbances. On the other hand, adaptive power system stabilizers have been proposed (Chen and Malik 1995; Cuk-Supriyadi et al. 2014; Karimi and Feliachi 2008; Kothan et al. 1996; Teh-Lu 1999; Wu and Malik 2006). These stabilizers provide better dynamic performance over a wide range of operating conditions, but they suffer from the major drawback of requiring parameter model identification, state observation and feedback gain computation 'on-line'.

This inadequacy is somewhat countered by the use of the merits of adaptive control and artificial intelligences techniques in promising design of adaptive fuzzy power system stabilizers (Bouchama and Harmas 2012; Elshafei et al. 2005; Hosseinzadeh and Kalam 1999; Hussein et al. 2010, 2009; Saoudi et al. 2008) and adaptive neural power system stabilizers (Fraile-Ardanuy and Zufiria 2007; Hosseini and Etemadi 2008; Liu et al. 2003; Radaideh et al. 2012; You et al. 2003). The main idea of adaptive fuzzy control is as follows: first construct a fuzzy model to describe the input/output behavior of the controlled system. A controller is designed based on the fuzzy model and then the adaptive laws are derived to adjust the parameters of the fuzzy modes on-line. However, these stabilizers do not make it possible to maintain good performances of continuation in the presence of external disturbances. On the other hand, robust control provides an effective approach to dealing with uncertainties introduced by variations of operating conditions. Among many techniques available in the control literature, sliding mode control has been reported as one of the most effective control methodologies for nonlinear power system applications

(Al-Duwaish and Al-Hamouz 2011; Bandal and Bandyopadhyay 2007; Bandal et al. 2005; Cao et al. 1994; Colbia-Vega et al. 2008; Fernandez-Vargas and Ledwich 2010; Ghazi et al. 2001; Huerta et al. 2010, 2011; Samarasinghe and Pahalawaththa 1997; Rashidi et al. 2003; Saoudi and Harmas 2014; Saoudi et al. 2011, 2008) in improving the power system stability due to its robust response characteristic.

In this chapter, a new indirect adaptive fuzzy sliding mode stabilizer (AFSMPSS) is designed for enhancing the damping of oscillations in nonlinear single and multi-machine power system using nonlinear models. The advantages application of the proposed PSS is to counteract the problem of variations in the system parameters, operating conditions, to improve the stability and robustness performance of the control systems. The nonlinear model of the power system is constructed with the differential equations with nonlinear parameters which are functions of the state of the system. Some of these parameters of nonlinear function are not known and others are not exact precise. i.e. it is not possible build a relatively exact mathematical model of the system. In order to design the proposed indirect AFSMPSS, the fuzzy logic system is used to approximate the unknown system function present in the model of power system. Moreover, the chattering phenomenon was eliminated due to the utilisation of proportional integral (PI) term control in the design of SMC. The optimal control gains are obtained via a particle swarm optimization (PSO) technique. Using Lyapunov stability theory, the adaptation laws are developed to make the fuzzy sliding mode controller adaptive and the PI parameters can be tuned on-line by adaptation law to take care of the changes due to the different operating conditions occurring in the power system and guarantee stability converge.

The performance of the newly designed controller is evaluated in a single machine infinite bus and two-area four machine power system under the different types of disturbances in comparison with the indirect adaptive fuzzy PSS. Simulation results show the effectiveness and robustness of the proposed stabilizer in enhancing damping power system oscillations under various disturbances. Moreover, it is superior in the comparison with other types of PSSs.

The rest of the chapter is organized as follows: In Sect. 2, an indirect adaptive fuzzy sliding mode control Based PSS Design for power system to enhance the transient stability of the system is presented. In Sect. 3, the proposed control design procedure is given. The optimal controller gains are obtained using a Particle Swarm Optimization (PSO) search technique is given with the procedure in Sect. 4. In Sect. 5, the simulation results that demonstrate the effectiveness of the proposed controller are presented and compared with those of the adaptive fuzzy controller and the conventional controllers using the single machine infinite bus and four machine two-area bench-mark test power systems. Conclusion is stated in Sect. 6.

2 Indirect Adaptive Fuzzy Sliding Mode Control Based PSS Design

2.1 Power System Model

In order to design the power system controller proposed in this paper, the dynamics model of generator can be expressed in a canonical form given in Slotine and Li (1991), this is obtained using the speed variation x_1 and instead of direct and quadrature voltages the accelerating power x_2 are used as a state variables, the system model of synchronous machine is represented in the following nonlinear state-space equations (Saoudi and Harmas 2014, Saoudi et al. 2011):

$$\begin{aligned} \dot{x}_1 &= a x_2 \\ a\dot{x}_2 &= f(x_1, x_2) + g(x_1, x_2)u \\ y &= x_1 \end{aligned} \tag{1}$$

where $a = -1/2H, x_1 = \Delta\omega = \omega - \omega_s$ and $x_2 = \Delta P = P_m - P_e$, H is the per unit machine inertia constant, ω is the rotor speed and ω_s is the synchronous speed are in per unit, P_m is the mechanical input power treated as a constant in the excitation controller design, i.e., it is assumed that the governor action is slow enough not to have any significant impact on the machine dynamics and P_e is the delivered electrical power. $\underline{x} = [x_1, x_2]^T \in R^2$ is a measurable state vector. The PSS output u represents the controlling supplementary signal to be designed and $y = \Delta\omega$ is the output state while f and g are nonlinear functions which are assumed to be unknown. (Eq. 1) represents the machine during a transient period after a major disturbance has occurred in the system. The design of the sliding mode control is presented in the following section.

2.2 Sliding Mode Control Design

The control objective is to force y in the system (Eq. 1) to track a given bounded desired trajectory y_d, under the constraint that all single involved must be bounded. Then the control objective (Slotine and Li 1991; Wang 1996) is determine a feedback control $u = u(\underline{x}|\underline{\theta})$ and an adaptation law for adjusting the parameters vector $\underline{\theta}$, such that:

The close loop system must be globally stable and robust in the sense that all variables $\underline{x}(t), \underline{\theta}(t)$ and $u(\underline{x}|\underline{\theta})$, must be uniformly bounded, i.e., $|\underline{x}| \leq M_x \leq \infty$, $|\underline{\theta}| \leq M_\theta \leq \infty$ and $|u| \leq M_\theta \leq \infty$ for all $t \geq 0$, where M_x, M_θ and M_u are parameters designer specified.

The traking error, $e = y - y_d$, should be as small as possible under the constraint in the previously objective.

The elaboration of an indirect adaptive fuzzy sliding mode controller is presented in the rest of this section (Saoudi and Harmas 2014; Saoudi et al. 2011) to achieve the above control objectives is discussed.

Let the tracking error be defined as:

$$\underline{e} = \underline{y} - \underline{y}_d = [e, \dot{e}]^T \tag{2}$$

and a sliding surface defined as:

$$s(\underline{e}) = k_1 e + \dot{e} = \underline{k}^T \underline{e} \tag{3}$$

where $\underline{k} = [k_1, 1]^T$ are the coefficients of the Hurwitzian polynomial $h(\lambda) = \lambda + k_1$. If the initial error vector $\underline{e}(0) = 0$, then the tracking problem can be considered as the state error vector $\underline{e}$ remaining on the sliding surface $s(\underline{e}) = 0$ for all $t > 0$. A sufficient condition to achieve this behavior is to select the control strategy such that:

$$\frac{1}{2}\frac{d}{dt}(S^2(\underline{e})) \le -\eta\,|s| \quad \eta \ge 0 \tag{4}$$

From Eqs. (3) and (4), we have

$$\dot{s} = k_1\dot{e} + f(\underline{x}) + g(\underline{x})u - \ddot{y}_d. \tag{5}$$

If f and g are known, we can easily construct the sliding mode control $u^* = u_{eq} - u_{sw}$:

$$u^* = \frac{1}{g(\underline{x})}\left[-k_1\dot{e} - f(\underline{x}) - \eta\,\mathrm{sgn}(s) + \ddot{y}_d\right] \tag{6}$$

$$u_{eq} = \frac{1}{g(\underline{x})}\left[-k_1\dot{e} - f(\underline{x}) + \ddot{y}_d\right] \tag{7}$$

$$u_{sw} = \frac{1}{g(\underline{x})}\left[\eta\,\mathrm{sgn}(s)\right] \tag{8}$$

However, power system parameters for nonlinear functions are not well known and imprecise; therefore it is difficult to implement the control law (Eq. 6) for unknown nonlinear system model. Not only f and g are unknown but the switching-type control term will cause chattering. An adaptive fuzzy sliding mode controller using fuzzy logic system and PI control term is proposed to solve these problems.

2.3 Fuzzy Logic System

The basis of the fuzzy logic systems (Wang 1993, 1996) consists of a collection of fuzzy IF-THEN rules:

$$R(l) : IFx_1\ is\ F_1^l\ and \ldots andx_n\ is\ F_n^l\ THEN\ y\ is\ G^l \tag{9}$$

By using the strategy of singleton fuzzification, product inference and center average defuzzification, the output value of the fuzzy system can be formulated

$$y\left(\underline{x}\right) = \frac{\sum_{l=1}^{M} \theta_l \left(\prod_{i=1}^{n} \mu_{F_i^l}\left(x_i\right)\right)}{\sum_{l=1}^{M}\left(\prod_{i=1}^{n} \mu_{F_i^l}\left(x_i\right)\right)} \tag{10}$$

where $\mu_{F_i^l}\left(x_i\right)$ is the membership function value of x_i in F_i^l, θ_l is the centre of gravity of the membership function of the output for the lth rule; (Eq. 10) can be rewritten as:

$$y(\underline{x}) = \sum_{l=1}^{M} \theta_l \xi_l(\underline{x}) = \underline{\theta}^T \underline{\xi}(\underline{x}) \tag{11}$$

where $\underline{\theta}_l = \left[\underline{\theta}_1 \ldots \underline{\theta}_M\right]^T$ and $\underline{\xi}(\underline{x}) = \left[\underline{\xi}_1(\underline{x}) \ldots \xi_M(\underline{x})\right]^T$ represents the fuzzy basis functions defined

$$\xi_l\left(\underline{x}\right) = \frac{\prod_{i=1}^{n} \mu_{F_i^l}\left(x_i\right)}{\sum_{l=1}^{M}\left(\prod_{i=1}^{n} \mu_{F_i^l}\left(x_i\right)\right)} \tag{12}$$

After this brief description, the following section explains the design of the adaptive fuzzy sliding mode control.

2.4 Indirect Adaptive Fuzzy Sliding Mode Control Design

If f and g were known, we could easily construct the sliding mode control u^* introduced in the previous section, however, f and g are not known, we thus replace $f(\underline{x}, t)$ and $g(\underline{x}, t)$ by the fuzzy estimates $\hat{f}(\underline{x}|\underline{\theta}_f)$, $\hat{g}(\underline{x}|\underline{\theta}_g)$ which are in the form of (Eq. 11) to which we append a proportional integral PI control term to suppress the chattering action. The inputs and output of the latter are defined as (Ho and Cheng 2009; Ho et al. 2009).

$$u_p = k_p h_1 + k_i h_2 \tag{13}$$

where $h_1 = s, h_2 = \int s dt, k_p$ and k_i are are PI control gains. Equation 13) can be rewritten as

$$\hat{p}(\underline{h}|\underline{\theta}_p) = \underline{\theta}_p^T \underline{\psi}(\underline{h}) \tag{14}$$

$\underline{\theta}_p = [k_p, k_i]^T \in R^2$ is an adjustable parameter vector, and $\underline{\psi}^T(\underline{h}) = [h_1, h_2] \in R^2$ is a regressive vector. We use fuzzy logic systems to approximate the unknown functions $f(\underline{x}), g(\underline{x})$ and design an adaptive PI control term eliminate chattering due to sliding mode control. Hence, the control law becomes:

$$u = \frac{1}{\hat{g}(\underline{x}|\underline{\theta}_g)}\left[-k_1\dot{e} - \hat{f}(\underline{x}|\underline{\theta}_f) - \hat{p}(\underline{h}|\underline{\theta}_p) + \ddot{y}_d\right] \tag{15}$$

$$\hat{f}(\underline{x}|\underline{\theta}_f) = \underline{\theta}_f^T \underline{\xi}(\underline{x}) \tag{16}$$

$$\hat{g}(\underline{x}|\underline{\theta}_g) = \underline{\theta}_g^T \underline{\xi}(\underline{x}) \tag{17}$$

In order to avoid the chattering problem, the switching term is replaced by a PI control action which changes continuously and will lead to smooth out of the chattering effect when the state is within a boundary layer $|s| < \Phi$. The control action is kept at the saturated value when the state is outside the boundary layer. Hence, we set $|\hat{p}(\underline{h}|\underline{\theta}_p)| = \eta$ when $|s| \geq \Phi$, where Φ is the thickness of the boundary layer.

Using the control law in (Eq. 15), then (Eq. 5) becomes:

$$\begin{aligned}\dot{s} &= k_1\dot{e} + f(\underline{x}, t) + g(\underline{x}, t)u - \ddot{y}_d \\ &= f(\underline{x}, t) - \hat{f}(\underline{x}|\underline{\theta}_f) + (g(\underline{x}, t) - \hat{g}(\underline{x}|\underline{\theta}_g))u - \hat{p}(\underline{h}|\underline{\theta}_p)\end{aligned} \tag{18}$$

The next task, is to replace $\hat{f}$ and $\hat{g}$ by fuzzy logic systems represented in (Eqs. 16 and 17), $\hat{p}$ is given by (Eq. 14) and to develop adequate adaptation laws for adjusting the parameters vector $\underline{\theta}_f, \underline{\theta}_g$ and $\underline{\theta}_p$ while seeking a zero tracking error. Using the procedure suggested in Hussein et al. (2009), Hussein et al. (2010), the parameter vectors of the fuzzy logic systems $\hat{f}(\underline{x}|\underline{\theta}_f)$ and $\hat{g}(\underline{x}|\underline{\theta}_g)$ will be adapted according to the following rules.

Theorem 1 *Consider the control problem of the nonlinear system* (Eq. 1). *If the control* (Eq. 15) *is used, the function $\hat{f}$, $\hat{g}$ and $\hat{p}$are estimated by* (Eqs. 16 and 17) *and* (Eq. 14), *the parameters vector $\underline{\theta}_f, \underline{\theta}_g$ and $\underline{\theta}_p$ are adjusted by the adaptive control law* (Eqs. 19–21), *the closed-loop system signals will be bounded and the tracking error will converge to zero asymptotically.*

$$\dot{\theta}_f = \gamma_1 s \underline{\xi}(\underline{x}) \tag{19}$$

$$\dot{\theta}_g = \gamma_2 s \underline{\xi}(\underline{x}) u \tag{20}$$

$$\dot{\theta}_p = \gamma_3 s \underline{\psi}(\underline{h}) \tag{21}$$

Proof Define the optimal parameters vector

$$\underline{\theta}_f^* = \arg \min_{\underline{\theta}_f \in \Omega_f} \left(\sup_{\underline{x} \in R^n} \left| \hat{f}(\underline{x}|\underline{\theta}_f) - f(\underline{x}, t) \right| \right) \tag{22}$$

$$\underline{\theta}_g^* = \arg \min_{\underline{\theta}_g \in \Omega g} \left(\sup_{\underline{x} \in R^n} \left| \hat{g}(\underline{x}|\underline{\theta}_g) - g(\underline{x}, t) \right| \right) \tag{23}$$

$$\underline{\theta}_p^* = \arg \min_{\underline{\theta}_p \in \Omega p} \left(\sup_{\underline{h} \in R^n} \left| \hat{p}(\underline{h}|\underline{\theta}_p) - u_{sw} \right| \right) \tag{24}$$

where Ω_f, Ω_g and Ω_p are constraint sets for $\underline{\theta}_f$, $\underline{\theta}_g$ and $\underline{\theta}_p$, respectively. Define the minimum approximation error:

$$\varepsilon = f(\underline{x}, t) - \hat{f}(\underline{x}|\underline{\theta}_f^*) + (g(\underline{x}, t) - \hat{g}(\underline{x}|\underline{\theta}_g^*))u. \tag{25}$$

Assumption 1 The parameters $\underline{\theta}_f$, $\underline{\theta}_g$ and $\underline{\theta}_p$ belong to the constraint sets Ω_f, Ω_g and Ω_p respectively, which are defined as

$$\Omega_f = \left\{ \underline{\theta}_f \in R^n : \left\| \underline{\theta}_f \right\| \le M_f \right\} \tag{26}$$

$$\Omega_g = \left\{ \underline{\theta}_g \in R^n : 0 < \zeta \le \left\| \underline{\theta}_g \right\| \le M_g \right\} \tag{27}$$

$$\Omega_p = \left\{ \underline{\theta}_p \in R^n : \left\| \underline{\theta}_p \right\| \le M_p \right\} \tag{28}$$

M_f, ζ, M_g and M_p are positive constants designer specified for estimated parameters' bounds. Assuming that fuzzy $\underline{\theta}_f$, $\underline{\theta}_g$ and PI control parameter $\underline{\theta}_p$ do not reach the boundaries.

So, (Eq. 18) can be written as

$$\dot{s} = \underline{\phi}_f^T \underline{\xi}(\underline{x}) + \underline{\phi}_g^T \underline{\xi}(\underline{x}) u + \underline{\theta}_p^T \underline{\psi}(\underline{h}) - \hat{p}(\underline{h}|\underline{\theta}_p^*) + \varepsilon \tag{29}$$

where $\underline{\phi}_f = \underline{\theta}_f^* - \underline{\theta}_f$, $\underline{\phi}_g = \underline{\theta}_g^* - \underline{\theta}_g$, $\underline{\phi}_p = \underline{\theta}_p^* - \underline{\theta}_p$.

Now let us consider the Lyapunov function candidate

$$V = \frac{1}{2}s^2 + \frac{1}{2\gamma_1}\underline{\phi}_f^T \underline{\phi}_f + \frac{1}{2\gamma_2}\underline{\phi}_g^T \underline{\phi}_g + \frac{1}{2\gamma_3}\underline{\phi}_p^T \underline{\phi}_p \tag{30}$$

The time derivative of V along the error trajectory (Eq. 29) is:

$$
\begin{aligned}
\dot{V} &= s\dot{s} + \frac{1}{\gamma_1}\underline{\phi}_f^T\dot{\underline{\phi}}_f + \frac{1}{\gamma_2}\underline{\phi}_g^T\dot{\underline{\phi}}_g + \frac{1}{\gamma_3}\underline{\phi}_p^T\dot{\underline{\phi}}_p \\
&= s(\underline{\phi}_f^T\underline{\xi}(\underline{x}) + \underline{\phi}_g^T\underline{\xi}(\underline{x})u - \hat{p}(\underline{h}|\underline{\theta}_p^*) + \varepsilon) + \frac{1}{\gamma_1}\phi_f^T\dot{\phi}_f + \frac{1}{\gamma_2}\phi_g^T\dot{\phi}_g + \frac{1}{\gamma_3}\underline{\phi}_p^T\dot{\underline{\phi}}_p \\
&= s\underline{\phi}_f^T\underline{\xi}(\underline{x}) + \frac{1}{\gamma_1}\phi_f^T\dot{\phi}_f + s\underline{\phi}_g^T\underline{\xi}(\underline{x})u + \frac{1}{\gamma_2}\phi_g^T\dot{\phi}_g + s\underline{\phi}_p^T\underline{\psi}(\underline{h}) + \frac{1}{\gamma_3}\underline{\phi}_p^T\dot{\underline{\phi}}_p - s\hat{p}(\underline{h}|\underline{\theta}_p^*) + s\varepsilon \qquad (31)\\
&= \frac{1}{\gamma_1}\phi_f^T(\gamma_1 s\underline{\xi}(\underline{x}) + \dot{\phi}_f) + \frac{1}{\gamma_2}\phi_g^T(\gamma_2 s\underline{\xi}(\underline{x})u + \dot{\phi}_g) + \frac{1}{\gamma_3}\underline{\phi}_p^T(s\underline{\psi}(\underline{h}) + \dot{\underline{\phi}}_p) - s\hat{p}(\underline{h}|\underline{\theta}_p^*) + s\varepsilon \\
&\le \frac{1}{\gamma_1}\phi_f^T(\gamma_1 s\underline{\xi}(\underline{x}) + \dot{\phi}_f) + \frac{1}{\gamma_2}\phi_g^T(\gamma_2 s\underline{\xi}(\underline{x})u + \dot{\phi}_g) + \frac{1}{\gamma_3}\underline{\phi}_p^T(s\underline{\psi}(\underline{h}) + \dot{\underline{\phi}}_p) - s\eta\,\mathrm{sgn}(s) + s\varepsilon \\
&< \frac{1}{\gamma_1}\phi_f^T(\gamma_1 s\underline{\xi}(\underline{x}) + \dot{\phi}_f) + \frac{1}{\gamma_2}\phi_g^T(\gamma_2 s\underline{\xi}(\underline{x})u + \dot{\phi}_g) + \frac{1}{\gamma_3}\underline{\phi}_p^T(s\underline{\psi}(\underline{h}) + \dot{\underline{\phi}}_p) - |s|\,\eta + s\varepsilon
\end{aligned}
$$

where $\dot{\underline{\phi}}_f = -\dot{\underline{\theta}}_f$, $\dot{\underline{\phi}}_g = -\dot{\underline{\theta}}_g$ and $\dot{\underline{\phi}}_p = -\dot{\underline{\theta}}_p$. Substitute (Eqs. 19–21) into (Eq. 31), then we have

$$\dot{V} \le s\varepsilon - |s|\,\eta \le 0 \qquad (32)$$

Since ε is being the minimum approximation error, (Eq. 32) is the best we can obtain. Therefore all signals in the system are bounded. Obviously, $e(t)$ will be bounded if $e(0)$ is bounded for all t. Since if the reference signal $\underline{y}_d$ is bounded, then system states $\underline{x}$ will be bounded. We need proving that $s \to 0$ as $t \to \infty$. Assuming that $|s| \le \eta_s$ then (Eq. 32) can be further simplified to

$$\dot{V} \le |s|\,|\varepsilon| - |s|\,\eta \le \eta_s\,|\varepsilon| - |s|\,\eta \qquad (33)$$

Integrating both sides of (33), we have

$$\int_0^t |s|d\tau \le \frac{1}{\eta}(|V(0)| + |V(t)|) + \frac{\eta_s}{\eta}\int_0^t |\varepsilon|d\tau \qquad (34)$$

then we have $s \in L_1$. From (Eq. 25), we know that s is bounded and every term in (Eq. 27) is bounded. Hence, $s, \dot{s} \in L_\infty$, use of Barbalat's lemma (Slotine and Li 1991). We have $s \to 0$ as $t \to \infty$, the system is stable and the error will asymptotically converge to zero.

3 Design Procedure

Let the inputs to the fuzzy logic system be $x_1 = \Delta\omega$ (speed variation), $x_2 = \Delta P$ (accelerating power), the procedure for designing an indirect AFSMPSS to damp low frequency oscillations in uncertain dynamic power systems can be summarized by the following steps:

3.1 Off-Line Initial Processing

- Use the PSO to search the control gains k_1 such that $k_1 e + \dot{e}$ is a hurwitzian polynomial and the values of PI control parameters k_p and k_i.
- Specify the learning coefficients $\gamma_1 = 2$, $\gamma_2 = 20$ and $\gamma_3 = 2$.

3.2 Initial Fuzzy Controller Construction

- Define m_i fuzzy sets $A_i^{l_i}$ for linguistic variable x_i, whose membership functions $\mu_{F_i^{li}}$ uniformly cover the corresponding universe of discourse, where $i = 1, 2$ and $l_i = 1, \ldots, m_i$. i.e. The input states $x_1 = \Delta\omega$, $x_2 = \Delta P$ and $m_1 = m_2 = 7$, the membership functions are selected Gaussian membership functions which are labelled Negative Big (NB), Negative Medium (NM), Negative Small (NS), Zero (ZR), Positive Small (PS), Positive Medium (PM), Positive Big (PB), linguistic variables respectively.
- Construct the fuzzy basis functions from the input membership functions
- Construct the fuzzy rule base of $\hat{g}(\underline{x}|\underline{\theta}_g)$, which consist of $m_1 \times m_2$ rules. Table 1 shows the fuzzy rules and forty nine initial parameter vector $\underline{\theta}_g$. Since there is enough information about $\hat{f}(\underline{x}|\underline{\theta}_f)$, the initial value of $\underline{\theta}_f$ is chosen to be zero.

$$R_g^{(l_1,l_2)} : \text{IF } x_1 \text{ is } A_1^{l1} \text{ and } x_2 \text{ is } A_2^{l2} \text{ THEN } \hat{g}(\underline{x}|\underline{\theta}_g) \text{ is } G^{(l_1,l_2)} \quad (35)$$

- Construct the fuzzy systems $\hat{f}(\underline{x}|\underline{\theta}_f) = \underline{\theta}_f^T \underline{\xi}(\underline{x})$ and $\hat{g}(\underline{x}|\underline{\theta}_g) = \underline{\theta}_g^T \underline{\xi}(\underline{x})$.

3.3 On-Line Adaptation

- Apply the feedback control (Eq. 15) as power system stabilizer to damping of the oscillations and improvement of the stability in the power system (Eq. 1).
- Use the adaptive laws (Eqs. 19–21) to adjust the parameters $\underline{\theta}_f$, $\underline{\theta}_g$ and $\underline{\theta}_p$.

The simplified schematic diagram of the proposed power system stabilizer and the interconnection of these techniques are illustrated in Fig. 1.

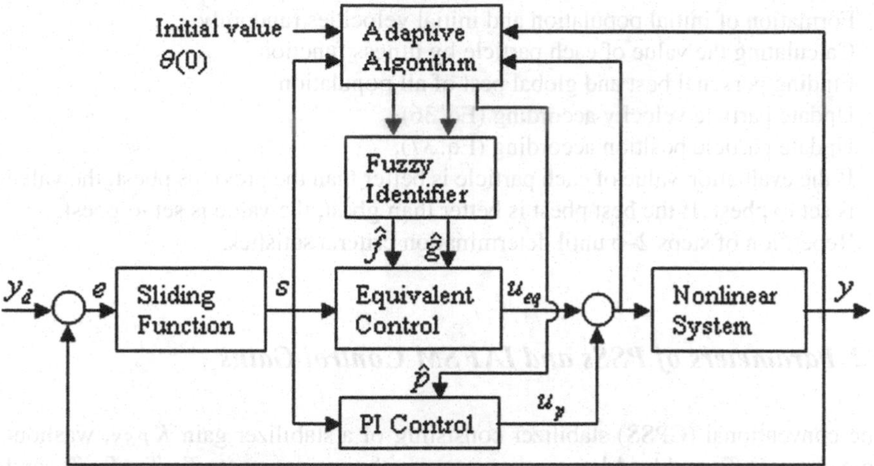

Fig. 1 The proposed indirect adaptive fuzzy sliding mode PSS

4 Optimal Parameters Settings of Controllers Gains

4.1 Overview of Particle Swarm Optimization

Similar to evolutionary algorithms, the particle swarm is one of the optimization techniques process is stochastic in nature. It is developed by Eberhart (Kennedy and Eberhart 1995, 2001). PSO is initialized with a population of candidate solutions. This population is called a swarm. Each candidate solution in PSO is called a particle. Each particle is treated as a point in the dimensional problem space. The *i*-th particle is represented as position vector $x_i = (x_{i1}, x_{i2}, \ldots x_{id})$ in d-dimensional space. The movement of this particle is specified by the velocity vector $v_i = (v_{i1}, v_{i2}, \ldots v_{id})$. The fitness of each particle can be evaluated according to the objective function of optimization problem. The personal best position found during the search by the *i*-th particle memory of the best position as $p_i = (p_{i1}, p_{i2}, \ldots p_{id})$. The position of the best personal of the entire swarm is noted as the global best position $p_g = (p_{g1}, p_{g2}, \ldots p_{gd})$. The velocity and position of each particle are updated as follows:

$$v_{id} = w.v_{id} + c_1 rand(p_{id} - x_{id}) + c_2 rand(p_{gd} - x_{id}) \tag{36}$$

$$x_{id} = x_{id} + v_{id} \tag{37}$$

where c_1 and c_2 are positive constants, and rand are randomly generated numbers in the range [0, 1], w is a positive inertia parameter. The steps of the PSO algorithm are (Abido 2002; Al-Awami et al. 2007; Mostafa et al. 2012):

1. Formation of initial population and initial velocities randomly.
2. Calculating the value of each particle by fitness function.
3. Finding personal best and global best of all population
4. Update particle velocity according (Eq. 36).
5. Update particle position according (Eq. 37).
6. If the evaluation value of each particle is better than the previous pbest, the value is set to pbest. If the best pbest is better than gbest, the value is set to gbest.
7. Repetition of steps 2–6 until determination criteria satisfies.

4.2 Parameters of PSSs and IAFSM Control Gains

The conventional (CPSS) stabilizer consisting of a stabilizer gain K_{PSS}, washout time constant T_w and lead-lag compensators with time constants T_1, T_2, T_3, T_4, and a limiter is used for comparison. The stabilizer transfer function is given by:

$$U_{PSS} = K_{PSS}\left(\frac{sT_W}{1+sT_W}\right)\left(\frac{1+sT_1}{1+sT_2}\right)\left(\frac{1+sT_3}{1+sT_4}\right)\Delta\omega \tag{38}$$

In this structure, the washout time constants T_w and the time constants T_2 and T_4 are usually prespecified. The controller gains K_{pss}, the time constants T_1 and T_3 are to be determined.

In the proposed IAFSM controller, the gains k_1 of the sliding mode surface such that $k_1 e + \dot{e}$ is a hurwitzian polynomial and PI controller gains k_p and k_i, where the first is proportional and the second proportional integral of the surface, all of these gains are to be optimized.

4.3 Objective Function

The optimizing objective function is based on the integral time absolute error index of the speed deviation of the synchronous generator. This fitness function is defined by:

$$J = \sum_{i=2}^{n} \int_{t=0}^{t=t_1} t\,|\Delta\omega_{i-1}|dt \tag{39}$$

where t_1 the time is range of the simulation and $\Delta\omega_{i-1}$ is the speed deviation of the ith generator relative to the first generator.

The proposed approach employs the PSO to search for the optimal parameter settings of the given controllers. The control parameters to be tuned through the optimization algorithm are K_{pss}, T_1, T_3, k_1, k_p and k_i of each generator in the system,

Table 1 The optimal parameters of the controllers gains for single machine

PSS		*SMC*		
K_{PSS}	T_1	K_1	K_p	K_i
19.8341	0.2084	5	2.8962	14.2279

Table 2 The optimal parameters of the controllers gains for multi-machine

	PSS			*SMC*		
	K_{PSS}	T_1	T_3	k_1	K_p	K_i
Gen.1	19.4231	0.0343	4.0811	0.1301	0.05	3.3857
Gen.2	18.9118	0.0627	2.8886	0.0582	0.1104	3.0194
Gen.3	18.2757	0.0736	3.0712	0.0501	0.1278	2.7128
Gen.4	23.8428	0.0189	3.8402	0.1851	0.0945	3.0750

that to aim minimize the selected fitness objective function in order to improve the system response in terms of the settling time and overshoots.

5 Results and Discussions

In this study, we will investigate the performance of the proposed indirect AFSMPSS as it is applied to both single machine infinite-bus and multi-machine power systems models. For the purpose of optimization of (39), to evaluate the objective function, the system dynamic model considering a Three-phase fault is simulated. The objective function J attains a finite value since the deviation in rotor speed is regulated to zero and the obtained optimal parameters are shown in the Tables 1 and 2. The success of the proposed PSS, with the single-machine infinite-bus case, motivates us to test its capability on a multi-machine model. To assess the effectiveness and robustness of the proposed individual design approach controllers under different kind of disturbance conditions, The nonlinear simulation of the power system model is carried out under the following severe faults cases considered are:
Case 1: Three-phase fault short circuit.
Case 2: Step change in the reference terminal voltage.

The performance of the proposed indirect adaptive fuzzy sliding mode PSS is compared with the indirect adaptive fuzzy PSS, fixed parameter fuzzy PSS and PSO-optimized conventional PSS.

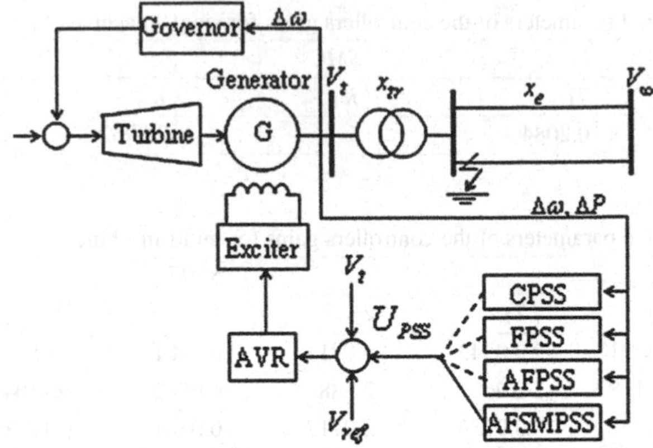

Fig. 2 Single machine infinite bus power system

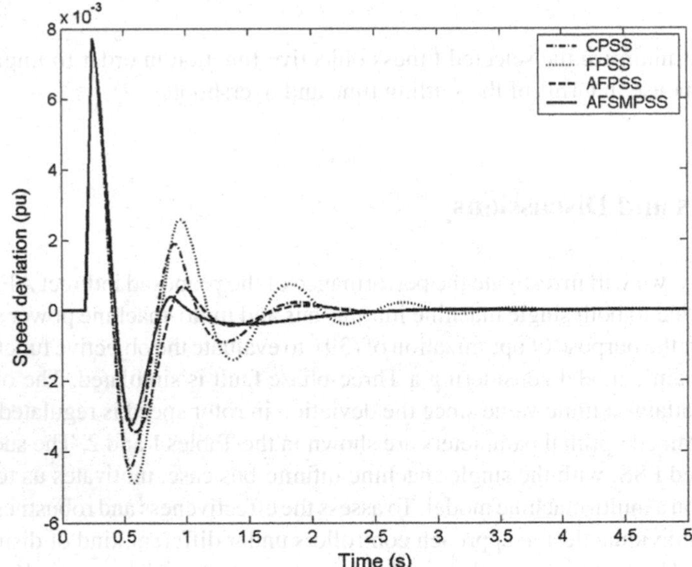

Fig. 3 Speed deviation response for a three-phase short-circuit fault disturbance

5.1 Application to the Single Machine Infinite-Bus Model

A nonlinear power system model consisting of a single machine connected to a infinite bus (SMIB) through a step-up transformer and double circuit of three phase transmission lines is chosen for time domain simulation studies. Details of the system data and the dynamics model of generator are given in Kundur (1994), Sauer and Pai (1998). A diagram representation of the power system is shown in Fig. 2.

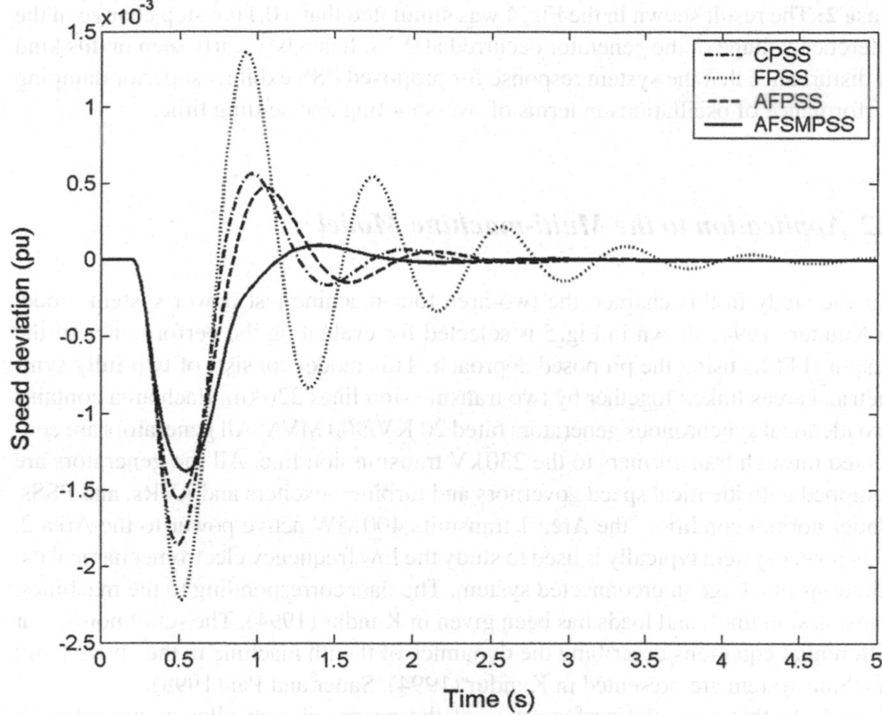

Fig. 4 Speed deviation response for a step change voltage reference disturbance

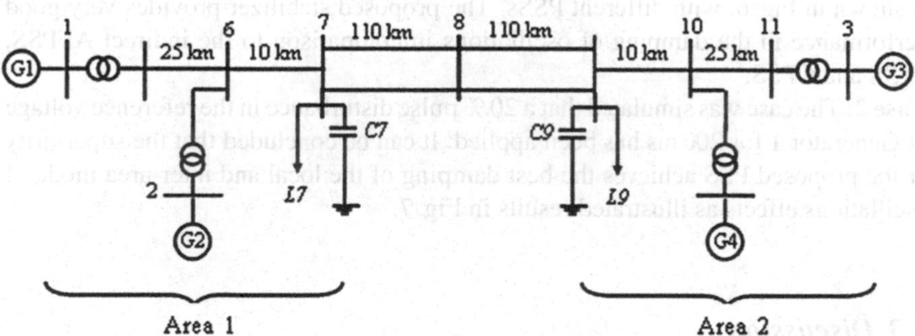

Fig. 5 Two area four machine test power system

Case 1: Figure 3, shows system response under a three-phase fault occurring at t = 0.2 s with a duration of 0.06 s, for the fourth different controllers when the system is simulated. It is obvious that indirect AFPSS has better damping of the speed deviation than CPSS and FPSS; while the proposed indirect AFSMPSS has the best damping of low frequency oscillations.

Case 2: The result shown in the Fig. 4 was simulated that a 0.1 p.u step change in the reference voltage of the generator occurred at 0.2 s. It can be clearly seen in this kind of disturbance that the system response for proposed PSS exhibits superior damping performance of oscillations in terms of overshooting and settling time.

5.2 Application to the Multi-machine Model

For the study in this chapter, the two-area four-machine test power system model in Kundur (1994) shown in Fig. 5 is selected for evaluating the performance of the designed PSSs using the proposed approach. This model consists of two fully symmetrical areas linked together by two transmission lines 220 km. Each area contains two identical synchronous generators rated 20 KV/900 MVA. All generators are connected through transformers to the 230 kV transmission line. All the generators are equipped with identical speed governors and turbines, exciters and AVRs, and PSSs. Under normal condition, the Area 1 transmits 400 MW active power to the Area 2. This power system typically is used to study the low frequency electromechanical oscillations of a large interconnected system. The data corresponding to the machines, transmission lines, and loads has been given in Kundur (1994). The set of non-linear differential equations describing the dynamics of the ith machine in the above multi machine system are presented in Kundur (1994), Sauer and Pai (1998).
Case 1: In this case, the performance of the proposed controller is evaluated by applying a six-cycle three-phase fault short circuit at the middle of one of the transmission lines between bus-7 and bus-8. The local and inter-area mode of oscillations is shown in Fig. 6, with different PSSs. The proposed stabilizer provides very good performance in the damping of oscillations in comparison to the indirect AFPSS, FPSS and CPSS.
Case 2: The case was simulated that a 20 % pulse disturbance in the reference voltage of Generator 1 for 200 ms has been applied. It can be concluded that the superiority of the proposed PSS achieves the best damping of the local and inter-area mode of oscillations effects as illustrated results in Fig. 7.

5.3 Discussion

The proposed indirect AFSMPSS is applied to both single machine infinite-bus and multi-machine power systems models. For evaluating the performance, severe fault disturbance and a voltage deviation are considered; In the first case test, three phase fault to ground short circuit type is considered and the system response is compared with the proposed controller and those obtained using a PSO optimized conventional (CPSS), a fuzzy power system stabilizer (FPSS) and an indirect adaptive fuzzy power system stabilizer (AFPSS). It is evident from the results in Fig. 3, for single machine infinite bus and Fig. 6 for the interconnection multi-machine power systems that the

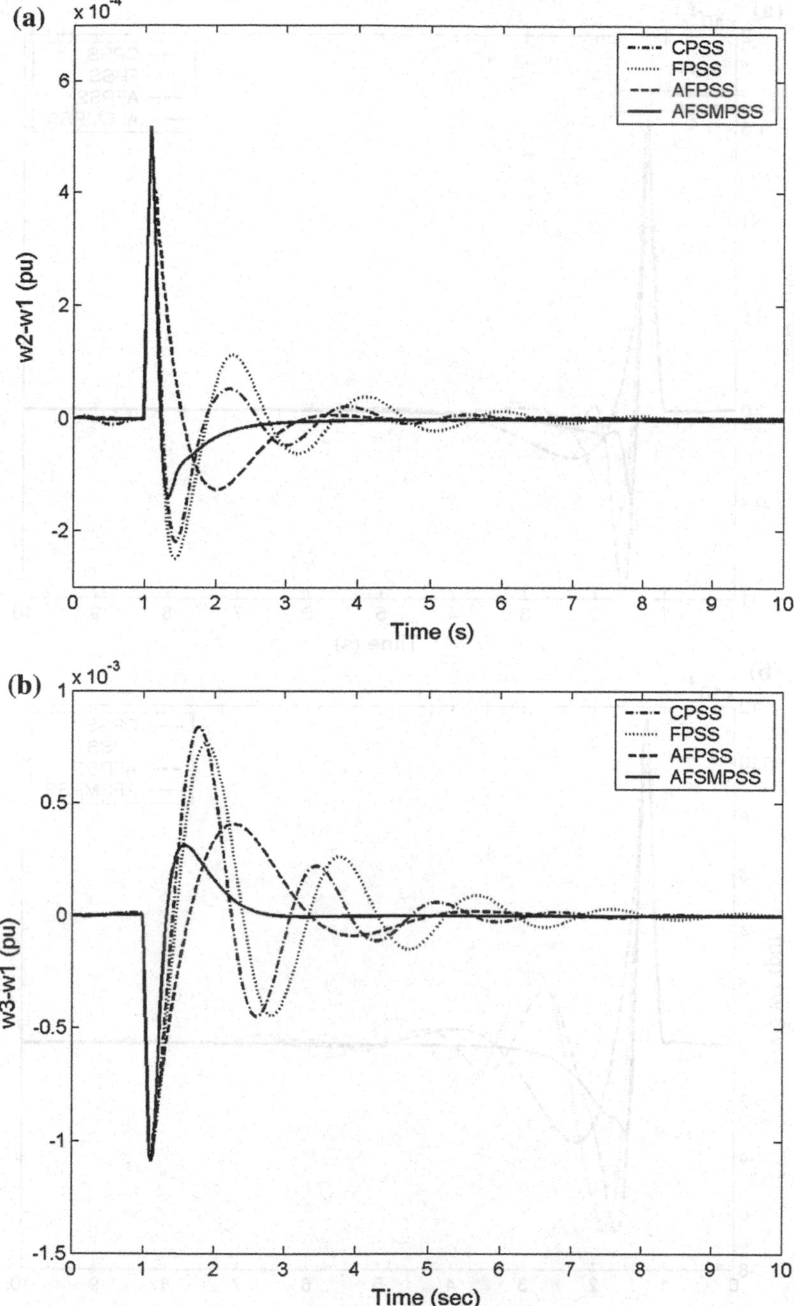

Fig. 6 System response to three-phase fault at the middle of one tie line applied between buses 7 and 8: **a** Local mode of oscillations, **b** Inter-area mode of oscillations

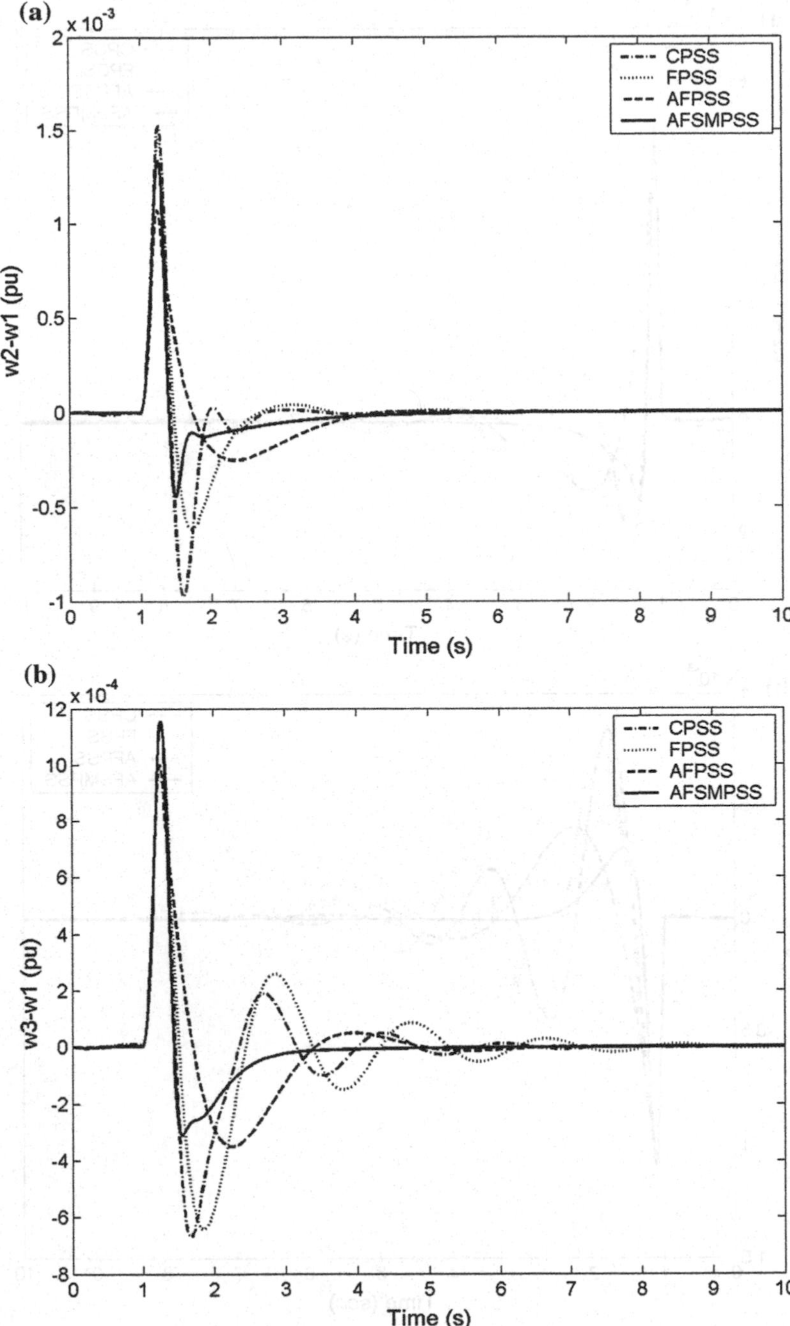

Fig. 7 System response due to step change in voltage reference of generator 1: **a** Local mode of oscillations, **b** Inter-area mode of oscillations

damping of the low frequency oscillations in both stabilizers the conventional CPSS and fixed parameter FPSS requires more time and has more oscillations before the speed deviation response is stabilized. The adaptive fuzzy controller improves the damping of oscillations due to the self-learning capability in the change of operating conditions. However, the superiority performance is clear with the proposed controller. The proposed controller provides significantly better damping enhancement in the power system oscillations. It is possible to observe that the overshoot and settling time are reduced as well, this in the presence of external disturbance.

In the second test case, the performance of designed controllers was evaluated in the presence of a step disturbance injected in reference voltage of the generator. Figures 4 and 6 in both power systems, the system response with the conventional CPSS and fuzzy FPSS have again more oscillations and large time to stabilize the systems. In comparison with the adaptive fuzzy AFPSS, the robustness is achieved by the designed AFSMPSS stabilizer than the AFPSS under generators parameters variations with a quite good damping performance.

To further assess the performance and the effectiveness of the proposed PSS, performance index $\left(J_P = \sum t\,|\Delta\omega_{i-1}|\right)$ is used to compare between the different PSSs considered. It is worth mentioning that the lower value of this index, the better is the system response in terms of overshooting and time-domain characteristics. Tables 3 and 4 show the values of performance index for all cases of disturbances. So, the proposed indirect AFSMPSS has clearly improved the system performance by reducing the speed deviations under the following types of disturbances: three-phase fault short circuit and step change in the reference voltage of generator which are all rejected by the proposed stabilizer.

6 Conclusion

A new an indirect adaptive fuzzy sliding mode power system stabilizer for a single machine infinite bus and multi-machine power system to damp oscillations has been proposed in this paper, based on the fuzzy logic system to approximate the unknown system function present in the model of power system and enhanced by a PI term controller that eliminates chattering in the control signal. Controller gains are tuned using PSO technique. An adaptation algorithm was derived based on the Lyapunov's direct method to cause the system follow a desired response. The effectiveness of the proposed design stabilizers have been tested on a single machine infinite bus and multi-machine power system under different system disturbances. The nonlinear time domain simulation results show the robustness performance of the proposed stabilizer and their ability to provide good quality damping of low frequency oscillations. Moreover, this controller exhibit better performance to damp the multi-machine power system with local and inter area modes of oscillations and improve greatly the system stability compared to the other power system stabilizers. As a future work, we intend to use a Type-2 fuzzy system to approximate system dynamics and design nonlinear decentralized controllers.

Table 3 The performance index for single machine

	CPSS	*FPSS*	*AFPSS*	*AFSMPSS*
Case 1	0.18506	0.32192	0.14065	0.13138
Case 2	0.060966	0.18476	0.055257	0.053798

Table 4 The performance index for multi-machine

		CPSS	*FPSS*	*AFPSS*	*AFSMPSS*
Case 1	Gen_{2-1}	0.044367	0.076493	0.044169	0.032748
	Gen_{3-1}	0.27498	0.38126	0.19403	0.071122
Case 2	Gen_{2-1}	0.10415	0.130625	0.13793	0.097366
	Gen_{3-1}	0.16238	0.258294	0.154403	0.083119

References

Abido, M.A.: Optimal design of power system stabilizers using particle Swarm optimization. IEEE Trans. Energy Convers. **17**(3), 406–413 (2002)

Abido, M.A., Abdel-Magid, Y.L.: Adaptive tuning of power system stabilizers using radial basis function networks. Electric Power Syst. Res. **49**(1), 21–29 (1999)

Al-Awami, A.T., Abdel-Magid, Y.L., Abido, M.A.: A particle-swarm-based approach of power system stability enhancement with unified power flow controller. Intern. J. Elect. Power Energy Syst. **29**(3), 251–259 (2007)

Al-Duwaish, H.N., Al-Hamouz, M.: A neural network based adaptive sliding mode controller: application to a power system stabilizer. Energy Convers. Manage. **52**(2), 1533–1538 (2011)

Anderson, P.M., Fouad, A.A.: Power System Control Stability. Iowa State University Press, Iowa, USA (1977)

Bandal, V., Bandyopadhyay, B.: Robust decentralised output feedback liding mode control technique-based power system stabiliser (PSS) for multimachine power system. IET Control Theory Appl. **1**(5), 1512–1522 (2007)

Bandal, V., Bandyopadhyay, B., Kulkarni, A. M.: Output feedback fuzzy sliding mode control technique based power system stabilizer (PSS) for single machine infinite bus (SMIB) system. In: IEEE International Conference on Industrial Technology, pp. 341–346 (2005)

Bhati, P.S., Gupta, R.: Robust fuzzy logic power system stabilizer based on evolution and learning. Intern. J. Electr. Power Energy Syst. **53**, 357–366 (2013)

Bouchama, Z., Harmas, M.N.: Optimal Robust aaptive fuzzy synergetic power system stabilizer design. Electric Power Syst. Res. **83**(1), 170–175 (2012)

Cao, Y., Jiang, L., Cheng, S., Chen, D., Malik, O.P., Hope, G.S.: A nonlinear variable structure stabilizer for power system stability. IEEE Trans. Energy Convers. **9**(3), 489–495 (1994)

Changaroon, B., Srivastava, S.C., Thukaram, D.: A neural network based power system stabilizer suitable for on-line training-a practical case study for EGAT system. IEEE Trans. Energy Convers. **15**(1), 103–109 (2000)

Chen, G.P., Malik, O.P.: Tracking constrained adaptive power system stabilizer. IEE Proc. Gener. Transm. Distrib. **142**(2), 149–156 (1995)

Cuk-Supriyadi, A.N., Takano, H., Murata, J., Goda, T.: Adaptive Robust PSS to enhance stabilization of interconnected power systems with high renewable energy penetration. Renew. Energy **63**, 767–774 (2014)

Colbia-Vega, A., León-Morales, J.D., Fridman, L., Salas-Peña, O., Mata- Jiménez, M.T.: Robust excitation control design using sliding-mode technique for multi-machine power systems. Electric Power Syst. Res. **78**(9), 1627–1634 (2008)

Demirören. Automatic generation control for power system with SMES by using neural network controller. Electric Power Compon. Syst. **31**(1), 1–25 (2003)

El-Metwally, K.A., Hancock, G.C., Malik, O.P.: Implementation of a fuzzy logic PSS using a micro-controller and experimental test results. IEEE Trans. Energy Convers. **11**(1), 91–96 (1996)

El-Metwally, K.A., Malik, O.P.: Fuzzy logic power system stabilizer. IEE Proc. Gen. Trans. Distrib. **142**(3), 277–281 (1995)

Elshafei, A.L., El-Metwally, K.A., Shaltout, A.A.: A variable-structure adaptive fuzzy-logic stabilizer for single and multi-machine power systems. Control Eng. Pract. **13**(4), 413–423 (2005)

Fernandez-Vargas, J., Ledwich, G.: Variable structure control for power systems stabilization. Intern. J. Electr. Power Energy Syst. **32**(2), 101–107 (2010)

Fraile-Ardanuy, J., Zufiria, P.J.: Design and comparison of adaptive power system stabilizers based on neural fuzzy networks and genetic algorithms. Neurocomputing **70**(16—18), 2902–2912 (2007)

Ghazi, R., Azemi, A., Badakhshan, K.P.: Adaptive fuzzy sliding mode control of SVC and TCSC for improving the dynamic performance of power systems. In: Proceedings of Seventh International Conference on AC-DC Power Transmission, pp. 333–337 (2001)

Hiyama, T.: Real time control of micro-machine system using micro-computer based fuzzy logic power system stabilizer. IEEE Trans. Energy Convers. **9**(4), 724–731 (1994)

Hassan, M.A., Malik, O.P., Hope, G.S.: A fuzzy logic based stabilizer for a synchronous machine. IEEE Trans. Energy Convers. **6**(3), 407–413 (1991)

Hosseinzadeh, N., Kalam, A.: A direct adaptive fuzzy power system stabilizer. IEEE Trans. Energy Convers. **14**(4), 1564–1571 (1999)

Huerta, H., Loukianov, A.G., Cañedo, J.M.: Decentralized sliding mode block control of multi-machine power systems. Intern. J. Electr. Power Energy Syst. **32**(1), 1–11 (2010)

Huerta, H., Loukianov, A.G., Cañedo, J.M.: Robust multi-machine power systems control via high order sliding modes. Electric Power Syst. Res. **81**(7), 1602–1609 (2011)

Hussein, T., Elshafei, A. L., Bahgat, A.: design of a hierarchical fuzzy logic PSS for a multi-machine power system. In Proceedings of Mediterranean Conference on Control and Automation, pp. 1–6 (2007)

Hussein, T., Saad, M.S., Elshafei, A.L., Bahgat, A.: Damping inter-area modes of oscillation using an adaptive fuzzy power system stabilizer. Electric Power Syst. Res. **80**(12), 1428–1436 (2010)

Hussein, T., Saad, M.S., Elshafei, A.L., Bahgat, A.: Robust adaptive fuzzy logic power system stabilizer. Expert Syst. Appl. **36**(10), 12104–12112 (2009)

Ho, H.F., Cheng, K.W.E.: Position control of induction motor using indirect adaptive fuzzy sliding mode control. In Proceedings of Third International Conference on Power Electronics Systems and Applications, pp. 1–5, (2009)

Ho, H.F., Wong, Y.K., Rad, A.B.: Adaptive fuzzy sliding mode control with chattering elimination for nonlinear SISO systems. Simul. Model. Pract. Theory **17**(7), 1199–1210 (2009)

Hosseini, S.H., Etemadi, A.H.: Adaptive neuro-fuzzy inference system based automatic generation control. Electric Power Syst. Res. **78**(7), 1230–1239 (2008)

Karimi, A., Feliachi, A.: Decentralized adaptive backstepping control of electric power systems. Electric Power Syst. Res. **18**(3), 484–493 (2008)

Kennedy, J., Eberhart, R.C.: Particle swarm optimization. In: IEEE Proceedings of International Conference on Neural Networks, pp. 1942–1948 (1995)

Kennedy, J., Eberhart, R.C.: Swarm Intelligence. Morgan Kaufmann Publishers, San Francisco (2001)

Klein, M., Rogers, G.J., Kundur, P.: A fundamental study of inter-area oscillations in power systems. IEEE Trans. Power Syst. **6**(3), 914–921 (1991)

Kundur, P.: Power System Stability and Control. McGrew Hill, New York (1994)

Kundur, P., Klein, M., Rogers, G.J., Zywno, M.S.: Application of power system stabilizers for enhancement of overall system stability. IEEE Trans. Power Syst. **4**(2), 614–626 (1989)
Kothan, M.L., Bhattacharya, K., Nanda, J.: Adaptive power system stabilizer based on pole shifting technique. IEE Proc. Gener. Transm. Distrib. **143**(1), 96–98 (1996)
Larsen, E.V., Swann, D.A.: Applying power system stabilizers Part-I: general concepts. IEEE Trans. Power Appl. Syst. **100**(6), 3017–3024 (1981)
Lin, Y.J.: Proportional plus derivative output feedback based fuzzy logic power system stabilizer. Intern. J. Electr. Power Energy Syst. **44**(1), 301–307 (2013)
Liu, W., Venayagamoorthy, G.K., Wunsch, D.C.: Design of an adaptive neural network based power system stabilizer. Neural Netw. **16**(5—-6), 891–898 (2003)
Mostafa, H.E., El-Sharkawy, M.A., Emary, A.A., Yassin, K.: Design and allocation of power system stabilizers using the particle swarm optimization technique for an interconnected power system. Intern. J. Electr. Power Energy Syst. **34**(1), 57–65 (2012)
Radaideh, S.M., Nejdawi, I.M., Mushtaha, M.H.: Design of power system stabilizers using two level fuzzy and adaptive neuro-fuzzy inference systems. Intern. J. Electr. Power Energy Syst. **35**(1), 47–56 (2012)
Rashidi, F., Rashidi, M., Amiri, H.: An adaptive fuzzy sliding mode control for power systems stabilizer. In: IEEE Proceedings Twenty-nine Annual Conference of the Industrial Electronics Society, pp 626–630 (2003)
Samarasinghe, V.G.D.C., Pahalawaththa, N.C.: Stabilization of a multi-machine power system using nonlinear robust variable structure control. Electric Power Syst. Res. **43**(1), 11–17 (1997)
Saoudi, K., Bouchama, Z., Harmas, M.N., Zehar, K.: Indirect adaptive fuzzy power system stabilizer. In: AIP Proceedings of First Mediterranean Conference on Intelligent Systems and Automation, pp. 512–515 (2008)
Saoudi, K., Harmas, M.N.: Enhanced design of an indirect adaptive fuzzy sliding mode power system stabilizer for multi-machine power systems. Intern. J. Electr. Power Energy Syst. **54**(1), 425–431 (2014)
Saoudi, K., Harmas, M.N., Bouchama, Z.: Design of a robust and indirect adaptive fuzzy power system stabilizer using particle swarm optimisation. Recovery utilization environment effects. Energy Sources Part A **36**(15), 1670–1680 (2011)
Saoudi, K., Harmas, M.N., Bouchama, Z.: Design and analysis of an indirect adaptive fuzzy sliding mode power system stabilizer. In: Proceedings of Second International Conference on Electrical and Electronics Engineering, pages 96–100 (2008)
Sauer, P.W., Pai, M.A.: Power System Dynamics and Stability. Prentice-Hall, Englewood Cliffs (1998)
Slotine, J.E., Li, W.P.: Applied Nonlinear Control. Prentice-Hall, Englewood Cliffs (1991)
Teh-Lu, L.: Design of an adaptive nonlinear controller to improve stabilization of a power system. Electr. Power Energy Syst. **21**(6), 433–441 (1999)
You, R., Hassan, J.E., Nehrir, M.H.: An online adaptive neuro-fuzzy power system stabilizer for multi-machine systems. IEEE Trans. Power Syst. **18**(1), 128–135 (2003)
Wang, L.X.: Stable adaptive fuzzy control of nonlinear system. IEEE Trans. Fuzzy Syst. **1**(2), 146–155 (1993)
Wang, L.X.: Stable adaptive fuzzy controllers with application to inverted pendulum tracking. IEEE Trans. Syst. Man Cybern. Part b **26**(5), 677–691 (1996)
Wu, B., Malik, O.P.: Multivariable adaptive control of synchronous machines in a multi-machine power system. IEEE Trans. Power Syst. **21**(4), 1772–1781 (2006)
Zeynelgil, H.L., Demiroren, A., Sengor, N.S.: The Application of ANN technique to automatic generation control for multi-area power system. Intern. J. Electr. Power Energy Syst. **24**(5), 345–354 (2002)
Zhang, Y., Malik, O.P., Chen, G.P.: Artificial neural network power system stabilizers in multi-machine power system environment. IEEE Trans. Energy Convers. **10**(1), 147–155 (1995)

Higher Order Sliding Mode Control of Uncertain Robot Manipulators

Neila Mezghani Ben Romdhane and Tarak Damak

Abstract This chapter deals with the tracking problem of robot manipulators. These systems are described by highly nonlinear and coupled equations. Higher order sliding mode controllers are then proposed to ensure stability and robustness of uncertain robot manipulators. The motivation for using high order sliding mode mainly relies on its appreciable features, such as high precision and elimination of chattering in addition that ensures the same performance of conventional sliding mode like robustness. In this chapter we propose two high order sliding mode controllers. The first guarantees a continuous control eliminating the chattering phenomenon. Instead of a regular control input, the derivative of the control input is used in the proposed control law. The discontinuity in the controller is made to act on the time derivative of the control input. The actual control signal obtained by integrating the derivative control signal is smooth and chattering free. The second controller is an adaptive version of high order sliding mode controller. The goal is to obtain a robust high order sliding mode adaptive gain control law to respect to uncertainties and perturbations without the knowledge of uncertainties/perturbations bound. The proposed controller ensures robustness, precision and smoothness of the control signal. The stability and the robustness of the proposed controllers can be easily verified by using the classical Lyapunov criterion. The proposed controllers are tested to a three-degree-of-freedom robot to prove their effectiveness.

1 Introduction

Control under uncertainty conditions is one of the main topics of the modern control theory. Among the existing control techniques (Haddad and Hayakawa 2002; Zhou et al. 2004; Yan et al. 2005), sliding mode control (Kachroo and Tomizuka 1996;

N. Mezghani Ben Romdhane (✉) · T. Damak
Laboratory of Sciences and Techniques of Automatic Control and Computer Engineering (Lab-STA), National Engineering School of Sfax (ENIS), Sfax, Tunisia
e-mail: neilamezghani@yahoo.fr

T. Damak
e-mail: tarak.damak@enis.rnu.tn

A.T. Azar and Q. Zhu (eds.), *Advances and Applications in Sliding Mode Control systems*, Studies in Computational Intelligence 576, DOI 10.1007/978-3-319-11173-5_12

Lu and Spurgeon 1997) is a powerful method to control nonlinear systems having uncertainties and disturbances. The control laws are designed so that the systems trajectory always reaches the sliding surface (Perruquetti and Barbot 2002). This is known as the reaching phase. Once on the sliding surface, the control structure is changed discontinuously to maintain the system on the sliding surface. At this stage, the system is in the sliding phase. In this phase the system becomes totally insensitive to parametric uncertainty and external disturbances. The control law may be linear or nonlinear during the whole or parts of the control mission. Its structure changes according to a preselected switching logic. The switch in the control structure depends on the instantaneous values of the systems state along the trajectory. The high frequency switching of the control causes the so-called chattering phenomenon which is the main drawback of the sliding mode control. This phenomenon is extremely dangerous to the actuator of electromechanical systems.

Several approaches are proposed to eliminate chattering. One such is to replace the sign function in a small area of the surface by a smooth approximation, which is the so-called boundary layer control (Kachroo and Tomizuka 1996). Then the chattering is reduced but accuracy and robustness are deteriorated. Another technique uses the observer design. This approach exploits a localization of the high frequency phenomenon in the feedback loop by introducing a discontinuous feedback control loop which is closed through an asymptotic observer of the plant (Young et al. 1999). Consequently, it suppresses the high frequency oscillations of the control input (Young et al. 1999). Recently, new approach has been proposed called higher order sliding mode (Levant 2003; Plestan et al. 2008; Rhif 2012; Kamal and Bandyopadlyay 2012; Beltran et al. 2009; Abouissa et al. 2013; Msaddek et al. 2013). Instead of influencing the first sliding surface time derivative, the sign function is acting on its higher time derivative. Keeping the main advantage of standard sliding mode control, the chattering effect is eliminated and higher order precision is provided (Perruquetti and Barbot 2002). In the case of γ th order sliding mode control, the objective is to keep the sliding variable and its $\gamma - 1$ first time derivatives to zero through discontinuous function acting on the time γ th derivative of the sliding variable.

Several second order sliding modes control algorithms are introduced such as twisting and super-twisting controllers, the suboptimal control algorithm, the control algorithm which prescribed convergence law and the quasi-continuous control algorithm (Zhao et al. 2013). Many papers are available in the case of second order sliding mode control (Hamerlain et al. 2005; Boiko et al. 2006; Bartolini et al. 2000; Zhang et al. 2013). Arbitrary order sliding mode controllers have recently been proposed in (Laghrouche et al. 2006, 2007; Levant 2001, 2005a; Defoort et al. 2009). In 2001, the first arbitrary order sliding mode controller was proposed (Levant 2001) by tuning only one gain parameter. Such controller allowed solving the finite-time output stabilization and exact disturbance compensation problem for an output with an arbitrary relative degree. There, its finite time convergence is proved by means of geometrical (point-to-point transformation) method. However, the convergence rate cannot be arbitrary selected. The main problem of the algorithms in (Levant 2001, 2005a, b) is parameter adjustment. Indeed, there is no explicit condition for the gain tuning. Therefore, the convergence cannot easily be made arbitrary fast or slow. The approach given in (Laghrouche et al. 2006) proposes higher order sliding mode based on linear quadratic approach. In spite of their advantages (constructive approach, practical applicability), its major drawback is that the higher order sliding mode control is only practical. The system trajectory reaches the small neighborhood of the origin in finite time. Similar type of approach is used in (Plestan et al. 2008). Based on the information of initial and final values for each state variable for the control input the higher order controller is designed. In (Laghrouche et al. 2007), the authors use the integral sliding mode control and guarantee the establishment of a higher order sliding mode. The advantages of this algorithm are easy to implement and guarantee the robustness of the system during the entire response. But it directly depends on the initial conditions of the system and complex off-line computations are needed before starting the control action. In 2007, a new type of arbitrary-order controller (Levant 2007), which is γ th-sliding homogeneous, controller was proposed. Considering all the above mentioned drawbacks, in 2009 (Defoort et al. 2009) a new proposal of higher order sliding mode came into existence, which was based on combined approach of geometrical homogeneity based linear controller (Bhat and Bernstein 2005) and classical sliding mode technique. Since this controller is based on geometrical homogeneity principle, it is again not possible to calculate exact time of convergence. Also, the control of (Defoort et al. 2009) suffers from the undesired phenomenon of chattering. In this chapter, another high order sliding mode control is presented assuring the elimination of the chattering.

This chapter proposes two higher order sliding mode controls applied to robotic manipulator in uncertainty conditions. The first proposed controller is inspired from classical sliding mode control. The main attributes of this controller are robustness and precision which are the basic properties of a higher order sliding mode controller. Moreover, the chattering phenomenon, in the control input, is eliminated. Indeed, the discontinuity is used in the derivative of the control, instead in the control. The second controller is an adaptive high order sliding mode controller. This controller ensures robustness and precision. The unknown upper bound of uncertainties is estimated using an adaptive tuning law. Consequently, prior knowledge of the upper bound of the system uncertainties is not required. Moreover, the chattering phenomenon, in the control input, is eliminated.

The outline of this paper is as follows. Section 2 presents the state of the art of high order sliding mode control. In Section 3, the second order sliding mode controller is designed for uncertain robot manipulator. The controller eliminates the chattering in the control input. Section 4 presents the adaptive second order sliding mode control of robot manipulators. The Section 5 presents simulation results that demonstrate the efficiency and advantages of the proposed controller. The discussion is presented in Section 6. Section 7 concludes the paper.

2 State of the Art of High Order Sliding Mode Control

Control of robotic system is vital due to wide range of their applications because this system is multi-input, multi-output, nonlinear and uncertain. Consequently, it is difficult to design accurately mathematical models for multiple degrees of freedoms robot manipulators. Therefore, strong mathematical tools used in new control methodologies to design a controller with acceptable performance. As it is obvious stability is the minimum requirement in any control system, however the proof of stability is not trivial especially in the case of nonlinear systems. One of the best nonlinear robust to control of robot manipulator is sliding mode controller.

As known to all, the sliding mode control with the strong robustness for internal parameters and external disturbances. In addition, the appropriate sliding surface can be selected to reduce order for control system. However, due to the chattering phenomena of sliding mode control, the high frequency oscillation of control system brings challenge for the application of sliding mode control. On the other hand, the choice of sliding surface strictly requires system relative degree to equal to1, which limits the choice of sliding surface.

In order to solve the above problems, this chapter focuses on a new type of sliding mode control, that is, higher order sliding mode control. The technology not only retains advantage of strong robustness in the traditional sliding mode control, but also enables discontinuous items transmit into the first order or higher order sliding mode derivative to eliminate the chattering. Besides, the design of the controller no longer must require relative degree to be 1. Therefore, it is greatly simplified to design parameters of sliding mode surface.

In recent years, because arbitrary order sliding mode control technique not only retains the traditional sliding mode control simple structure with strong robustness, but also eliminates the chattering phenomenon in the traditional sliding mode, at the same time, gets rid of the constraints of system relative degree. Therefore theoretical research and engineering applications has caused widespread concern and has been constant development.

Without losing generality, considering a state equation of single input nonlinear system as

$$\begin{aligned} \dot{x} &= f(x) + g(x)u \\ y &= s(x, t) \end{aligned} \tag{1}$$

where, $x \in R^n$ is system state variable, t is time, u is control input. Here, $f(x)$, $g(x)$ and $s(x, t)$ are smooth functions. The control objective is making output function $s \equiv 0$.

Differentiate the output variables continuously, we can get every order derivative of. According to the conception of system relative degree, there are two conditions.

i. Relative degree $r = 1$, if and only if $\frac{\partial \dot{s}}{\partial s} \neq 0$,
ii. Relative degree $r \geq 2$, if $\frac{\partial s^{(i)}}{\partial s} = 0\,(i = 1, 2, \ldots r - 1)$, and $\frac{\partial s^{(r)}}{\partial s} \neq 0$.

In arbitrary order sliding mode control, its core idea is the discrete function acts on higher order sliding mode surface, making

$$s(x,t) = \dot{s}(x,t) = \ddot{s}(x,t) = \cdots = s^{(r-1)}(x,t) = 0 \quad (2)$$

Suppose the relative degree of system (1) equals to r, generally speaking, when the control input u first time appears in r-order derivative of s, that is $\frac{\partial s^{(r)}}{\partial u} \neq 0$, then we take r-order derivative of s for output of system (1), $s, \dot{s}, \ddot{s}, \ldots, s^{(r-1)}$can be obtained. They are continuous functions for all x and t. However, corresponding discrete control law u acts on $s^{(r)}$. Selecting a new local coordinate, then

$$y = (y_1, y_2, \ldots y_r) = (s, \dot{s}, \ddot{s}, \ldots, s^{(r-1)}) \quad (3)$$

So, the following expression can be obtained

$$s^{(r)} = a(y,t) + b(y,t)u, \quad b(y,t) \neq 0 \quad (4)$$

Therefore, high order sliding mode control is transformed to stability of r th order dynamic system (2), (4). Through the Lie derivative calculation, it is very easy to verify that

$$\begin{aligned} b &= L_g L_f^{r-1} s = \frac{ds^{(r)}}{du} \\ a &= L_f^r s \end{aligned} \quad (5)$$

Suppose $\eta = (y_{r+1}, y_{r+2}, \ldots, y_n)$, then

$$\eta = \xi(t, s, \dot{s}, \ldots, s^{(r-1)}, \eta) + \chi(t, s, \dot{s}, \ldots, s^{(r-1)}, \eta)u \quad (6)$$

Now, Eqs. (3), (4) and (6) are transformed to Isidori-Brunowsky canonical form. The sliding mode equivalent control is $u_{eq} = \frac{a(y,t)}{b(y,t)}$ (Utkin 1992). At present, the aim of control is to design a discrete feedback control $u = U(x,t)$, so that new system converges into origin on the order sliding mode surface within limited time. Therefore, in Eq. (4), both $a(y,t)$ and $b(y,t)$ are bounded function. There are positive constants K_m, K_M and C so that

$$\begin{aligned} 0 < K_m < b(y,t) < K_M \\ a(y,t) < C \end{aligned} \quad (7)$$

Theorem 1 (Levant 1998, 2003) *Suppose the relative degree of nonlinear system* (1) *to output function* $s(x,t)$ *is* r, *and satisfying the condition* (7), the arbitrary order sliding mode controller has following expression

$$u = -\alpha\ sgn(\psi_{r-1,r}(s, \dot{s}, \ldots, s^{(r-1)})) \tag{8}$$

where

$$\begin{aligned}
\psi_{0,r} &= s \\
\psi_{1,r} &= \dot{s} + \beta_1 N_{1,r} sgn(s) \\
\psi_{i,r} &= s^{(r)} + \beta_i N_{i,r} sgn(\psi_{i-1,r}), \quad i = 1, \ldots, r-1 \\
N_{1,r} &= |s|^{(r-1)/r} \\
N_{i,r} &= (|s|^{p/r} + |\dot{s}|^{p/(r-1)} + \cdots + |s^{(i-1)}|^{p/(r-i-1)})^{(r-i)/p} \quad i = 1, \ldots, r-1 \\
N_{r-1,r} &= (|s|^{\frac{p}{r}} + |\dot{s}|^{\frac{p}{r-1}} + \cdots + |s^{(s^{(r-2)})}|^{p/2})^{1/p}
\end{aligned} \tag{9}$$

Properly choose positive parameters $\beta_1, \beta_2, \ldots, \beta_{r-1}$, the system converges into origin on the r order sliding mode surface within limited time. Finally, when $s \equiv 0$, it achieves control object. The choice of positive parameters $\beta_1, \beta_2, \ldots, \beta_{r-1}$ is not unique. Here, $r \leq 4$ order sliding mode controller is given, which is also tested.

$$\begin{aligned}
&1.u = -\alpha\ sgn(s) \\
&2.u = -\alpha\ sgn(\dot{s} + |s|^{1/2} sgn(s)) \\
&3.u = -\alpha\ sgn(\ddot{s} + 2(|\dot{s}|^3 + |s|^{02})^{1/6} sgn(\dot{s} + |s|^{2/3} sgn(s))) \\
&4.u = -\alpha\ sgn\{\dddot{s} + 3[|\ddot{s}|^6 + |\dot{s}|^4 + |s|^3]^{\frac{1}{12}} sgn[\ddot{s} + (|\dot{s}|^4 + |s|^3)^{1/6} \\
&\qquad sgn(\dot{s} + 0.5|s|^{\frac{3}{4}} sgn(s))]\} \\
&\qquad \vdots
\end{aligned} \tag{10}$$

From the above Eq. (10) we can also see that, when $r = 1$, the controller is traditional relay sliding mode control; when $r = 2$, in fact, the controller is super twisting algorithm of second order sliding mode.

To get the differentiation of a given signal is always essential in automatic control systems. We often need derivative a variable or function. So there are a lot of numerical algorithms for this issue. The same situation also appears in the design of high order sliding mode controller (10) that needs to calculate the derivative values of sliding mode variable.

Presentation above in the previous has been explained in detail the principles of high order sliding mode control and sliding mode controller design method. This part focuses on how to take use of high order sliding mode technique to solve the differentiation of a given signal or variable function. And their simulation results are verified.

Suppose given signal is $f(t)$, now set a dynamic system as

$$\dot{x} = u \tag{11}$$

The control objectiveis to make the variable x follow given signal $f(t)$, that is

$$x = f(t) \tag{12}$$

Therefore, sliding mode surface is selected as

$$s = x - f(t) \tag{13}$$

At this moment, according to the principle of sliding mode control, a proper controller is designed. When the system enter into sliding mode, $s = x - f(t) = 0$. Derivative of sliding mode surface (13),

$$\dot{s} = \dot{x} - \dot{f}(t) = u - \dot{f}(t) \tag{14}$$

Because control input first time appears in the derivative of sliding mode surface s, the relative degree of system is $r = 1$. It satisfies the requirement about relative degree of second order sliding mode. So the super twisting algorithm (Fridman and Levant 2002) is adopted. Thus,

$$\begin{aligned} u &= -\lambda|x - f(t)|^{1/2} sgn(x - f(t)) + u_1 \\ \dot{u}_1 &= -\alpha sgn(x - f(t)) \end{aligned} \tag{15}$$

where, $\lambda > 0$, $\alpha > 0$ are positive constant. Definite a function as $\Theta(\alpha, \lambda, C) = |\Psi(t)|$, C is Lipschitz constant about derivative of $f(t)$. $(\Sigma t, \Psi(t))$is the solution of equation of (16), the initial value are $\Sigma(0) = 0$, $\Psi(0) = 1$

$$\begin{aligned} \dot{\Sigma} &= -|\Sigma|^{1/2} + \Psi \\ \dot{\Psi} &= \begin{cases} -\frac{1}{\lambda^2}(a - C), & -|\Sigma|^{1/2} + \Psi > 0 \\ -\frac{1}{\lambda^2}(a + C), & -|\Sigma|^{1/2} + \Psi \le 0 \end{cases} \end{aligned} \tag{16}$$

Theorem 2 (Levant 1998) *Let* $\alpha > C > 0$, $\lambda > 0$ *function* $\Theta(\alpha, \lambda, C) < 1$.

Then, provided $f(t)$ *has a derivative with Lipschitz's constant* C, *the equality* $u = \dot{f}(t)$ *is fulfilled identically after finite time transient process. And the smaller value of* Θ *assume, faster convergence; If* $\Theta(\alpha, \lambda, C) > 1$, *control input* u *will not converge into* $\dot{f}(t)$. *Observer parameters should meet the following sufficient condition for convergence of the second order sliding mode control,*

$$\begin{aligned} &\alpha > C \\ &\lambda^2 \ge 4C\frac{\alpha + C}{\alpha - C} \end{aligned} \tag{17}$$

According to the principle of second order sliding mode, after a finite time, the system will converge into the origin, that is,

$$s(x,t) = \dot{s}(x,t) = 0 \tag{18}$$

Then,

$$u = \dot{f}(t) \tag{19}$$

Now, observer input is the estimation of derivative of given signal $f(t)$. Using a sliding mode controller achieve differentiation of variable function.

Let input signal be presented in the form $f(t) = f_0(t) + n(t)$, where $f_0(t)$ is a differentiable base signal $f_0(t)$, has a derivative with Lipschitz 's constant $C > 0$, and $n(t)$ is a noise, $|n(t)| < \varepsilon$. Then, there exists such a constant $b > 0$ depend on $(\alpha - C)/\lambda^2$ and $(\alpha + C)/\lambda^2$ that after a finite time, the inequality $|u(t) - \dot{f}_0(t)| < \lambda b \varepsilon^{1/2}$ holds (Levant 1998).

Through the first order sliding mode differentiator description of the working principle, it will naturally think, whether can design a sliding mode differentiator to obtain the arbitrary order derivative of given signal. Well, the design of high order sliding mode controller (10) needs to know all sliding mode variables and their corresponding differentiation.

Theorem 3 *Design an arbitrary order sliding mode differentiator, which can be used to estimate the derivative value of sliding mode variables, so as to achieve a simplified numerical differential purpose as following.*

$$\begin{aligned}
\dot{z}_0 &= v_0 \\
v_0 &= -\lambda_0 |z_0 - f(t)|^{n/(n+1)} sgn(z_0 - f(t)) + z_1 \\
\dot{z}_1 &= v_1 \\
v_1 &= -\lambda_1 |z_1 - v_0|^{(n-1)/n} sgn(z_1 - v_0) + z_2 \\
&\vdots \\
\dot{z}_{n-1} &= v_{n-1} \\
v_{n-1} &= -\lambda_{n-1} |z_{n-1} - v_{n-2}|^{1)/2} sgn(z_{n-1} - v_{n-2}) + z_n \\
\dot{z}_n &= -\lambda_n sgn(z_n - v_{n-1})
\end{aligned} \tag{20}$$

The same with first order sliding mode differentiator, suppose given signal is $f(t), t \in [0, \infty)$. *It has been known that the n order derivative of* $f(t)$ *has Lipschitz constant, recorded as* $L > 0$. *Now, the object of sliding mode differentiator is estimating the value of* $f'(t), f''(t), \ldots, f^n(t)$,*in real time.*

Arbitrary order sliding mode differentiator has the following recursive form as Eq. (20).

It can be verified, when $n = 1$, it is first order differentiator. Suppose $f_0(t)$ is basic value of given signal $f(t)$, $\delta(t)$ is uncertain part, but bounded, satisfying $|\delta(t)| < \varepsilon$, then $f(t) = f_0 + \delta(t)$.

Theorem 4 (Levant 2003) *If properly choose parameter* $\lambda_i (0 \le i \le n)$*, the following equalities are true in the absence of input noise after a finite time of a transient process.*

$$z_0 = f_0(t)$$
$$z_i = v_i = f_0^{(i)}(t), \ i = 1, \ldots, n \tag{21}$$

The Theorem 4 illustrates that arbitrary order sliding mode differentiator can use differentiation $z_i (0 \le i \le n)$, to estimate any order derivative of input $f(t)$ function online within limited time.

Theorem 5 (Levant 2003) *Let the input noise satisfy the inequality* $\delta(t) = |f(t) - f_0(t)| \le \varepsilon$*. Then the following inequality are established in finite time for some positive constants* μ_i, τ_i*, depending exclusively on the parameters of the differentiator.*

$$|z_i - f_0^i(t)| \le \mu_i \varepsilon^{(n-i+1)/(n+1)} \quad i = 0, \ldots, n$$
$$|v_i - f_0^{i+1}(t)| \le \tau_i \varepsilon^{(n-1)/(n+1)} \quad i = 0, \ldots, n-1 \tag{22}$$

By Theorem 5, we can see that the arbitrary order sliding mode differentiator has robustness.

The arbitrary order sliding mode differentiator can accurately estimate any order derivative of a given input. If this differentiator can be used in high order sliding mode controller (10), any order derivative of sliding mode variable can be accurately estimated avoiding the complicated calculation, which greatly simplifies the controller design. Adopting the differentiator, consider $s(t)$ in high order sliding mode controller as given input for differentiator. Then the output of differentiator $z_i (0 \le i \le n)$ can substitute any order derivative of $s(t)$, that is

$$z_0 = s$$
$$z_i = s^{(i)} \ i = 1, \ldots, n \tag{23}$$

The sliding mode controller (8) can written be

$$u = -\alpha \ sgn(\psi_{r-1,r}(z_0, z_1, \ldots, z_{(r-1)})) \tag{24}$$

The expression from this controller can also be clearly seen, with high order sliding mode differentiator, the differentiation of arbitrary order sliding mode variable will not be difficult to solve, which makes the high order sliding mode controller design has been simplified greatly.

3 Second Order Sliding Mode Control of Robot Manipulator

Robots manipulators are well-known as nonlinear systems including strong coupling between their dynamics. These characteristics, in company with: (1) structured uncertainties caused by model imprecision of link parameters, payload variation, etc., and (2) unstructured uncertainties produced by un-modeled dynamics such as nonlinear friction and external disturbances make the motion control of rigid-link manipulator a complicated problem. Many controllers are proposed in the literature to solve this problem like computed torque control, adaptive control and sliding mode control.

The sliding mode control is known to be a robust approach to solve the control problems of nonlinear systems. Robustness properties against various kinds of uncertainties such as parameter perturbations and external disturbances can be guaranteed. However, this control strategy has a main drawback: the well-known chattering phenomenon. In order to reduce the chattering, the sign function can be replaced by a smooth approximation. However, this technique induces deterioration in accuracy and robustness. In last decade, another approach called higher order sliding mode has been proposed and developed.it is the generalization of classical sliding mode control and can be applied to control systems with arbitrary relative degree respecting to the considered output. In High order sliding mode control, the main objective is to obtain a finite time convergence in the non-empty manifold $S = \{x \in X | s = \dot{s} = \ddot{s} = \cdots = s^{(r-1)} = 0\}$ by acting discontinuously on high order derivatives of the sliding variable . Advantageous properties of high order sliding mode control are: the chattering effect is eliminated, higher order precision is provided whereas all the qualities of standard sliding mode are kept, and control law is not limited by relative degree of the output.

3.1 Robot Manipulator Model

According to the Lagrange theory (Artega and Kelly 2004), the dynamic equation of -joint robot manipulator can be described by

$$M(q)\ddot{q} + C(q, \dot{q}) + G(q) = \tau + d(t) \tag{25}$$

where $q \in R^n$ is the vector of joint angles, $M(q) \in R^{n\times n}$ is the inertia matrix, $C(q, \dot{q}) \in R^n$ is the Coriolis and Centrifugal terms, $G(q) \in R^n$ is the gravitational torque, $\tau \in R^n$ is the vector of the torque produced by actuators, and $d(t) \in R^n$ is the vector of bounded input disturbance, $\|d\|(t) < d_1$ where $d_1 > 0$.

Assuming that the system described by (1) has parts which are known $M_0(q)$, $C_0(q, \dot{q})$, $G_{0(q)}$ and unknwown $\Delta M(q)$, $\Delta C(q, \dot{q})$, $\Delta G(q)$, then

$$M(q) = M_0(q) + \Delta M(q) \tag{26}$$

$$C(q,\dot{q}) = C_0(q,\dot{q}) + \Delta C(q,\dot{q}) \tag{27}$$

$$G(q) = G_0(q) + \Delta G(q) \tag{28}$$

From (26)–(28), (25) can be written in the following form

$$M_0(q)\ddot{q} + C_0(q,\dot{q}) + G_0(q) = \tau + \rho(t) \tag{29}$$

where $\rho(t) = -\Delta M(q)\ddot{q} - \Delta C(q,\dot{q}) - \Delta G(q) + d(t)$.

The control objective is to ensure the tracking of the angular position to the desired position in finite time, with robustness and without chattering.

Consider the robot manipulator model and define the desired trajectory as

$$Q_d(t) = [q_d(t) \quad \dot{q_d}(t)]^T \tag{30}$$

where $q_d(t) \in R^n$ is the vector of desired joint angular and $\dot{q_d}(t)$ is the vector of desired angular velocities.
Define the tracking error vector as

$$e = \begin{pmatrix} q - q_d(t) \\ \dot{q} - \dot{q}_d(t) \end{pmatrix} = \begin{pmatrix} e_1 \\ e_2 \end{pmatrix} \tag{31}$$

The matrix form corresponding to the robot model (25), without disturbance, is

$$\dot{e} = Ae + F(q,\dot{q}) + B(q)\tau = f(e,\tau) \tag{32}$$

where

$$A = \begin{pmatrix} 0 \; I_n \\ 0 \; 0 \end{pmatrix}, F(q,\dot{q}) = \begin{pmatrix} 0 \\ -\ddot{q}_d(t) - M(q)^{-1}(C(q,\dot{q}) + G(q)) \end{pmatrix}$$

$$B(q) = \begin{pmatrix} 0 \\ M(q)^{-1} \end{pmatrix}$$

3.2 Second Order Sliding Mode Control

A sliding surface is chosen for the system (32), in the following form

$$S = C_e \tag{33}$$

Such that $C = (C' \quad I_n)$ and $C' = diag(c_1, c_2, \ldots, c_n)$.

Define the new system formed by $y_1 = S$ and $y_2 = \dot{S}$, then

$$\begin{cases} \dot{y_1} = y_2 \\ \dot{y_2} = \varphi(e) + \psi(e)\dot{\tau} \end{cases} \tag{34}$$

where $\varphi(e) = C\frac{\partial f(e,\tau)}{\partial e}\dot{e}$ and $\psi(e) = C\frac{\partial f(e,\tau)}{\partial \tau}$.
In (9), the time derivative of the control input $\dot{\tau}$ would be designed to act on the higher order derivative of the sliding surface. Hence, instead of the actual control τ, the time derivative control, $\dot{\tau}$ would be used as the control input. The new control would be designed as a discontinuous signal, but its integral (the actual control τ) would be continuous thereby eliminating the high frequency chattering.

Matrices $\varphi(e)$ and $\psi(e)$, in (34), consist of nominal parts $\bar{\varphi}(e)$ and $\bar{\psi}(e)$ which are known apriori and uncertain parts $\Delta\varphi(e)$ and $\Delta\psi(e)$ which are unknown and we suppose that are bounded. Thus we have

$$\begin{cases} \varphi(e) = \bar{\varphi}(e) + \Delta\varphi(e) \\ \psi(e) = \bar{\psi}(e) + \Delta\psi(e) \end{cases} \tag{35}$$

Using (35), the rth order sliding mode system can be written as

$$\begin{cases} \dot{y_1} = y_2 \\ \dot{y_2} = \bar{\varphi}(e) + \bar{\psi}(e)\dot{\tau} + \Delta P(e,t) \end{cases} \tag{36}$$

where $\Delta P(e,t) = \Delta\varphi(e) + \Delta\psi(e)\dot{\tau}$ include all uncertain parameters and external disturbance.

To determine a high order sliding mode control, a novel surface is defined for the system (36) as

$$\sigma = y_2 + Dy_1 \tag{37}$$

where $D = diag(D_i), i = 1, \ldots, n$, such that σ satisfy

$$\dot{\sigma} = -N(\sigma + Wsign(\sigma)) \tag{38}$$

where $N = diag(N_i)$ and $W = diag(W_i)$, $N_i > 0$, $W_i > 0, i = 1, \ldots, n$.

Differentiating (37) and using (36) and (38), the derivative of the control is expressed as

$$\dot{\tau} = -\bar{\psi}(e)^{-1}(\bar{\varphi}(e) + Dy_2 + N(\sigma + Wsign(\sigma))) \tag{39}$$

where

$$N_i W_i > |\Delta P_i(e,t)| \tag{40}$$

Theorem 6 *Consider the robot model* (32)*, if the gains N_i and W_i fulfill the condition* (40) *the control law* (39) *ensures the establishment of the 2nd order sliding mode* *in the sliding surface S, i.e. the trajectory of the system converges asymptotically to zero.*

Proof A Lyapunov function V is selected as

$$V = \frac{1}{2}\sigma^2\sigma \tag{41}$$

Differentiating (41) and using (37) and (36), one obtain

$$\dot{V} = \sigma^T\dot{\sigma} = \bar{\varphi}(e)\dot{\tau} + \Delta P(e,t) + Dy_2 \tag{42}$$

Substituting (39) and simplifying, then

$$\begin{aligned}\dot{V} &= \sigma^T(-N(\sigma + Wsign(\sigma)) + \Delta P(e,t)) \\ &\leq -||\sigma^T N\sigma|| - \sum_{i=1}^{n}\sigma_i N_i W_i sing(\sigma_i) + \sum_{i=1}^{n}\sigma_i \Delta P_i(e,t) \\ &\leq -||\sigma^T N\sigma|| - \sum_{i=1}^{n}|\sigma_i|(N_i W_i - |\Delta P_i(e,t)|)\end{aligned} \tag{43}$$

Then, using (40) yields. $\dot{V} < 0$

Therefore, asymptotic convergence to a domain is guaranteed from any initial condition.

As is evident from (39), $\dot{\tau}$ is discontinuous but after integration it yields a continuous control law τ. Hence, the undesirable high frequency chattering of the control signal is alleviated.

4 Adaptive Second Sliding Mode Control of Robot Manipulator

In practice, the upper bound of the system uncertainty is often unknown in advance and hence the components of the vector uncertainty $|\Delta p_i|$ are difficult to find. Therefore, an adaptive tuning law is used to estimate W_i. Then the control law (39) can be written as

$$\dot{\tau} = -\bar{\psi}(e)^{-1}(\bar{\varphi}(e) + Dy_2 + N\sigma + \widehat{W}_1 sign(\sigma)) \tag{44}$$

where $\widehat{W}_1 = diag(\widehat{W}_{1i}), i = 1, \ldots, n$, $\widehat{W}_{1i}$ is the estimate of W_i.

Defining the adaption error as $\tilde{W}_1 = \widehat{W}_1 - W_1$. The adaptation law, to estimate $\hat{W}_{1i}$, is inspired from the adaptive conventional sliding mode control of (Plestan et al. 2010)

$$\dot{\widehat{W}}_{1i} = \begin{cases} \gamma_i|\sigma_i|sign(|\sigma_i| - \varepsilon) & if \;\; \widehat{W}_{1i} > \mu \\ \mu & if \;\; \widehat{W}_{1i} < \mu \end{cases} \tag{45}$$

Lemma (Plestan et al. 2010)
For the nonlinear uncertain system (36) with the sliding variable σ dynamics (37) controlled by (44), (45) the gain $\widehat{W}_{1i}$ has an upper-bound, i.e. there exists a positive constant W_{1i}^* so that

$$\widehat{W}_{1i} \leq W_{1i}^*, \quad \forall t \tag{46}$$

Theorem 7 *If the control law* (44), *with the adaptation law* (45), *is applied to the nonlinear uncertain system defined by* (7), *the error converges to zero in finite time.*

Proof Consider the following Lyapunov function

$$V = \frac{1}{2}\sigma^T\sigma + \frac{1}{2}\sum_1^n \frac{1}{\gamma_i}(\widehat{W}_{1i} - W_{1i}^*)^2 \tag{47}$$

Differencing V and using the adaptation law (45) for $\widehat{W}_{1i} > \mu$, we obtain

$$\begin{aligned}
\dot{V} &= \sigma^T\dot{\sigma} + \sum_{i=1}^n \frac{1}{\gamma i}(\widehat{W}_{1i} - W_{1i}^*)\dot{\widehat{W}}_{1i} \\
&= \sigma^T\dot{\sigma} + \sum_{i=1}^n (\widehat{W}_{1i} - W_{1i}^*)|\sigma_i| sign(|\sigma_i| - \varepsilon) \\
&= \sigma^T(\dot{y}_2 + D\dot{y}_1) + \sum_1^n (\widehat{W}_{1i} - W_{1i}^*)|\sigma_i| sign(|\sigma_i| - \varepsilon) \\
&= \sigma^T(\bar{\varphi}(e) + \bar{\psi}(e)^{-1}\dot{\tau} + \Delta P(e,t)) + \sum_{i=1}^n (\widehat{W}_{1i} - W_{1i}^*)|\sigma_i| sign(|\sigma_i| - \varepsilon)
\end{aligned} \tag{48}$$

Substituting $\dot{\tau}$ by the expression defined by (44) and simplifying, we obtain

$$\begin{aligned}
\dot{V} &= -\sigma^T N\sigma - \sigma^T\widehat{W}_1 sign(\sigma) + \sum_{i=1}^n (\widehat{W}_{1i} - W_{1i}^*)|\sigma_i| sign(|\sigma_i| - \varepsilon) \\
&= -\sum_{i=1}^n N_i\sigma_i^2 - \sum_{i=1}^n \widehat{W}_{1i}|\sigma_i| + \sum_{i=1}^n (\widehat{W}_{1i} - W_{1i}^*)|\sigma_i| sign(|\sigma_i| - \varepsilon) \\
&= -\sum_{i=1}^n N_i\sigma_i^2 - \sum_{i=1}^n \widehat{W}_{1i}|\sigma_i| - \sum_{i=1}^n W_{1i}^*|\sigma_i| + \sum_{i=1}^n \widehat{W}_{1i}^*|\sigma_i| \\
&\quad + \sum_{i=1}^n (\widehat{W}_{1i} - W_{1i}^*)|\sigma_i| sign(|\sigma_i| - \varepsilon) \\
&= -\sum_{i=1}^n N_i\sigma_i^2 - \sum_{i=1}^n W_{1i}^*|\sigma_i| - \sum_{i=1}^n (\widehat{W}_{1i} - W_{1i}^*)|\sigma_i|
\end{aligned}$$

$$+\sum_{i=1}^{n}(\widehat{W}_{1i}-W_{1i}^{*})|\sigma_i|sign(|\sigma_i|-\varepsilon)$$

$$=-\sum_{i=1}^{n}N_i\sigma_i^2-\sum_{i=1}^{n}W_{1i}^{*}|\sigma_i|+\sum_{i=1}^{n}(\widehat{W}_{1i}-W_{1i}^{*})|\sigma_i|(-1+sign|(\sigma_i|-\varepsilon)) \quad (49)$$

From Lemma, one has always $\widehat{W}_{1i}-W_{1i}^{*}<0$ for all $t>0$. It yields

$$\dot{V}<0 \quad (50)$$

For $\widehat{W}_{1i}<\mu$ it is easy to proof that $\dot{V}<0$.

Therefore, finite time convergence to a domain $S=0$ is guaranteed from any initial condition.

As is evident from (44), is discontinuous but integration $\dot{\tau}$ of yield a continuous control law τ. Hence, the undesirable high frequency chattering of the control signal is alleviated.

5 Simulation Results

The proposed higher order sliding mode control is applied to a three degree freedom robot manipulator. The model of this robot is simulated by using MATLAB Simulink platform with fixed step size of 0.001.

The robot model is defined by the following equation (Mezghani Ben Romdhane and Damak 2011)

$$\begin{pmatrix}M_{11} & M_{12} & M_{13}\\ M_{12} & M_{22} & M_{23}\\ M_{13} & M_{32} & M_{33}\end{pmatrix}\begin{pmatrix}\ddot{q}_1\\ \ddot{q}_2\\ \ddot{q}_3\end{pmatrix}+\begin{pmatrix}C_1\\ C_2\\ C_3\end{pmatrix}+\begin{pmatrix}G_1\\ G_2\\ G_3\end{pmatrix}=\begin{pmatrix}\tau_1\\ \tau_2\\ \tau_3\end{pmatrix}+\begin{pmatrix}d_1\\ d_2\\ d_3\end{pmatrix}$$

where:

$$\begin{aligned}
M_{11}&=2b_1cosq_2+2b_2cos(q_2+q_3)+2b_3cosq_3+a1\\
M_{12}&=b_1cosq_2+b_2cos(q_2+q_3)+2b_3cosq_3+a2\\
M_{22}&=a_2+2b_3cosq_3\\
M_{13}&=b_2cos(q_2+q_3)+b_3cosq_3+a3\\
M_{23}&=a_3+b_3cosq_3\\
M_{23}&=a_3
\end{aligned}$$

$$a_1 = J_1 + m_1L_{c1}^2 + J_2 + m_2(L_1^2 + L_{c2}^2) + J_3 + m_3(L_1^2 + L_2^2 + L_{c3}^2)$$
$$a_2 = J_2 + m_2L_{c2}^2 + m_3(L_2^2 + L_{c3}^2)$$
$$a_3 = J_3 + m_3L_{c3}^2$$

$$C_1 = -b_1\dot{q}_2(2\dot{q}_1 + \dot{q}_2)sinq_2 - b_2(2\dot{q}_1 + \dot{q}_2 + \dot{q}_3)(\dot{q}_2 + \dot{q}_3)sin(q_2 + q_3) - b_3\dot{q}_3(2\dot{q}_1 + \dot{q}_2 + \dot{q}_3)sinq_3$$
$$C_2 = -b_1\dot{q}_1^2 sinq_2 + b_2\dot{q}_1^2 sin(q_2 + q_3) - b_2(2\dot{q}_1 + \dot{q}_2 + \dot{q}_3) + \dot{q}_3 sinq_3$$
$$C_3 = -b_2\dot{q}_1^2 sin(q_2 + q_3) + b_3(\dot{q}_1 + \dot{q}_2)^2 sinq_3$$

$$b_1 = m_2L_1L_{c2} + m_3L_1L_2$$
$$b_2 = m_3L_1L_{c3}$$
$$b_3 = m_3L_2L_{c3}$$

$$G_1 = k_1cosq_1 + k_2cos(q_1 + q_2) + k_3cos(q_1 + q_2 + q_3)$$
$$G_2 = k_1cos(q_1 + q_2) + k_3cos(q_1 + q_2 + q_3)$$
$$G_3 = k_3cos(q_1 + q_2 + q_3)$$

$$k_1 = (m_1L_{c1} + m_2L_1 + m_3L_1)g$$
$$k_2 = (m_2L_{c2} + m_3L_2)g$$
$$k_3 = m_3L_{c3}g$$

The nominal values of m_1 , m_2 and m_3 are assumed to be (Mezghani Ben Romdhane and Damak 2011)

$$m_{10} = 0.5\,\text{Kg},\ m_{20} = 1\,\text{Kg},\ m_{30} = 0.2\,\text{Kg}$$

and the other system parameters are assumed to be known (Mezghani Ben Romdhane and Damak 2011)

$$J_1 = 0.12\,\text{Kg}\,\text{m}^2 \quad L_1 = 0.5\,\text{m}$$
$$J_2 = 0.25\,\text{Kg}\,\text{m}^2 \quad L_2 = 0.5\,\text{m}$$
$$J_3 = 0.3\,\text{Kg}\,\text{m}^2 \quad L_1 = 0.25\,\text{m}$$
$$L_{c2} = 0.35\,\text{m} \quad L_{c3} = 0.15\,\text{m}$$
$$g = 9.81\,\text{m/s}^2$$

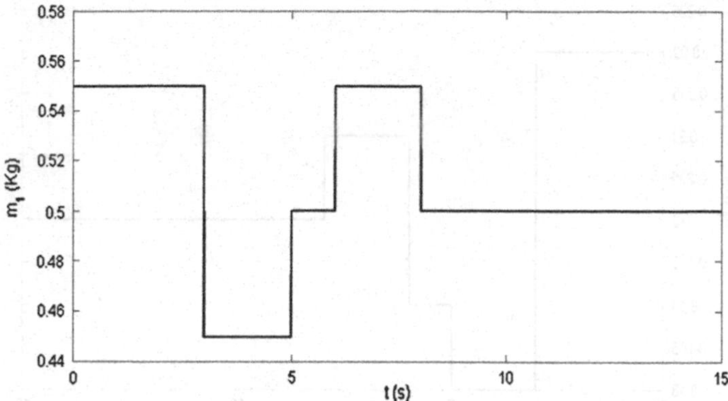

Fig. 1 Variation of the mass m_1

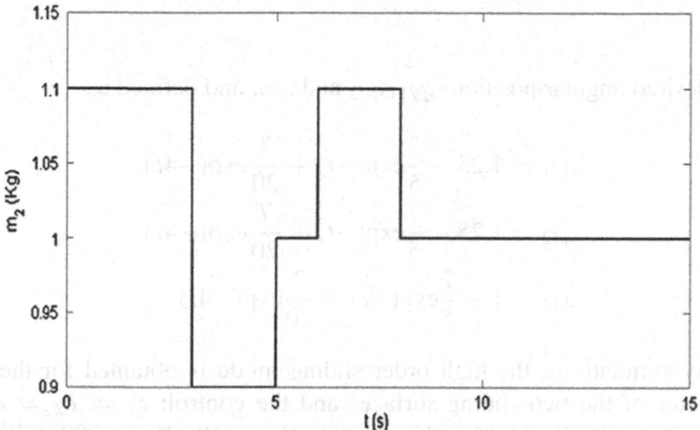

Fig. 2 Variation of the mass m_2

We suppose that we have an uncertainty on masses of the order $\pm$ 10 % (Figs. 1, 2 and 3), and the disturbance vector is $d(t) = [d_1(t)\ d_2(t)\ d_3(t)]^T$ where

$$d_1(t) = 0.2\sin(3t) + 0.02\sin(26\pi t)$$
$$d_2(t) = 0.1\sin(3t) + 0.01\sin(26\pi t)$$
$$d_3(t) = 0.1\sin(3t) + 0.01\sin(26\pi t)$$

The control objective is to design a robust control law such that the angular positions q_1, q_2 and q_3 and evolved from the following initial conditions

$$[q_1(0)\ q_2(0)\ q_3(0)]^T = [-0.2\ \ -0.2\ \ -0.4]^T$$
$$[\dot{q}_1(0)\ \dot{q}_2(0)\ \dot{q}_3(0)]^T = [0\ 0\ 0]^T$$

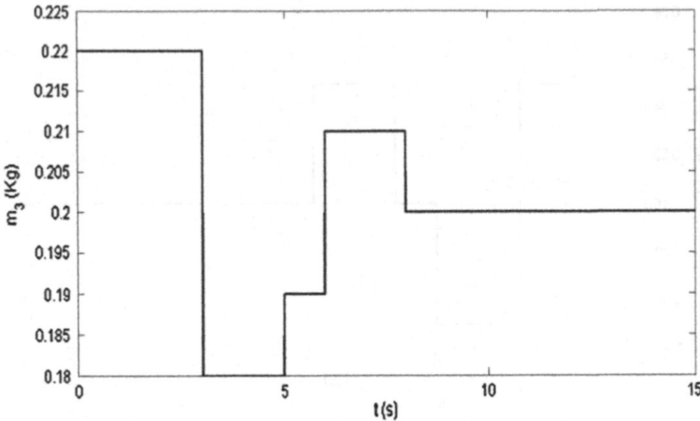

Fig. 3 Variation of the mass m_3

track the desired angular positions q_{d1}, q_{d2} and q_{d3}, and defined as

$$q_{d1} = 1.25 - \frac{7}{5}\exp(-t) + \frac{7}{20}\exp(-4t)$$
$$q_{d2} = 1.25 - \frac{7}{5}\exp(-t) + \frac{7}{20}\exp(-4t)$$
$$q_{d3} = 1 - \frac{7}{5}\exp(-t) + \frac{7}{20}\exp(-4t)$$

After many simulations, the high order sliding mode is obtained for the following parameter of the two sliding surfaces and the control: $c_1 = c_2 = c_3 = 2$, $D_1 = 10$, $D_2 = 8$, $D_3 = 15$, $N_1 = 100$, $N_2 = 10$, $N_3 = 700$, $W_1 = 10$, $W_2 = 60$, $W_3 = 1$.

Figures 4, 5 and 6 show the tracking error, the control input, the sliding surface S and the state trajectory of each joint obtained by using the proposed high order sliding mode controller. It is obvious that the proposed controller ensures finite time convergence of tracking error of three joint and robustness. From control signal it is clear that the control input has a negligible chattering especially in beginning, then it is smooth having no chattering. A second order sliding mode is achieved on the sliding surface S and its components reach zero in finite time. It is also chatterless. The state trajectory of the system evolves without chattering.

The results of the sliding surface σ are presented in Fig. 7. The three sliding surface converge to zero in finite time.

The sliding variable σ converges to zero in finite time. A first order sliding mode control is then established on this surface. Because the discontinuity act on the first derivative of σ, their components present the chattering phenomenon.

The proposed adaptive high order sliding mode control is also tested to the three degree of freedom robot in the same conditions.

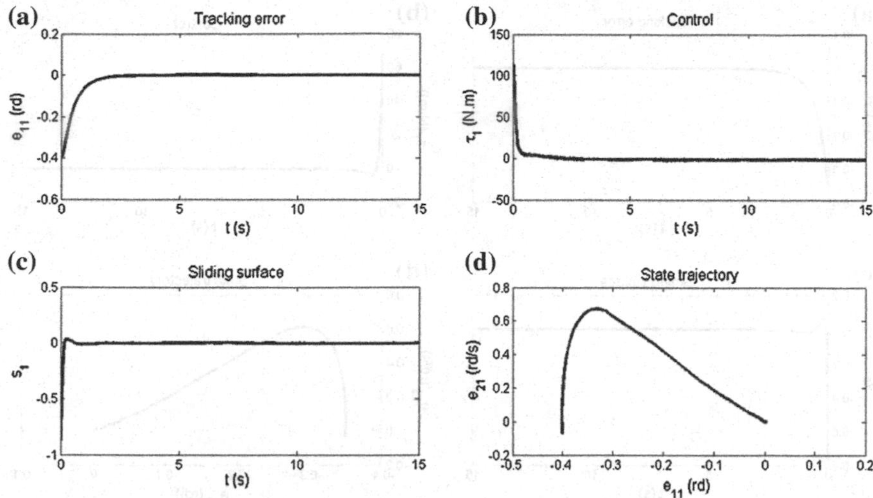

Fig. 4 Tracking of the first joint

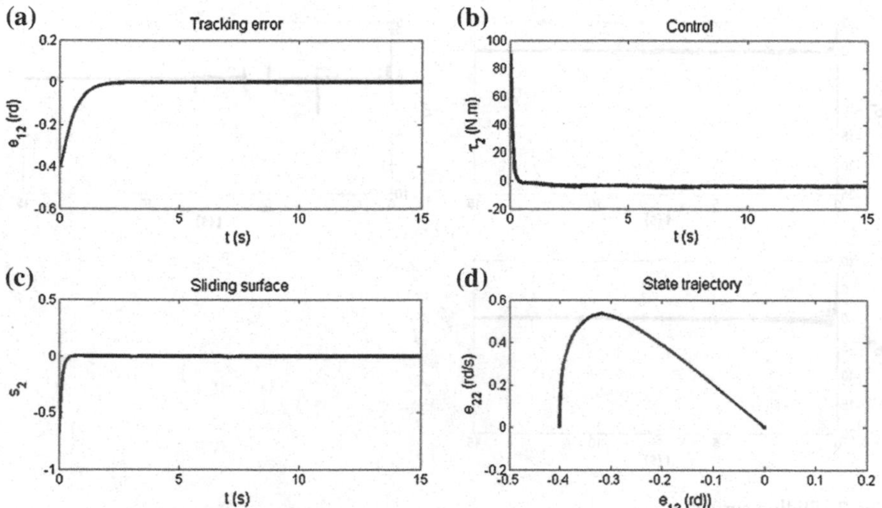

Fig. 5 Tracking of the second joint

The parameters of the two sliding surfaces, the control and the adaptation law are: $c_1 = c_2 = c_3 = 2$, $N_1 = 280$, $N_2 = 250$, $N_3 = 150$, $D_1 = D_2 = D_3 = 15$, $\gamma_1 = 20$, $\gamma_2 = 10$, $\gamma_3 = 20$, $\varepsilon = 0.01$ et $\mu = 0.001$.

Figures 8, 9, 10 show the tracking error, the control input, the sliding surface and the state trajectory of each joint obtained by using the proposed adaptive high order sliding mode controller. It is obvious that the proposed controller ensures finite time convergence of tracking error of three joint and robustness. Because the

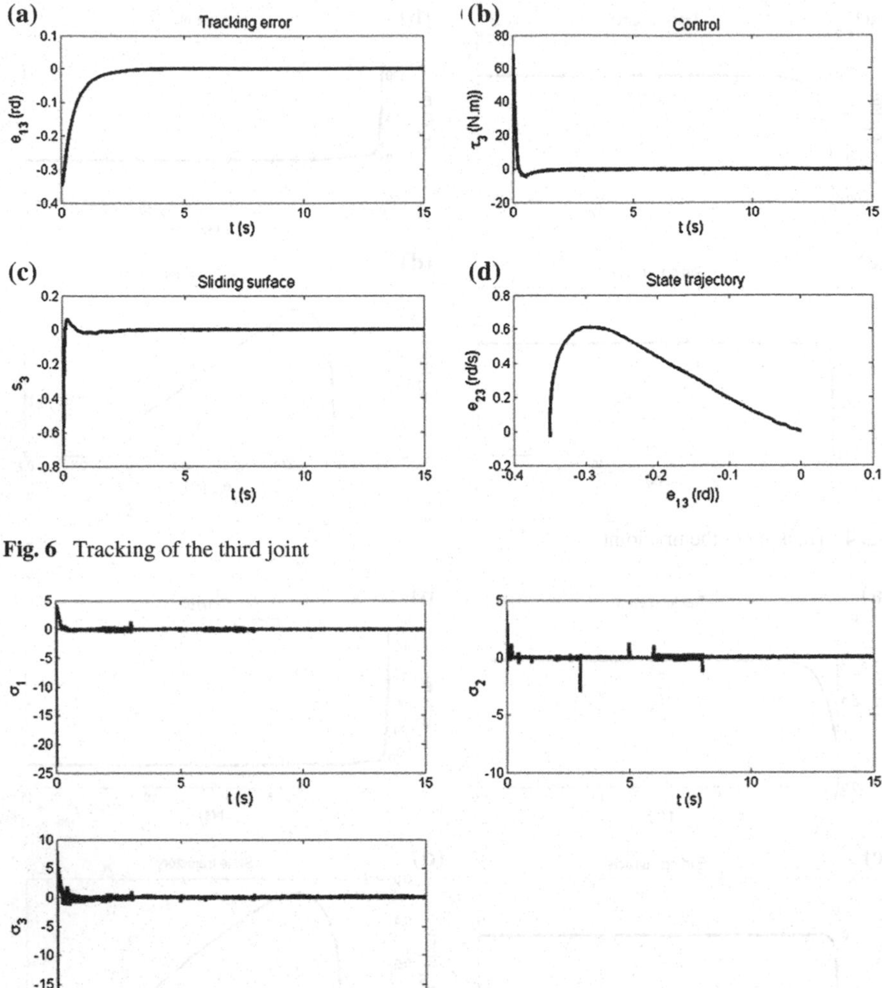

Fig. 6 Tracking of the third joint

Fig. 7 Sliding surface σ

discontinuity is in the derivative of the control the signal control is smooth having no chattering. A conventional sliding mode is ensured on the sliding mode surface σ (Fig. 11); consequently, a second order sliding mode is guaranteed on the sliding surface S_i, $i = 1, 2, 3$, that reach zero in finite time. The state trajectories evolve without chattering for the three joints.

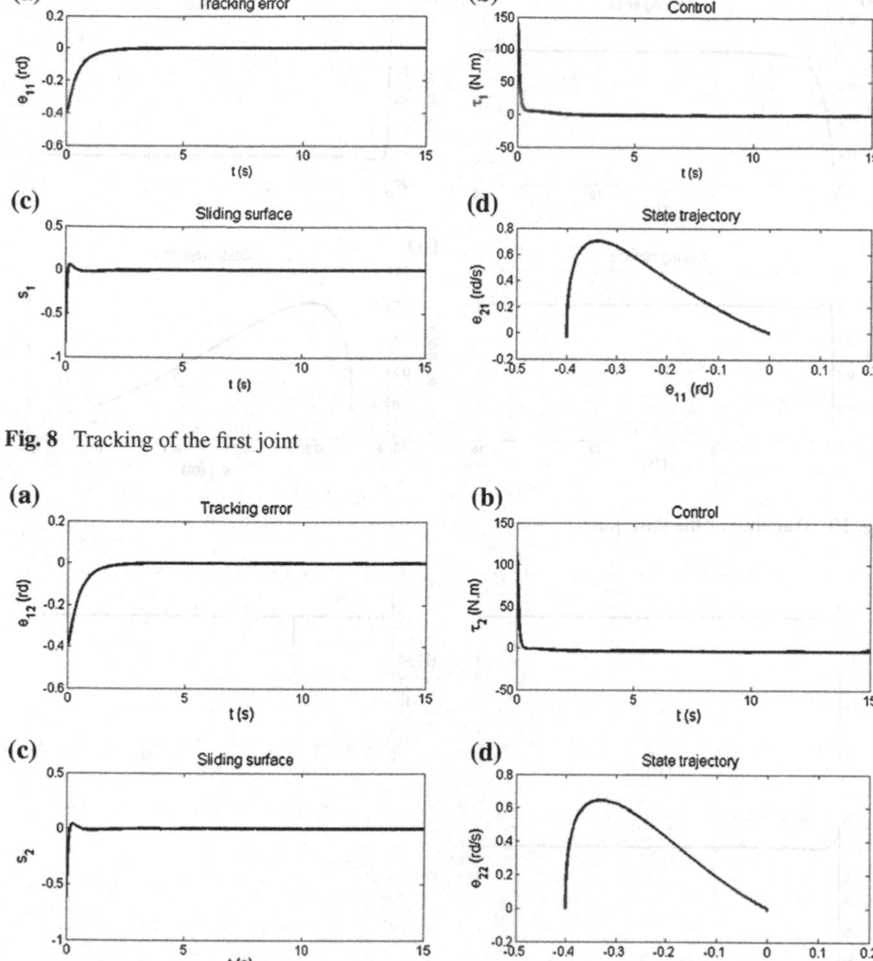

Fig. 8 Tracking of the first joint

Fig. 9 Tracking of the second joint

6 Discussion

In this chapter we have deal with the problem of robot manipulator control in presence of uncertainty. We have distinguished two types of uncertainties:structured uncertainties caused by model imprecision of link parameters, and unstructured uncertainties produced by un-modeled dynamics such as external disturbances. Two main techniques are suitable for the control of uncertain systems: the sliding mode control in particular higher order sliding mode control and adaptive control.

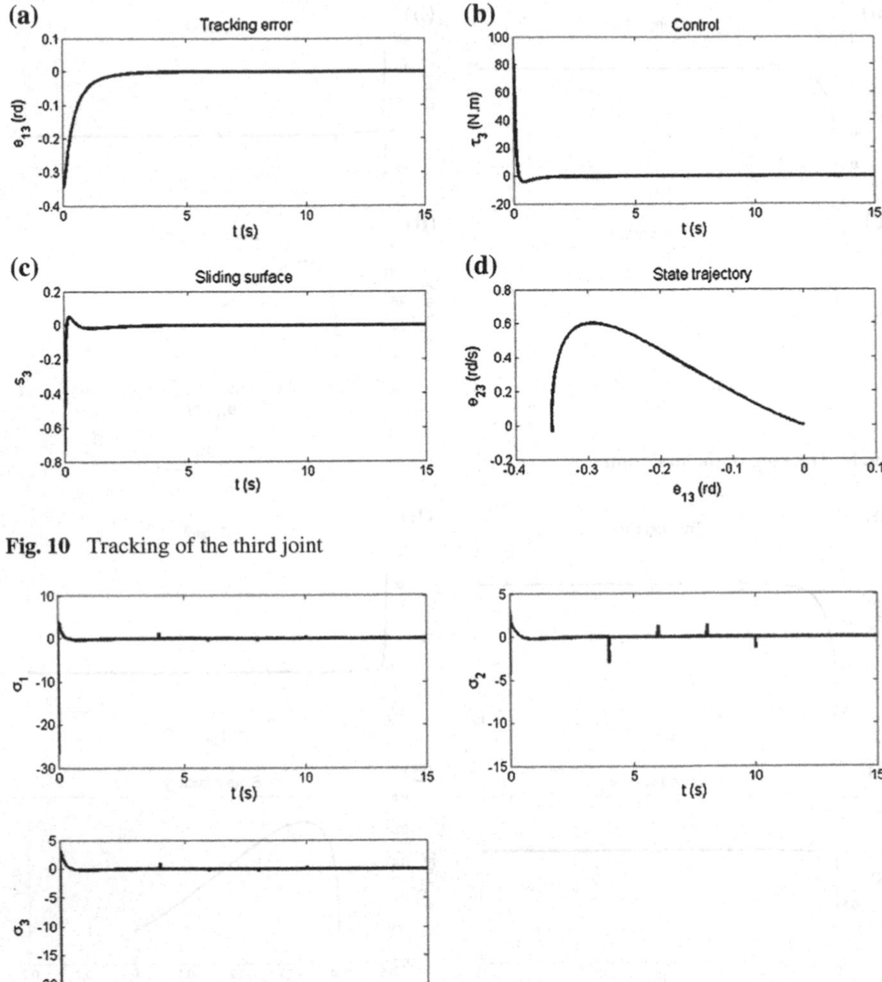

Fig. 10 Tracking of the third joint

Fig. 11 Sliding surface σ

First we have proposed a second order sliding mode control which is inspired from the conventional sliding mode control. The discontinuity, instead of acting on the first time derivative of the sliding variable, is acting on the second derivative of this variable. And the sign function is used in the first derivative of the control law then the control is obtained by integration. This latter is smooth compared to the proposed control of (Defoort et al. 2009) where the sign function is used in the control.This control has been tested to a three degree of freedom robot in presence of uncertainties. The simulation results show the robustness and precision of this con-

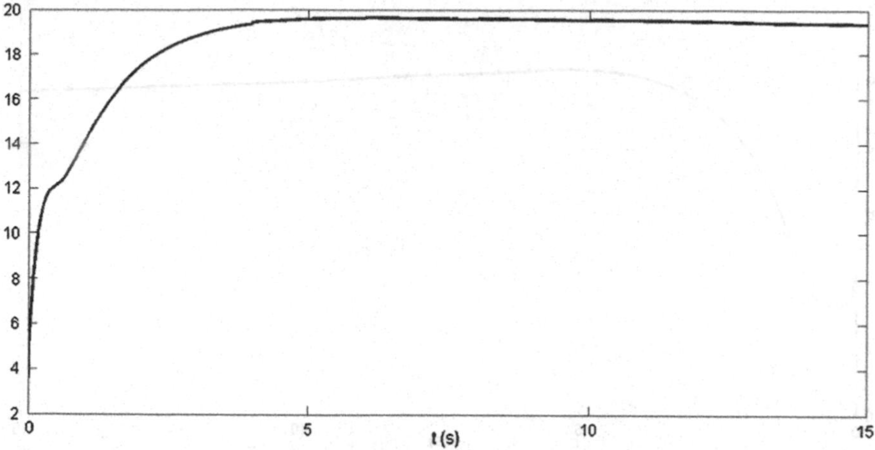

Fig. 12 Estimated parameter $\widehat{W}_{11}$

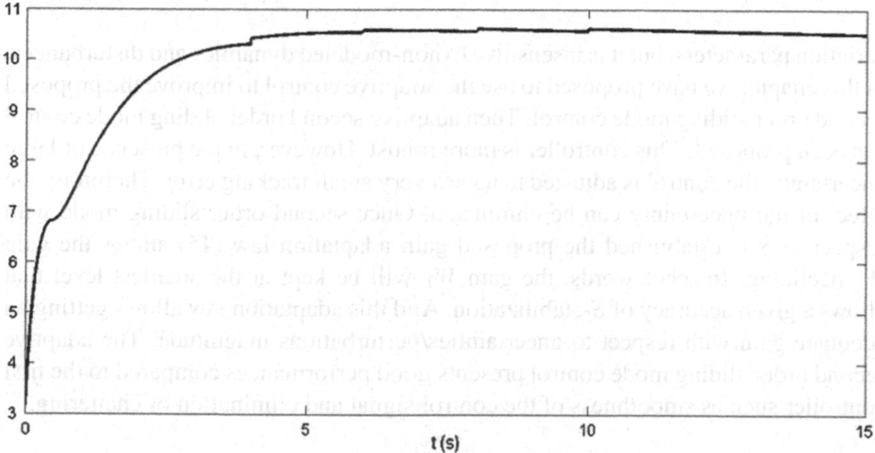

Fig. 13 Estimated parameter $\widehat{W}_{12}$

troller. The control signal is continuous with presence of negligible commutations. But this controller depends on the upper bound of uncertainties.

In order to guarantee the stability of the high order sliding mode control the upper bound of uncertainty must be estimated. Unfortunately, because of the complexity of the structure of the uncertainties of robot manipulators, it is difficult to estimate this bound. Then high control gain can be used when the upper bound of uncertainties is unknown. But this gain can cause the chattering phenomenon at the sliding surface and accordingly the deterioration of the system performances.

Adaptive control uses a parametric adaptation law providing an estimate of the unknown parameters at each instant. This control is applicable to a wide range of

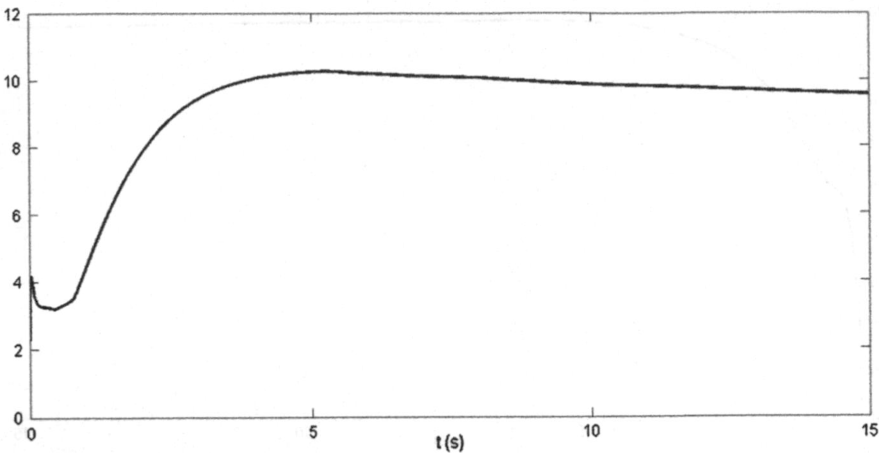

Fig. 14 Estimated parameter $\widehat{W}_{13}$

variation parameters, but it is insensitive to non-modeled dynamics and disturbances. In this chapter we have proposed to use the adaptive control to improve the proposed second order sliding mode control. Then adaptive second order sliding mode control has been proposed. This controller is more robust. However, in the presence of large uncertainty, the control is adjusted to have a very small tracking error. Therefore, the effect of the uncertainty can be eliminated. Once second order sliding mode with respect to S is established the proposed gain adaptation law (45) allows the gain W_1 declining. In other words, the gain W_1 will be kept at the smallest level that allows a given accuracy of S-stabilization. And this adaptation law allows getting an adequate gain with respect to uncertainties/perturbations magnitude. The adaptive second order sliding mode control presents good performances compared to the first controller such as smoothness of the control signal and elimination of chattering.

7 Conclusion

In this paper, we have presented the design of the two robust high order sliding mode control for the tracking problem of rigid robot manipulators. The first is inspired from classical sliding mode control. The main feature of this controller is assuring a smooth high order sliding mode control. The time derivative of the control acts on the second derivative of the sliding surface. Therefore the obtained control law is continuous and robust. The proposed controller guarantees a finite time convergence of the tracking error. Also this controller has eliminated the chattering phenomenon without losing robustness property and precision. To improve this controller and in order to guarantee tracking with more robustness an adaptive high order sliding mode control is proposed. The control gain is adjusted using an adaptive law to make the system robust to uncertainties and perturbations without knowledge of their upper bound. Hence, the proposed controller is highly suitable for practical applications. The stability of the controlled system is proved by using Lyapunov stability criterion.Simulation results demonstrate the efficacy and advantage of the proposed controllers. These controllers can be improved by using high order differentiator to estimate the successive time derivative of the sliding variable. And the combination of integral sliding mode control and adaptive high order sliding mode control can be used in order to eliminate the reaching phase making the control more robust.

References

Abouissa, H., Dryankova, V., Jolly, D.: Real-time ramp metering: high-order sliding mode control and differential flatness concept. Int. J. Autom. Control Eng. **2**(4), 151–160 (2013)

Artega, M.-A., Kelly, R.: Robot control without velocity measeurements: new theory and experimental results. IEEE Trans. Robot. Autom. **20**(2), 297–308 (2004)

Bartolini, G., Ferrara, A., Usai, E., Utkin, V.I.: On multi-input chattering-free second-order sliding mode control. IEEE Trans. Autom. Control **45**(9), 1711–1717 (2000)

Beltran, B., Ahmed-Ali, T.: High-order sliding-mode controlof variable-speed wind turbines. IEEE Trans. Ind. Electron. **56**(9), 3314–3320 (2009)

Bhat, S., Bernstein, D.: Geometric homogenity with applications to finite time stability. Math. Control signals Syst. **17**(2), 101–127 (2005)

Boiko, I., Fridman, L., Iriarte, R., Pisano, A., Usai, E.: Parameter tuning of second-order sliding mode controllers for linear plants with dynamic actuators. Automatica **42**(5), 833–839 (2006)

Defoort, M., Floquet, T., Kokosy, A., Perruquetti, W.: A novel higher order sliding mode control scheme. Syst. Control Lett. **58**(2), 102–108 (2009)

Fridman, L., Levant, A.: High order sliding mode. In: Perruquetti, W., Barbot, J.P. (eds.) Sliding Mode Control In Engineering, pp. 53–102. Marcel Dekker, New York (2002)

Haddad, W.H., Hayakawa, T.: Direct adaptive controlfor non-linear uncertain systems with exogenous disturbances. Int. J. Adapt. Control **16**(2), 151–172 (2002)

Hamerlain, F., Achour, K., Floquet, T., Perruquetti, W.: Higher order sliding mode control of wheeled mobile robots in the presence of sliding effects. In: Proceedings of the 44th IEEE Conference on Decision and Control, and the European Control Conference pp. 1959–1963 (2005)

Han, X., Fridman, E., Spurgeon, S.K.: Sliding-mode control of uncertain systems in presence of unmatched disturbances with applications. Int. J. Control **83**(12), 2413–2426 (2010)

Kachroo, P., Tomizuka, M.: Chattering reduction and error convergence in the sliding-mode control of a class of nonlinear systems. IEEE Trans. Autom. Control **41**(7), 1063–1068 (1996)

Kamal, S., Bandyopadlyay, B.: Arbitrary higher order sliding mode control based on control Lyapunov approach. In: 12th IEEE Workshop on Variable Structure Systems pp. 446-450 VSS (2012)

Laghrouche, S., Smaoui, M., Plestan, F., Brun, X.: Higher order sliding modec ontrol based on optimal approach of an electropneumatic actuator. Int. J. Control **79**(2), 119–131 (2006)

Laghrouche, S., Plestan, F., Glumineau, A.: Higher order sliding mode control based on integral sliding mode. Automatica **43**(3), 531–537 (2007)

Levant, A.: Robust exact differentiation via sliding mode technique. Automatica **34**(3), 379–384 (1998)

Levant, A.: Control Universal SISO sliding mode controllers with finite time convergence. IEEE Trans. Autom **46**(9), 1447–1451 (2001)

Levant, A.: Higher-order sliding modes, differentiation and output-feedback control. Int. J. Control **76**(9/10), 924–941 (2003)

Levant, A.: Homogenity approach to high_order sliding mode design. Automatica **41**(5), 823–830 (2005a)

Levant, A.: Quasi-continuous high-order sliding-mode controllers. IEEE Trans. Autom. control **50**(9), 1447–1451 (2005b)

Levant, A.: Finite-time stabilization of uncertain SISO system. In: Proceedings of the 46th IEEE Conference on DecisionandControl, pp. 1728–1733 (2007)

Lu, X.-Y., Spurgeon, S.K.: Robust sliding mode control of uncertain nonlinear systems. Syst. Control Lett. **32**(2), 75–90 (1997)

Mezghani Ben Romdhane, N.: Control of robotic manipulator with integral sliding mode. J. Autom. Syst. Eng. **5**(4), 185–198 (2011)

Moreno, J.A.: Lyapunov approach for analysis and design of second order sliding mode alghorithms. In: 11th IEEE Workshop on Variable Structure Systems Plenaries andSemiplenaries pp. 121–149, Springer (2010)

Msaddek, A., Gaaloul, A., Msahli, F.: A novel higher order sliding mode control: Application to an induction motor. In: International Conference on Control, Engineering and Information Technology, pp. 13–21. CEIT (2013)

Perruquetti, W., Barbot, J.P.: Sliding Mode Control in Engineering. Marcel Dekker, New York (2002)

Plestan, F., Glumineau, A., Laghrouche, S.: A new algorithm for high-order sliding mode control. Int. J. Robust Nonlinear Control **18**(4/5), 441–453 (2008)

Plestan, F., Shtessel, Y., Bregeault, V., Poznyak, A.: New methodologies for adaptive sliding mode control. Int. J. Control **83**(9), 1907–1919 (2010)

Polyakov, A.E., Poznyak, A.S.: Method of Lyapunov functions for systems with higher order sliding modes. Autom. Remote Control **72**(5), 944–963 (2011)

Rhif, A.: A High order sliding mode control with PID sliding surface: simulation on a torpedo. Int. J. Inf. Technol. Control Autom. **2**(1), 1–13 (2012)

Tsai, Y.-W., Mai, K.-H., Shyu, K.-K.: Sliding mode control for unmatched uncertain systems with totally invariant property and exponential stability. J. Chin. Inst. Eng. **29**(1), 179–183 (2006)

Utkin, V.: Slidingmodes in controlandoptimization. Springer, Berlin (1992)

Yan, F., Wang Z., Hung Y.S., Shu H.: Mixed filteringforuncertainsystemswith regional pole assignment. IEEE Trans. Aerospace Electr. Syst **41**(2), 438–448 (2005)

Young, K.D., Utkin, V.I., Ozguner, U.: A control engineer's guide to sliding mode control. IEEE Trans. Control Syst. Technol **7**(3), 328–342 (1999).

Zhao, Y.-X., Wu, T., Li, G.: A second-order slidingmodecontroller design for spacecraft tracking control. Math. Probl. Eng. **2013**, 1–9 (2013)

Zhang, R., Sun, C., Zhang, J., Zhou, Y.: Second-order terminal sliding mode control for hypersonic vehicle in cruising flight with sliding mode disturbance observer. J. Control Theory Appl. **11**(2), 299–305 (2013)

Zhou, J., Wen, C., Zhang, Y.: Adaptive backstepping control of a class of uncertain nonlinear systems with unknown backlash-like hysteresis. IEEE Trans. Autom. Control **49**(10), 1751–1757 (2004)

Generalized H_2 Sliding Mode Control for a Class of (TS) Fuzzy Descriptor Systems with Time-Varying Delay and Nonlinear Perturbations

Mourad Kchaou and Ahmed Toumi

Abstract This chapter considers the development of robust performance control based-on integral sliding-mode for descriptor system with nonlinearities and perturbations which consist on external disturbances and model uncertainties of great possibility time-varying manner. Sliding-mode control (SMC) is one of robust control methodologies that deal with both linear and nonlinear systems. The most distinguishing feature of (SMC) is its robustness as well as in the case of invariant control systems. Loosely speaking, the term "invariant" means that the system is completely insensitive to parametric uncertainty and external disturbances. Another type of advanced sliding mode control law is "integral sliding mode". The integral sliding mode control differs from the sliding mode control by the use of an integration term in the sliding variable (surface) design in addition to the linear term. In this work the problem of sliding mode control (SMC) for a class of uncertain (TS) fuzzy descriptor systems with time-varying delay is studied. An integral-type sliding function is proposed and a delay-dependent criterion is developed in terms of linear matrix inequality (LMI), which ensures the sliding mode dynamics to be robustly admissible with generalized H_2 disturbance rejection level. Moreover, a SMC law is established to satisfy the reaching condition of the specified sliding surface for all admissible uncertainties and time-varying delay. The developed results are tested on two representative examples to illustrate the theoretical developments.

1 Introduction

Mathematically, a descriptor system model is formulated as a set of coupled differential and algebraic equations, which include information on both static and dynamic constraints of a real plant (Dai 1989). Descriptor systems have been widely studied in the past two decades. This is, in part, because of descriptor form is useful to represent and to handle systems such as mechanical systems, electric circuits, interconnected

M. Kchaou (✉) · A. Toumi
Laboratory of Sciences and Techniques of Automatic Control and Computer Engineering (Lab-STA), Sfax, Tunisia
e-mail: mouradkchaou@gmail.com

A.T. Azar and Q. Zhu (eds.), *Advances and Applications in Sliding Mode Control systems*, Studies in Computational Intelligence 576, DOI 10.1007/978-3-319-11173-5_13

systems, social and economic, and biological systems, and in part, because of additional challenges that these systems present. For example, these systems may possess impulse behavior, which is quite different from the standard state-space systems (Buzurovic and Debeljkovic 2010; Duan 2010; Muller 1997).

It is well known that in many physical, industrial and engineering systems, delays occur due to the finite capabilities of information processing and data transmission among various parts of the system. Delays could arise as well from inherent physical phenomena like mass transport flow or recycling (Gu et al. 2003). Since delay cannot be ignored, it offers many open research topics because it is often major sources of undesirable system transient response or even instability and degradation in control performance. Design procedures of desired controllers, including state feedback controllers, filters, and output feedback controllers, have been developed to deal with time delay systems (Feng et al. 2005; Ma et al. 2008; Wu et al. 2009b; Xu et al. 2003; Zhang et al. 2009; Zhou and Zheng 2009; Zhou and Fang 2009).

On the other hand, there are a lot of delay systems can be described by descriptor representation. We call such systems: descriptor systems with time delay. It should be pointed out that the stability problem for descriptor time-delay systems is much more complicated than that for regular systems, because it is necessary to ensure that they are not only stable, but also regular and impulse free. Many problems for the class of descriptor delayed systems either in the continuous-time or discrete-time have been tackled, and interesting results have been reported in the literature (Boukas 2007; Kchaou et al. 2013, 2014; Li et al. 2008; Ma et al. 2007; Wu et al. 2009a; Xu et al. 2002; Yang and Zhang 2005).

Because most physical systems and processes in the real world are nonlinear, many research efforts have been devoted to seeking an effective means of controlling nonlinear systems. Among the many developments, there are growing interests in the fuzzy control of complex nonlinear systems (Azar 2010, 2012; Takagi and Sugeno 1985). This is because of their capability to approximate very complex nonlinear dynamics in a very natural manner. Recently, the (TS) fuzzy model (Takagi and Sugeno 1985) has been extended to nonlinear descriptor systems with time delay and various problems of analysis and synthesis have been studied (Gassara et al. 2013; Kchaou et al. 2011; Lin et al. 2006; Su et al. 2009; Tian and Zhang 2008; Zhang et al. 2009). We must point out that considerable attention and effort have been paid to the challenging issue of analysis and synthesis of practical engineering systems using the (TS) descriptor fuzzy approach. In (Schulte and Guelton 2006), the use of this representation is justified to model and control design for a SCARA robot. This model, is adopted in Aguilera-gonzalez et al. (2013) to regulate the intake and exhaust manifold pressures for four-cylinder diesel engine with Exhaust Gas Recirculation. The guaranteed cost control for a practical overhead crane is investigated in Chen et al. (2009). The dynamics of the system is exactly transformed into a fuzzy descriptor model and then a fuzzy controller is designed under input/state constraints.

As the dual of the robust control problem, generalized H_2 $(L_2 - L_\infty)$ control for dynamic systems has been extensively investigated. Generalized H_2 performance has been well recognized to be most appropriate for systems with noise input, whose stochastic information is not precisely known. The objective of this problem is to design a controller such that the resulting closed-loop system is stable and ensures that the peak value of the controlled output is often required to be within a certain range (Kchaou et al. 2011; Li et al. 2008; Wu and Wang 2008)

Due to its attractive features such as easy realization, fast response, good transient response and invariance to matched uncertainties, sliding-mode control (SMC), as the most popular kind of variable-structure control (VSC), has become a very focused topic for control engineers and it has been successfully applied to solving many practical control problems (Chang 2012; Ding et al. 2011; Li et al. 2007; Wu et al. 2008). In SMC the design cycle consists of two stages. First, a sliding surface is designed such that, when the system trajectories are restricted to the sliding surface, the system meets the control objectives. During the second stage, a (possibly discontinuous) control is designed to drive and constrain the system trajectories to the sliding surface, irrespectively of the disturbances acting on the system.

In contrast with conventional SMC, the system motion under integral sliding mode has a dimension, which equals to that of the state space of the system. The main advantages of using the integral sliding surface are that, once the system is in the sliding mode, the effect of matched perturbation can be completely eliminated and the robust stability problem of the closed-loop system becomes a standard feedback controller design problem for a system with mismatched uncertainty and disturbance (Chang 2012). Based on integral SMC strategy, many results have been developed to control a various class of systems, such as, uncertain time-delay systems, stochastic systems, and Markovian jump systems (Gao and Wu 2007; Seuret et al. 2009; Wu and Ho 2010; Wu and Zheng 2009). However, to the author's knowledge, there is little related results reported on SMC of (TS) descriptor systems (Han et al. 2012).

In this chapter, we will investigate the integral SMC for (TS) fuzzy descriptor systems subject to time-varying delay, mismatched norm-bounded uncertainties, disturbances and matched nonlinear perturbation. We will design an appropriate integral sliding surface function where the singular matrix E is taken into account. Since mismatch disturbances cannot be eliminated completely once a system is in the sliding mode, the generalized H_2 disturbance attenuation technique can reduce the effect of the disturbance acting on a system to an acceptable level. In this work, we address the following issues of sliding-mode control for delayed (TS) fuzzy descriptor systems:

1. we design a suitable integral sliding surface function by taking the singular matrix into account, thus the resulting sliding mode dynamics is a full-order descriptor time-delay system,
2. we derive sufficient LMI conditions under which the robust admissibility of the sliding mode dynamics with generalized H_2 performance is guaranteed,
3. we synthesize a SMC law to drive the system trajectories onto the predefined switching surface.

This chapter is organized as follows. The description of (TS) fuzzy descriptor systems with time-varying delay and some preliminaries are given in Sect. 2. Section 3 is divided into four parts. In Sect. 3.1, the integral sliding surface is designed. A delay-dependent sufficient condition that ensure for sliding mode dynamics to be admissible with generalized H_2 performance is developed in Sect. 3.2. Section 3.3 is devoted to how to compute gain K_i in the switching surface function such that the sliding mode dynamics is robustly admissible with generalized H_2 performance. The synthesizing of SMC law, by which the trajectories of the fuzzy descriptor system can be driven onto the pre-specified switching surface, is presented in Sect. 3.4. The feasibility of the proposed method is illustrated in Sect. 4 with two numerical examples. Conclusions are given in Sect. 5.

Notations Throughout this paper, $X \in \mathbb{R}^n$ denotes the n—dimensional Euclidean space, while $X \in \mathbb{R}^{n\times m}$ refers to the set of all $n \times m$ real matrices. Notation $X > 0$ (respectively, $X \geq 0$) means that matrix X is real symmetric positive definite (respectively, positive semi-definite). If not explicitly stated, all matrices are assumed to have compatible dimensions for algebraic operations. L_2 is the space of integral vector over $[0, \infty)$. The L_2 and L_∞-norm over $[0, \infty)$ are defined as $||g||_2^2 = \int_0^\infty g^T(t)g(t)dt$, and $||f||_\infty = \sup_t |f(t)|$. Symbol $(*)$ stands for matrix block induced by symmetry, sym(X) stands for $X + X^T$.

2 System Description and Preliminaries

The (TS) dynamic model is a class of fuzzy systems described by fuzzy IF-THEN rules, which locally represent linear input-output relations of nonlinear systems.

A continuous fuzzy descriptor model with time delay and parametric uncertainties can be described by:

$\mathbf{R}_i$: If $\theta_1(t)$ is F_1^i and If $\theta_2(t)$ is $F_2^i \cdots$ If $\theta_s(t)$ is F_s^i, Then

$$\begin{cases} E\dot{x}(t) = A_i(t)x(t) + A_{hi}(t)x(t - h(t)) + B_i\big(u(t) + f_i(t, x(t))\big) + B_{wi}(t)w(t) \\ z(t) = C_i x(t) \quad i = 1, 2, \ldots, r \\ x(t) = \varphi(t), \ t \in [-h_M, 0]. \end{cases} \tag{1}$$

where $x(t) \in \mathbb{R}^n$ is the state, $u(t) \in \mathbb{R}^m$ is the control input, $w(t) \in \mathbb{R}^w$ is the external disturbance input, $f_i(t, x(t))$ represents the system nonlinearity and any model uncertainties in the system including external disturbances, $z(t) \in \mathbb{R}^s$ is the controlled output, F_j^i $(j = 1 \ldots s)$ are fuzzy sets, $\theta(t) = [\theta_1(t), \ldots, \theta_s(t)]$ is the premise variable vector. It is assumed that the premise variables do not depend on the input variables $u(t)$, which is needed to avoid a complicated defuzzification process of fuzzy controllers. Delay $h(t)$ is time-varying and satisfies

$$0 \leq h(t) \leq h_M, \quad \dot{h}(t) \leq h_d. \tag{2}$$

where h_M is constant representing the bounds of the delay, h_d is a positive constant. $\varphi(t)$ is a compatible vector-valued initial function in $[-h_M, 0]$ representing the initial condition of the system.

The system disturbance, $w(t)$, is assumed to belong to $L_2[0, \infty)$, that is, $\int_0^\infty w^T(t) w(t)dt < \infty$. This implies that the disturbance has finite energy. Matrix $E \in \mathbb{R}^{n\times n}$ may be singular with rank(E) = q $\leq$ n. $A_i(t) = A_i + \Delta A_i(t)$, $A_{hi}(t) = A_{hi} + \Delta A_{hi}(t)$ and $B_{wi}(t) = B_{wi} + \Delta B_{wi}(t)$ are time-varying system matrices. A_i, A_{hi}, B_i, B_{wi} and C_i are constant matrices with appropriate dimensions. The overall fuzzy model is inferred as follows:

$$\begin{cases} E\dot{x}(t) = \sum_{i=1}^{r} \mu_i(\theta(t))\left\{A_i(t)x(t) + A_{hi}(t)x(t-h(t)) + B_i\left(u(t) + f_i(t, x(t))\right) + B_{wi}(t)w(t)\right\} \\ z(t) = \sum_{i=1}^{r} \mu_i(\theta(t))C_i x(t) \end{cases} \tag{3}$$

where $\mu_i(\theta(t))$ is the normalized membership function defined by

$$\mu_i(\theta(t)) = \frac{\prod_{j=1}^{s} F_j^i(\theta_j(t))}{\sum_{i=1}^{r}\prod_{j=1}^{s} F_j^i(\theta_j(t))}, \quad i = 1, 2, \ldots, r$$

and $F_j^i(\theta_j(t))$ represents the membership degrees of $\theta_j(t)$ in fuzzy set F_j^i. Note that normalized membership $\mu_i(\theta(t))$ satisfies

$$\mu_i(\theta(t)) \geq 0, \; i = 1, 2, \ldots, r \quad \sum_{i=1}^{r} \mu_i(\theta(t)) = 1. \tag{4}$$

Without loss of generality, we introduce the following assumption for technical convenience. For brevity, we use in the sequel the following notation where μ_i stands for $\mu_i(\theta(t))$.

1. $\Delta A_i(t)$, $\Delta A_{hi}(t)$ and $\Delta B_{wi}(t)$ are the unmatched uncertainties satisfying

$$[\Delta A_i(t)\Delta A_{hi}(t)\Delta B_{wi}(t)] = M_i F(t)[N_i N_{hi} N_{wi}], \tag{5}$$

   where M_i, N_i, N_{hi} and N_{wi} are known real constant matrices and $F(t)$ is unknown time-varying matrix function satisfying $F^T(t)F(t) \leq I$.
2. Matrices $B_i, i = 1, 2, \ldots, r$ are assumed to satisfy $B_1 = B_2, \ldots, B_r = B$.
3. Matched nonlinearities $f_i(x)$ satisfies the inequality

$$f_i(x) \leq \eta_i(x) \tag{6}$$

   where $\eta_i(x)$ is positive known vector-valued function.
4. Exogenous signal, $w(t)$ is bounded.

First of all, we recall some definitions.

Consider an unforced linear descriptor system with delay described by:

$$\begin{aligned} E\dot{x}(t) &= Ax(t) + A_h x(t - h(t)), \quad 0 \le h(t) \le h_M \\ x(t) &= \varphi(t), \ t \in [-h_M, 0]. \end{aligned} \tag{7}$$

Definition 1 (Dai 1989)

1. System (7) is said to be regular if $det\Big(sE - A\Big) \neq 0$.
2. System (7) is said to be impulse free if $\deg\Big(det\Big(sE - A\Big)\Big) = \text{rank(E)}$.
3. System (7) is said to be admissible if it is regular, impulse-free and stable.

Descriptor time-delay system (7) may have an impulsive solution, however, the regularity and non-impulse of (E, A) guarantee the existence and uniqueness of impulse-free solution to (7) on $[0, \infty)$.

Definition 2 (Xu et al. 2002) The descriptor delay system (7) is said to be regular and impulse free if the pair (E, A) is regular and impulse free. System (7) is said asymptotically stable, if for any $\varepsilon > 0$ there exists a scalar $\delta(\varepsilon) > 0$ such that for any compatible initial condition, $\phi(t)$ with $\sup_{-h_M < t \le 0} \|\phi(t)\| < \delta(\varepsilon)$, the solution $x(t)$ of (7) satisfies $\|x(t)\| < \varepsilon$ for $t \le 0$ and $\lim_{t \to 0} x(t) = 0$

Definition 3 Descriptor system (7) is said to be asymptotically stable with generalized H_2 performance if the open-loop system is asymptotically stable and under the zero initial condition, the L_2-L_∞ norm of the open-loop transfer function $T_{zw}(s)$ from external disturbance $w(t)$ to controlled output $z(t)$ satisfies

$$||T_{zw}(s)||_{L_2 - L_\infty} = \sup_{0 \neq w(t) \in L_2} \frac{||z(t)||_\infty}{||w(t)||_2} < \gamma \tag{8}$$

where γ is a given positive scalar.

Lemma 1 (Fridman 2000) *If a functional* $V : C_n[-\tau, 0] \to \mathbb{R}$ *is continous and* $x(t, \Phi)$ *is a solution to* (7)*, we define*

$$\dot{V}(\Phi) = \lim_{h \to 0^+} \sup \frac{1}{h}\Big(V(x(t + h, \Phi) - V(\phi))\Big)$$

Denote the system parameters of (7) *as*

$$(E,\ A\ , A_h) = \left(\begin{bmatrix} I_q & 0 \\ 0 & 0 \end{bmatrix}, \begin{bmatrix} A_{11} & A_{12} \\ A_{21} & A_{22} \end{bmatrix}, \begin{bmatrix} A_{h11} & A_{h12} \\ A_{h21} & A_{h22} \end{bmatrix}\right)$$

Assume that the descriptor system (7) *is regular and impulse-free,* A_{22} *is invertible and* $\rho(A_{22}^{-1} A_{h22}) < 1$*. Then, system* (7) *is stable if there exists positive numbers* $\alpha,\ \mu,\ \nu$ *and a continuous function,* $V : C_n[-\tau, 0] \to \mathbb{R}$*, such that*

$$\mu\|\Phi_1(0)\|^2 \leq V(\Phi) \leq \nu\|\Phi\|^2,$$
$$\dot{V}(x_t) \leq -\alpha\|x_t\|^2$$

where $x_t = x(t+\theta)$ *with* $\theta \in [-\tau, 0]$ *and* $\Phi = \left[\Phi_1^T \ \Phi_2^T\right]^T$ *with* $\Phi_1 \in \mathbb{R}^q$

Lemma 2 (Gu et al. 2003) *For any constant matrix* $M > 0$*, any scalar* h_m *and* h_M *with* $0 < h_m < h_M$*, and vector function* $x(t) : [-h_M, -h_m] \to \mathbb{R}^n$ *such that the integrals concerned as well defined, then the following holds*

$$-(h_M - h_m)\int_{t-h_M}^{t-h_m} x^T(s)Mx(s)ds \leq -\int_{t-h_M}^{t-h_m} x^T(s)dsM\int_{t-h_M}^{t-h_m} x(s)ds$$

Lemma 3 (Peterson 1987) *Let* M *and* N *be real matrices with appropriate dimensions. Then, for any* Δ *matrix satisfying* $\Delta^T\Delta \leq I$ *and scalar* $\varepsilon > 0$,

$$\mathrm{sym}(M\Delta N) \leq \varepsilon MM^T + \varepsilon^{-1}N^TN \tag{9}$$

3 Integral Sliding Mode Controller Design

SMC design involves two basic steps. The first one is to design an appropriate switching surface such the sliding mode dynamics restricted to the surface is admissible with generalized H_2 disturbance rejection level γ. In the second step a SMC law is synthesized to guarantee that the sliding mode is reached and the system states maintain in the sliding mode thereafter.

3.1 Integral Sliding Mode Surface

The integral sliding-mode control completely eliminating the matched-type nonlinearities and uncertainties of (3) while keeping $s = 0$. In this work, the following integral sliding surface is considered:

$$s(x,t) = \mathcal{M}Ex(t) - \mathcal{M}\Big(Ex_0 + \int_0^t \sum_{i=1}^{r}\mu_i\Big\{(A_i + BK_i)x(\theta) + A_{hi}x(\theta - h(\theta))d\theta\Big\}\Big) \tag{10}$$

where $K_i \in \mathbb{R}^{m\times n}$ is real matrix to be designed and $\mathcal{M} \in \mathbb{R}^{m\times n}$ is constant matrix satisfying $\mathcal{M}B$ is nonsingular. According to SMC theory, when the system trajectories reach onto the sliding surface, it follows that $s(x,t) = 0$ and $\dot{s}(x,t) = 0$. Therefore, from $\dot{s}(x,t) = 0$, the equivalent control law can be established as

$$u_s = -(\mathcal{M}B)^{-1}\mathcal{M}\sum_{i=1}^{r}\mu_i\Big\{\big(\Delta A_i(t) + BK_i\big)x(t) + \Delta A_{hi}(t)x(t-h(t))) + B_{wi}(t)w(t)\Big\}$$
$$-\sum_{i=1}^{r}\mu_i f_i(x(t)) \tag{11}$$

Substituting (11) into (3), we obtain the following sliding mode dynamics:

$$\begin{cases} E\dot{x}(t) = \sum_{i=1}^{r}\mu_i\Big\{\overline{A}_i(t)x(t) + \overline{A}_{hi}(t)x(t-h(t)) + \overline{B}_{wi}(t)w(t)\Big\} \\ z(t) = \sum_{i=1}^{r}\mu_i C_i x(t) \end{cases} \tag{12}$$

where $\overline{\mathcal{M}} = I - B(\mathcal{M}B)^{-1}\mathcal{M}$ and

$$\begin{aligned} &\overline{A}_i(t) = \overline{A}_i + \Delta\overline{A}_i(t), \qquad \overline{A}_i = A_i + BK_i, \quad \overline{A}_{hi}(t) = A_{hi} + \Delta\overline{A}_{hi}(t),\\ &\overline{B}_{wi}(t) = \overline{B}_{wi} + \Delta\overline{B}_{wi}(t), \quad \overline{B}_{wi} = \overline{\mathcal{M}}B_{wi}, \qquad \overline{M}_i = \overline{\mathcal{M}}M_i,\\ &\big[\Delta\overline{A}_i(t)\ \Delta\overline{A}_{hi}(t)\ \Delta\overline{B}_{wi}(t)\big] = \overline{M}_i F(t)\big[N_i\ N_{hi}\ N_{wi}\big]. \end{aligned} \tag{13}$$

3.2 Sliding Mode Dynamics Generalized H_2 Analysis

In this subsection, we develop a delay-dependent sufficient condition which ensures the admissibility of the sliding mode dynamics (12) with generalized H_2 performance.

3.2.1 Nominal Case

In what follows, we are presenting a delay-dependent sufficient condition such that the nominal case of (12) (that is, $\Delta\overline{A}_i(t) = 0$, $\Delta\overline{A}_{hi}(t) = 0$ and $\Delta\overline{B}_{wi}(t) = 0$) is admissible with generalized H_2 performance.
For brevity, we use in the sequel the following notation:

$$\mathbb{A} = \sum_{i=1}^{r}\mu_i\overline{A}_i \quad \mathbb{A}_h = \sum_{i=1}^{r}\mu_i A_{hi} \quad \mathbb{B}_w = \sum_{i=1}^{r}\mu_i\overline{B}_{wi} \quad \mathbb{C} = \sum_{i=1}^{r}\mu_i C_i. \tag{14}$$

we obtain

$$\begin{cases} E\dot{x}(t) = \mathbb{A}x(t) + \mathbb{A}_h x(t-h(t)) + \mathbb{B}_w w(t) \\ z(t) = \mathbb{C}x(t) \end{cases} \tag{15}$$

Theorem 1 *Let γ, h_M and h_d given positive scalars. Then fuzzy descriptor system (12) is regular, impulse free and asymptotically stable with generalized H_2 norm bound γ, if there exist a non-singular matrix P, some matrices $Q_1 > 0$, $Q_2 > 0$, $S > 0$, with appropriate dimensions such that the following set of LMIs holds:*

$$E^T P = P^T E \geq 0 \tag{16}$$

$$\begin{bmatrix} \Phi_i & \mathbf{B}_{wi} & \sqrt{h_M}\mathbf{A}_i S \\ * & -\gamma I & \sqrt{h_M}\,\overline{B}_{wi}^T S \\ * & * & -S \end{bmatrix} < 0 \tag{17}$$

$$\begin{bmatrix} -E^T P & C_i^T \\ * & -\gamma I \end{bmatrix} < 0, \tag{18}$$

where

$$\Phi_i = \begin{bmatrix} \Phi_{11i} & P^T A_{hi} + \frac{1}{h_M}E^T SE & 0 \\ * & -(1-h_d)Q_1 - \frac{2}{h_M}E^T SE & \frac{1}{h_M}E^T SE \\ * & * & -Q_2 - \frac{1}{h_M}E^T SE \end{bmatrix}$$

$$\Phi_{11i} = Q_1 + Q_2 + \mathrm{sym}(P^T\overline{A}_i) - \frac{1}{h_M}E^T SE$$

$$\mathbf{A}_i = \left[\overline{A}_i \; A_{hi} \; 0\right]^T, \; \mathbf{B}_{wi} = \left[\overline{B}_{wi}^T P \; 0 \; 0\right]^T$$

Proof The proof of this theorem is divided into two parts. The first one is concerned with the regularity and the impulse-free characterizations, and the second one treats the stability property of system (15).
Since rank(E) $=$ q $\leq$ n, there always exist two non singular matrices M and $N \in \mathbb{R}^{n\times n}$ such that

$$\overline{E} = MEN = \begin{bmatrix} I_q & 0 \\ 0 & 0 \end{bmatrix} \tag{19}$$

Set

$$\overline{\mathbb{A}} = M\mathbb{A}N = \begin{bmatrix} \overline{\mathbb{A}}_{11} & \overline{\mathbb{A}}_{12} \\ \overline{\mathbb{A}}_{21} & \overline{\mathbb{A}}_{22} \end{bmatrix}, \qquad \overline{\mathbb{A}}_h = M\mathbb{A}_h N = \begin{bmatrix} \overline{\mathbb{A}}_{h11} & \overline{\mathbb{A}}_{h12} \\ \overline{\mathbb{A}}_{h21} & \overline{\mathbb{A}}_{h22} \end{bmatrix},$$
$$\overline{P} = M^{-T} P N = \begin{bmatrix} \overline{P}_{11} & \overline{P}_{12} \\ \overline{P}_{21} & \overline{P}_{22} \end{bmatrix}. \tag{20}$$

Using the fact that $\overline{P}$ is non-singular, it is easy to see from (16) and (20) that $\overline{P}_{11} > 0$, $\overline{P}_{12} = 0$ and $\overline{P}_{22}$ is also non-singular.
From (17), it is easy to verify that $\Phi_{11i} < 0$, and thus

$$\mathrm{sym}(P^T \mathbb{A}) - \frac{1}{h_M} E^T S E < 0 \tag{21}$$

Pre- and post-multiplying (21) by N^T and N, respectively, we obtain

$$\begin{bmatrix} \star & \star \\ \star & \mathrm{sym}(\overline{P}_{22}^T \overline{\mathbb{A}}_{22}) \end{bmatrix} < 0 \tag{22}$$

where $\star$ will not be used in the following development. Hence, we can deduce that $\overline{\mathbb{A}}_{22}$ is non-singular. Therefore, according to Definition 1, singular time-delay system (15) is regular and impulse free for any time-delay $h(t)$ satisfying (2).
From (17), we conclude that

$$\begin{bmatrix} \mathrm{sym}(P^T \mathbb{A}) + Q_1 + Q_2 - \frac{1}{h_M} E^T S E & P^T \mathbb{A}_h + \frac{1}{h_M} E^T S E \\ * & -(1 - h_d) Q_1 - \frac{2}{h_M} E^T S E \end{bmatrix} < 0$$

Pre- and post-multiplying the above inequality by

$$\begin{bmatrix} 0 & I & 0 & 0 \\ 0 & 0 & 0 & I \end{bmatrix}$$

and it transpose, respectively, we obtain

$$\begin{bmatrix} \mathrm{sym}\left(\overline{P}_{22}^T \overline{\mathbb{A}}_{22}\right) + Q_1^{22} + Q_2^{22} & P_{22}^T \overline{\mathbb{A}}_{h22} \\ * & -(1 - h_d) Q_1^{22} \end{bmatrix} < 0$$

which implies that

$$\begin{bmatrix} \mathrm{sym}\left(\overline{P}_{22}^T \overline{\mathbb{A}}_{22}\right) + Q_1^{22} & P_{22}^T \overline{\mathbb{A}}_{h22} \\ * & -(1 - h_d) Q_1^{22} \end{bmatrix} < 0$$

It follows from (1,1)-block that $\overline{\mathbb{A}}_{22}$ is invertible. Then, pre- and post multiplying the above inequality by $\left[-\left(\overline{\mathbb{A}}_{h22}\right)^T\left(\overline{\mathbb{A}}_{22}\right)^{-T}\ I\right] < 0$ and its transpose, respectively, yields

$$\left(\left(\overline{\mathbb{A}}_{22}\right)^{-1}\left(\overline{\mathbb{A}}_{h22}\right)\right)^T Q_1^{22}\left(\left(\overline{\mathbb{A}}_{22}\right)^{-1}\left(\overline{\mathbb{A}}_{h22}\right)\right) - (1-h_d)Q_1^{22} < 0$$

which shows that $\rho\left(\left(\left(\overline{\mathbb{A}}_{22}\right)^{-1}\left(\overline{\mathbb{A}}_{h22}\right)\right)\right) < 1$ holds.

Now, let us choose the following Lyapunov-Krasovskii functional as follows:

$$\begin{aligned}
V(t) &= V_1(t) + V_2(t) + V_3(t)\\
V_1(t) &= x^T(t)E^T Px(t)\\
V_2(t) &= \int_{t-h(t)}^{t} x^T(s)Q_1x(s)ds + \int_{t-h_M}^{t} x^T(s)Q_2x(s)ds\\
V_3(t) &= \int_{-h_M}^{0}\int_{t+\theta}^{t} \dot{x}^T(s)E^T SE\dot{x}(s)dsd\theta
\end{aligned} \tag{23}$$

The derivative along the trajectories of (12) is expressed as

$$\begin{aligned}
\dot{V}_1(t) &= 2x^T(t)P^T\left(\mathbb{A}x(t) + \mathbb{A}_h x(t-h(t))\right)\\
\dot{V}_2(t) &= x^T(t)Q_1x(t) - (1-\dot{h}(t))x^T(t-h(t))Q_1x(t-h(t)) + x^T(t)Q_2x(t)\\
&\quad - x^T(t-h_M)Q_2x(t-h_M)\\
&\le x^T(t)(Q_1+Q_2)x(t) - (1-h_d)x^T(t-h(t))Q_1x(t-h(t))\\
&\quad - x^T(t-h_M)Q_2x(t-h_M)\\
\dot{V}_3(t) &= h_M\dot{x}^T(t)E^T SE\dot{x}(t) - \int_{t-h_M}^{t} \dot{x}^T(s)E^T SE\dot{x}(s)ds\\
&= h_M\dot{x}^T(t)E^T SE\dot{x}(t) - \int_{t-h_M}^{t-h(t)} \dot{x}^T(s)E^T SE\dot{x}(s)ds - \int_{t-h(t)}^{t} \dot{x}^T(s)E^T SE\dot{x}(s)ds
\end{aligned} \tag{24}$$

According to Lemma 2, we develop

$$\dot{V}_3(t) \le -\frac{1}{h_M}\gamma_1 E^T SE\gamma_1 - \frac{1}{h_M}\gamma_2 E^T SE\gamma_2 \tag{25}$$

where $\gamma_1 = x(t - h(t)) - x(t - h_M)$ and $\gamma_2 = x(t) - x(t - h(t))$.
Considering (25), we get from (24)

$$\dot{V}(t) \leq \xi^T(t)\bar{\Phi}\xi(t) \tag{26}$$

where $\xi(t) = \left[x^T(t)\ x^T(t - h(t))\ x^T(t - h_M)\right]^T$.
From (17), it is easy to see that $\Phi_i < 0$. Then we have

$$\bar{\Phi} = \sum_{i=1}^{r} \mu_i \Phi_i < 0 \tag{27}$$

Hence, $\dot{V}(t) \leq -\lambda_{min}(-\bar{\Phi})\|\xi_1(t)\|^2$ which implies that nominal singular system (12), with $w(t) = 0$, is asymptotically stable.
Next the the generalized H_2 performance of system (12) is established under zero initial conditions.
From (17) we can obtain

$$\Psi_i = \begin{bmatrix} \Phi_i & \mathbf{B}_{wi} \\ * & -\gamma I \end{bmatrix} + \frac{1}{h_M}\begin{bmatrix} \mathbf{A}_i \\ \overline{B}_{wi}^T \end{bmatrix} S \begin{bmatrix} \mathbf{A}_i \\ \overline{B}_{wi}^T \end{bmatrix}^T < 0 \tag{28}$$

Consider the following performance index:

$$J_0 = V(t) - \gamma \int_0^t w^T(s)w(s)ds \tag{29}$$

where $V(t)$ is defined as in (23). For any non-zero $w(s) \in L_2$, $t > 0$ and zero initial state condition $\varphi(t) = 0$, $t \in [-h_M, 0]$, it is not difficult to achieve

$$J_0 = V(t) - V(0) - \gamma \int_0^t w^T(s)w(s)ds = \int_0^t \dot{V}(x(s)) - \gamma w^T(s)w(s)ds \tag{30}$$

Define $\zeta(t) = \left[\xi^T(t)\ w^T(t)\right]^T$. According to the aforementioned method we get

$$\dot{V}(x(t)) - \gamma w^T(t)w(t) \leq \sum_{i=1}^{r} h_i \zeta^T(t)\Psi_i\zeta(t) \tag{31}$$

From (28) to (31), we conclude that $J_0 < 0$.
Therefore, we can obtain the following inequality

$$x^T(t)E^T Px(t) \leq V(t) < \gamma \int_0^t w^T(s)w(s)ds \tag{32}$$

On the other hand, from (18) it yields $\gamma^{-1}C_i^T C_i - E^T P < 0$ which, in turn, leads to

$$\begin{aligned} z^T(t)z(t) &\leq \gamma x^T(t)E^T Px(t) \leq \gamma V(t) \\ &< \gamma^2 \int_0^t w^T(s)w(s)ds \leq \gamma^2 \int_0^\infty w^T(s)w(s)ds \end{aligned} \tag{33}$$

Taking the maximum value of $||z(t)||_\infty^2$, we have $||z(t)||_\infty^2 < \gamma^2 ||w(t)||_2^2$ for any $0 \neq w(t) \in L_2$ which means that system (12) is delay-dependent asymptotically stable with generalized H_2 norm bound γ. This completes the proof. □

3.2.2 Uncertain Case

Based on Theorem 1, we develop a delay-dependent criterion such that system (12) with norm-bounded parameter uncertainties described in (13) is robustly admissible with generalized H_2 performance.

Theorem 2 *Let γ, h_M and h_d given positive scalars. Then fuzzy descriptor system (12) is regular, impulse free and asymptotically stable with generalized H_2 norm bound γ, if there exist a non-singular matrix P, some matrices $Q_1 > 0$, $Q_2 > 0$, $S > 0$, with appropriate dimensions and positive scalars ε_i $(i = 1, \ldots, r)$ such that the following set of LMIs holds:*

$$E^T P = P^T E \geq 0 \tag{34}$$

$$\begin{bmatrix} \Phi_i & \mathbf{B}_{wi} & \sqrt{h_M}\mathbf{A}_i S & \mathbf{M}_i & \varepsilon_i \mathbf{N}_i \\ * & -\gamma I & \sqrt{h_M}\,\overline{B}_{wi}^T S & 0 & \varepsilon_i N_{wi} \\ * & * & -S & 0 & 0 \\ * & * & * & -\varepsilon_i I & 0 \\ * & * & * & * & -\varepsilon_i I \end{bmatrix} < 0 \tag{35}$$

$$\begin{bmatrix} -E^T P & C_i^T \\ * & -\gamma I \end{bmatrix} < 0, \tag{36}$$

where

$$\mathbf{N}_i = \left[N_i^T \; N_{hi}^T \; 0\right]^T, \; \mathbf{M}_i = \left[\overline{M}_i^T P \; 0 \; 0\right]^T$$

Proof Replacing $\overline{A}_i$, A_{hi} and $\overline{B}_{wi}$ by $\overline{A}_i(t)$, $\overline{A}_{hi}(t)$ and $\overline{B}_{wi}(t)$ in (16), respectively, we can conclude that the uncertain sliding mode dynamics (12) is admissible with generalized H_2 performance.

$$\Psi_i + sym(\mathbf{N}_i F(t)\mathbf{M}_i^T) < 0 \tag{37}$$

Then, according to Lemma 3, (35) holds using the Schur complement. □

3.3 Sliding Mode Dynamics Generalized H_2 Synthesis

Using the previous results we focus on this section to determine gain K_i in the switching surface function in (10) such that sliding mode dynamics (12) is robustly admissible with generalized H_2 performance.

Theorem 3 *Let h_M, h_d, γ and σ given positive scalars. Then, sliding mode dynamics (12) is robustly admissible with H_2 performance γ, for any delay $h(t)$ satisfying (2), if there exist a non-singular matrix X, symmetric positive-definite matrices $\tilde{Q}_1$, $\tilde{Q}_1$, $\tilde{S}$ and some positive scalars ε_i, $i = 1, \ldots, r$ such that the following LMIs hold:*

$$EX = X^T E^T \geq 0 \tag{38}$$

$$\Upsilon_i = \begin{bmatrix} \tilde{\Phi}_i & \tilde{\mathbf{B}}_{wi} & \sqrt{h_M}\tilde{\mathbf{A}}_i & \tilde{\mathbf{M}}_i & \varepsilon_i \tilde{\mathbf{N}}_i \\ * & -\gamma I & \sqrt{h_M}\, B_{wi}^T & 0 & \varepsilon_i N_{wi}^T \\ * & * & \sigma^2 \tilde{S} - \sigma \,\mathrm{sym}(X) & 0 & 0 \\ * & * & * & -\varepsilon_i I & 0 \\ * & * & * & * & -\varepsilon_i I \end{bmatrix} < 0 \tag{39}$$

$$\Gamma_i = \begin{bmatrix} -X^T E^T & \tilde{C}_i^T \\ * & -\gamma I \end{bmatrix} < 0, \qquad i, j = 1, \ldots, r \tag{40}$$

where

$$\Phi_i = \begin{bmatrix} \Phi_{1i} & \overline{A}_{hi} X + E\tilde{S}E^T & 0 \\ * & -(1-h_d)\tilde{Q}_1 - 2E\tilde{S}E^T & E\tilde{S}E^T \\ * & * & -\tilde{Q}_2 - E\tilde{S}E^T \end{bmatrix} \tag{41}$$

$$\Phi_{1i} = \mathrm{sym}(A_i X + BF_i) + \tilde{Q}_1 + \tilde{Q}_2 - \frac{1}{h_M} E\tilde{S}E^T, \qquad \tilde{C}_i = C_i X$$

$$\begin{aligned} \tilde{\mathbf{A}}_i &= \begin{bmatrix} A_i X + BF_i & A_{hi} X & 0 \end{bmatrix}^T, & \tilde{\mathbf{B}}_{wi} &= \begin{bmatrix} \overline{B}_{wi}^T & 0 & 0 \end{bmatrix}^T \\ \tilde{\mathbf{M}}_i &= \begin{bmatrix} \overline{M}_i^T & 0 & 0 \end{bmatrix}^T, & \tilde{\mathbf{N}}_i &= \begin{bmatrix} N_i X & N_{hi} X & 0 \end{bmatrix}^T. \end{aligned} \tag{42}$$

The stabilising controller gains are given by $K_i = F_i X^{-1}$

Proof Under the conditions of Theorem 3, a feasible solution satisfies condition $-\sigma \operatorname{sym}(X) + \sigma^2 \tilde{S} < 0$ which implies that X is nonsingular.
On another hand, we note for any $\sigma > 0$ that

$$0 \leq (X - \sigma\tilde{S})^T \tilde{S}^{-1}(X - \sigma\tilde{S}) = X^T \tilde{S}^{-1} X - \sigma \operatorname{sym}(X) + \sigma^2 \tilde{S} \tag{43}$$

which implies that

$$-X^T \tilde{S}^{-1} X \leq -\sigma \operatorname{sym}(X) + \sigma^2 \tilde{S} \tag{44}$$

Let $P = X^{-1}$, $\tilde{Q}_l = X^T Q_l X$ $(l = 1, 2)$ and $\tilde{S} = X^T S X$ and $Y_i = K_i X$ $(i = 1, \ldots, r)$.
Considering (44) and checking a congruence transformation to (38), (39) and (40) by P, $diag\Big\{P, P, I, I, I, I, I\Big\}$ and $diag\Big\{P, I\Big\}$, respectively, inequalities (34), (35) and (36) hold. □

3.4 SMC Law Synthesis

Now, we are in position to synthesize a SMC law, by which the trajectories of the fuzzy descriptor system with delay in (3) can be driven onto the pre-specified switching surface $s(t) = 0$ and then are maintained there for all subsequent time.

Theorem 4 *Consider the uncertain descriptor time-delay system* (3). *Suppose that the switching surface function is given by* (10), *then the trajectories of system* (3) *can be driven onto the switching surface* $s(t) = 0$ *by the following SMC law:*

$$u(t) = \sum_{i=1}^{r} \mu_i \left(K_i x(t) - \alpha_i \frac{s(t)}{\|s(t)\|}\right) \tag{45}$$

where

$$\begin{aligned}\alpha_i = \lambda + \eta_i(x) + \|\big(\mathcal{M}B\big)^{-1} \mathcal{M} M_i\| \Big(\|N_i x(t)\| + \|N_{hi} x(t - h(t))\| + \|N_{wi} w(t)\|\Big) \\ + \|\big(\mathcal{M}B\big)^{-1} \mathcal{M} B_{wi}\| \|w(t)\|\end{aligned} \tag{46}$$

Proof Without loss of generality, we can choose $\mathcal{M} = B^T X_0$, where X_0 is positive definite matrix. So $\mathcal{M}B = B^T X_0 B$ is nonsingular. Consider the following Lyapunov function:

$$V_s(t) = \frac{1}{2} s^T(t) \big(\mathcal{M}B\big)^{-1} s(t) \tag{47}$$

According to (10), we obtain

$$\dot{s}(t) = \mathcal{M}\sum_{i=1}^{r}\mu_i\Big\{\big(\Delta A_i(t) - BK_i\big)x(t) + \Delta A_{hi}(t)x(t-h(t))) + B_{wi}(t)w(t) + B\Big(u(t) + f_i(x(t))\Big)\Big\} \tag{48}$$

Considering (48), the derivative of $V_s(t)$ is expressed as

$$\begin{aligned}\dot{V}_s(t) &= s^T(t)\big(\mathcal{M}B\big)^{-1}\dot{s}(t)\\ &= s^T(t)\big(\mathcal{M}B\big)^{-1}\mathcal{M}\sum_{i=1}^{r}\mu_i\Big\{\Delta A_i(t)x(t) + \Delta A_{hi}(t)x(t-h(t))) + B_{wi}(t)w(t)\Big\}\\ &\quad + s^T(t)\Big(u(t) + \sum_{i=1}^{r}\mu_i\big(f_i(x(t)) - K_i x(t)\big)\Big)\\ &\leq \|s(t)\|\sum_{i=1}^{r}\mu_i\Big\{\|\big(\mathcal{M}B\big)^{-1}\mathcal{M}M_i\|\Big(\|N_i x(t)\| + \|N_{hi}x(t-h(t))\| + \|N_{wi}w(t)\|\Big)\\ &\quad + \|\big(\mathcal{M}B\big)^{-1}\mathcal{M}B_{wi}\|\|w(t)\| + \eta_i(x)\Big\} + s^T(t)\Big(u(t) - \sum_{i=1}^{r}\mu_i K_i x(t)\Big)\end{aligned} \tag{49}$$

Substituting (45) into (49), we get

$$\dot{V}_s(t) = -\lambda\|s(t)\| < 0, \quad \forall\|s(t)\| \neq 0 \tag{50}$$

Then the system trajectories converges to the predefined sliding surface and is restricted to the surface for all subsequent time, thereby completing the proof. □

Remark 1 One popular solution to eliminate chattering is to approximate discontinuous function $\dfrac{s(t)}{\|s(t)\|}$ by some continuous and smooth functions. For example, it could be replaced by $\dfrac{s(t)}{\epsilon + \|s(t)\|}$, where ϵ is a small positive scalar value.
However, due to the consequence of disturbances, the smooth control function cannot provide finite-time convergence of the sliding variable to zero and the state variables converge to domains in a vicinity of the origin. This smooth control law is known as quasi-sliding mode control.

4 Numerical Examples

Example 1 To illustrate the merit and effectiveness of our results, we consider the following nonlinear time delay system borrowed from (Zhang et al. 2009)

$$\begin{aligned}\Big(1+(a+\delta a(t))\cos(\theta(t))\Big)\ddot{\theta}(t) =& -b\dot{\theta}^3(t)+c\theta(t)+(c_h+\delta c_h(t))\theta(t-h(t))\\ &+d\big(u(t)+f_i(t,x(t))\big)+ew(t)\end{aligned} \tag{51}$$

where the range of $\dot{\theta}(t)$ is assumed to satisfy $|\dot{\theta}(t)| < \Phi$, $\Phi = 2$, $c_h = 0.8$, $u(t)$ is the control input and $w(t)$ is the disturbance input. For simulation purposes, we set $a = 1, b = e = 1, c = 1$ and $d = 1$.
Defining $x(t) = \big[\theta(t)\ \dot{\theta}(t)\ \ddot{\theta}(t)\big]$. System (51) can expressed as

$$\begin{bmatrix}1&0&0\\0&1&0\\0&0&0\end{bmatrix}\dot{x}(t) = \begin{bmatrix}0&1&0\\0&0&1\\c&-bx_2^2&-1-a(t)cos(x_1)\end{bmatrix}x(t)+\begin{bmatrix}0&0&0\\0&0&0\\c_h(t)&0&0\end{bmatrix}x(t-h(t))+\begin{bmatrix}0\\0\\d\end{bmatrix}$$
$$+d\big(u(t)+f_i(t,x(t))\big)+\begin{bmatrix}0\\0\\e\end{bmatrix}w(t) \tag{52}$$

Based on the sector nonlinearity concept (Tanaka and Wang 2001), time-delay system (51) can be expressed exactly by the following (TS) fuzzy descriptor model (Zhang et al. 2009):

$$\begin{aligned}E\dot{x}(t) &= \sum_{i=1}^{3}\mu_i\Big\{A_i(t)x(t)+A_{hi}(t)x(t-h(t))+B\big(u(t)+f_i(t,x(t))\big)+B_{wi}(t)w(t)\Big\}\\ z(t) &= \sum_{i=1}^{3}\mu_iC_ix(t)\end{aligned} \tag{53}$$

where

$$E=\begin{bmatrix}1&0&0\\0&1&0\\0&0&0\end{bmatrix},\quad A_1=\begin{bmatrix}0&1&0\\0&0&1\\c&-b(\Phi^2+2)&a-1\end{bmatrix},\quad A_2=\begin{bmatrix}0&1&0\\0&0&1\\c&0&-a-1-a\Phi^2\end{bmatrix},$$

$$A_3=\begin{bmatrix}0&1&0\\0&0&1\\c&0&a-1\end{bmatrix},\quad A_{hi}=\begin{bmatrix}0&0&0\\0&0&0\\c_h&0&0\end{bmatrix},\quad B_i=\begin{bmatrix}0\\0\\d\end{bmatrix},$$

$$B_{wi}=\begin{bmatrix}0\\0\\e\end{bmatrix},\quad C_i=\big[0.5\ 0\ 0\big],\ i=1,2,3.$$

$$\mu_1=\frac{x_2^2(t)}{\Phi^2+2},\quad \mu_2=\frac{1+\cos(x_1(t))}{\Phi^2+2},\quad \mu_3=\frac{\Phi^2-x_2^2(t)+1-\cos(x_1(t))}{\Phi^2+2}$$

Assume that $\delta a(t) = \alpha\Delta(t)a$ and $\delta c_h(t) = \alpha\Delta(t)c_h$. The uncertain matrices in (5) can be described as

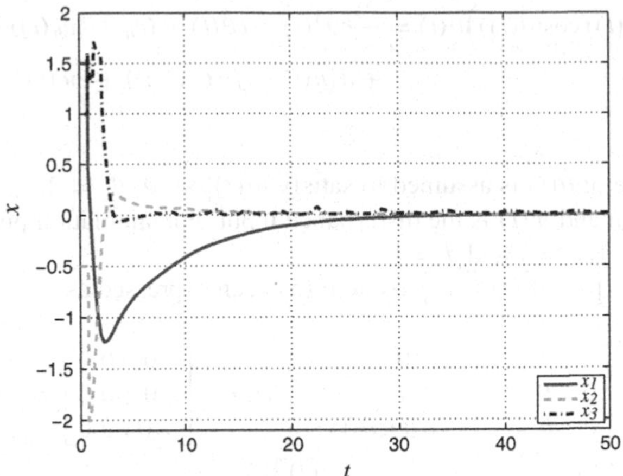

Fig. 1 States of the closed-loop system

$$M_i = \begin{bmatrix} 0 \\ 0 \\ \alpha \end{bmatrix}, \quad N_{1,3} = \begin{bmatrix} 0 \ 0 \ a \end{bmatrix}, \quad N_2 = \begin{bmatrix} 0 \ 0 \ -a(\Phi^2 + 1) \end{bmatrix}$$

For computational simplicity, set $\mathcal{M} = \begin{bmatrix} 0.3 \ 0.2 \ 1 \end{bmatrix}$. Then $\mathcal{M}B = 1$ is nonsingular. Set $\alpha = 0.25$, $\gamma = 0.1$, $\sigma = 1.1$ and time-varying delay $h(t) = 1.2 + 0.1sin(t)$. A straightforward calculation gives $h_M = 1.3$ and $h_d = 0.1$.

Our aim is to design an SMC law $u(t)$ as given in (45) such that the closed-loop system is robustly stable with generalized H_2 performance.
For aforementioned parameters, Theorem 3 produces a feasible solution to the corresponding LMIs with the following matrices

$$X = \begin{bmatrix} 0.088069 & -0.04124 & 0 \\ -0.04124 & 0.03636 & 0 \\ 0.011769 & -0.030176 & 0.037596 \end{bmatrix}, \quad K_1 = \begin{bmatrix} -2.2383 \ 3.78 \ -2.0092 \end{bmatrix},$$
$$K_2 = \begin{bmatrix} -2.2206 \ -2.2258 \ 4.1739 \end{bmatrix}, \quad K_3 = \begin{bmatrix} -2.2383 \ -2.22 \ -2.0092 \end{bmatrix}, \tag{54}$$

In addition, we take for simulation purpose $f_i(t, x(t)) = 0.75sin(x_1(t))x_1(t)$, $(i = 1, 2, 3)$, the exogenous input $w(t) = \dfrac{0.5}{1 + t^2}$ and the uncertain matrix function $\Delta(t) = sin(t)$. By setting $\lambda = 0.5$, the SMC law can be designed according to (45)–(46) and the simulation results are depicted in Figs. 1, 2 and 3 with initial condition $x(0) = \begin{bmatrix} 1.5 \ -0.5 \ 1 \end{bmatrix}^T$. To prevent the control signals from chattering, we replace sign $\frac{s(t)}{\|s(t)\|}$ with $\frac{s(t)}{0.05+\|s(t)\|}$.

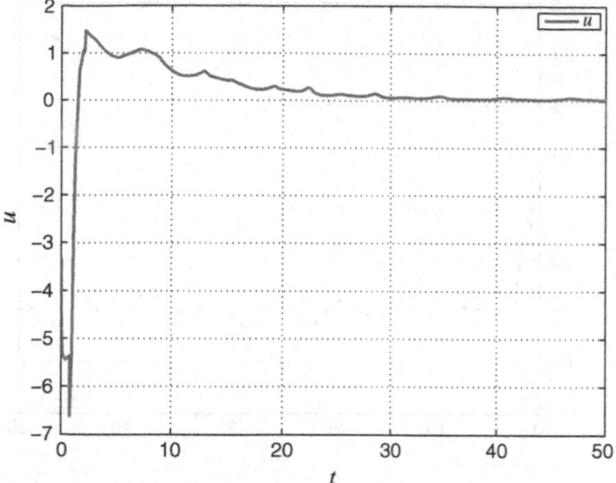

Fig. 2 Control input $u(t)$

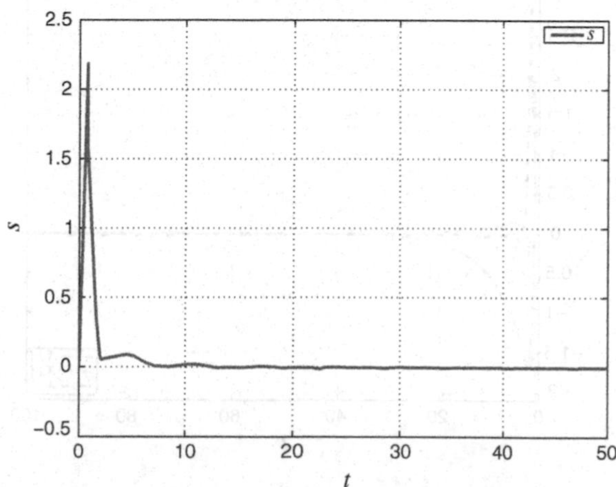

Fig. 3 Switching surface function $s(t)$

Figure 1 plots the evolution of the system states and Fig. 2 depicts the control input vector. The response of $s(t)$ is given in Fig. 3. It is observed from Fig. 1 that the state trajectories of the system all converge to the origin quickly. The system can be stabilized by the proposed method and the reaching motion satisfies the sliding reaching condition in spite of the time-varying delay, uncertainties and matched input. Figure 4 shows the state trajectories of the closed-loop system without sliding mode term. From this figure, we can see the effectiveness of the sliding mode term, which is used to compensate the effect of unknown input.

Now, assume that $f_i(t, x(t)) = 0$ and for time $t \geq 70$ s model parameters a, c_h and also c abruptly change. For $\alpha = 0.4$ and $\Delta(t) = 0.8 + 0.2sin(t)$, applying control law (45) to the system, one gets the state trajectory evolutions shown in Fig. 5.

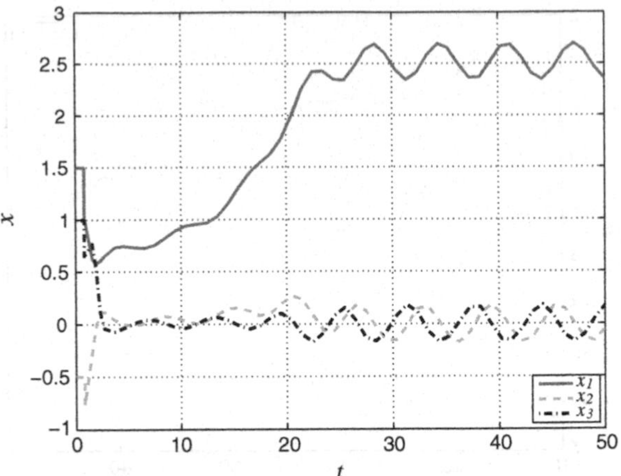

Fig. 4 States without a sliding mode term

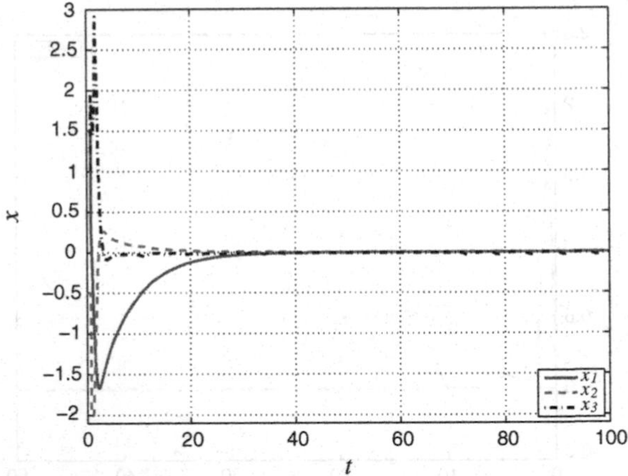

Fig. 5 State trajectories for $\alpha = 0.4$

However, when the control law is applied to the system without sliding mode term, the stability of the uncertain system can be degraded and the poor performance is shown in Fig. 6. It is clear that the proposed SMC scheme effectively eliminates effects of parameter uncertainties and guarantees the asymptotic stability of the closed-loop systems. Regarding these results, we conclude that the proposed SMC law yields a good performance and stabilizes the nonlinear system with time varying delay, unknown parameters uncertainties and nonlinear input.

Example 2 Consider a single-link manipulator of mass m and languor l, which is equipped with a DC motor as shown in Fig. 7. α is the angle of arm rotation. The dynamic of the system is given by the third-order equation (Manceur et al. 2012).

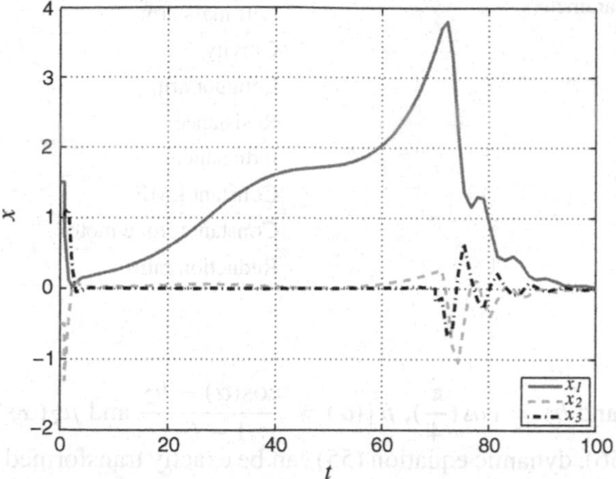

Fig. 6 State trajectories without a sliding mode term for $\alpha = 0.4$

Fig. 7 single-link manipulator

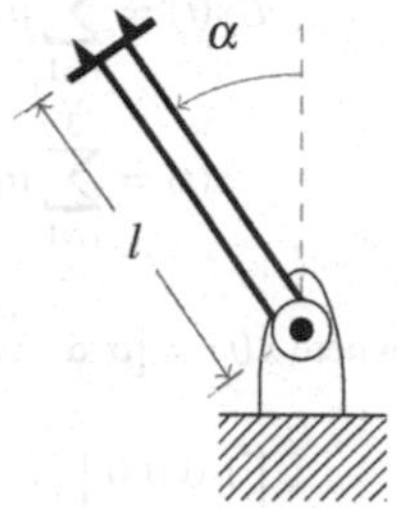

$$\begin{cases} \alpha^{(3)} = f_0(\alpha, \dot{\alpha}, \alpha(t-h(t))) + g_0(\alpha, \dot{\alpha})u + w \\ f_0(\alpha, \dot{\alpha}, \alpha(t-h(t))) = -\frac{R}{L}\ddot{\alpha} - \left(\frac{K_b N^2 K_t}{ml^2 L} + \frac{g}{l}cos(\alpha)\right)\dot{\alpha} - \frac{Rg}{lL}sin(\alpha) - 0.1\frac{Rg}{lL}\alpha(t-h(t)) \\ g_0(\alpha, \dot{\alpha}) = \frac{K_t N}{ml^2 L} \end{cases} \tag{55}$$

where $\alpha^{(3)}$, $\ddot{\alpha}$, and $\dot{\alpha}$ are the time derivatives of the angle α. w represents the unknown external disturbance. g, L, R, N, K_b, and K_t are, respectively, the gravity and the motor parameters whose signification is given in Table 1.

By the sector nonlinearity concept, $cos(\alpha)$, under the constraint $|\alpha| \leq \frac{\pi}{4}$, can be exactly represented as

$$cos(\alpha) = \sum_{i=1}^{2} \mu_i(\alpha) b_i \tag{56}$$

Table 1 Model parameters

Left mass arm	$m = 2\,\text{kg}$
Gravity	$g = 9.8\,\text{m/s}^2$
Longuor arm	$l = 0.5\,\text{m}$
Resistance	$R = 1.5\,\Omega$
Inductance	$L = 0.15\,\text{H}$
Constant EMF	$K_b = 0.2$
Constant torque motor	$K_t = 0.3$
Reduction ratio	$N = 60$

where $b_1 = 1$ and $b_2 = cos(\frac{\pi}{4})$, $\mu_1(\alpha) = \dfrac{\cos(\alpha) - b_2}{1 - b_2}$ and $\mu_2(\alpha) = 1 - \mu_1(\alpha)$. According to (56), dynamic equation (55) can be exactly transformed into following fuzzy descriptor model.

$$
\begin{aligned}
E\dot{x}(t) &= \sum_{i=1}^{2} \mu_i \Big\{A_i x(t) + A_{hi} x(t - h(t)) + B\big(u + f(\alpha)\big) + B_{wi} w(t)\Big\} \\
z(t) &= \sum_{i=1}^{3} \mu_i C_i x(t)
\end{aligned}
\tag{57}
$$

where $x(t) = \big[\alpha\ \dot{\alpha}\ \ddot{\alpha}\ \alpha^{(3)}\big]$ and

$$
E = \begin{bmatrix} 1\,0\,0\,0 \\ 0\,1\,0\,0 \\ 0\,0\,1\,0 \\ 0\,0\,0\,0 \end{bmatrix}, \quad
A_1 = \begin{bmatrix} 0 & 1 & 0\,0 \\ 0 & 0 & 1\,0 \\ 0 & 0 & 0\,1 \\ 0 & (b + cb_1) & a\ 1 \end{bmatrix}, \quad
A_2 = \begin{bmatrix} 0 & 1 & 0\,0 \\ 0 & 0 & 1\,0 \\ 0 & 0 & 0\,1 \\ 0 & (b + cb_2) & a\ 1 \end{bmatrix}, \quad
B_i = \begin{bmatrix} 0 \\ 0 \\ 0 \\ d \end{bmatrix}
$$

$$
A_{hi} = \begin{bmatrix} 0 & 0\,0\,0 \\ 0 & 0\,0\,0 \\ 0 & 0\,0\,0 \\ 0.1e & 0\,0\,0 \end{bmatrix}, \quad
B_{wi} = \begin{bmatrix} 0 \\ 0 \\ 0 \\ -1 \end{bmatrix}, \quad
C_i = \begin{bmatrix} 0.01\ 0\ 0\ 0 \\ 0\ \ 1\ 0\ 0 \end{bmatrix}.
$$

$$
a = \frac{R}{L}, \quad b = \frac{K_b N^2 K_t}{ml^2 L}, \quad c = \frac{g}{l}, \quad d = -\frac{K_t N}{ml^2 L}, \quad e = \frac{Rg}{lL}, \quad f(\alpha) = \frac{e}{d} sin(\alpha).
$$

Remark 2 From single-link manipulator dynamics (55), there are two nonlinear terms $cos(\alpha)$ and $sin(\alpha)$ Therefore it needs 2^2 rules for traditional (TS) fuzzy system to represent this system. To reduce the number of rules and thus the LMI number, we consider only the first term to represent the (TS) model, whereas, the second is regarded as nonlinear input.

For simulation purpose, we set $\mathcal{M} = \big[0.1\ 0.1\ {-0.1}\ 1/d\big]$, $\sigma = 0.9$, time-varying delay $h(t) = 0.2 + 0.1sin(t)$, exogenous disturbance $w = 4e^{-0.3t} sin(t)$ and initial condition $x(0) = \Big[-\frac{\pi}{8}\ 0.1\ 0\ 0\Big]^T$. Theorem 3 produces a feasible solution with the

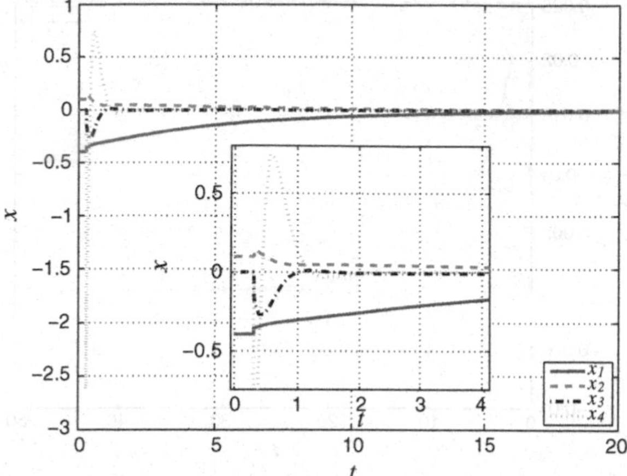

Fig. 8 States of the closed-loop system

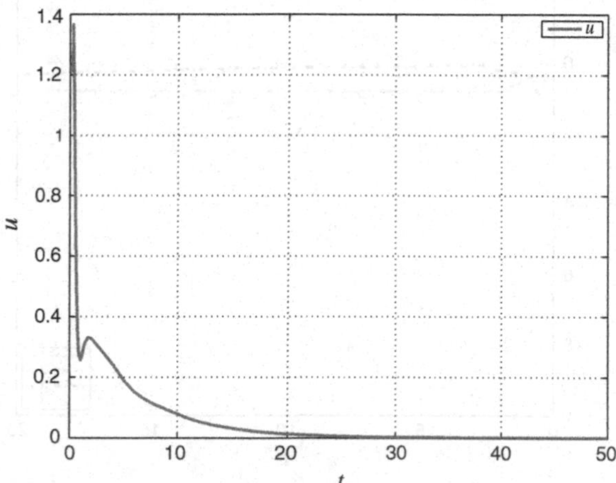

Fig. 9 Control response

following gains matrices

$$\begin{aligned} K_1 &= \left[0.20635\ 12.869\ 0.30169\ 0.040583\right], \\ K_2 &= \left[0.20635\ 12.845\ 0.30169\ 0.040583\right]. \end{aligned} \tag{58}$$

By Letting $\lambda = 0.55$ and $\varepsilon = 0.05$ the SMC law can be designed according to (45)–(46) and the simulation results are depicted in Figs. 8, 9 and 10.

It is seen that the proposed sliding-mode controller can effectively cope with the effect of time-delay and input non-linearity, and ensure the global asymptotic stability of the overall closed-loop system. Although there are some notable variations in the

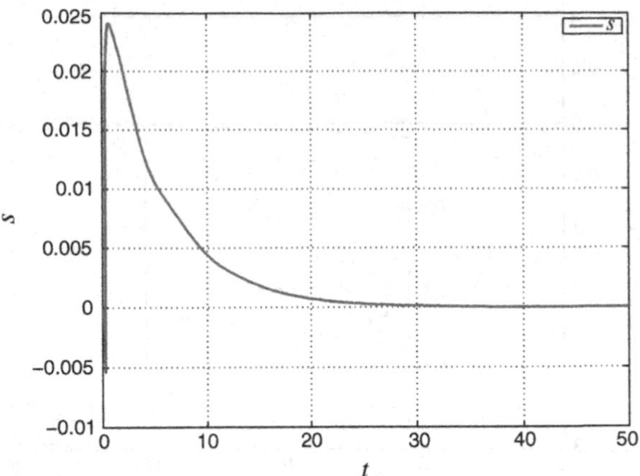

Fig. 10 Surface response

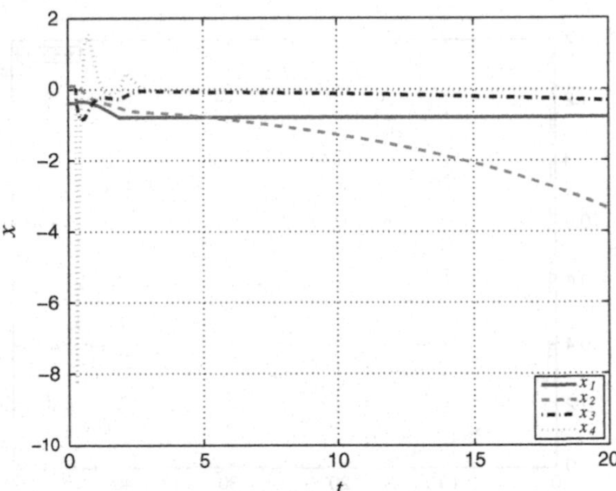

Fig. 11 States response without SMC term

curves, which are caused by the effect of exogenous disturbance, these undesirable effects are effectively attenuated. Figure 11 charts the state trajectories when the control law is applied without sliding mode term. Thus, the proposed SMC scheme effectively eliminates the effect of input nonlinearity,

5 Conclusion

Complete results have been developed for robust control of a class of continuous (TS) fuzzy descriptor systems with time-varying delay based-on integral sliding-mode control in the presence of nonlinearities, external disturbances, and parameters uncertainties. A new sliding function is proposed and a delay-dependent sufficient condition is derived to guarantee that sliding mode dynamics is robustly admissible with generalized H_2 disturbance rejection level. The integral sliding mode control aims to eliminate the reaching phase in normal sliding mode in order to improve the robustness of the system. Moreover, a SMC control law is designed such that the reaching condition is satisfied and the chattering can be reduced. The existence and the effectiveness of theoretical developments has been verified by two numerical examples.

The future work could be associated with the following directions:

- extend the method to systems with multiple state delays.
- using observer-based on SMC should be considered for delayed descriptor systems.
- propose a higher-order sliding-mode control for nonlinear descriptor systems,
- for numerical control, discrete-time sliding mode control should be developed for various classes of discrete time descriptor systems.

References

Aguilera-gonzalez, A., Bosche, J., El-Hajjaji, A., Abidi, I.: Feedback design air-path control on a diesel engine based on Takagi-Sugeno fuzzy descriptor systems. In: 21st Mediterranean Conference, pp. 1544–1549 (20013)

Azar, A.: Fuzzy Systems. IN-TECH, Vienna (2010)

Azar, A.: Overview of type-2 fuzzy logic systems. Int. J. Fuzzy Syst. Appl. **2**(4), 1–28 (2012)

Boukas, E.: Singular linear systems with delay: H_∞ stabilization. Optim. Control Appl. Methods **28**(4), 259–274 (2007)

Buzurovic, I., Debeljkovic, D.: Contact problem and controllability for singular systems in biomedical robotics. Int. J. Inform. Syst. Sci. **6**(2), 128–141 (2010)

Chang, J.: Dynamic output feedback integral sliding mode control design for uncertain systems. Int. J. Robust Nonlinear Control **22**(8), 841–857 (2012)

Chen, Y., Wang, W., Chang, C.: Guaranteed cost control for an overhead crane with practical constraints: Fuzzy descriptor system approach. Eng. Appl. Artif. Intell. **22**(4–5), 639–645 (2009)

Dai, L.: Singular Control Systems. Lecture Notes in Control and Information Sciences, vol. 118. Springer, New York (1989)

Ding, Y., Zhu, H., Zhong, S.: Exponential stabilization using sliding mode control for singular systems with time-varying delays and nonlinear perturbations. Commun. Nonlinear Sci. Numer. Simul. **16**(10), 4099–4107 (2011)

Duan, G.: Analysis and Design of Descriptor Linear Systems. Springer, New York (2010)

Feng, J., Cui, P., Cheng, Z.: H_∞ output feedback control for descriptor systems with delayed states. J. Control Theory Appl. **4**(4), 342–347 (2005)

Fridman, E.: Stability of linear descriptor systems with delay: a Lyapunov-based approach. J. Math. Anal. Appl. **273**(1), 24–44 (2000)

Gao, L., Wu, Y.: An LMI-based variable structure control for a class of uncertain singular markov switched systems. J. Control Theory Appl. **5**(4), 415–419 (2007)

Gassara, H., Hajjaji, A.E., Kchaou, M., Chaabane, M.: Observer based $(Q, V, R) - \alpha$ -dissipative control for TS fuzzy descriptor systems with time delay. J. Franklin Inst. **351**(1), 187–206 (2013)

Gu, K., Kharitonov, V., Chen, J.: Stability of Time-Delay Systems. Springer, Berlin (2003)

Han, C., Zhang, G., Wu, L., Zeng, Q.: Sliding mode control of TS fuzzy descriptor systems with time-delay. J. Franklin Inst. **349**(4), 1430–1444 (2012)

Kchaou, M., Souissi, M., Toumi, A.: Delay-dependent stability and robust $L_2 - L_\infty$ control for a class of fuzzy descriptor systems with time-varying delay. Int. J. Robust Nonlinear Control **23**(3), 284–304 (2011)

Kchaou, M., Tadeo, F., Chaabane, M.: A partitioning approach for H_∞ control of singular time-delay systems. Optim. Control Appl. Methods **34**(4), 472–486 (2013)

Kchaou, M., Tadeo, F., Chaabane, M., Toumi, A.: Delay-dependent robust observer-based control for discrete-time uncertain singular systems with interval time-varying state delay. Int. J. Control Autom. Syst. **12**(1), 12–22 (2014)

Li, L., Jia, Y., Dub, J., Yuan, S.: Robust $L_2 - L\infty$ control for uncertain singular systems with time-varying delay. Prog. Nat. Sci. **18**(8), 1015–1021 (2008)

Li, T., Kuo, C., Guo, N.: Design of an EP-based fuzzy sliding-mode control for a magnetic ball suspension system. Chaos Solitons Fractals **33**(5), 1523–1531 (2007)

Lin, C., Wang, Q., Lee, T.: Stability and stabilization of a class of fuzzy time-delay descriptor systems. IEEE Trans. Fuzzy Syst. **14**(4), 542–551 (2006)

Ma, S., Cheng, Z., Zhang, C.: Delay-dependent robust stability and stabilisation for uncertain discrete singular systems with time-varying delays. IET Control Theory Appl. **1**(4), 1086–1095 (2007)

Ma, S., Zhang, C., Cheng, Z.: Delay-dependent robust H_∞ control for uncertain discrete-time singular systems with time-delays. J. Comput. Appl. Math. **217**(15), 194–211 (2008)

Manceur, M., Essounbouli, N., Hamzaoui, A.: Second-order sliding fuzzy interval type-2 control for an uncertain system with real application. IEEE Trans. Fuzzy Syst. **20**(2), 262–275 (2012)

Muller, P.: Linear mechanical descriptor systems: identification, analysis and design. In: IFAC Conference on Control of Independent Systems, pp. 501–506. Belfort, France (1997)

Peterson, I.: A stabilization algorithm for a class uncertain linear systems. Syst. Control Lett. **8**(4), 351–357 (1987)

Schulte, H., Guelton, K.: Modelling and simulation of two-link robot manipulators based on Takagi Sugeno fuzzy descriptor systems. In: International Conference on Industrial Technology, ICIT, pp. 2692–2697, Mumbai. IEEE (2006)

Seuret, A., Edwards, C., Spurgeon, S.: Static output feedback sliding mode control design via an artificial stabilizing delay. IEEE Trans. Autom. Control **54**(2), 256–266 (2009)

Su, X., Zheng, W., Zhang, N.: Optimal control for TS fuzzy descriptor systems with time domain hard constraints. Int. J. Inf. Syst. Sci. **5**(3–4), 447–456 (2009)

Takagi, T., Sugeno, M.: Fuzzy identification of systems an dits application to modelling and control. Trans. Syst. Man Cybern. **15**(1), 116–132 (1985)

Tanaka, K., Wang, H.: Fuzzy Control Systems Design and Analysis. A Linear Matrix Inequality Approach. Wiley, New York (2001)

Tian, W., Zhang, H.: Optimal guaranteed cost control for fuzzy descriptor systems with time-varying delay. J. Syst. Eng. Electron. **19**(3), 584–591 (2008)

Wu, J., Weng, Z., Tian, Z., Shi, S.: Fault tolerant control for uncertain time-delay systems based on sliding mode control. Kybernetika **44**(5), 617–632 (2008)

Wu, L., Ho, D.: Sliding mode control of singular stochastic hybrid systems. Automatica **46**(4), 779–783 (2010)

Wu, L., Wang, Z.: Robust $L_2 - L\infty$ control of uncertain differential linear repetitive processes. Syst. Control Lett. **57**(5), 425–435 (2008)

Wu, L., Zheng, W.: Passivity-based sliding mode control of uncertain singular time-delay systems. Automatica **45**(9), 2120–2127 (2009)

Wu, Z., Su, H., Chu, J.: Delay-dependent $L_2 - L_\infty$ filter for singular time-delay systems. Acta Automatica Sinica **35**(9), 1226–1229 (2009a)

Wu, Z., Su, H., Chu, J.: Improved results on delay-dependent H_∞ control for singular time-delay systems. Acta Automatica Sinica **35**(9), 1226–1229 (2009b)

Xu, S., Dooren, P.V., Stefan, R., Lam, J.: Robust stability and stabilization for singular systems with state delay and parameter uncertainty. IEEE Trans. Autom. Control **47**(7), 1122–1128 (2002)

Xu, S., Lam, J., Zou, Y.: H_∞ filtering for singular systems. IEEE Trans. Autom. Control **48**(12), 2217–2222 (2003)

Yang, F., Zhang, Q.: Delay-dependent H_∞ control for linear descriptor systems with delay in state. J. Control Theory Appl. **3**(1), 76–84 (2005)

Zhang, H., Shenb, Y., Feng, G.: Delay-dependent stabilityand H_∞ control for a class of fuzzy descriptor systems with time-delay. Fuzzy Sets Syst. **160**(12), 1689–1707 (2009)

Zhou, S., Zheng, W.: Robust H_∞ control of delayed singular systems with linear fractional parametric uncertainties. J. Franklin Inst. **346**(2), 147–158 (2009)

Zhou, W., Fang, J.: Delay-dependent robust H_∞ admissibility and stabilization for uncertain singular system with markovian jumping parameters. Circ. Syst. Sig. Process. **28**(3), 433–450 (2009)

Rigid Spacecraft Fault-Tolerant Control Using Adaptive Fast Terminal Sliding Mode

Pyare Mohan Tiwari, S. Janardhanan and Mashuq un-Nabi

Abstract In addition to the robustness against inertia uncertainty and external disturbances, the efficient and quick fault-tolerant property is expected by the on-board attitude controller for any spacecraft mission. In comparison to the active fault tolerant control methods, the passive fault-tolerant methods are simpler and require less computation time and power. The finite-time sliding mode using the terminal sliding mode has been proven the efficacy to address the attitude control related issues, but in most of the cases, fault-tolerant issues were not taken into account. The objective of the chapter here is to propose a passive fault-tolerant control by using the finite-time sliding mode control. Firstly, an extensive review has been given to discuss the application of terminal sliding mode and its variants for the attitude control problem. Then, in control design, a non-singular fast terminal sliding mode has been integrated together with the adaptive control, and an adaptive non-singular fast terminal sliding mode control has been designed. In most of the finite time fault-tolerant designed using terminal sliding modes, the controllers gains are remain to constant; which can be cause for chattering. Therefore, to limit the chattering effect, and to avoid the need of upper bounds of uncertainty and external disturbances, adaptive estimate laws have been designed to estimate the controller's gains. Finite time stability has been analyzed by the Lyapunov theorem. Further, to show the fault-tolerance effectiveness of the proposed control law in attitude stabilization and tracking, various simulation results have been presented. The proposed control law is quick, and robust enough to negate the effects of external disturbances, mass inertia uncertainty, and actuator faults.

P.M. Tiwari (✉)
Amity University, Noida, India
e-mail: pmtiwari@amity.edu

S. Janardhanan · Mashuq-un-Nabi
Indian Institute of Technology, New Delhi, India
e-mail: janas@ee.iitd.ac.in

Mashuq-un-Nabi
e-mail: mnabi@ee.iitd.ac.in

A.T. Azar and Q. Zhu (eds.), *Advances and Applications in Sliding Mode Control systems*, Studies in Computational Intelligence 576, DOI 10.1007/978-3-319-11173-5_14

1 Introduction

Attitude control system (ACS) is an important module in the spacecraft mission design, and in the success of mission, the ACS design plays a vital role. To maintain the efficient performance of ACS, the on-board attitude controller should show the robustness against inertia uncertainty, external disturbances, and actuator fault; and additionally, it is also expected by the attitude controller to ensure the proper attitude stabilization or attitude tracking error reduction, in finite-time.

Sliding mode control (SMC) has considerably used to provide the solution for many non-linear problems (Utkin 1977; John et al. 1993). In this series, in the eighties, SMC application started for the spacecraft attitude control (Vadali 1986). In this continuation , recently, some other works for the rigid spacecraft attitude control have been reported by using the SMC (Yeh 2010; Lu et al. 2013). In these applications of SMC for attitude control design, the sliding surface is of linear structure. The major limitation of SMC is the asymptotic convergence of the system states to equilibrium, and it is due to the linear sliding surface. In conclusion, the SMC attitude control will control the attitude in infinite time.

In the eighties, a new and interesting theory the finite time control (FTC) has been developed (Haimo 1986). In the FTC, it is possible that the system states converge to the respective equilibrium in finite time. Inspired by the FTC theory, researchers have developed the terminal sliding mode (TSM) theory (Venkataraman 1991; Yu and Man 1996; Man and Yu 1997; Tang 1998). In TSM, contrary to the SMC, the sliding surface is the non-linear combination of system states; which ensures the finite time convergence to equilibrium. The application of TSM theory to design the spacecraft attitude control first appeared in Erdong and Zhaowei (2008). The originally proposed TSM suffers with the two drawbacks: one is the singularity in control for some initial condition, and the other is the slower convergence speed when the system states start remotely from the equilibrium. Hence, schemes NTSM (Feng et al. 2002) and FTSM (Yu and Man 2002) have been developed to solve the problem of singularity and convergence speed, respectively. By using the NTSM and FTSM, attitude control laws have been designed in Ding et al. (2009), Li et al. (2011), Lu and Xia (2013) and Tiwari et al. (2010), Zou and Kumar (2011), respectively. To design a control that solves the singularity and the finite time convergence together, non-singular fast terminal sliding mode (NSFTSM) has been developed in Yang et al. (2011). Inspired by NSFTSM Yang et al. (2011), for the attitude stabilization and tracking cases, control laws have been presented in Tiwari et al. (2012) and Tiwari et al. (2014), respectively. In these all the afore-mentioned finite time attitude control references, the actuator fault condition has not been taken into account.

Through the technological advancement, tremendous improvements have been made in the attitude actuators design and their implementation techniques. However, to design a fully autonomous space mission, it is important that the on-board ACS should be able to defeat the actuator fault in finite-time with high speed and efficacy. It is worth mentioning that fault tolerance should be done in finite- time with high speed; otherwise in some specific space missions designed specially for military applications, the ACS may not be able to maintain the control performance all round. Mainly, the fault tolerant control are categorized in two methods, the active and passive. The active fault tolerant control methods are equipped with fault diagnosis and detection (FDD). The FDD role is to detect and identify the actuator faults; and then to re-configure the controllers to compensate the faults effects. Therefore, obviously that the series of computation and decision steps are required in active fault tolerant method. So, the ACS with active fault tolerant method require more time to complete the online computations; but the long computation time could delay the timely control action, and the control performance may deteriorated to the level that can lead to the catastrophic failures. In contrary to the active methods, the passive fault tolerant control is equipped with the only one controller, and with this controller, both the lumped uncertainty and the actuator faults and saturation are handled. Numerous efforts by taking different control techniques have been reported for the design of passive fault tolerant controllers (Bustan et al. 2013) (and other references mentioned in Bustan et al.).

Inspired by the finite time convergence property of TSM and its variants (NTSM, FTSM, NFTSM), recently, they have been introduced as a qualified passive fault tolerant method for the spacecraft attitude. In Hu et al. (2012), TSM has been applied to compensate the effects of actuator effectiveness loss, inertia uncertainty, and external disturbances. However, the chosen sliding surface suffers with the same limitations as with the originally proposed TSM. The NTSM based fault tolerant control appeared in Hu et al. (2013), Lu et al. (2013). In Hu et al. (2013), finite-time attitude stabilization law under actuator misalignment is addressed. In reference Lu et al. (2013), attitude tracking performance is checked under the actuator fault and effectiveness loss. In these references, the controller gains are remain constant; and gains values are linked with the upper bounds of uncertainty and external disturbance. More than that, to enhance the fault tolerant control quality, recently, the FTSM control and the adaptive control appeared together. In Hu et al. (2012), authors developed the adaptive law updated finite-time controller using FTSM, and applied for the reaction wheel fault tolerance. However, the control law may suffer with limitations of singularity and unbound increment in control gains estimate. In this series, authors of Xiao et al. (2013) have developed the attitude tracking compensation controller, and shown the performance under actuator fault, actuator misalignment, and external disturbances. Though, the recommended controller may cause the singularity problem. In Zhang et al. (2013), by using FTSM, authors have developed the finite-time fault tolerant control; it is shown that together the nominal controller and the adaptive compensation control is successfully accomplished the attitude tracking in the presence of actuator fault and actuator misalignment. In this work also, while discussing the stability proof, the error quaternion vector $e \neq 0$ is considered, but this is not the case always possible. For example, if one of the error quaternion will start or attain value zero, condition $||e^{r-1}|| \leq \ell_3$ will not be fulfilled for $r \in (0, 1)$.

It is noticed that by using TSM and its variants, finite time fault tolerant control is in its early stage; and in the selection of sliding surface, method to decide the controller's gains, and consideration of the different types of faults, are the major areas of improvements in proposing the solution. Our endeavor here is to develop a control law for the control of rigid spacecraft in the presence of actuator fault, actuator effectiveness loss, external disturbances, uncertain mass inertia parameters. In the control development, the nominal control component is derived by using a non-singular fast terminal sliding mode (NSFTSM) surface. Additionally, to negate the actuator fault and external disturbances as well inertia uncertainty, the nominal controller is supported by the adaptive control component. The closed loop finite time stability has been proved using the Lyapunov stability theory.

The structure of the chapter is as follows: The rigid spacecraft attitude mathematical modeling for stabilization and tracking are discussed in Sect. 2. In Sect. 3 of the chapter discusses the control objective and the proposed fault -tolerant control design with the finite-time stability proof. In Sect. 4, simulation results are illustrated with extensive discussion. Finally, conclusion is given in Sect. 5.

2 System Description

In any space mission, attitude stabilization and tracking are the main aim of ACS. This section discusses the mathematical model for the attitude stabilization as well as for the attitude tracking of a rigid spacecraft. Unit quaternion, due to non-trigonometric expression and non-singularity computation (Wertz 1978), are extensively used parameter to represent the kinematics of a rigid spacecraft; that is why the attitude kinematics is described here by using the unit quaternion.

2.1 Mathematical Model for Attitude Stabilization

Rigid spacecraft attitude control is described by the kinematics and dynamics equations (Wertz 1978). The attitude kinematics representation using the unit quaternion is given as follows.

$$\bar{q} = [q_v^T \quad q_4]^T. \tag{1}$$

where $q_v = [q_1 \quad q_2 \quad q_3]^T = \varepsilon \sin\frac{\theta}{2}$ and $q_4 = \cos\frac{\theta}{2}$ are the vector and the scalar components of the unit quaternion respectively, where $\theta \in \Re$ is the rotation angle about the eigen axis, which is given by the unit vector $\varepsilon = \left[\varepsilon_1 \; \varepsilon_2 \; \varepsilon_3\right]^T$. The scalar and the vector components of unit quaternion satisfies the constraint

$$q_v^T q_v + q_4^2 = 1. \tag{2}$$

The kinematics equations are given as

$$\dot{q_v} = \frac{1}{2}(q_4 I_3 + q_v{}^{\times})\omega$$

$$\dot{q}_4 = -\frac{1}{2} q_v^T \omega, \tag{3}$$

where I_3 is the 3×3 identity matrix, $\omega \in \Re^3$ is the the body angular velocity vector measured with respect to the inertial frame expressed in the body frame. The notation $q_v^\times$ represents the following skew-symmetric matrix generated by the vector $q_v = [q_1 \quad q_2 \quad q_3]^T$

$$q_v{}^\times = \begin{bmatrix} 0 & -q_3 & q_2 \\ q_3 & 0 & -q_1 \\ -q_2 & q_1 & 0 \end{bmatrix}$$

The dynamics equation for a rigid body spacecraft is given by

$$\dot{\omega} = J^{-1}(-\omega^\times J\omega + D\ \Delta(t)\ u(t) + d(t)) \tag{4}$$

where $J = J_0 + \delta J$ is the inertia matrix of spacecraft, where $J_0 \in \Re^{3\times3}$ and $\delta J \in \Re^{3\times3}$ are the nominal component and the uncertain components respectively, $u(t) \in \Re^n$ is the control input generated by n actuators, $D \in R^{3\times n}$ is the actuator distribution matrix, and $d(t) \in \Re^3$ is the bounded external disturbance torque acting on the body, $\Delta(t) = diag(\Delta_1,\ \Delta_2,\ \Delta_3, ...,\ \Delta_n) \in \Re^{n\times n}$ is the actuator effectiveness matrix with $0 \le \Delta_i \le 1$. Important to note that $\Delta_i = 1$ means that particular actuator is fully healthy, and $\Delta_i = 0$ means particular actuator is lost its strength completely, and if $\Delta_i < 1$, then particular actuator is partially functioning. The notation $\omega^\times$ is a skew-symmetric matrix generated by ω.

2.2 Mathematical Model for Attitude Tracking

To define the attitude kinematics and dynamics equation for tracking control problem, the relative attitude error between reference frame and a desired reference frame is required to be established. The error quaternion $q_e = [q_{ev}^T,\ q_{e4}]^T \in \Re \times \Re^3$ and the angular velocity error $\omega_e \in \Re^3$ are measured from body fixed reference frame to the desired reference frame, and the defining equations are as follows

$$\begin{aligned} q_{ev} &= q_{d4} q_v - q_{dv}^\times q_v - q_4 q_{dv} \\ q_{e4} &= q_{dv}^T q_v + q_4 q_{dv} \\ \omega_e &= \omega - C\omega_d, \end{aligned} \tag{5}$$

where $q_{ev} = [q_{e1} \quad q_{e2} \quad q_{e3}]^T$ and q_{e4} are the vector and scalar components of the error quaternion, respectively, $q_{dv} = [q_{d1} \quad q_{d2} \quad q_{d3}]^T \in \Re^3$, $q_{d4} \in \Re$, and $\omega_d = [\omega_{d1} \quad \omega_{d2} \quad \omega_{d3}]^T \in \Re^3$ are the desired attitude frame vector quaternion, scalar quaternion, and angular velocity, respectively. Both q_e and q_d =

$[q_{d1} \quad q_{d2} \quad q_{d3} \quad q_{d4}]^T$ satisfy the constraint $q_{ev}^T q_{ev} + q_{e4}^2 = 1$ and $q_{dv}^T q_{dv} + q_{d4}^2 = 1$; respectively. $C = (q_{e4}^2 - 2q_{ev}^T)I + 2q_{ev}q_{ev}^T - 2q_{e4}q_{ev}^\times \in \Re^{3\times 3}$ with $||C|| = 1$ and $\dot{C} = -\omega^\times C$ represents the rotation matrix between body fixed reference frame and desired reference frame.

Then, using (5), the attitude kinematics and the dynamics equation for the tracking problem could be written as

$$\begin{aligned} \dot{q}_{ev} &= \frac{1}{2}(q_{e4}I + q_{ev}^\times)\omega_e \\ \dot{q}_{e4} &= -\frac{1}{2}q_{ev}^T\omega_e \end{aligned} \tag{6}$$

$$\dot{\omega_e} = J^{-1}\left((-\omega_e + C\omega_d)^\times J(\omega_e + C\omega_d) + J(\omega_e^\times C\omega_d - C\dot{\omega_d}) + D\,\Delta(t)\,u(t) + d(t)\right). \tag{7}$$

3 Fault Tolerant Control Design

In this section, first the control objective is defined, and then the control design method is developed by using a non-singular fast terminal sliding mode. To compensate the effects of actuator fault, external disturbances, and inertia matrix uncertainty, an adaptive control component is applied with nominal control. The proposed fault-tolerant control design will be completed in following three steps

1. Selection of sliding surface
2. Control structure
3. Stability proof both in reaching phase and in sliding phase.

3.1 Control Objective

The control objective is to design a robust fault tolerant controller that to ensure the finite time attitude control in presence of external disturbance, inertia uncertainty, loss of actuator effectiveness, and any fault. Mathematical representation for the control objective is

$$\begin{cases} \lim\limits_{t\to t_f}(q_v - q_d) = 0 \\ \lim\limits_{t\to t_f} \Omega = 0, \end{cases} \tag{8}$$

where $\Omega = (\omega - \omega_d) \in \Re^3$.

In the fore coming control design to achieve the afore-mentioned control objective, the following assumptions are made.

Assumption 1 The body frame quaternion q and angular velocity ω are measurable, and available for feedback.

Assumption 2 The desired reference attitude frame angular velocity ω_d and its first time derivative $\dot{\omega_d}$ are bounded.

Assumption 3 Spacecraft mass inertia matrix nominal component J_0 and the uncertain component δJ are bounded; though the bound limits are not known in advance.

Assumption 4 The control input is not unlimited, and constrained by the limit $u(t) \leq u_{max}$.

Assumption 5 External disturbance $d(t)$ is bounded, and the bound limit is not known in advance.

3.2 Control Design

The detailed design steps are as under

Step 1: Sliding surface design

In contrary to the published finite-time fault tolerant control, here the sliding surface is chosen that to avoid the singularity and to get quick convergence speed. Therefore, using the angular velocity error and the quaternion error information the sliding surface selected is

$$\sigma_e = sig^{\alpha}(\omega_e) + M_1 sig^{\alpha}(q_{ev}) + M_2(q_{ev}) \tag{9}$$

here, $\sigma_e \in \Re^3$ is the sliding surface chosen, $\alpha \in (1, 2)$, $M_1 = diag(m_{11}, m_{12}, m_{13})$, $M_2 = diag(m_{21}, m_{22}, m_{23})$ with $m_{ij} \in \Re$ for $i = 1, 2$ and $j = 1, 2, 3$, and for any vector $y \in \Re^3$, the notation $sig^{\alpha}(y) = [|y_1|^{\alpha} sign(y_1) \quad |y_2|^{\alpha} sign(y_2) \quad |y_3|^{\alpha} sign(y_3)]$.

Now, evaluating

$$\begin{aligned} J\dot{\sigma_e} &= \alpha diag(|\omega_e|^{\alpha-1}) J\dot{\omega_e} \\ &\quad + \frac{J}{2}(M_1 \alpha diag(|q_{ev}|)^{\alpha-1} + M_2)(q_{e4} I_3 + q_{ev}^{\times})\omega_e \end{aligned} \tag{10}$$

Applying (7) in (10), we have

$$\begin{aligned} J\dot{\sigma_e} = \alpha diag(|\omega_e|^{\alpha-1})\Big(& (-\omega_e + C\omega_d)^{\times} J_o(\omega_e + C\omega_d) + J_o(\omega_e^{\times} C\omega_d - C\dot{\omega_d}) \\ & + D\,\Delta(t)\, u(t) + d(t) + L(q_e, \omega_e, \omega_d, \dot{\omega}_d, \delta J)\Big) + \frac{J_o}{2}(M_1 \alpha diag(|q_{ev}|)^{\alpha-1} \\ & + M_2)(q_{e4} I_3 + q_{ev}^{\times})\omega_e \end{aligned} \tag{11}$$

where $L(q_e, \omega_e, \omega_d, \dot{\omega}_d, \delta J) = (-\omega_e + C\omega_d)^{\times}\delta J(\omega_e + C\omega_d) + \delta J(\omega_e^{\times} C\omega_d - C\dot{\omega}_d) + \frac{\delta J}{2\alpha}(M_1 \alpha diag(|q_{ev}|)^{\alpha-1} + M_2)(q_{e4}I_3 + q_{ev}^{\times})sig^{2-\alpha}(\omega_e)$ represents the uncertain terms due to inertia matrix uncertainty.

Step 2: Control structure

To achieve the desired control objective (8), the proposed control structure is given as follows

$$u(t) = u_{nom}(t) + u_{ada}(t) \tag{12}$$

where

$$u_{nom}(t) = D^{\dagger}\Big((\omega_e + C\omega_d)^{\times} J_o(\omega_e + C\omega_d) - J_o(\omega_e^{\times} C\omega_d - C\dot{\omega}_d) \\ - \frac{J_o}{2\alpha}(M_1 \alpha diag(|q_{ev}|^{\alpha-1}) + M_2)(q_{e4}I_3 + q_{ev}^{\times})sig^{2-\alpha}\omega_e\Big) \tag{13}$$

and

$$u_{ada}(t) = \begin{cases} D^{\dagger}(-\hat{k}_1\sigma_e - \hat{k}_2 sig^{\gamma}(\sigma_e)), & \text{if } ||\sigma_e|| \geq \varepsilon \\ 0, & \text{if } ||\sigma_e|| < \varepsilon, \end{cases} \tag{14}$$

where $D^{\dagger} = D^T(D\,D^T)^{-1}$ is a right-pseudo inverse of actuator distribution matrix D, $\hat{k}_1$ and $\hat{k}_2$ are the estimates of controller gains k_1 and k_2, respectively.

The adaptive estimate laws proposed here are as follows

$$\dot{\hat{k}}_1 = \begin{cases} \alpha\,\eta||\omega_e^{\alpha-1}||_{\infty}\,||\sigma_e||^2, & \text{if } ||\sigma_e|| \geq \varepsilon \\ 0, & \text{if } ||\sigma_e|| < \varepsilon, \end{cases} \tag{15}$$

$$\dot{\hat{k}}_2 = \begin{cases} \alpha\,\theta||\omega_e^{\alpha-1}||_{\infty}\,||\sigma_e||_{\gamma+1}^{\gamma+1}, & \text{if } ||\sigma_e|| \geq \varepsilon \\ 0, & \text{if } ||\sigma_e|| < \varepsilon, \end{cases} \tag{16}$$

where, $\eta \in \Re$, $\theta \in \Re$, $\varepsilon \in \Re$ are the design parameters, and $||\sigma_e||_{\gamma+1}^{\gamma+1} = [|\sigma_{e1}|^{\gamma+1} + |\sigma_{e2}|^{\gamma+1} + |\sigma_{e3}|^{\gamma+1}]^{\gamma+1}$.

Remark 1 The nominal component of the proposed controller is evaluated by applying the invariance principle. Obviously, the nominal control expression, (13) have two terms with the fractional power, but both powers are nonnegative and hence there is no point of singularity.

Remark 2 To negate the effects of inertia uncertainty and external disturbance, and to accelerate the convergence speed in reaching phase, an adaptive control component is added with the nominal control. In most of the published fault tolerant control, the controller's gains are static, and are decided on the basis of uncertainty and disturbance upper bounds, but practically it is difficult to know the bounds in advance.

Therefore, here to estimate the gains values, adaptive estimate laws are proposed. From (15) and (16), it is notable that the adaptive estimate laws depend on only the system states, and not to the uncertainty, disturbance, and actuator faults. Further, to reject the possibility of unbound growth in the controller's gains, dead zone technique has been applied. It is evident from (14), (15), and (16), that both the adaptive control and the adaptive gains will not be updated once the attitude states reach into the desired boundary.

Remark 3 In the proposed adaptive laws, parameters η and θ are working to regulate the estimate speed. Higher the values of η and θ, higher the convergence speed for $\sigma_e = 0$ is obtained.

Step 3: Stability Analysis

The finite time stability check of closed loop system (6)–(7) is completed in two steps. In first step, the reaching phase stability is proved, and in second step the sliding phase stability is proved. Before the stability discussion, the following lemma, discussing the finite-time stability is useful.

Lemma 1 (Yu et al. 2005) *An extended Lyapunov description of finite-time stability with faster finite time convergence can be given as*

$$\dot{V}(x) + \lambda_1 V(x) + \lambda_2 V^m(x) \leq 0 \tag{17}$$

and the convergence time can be given as

$$t \leq \frac{1}{\lambda_1(1-m)} ln \frac{\lambda_1 V^{1-m}(x_0) + \lambda_2}{\lambda_2} \tag{18}$$

where $\lambda_1 > 0$, $\lambda_2 > 0$, *and* $m \in (0, 1)$.

Reaching phase stability

Theorem 1 *With the controller* (12), *the attitude states of* (6)–(7) *will be able to reach the neighborhood of* $\sigma_e = 0$ *in finite time.*

Proof Select the Lyapunov function

$$V_1 = \frac{1}{2}\sigma_e^T J \sigma_e + \frac{1}{2\eta}\bar{k}_1^2 + \frac{1}{2\theta}\bar{k}_2^2 \tag{19}$$

where, $\bar{k}_1 = \hat{k}_1 - k_1$ and $\bar{k}_2 = \hat{k}_2 - k_2$, and V_1 satisfies

$$\frac{1}{2}\lambda_{min}(J)||\sigma_e||^2 \leq \frac{1}{2}\lambda_{max}(J)||\sigma_e||^2 \tag{20}$$

Taking the first time derivative V_1, and applying (11) leads to

$$\dot{V}_1 = \sigma_e^T \Bigg(\alpha\, diag(|\omega_e|^{\alpha-1}) \Big((-\omega_e + C\omega_d)^\times J_o(\omega_e + C\omega_d) + J_o(\omega_e^\times C\omega_d - C\dot{\omega_d}) + D\, u(t) + D\, (\Delta - I_4)u(t) + d(t) + L(q_e, \omega_e, \omega_d, \dot{\omega}_d, \delta J) \Big) \Bigg) + \frac{1}{\eta}(\hat{k}_1 - k_1)\dot{\hat{k}}_1 + \frac{1}{\theta}(\hat{k}_2 - k_2)\dot{\hat{k}}_2 \tag{21}$$

Further, defining $\bar{L} = D\, (\Delta - I_4)u(t) + d(t) + L(q_e, \omega_e, \omega_d, \dot{\omega}_d, \delta J)$, and then it could be written that

$$||\bar{L}|| \le ||\bar{L}||_1 \le ||D\, (\Delta - I_4)u(t)||_1 + ||d(t)||_1 + ||L||_1 \tag{22}$$

and then substituting (12)–(16), (22); and (21) yields

$$\begin{aligned}
\dot{V}_1 &= \sigma_e^T \alpha diag(|\omega_e|^{\alpha-1}) \Big(- \hat{k}_1 \sigma_e - \hat{k}_2 sig^\gamma(\sigma_e) + \bar{L} \Big) \\
&\quad + \alpha(\hat{k}_1 - k_1)||\, |\omega_e^{\alpha-1}|\, ||_\infty\, ||\sigma_e||^2 + \alpha(\hat{k}_2 - k_2)||\omega_e^{\alpha-1}||_\infty ||\,||\sigma_e||_{\gamma+1}^{\gamma+1} \\
&\le -\alpha\hat{k}_1 |||\omega_e|^{\alpha-1}||_\infty\, ||\sigma_e||^2 - \alpha\hat{k}_2 |||\omega_e|^{\alpha-1}||_\infty\, ||\sigma_e||_{\gamma+1}^{\gamma+1} + \alpha||\bar{L}||\; |||\omega_e|^{\alpha-1}||_\infty ||\sigma_e^T|| \\
&\quad + \alpha(\hat{k}_1 - k_1)||\omega_e^{\alpha-1}||_\infty ||\sigma_e||^2 + \alpha(\hat{k}_2 - k_2)||\omega_e^{\alpha-1}||_\infty ||\; ||\sigma_e||_{\gamma+1}^{\gamma+1} \\
&\le -\alpha k_1 |||\omega_e|^{\alpha-1}||_\infty ||||\sigma_e||^2 - \alpha k_2 |||\omega_e|^{\alpha-1}||_\infty \;||\; ||\sigma_e||_{\gamma+1}^{\gamma+1} \\
&\quad + \alpha||\bar{L}||\; |||\omega_e|^{\alpha-1}||_\infty\; ||\sigma_e^T||
\end{aligned} \tag{23}$$

In reaching phase ($\sigma_e \neq 0$), it is easy to show that $\omega_e = 0$ is not an attractor (Appendix). Therefore, (23) can be rewritten in the following two forms:

$$\dot{V}_1 \le -\frac{2\alpha |||\omega_e|^{\alpha-1}||_\infty}{\lambda_{max}(J)} \left(k_1 - \frac{||\bar{L}||}{||\sigma_e||} \right) V_1 - \left(\frac{2}{\lambda_{max}(J)} \right)^{\frac{\gamma+1}{2}} \alpha k_2 |||\omega_e|^{\alpha-1}||_\infty V_1^{\frac{\gamma+1}{2}} \tag{24}$$

$$\dot{V}_1 \le -\frac{2\alpha k_1 |||\omega_e|^{\alpha-1}||_\infty}{\lambda_{max}(J)} V_1 - \left(\frac{2}{\lambda_{max}(J)} \right)^{\frac{\gamma+1}{2}} \alpha |||\omega_e|^{\alpha-1}||_\infty \left(k_2 - \frac{||\bar{L}||\; ||\sigma_e||}{||\sigma_e||_{\gamma+1}^{\gamma+1}} \right) V_1^{\frac{\gamma+1}{2}} \tag{25}$$

The stability analysis of (24) and (25) are completed in two scenario.

1. **Scenario1** ($d(t) = 0$), $\delta J = 0$, $\Delta = I_4$)
   In this scenario, both (24) and (25) are simplified to $\dot{V}_1 + \Gamma_1 V_1 + \Gamma_2 V_1^{\frac{\gamma+1}{2}}$, where $\Gamma_1 = \frac{2\alpha k_1 |||\omega_e|^{\alpha-1}||_\infty}{\lambda_{max}(J)}$, $\Gamma_2 = \left(\frac{2}{\lambda_{max}(J)} \right)^{\frac{\gamma+1}{2}} \alpha k_2 |||\omega_e|^{\alpha-1}||_\infty$, and therefore,

convergence to $\sigma_e = 0$ is ensured in finite time

$$t_1 \le \frac{1}{\Gamma_1(1-\gamma)/2} \ln \frac{\Gamma_1 V_2^{(1-\gamma)/2}(\sigma_e(0)) + \Gamma_2}{\Gamma_2}. \tag{26}$$

2. **Scenario2** ($d(t) \neq 0$), $\delta J \neq 0$, $\Delta \neq I_4$)
Following the analysis given in Yu et al. (2005), from (24) and (25) it is obvious that, if $k_1 - \frac{\|\bar{L}\|}{\|\sigma_e\|} > 0$ and $k_2 - \frac{\|\bar{L}\| \|\sigma_e\|}{\|\sigma_e\|_{\gamma+1}^{\gamma+1}} > 0$ is ensured, then (24) and (25) structure will take the faster finite time stability condition (17) of Lemma 1, and the region $\|\sigma_e\| < \frac{\|\bar{L}\|}{k_1}$ and $\|\sigma_e\| < \left(\frac{\|\bar{L}\|}{k_2}\right)^{1/\gamma}$ will be achieved in finite time, respectively. This completes the proof.

Sliding phase stability

Theorem 2 *After the attitude trajectory of system* (6)–(7) *reach to the neighborhood of* $\sigma_e = 0$, *the tracking error in attitude states will converge to zero in finite time.*

Proof Once the attitude states of (6)–(7) reach to $\sigma_e = 0$, we have

$$\begin{aligned} sig^{\alpha}(\omega_e) + M_1\, sig^{\alpha}(q_{ev}) + M_2\, q_{ev} = 0 \\ \omega_e \le -M_1^{\frac{1}{\alpha}} q_{ev} - M_1^{\frac{1}{\alpha}}\, sig^{\frac{1}{\alpha}} q_{ev} \end{aligned} \tag{27}$$

Define another Lyapunov function

$$V_2 = q_{ev}^T q_{ev}. \tag{28}$$

Evaluating the first time derivative of (28), give

$$\begin{aligned} \dot{V}_2 &= 2q_{ev}^T \dot{q}_{ev} \\ &= q_{ev}^T (q_{e4} I + q_{ev}^{\times})\omega_e. \end{aligned} \tag{29}$$

Applying the inequality (27), and using the fact $\|(q_{e4}I + q_{ev}^{\times})\| \le 1$, the above expression results

$$\begin{aligned} \dot{V}_2 &\le q_{ev}^T(-M_1^{1/\alpha}(q_{ev}) - M_2^{1/\alpha} sig^{1/\alpha}(q_{ev})) \\ &\le -M_1^{1/\alpha} V_2 - M_2^{1/\alpha} V_2^{(\alpha+1)/2\alpha} \\ &\le -\lambda_1 V_2 - \lambda_2 V_2^{(\alpha+1)/2\alpha} \end{aligned} \tag{30}$$

where λ_1 and λ_2 are the minimum eigen values of $M_1^{1/\alpha}$ and $M_2^{1/\alpha}$, respectively. Therefore,

$$\dot{V}_2 + \lambda_1 V_2 + \lambda_2 V_2^{(\alpha+1)/2\alpha} \le 0 \tag{31}$$

The above Eq. (31) satisfies the finite time stability criteria (17), and hence, once the attitude trajectory falls on to the sliding surface, then the quaternion error will converge to zero in finite time

$$t_2 \leq \frac{1}{\lambda_1(\alpha-1)/2\alpha} \ln \frac{\lambda_1 V_2^{(\alpha-1)/2\alpha}(\sigma_e(t_1)) + \lambda_2}{\lambda_2}, \tag{32}$$

where t_1 is the time to cross the reaching phase and to enter into the neighborhood region of $\sigma_e = 0$. Subsequently, by (6), it is proved that $\omega_e = 0$. This completes the proof.

Remark 4 From (32), it is obvious that M_1 and M_2 both have significant effect on the convergence speed. However, the nominal control component (13) is also linked with M_1 and M_2. Therefore, it is desired that while deciding M_1 and M_2, both the convergence speed and the control input level should be monitored.

4 Simulation and Result Discussion

In this section, to verify the effectiveness of the proposed control method, simulations are conducted and the outcomes are presented with extensive discussion. The proposed controller effectiveness demonstration completes in two steps. In the first step, a small spacecraft attitude stabilization performance for the pure condition ($d(t) = 0$, $\delta J = 0, \Delta = 0$), and the practical condition ($d(t) \neq 0, \delta J \neq 0, \Delta \neq 0$) is checked. In the second step, a practical spacecraft is taken, and its attitude tracking performance is demonstrated for practical conditions. Other than the structural difference and type of control, the both simulation steps differs on certain other grounds. In stabilization example, number of actuators are four, and in tracking example three actuators are applied. Additionally, in the tracking example, both the additive fault and actuator effectiveness loss are applied.

4.1 Step 1: Spacecraft Attitude Stabilization

The considered spacecraft parameters are referred from Hu et al. (2012). The nominal component and the uncertain component of spacecraft inertia matrix are $J_o = [20\,0\,0.9;\ 0\,17\,0;\ 0.9\,0\,15]$ and $\delta J = diag[3,\ 2,\ 1][1 + e^{-0.1t} + 2\Upsilon(t-10) - 4\Upsilon(t-20)]$, respectively, where $\Upsilon(x) = 1$ for $x \geq 0$, else $\Upsilon(x) = 0$. The spacecraft attitude is controlled with four reaction wheels with torque constraint limit $|u(t)| \leq 0.2$ N-m; the distribution matrix for reaction wheels is

$$D = \begin{bmatrix} -1 & 0 & 0 & 1/\sqrt{3} \\ 0 & -1 & 0 & 1/\sqrt{3} \\ 0 & 0 & -1 & 1/\sqrt{3} \end{bmatrix}.$$

For the simulations, rigid spacecraft mathematical model discussed in (6)–(7) is used. The initial conditions of the body frame quaternion and the body frame angular velocity are taken as $q_v(0) = [0.3 \;\; -0.2 \;\; -0.3 \; 0.8832]^T$ and $\omega(0) = [0\ 0\ 0]^T$, respectively. The desired frame is characterized with $q_d(0) = [0\,0\,0\,1]^T$ and $\omega_d(0) = [0\ 0\ 0]^T$.

Further, to check the robustness against external disturbances, and to investigate the proposed controller effectiveness against actuator faults, the following mathematical model for the external disturbance and the actuator fault, respectively, are taken (Hu et al. 2012).

External disturbance:

$$d(t) = (||\omega||^2 + 0.005)[sin\ 0.8t\ cos\ 0.5t\ cos\ 0.3t]^T\, N - m \tag{33}$$

Actuator effectiveness:

$$\Delta_1 = \begin{cases} 1, & \text{if } t \le 2.4\,\text{s.} \\ 0.45 + 0.15\ rand(t_i) + 0.1 sin(0.5t + \pi/3), & \text{if } t > 2.4\,\text{s.} \end{cases}$$

$$\Delta_2 = \begin{cases} 1, & \text{if } t \le 5.0\,\text{s.} \\ 0.50 + 0.15\ rand(t_i) + 0.1 sin(0.5t + 2\pi/3), & \text{if } t > 5.0\,\text{s.} \end{cases}$$

$$\Delta_3 = \begin{cases} 1, & \text{if } t \le 10.0\,\text{s.} \\ 0.40 + 0.15\ rand(t_i) + 0.1 sin(0.5t + \pi), & \text{if } t > 10.0\,\text{s.} \end{cases}$$

$$\Delta_4 = \begin{cases} 1, & \text{if } t \le 15.0\,\text{s.} \\ 0, & \text{if } t > 15.0\,\text{s.} \end{cases} \tag{34}$$

The controller settings used for the stabilization are mentioned in Table 1.

4.1.1 Interpretation and Discussion with Comparative Comments on Stabilization Performance

Case 1: Stabilization performance for pure condition

For the pure condition, simulation results are shown in Figs. 1, 2 and 3. The top and bottom frames of Fig. 1a illustrate that the error quaternion and the angular velocity error vectors, respectively, reduces to zero in finite time. Figure 1b illustrates the steady precision of the error quaternion and angular velocity error vectors. Figure 2 depicts the time evolution of the NSFTSM vector and the control input; in which the bottom frame of Fig. 2a illustrates that the control input is continuous. The steady

Table 1 Controller parameters

Step 1	Attitude stabilization	$M_1 =$ $diag(0.080,\ 0.080,\ 0.080)$ $\alpha = 1.1$ $k_1(0) = 1.5$ $\eta = 30$	$M_2 =$ $diag(0.152,\ 0.152,\ 0.152)$ $\gamma = 0.47$ $k_2(0) = 0.20$ $\phi = 25$
Step 2	Attitude tracking	$M_1 =$ $diag(0.07,\ 0.07,\ 0.07)$ $\alpha = 1.1$ $k_1(0) = 0.80$ $\eta = 60$	$M_2 =$ $diag(2.90,\ 2.90,\ 2.90)$ $\gamma = 0.49$ $k_2(0) = 0.02$ $\phi = 56$

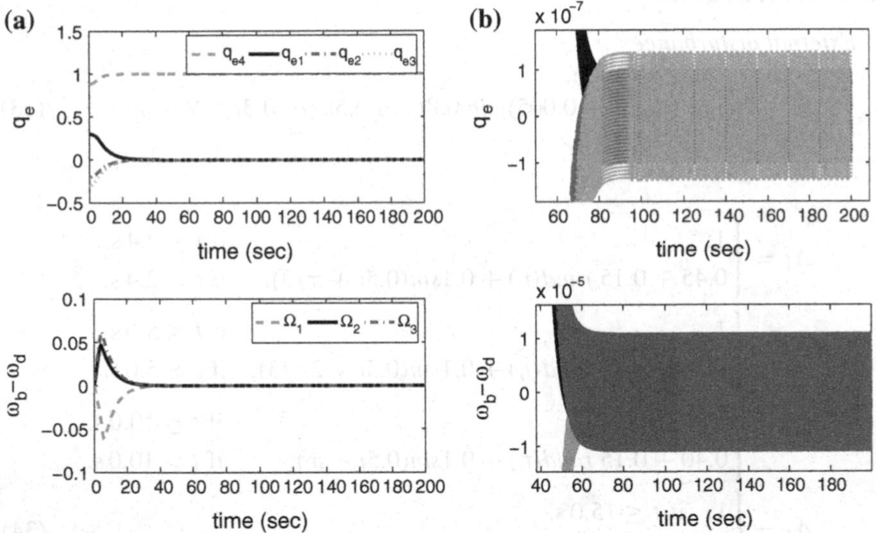

Fig. 1 Stabilization performance Case 1: quaternion and angular velocity pattern under pure condition. **a** Quaternion and angular velocity. **b** Steady precision

precision of NSFTSM vector and control input are illustrated in Fig. 2b. Figure 3 illustrates the estimates of control gains.

As illustrated in Figs. 1 and 2 and mentioned in Table 2, the proposed controller needs time 5.8 s to guarantee that the NSFTSM vector entered into the band $|\sigma_e| \leq 5e-3$; and further, total time 18.15 s is required to satisfy the condition $(|q_{ev}|, |\Omega|) \leq 2e-2$. Additionally, it is revealed that the steady precision for the error quaternion vector, the angular velocity error vector, and the NSFTSM vector are ensured in the range $|q_{ev}| \leq 1.35e-7$, $|\Omega| \leq 1.07e-5$, $|\sigma_e| < 3.41e-6$, respectively.

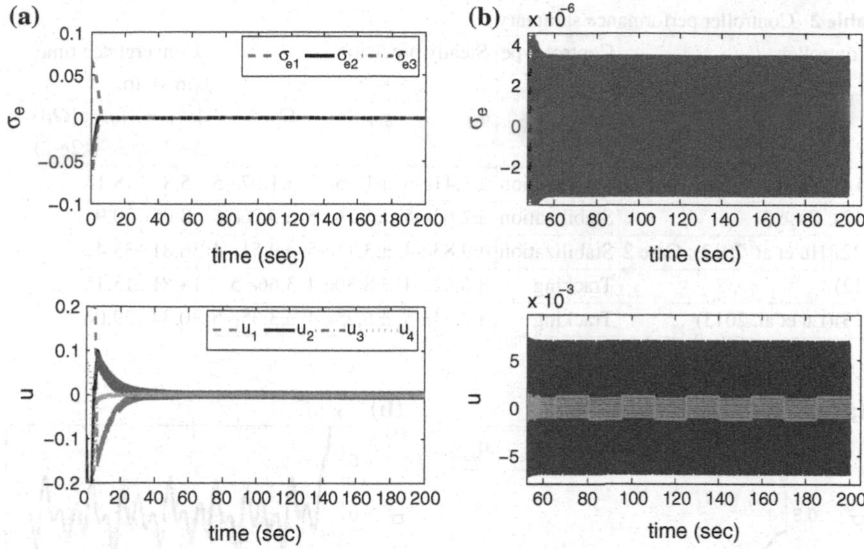

Fig. 2 Stabilization performance Case 1: sliding vectors and control input pattern under pure condition. **a** Sliding vectors and control input. **b** Steady precision

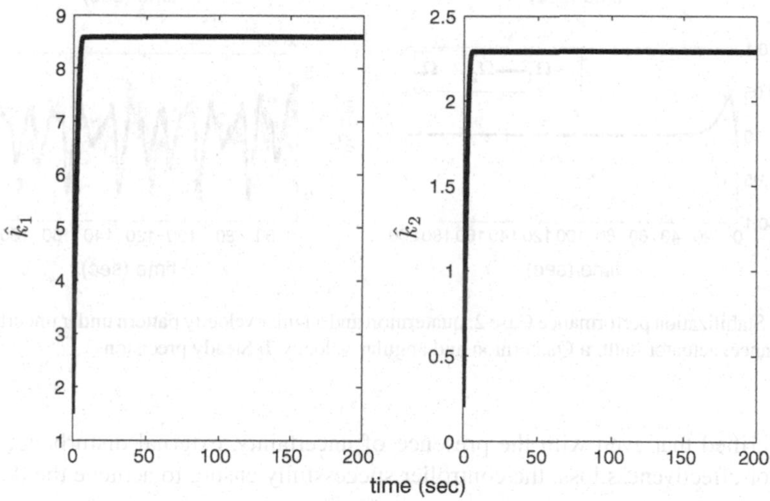

Fig. 3 Stabilization performance Case 1: estimate of control gains

Case 2: Stabilization performance for practical condition

Further, the spacecraft stabilization performance is verified with inertia uncertainty, external disturbance (33), and loss of actuator effectiveness (34). The simulation results are illustrated in Figs. 4 and 6. In Fig. 4a, the finite time convergence of the quaternion and the angular velocity to the desired level, respectively, are portrayed.

Table 2 Controller performance summary

Controller	Control type	Steady precision			Convergence time (in s) for	
		σ_e	q_{ev}	Ω	$\lvert\sigma_e\rvert <$ 5e-3	$(\lvert q_{ei}\rvert, \lvert\Omega_i\rvert) <$ (2e-2,2e-2)
(12): Case 1	Stabilization	±3.41*e*-6	±1.35*e*-7	±1.07*e*-5	5.8	18.15
(12): Case 2	Stabilization	±2.62*e*-5	±6.26*e*-5	±5.26*e*-5	8.05	20.9
(22)(Hu et al. 2012): Case 2	Stabilization	±9.83*e*-4	±3.11*e*-5	±9.51*e*-4	36.81	35.42
(12)	Tracking	±6.62*e*-4	±8.80*e*-4	3.66*e*-5	14.81	13.15
(13)(Lu et al. 2013)	Tracking	±2.74*e*-5	±6.25*e*-4	±3.35*e*-5	30.34	29.06

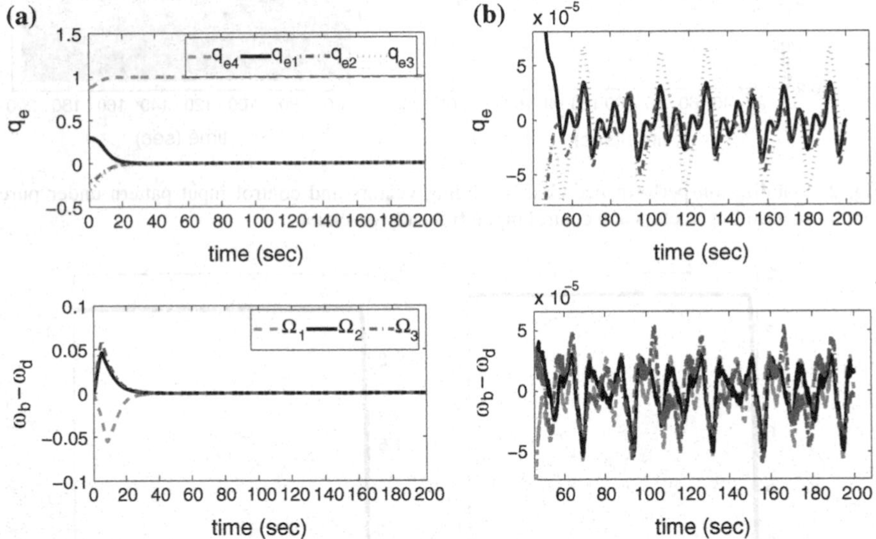

Fig. 4 Stabilization performance Case 2: quaternion and angular velocity pattern under uncertainty, disturbance, actuator fault. **a** Quaternion and angular velocity. **b** Steady precision

It is verified that even with the presence of uncertainty, external disturbance, and actuator effectiveness loss, the controller successfully ensure to achieve the desired objective.

As mentioned in Table 2, and illustrated in the top and the bottom frame of Fig. 4, $(|q_{ev}|, |\Omega|) \leq 2e - 2$ is achieved in 20.9 s. Figure 5 top frame exhibits the time evolution of NSFTSM vector, and it is verifiable that in time 8.05 s, $|\sigma_e| \leq 5e - 3$ is established. Figure 5 bottom frame illustrates that the control input pattern is within the imposed limitation $|u| \leq 0.2$ N-m; and no sign of chattering is appeared. The steady precision performance is shown in Figs. 4b and 5b. Figure 6 illustrates the estimates of control gains; as expected, the estimate of control gains are attained the higher values than their counterparts of pure condition.

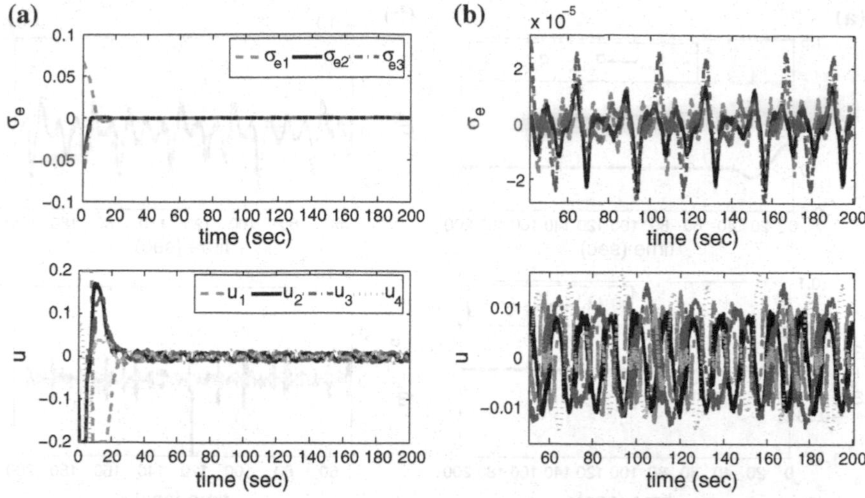

Fig. 5 Stabilization performance Case 2: sliding vectors and control input pattern under uncertainty, disturbance, actuator fault. **a** Sliding vectors and control input. **b** Steady precision

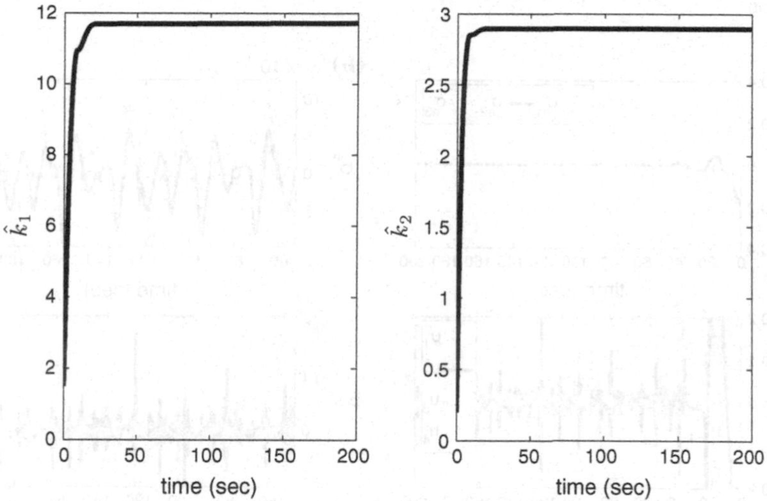

Fig. 6 Stabilization performance Case 2: estimate of control gains

Interestingly, in the pure conditions' control input patterns, the fourth reaction wheel output always satisfy $|u| < 0.2\,\text{N-m}$, and attain never to the maximum limit ($|u| = 0.2\,\text{N-m}$). In contrary, for the practical case, the all four reaction wheels output need to attain to the maximum limit. Additionally, in the practical case, the reaction wheels maximum control output is to be required to apply for the longer duration.

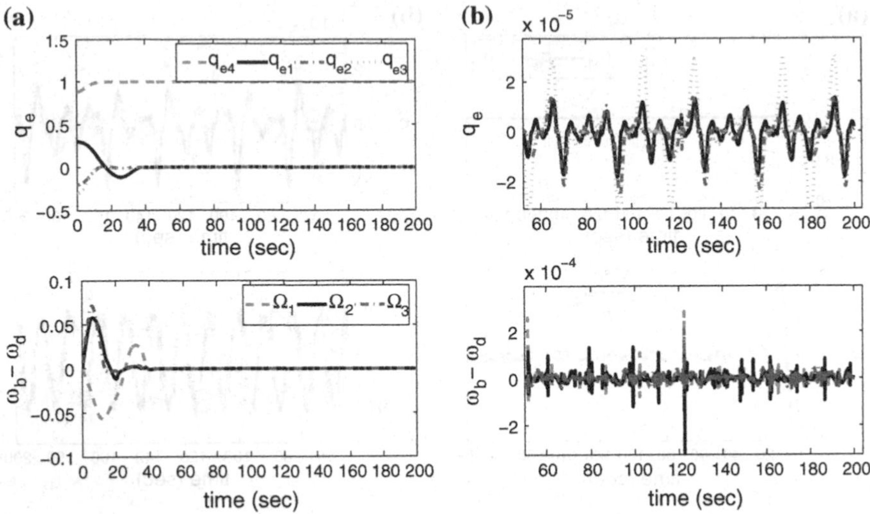

Fig. 7 Controller (22) (Hu et al. 2012) stabilization performance: quaternion and angular velocity pattern under uncertainty, disturbance, actuator fault. **a** Quaternion and angular velocity. **b** Steady precision

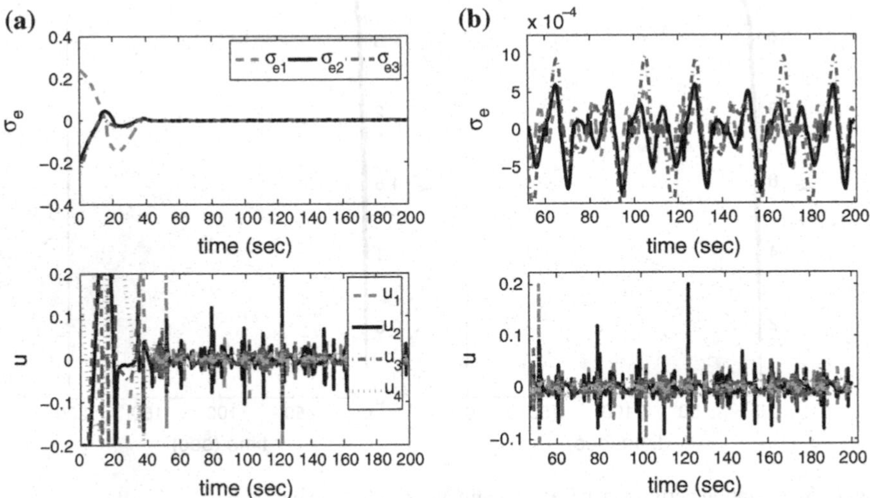

Fig. 8 Controller (22) (Hu et al. 2012) stabilization performance: sliding vectors and control input pattern under uncertainty, disturbance, actuator fault. **a** Sliding vectors and control input. **b** Steady precision

For the comparative analysis, the proposed controller stabilization performance is compared with the controller (22) (Hu et al. 2012). Simulation results to the similar condition are shown in Figs. 7 and 8. The salient points of the proposed controller are the non-singularity, faster convergence speed, and the adaptive law estimated gains. Particularly, the proposed controller ensured to satisfy the criterion $|\sigma_e| \leq 5e\text{-}3$ and $(|q_{ei}|, |\Omega_i|) < (2e\text{-}2, 2e\text{-}2)$ for $i = 1, 2, 3$ in time 8.05 and 20.9 s. respectively; but the controller (22) (Hu et al. 2012) takes 36.81 and 35.42 s. respectively, to satisfy the similar criterion. The steady precision level are almost same for both controller, but the the proposed controller is more quicker in fault tolerant, and it doesn't demand the uncertainty or disturbance bounds to decide the controller's gains.

4.2 Step 2: Spacecraft Attitude Tracking

Further, to examine the tracking performance of controller (12), simulations are conducted on the spacecraft model mentioned in Lu et al. (2013). The nominal component and the uncertain component of inertia matrix are $J_o = [800.027\ 0\ 0;\ 0\ 839.93\ 0;\ 0\ 0\ 289.93]$ and $\delta J = diag[100,\ 100,\ 50]$, respectively. In contrast to stabilization Step, in tracking case the spacecraft attitude is controlled with three reaction wheels only, and the wheels are constrained with torque limit $u(t) = 30$ N-m. The initial conditions of the body frame quaternion and angular velocity are the same as it taken in Step 1 simulations. The initial conditions for the desired frame quaternion and angular velocity are $q_d(0) = [0\ 0\ 0\ 1]^T$ and $\omega_d(t) = 0.05[sin(\pi t \backslash 100)\ sin(2\pi t \backslash 100)\ sin(3\pi t \backslash 100)]^T$, respectively. In this simulation step, in addition to the actuator effectiveness loss the additive fault possibility is also included, and therefore the dynamics of rigid spacecraft is modified to

$$J\dot{\omega}_e = (-\omega_e + C\omega_d)^{\times} J(\omega_e + C\omega_d) + J(\omega_e^{\times} C\omega_d - C\dot{\omega}_d) + (D\,\Delta(t)\,u(t) + E(t)) + d(t). \tag{35}$$

where $E(t) = [e_1\ \ e_2\ \ e_3]^T$ is the additive fault.

The mathematical model considered for the external disturbance, the actuator effectiveness loss, and the additive fault are as follows

Disturbance:

$$d(t) = [0.1\ sin\ 0.1t\ 0.2\ sin\ 0.2t\ 0.3\ sin\ 0.3t]^T \tag{36}$$

Actuator Effectiveness:

$$\Delta_1 = \begin{cases} 1, & \text{if } t \leq 10\,\text{s.} \\ 0.75 + 0.1 sin(0.5t + \pi/3), & \text{if } t > 10\,\text{s.} \end{cases}$$

$$\Delta_2 = \begin{cases} 1, & \text{if } t \leq 10\,\text{s.} \\ 0.75 + 0.1 sin(0.5t + 2\pi/3), & \text{if } t > 10\,\text{s.} \end{cases}$$

$$\Delta_3 = \begin{cases} 1, & \text{if } t \leq 10\,\text{s.} \\ 0.75 + 0.1 sin(0.5t + \pi), & \text{if } t > 10\,\text{s.} \end{cases} \quad (37)$$

Additive fault:

$$e_i = \begin{cases} 0, & \text{if } t < 15\,\text{s.} \\ 0.1 + 0.05 sin(0.5\pi t), & \text{if } t \geq 15\,\text{s.} \\ \text{for } i = 1, 2, 3 \end{cases}$$

4.2.1 Interpretation and Discussion with Comparative Comments on Tracking Performance

By applying the afore-mentioned external disturbance, inertia uncertainty, actuator effectiveness loss, and additive fault, the simulation results for the attitude tracking are shown in Figs. 9 10, 11, and 12.

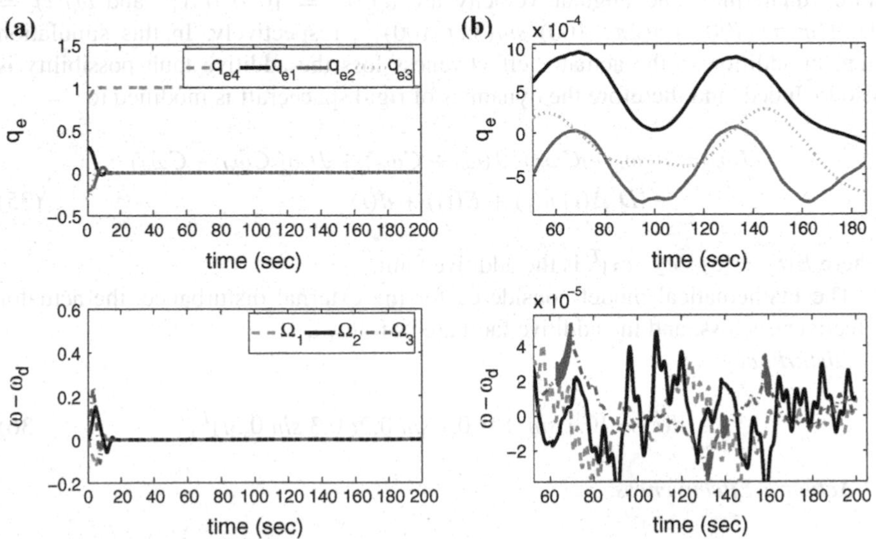

Fig. 9 Tracking performance: quaternion and angular velocity pattern under uncertainty, disturbance, actuator fault. **a** Quaternion and angular velocity. **b** Steady precision

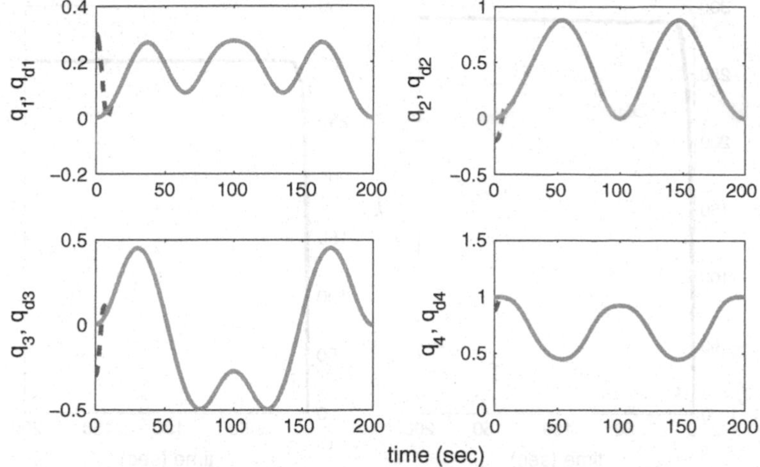

Fig. 10 Tracking performance: quaternion tracking pattern

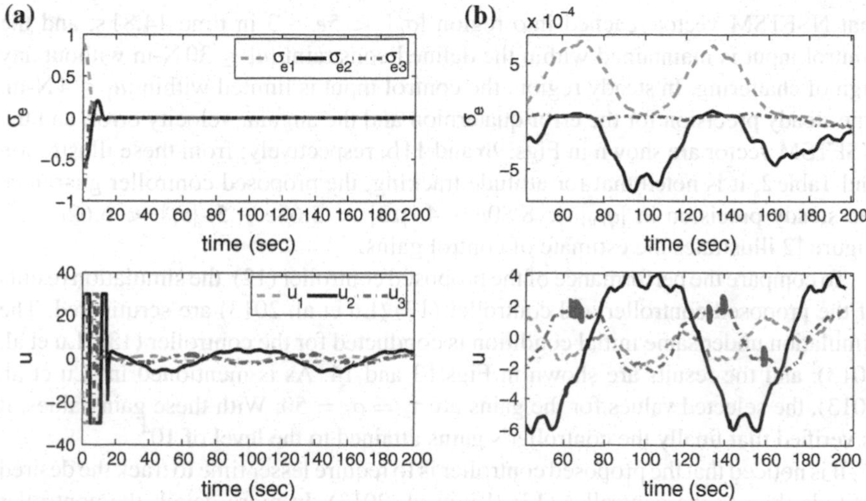

Fig. 11 Tracking performance: sliding vectors and control input pattern under uncertainty, disturbance, actuator fault. **a** Sliding vectors and control input. **b** Steady precision

The error quaternion and the angular velocity tracking errors are portrayed in Fig. 9, it illustrates that the controller is successfully negates the odd effects, and ensures the tracking performance in finite time. Additionally, as mentioned in Table 2, and illustrated in the top and the bottom frame of Fig. 9a, $(|q_{ev}|, |\Omega|) \leq 2e-2$ is achieved in 13.15 s. For the better lucidity, the quaternion tracking pattern is also shown in Fig. 10, and this also approves the attitude tracking performance. The NSFTSM and the control input time evolution are depicted in Fig. 11, it illustrates

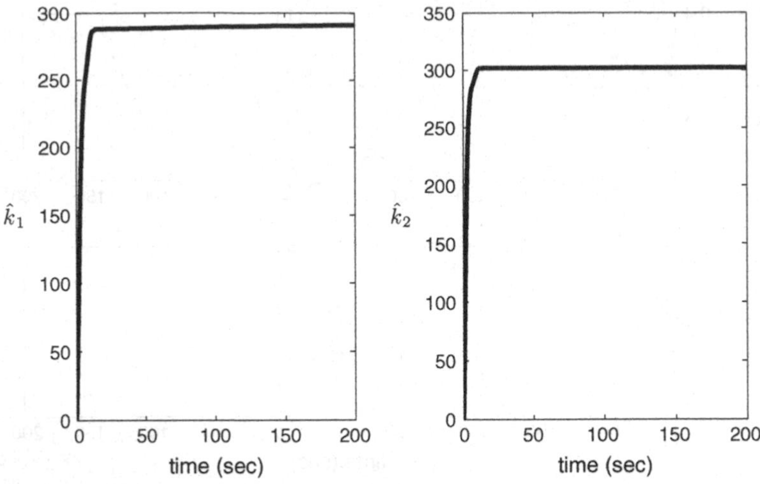

Fig. 12 Tracking performance: estimate of control gains

that NSFTSM vector reached into region $|\sigma_e| < 5e - 3$ in time 14.81 s, and the control input is maintained within the defined constraint $|u| \leq 30$ N-m without any sign of chattering. In steady region, the control input is limited within $|u| \leq 4$ N-m. The steady precision for the error quaternion and the angular velocity error, and the NSFTSM vector are shown in Figs. 9b and 11b, respectively; from these illustration and Table 2, it is noted that for attitude tracking, the proposed controller guarantee the steady precision in $|q_{ev}| < 8.80e - 4$, $|\Omega| < 3.66e - 5$, $|\sigma_e| < 6.62e - 5$. Figure 12 illustrates the estimate of control gains.

To compare the performance of the proposed controller (12), the simulation results of the proposed controller and controller (13) (Lu et al. 2013) are scrutinized. The simulation under same initial condition is conducted for the controller (13) (Lu et al. 2013); and the results are shown in Figs.13 and 14. As is mentioned in (Lu et al. 2013), the selected values for the gains are $\tau_i = \sigma_i = 50$. With these gain values, it is verified that finally the controller's gains attained to the level of 10^4.

It is noticed that the proposed controller is to require lesser time to track the desired attitude than to the controller (13) (Lu et al. 2013). In more detail, the controller (13) (Lu et al. 2013) took 30.34 and 29.06 s. to satisfy the criterion $|\sigma_e| \leq 5e\text{-}3$ and $(|q_{ei}|, |\Omega_i|) < (2e\text{-}2, 2e\text{-}2)$ for $i = 1, 2, 3$, respectively; in contrary, to satisfy the same criterion, the proposed controller (12) is demanded 14.81 and 13.15 s, respectively. Though, the steady precision for error quaternion and angular velocity error is slightly lower than to the controller (13) (Lu et al. 2013)(*Refer to Table* 2), yet it is acceptable and comes in high precision range. Additionally, in contrary to the controller (13) (Lu et al. 2013), the proposed controller's gains are not selected on any conservative approach, and in fact are being estimated with the proposed adaptive law; hence, even if any unwanted and unaccounted external disturbance and uncertainty surfaced, then also the proposed controller is equipped with adaptive laws to overcome its effect.

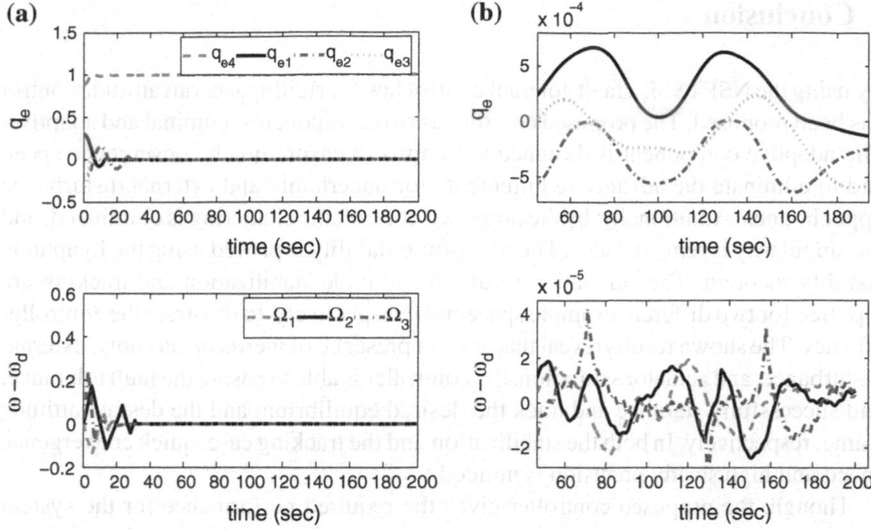

Fig. 13 Controller (13) (Lu et al. 2013) tracking performance: quaternion and angular velocity pattern under uncertainty, disturbance, actuator fault. **a** Quaternion and angular velocity. **b** Steady precision

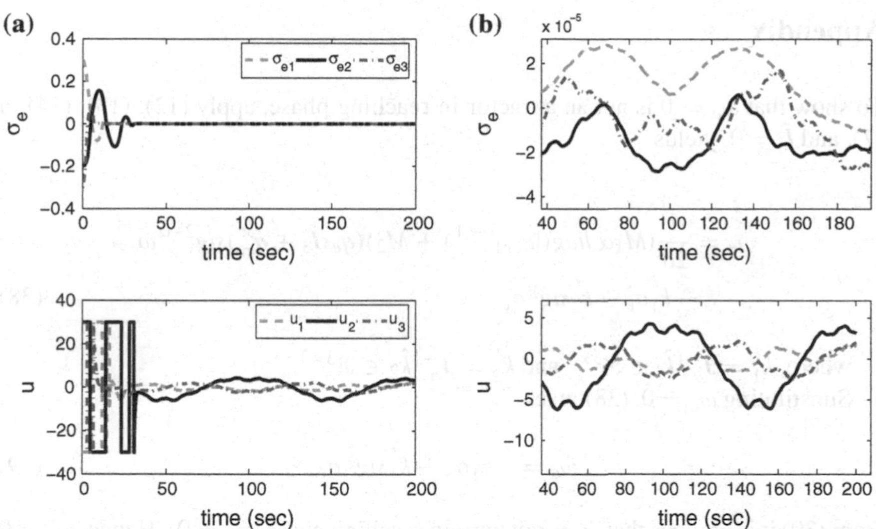

Fig. 14 Controller (13) (Lu et al. 2013) tracking performance: sliding vectors and control input pattern under uncertainty, disturbance, actuator fault. **a** Sliding vectors and control input. **b** Steady precision

5 Conclusion

By using the NSFTSM, a fault-tolerant control law for rigid spacecraft attitude control has been proposed. The proposed control has two components, nominal and adaptive. The adaptive component is designed with aims to ensure quick convergence speed and to eliminate the advance requirements for uncertainty and external disturbance upper bounds. Additionally, by the proposed control, the chattering is eliminated, and the singularity is removed also. The finite-time stability is proved using the Lyapunov stability theorem. The simulation results for attitude stabilization and tracking are reported for two different example spacecraft, respectively, to illustrate the controller efficacy. The shown results reveal that even in presence of inertia uncertainty, external disturbance, and actuator saturation, the controller is able to ensure the fault tolerance, and successfully stabilize and track the desired equilibrium and the desired attitude frame, respectively. In both the stabilization and the tracking case, quick convergence speed and high steady precision is noticed.

Though, the proposed controller gives the required performance for the system considered; in the future control design, to give the most practical solution, actuator dynamics and spacecraft structure flexibility may be include to.

Appendix

To show that $\omega_e = 0$ is not an attractor in reaching phase, apply (12), (13), (14) in (7), and $\bar{L} = 0$, yields

$$\begin{aligned}\dot{\omega}_e = &\frac{1}{2\alpha}(M_1 \alpha diag(|q_{ev}|^{\alpha-1}) + M_2)(q_{e4}I_3 + q_{ev}^{\times})sig^{2-\alpha}\omega_e \\ &-k_1^{'}\sigma_e - k_2^{'}sig^{\gamma}\sigma_e\end{aligned} \quad (38)$$

where, $k_1^{'} = J_o^{-1}\hat{k}_1 \in \Re^{3\times 3}$ and $k_2^{'} = J_o^{-1}\hat{k}_2 \in \Re^{3\times 3}$.
Substituting $\omega_e = 0$, (38) gives

$$\dot{\omega}_e = -k_1^{'}\sigma_e - k_2^{'}sig^{\gamma}\sigma_e, \quad (39)$$

from (39) it is obvious that $\dot{\omega}_e$ is not zero in reaching phase ($\sigma_e \neq 0$). Hence, $\omega_e = 0$ is not an attractor in reaching phase.

References

Bustan, D., Sani, S.K.H., Pariz, N.: Adaptive fault-tolerant spacecraft attitude control design with transient response Control. IEEE/ASME Trans. Mechatron. **19**(4), 1404–1411 (2013)

Ding, S., Li, S.: Stabilization of the attitude of a rigid spacecraft with external disturbances using finite-time control techniques. Aerosp. Sci. Technol. **13**(4–5), 256–265 (2009)

Erdong, J., Zhaowei, S.: Robust controllers design with finite time convergence for rigid spacecraft attitude tracking control. Aerosp. Sci. Technol. **12**(4), 324–330 (2008)

Feng, Y., Yu, X.H., Man, Z.: Non-singular terminal sliding mode control of rigid manipulator. Automatica **38**(12), 2159–2167 (2002)

Haimo, V.T.: Finite time controllers. SIAM J. Control Optim **24**(4), 760–770 (1986)

Hung, J.Y., Gao, W., Hung, J.C.: Variable structure control: a survey. IEEE. Trans. Ind. Electron. **40**(1), 1–12 (1993)

Hu, Q., Huo, X., Xiao, B., Zhang, Z.: Robust finite-time control for spacecraft attitude stabilization under actuator fault. Proc. Inst. Mech. Eng. Part I: J. Syst. Control Eng. **226**(3), 416–428 (2012)

Hu, Q., Xing, Huo, Xiao, B.: Reaction wheel fault tolerant control for spacecraft attitude stabilization with finite time convergence. Int. J. Robust Nonlinear Control **23**(15), 1737–1752 (2012)

Hu, Q., Li, B., Zhang, Aihua: Robust finite-time control allocation in spacecraft attitude stabilization under actuator misalignment. Nonlinear Dyn. **73**(1–2), 53–71 (2013)

Li, S., Wang, Z., Fei, S.: Comments on paper: Robust controllers design with finite time convergence for rigid spacecraft attitude tracking control. Aerosp. Sci. Technol. **15**(3), 193–195 (2011)

Lu, K.,Xia, Y.: Finite-time attitude stabilization for rigid spacecraft. Intern. J. Robust Nonlinear Control (2013). doi:10.1002/rnc.3071

Lu, K., Xia, Y., Fu, M.: Controller design for rigid spacecraft attitude tracking with actuator saturation. Inf. Sci. **220**, 343–366 (2013)

Lu, Kunfeng, Xia, Y., Fu, M.: Finite-time fault-tolerant control for rigid spacecraft with actuator saturations. IET Control Theory Appl. **7**(11), 1529–1539 (2013)

Man, Z., Yu, X.H.: Terminal sliding mode control of MIMO linear systems. IEEE. Trans. on Circuits Syst. **44**(11), 1065–1070 (1997)

Tang, Y.: Terminal sliding mode control of rigid robots. Automatica **34**(1), 51–56 (1998)

Tiwari, P.M., Janardhanan, S., Nabi, M.: A finite time convergent continuous time sliding mode controller for spacecraft attitude control. The 2010 IEEE International Workshop on Variable Structure Systems, 26–28 June 2010, Mexico City, pp. 399–403 (2010). doi:10.1109/VSS.2010.5544630

Tiwari, P.M., Janardhanan, S., Nabi, M.: Spacecraft attitude control using non-singular finite time convergence fast terminal sliding mode. Intern. J. Instrum. Technol. **1**(2), 124–142 (2012)

Tiwari, P.M., Janardhanan, S., Nabi, M.: Rigid spacecraft attitude tracking using finite time sliding mode control. In: The 2014 International Conference on Advances in Control and Optimization of Dynamical Systems, 13–15 March 2014, India, pp. 263–270, (2014). doi:10.3182/20140313-3-IN-3024.00168

Utkin, V.I.: Variable structure systems with sliding modes. IEEE Trans. Autom. Control **22**(2), 212–222 (1977)

Venkataraman, S.T., Gulati, S.: Terminal sliding modes: A new approach. The 1991 International Conference on Advanced Robotics, 19–22 June 1991, Italy, pp. 443–448, (1991). doi:10.1109/ICAR.1991.240613

Vadali, S.R.: Variable-structure control of spacecraft large-Angle Maneuvers. J. Guidance **9**(2), 235–239 (1986)

Wertz, W.: Spacecraft Attitude Determination and Control. In: J. Wertz (ed.), Academic Publishers, New York (1978)

Xiao, B., Hu, Q., Wang, D., Poh, E.K.: Attitude tracking control of rigid spacecrafts with actuator misalignment and fault. IEEE Trans. Control System Technol. **21**(6), 2360–2366 (2013)

Yeh, F.K.: Sliding-mode adaptive attitude controller design for spacecrafts with thrusters. IET Control Theory Appl. **4**(7), 1254–1264 (2010)

Yu, X.H., Man, Z.: On finite time mechanism: Terminal sliding modes. In: The 1996 IEEE International Workshop on Variable Structure Systems, 5–6 Dec 1996, Tokyo, pp. 164–167, (1996). doi:10.1109/VSS.1996.578596

Yu, X.H., Man, Z.: Fast terminal sliding mode control for nonlinear dynamical systems. IEEE. Trans. Circuits Syst. I: Fundam. Theory Appl. **49**(2), 261–264 (2002)

Yu, S., Yu, X.H., Shirinzadeh, B., Man, Z.: Continuous finite-time control for Robotic manipulator with terminal sliding mode. Automatica **41**(11), 1957–1964 (2005)

Yang, L., Yang, J.: Nonsingular fast terminal sliding mode control for nonlinear dynamical systems. Intern. J. Robust Nonlinear Control **21**(16), 1865–1879 (2011)

Zou, A.-M., Kumar, K.D.: Finite-time attitude tracking control for spacecraft using terminal sliding mode and chebyshev neural network. IEEE. Trans. Syst. Man Cybern. **41**(4), 950–963 (2011)

Zhang, A., Hu, Q., Friswell, M.: Finite-time fault tolerant attitude control for over-activated spacecraft subject to actuator misalignment and faults. IET Control Theory Appl. **7**(16), 2007–2020 (2013)

Sliding Modes for Fault Tolerant Control

Hemza Mekki, Djamel Boukhetala and Ahmad Taher Azar

Abstract As modern technological systems increase in complexity, their corresponding control systems become more and more sophisticated. In order to increase the reliability, which is crucial topic in industrial applications. The main focus of this chapter will be on the design of fault tolerant control (FTC) strategy. Therefore, FTC has found extensive applications in multiple domains including mechanical engineering, electrical engineering, control engineering, biomedical engineering, and micro-engineering. This chapter gives a brief overview in the field of FTC (definitions, practical requirements and classification). On the other hand, give a brief introduction to the concept of sliding mode control and examine its properties. Sliding surface design and tracking requirements are also discussed. In many ways, this chapter demonstrates the true theoretical and applications depth to which the sliding mode control paradigm has been developed today in the fields of FTC. Also, highlights the benefits and give discussions of some FTC approaches based SMC. At the end, in order to introduce the concept and to prove the effectiveness of the proposed approach a permanent magnet synchronous motor (PMSM) systems case study will be presented.

H. Mekki (✉)
Department of Electrical Engineering, University of M'sila, M'sila, Algeria
e-mail: hamza.mekki@g.enp.edu.dz

D. Boukhetala
Process Control Laboratory, Department of Automatic,
Ecole Nationale Polytechnique (ENP), Algiers, Algeria
e-mail: djamel.boukhetala@g.enp.edu.dz

A.T. Azar
Faculty of Computers and Information, Department of Scientific Computing,
Benha University, Benha, Egypt
e-mail: ahmad_t_azar@ieee.org

A.T. Azar and Q. Zhu (eds.), *Advances and Applications in Sliding Mode Control systems*,
Studies in Computational Intelligence 576, DOI 10.1007/978-3-319-11173-5_15

1 Introduction

To meet the market requirements in terms of functionality, cost and flexibility, automated systems have long pushed to ever-increasing modularity, which passes including reuse many components developed independently. This high degree of reuse leads to increasingly complex architectures incorporating heterogeneous elements in multifaceted systems, which increases the risk of a fault.

Currently, when a faults occurs in the level of complex systems such as nuclear central, aircraft system and railway system. The most research into reconfigurable systems control focuses on faults detection and isolation (FDI) where the fundamental purpose of a FDI scheme is to generate an alarm when a fault occurs and to pin-point the source (Patton et al. 1989). The FDI is very important phase but it's not sufficient to ensure safe operation and guarantee the good performance. Where, it is important for these complex systems to be kept stable with an acceptable closed loop control performance. Ideally, in these applications where continuity of operation is a key feature, the closed loop system should be capable of maintaining its pre-specified performance in terms of quality, safety, and stability despite the presence of faults (Patton 1997). This procedure is rendered possible thanks to the fault tolerant control.

The common Fault tolerant control definition can be found in Zhang and Jiang (2008) where authors states: ...FTCS are control systems which possess the ability to accommodate component failures automatically. They are capable of maintaining overall system stability and acceptable performance in the event of such failures. In other words, a closed-loop control system which can tolerate component malfunctions, while maintaining desirable performance and stability properties is said to be a fault-tolerant control system...

In the last decade where the automation has become more complex, fault tolerance has become an increasingly interesting topic. Fault tolerance is no longer limited to high-end systems but also to railway (Bennett et al. 1999), and automobile applications (Benbouzid et al. 2007). While Jones (2005) gives a survey on reconfiguration methods used specifically for FTC in flight control applications. FTC becomes an important means to increase the reliability, availability, and continuous operation of electromechanical systems among the automotive ones.

In general, the FTC approaches can be classified into two types: the passive approach and the active approach. The survey papers (Patton 1997), (Prashant et al. 2013) and (Zhang and Jiang 2008) review the concepts and the state of the art in the field of FTCs. comparative study between these two approaches have been reported in Jiang and Xiang (2012). Then, more discussion of these FTC strategies is provided in the following next chapter subsections.

In high performance nonlinear systems, a conventional state feedback such as the proportional and integral controller (PI) can be very limited and cause the system to unwanted behaviors or instability. To overcome such drawbacks, the method presented in this chapter, uses Sliding Mode Control (SMC) to develop a robust controller which adaptively handles input magnitude and rate constraints. We first give an overview of SMC before going through the particulars of its use in reconfiguration.

Sliding mode control (SMC) theory was introduced for the first time the context of the variable structure system (VSS). It was developed for electrome-chanical systems by Utkin see Utkin et al. (1999). It has become so popular that now it represents this class of control systems. Even through, in its early stage of development, the SMC theory was over-looked because of the development in the famous linear control theory, during the last 20 years it has shown to be a very effective control method (Edwards and Spurgeon 1998). Various books on SMC have also been published recently Utkin (1992), Edwards and Spurgeon (1998), Fridman et al. (2011).

During the last two decades, new techniques for fault tolerant control (FTC) have been developed for the specific purpose of maintaining relatively system performance and stability. Development of FTC systems has also provided us a systematic way to design and implement various control methods; the most FTC implemented control methods based sliding mode control. Some application, of SMC in the field of FTC can be found in these papers Benbouzid et al. (2007), Alwi et al. (2011), Alwi and Edwards (2008), Fekih (2008), Mekki et al. (2013).

In this chapter in order to facilitate quantitative assessment and to prove the effectiveness of the proposed approach, a permanent magnet synchronous motor (PMSM) system is selected as case study. Nowadays PMSM drive is widely used in the industry applications due to their high efficiency and high power/torque density (Teng et al. 2012). These motors are used in many applications such as traction with variable speeds in transportation as presented in Erginer and Sarul (2013). In order to introduce the concept and to prove the effectiveness of the proposed approach a PMS motor systems case study will be presented.

In this chapter, an introduction to fault tolerant control (FTC) and sliding modes approach will be presented. The chapter will start with some definitions of faults and failures which can occur in systems and describe some practical requirements of FTC systems. Then, classification of different of FTC types will be presented and discussed. Later, the sliding modes approach details will be given as well as a brief overview. Also, some discussions on the benefits and motivation for sliding mode techniques in the fields of FTC will be presented. At the end, this chapter wills give a survey on sliding modes methods used specifically for FTC in PMSM systems case study, in order to prove the high precision of the proposed control techniques in healthy and during the faulty condition, some simulation results will be presented and discussed.

2 Fault Tolerant Control Methods

This section will discussion and present a brief fault tolerant control (FTC) strategies overview (definitions, requirements and classifications). In order to develop this area further the terms fault and failure need to be defined in the context of uncertain systems. This will also enable the concept of fault tolerant control (FTC) to be specified in terms of faults and/or failures later in this chapter.

2.1 Faults and Failures

Compressive area about faults and failures topics and more detail can be found in Alwi et al. (2011) book. The faults and failures definitions provided in Alwi et al., book is in compliance with the definitions given by the IFAC SAFEPROCESS technical committee as given in Isermann and Balle (1997). Where the IFAC technical committee, makes the following definitions:

Fault: *an unpermitted deviation of at least one characteristic property or parameter of the system from the acceptable/usual/standard condition.*

Failure: *a permanent interruption of a system's ability to perform a required function under specified operating conditions.*

Faults in the components of controlled systems may lead to total system failure, depending on the precise conditions, the criticality of the fault, etc. and if appropriate action is not taken (Isermann and Balle 1997).

On the other hand, a Failure describes the condition when the system is no longer performing the required function i.e. the system function involving the faulty component may have failed (Klinkhieo 2009).

Clearly, a failure is a condition which is much more severe than a fault. When a fault occurs in an actuator for example, the actuator is still usable but may have a slower response or become less effective. But when a failure occurs, a totally different actuator is needed to be able to produce the desired effect (Alwi et al. 2011).

2.2 Practical Requirements for FTC

Due to the large and rapid development experienced by the industrial world, the manufacturing processes are becoming more complex and sophisticated. Therefore, the increase in reliability, availability and dependability, is at present, one of the major concerns of industry.

In several complex systems; such as nuclear central (see Fig. 1) and/or aircraft systems (see Fig. 2). The faults detection and reconstruction phase is needed but will not sufficient to ensure safe operation and guarantee the good performance. When the fault appear in these systems it's essential to change the control law in real time to maintain stability of the system and ensure an acceptable operating in degraded mode. Thus, it is necessary to associate to FDD a fault tolerant control unit. Figures 1 and 2 shows some examples that faults can lead to serious accidents also show the importance of fault information and fault tolerant control.

In September 2004 the Group for Aeronautical Research and Technology in Europe (GARTEUR) launch a project a Flight Mechanics Action Group FM-AG16: Fault Tolerant Control. The project has started from and is expected to be finished in September 2007. The AG16 group developed further research on fault-tolerant flight control and demonstrated the value of using FTC methods to reduce the probability

Fig. 1 Fukushima Daiichi nuclear accident

of accident. The goal was to apply a number of FDD and FTC algorithms within a realistic failure scenario. Currently, the project has been participated by the several leading universities in Europe (such as Cambridge University, Leicester University, Hull University and Brunel University in UK; Delft University of Technology in the Netherlands; University of Lille 1 and University of Bordeaux in France),

Figure 1 illustrates the Fukushima Daiichi nuclear accident on March 11, 2012. This figure represents an aerial view of Fukushima Daiichi nuclear central before and one year after disaster (www.maxisciences.com). Evacuation order was issued for an initial range of 3 Km from the periphery of the reactor and included on the 5,800 people living within this range. He also advised people living within 10 Km of the plant to remain in their homes. Later, the evacuation order covered all of the population within 10 km. Today, it is completely useless and should be dismantled.

Figure 2 illustrates Hudson River flight accident on January 15, 2009. The downed US Airways 1549 floating on the Hudson River in the west side of Manhattan, The flight lasted 5 min 8 s after takeoff. No loss of life in this accident (www.celebitchy.com).

Fig. 2 Hudson River flight accident

2.3 Definitions of FTC Systems

Fault tolerance is an issue that has been addressed by many authors. During the last two decades there has been a substantial literature on the subject of FTC according to reviews, survey papers and books (e.g. Patton 1997; Blanke et al. 2001; Zhang and Jiang 2008; Noura et al. 2009; Alwi et al. 2011; Prashant et al. 2013), which give the state of the art and perspectives in the field of control reconfiguration in FTC. As discussed above, approaches to FTC are motivated only by a particular application. For example, safety in flight control, efficiency and quality improvements in industrial processes, etc.

Patton in Patton (1997) stated in his survey that: *...Most often, the main requirement is that the system should maintain some "acceptable" level of performance or degrade gracefully, subsequent to a malfunction. When it is proved that this can be achieved the fault-tolerant approach becomes acceptable to systems and control engineers...*

Noura Hassan in Noura et al. (2000) stated that: *...Fault-tolerant control systems are characterized here by their capabilities, alter fault occurrence, to recover per-*

formance close to the nominal desired performance. In addition, their ability to react successfully (stably) during a transient period between the fault occurrence and the performance recovery is an important feature. Accommodation capability of a control system depends on many factors such as the severity of the failure, the robustness of the nominal system, and the actuators' redundancy...

2.4 Classification of FTC Systems

Generally, in Patton (1997) and Zhang and Jiang (2008), authors classify FTCS into two major groups: active fault tolerant control systems (AFTCS) and passive fault tolerant control systems (PFTCS) as presented in Fig. 3. These two approaches use different design methodologies for the same control objective. The survey books Noura et al. (2009) and Prashant et al. (2013) review the design and practical applications of FTCS, recent advances and comparative study between these two FTCS approaches are available in Alwi et al. (2011) and Jiang and Xiang (2012). A Table presents a brief comparison of the FTC methods can be find in Jones (2005). The taxonomy of active and passive FTC methods adapted from Patton (1997) is illustrate in Fig. 3 healthy

2.4.1 Active Fault Tolerant Control

In the active approach, as presented in Alwi and Edwards (2008), Alwi et al. (2011), Benbouzid et al. (2007), Zhang and Jiang (2008), Tabbache et al. (2013)), a new control system is redesigned using desirable properties of performance and robustness

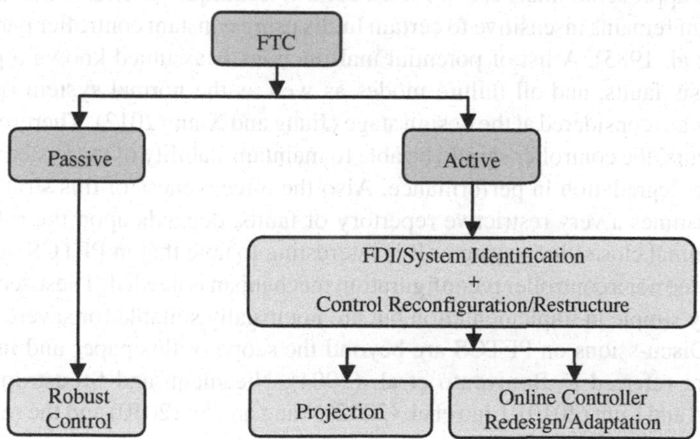

Fig. 3 Decomposition of fault tolerant control

that were important in the original system, but with the reduced capability of the impaired system in mind (Patton 1997). AFTCS react to the system component failures actively by reconfiguring control actions so that the stability and acceptable performance of the entire system can be maintained. In certain circumstances, degraded performance may have to be accepted (Zhang and Jiang 2008). Typically active FTCS consists of an FDD scheme to provide the fault or failure information, a reconfigurable controller, and a controller reconfiguration mechanism. These three units have to work in harmony to complete successful control tasks (Jiang and Xiang 2012).

Active approaches are divided into two main types of methods: projection-based methods and on-line automatic controller redesign methods. The latter involves the calculation of new controller parameters in response to a control impairment. This is often referred to as reconfigurable control (Patton 1997). In projection-based methods, controllers are designed a priori for all possible faults/failures that might occur in the system. The projected controller will only be active when the corresponding fault/failure occur see Alwi et al. (2011). A discussion of these different strategies is provided in Jones (2005) and Alwi et al. (2011).

2.4.2 Passive Fault Tolerant Control

A closed-loop system can have limited fault-tolerance by means of a carefully chosen feedback design, taking care of effects of both faults and system uncertain-ties. Such a system is sometimes called a passive fault-tolerant control system (Eterno et al. 1985; Stengel 1991). Although there are systems in which a specially fixed controller can compensate for the effects of certain faults, usually information about the fault nature and location is required before the controller is able to react to the fault (Patton 1997).

Passive approaches make use of robust control techniques to ensure that a closed-loop system remains insensitive to certain faults using constant controller parameters (Eterno et al. 1985). A list of potential malfunctions is assumed known a priori as design basis faults, and all failure modes as well as the normal system operating conditions are considered at the design stage (Jiang and Xiang 2012). Therefore when a fault occurs, the controller should be able to maintain stability of the system with an acceptable degradation in performance. Also the effectiveness of this strategy, that usually assumes a very restrictive repertory of faults, depends upon the robustness of the nominal closed-loop system. It is interesting to note that in PFTCS neither an FDD scheme nor a controller reconfiguration mechanism is needed. These techniques are usually simple in implementation but are not usually suitable for severe cases of failures. Discu-ssions on PFTCS are beyond the scope of this paper and interested readers are referred to Bonivento et al. (2004), Nieamann and Stoustrup (2005), Benosman and Lum (2010), Liao et al. (2002), Yang and Ye (2010) and the references therein for recent development.

3 Sliding Mode Control in Fields of FTC

The main advantage of the SMC over the other nonlinear control laws is its robustness to external disturbances, model uncertainties, and variations in system parameters. Sliding mode techniques have good robustness and are completely insensitive to so-called matched uncertainty (Edwards and Spurgeon 1998; Utkin 1992). It has been shown that sliding mode techniques can be used to deal with both structural and unmatched uncertainty.

In the last few decades, various techniques based fault tolerant control approaches have been developed for the specific purpose of maintaining relatively system performance and stability. Development of FTC systems has also provided us a systematic way to design and implement various control methods; the most FTC implemented control methods based sliding mode control. Therefore, the application of sliding mode techniques for FTC offers good potential. The comprehensive survey papers Härkegard and Glad (2005), Shin et al. (2005), Benbouzid et al. (2007), Alwi et al. (2011), Alwi and Edwards (2008), Fekih (2008), Xiao et al. (2012), Mekki et al. (2013), Gouichiche et al. (2013), Hamayun et al. (2013), Cortés-Romero et al. (2013) provides an overview of SMC based active and passive FTC approaches, and many results have been established.

3.1 Sliding Mode Control in Active FTC

3.1.1 Projection Based Approach

Early publications focused on so-called projection methods as presented in Benbouzid et al. (2007), Fekih (2008), Gouichiche et al. (2013), Alwi et al. (2011). In this case all expected failure scenarios are enumerated during failure modes and fault models constructed which cover each situation. Whereby if a particular fault was identified and detected, a corresponding control law from a pre-computed and prespecified set of controllers, projection method select and switches online to the pre-computed control law corresponding to the current failure situation.

In Fekih (2008) and Gouichiche et al. (2013) authors propos a FTC based projection method which need a switching bloc, to switch between tow control strategies Field Oriented Control for the healthy condition and SMC for the faulty condition . In our case the proposed FTC structures don't need a switching bloc and used only one control strategy (SMC).

The fault tolerant architecture proposed in Fekih (2008), Gouichiche et al. (2013) is illustrated in Fig. 4 In this diagram, each technique is being used where it is most advantageous which guarantees the achievement of control objectives under any considered conditions.

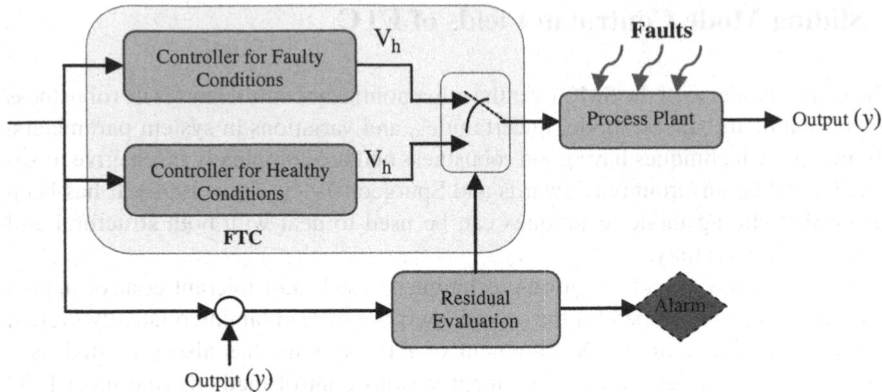

Fig. 4 Bloc diagram of FTC based projection methods (figure adapted from: Fekih 2008)

3.1.2 Control Allocations

Control allocation (CA) has been heavily studied in relation to over-actuated systems and has received a great deal of attention in the literature for reconfigurable systems as it allows actuator failures to be handled without the need to modify the control law; for a survey see (Enns 1998). CA has the capability of redistributing the control command signals to the actuators especially during faults/failures. In CA, the controller is designed based on a 'virtual control' signal and the CA element will map the virtual control to the actual control demand to the actuators. The benefit here is that the controller design is independent of the CA unit (Alwi and Edwards 2008). Therefore, CA can be used in conjunction with any other controller design paradigm. Some of the recent work in this area can be found in these papers (Enns 1998; Härkegard and Glad 2005; Shin et al. 2005; Alwi and Edwards 2008; Hamayun et al. 2013).

A recent application of sliding mode controllers for active FTC is also presented. Here the inherent robustness properties of sliding modes to matched uncertainty are exploited. Although sliding mode controllers can cope easily with faults, they are not able to directly deal with failures—i.e. the total loss of an actuator. In order to overcome this problem, the integration and combination of sliding mode scheme with control allocation framework is considered in Shin et al. (2005), Alwi and Edwards (2008), Hamayun et al. (2013) where in these papers authors present a FTC using SMC with on-line control allocation also considered in the 11th chapter edited by Alwi and Edwards in this book Fridman et al. (2011). Whereby the effectiveness level of the actuators is used by the control allocation scheme to redistribute the control signals to the 'healthy' actuators when a fault occurs (Fig. 5).

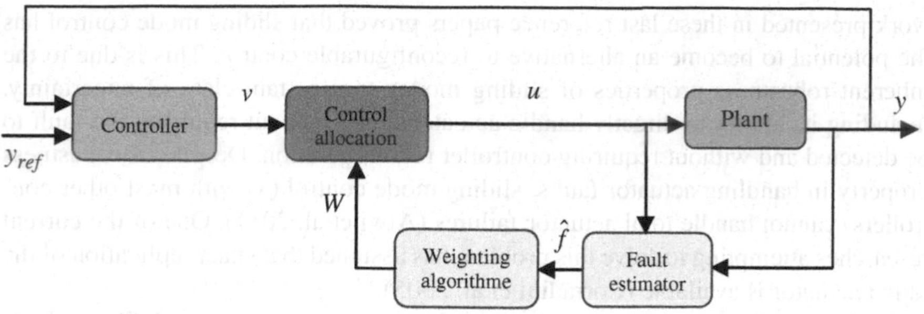

Fig. 5 Bloc diagram of FTC based Control Allocation Strategy (figure adapted from: Alwi and Edwards 2008)

3.2 Sliding Mode Control in Passive FTC

Robust controls have great potential for the development of simple, robust fault tolerant controllers. This controller is closely related to passive fault tolerant control systems (PFTCS). As shown in Fig. 6, in this case the controller is designed to be robust against uncertainty and disturbances during the stage design also to tolerate system component faults by using the system redundancies without any parameter adjustment or controller structure.

The robust characteristics of the sliding mode method provide a natural environment for the use of such methods on passive FTC schemes. This technique has been properly used in different control schemes and assisted by other effective control strategies which have shown proper performance under fault tolerant operations see for examples Xiao et al. (2012), Hamayun et al. (2013), Mekki et al. (2013). Over the past years, considerable attention has been paid to the design of for sliding mode controller assistance in order to overcome several issues like disturbance rejection as presented in Cortés-Romero et al. (2013) also to accommodating actuator failures as presented in Corradini et al. (2005).

The possibilities of exploiting the inherent robustness properties of SM for FTC previously been explored for several applications as flexible spacecraft (Xiao et al. 2012) and flight systems (Hess and Wells 2003; Shin et al. 2005; Jones 2005) the

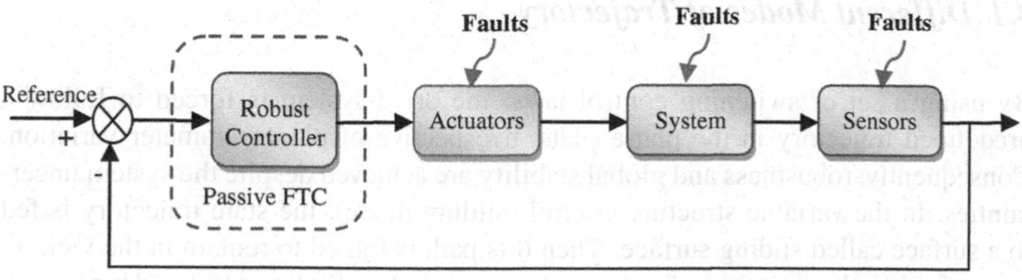

Fig. 6 Bloc diagram of Passive FTC Strategy

work presented in these last reference papers proved that sliding mode control has the potential to become an alternative to reconfigurable control. This is due to the inherent robustness properties of sliding modes to a certain class of uncertainty, including its ability to directly handle actuator faults without requiring the fault to be detected and without requiring controller reconfiguration. Despite its robustness property in handling actuator faults, sliding mode control (as with most other controllers) cannot handle total actuator failures (Alwi et al. 2011). One of the current researches attempting to solve this problem has assumed that exact replication of the failed actuator is available (Corradini et al. 2005).

In Jones (2005) author presents a brief comparison of current FTC methods resumed in table the same table is presented in Alwi et al. (2011). As can be seen from this table, that SMC method can handle partial loss of effectiveness of actuators but not complete loss also SMC assumes robust control can handle all forms of structural failures

4 Sliding Modes Control Methods

In the recent past years, the SMC strategies have received worldwide interest, and many theoretical studies and application researches are reported recently in Hung et al. (1993), Edwards and Spurgeon (1998), Utkin et al. (1999), Fridman et al. (2011). Even through, in its early stage of development, the SMC theory was overlooked because of the development in the famous linear control theory, during the last 20 years it has shown to be a very effective control method (Edwards and Spurgeon 1998). It is known that the SMC can offer such properties as insensi-tivity to parameters variations, external disturbance rejection, fast dynamic response, and simplicity of design and implementation. In this ways, this chapter demonstrates the true theoretical and applications depth to which the sliding mode control paradigm has been developed today in the field of FTC. SMC has three very strong themes: control design, theoretical extensions and industrial applications (Alwi et al. 2011). For sliding mode control, this section gives brief introduction to understands of the design method.

4.1 Different Modes of Trajectory

By using a set of switching control laws, the drive system is forced to follow a predefined trajectory in the phase plane irrespective of plant parameter variation. Consequently, robustness and global stability are achieved despite the system uncertainties. In the variable structure control (sliding mode), the state trajectory is fed to a surface called sliding surface. Then this path is forced to remain in the vicinity thereof using the switching function. As presented in Slotine (1984), Hung et al. (1993), the trajectory in the phase plane consists of three distinct parts; see Fig. 7.

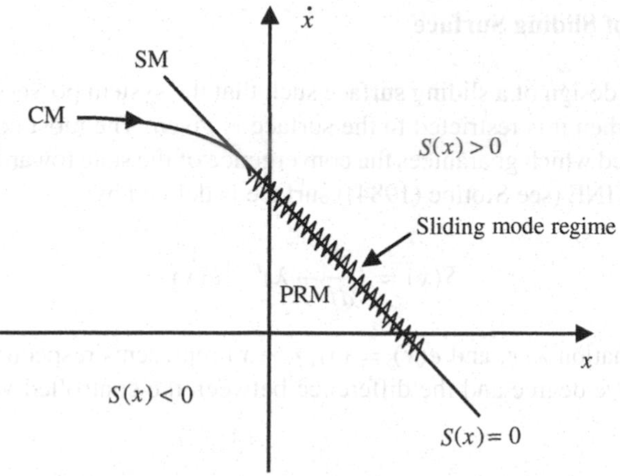

Fig. 7 Different modes of trajectory in the phase plane

4.1.1 Convergence Mode (CM)

This is the mode in which the controlled variable moves from any starting point in the phase plane, and tends towards the switching surface $S(x) = 0$ and reaches in a finite time. This mode is characterized by the control law and the convergence criterion.

4.1.2 Sliding Mode (SM)

This is the mode in which the state variable is reached and the sliding surface tends to cause the phase plane. The dynamics of this mode is characterized by the choice of the sliding surface $S(x) = 0$.

4.1.3 Permanent Regime Mode (PRM)

This mode is added to the study of the response of the system around the equilibrium point. It characterizes the quality and performance of the control (Hung et al. 1993).

4.2 Design of Sliding Mode Control

The design of the sliding mode control takes into account the problems of stability and good performance in its approach. In general, for this type of control three steps must be performed:

4.2.1 Design of Sliding Surface

In this step the design of a sliding surface such that the system possesses the desired performance when it is restricted to the surface is given. The most commonly used for the slip speed which guarantees the convergence of the state towards its reference given by SOLTINE (see Slotine (1984)) surface is defined by:

$$S(x) = (\frac{d}{dt} + \lambda)^{r-1} e(x) \tag{1}$$

In this first equation λ, r, and $e(x) = (x_{ref} - x)$ represents respectively a positive constant, relative degree and the difference between the controlled variable and its reference.

4.2.2 Convergence Conditions and Existence

The conditions of existence and convergence criteria are different dynamics that allow the system to converge to the sliding surface and stay there regardless of the disturbance. There are two considerations to ensure convergence of the user: (i) the direct switching function, which is the first condition of convergence, it is proposed and studied by EMILYANOV and UTKIN. This is to give the surface a convergent dynamic to zero. It is given by:

$$S(x)\dot{S}(x) < 0 \tag{2}$$

(ii) the LYAPUNOV Function, which is the second condition of convergence. The Lyapunov function is a positive scalar function ($V(x) > 0$) for the system state variables of the. We define the Lyapunov function as follows:

$$V(x) = \frac{1}{2} S^2(x) \tag{3}$$

Then the derivative of this function is given by:

$$\dot{V}(x) = S(x)\dot{S}(x) \tag{4}$$

For the LYAPUNOV function $V(x)$ can decrease and converge to zero, just to ensure that its derivative is negative. This is checked only if condition (2) holds.

4.2.3 Design of a Variable Structure Control Law

Obtaining a sliding regime implies a discontinuous control. The sliding surface should be attractive to both sides. Therefore, if this is essential discontinuous control, it does not prevent a continued part it is added. The party continues in effect lead to reduce as much as we want the amplitude of the discontinuous portion. In the presence of a disturbance, the discontinuous portion is essentially to check the terms of attractiveness.

In this case, the structure of a sliding mode controller consists of two parts: (i) switching control U_n; (ii) equivalent control U_{eq}. To force the system to follow the path imposed just now to make $S = 0$ attractive. For this, a switching control U_n will be added to the equivalent control U_{eq} in the form:

$$U = U_{eq} + U_n \tag{5}$$

Necessary for the system states provided, follow the defined surfaces sliding trajectory is what brings us to define the equivalent command. While the control law which ensures the attractiveness is given by:

$$U_n = -k\, sign(S) \tag{6}$$

Then the condition is always performed, which proves that $S = 0$ is attractive and invariant, despite the disruptions.

Remark 1 It's important to notice that the most work's as presented in Edwards and Spurgeon (1998), Yan and Edwards (2007), Alwi et al. (2011) consider that the sliding mode approach consists of two steps: (i) the design of a sliding surface such that the system possesses the desired performance when it is restricted to the surface; (ii) the design of a variable structure control law which drives the system trajectories to the sliding surface in finite time and maintains a sliding motion on it thereafter. In these cases the conditions of existence and convergence step will be associated to the design of a variable structure control law.

5 Case Study: PMSM Example

The method described above will now be demonstrated using a permanent magnet synchronous motor (PMSM) as application example. Where; permanent magnet synchronous motor is nowadays widely used in the industry applications due to their high efficiency and high power/torque density (Teng et al. 2012). These motors are used in many applications such as traction with variable speeds in transportation as presented in Erginer and Sarul (2013).

5.1 PMSM Healthy Model

The setting in the state form of the PMSM model allows the simulation of this latter. In the rotor rotating $(d-q)$ reference frame, the PMSM stator current model is described as follows (Tang 1997):

$$\begin{cases} \dot{x} = f(x) + Bu + DT_L \\ x = [x_1 \quad x_2 \quad x_3]^T = \left[i_d \; i_q \; \omega_r\right]^T \\ u = \begin{bmatrix} u_d = V_{sd} \\ u_q = V_{sq} \end{bmatrix}; \quad B = \begin{bmatrix} b_1 & 0 & 0 \\ 0 & b_2 & 0 \end{bmatrix}^T \\ D = \begin{bmatrix} 0 & 0 & d \end{bmatrix}^T \end{cases} \tag{7}$$

With the following expression of field vector $f(x)$:

$$\begin{cases} f_1(x) = a_1 x_1 + a_2 x_2 x_3 \\ f_2(x) = a_3 x_2 + a_4 x_3 + a_5 x_1 x_3 \\ f_3(x) = a_6 x_2 + a_7 x_3 + a_8 x_1 x_2 \end{cases} \tag{8}$$

The components of this vector are expressed according to the PMSM parameters as follows:

$$\begin{cases} a_1 = -\frac{R_s}{L_d};\, a_2 = \frac{L_q}{L_d};\, a_3 = -\frac{R_s}{L_q} a_4 = -\frac{\varphi_f}{L_d};\, a_5 = \frac{L_d}{L_q};\, a_6 = -\frac{n_p^2 \varphi_f}{j} \\ a_7 = -\frac{f}{J};\, a_8 = \frac{n_p^2 \varphi_f}{j}(L_d - L_q);\, d = -\frac{n_p}{J};\, b_1 = \frac{1}{L_d};\, b_2 = -\frac{1}{L_q} \end{cases}$$

where: i_d, i_q stator current; V_d, V_q stator voltage; L_d, L_q stator inductance; R_s stator resistance; φ_f rotor permanent magnet flux. ω_r rotor speed, f viscous friction coefficient, L_T load torque, J moment of Inertia. As presented in the appendix (see Table 1) we take in this paper in PMSM with smooth poles $L_d = L_q = L$ in this case $(a_g = 0)$

The use of the classical controllers such as the proportional and integral controller (PI) is insufficient to provide good speed tracking performance. To overcome these problems, a robust controller based SMC approach is proposed.

5.2 Sliding Mode Control Design

Sliding mode control (SMC) is an effective control strategy in modern theory of strong control robustness and simple implementation. Due to its order reduction, disturbance rejection, strong robustness and simple operation using the power converter is one of the possible methods of control for electromechanical systems (Utkin et al. 1999). As presented in Ezzat (2011) and Attou et al. (2013) the application of the sliding mode control strategy to PMSM in this case is divided into two steps.

First we take the following equilibrium surface:

$$\begin{pmatrix} S_1 = e_w = x_3^* - x_3 \\ S_2 = e_q = x_2^* - x_2 \\ S_3 = e_d = x_1^* - x_1 \end{pmatrix} \tag{9}$$

5.2.1 Speed Regulator

The condition necessary for the system states follow the trajectory defined by the sliding surfaces is $S_1 = 0$ which brings back us to define the speed equivalent control in the following way:

$$(S_1 = 0) \Rightarrow \dot{S}_1 = \dot{x}_3 - \dot{x}_3^* = 0 \tag{10}$$

With: x_3 and x_3^* represents the real and reference speed. In this case we get:

$$x_2^d = \frac{1}{a_6}(-a_7 x_3 - dT_L + \dot{x}_3^*) \tag{11}$$

From 6 the control law which ensures the attractivity is given by:

$$i_d^n = -k_1 sign(S_1) \tag{12}$$

where: k_1 are positive constant. Then from (5), (11) and (12) we get:

$$x_2^* = \frac{1}{a_6}(-a_7 x_3 - dT_L + \dot{x}_3^*) - k_1 sign(S_1) \tag{13}$$

5.2.2 Currents Regulator

The condition necessary for the system states follow the trajectory defined by the sliding surfaces is $S_2 = 0$ and $S_3 = 0$ which brings back us to define the direct and quadratic currents equivalents control in the following way:

$$\begin{cases} S_2 = 0 \\ S_3 = 0 \end{cases} \Rightarrow \begin{cases} \dot{S}_2 = \dot{x}_2^* - \dot{x}_2 = 0 \\ \dot{S}_3 = \dot{x}_1^* - \dot{x}_1 = 0 \end{cases} \tag{14}$$

According to the derivative of the direct and quadratic currents surfaces we can generate the tension given as fellow:

$$\begin{cases} u_q^{eq} = V_{sq}^{eq} = \frac{1}{b}\left(\dot{x}_2^* - a_3 x_2 - a_4 x_3 - a_5 x_1 x_3\right) \\ u_d^{eq} = V_{sd}^{eq} = \frac{1}{b}\left(\dot{x}_1^* - a_1 x_1 - a_2 x_2 x_3\right) \end{cases} \tag{15}$$

From (6) the attractive control law is ensured and given by:

$$\begin{cases} V_{sq}^{n} = -k_2 sign(S_2) \\ V_{sd}^{n} = -k_3 sign(S_3) \end{cases} \tag{16}$$

With k_2 and k_3 are positive constants. Finely from (5) we get:

$$\begin{cases} u_q^{nom} = \frac{1}{b}\left(\dot{x}_2^* - a_3x_2 - a_4x_3 - a_5x_1x_3\right) - k_2 sign(S_2) \\ u_d^{nom} = \frac{1}{b}\left(\dot{x}_1^* - a_1x_1 - a_2x_2x_3\right) - k_3 sign(S_3) \end{cases} \tag{17}$$

5.2.3 Stability of the Closed Loop

The control objective in this case is to force the PMSM speed ($\omega_r = x_3$) to follow its reference x_3^* and maintain in the same time the direct current ($i_d = x_1$) to zero under load torque disturbance. Let e_d, e_q and e_ω be the tracking errors of the currents and the speed then the dynamic of the tracking errors are given by:

$$\begin{cases} \dot{e}_d = a_1x_1 + a_2x_2x_3 + b_1V_{sd} - \dot{x}_1^* \\ \dot{e}_q = a_3x_2 + a_4x_3 + a_5x_1x_3 + b_2V_{sq} - \dot{x}_2^* \\ \dot{e}_\omega = a_6x_2 + a_7x_3 + dT_L - \dot{x}_3^* \end{cases} \tag{18}$$

By taking $k_1 = \frac{k_\omega}{a_6}$ in (13), from this equation and $\dot{e}_\omega$given in (18) we get:

$$\dot{e}_\omega = -k_\omega sign e_\omega \tag{19}$$

By taking $k_2 = \frac{k_q}{b_2}$ and $k_3 = \frac{k_d}{b_1}$ in (17) from this new equation, $\dot{e}_q$ and $\dot{e}_d$ given in (18) we get:

$$\begin{cases} \dot{e}_q = -k_q sign e_q \\ \dot{e}_d = -k_d sign e_d \end{cases} \tag{20}$$

Consider the following Lyapunov function:

$$V = \frac{1}{2}e_d^2 + \frac{1}{2}e_q^2 + \frac{1}{2}e_\omega^2 \tag{21}$$

The derivative of V with respect to time is:

$$\dot{V} = e_d(-k_d sign e_d) + e_q(-k_q sign e_q) + e_\omega(-k_\omega sign e_\omega) \tag{22}$$

Remark 2 We have at $t \to \infty e_i \to 0$ and $sign\, e_i \to 0$ then we take $sign e_i = e_i$ where $e_i = e_d, e_q, e_\omega$ then the derivative of the Lyapunov function given by (22) becomes:

$$\dot{V} = -k_d e_d^2 - k_q e_q^2 - k_\omega e_\omega^2 \tag{23}$$

Finely From (23) we see that ($\dot{V} \le 0$) the derivative of the complete Lyapunov function be negative definite this implies that all the error variables are globally uniformly bounded and maintain the system closed loop performance in presence of load torque disturbances.

5.3 Fault Tolerant Control Design

5.3.1 PMSM Faulty Model

In this section we briefly review how the PMSM model will be modifies in presence of faults which can be both of mechanical and electrical nature. The faults dealt with in this paper can be summarized in the class of Stator asymmetries, mainly due to static eccentricity as presented in Akar and Çankaya (2009) and Ebrahimi et al. (2009). Following the theory in Vas (1994), it turns out that the presence of stator faults generates asymmetries in the PMSM, yielding some slot harmonics (sinusoidal components) in the stator currents (see Akar and Çankaya 2009; Ebrahimi et al. 2009). Then the currents will be modified as fellow:

$$\begin{cases} i_d \to i_d + A\sin(\omega_1 t + \varphi) \\ i_q \to i_q + A\cos(\omega_1 t + \varphi) \end{cases} \tag{24}$$

where i_d and i_q denote the stator currents in the $(d-q)$ reference frame. The pulsations ω_1 of the harmonic components depend on the kind of fault (due to the stator asymmetries). The amplitude A and the phases φ are unknown; they depend on the stator faults entity. The sinusoidal components generated by the presence of the stator faults can be modeled by the following exosystem presented in Bonivento et al. (2004), Mekki et al. (2013):

$$\dot{w} = S(\omega_1)\cdot w \quad w \in \Re^{2n_f} \tag{25}$$

With: $S(\omega_1)$ is the vector of the pulsations.

$$S(\omega_1) = \begin{pmatrix} 0 & \omega_1 \\ -\omega_1 & 0 \end{pmatrix} \tag{26}$$

where ω_1 the pulsation of the harmonic is generated by the stator faults; the amplitudes and the phases of the harmonics are unknown; they depend on the initial state $w(0)$ of the exosystem. Then, the additive sinusoidal terms in (24) can be as a suitable combination of the exosystem state, i.e.:

$$\begin{cases} \dot{i}_d \to \dot{i}_d + Q_d w \\ \dot{i}_q \to \dot{i}_q + Q_q w \end{cases} \tag{27}$$

where: $\begin{cases} Q_d = (1\,0\,1\,0\,\ldots\,1\,0) \\ Q_q = (0\,1\,0\,1\,\ldots\,0\,1) \end{cases}$

Recalling the current dynamics in the un-faulty operative condition reported in the previous section, a simple computation shows that, once the perturbing terms $Q_d w$ and $Q_d w$ are added, by deriving (27) the $(i_d - i_q)$ modify as:

$$\begin{cases} \frac{di_d}{dt} = \dot{x}_1 = a_1 x_1 + a_2 x_2 x_3 + b_1 u_d - (a_1 Q_d + a_2 Q_q x_3 + Q_d S)w \\ \frac{di_q}{dt} = \dot{x}_2 = a_3 x_2 + a_4 x_3 + a_5 x_1 x_3 + b_2 u_q - (a_3 Q_d + a_4 Q_q x_3 + Q_d S)w \end{cases} \tag{28}$$

Bearing in mind the dynamics of the stator currents in the healthy operative conditions, it is also simple to get the PMSM dynamics after the occurrence of a fault. As a matter of fact, taking Eq. (28) it is readily seen that the PMSM healthy model given by (7) and (8) in faulty condition (presence of stator faults) will be given by:

$$\begin{cases} \dot{x}_1 = a_1 x_1 + a_2 x_2 x_3 + b_1 u_d + \Gamma_d w \\ \dot{x}_2 = a_3 x_2 + a_4 x_3 + a_5 x_1 x_3 + b_2 u_q + \Gamma_q w \\ \dot{x}_3 = a_6 x_2 + a_7 x_3 + dT_L \end{cases} \tag{29}$$

With: $\Gamma(w) = -\begin{pmatrix} \Gamma_d \\ \Gamma_q \end{pmatrix} w \quad \begin{cases} \Gamma_d = -(a_1 Q_d + a_2 Q_q x_3 + Q_d S) \\ \Gamma_q = -(a_3 Q_d + a_5 Q_q x_3 + Q_d S) \end{cases}$

Finely, in the presence of stator faults the PMSM model becomes:

$$\dot{x} = f(x) + Bu + dT_L + \Gamma(w) \tag{30}$$

5.3.2 Control Reconfiguration

The principal of the new FTC control law will be expressed by (31) in this section the total control law is dividing to two control laws: (i) the nominal control u_{nom} resulting from the SMC law presented in Sect. 5.2 by Eq. (17); (ii) the additive control u_{ad} which is added to the control and setting to compensate the faults effect. This additive control results from the internal model whose role is to reproduce the signal representing the faults effect.

$$\begin{cases} u = u_{nom} + u_{ad} \\ u = \begin{bmatrix} u_d \\ u_q \end{bmatrix} = \begin{bmatrix} u_d^{nom} \\ u_q^{nom} \end{bmatrix} + \begin{bmatrix} u_d^{ad} \\ u_q^{ad} \end{bmatrix} \end{cases} \tag{31}$$

The instantaneous difference between the state derivative of the system and the reference becomes:

$$\dot{\tilde{x}} = \begin{bmatrix} \dot{e}_d \\ \dot{e}_q \\ \dot{e}_\omega \end{bmatrix} = \begin{bmatrix} \dot{x}_1 \\ \dot{x}_2 \\ \dot{x}_3 \end{bmatrix} - \begin{bmatrix} \dot{x}_1^* \\ \dot{x}_2^* \\ \dot{x}_3^* \end{bmatrix} = \begin{cases} f_1(x) + b_1 u_d - \dot{x}_1^* - \Gamma_d w \\ f_2(x) + b_2 u_q - \dot{x}_2^* - \Gamma_q w \\ f_2(x) - \dot{x}_3^* \end{cases} \tag{32}$$

From the SMC law given by (17) and the FTC law given by (31) after rep-lacing in (32) we get:

$$\dot{\tilde{x}} = \begin{bmatrix} \dot{e}_d \\ \dot{e}_q \\ \dot{e}_\omega \end{bmatrix} = \begin{cases} -k_d sign\, e_d + b_1 u_d^{ad} - \Gamma_d w \\ -k_q sign\, e_q + b_2 u_q^{ad} - \Gamma_q w \\ -k_\omega sign\, e_\omega \end{cases} \tag{33}$$

In the third equation if $e_\omega \to 0 \Rightarrow \dot{e}_\omega \to 0$ Let us notice that the first two equations do not depend on the variable e_ω. In the continuation, for the determination of u_{ad} let us consider the new variables ($\tilde{x}_1 = e_d$ and $\tilde{x}_2 = e_q$), whose dynamics results from Remark 2 and from (33) as follow:

$$\begin{cases} \dot{w} = S(\varpi) \cdot w \\ \dot{\tilde{x}} = \begin{bmatrix} \dot{\tilde{x}}_1 \\ \dot{\tilde{x}}_1 \end{bmatrix} = \begin{cases} -k_3 \tilde{x}_1 + b_1 u_d^{ad} - \Gamma_d w \\ -k_4 \tilde{x}_2 + b_2 u_q^{ad} - \Gamma_q w \end{cases} \end{cases} \tag{34}$$

Then we can write the system (34) in a matrix form:

$$\dot{\tilde{x}} = H(\tilde{x}) + \tilde{B} \cdot u_{ad} - \Gamma \cdot w \tag{35}$$

$$\begin{cases} H(\tilde{x}) = \tilde{A} \cdot \tilde{x} \; and \;\; \tilde{A} = \begin{bmatrix} -k_3 & 0 \\ 0 & -k_4 \end{bmatrix} \\ \tilde{B} = \begin{bmatrix} b_d & 0 \\ 0 & b_q \end{bmatrix} \quad and \;\; \Gamma = \begin{bmatrix} \Gamma_d & 0 \\ 0 & \Gamma_q \end{bmatrix} \end{cases} \tag{36}$$

In this case for the determination of the internal model we introduce a resent implicit fault tolerant control approach which does not rest on the resolution of the Sylvester equation problem proposed in Bonivento et al. (2004). To solve this problem we propose in this paper an internal based Lyapunov theory which takes the following form Mekki et al. (2013):

$$\begin{cases} \dot{\xi} = S(\varpi)\,\xi + N(\tilde{x}) \\ \dim(\xi) = \dim(w) = 2n_f \end{cases} \tag{37}$$

As presented in Mekki et al. (2013) the additive control law u_{ad} is chosen like:

$$u_{ad} = \tilde{B}^{-1}\Gamma\xi \tag{38}$$

Consider the systems (35) and the additive term given by (38) in this case we have:

$$\dot{\tilde{x}} = H(\tilde{x}) + \Gamma \cdot (\xi - w) \tag{39}$$

The new error variable is considered is $e = (\xi - w)$. Its derivative compared to time takes this form:

$$\dot{e} = \dot{\xi} - \dot{w} = S(\varpi)\,\xi + N(\tilde{x}) + S(\varpi)\,w \tag{40}$$

The equations describing the dynamics of the errors in closed loop are thus:

$$\begin{cases} \dot{\tilde{x}} = \tilde{A} \cdot \tilde{x} + \Gamma \cdot e \\ \dot{e} = S(\varpi)\, e + N(\tilde{x}) \end{cases} \tag{41}$$

It is necessary to find the expression of $N(\tilde{x})$ which cancels the observation error of the faults e and makes it possible at the same time to reject their effect for it cancels also $\tilde{x}$. That is to say the Lyapunov function of the system (41):

$$V = \frac{1}{2}\tilde{x}^T \cdot \tilde{x} + \frac{1}{2}e^T \cdot e \tag{42}$$

After develop of calculates $\dot{V}$ becomes:

$$\dot{V} = \tilde{x}^T \cdot \tilde{A} \cdot \tilde{x} + e^T \cdot \Gamma^T \cdot \tilde{x} + e^T \cdot N(\tilde{x}) \tag{43}$$

In this case the $N(\tilde{x})$ choice is given by:

$$N(\tilde{x}) = -\Gamma^T \tilde{x} \tag{44}$$

After replacing of (43) in (44). The derivative of Lyapunov function become:

$$\dot{V} = \tilde{x}^T \cdot \tilde{A} \cdot \tilde{x} \leq 0 \tag{45}$$

Finally the system (41) will be:

$$\begin{cases} \Gamma \cdot e = 0 \\ \dot{e} = S(\varpi)\, e \end{cases} \tag{46}$$

The objective of the control is achieved by adopting the procedure suggested and we able to compensate the faults effect on the system ($x \rightarrow 0$) and to reproduce ($e \rightarrow 0$) thanks to sliding mode control and internal model.

5.4 Simulation Results

In this section numerical simulations have been performed to test and validate the proposed FTC based SMC scheme. The PMSM parameters are given as follow: Power = 22 W; Voltage = 220 V; Frequency = 50 Hz; $n_P = 2$; $R_s = 3.4\Omega$; $L_d = 0.0121$ H; $L_q = 0.0121$ H ; $\varphi_f = 0.013$ H ; $f = 0.0005$IS. The speed and direct current references are fixed at 100 (*rad/s*) and zero (*A*) respectively, also a load torque disturbance $T_L = 0.05\ N.m$ is applied at t = 0.3 s. In Fig. 8 after association between PMSM and PWM inverter controlled by the SMC technique we start the simulation without any load torque, then at t = 0.3 s the application of a load torque equal to the

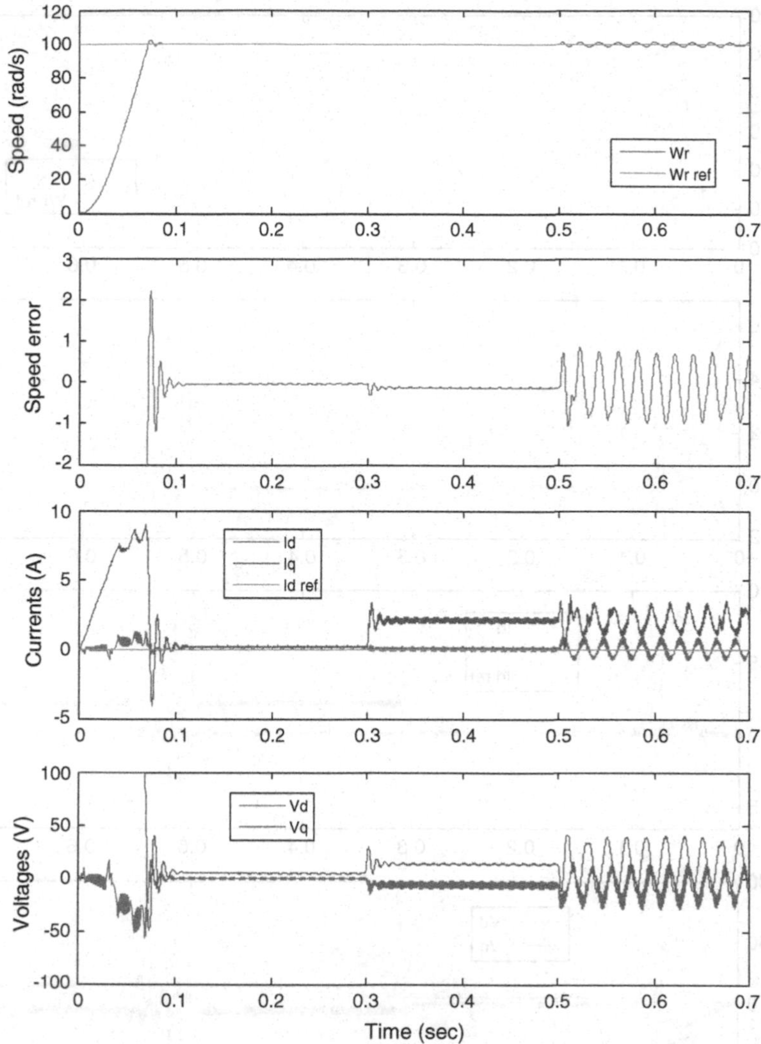

Fig. 8 SMC performance under load torque disturbance and stator fault

nominal torque ($T_L = 0.05N.m$) is presented, then the effect of stator fault will be introduced at t = 0.5 s. For Fig. 9 the same consideration is taken account but in this case we simulate the global closed loop system with the FTC based SMC approach.

From these simulations we can noticed that sliding mode controller (nominal controller) which we synthesized present a robustness compared to the load torque disturbance, but can't deal with total faults in this case it proves to be insufficient in the event of stator fault. This is checked by simulations represented above when the

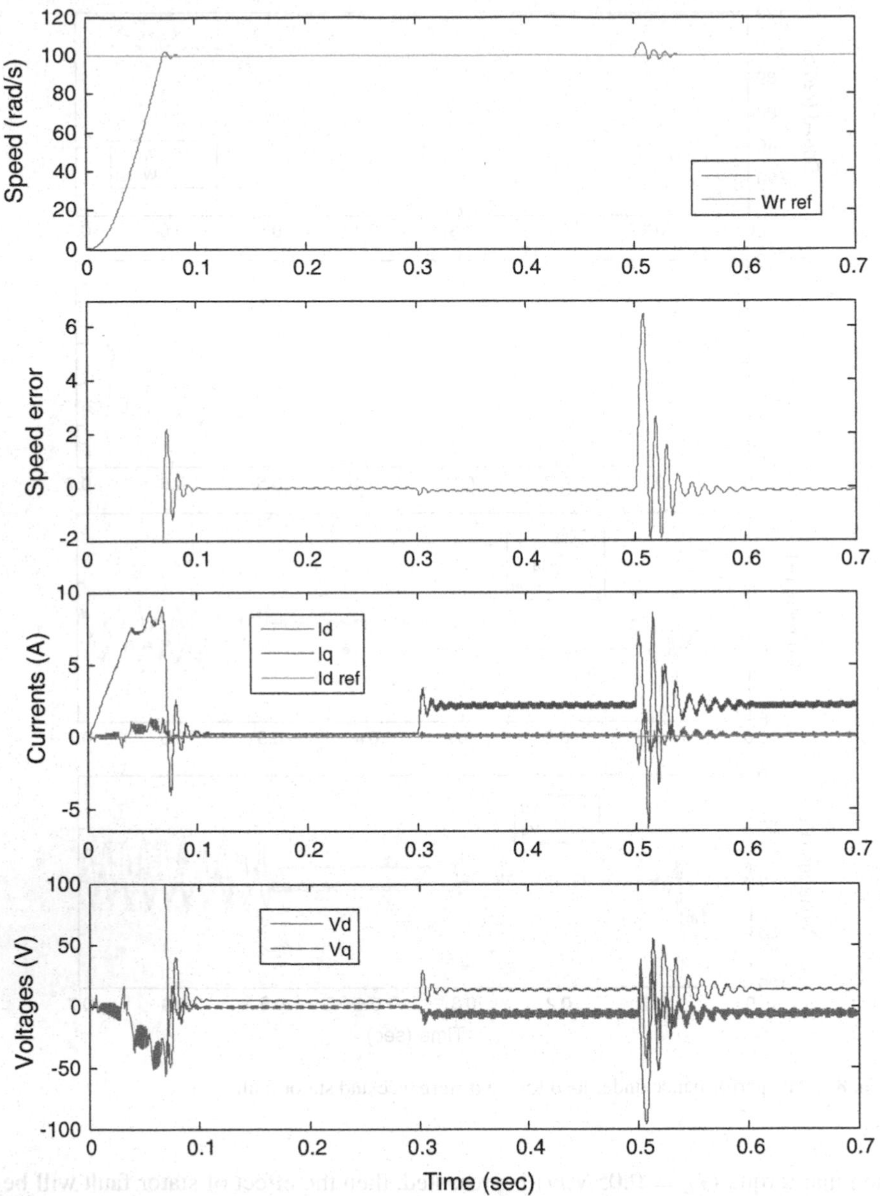

Fig. 9 FTC approach based SMC performance under load torque disturbance and stator fault

internal model is not active. Therefore, the FTC approach (when the internal model is active) which we synthesized rejects the effect of the load torque disturbances and also the stator faults effect.

The work presented in this section focuses on the concept of fault tolerant control approach based sliding mode control strategy for PMS Motors. In un-faulty condition the SM Controller permits to steer the direct current and the speed variables to their desired references and to reject the load torque disturbances, however the presence of stator fault degraded the performances of the PMSM. In order to compensate the faults effect a FTC approach can be designed starting with generating from the internal model, an additive term wish we add to the nominal control (SMC) to compensate the faults effect. The proposed approach is validated through the simulated results.

6 Conclusion

This chapter highlights the benefits of sliding modes when applied to the fields of fault tolerant control (FTC). The presented considerations in this chapter provide a strong motivation for the development of methods and strategies for the design of advanced fault tolerant control systems based sliding mode control that ensure an efficient and timely response to enhance fault recovery, prevent faults from propagating or developing into large class of failures, and reduce the risk. The presented chapter gives a general FTC strategies overview. Also this chapter briefly discuses and highlights the benefits of SMC strategy when applied to active and passive FTC. On the other hand, give a brief introduction to the concept of sliding mode control and examine its properties. Finely, a permanent magnet synchronous motor (PMSM) systems case study is presented in order to introduce the concept and to prove the effectiveness of the proposed approach. Where computer numerical simulations show the effectiveness of the proposed FTC based SMC scheme. We hope that this work will serve as a good starting point and a useful reference for researchers working on the development of new sliding modes based fault tolerant control approaches.

References

Akar, M., Çankaya, İ.: Diagnosis of static eccentricity fault in permanent magnet synchronous motor by on-line monitoring of motor current and voltage. J. Electr. Electron. Eng. **9**(2), 951–958 (2009)

Alwi, H., Edwards, C.: Fault tolerant control using sliding modes with online control allocation. Automatica **44**(7), 1859–1866 (2008)

Alwi, H., Edwards, C., Tan, C.P.: Fault Detection and Fault-Tolerant Control Using Sliding Modes. Springer, London (2011)

Attou, A., Massoum, A., Chiali, E.:Sliding mode control of a permanent magnets synchronous machine. In: 2013 Fourth International Conference on Power Engineering, Energy and Electrical Drives (POWERENG), pp. 115–119. Istanbul, 13–17 May 2013, doi:10.1109/PowerEng.6635591

Benbouzid, M.E.H., Diallo, D., Zeraoulia, M.: Advanced fault tolerant control of induction-motor drives for EV/HEV traction applications: from conventional to modern and intelligent control techniques. IEEE Trans. Veh. Technol. **56**(2), 519–528 (2007)

Bonivento, C., Isidori, A., Marconi, L., Paoli, A.: Implicit fault tolerant control: application to induction motors. Automatica **40**(3), 355–371 (2004)

Blanke, M., Staroswiecki, M., Wu, N. E.: Concepts and methods in fault-tolerant control. In: Proceedings of the 2001 American Control Conference, vol. 4, pp. 2606–2620. Arlington, VA, 25 Jun 2001–27 Jun 2001,doi:10.1109/ACC.2001.946264

Bennett, S.M., Patton, R.J., Daley, S.: Sensor fault-tolerant control of a rail traction drive. Control Eng. Pract. **7**(2), 217–225 (1999)

Benosman, M., Lum, K.Y.: Passive actuators' fault-tolerant control for affine nonlinear systems. IEEE Trans. Consum. Electron. **18**(1), 152–163 (2010)

Corradini, M.L., Orlando, G., Parlangeli, G.: A fault tolerant sliding mode controller for accommodating actuator failures. In: 44th IEEE Conference on Decision and Control, pp. 3091–3096. Spain, 12–15 December 2005, doi:10.1109/CDC.2005.1582636

Cortés-Romero, J., Rojas-Cubides, H., Coral-Enriquez, H., Sira-Ramírez, H., Luviano-Juárez, A.: Active disturbance rejection approach for robust fault-tolerant control via observer assisted sliding mode control. Mathematical Problems in Engineering, vol. 2013, ID 609523, pp. 12. (2013)

Ebrahimi, B.M., Faiz, J., Roshtkhari, M.J.: Static, dynamic and mixed eccentricity fault diagnosis in permanent magnet synchronous motors. IEEE Trans. Industr. Electron. **56**(11), 4727–4739 (2009)

Edwards, C., Spurgeon, S.K.: Sliding Mode Control: Theory and Applications. Taylor & Francis, London (1998)

Enns, D.: Control allocation approaches. In: AIAA Guidance, Navigation and Control, pp. 98–108. (1998)

Erginer, V., Sarul, M.H.: High Performance and Reliable Torque Control of Permanent Magnet Synchronous Motors in Electric Vehicle Applications. Elektronika ir Elektrotechnika **19**(7), 41–46 (2013)

Eterno, J.S., Looze, D.P., Weiss, J.L., Willsky, A.S.: Design issues for fault-tolerant restructurable aircraft control. In: Proceeding 24th IEEE Conference on Decision and Control, Fort Lauderdale, pp. 900–905. (1985), doi:10.1109/CDC.1985.268630

Ezzat, M.: Nonlinear control of a permanent magnet synchronous motor without mechanical sensor. Ph.D. Thesis, Central School Nantes, France (2011)

Fekih, A.: Effective fault tolerant control design for nonlinear systems: application to a class of motor control system. IET Control Theory Appl. **2**(9), 762–772 (2008)

Fridman, L., Moreno, J., Iriarte, R.: Sliding Modes after the First Decade of the 21st Century - State of the Art. Springer, Berlin (2011)

Gouichiche, A., Boucherit, S.M., Tadjine, M., Safa, A., Messlem, Y.: An improved stator winding fault tolerance architecture for vector control of induction motor: theory and experiment. Electr. Power Syst. Res. **104**(2013), 129–137 (2013)

Hamayun, M.T., Edwards, C., Alwi, H.: A fault tolerant control allocation scheme with output integral sliding modes. Automatica **49**(6), 1830–1837 (2013)

Härkegard, O., Glad, S.T.: Resolving actuator redundancy-optimal control versus control allocation. Automatica **41**(1), 137–144 (2005)

Hess, R.A., Wells, S.R.: Sliding mode control applied to reconfigurable flight control design. J. Guidance Control Dyn. **26**(3), 452–462 (2003)

Hung, J.Y., Gao, W.B., Hung, J.C.: Variable structure control: A survey. IEEE Trans. Industr. Electron. **40**(1), 2–22 (1993)

Isermann, R., Balle, P.: Trends in the application of model-based fault detection and diagnosis of technical processes. Control Eng. Pract. **5**(5), 709–719 (1997)

Jiang, J., Xiang, Y.: Fault-tolerant control systems: a comparative study between active and passive approaches. Annu. Rev. Control **36**(1), 60–72 (2012)

Jones, C.N.: Reconfigurable flight control: First year report. Cambridge University, Technical report (2005)

Klinkhieo, S.: On-line Estimation Approaches to Fault-Tolerant Control of Uncertain Systems. Ph.D. thesis, The University of Hull (2009)

Liao, F., Wang, J.L., Yang, G.H.: Reliable robust flight tracking control: An LMI approach. IEEE Trans. Control Syst. Technol. **10**(1), 76–89 (2002)

Mekki, H., Benzineb, O., Boukhetala, D., Tadjine, M.: Fault tolerant control based sliding mode application to induction motor. In: Proceeding Engineering and Technology, PET vol. 3, pp. 1–6. (2013)

Nieamann, H., Stoustrup, J.: Passive fault tolerant control of a double inverted pendulium-a case study. Control Eng. Pract. **13**(8), 1047–1059 (2005)

Noura, H., Sauter, D., Hamelin, F., Theilliol, D.: Fault tolerant control in dynamic systems application to a winding machine. IEEE Control Mag. **20**(1), 33–49 (2000)

Noura, H., Theilliol, D., Ponsart, J.C., Chamseddine, A.: Fault-Tolerant Control Systems Design and Practical Applications. Springer, London (2009)

Patton, R.J.: Fault tolerant control systems: the 1997 situation. In: Proceedings of the IFAC Safe process, pp. 1033–1055. Hull, UK (1997)

Patton, R.J., Frank, P.M., Clark, R.N.: Fault Diagnosis in Dynamic Systems: Theory and Application. Prentice-Hall, New York (1989)

Prashant, M., Jinfeng, L., Panagiotis, D.C.: Fault-Tolerant Process Control Methods and Applications. Springer, London (2013)

Shin, D., Moon, G., Kim, Y.: Design of reconfigurable flight control system using adaptive sliding mode control: actuator fault. Proc. Inst. Mech. Eng. Part G J. Aerosp. Eng. **219**(4), 321–328 (2005)

Slotine, J.J.: Sliding controller design for nonlinear system. I. J. C. **4**(2), 421–434 (1984)

Stengel, R.F.: Intelligent failure-tolerant control. IEEE Control Syst. **11**(4), 14–23 (1991)

Tabbache, B., Rizoug, N., Benbouzid, M.E.H., Kheloui, A.: A control reconfiguration strategy for post-sensor FTC in induction motor based EVs. IEEE Trans. Veh. Technol. **62**(3), 965–971 (2013)

Tang, R.Y.: Modern Permanent Magnet Synchronous Motor Theory and Design. Machinery Industry Press, Beijing (1997)

Teng, Q., Zhu, J., Wang, T., Lei, G.: Fault tolerant direct torque control of three-phase permanent magnet synchronous motors. WSEAS Trans. Syst. **11**(8), 465–476 (2012)

Utkin, V. (ed.): Sliding Modes in Control Optimization. Springer, Berlin (1992)

Utkin, V., Guldner, J., Shi, J.: Sliding Mode Control in Electromechanical Systems. Taylor & Francis, New York (1999)

Vas, P.: Parameter Estimation, Condition Monitoring and Diagnosis of Electrical Machines. Oxford Science Publications, Oxford (1994)

Xiao, B., Hu, Q., Zhang, Y.: Adaptive sliding mode fault tolerant attitude tracking control for flexible spacecraft under actuator saturation. IEEE Trans. Control Syst. Technol. **20**(6), 1605–1612 (2012)

Yan, X.G., Edwards, C.: Nonlinear robust fault reconstruction and estimation using a sliding mode observer. Automatica **43**(2007), 1605–1614 (2007)

Yang, G.H., Ye, D.: Reliable H1 control of linear systems with adaptive mechanism. IEEE Trans. Autom. Control **55**(1), 242–247 (2010)

Zhang, Y., Jiang, J.: Bibliographic review on reconfigurable fault-tolerant control systems. Annu. Rev. Control **32**(2), 229–252 (2008)

Transient Stability Enhancement of Power Systems Using Observer-Based Sliding Mode Control

M. Ouassaid, M. Maaroufi and M. Cherkaoui

Abstract The high complexity and nonlinearity of power systems, together with their almost continuously time-varying nature, have presented a big challenge for control engineers, for decades. The disadvantages of the linear controllers/models, such as being dependent on the operating condition, sensibility to the disturbance such as parametric variations or faults can be overcome by using appropriate nonlinear control techniques. Sliding-mode control technique has been extensively used when a robust control scheme is required. This chapter presents the transient stabilization with voltage regulation analysis of a synchronous power generator driven by steam turbine and connected to an infinite bus. The aim is to obtain high performance for the terminal voltage and the rotor speed simultaneously under a large sudden fault and a wide range of operating conditions. The methodology adopted is based on sliding mode control technique. First, a nonlinear sliding mode observer for the synchronous machine damper currents is proposed. Next, the control laws of the complete ninth order model of a power system, which takes into account the stator dynamics as well as the damper effects, are developed. They are shown to be asymptotically stable in the context of Lyapunov theory. Finally, the effectiveness of the proposed combined observer-controller for the transient stabilization and voltage regulation is demonstrated.

Nomenculture

v_d, v_q	Direct and quadrature axis stator terminal voltage components, respectively
v_{fd}	Excitation control input

M. Ouassaid (✉)
Cadi Ayyad University, Ecole Nationale des Sciences Appliqu ées, Safi, Morocco
e-mail: ouassaid@emi.ac.ma

M. Maaroufi · M. Cherkaoui
Mohammed V University, Ecole Mohammadia D'Ing énieurs, Rabat, Morocco
e-mail: maaroufi@emi.ac.ma

M. Cherkaoui
e-mail: cherkaoui@emi.ac.ma

A.T. Azar and Q. Zhu (eds.), *Advances and Applications in Sliding Mode Control systems*, Studies in Computational Intelligence 576, DOI 10.1007/978-3-319-11173-5_16

v_t	Terminal voltage
i_d, i_q	Direct and quadrature axis stator current components, respectively
i_{fd}	Field winding Current
i_{kd}, i_{kq}	Direct and quadrature axis damper winding current components, respectively
λ_d, λ_q	Direct and quadrature axis flux linkages, respectively
R_s	Stator resistance
R_{fd}	Field resistance
R_{kd}, R_{kq}	Damper winding resistances
L_d, L_q	Direct and quadrature self inductances, respectively
L_{fd}	Rotor self inductance
L_{kd}, L_{kq}	Direct and quadrature damper winding self inductances, respectively
L_{md}, L_{mq}	Direct and quadrature magnetizing inductances, respectively
ω	Angular speed of the generator
δ	Rotor angle of the generator
T_m	Mechanical torque
T_e	Electromagnetic torque
D	Damping constant
H	Inertia constant
a	Phase angle *of* infinite bus voltage
$V^{\propto}$	Infinite bus voltage
L_e	Inductance of the transmission line
R_e	Resistance of the transmission line

1 Introduction

Nowadays, electric power systems have evolved through continuing growth in interconnections, use of new technologies and controls. They are operating more and more closely to their limit stability in highly stressed conditions. To maintain a high degree of reliability and security, different forms of system instability must be considered in the design of controllers.

Stability is a condition of equilibrium between opposing forces. Depending on the network topology, system operating condition and the form of disturbance, different sets of opposing forces may experience sustained imbalance leading to different forms of instability. Figure 1 identifies the categories and subcategories of the power system stability problem. The classification of power system stability is generally based on the physical nature of the resulting mode of instability, the size of the disturbance considered, the devices, processes, and the time span (Kundur 1994).

The reliability of the power supply implies much more than merely being available. Ideally, the loads must be fed at constant voltage and frequency at all times. However, small or large disturbances such as power changes or short circuits may transpire. One of the most vital operation demands is maintaining good stability and transient

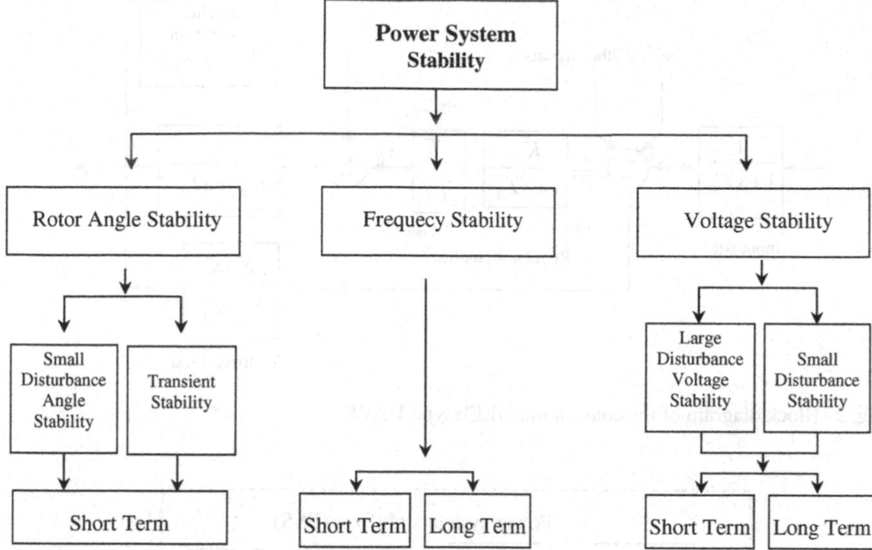

Fig. 1 Classification of power system stability

performance of the terminal voltage, rotor speed and the power transfer to the network (Guo et al. 2001; Jiawei et al. 2014). This requirement should be achieved by an adequate control of the system.

Traditionally, excitation controllers, which are mainly designed by using linear control theory, are used to regulate the terminal voltage at a specified value and ensure the stability under small and large disturbances. The principal conventional excitation controller is the automatic voltage regulator (AVR). Many different AVR models have been developed to represent the various types used in a power system. The IEEE defined several AVR types, the main one of which (Type 1) is shown in Fig. 2. The modern AVR employing conventional, fixed parameter compensators, whilst capable of providing good steady state voltage regulation and fast dynamic response to disturbances, suffers from considerable variations in voltage control performance as the generator operating change. Several forms of adaptive control have been investigated to address the problem of performance variation (Ghazizadeh and Hughes 1998).

Adversely, the generator automatic voltage regulator which reacts only to the voltage error weakens the damping introduced by damper windings. This detrimental effect of the AVR can be compensated using supplementary control loop which is the power system stabilizer (PSS). The structure of the PSS is given in Fig. 3. These stabilizers introduced additional system damping signals derived from the machine speed or power through the excitation system in order to improve the damping of power swings (Ghandakly and Farhoud 1992). Conventional fixed parameter stabilizers work reasonably well over medium range of operating conditions. However may diminish as the generator load changes or the network configuration is altered

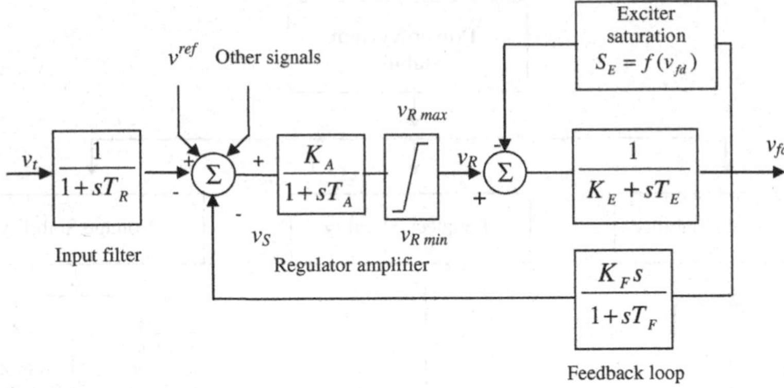

Fig. 2 Block diagram of the conventional IEEE type 1 AVR

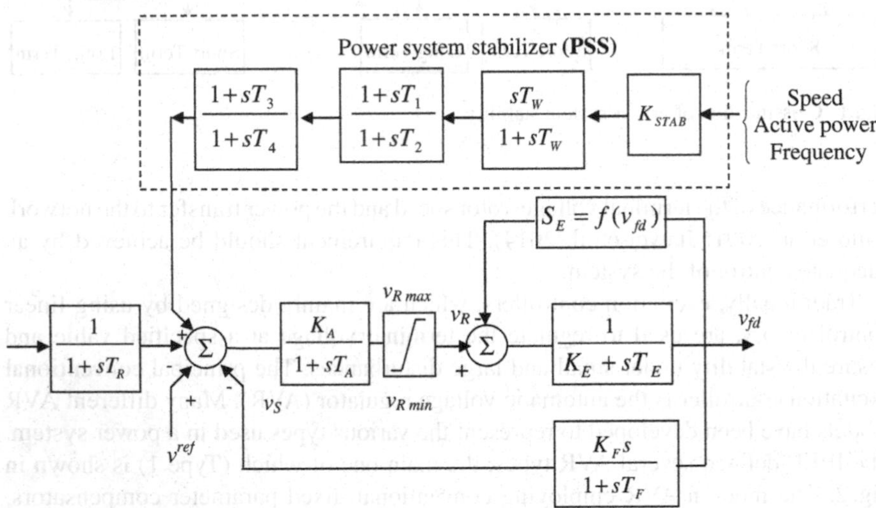

Fig. 3 Block diagram of the AVR+PSS structure

by faults or other switching conditions which lead to deterioration in the stabilizer performance. Remarkable efforts have been devoted to the design of appropriate PSS; various methods, such as root locus, eigenvalue techniques, pole placement, adaptive control, etc. have been used. But in all these methods, model uncertainties cannot be considered explicitly at the design stage (Zhao and Jiang 1995).

To deal with a high complexity and nonlinearity of power systems, together with their almost continuously time-varying nature, different techniques have been investigated in aim to:

- Tackle the problem of transient stability by considering nonlinear models of power systems.

- Overcome the drawbacks of the linear controllers via design of nonlinear controllers.

The main features of those controllers are summarized as follows:

- Independence of the equilibrium point and taking into account the important nonlinearities of the power system model.
- *Robustness* The designed controller must be insensible to all kinds of perturbations such as parametric variations or faults and the non-modeled dynamics.
- *Dynamics performance and Tracking* Terminal voltage, rotor speed and rotor angle converge to their references with accuracy and rapidity.
- *Enhancing the transient stability* Damping of all types of oscillations (local and inter area).

Several control approaches have been applied. As a summary, the main strategies are outlined as follows:

1.1 Feedback linearization

The essence of this technique is to first transform a nonlinear system into a linear on by a nonlinear feedback, and then uses the well-known linear design techniques to complete the controller design (Isidori 1995). Nevertheless these control designs require the exact cancellation of nonlinear terms. With parametric uncertainties present in the system, the cancellation is no longer applied. This constitutes an important drawback in the implementation of such controllers in the presence of model uncertainties and/or external disturbances, thus affecting the robustness of the closed loop system (Gao et al. 1992; King et al. 1994). Several adaptive versions of the feedback linearizing controls are then developed in (Jain et al. 1994; Tan and Wang 1998).

1.2 Passivity based control

The control based on the passivity has been the subject of several investigations. The aim of the method is to make the system passive closed loop (Byrnes et al. 1991; Kokotovic 1992; Ortega et al. 1998). This approach is limited to physical systems described by equations of motion Euler-Lagrange. The major problem with this approach is that the performance of the closed loop system depends on the knowledge of the model parameters used to define terms of energy dissipation. Therefore, the performance is not satisfactory if terms of energy, which are used to ensure the asymptotic stability of the controlled system dissipation, are used to ensure the passivity for all operating conditions (Nickllasson et al. 1997). References (Ortega et al. 1998) and (Galaz et al. 2003) present an application of this technique.

1.3 Robust control

To cope with parametric uncertainties in power systems, many robust voltage regulators have been proposed using the theory of linear robust control such as H ∞ (Ahmed et al. 1996) and the L ∞ stability theory (Guo et al. 2001; Jiawei et al. 2014). In (Ohtsuk 1992), several types of uncertainties and changes in variables are taken into account in the design of H ∞ controller. The maximum effects of these disturbances are minimized. The use of this type of control for electric power system is investigated in (Xi et al. 2002) and (Wang et al. 2003). The disadvantage of these regulators is excessive gain values, which makes it difficult their practical achievements.

1.4 Adaptive control

It should be noted that the model of a process, even relatively complex, is never perfect. This type of approach applies to systems whose dynamics are known but whose parameters are poorly identified or unknown or even slowly varying in time (Astrom and Wittenmark 1995). The weakness of this type of controller resides essentially in the fact that the dynamics of the estimator is not considered in the design process. The relatively slow convergence of the adaptation may result in some cases irreversible instability of the loop (Narendra and Balakrishnan 1997). In (Khorrami et al. 1994; Ghandakly and Dai 2000; Shen et al. 2003; Jiao et al. 2005; Wu and Malik 2006), regulators of power system are based on adaptive control.

1.5 Backstepping technique

This approach widely detailed by Krstic and Kokotovic Kanellakopolus in (Krstić et al. 1995) provides solutions to the aforementioned problems. Indeed, the backstepping, whose basic idea is to synthesize the control law in a recursive manner, is less restrictive compared to the control non-linear state feedback which cancels the nonlinearities that might be useful. Unlike the adaptive controllers, based on certain equivalence, which separate the design of the controller and the terms of adaptation, adaptive backstepping has emerged as an alternative. In adaptive backstepping, the control law takes into account the dynamic adaptation. These last two and the Lyapunov function which guarantees the stability and performance of the overall system are designed simultaneously. This technique has been successfully applied for power system in (Karimi and Feliachi 2008; Ouassaid et al. 2008, 2010).

1.6 Intelligent control

New approaches have been proposed for power stability such as fuzzy logic control (Mrad et al. 2000; Abbadi et al. 2013), neurocontrol (Shamsollahi and Malik 1997; Park et al. 2003; Venayagamoorthy et al. 2003; Mohagheghi et al. 2007) and algorithm genetic (Alkhatib and Duveau 2013). Combinations of the above techniques are also proposed in order to exploit the advantages of each method. These solutions are efficient, but they increase the cost and complexity of the control system (Segal et al. 2000; Wang 2013).

1.7 Sliding mode control

This method is a very interesting technique. It dates back to the 70s with the work of Utkin (Utkin 1977). It is a robust control to the parametric uncertainties and neglected dynamics. Nevertheless, the problems of chattering inherent in this type of discontinuous control appear quickly. Note that the chattering may excite the high-frequency dynamics neglected sometimes leading to instability. Methods to reduce this phenomenon have been developed (Slotine and Li 1991). This technique was applied to electric power systems in (Morales et al. 2001; Colbia-Vega et al. 2008; Huerta et al. 2010).

Almost all the mentioned above controllers for EPS consider reduced order models, taking into account the generator mechanical dynamics only. In the most of those studies, the nonlinear model used was a reduced third order model of the synchronous machine. In (Loukianov et al. 2004; Cabrera-Vazquez et al. 2007) sliding mode controllers for infinite machine bus systems have been designed considering the mechanical rotor, and electrical stator dynamics. Likewise, In (Akhrif et al. 1999), the feedback linearization technique was used to improve the system's stability and to obtain good post-fault voltage regulation. It is based on a 7 order model of the synchronous machine which takes into account the damper windings effects. However the authors assumed that the damper currents are available for measurement. In fact, the technology for direct damper current measurement is not fully developed yet. Because, damper windings are metal bars placed in slots in the pole faces and connected together at each end.

Thanks to the mentioned assumption, implementation of a controller based on a complete 7th order model of power synchronous machine requires information about the entire states of the power system. As a result, the estimation problem of damper currents of synchronous generator arises. For this purpose, a nonlinear observer for damper currents is developed, based on the sliding mode technique (Ouassaid et al. 2012).

The rest of this chapter is organized as follows: In Sect. 2, a mathematical model of a power system is introduced. It is based on a detailed 9th order model of a system which consists of a steam turbine and Single Machine Infinite Bus (SMIB) and takes into account the stator dynamics as well as the damper winding effects and practical limitation on controls. A nonlinear observer for damper winding currents is developed in Sect. 3. Then, in Sect. 4, a sliding mode controller is constructed based on a time-varying sliding surface to control the rotor speed and terminal voltage, simultaneously, in order to enhance the transient stability and to ensure good post-fault voltage regulation for power system. Section 5 presents a number of numerical simulations results of the proposed observer-based nonlinear controller. Finally, conclusions are given in Sect. 6.

2 Power System Model

The system to be controlled is shown in Fig. 4. It consists of synchronous generator driven by steam turbine and connected to an infinite bus via a transmission line. The synchronous generator is described by a 7th order nonlinear mathematical model which comprises three stator windings, one field winding and two damper windings.

The synchronous machine equations in terms of the Park's d-q axis are expressed (Fig. 5) as follows (Cheng and Hsu 1992; Anderson and Fouad 1994):

Armature windings

$$v_d = -R_s i_d - \omega\lambda_q + \frac{d\lambda_d}{dt} \quad (1)$$

$$v_q = -R_s i_q + \omega\lambda_d + \frac{d\lambda_q}{dt} \quad (2)$$

where

$$\lambda_d = -L_d i_d + L_{md}(i_{fd} + i_{kd}) \quad (3)$$

$$\lambda_q = -L_q i_q + L_{mq} i_{kq} \quad (4)$$

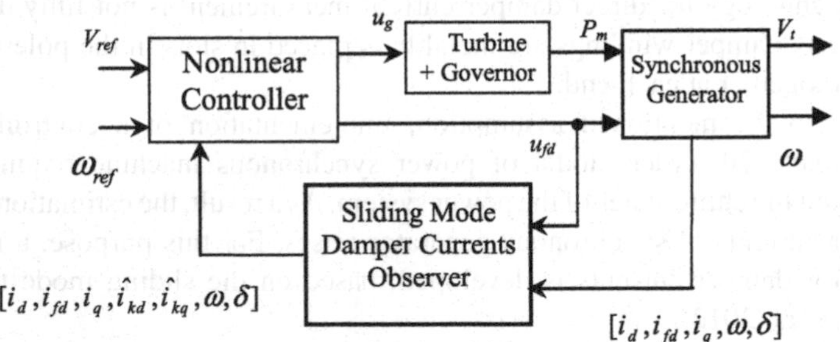

Fig. 4 Block diagram of the power system with observer based-controller

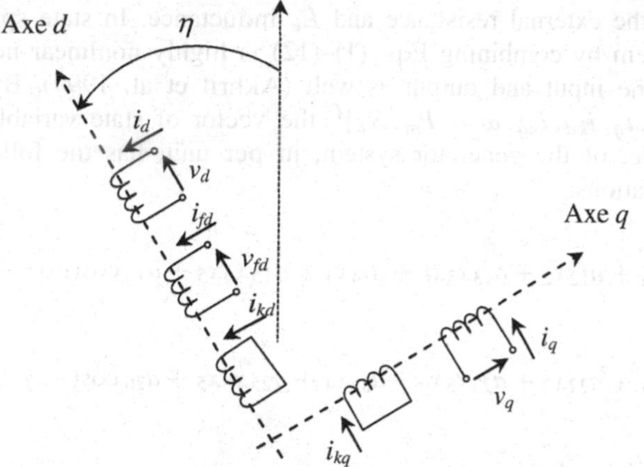

Fig. 5 Synchronous machine in Park's d-q axis

Field winding

$$v_{fd} = R_s i_{fd} - L_{md}\frac{di_d}{dt} + L_{fd}\frac{di_{fd}}{dt} + L_{md}\frac{di_{kd}}{dt} \tag{5}$$

Damper windings

$$0 = R_{kd}i_{kd} - L_{md}\frac{di_d}{dt} + L_{md}\frac{di_{fd}}{dt} + L_{kd}\frac{di_{kd}}{dt} \tag{6}$$

$$0 = R_{kq}i_{kq} - L_{mq}\frac{di_d}{dt} + L_{kq}\frac{di_{kq}}{dt} \tag{7}$$

Mechanical equations

$$\frac{d\delta}{dt} = \omega - 1 \tag{8}$$

$$2H\frac{d\omega}{dt} = T_m - T_e - D\omega \tag{9}$$

The electromagnetic torque is

$$T_e = \left(L_q - L_d\right) i_d i_q + L_{mfd} i_{fd} i_q + L_{md} i_{kd} i_q - L_{mq} i_d i_{kq} \tag{10}$$

The equation of transmission network in the Park's coordinates is

$$v_d = R_e i_d + L_e\frac{di_d}{dt} - \omega L_e i_q + V^{\infty}\cos(\delta - a) \tag{11}$$

$$v_q = R_e i_q + L_e\frac{di_q}{dt} + \omega L_e i_d - V^{\infty}\sin(\delta - a) \tag{12}$$

where R_e is the external resistance and L_e inductance. In state space form, the resulting system by combining Eqs. (1)–(12) is highly nonlinear not only in the state but in the input and output as well (Akhrif et al. 1999). By considering $x = [i_d, i_{fd}, i_q, i_{kd}, i_{kq}, \omega, \delta, P_m, X_e]^T$ the vector of state variables, the mathematical model of the generator system, in per unit, has the following form: Electrical equations:

$$\frac{dx_1}{dt} = x_1 a_{11} + a_{12}x_2 + a_{13}x_3x_6 + a_{14}x_4 + a_{15}x_6x_5 + a_{16}\cos(-x_7+\sigma) + b_1 u_{fd} \tag{13}$$

$$\frac{dx_2}{dt} = a_{21}x_1 + a_{22}x_2 + a_{23}x_3x_6 + a_{24}x_4 + a_{25}x_6x_5 + a_{26}\cos(-x_7+\sigma) + b_2 u_{fd} \tag{14}$$

$$\frac{dx_3}{dt} = a_{31}x_1x_6 + a_{32}x_2x_6 + a_{33}x_3 + a_{34}x_4x_6 + a_{35}x_5 + a_{36}\sin(-x_7+\sigma) \tag{15}$$

$$\frac{dx_4}{dt} = a_{41}x_1 + a_{42}x_2 + a_{43}x_3x_6 + a_{44}x_4 + a_{45}x_6x_5 + a_{46}\cos(-x_7+\sigma) + b_3 u_{fd} \tag{16}$$

$$\frac{dx_5}{dt} = a_{51}x_1x_6 + a_{52}x_2x_6 + a_{53}x_3 + a_{54}x_4x_6 + a_{55}x_5 + a_{56}\sin(-x_7+\sigma) \tag{17}$$

Mechanical equations:

$$\frac{dx_6}{dt} = a_{61}x_6 + a_{62}\frac{x_8}{x_6} - a_{62}T_e \tag{18}$$

$$\frac{dx_7}{dt} = \omega_R(x_6 - 1) \tag{19}$$

Turbine dynamics (Hill and Wang 2000):

$$\frac{dx_8}{dt} = a_{81}x_8 + a_{82}x_9 \tag{20}$$

Turbine valve control (Hill and Wang 2000):

$$\frac{dx_9}{dt} = a_{91}x_9 + a_{92}x_6 + b_4 u_g \tag{21}$$

where, u_{fd} the excitation control input, u_g the input power of control system. The parameters a_{ij} and b_i are described as follow

$$\begin{aligned}
a_{11} &= -(R_s+R_e)(L_{fd}L_{kd} - L_{md}^2)\omega_R D_d^{-1} & a_{41} &= -(R_s+R_e)(L_{fd}L_{md} - L_{md}^2)\omega_R D_d^{-1}\\
a_{12} &= -R_{fd}(L_{mq}L_{kd} - L_{md}^2)\omega_R D_d^{-1} & a_{42} &= R_{fd}((L_d+L_e)L_{md} - L_{md}^2)\omega_R D_d^{-1}\\
a_{13} &= (L_q+L_e)(L_{md}L_{kd} - L_{md}^2)\omega_R D_d^{-1} & a_{45} &= -L_{md}(L_{mq}.L_{fd} - L_{md}^2)\omega_R D_d^{-1}\\
a_{15} &= -L_{mq}(L_{fd}L_{kd} - L_{md}^2)\omega_R D_d^{-1} & a_{43} &= (L_q+L_e)(L_{md}L_d - L_{md}^2)\omega_R D_d^{-1}\\
a_{14} &= R_{kd}((L_d+L_e)L_{md} - L_{md}^2)\omega_R D_d^{-1} & a_{44} &= -R_{kd}((L_d+L_e)L_{fd} - L_{md}^2)\omega_R D_d^{-1}
\end{aligned}$$

$$a_{16} = -V^{\infty}((L_d + L_e)L_{kd} - L_{md}^2)\omega_R D_d^{-1} \quad a_{46} = -V^{\infty}(L_{md}.L_{fd} + L_{md}^2)\omega_R D_d^{-1}$$
$$b_1 = (L_{md}L_{kd} - L_{md}^2)\omega_R D_d^{-1} \quad a_{51} = -(L_d + L_e)L_{mq}\omega_R D_q^{-1}$$
$$a_{21} = -(R_s + R_e)(L_{md}L_{kd} - L_{md}^2)\omega_R D_d^{-1} \quad a_{52} = L_{md}L_{mq}\omega_R D_q^{-1}$$
$$a_{22} = -R_{fd}((L_d + L_e)L_{kd} - L_{md}^2)\omega_R D_d^{-1} \quad a_{53} = -(R_s + R_e)L_{mq}\omega_R D_q^{-1}$$
$$a_{23} = (L_q + L_e)(L_{md}L_{kd} - L_{md}^2)\omega_R D_d^{-1} \quad a_{54} = L_{md}L_{mq}\omega_R D_q^{-1}$$
$$a_{24} = R_{kd}((L_d + L_e)L_{md} - L_{md}^2)\omega_R D_d^{-1} \quad a_{55} = -R_{kq}(L_q + L_e)\omega_R D_q^{-1}$$
$$a_{25} = -L_{mq}(L_{md}L_{kd} - L_{md}^2)\omega_R D_d^{-1} \quad a_{56} = -V^{\infty}L_{mq}\omega_R D_d^{-1}$$
$$a_{26} = -V^{\infty}(L_{md}L_{kd} - L_{md}^2)\omega_R D_d^{-1} \quad a_{61} = -D(2H)^{-1}$$
$$b_2 = ((L_d + L_{fd})L_{kd} - L_{md}^2)\omega_R D_d^{-1} \quad a_{62} = (2H)^{-1}$$
$$a_{31} = -(L_d + L_e)L_{kq}\omega_R D_q^{-1} \quad a_{81} = -(T_m)^{-1}$$
$$a_{32} = L_{md}L_{kq}\omega_R D_q^{-1} \quad a_{82} = K_m(T_m)^{-1}$$
$$a_{33} = -(R_s + R_e)L_{kq}\omega_R D_q^{-1} \quad a_{91} = -(T_g)^{-1}$$
$$a_{34} = L_{md}L_{kq}\omega_R D_q^{-1} \quad a_{92} = -K_g(T_g R\omega_R)^{-1}$$
$$a_{35} = -L_{mq}.R_{kq}\omega_R D_q^{-1} \quad b_4 = K_g(T_g)^{-1}$$
$$a_{36} = V^{\infty}L_{kq}\omega_R D_q^{-1}$$
$$b_3 = ((L_d + L_e)L_{md} - L_{md}^2)\omega_R D_d^{-1}$$

Here we have denoted

$$D_d = (L_d + L_e)L_{fd}L_{kd} - L_{md}^2(L_d + L_{fd} + L_{kd}) + 2L_{md}^3$$
$$D_q = (L_q + L_e)L_{kq} - L_{mq}^2$$

The machine terminal voltage is calculated from Park components v_d and v_q as follows (Anderson and Fouad 1994; Akhrif et al. 1999):

$$v_t = \left(v_d^2 + v_q^2\right)^{\frac{1}{2}} \tag{22}$$

with

$$v_d = c_{11}x_1 + c_{12}x_2 + c_{13}x_3x_6 + c_{14}x_4 + c_{15}x_5x_6 + c_{16}\cos(-x_7 + \sigma) + c_{17}u_{fd} \tag{23}$$

$$v_q = c_{21}x_1x_6 + c_{22}x_2x_6 + c_{23}x_3 + c_{24}x_4x_6 + c_{25}x_5 + c_{26}\sin(-x_7 + \sigma) \tag{24}$$

where c_{ij} are coefficients which depend on the coefficients a_{ij}, on the infinite bus phase voltage V^{∞} and the transmission line parameters R_e and L_e. They are described as follow

$$c_{11} = R_e + a_{11}L_e\omega_R^{-1} \quad c_{17} = b_1 L_e\omega_R^{-1}$$
$$c_{12} = a_{12}L_e\omega_R^{-1} \quad c_{21} = L_e + a_{31}L_e\omega_R^{-1}$$
$$c_{13} = L_e(a_{13}\omega_R^{-1} - 1) \quad c_{22} = a_{32}L_e\omega_R^{-1}$$
$$c_{14} = a_{14}L_e\omega_R^{-1} \quad c_{23} = a_{33}L_e\omega_R^{-1} + R_e$$
$$c_{15} = a_{15}L_e\omega_R^{-1} \quad c_{24} = a_{34}L_e\omega_R^{-1}$$

$$c_{16} = V^{\infty} + a_{16} L_e \omega_R^{-1} \qquad c_{25} = a_{35} L_e \omega_R^{-1}$$
$$c_{26} = V^{\infty} + a_{36} L_e \omega_R^{-1}$$

Available states for synchronous generator are the stator phase currents i_d and i_q, voltages at the terminals of the machine v_d and v_q, field current i_{fd}. It is also assumed that the angular speed ω and the power angle δ are available for measurement (De Mello 1994). In the next section the construction an observer of the damper currents i_{kd} and i_{kq} will be given.

3 Sliding Mode Observer for the Damper Winding Currents

The state space representation of the electrical dynamics of the power system model (13)–(17) is given as

$$\frac{d}{dt}\begin{bmatrix} x_1 \\ x_2 \\ x_3 \end{bmatrix} = F_{11}\begin{bmatrix} x_1 \\ x_2 \\ x_3 \end{bmatrix} + F_{12}\begin{bmatrix} x_4 \\ x_5 \end{bmatrix} + \begin{bmatrix} b_1 \\ b_2 \\ 0 \end{bmatrix} u_{fd} + H_1(t) \tag{25}$$

$$\frac{d}{dt}\begin{bmatrix} x_4 \\ x_5 \end{bmatrix} = F_{21}\begin{bmatrix} x_1 \\ x_2 \\ x_3 \end{bmatrix} + F_{22}\begin{bmatrix} x_4 \\ x_5 \end{bmatrix} + \begin{bmatrix} b_3 \\ 0 \end{bmatrix} u_{fd} + H_2(t) \tag{26}$$

where

$$H_1(t) = [a_{16}\cos(-x_7+\sigma), a_{26}\cos(-x_7+\sigma), a_{36}\sin(-x_7+\sigma)]^T$$

$$H_2(t) = [a_{46}\cos(-x_7+\sigma), a_{56}\sin(-x_7+\sigma)]^T$$

$$F_{11} = \begin{bmatrix} a_{11} & a_{12} & a_{13}x_6 \\ a_{21} & a_{22} & a_{23}x_6 \\ a_{31}x_6 & a_{32}x_6 & a_{33} \end{bmatrix}$$

$$F_{21} = \begin{bmatrix} a_{41} & a_{42} & a_{43}x_6 \\ a_{51}x_6 & a_{52}x_6 & a_{53} \end{bmatrix}$$

$$F_{12} = \begin{bmatrix} a_{14} & a_{15}x_6 \\ a_{24} & a_{25}x_6 \\ a_{34}x_6 & a_{35} \end{bmatrix}$$

$$F_{22} = \begin{bmatrix} a_{44} & a_{45}x_6 \\ a_{54}x_6 & a_{55} \end{bmatrix}$$

Considering the switching surface S as follows

$$S(t) = \begin{bmatrix} \hat{x}_1 - x_1 \\ \hat{x}_2 - x_2 \\ \hat{x}_3 - x_3 \end{bmatrix} \equiv \begin{bmatrix} z_1 \\ z_2 \\ z_3 \end{bmatrix} = 0 \tag{27}$$

Hence, a sliding mode observer for (25) is defined as

$$\frac{d}{dt}\begin{bmatrix} \hat{x}_1 \\ \hat{x}_2 \\ \hat{x}_3 \end{bmatrix} = F_{11}\begin{bmatrix} \hat{x}_1 \\ \hat{x}_2 \\ \hat{x}_3 \end{bmatrix} + F_{12}\begin{bmatrix} \hat{x}_4 \\ \hat{x}_5 \end{bmatrix} + \begin{bmatrix} b_1 \\ b_2 \\ 0 \end{bmatrix} u_{fd} + H_1(t) + K\begin{bmatrix} \mathrm{sgn}(\hat{x}_1 - x_1) \\ \mathrm{sgn}(\hat{x}_2 - x_2) \\ \mathrm{sgn}(\hat{x}_3 - x_3) \end{bmatrix} \tag{28}$$

where $\hat{x}_1, \hat{x}_2$ and $\hat{x}_3$ are the observed values of i_d, i_{fd} and i_q, K is the switching gain, and sgn is the sign function.

Furthermore, the damper current observer is given from (26) as

$$\frac{d}{dt}\begin{bmatrix} \hat{x}_4 \\ \hat{x}_5 \end{bmatrix} = F_{21}\begin{bmatrix} \hat{x}_1 \\ \hat{x}_2 \\ \hat{x}_3 \end{bmatrix} + F_{22}\begin{bmatrix} \hat{x}_4 \\ \hat{x}_5 \end{bmatrix} + \begin{bmatrix} b_3 \\ 0 \end{bmatrix} u_{fd} + H_2(t) \tag{29}$$

where $\hat{x}_4$ and $\hat{x}_5$ are the observed values of i_{kd} and i_{kq}.

Subtracting (25) from (28), the error dynamics can be written in the following form

$$\frac{d}{dt}\begin{bmatrix} z_1 \\ z_2 \\ z_3 \end{bmatrix} = F_{11}\begin{bmatrix} z_1 \\ z_2 \\ z_3 \end{bmatrix} + F_{12}\begin{bmatrix} \tilde{x}_4 \\ \tilde{x}_5 \end{bmatrix} + K\begin{bmatrix} \mathrm{sgn} z_1 \\ \mathrm{sgn} z_2 \\ \mathrm{sgn} z_3 \end{bmatrix} \tag{30}$$

where $\tilde{x}_4$and $\tilde{x}_5$ are the estimation errors of the damper currents x_4 and x_5.

The switching gain is defined as

$$K = \min\left\{ \begin{array}{l} -a_{11}\left|z_1\right| - (a_{12}z_2 + a_{13}\omega z_3 + a_{14}\tilde{x}_4 + a_{15}\omega\tilde{x}_5)\,\mathrm{sgn} z_1 \\ -a_{22}\left|z_2\right| - (a_{21}z_1 + a_{23}\omega z_3 + a_{24}\tilde{x}_4 + a_{25}\omega\tilde{x}_5)\,\mathrm{sgn} z_2 \\ -a_{33}\left|z_3\right| - (a_{31}\omega z_1 + a_{32}\omega z_2 + a_{34}\omega\tilde{x}_4 + a_{35}\tilde{x}_5)\,\mathrm{sgn} z_3 \end{array} \right\} - \xi \tag{31}$$

where ξ is a positive small value.

Theorem 1 *The globally asymptotic stability of* (30) *is guaranteed, if the switching gain is given by* (31).

Proof The stability of the overall structure is guaranteed through the stability of the direct axis and quadrature axis currents x_1, x_2, and field current x_3 observer. The

Lyapunov function of the sliding mode observer for damper currents is chosen as

$$V_{obs} = \frac{1}{2} S^T \Gamma S \tag{32}$$

where Γ is an identity positive matrix. Consequently, the derivative of the Lyapunov function is

$$\begin{aligned} \frac{dV_{obs}}{dt} &= S^T \Gamma \frac{dS}{dt} \\ &= \begin{bmatrix} z_1 \\ z_2 \\ z_3 \end{bmatrix}^T \Gamma \left(F_{11} \begin{bmatrix} z_1 \\ z_2 \\ z_3 \end{bmatrix} + F_{12} \begin{bmatrix} \tilde{x}_4 \\ \tilde{x}_5 \end{bmatrix} + K \begin{bmatrix} \mathrm{sgn} z_1 \\ \mathrm{sgn} z_2 \\ \mathrm{sgn} z_3 \end{bmatrix} \right) \\ &= G_1 + G_2 + G_3 \end{aligned} \tag{33}$$

where

$$\begin{aligned} G_1 &= a_{11} z_1^2 + a_{12} z_1 z_2 + a_{13} \omega z_1 z_3 + a_{14} z_1 \tilde{x}_4 + a_{15} \omega z_1 \tilde{x}_5 + K \left| z_1 \right| \\ G_2 &= a_{21} z_1 z_2 + a_{22} z_2^2 + a_{23} \omega z_2 z_3 + a_{24} z_2 \tilde{x}_4 + a_{25} \omega z_2 \tilde{x}_5 + K \left| z_2 \right| \\ G_3 &= a_{31} \omega z_1 z_3 + a_{32} \omega z_2 z_3 + a_{33} z_3^2 + a_{34} \omega z_3 \tilde{x}_4 + a_{35} z_3 \tilde{x}_5 + K \left| z_3 \right| \end{aligned}$$

Using the designed switching gain in (31), both G_1, G_2 and G_3 are negatives. Therefore, $\dot{V}_{obs}$ is a negative definite, and the sliding mode condition is satisfied (Slotine and Li 1991). Furthermore the global asymptotic stability of the observer is guaranteed.

According to (31) by a proper selection of ξ, the influence of parametric uncertainties of the SMIB can be much reduced. The switching gain must large enough to satisfy the reaching condition of sliding mode. Hence the estimation error is confined into the sliding hyperplane

$$\frac{d}{dt} \begin{bmatrix} z_1 \\ z_2 \\ z_3 \end{bmatrix} = \begin{bmatrix} z_1 \\ z_2 \\ z_3 \end{bmatrix} = 0 \tag{34}$$

Nevertheless, if the switching gain is too large, the chattering noise may lead to estimation errors. To avoid the chattering phenomena, the sign function is replaced by the following continuous function

$$\frac{S(t)}{|S(t)| + \varsigma_1}$$

where ς_1 is a positive constant.

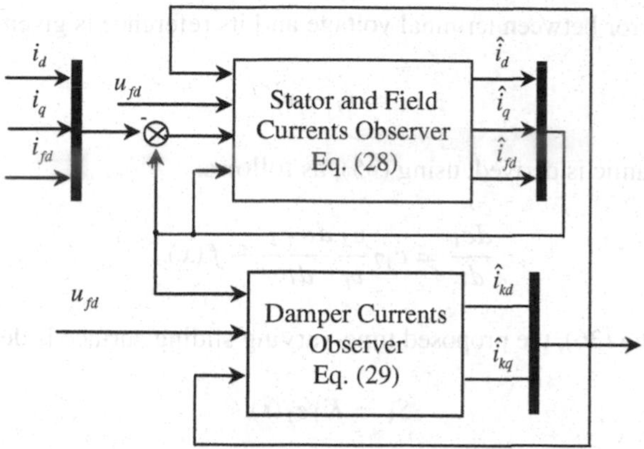

Fig. 6 Block diagram of the sliding mode damper currents observer

4 Design of Sliding Mode Controllers

This section deals with a procedure for the design of power system controllers, in order to improve the system's stability and damping properties under large disturbances and variation in operating points. The first objective is the terminal voltage regulation.

The dynamic of the terminal voltage (35), is obtained through the time derivative of (22) using (23) and (24) where the damper currents are replaced by the observer (29)

$$\begin{aligned}\frac{dv_t}{dt} &= \frac{1}{v_t}\left(v_d\frac{dv_d}{dt} + v_q\frac{dv_q}{dt}\right) \\ &= \frac{v_q}{v_t}\frac{dv_q}{dt} + c_{17}\frac{v_d}{v_t}\frac{du_{fd}}{dt} \\ &+ \frac{v_d}{v_t}\left[\begin{aligned} &c_{11}\frac{dx_1}{dt} + c_{12}\frac{dx_2}{dt} + c_{13}x_6\frac{dx_3}{dt} + c_{13}x_3\frac{dx_6}{dt} + c_{14}\frac{d\hat{x}_4}{dt} \\ &+c_{15}x_6\frac{d\hat{x}_5}{dt} + c_{15}\hat{x}_5\frac{dx_6}{dt} + c_{16}\frac{dx_7}{dt}\sin(-x_7+\sigma)\end{aligned}\right] \\ &= c_{17}\frac{v_d}{v_t}\frac{du_{fd}}{dt} + f(x)\end{aligned} \tag{35}$$

where

$$f(x) = \frac{v_q}{v_t}\frac{dv_q}{dt} + \frac{v_d}{v_t}\left[\begin{aligned} &c_{11}\frac{dx_1}{dt} + c_{12}\frac{dx_2}{dt} + c_{13}x_6\frac{dx_3}{dt} + c_{13}x_3\frac{dx_6}{dt} + c_{14}\frac{d\hat{x}_4}{dt} \\ &+c_{15}\hat{x}_5\frac{dx_6}{dt} + c_{15}x_6\frac{d\hat{x}_5}{dt} + c_{16}\frac{dx_7}{dt}\sin(-x_7+\sigma)\end{aligned}\right]$$

The tracking error between terminal voltage and its reference is given as

$$e_1 = v_t - v_t^{ref} \tag{36}$$

Hence, its dynamic is derived, using (35), as follows:

$$\frac{de_1}{dt} = c_{17}\frac{v_d}{v_t}\frac{du_{fd}}{dt} + f(x) \tag{37}$$

According to the (36), the proposed time-varying sliding surface is defined by

$$S_1 = K_1 e_1(t) \tag{38}$$

where K_1 is a positive constant feedback gain. The next step is to design a control input which satisfies the sliding mode existence law. The control input have the following structure

$$u(t) = u_{eq}(t) + u_n(t) \tag{39}$$

where $u_{eq}(t)$ is an equivalent control-input that determines the system's behavior on the sliding surface and $u_n(t)$ is a non-linear switching input, which drives the state to the sliding surface and maintains it on the sliding surface despite the presence of the parameter variations and disturbances. The equivalent control-input is obtained from the invariance condition and is given by the following condition (Utkin et al. 1999):

$$S_1 = 0 \text{ and } \frac{dS_1}{dt} = 0 \Rightarrow u(t) = u_{eq}(t)$$

From the above equation

$$\dot{S}_1 = K_1 c_{17}\frac{v_d}{v_t}\frac{du_{fd}}{dt} + K_1 f(x) = 0 \tag{40}$$

Therefore, the equivalent control-input is given as

$$u_{eq}(t) = -\frac{v_t}{c_{17}v_d} f(x) \tag{41}$$

By choosing the nonlinear switching input $u_n(t)$ as follows

$$u_n(t) = -\alpha_1 \frac{v_t}{c_{17}v_d}\text{sgn}(e_1) \tag{42}$$

where α_1 is a positive constant. The control input is derived from (39), (41) and (42) as follows:

$$u(t) = \frac{du_{fd}}{dt} = -\frac{v_t}{c_{17}v_d}\left(f(x) + \alpha_1 \mathrm{sgn}(e_1)\right) \tag{43}$$

Using the proposed control law (43), the reachability of sliding mode control of (37) is guaranteed.

Now, the attention is focused to the rotor speed tracking objective. The sliding mode-based rotor speed control methodology consists of three steps

Step 1: The rotor speed error is

$$e_2 = x_6 - \omega^{ref} \tag{44}$$

where $\omega^{ref} = 1$ p.u. is the desired trajectory. The sliding surface is selected as follow

$$S_2 = K_2 e_2(t) \tag{45}$$

where K_2 is a positive constant. By using (44) and (18), the derivative of the sliding surface (45) is calculated as:

$$\frac{dS_2}{dt} = K_2\left(a_{61}x_6 + a_{62}x_8/x_6 - a_{62}T_e\right) \tag{46}$$

The x_8 can be viewed as a virtual control in the above equation. To ensure the Lyapunov stability criteria i.e. $\frac{dS_2}{dt}S_2 \prec 0$ we define the nonlinear control input x^*_{8eq} as

$$x^*_{8eq} = \frac{x_6}{a_{62}}\left(a_{62}T_e - a_{61}x_6\right) \tag{47}$$

The nonlinear switching input x^*_{8n} can be chosen as follows

$$x^*_{8n} = -\alpha_2 \frac{x_6}{a_{62}} \mathrm{sgn}(e_2) \tag{48}$$

where α_2 is a positive constant.
Then, the stabilizing function of the mechanical power is obtained as

$$x^*_8 = \frac{x_6}{a_{62}}\left(a_{62}T_e - a_{61}x_6 - \alpha_2 \mathrm{sgn}(e_2)\right) \tag{49}$$

When a fault occurs, large currents and torques are produced. This electrical perturbation may destabilize the operating conditions. Hence, it becomes necessary to account for these uncertainties by designing a higher performance controller.

In (49), as electromagnetic load T_e is unknown, when fault occurs, it has to be estimated adaptively. Thus, let us define

$$\hat{x}_8^* = \frac{x_6}{a_{62}}\left(a_{62}\hat{T}_e - a_{61}x_6 - \alpha_2 \text{sgn}(e_2)\right) \tag{50}$$

where $\widehat{T_e}$ is the estimated value of the electromagnetic load which should be determined later. Substituting (50) in (46), the rotor speed sliding surface dynamics becomes

$$\frac{dS_2}{dt} = K_2\left(-\alpha_2 \text{sgn}(e_2) - a_{62}\widetilde{T_e}\right) \tag{51}$$

where $\widetilde{T_e} = T_e - \widehat{T_e}$ is the estimation error of electromagnetic load.

Step 2: Since the mechanical power x_8 is not our control input, the stabilizing error between x_8 and its desired trajectory x_8^* is defined as

$$e_3 = x_8^* - x_8 \tag{52}$$

To stabilize the mechanical power x_8, the new sliding surface is selected as

$$S_3 = K_3 e_3(t) \tag{53}$$

where K_3 is a positive constant. The derivative of S_3 using (52) and (20) is given as

$$\frac{dS_3}{dt} = K_3\left(a_{81}x_8 + a_{82}x_9 - \frac{dx_8^*}{dt}\right) \tag{54}$$

By considering the steam valve opening x_9 as a second virtual control, the equivalent control x_{9eq}^* is obtained as the solution of the equation $\frac{dS_3(t)}{dt} = 0$.

$$x_{9eq}^* = \frac{1}{a_{82}}\left(\frac{dx_8^*}{dt} - a_{81}x_8\right) \tag{55}$$

As a result, the stabilizing function of the steam valve opening x_9^* the mechanical power is computed as

$$x_9^* = \frac{1}{a_{82}}\left(\frac{dx_8^*}{dt} - a_{81}x_8 - \alpha_3 \text{sgn}(e_3)\right) \tag{56}$$

where α_3 is a positive constant. Substituting (56) in (54), the steam valve opening sliding surface dynamics becomes

$$\frac{dS_3}{dt} = -\alpha_3 K_3 \text{sgn}(e_3) \tag{57}$$

Step 3: Finally, the steam valve opening error is defined as

$$e_4 = x_9 - x_9^* \tag{58}$$

By defining a sliding surface $S_4(t) = K_4 e_4(t)$, the derivative of S_4 is calculated by time-differentiation of (58) and using (21)

$$\frac{dS_4}{dt} = K_4\left(a_{91}x_9 + a_{92}x_6 + b_4 u_g - \frac{dx_9^*}{dt}\right) \tag{59}$$

To assure the reaching condition $\frac{dS_4}{dt}S_4 \prec 0$, the equivalent control $u_{geq}(t)$ is obtained as

$$u_{geq} = \frac{1}{b_4}\left(\frac{dx_9^*}{dt} - a_{91}x_9 - a_{92}x_6\right) \tag{60}$$

Subsequently, the control law is written as

$$u_g = \frac{1}{b_4}\left(\frac{dx_9^*}{dt} - a_{91}x_9 - a_{92}x_6 - \alpha_4 \text{sgn}(e_4)\right) \tag{61}$$

Theorem 2 *The dynamic sliding mode control laws* (43) *and* (61) *with stabilizing functions* (50) *and* (56) *when applied to the single machine infinite power system, guarantee the asymptotic convergence of the outputs v_t and $x_6 = \omega$ to their desired values v_{tref} and ω_{ref} =1, respectively.*

Proof Consider the following positive definite Lyapunov function

$$V_{con} = \frac{1}{2}S_1^2 + \frac{1}{2}S_2^2 + \frac{1}{2}S_3^2 + \frac{1}{2}S_4^2 + \frac{1}{2\mu}\tilde{T}_e^2 \tag{62}$$

By considering (40), (51), (57) and (59), the derivative of (62) can be derived as follows:

$$\begin{aligned}
\dot{V}_{con} &= \frac{dS_1}{dt}S_1 + \frac{dS_2}{dt}S_2 + \frac{dS_3}{dt}S_3 + \frac{dS_4}{dt}S_4 + \tilde{T}_e\frac{1}{\mu}\frac{d\tilde{T}_e}{dt} \\
&= K_1 c_{17}\frac{v_d}{v_t}\frac{du_{fd}}{dt} + K_1 f(x) + K_2\left(-\alpha_2 \text{sgn}(e_2) - a_{62}\tilde{T}_e\right) \\
&\quad - \alpha_3 K_3 \text{sgn}(e_3) + K_4\left(a_{91}x_9 + a_{92}x_6 + b_4 u_g - \frac{dx_9^*}{dt}\right) + \tilde{T}_e\frac{1}{\mu}\frac{d\tilde{T}_e}{dt}
\end{aligned} \tag{63}$$

Substituting the control laws (43) and (61) in (63) gives

$$\begin{aligned}\dot{V}_{con} &= -\alpha_1 K_1^2 e_1 \mathrm{sgn}(e_1) - \alpha_2 K_2^2 e_2 \mathrm{sgn}(e_2) - \alpha_3 K_3^2 e_3 \mathrm{sgn}(e_3) \\ &\quad - \alpha_4 K_4^2 e_4 \mathrm{sgn}(e_3) - K_2^2 a_{62} \tilde{T}_e e_2 + \tilde{T}_e \frac{1}{\mu}\frac{d\tilde{T}_e}{dt} \\ &= -\alpha_1 K_1^2 |e_1| - \alpha_2 K_2^2 |e_2| - \alpha_3 K_3^2 |e_3| - \alpha_4 K_4^2 |e_4| \\ &\quad + \left(\frac{1}{\mu}\frac{d\tilde{T}_e}{dt} - K_2^2 a_{62} e_2\right)\tilde{T}_e\end{aligned} \tag{64}$$

By choosing the adaptive law (65), the time derivative of V_{con} is strictly negative.

$$\frac{d\tilde{T}_e}{dt} = \mu a_{62} K_2^2 e_2 \tag{65}$$

Thus

$$\begin{aligned}\frac{dV_{con}}{dt} &= -\alpha_1 K_1^2 |e_1| - \alpha_2 K_2^2 |e_2| - \alpha_3 K_3^2 |e_3| - \alpha_4 K_4^2 |e_4| \\ &= -\sum_{i=1}^{4} \alpha_i K_i^2 |e_i| < 0\end{aligned} \tag{66}$$

From the above analysis, it is evident that the reaching condition of sliding mode is guaranteed.

Remark In order to eliminate the chattering, the discontinuous control components in (43), (50), (56) and (61) can be replaced by a smooth sliding mode component to yield

$$\begin{aligned}\frac{du_{fd}}{dt} &= -\frac{v_t}{c_{17} v_d}\left(f(x) + \alpha_1 \frac{S_1(t)}{|S_1(t)| + \tau_2}\right) \\ x_8^* &= \frac{x_6}{a_{62}}\left(a_{62} T_e - a_{61} x_6 - \alpha_2 \frac{S_2(t)}{|S_2(t)| + \tau_3}\right) \\ x_9^* &= \frac{1}{a_{82}}\left(\frac{dx_8^*}{dt} - a_{81} x_8 - \alpha_3 \frac{S_3(t)}{|S_3(t)| + \tau_4}\right) \\ u_g &= \frac{1}{b_4}\left(\frac{dx_9^*}{dt} - a_{91} x_9 - a_{92} x_6 - \alpha_4 \frac{S_4(t)}{|S_4(t)| + \tau_5}\right)\end{aligned}$$

where $\tau_i \succ 0$ is a small constant. This modification creates a small boundary layer around the switching surface in which the system trajectory remains. Therefore, the chattering problem can be reduced significantly (Utkin et al. 1999).

5 Validation and Discussion

To verify the effectiveness of the developed observer based-controller, some simulation works are carried out for the power system under severe disturbance conditions which cause significant deviation in generator loading. Also, different operating points load are considered. The performance of the nonlinear controller was tested on the complete 9th order model of SMIB power system (202 MVA, 13.7 KV), including all kinds of nonlinearities such as exciter ceilings, control signal limiters, etc. and speed regulator. The parameter values used in the simulation are given in the Tables 1, 2 and 3. The physical limits of the plant are

$$\max \left| v_{fd} \right| = 10\,\text{p.u.}, \text{ and } 0 \le X_e(t) \le 1$$

The system configuration is presented as shown in Fig. 7. The proposed sliding mode observer is implemented based on the scheme shown in Fig. 6.

In order to verify the stability and asymptotic tracking performance of the proposed control system, a symmetrical three-phase short circuit occurs closer to the generator bus, at t = 10 s and removed by opening the barkers of the faulted line at t = 10.1 s. The operating points considered are $P_m = 0.6$ p.u. and $P_m = 0.9$ p.u. The

Table 1 Parameters of the transmission line in p.u.

Paramseter	Value
L_e, inductance of the transmission line	0.4
R_e, resistance of the transmission line	0.02

Table 2 Parameters of the synchronous generator in p.u

Parameter	Value
R_s, stator resistance	$1.096\ 10^{-3}$
R_{fd}, field resistance	$7.42\ 10^{-4}$
R_{kd}, direct damper winding resistance	$13.1\ 10^{-3}$
R_{kq}, quadrature damper winding resistance	$54\ 10^{-3}$
L_d, direct self-inductance	1.700
L_q quadrature self-inductances	1.640
L_{fd}, rotor self inductance	1.650
L_{kd}, direct damper winding self inductance	1.605
L_{kq}, quadrature damper winding self inductance	1.526
L_{md}, direct magnetizing inductance	1.550
L_{mq}, quadrature magnetizing inductance	1.490
$V^{\propto}$, infinite bus voltage	1
D, damping constant	0
H, inertia constant	2.37 s

Table 3 Parameters of the steam turbine and speed governor

Parameter	Value
T_t, time constant of the turbine	0.35 s
K_t, gain of the turbine	1
R regulation constant of the system	0.05
T_g, time constant of the speed governor	0.2 s
K_g, gain of the speed governor	1

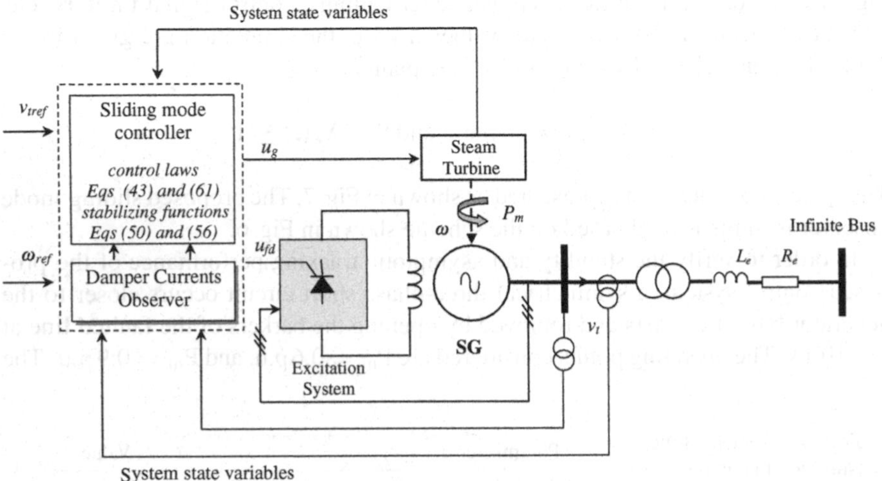

Fig. 7 Control system configuration

simulated results are given in Figs. 8 and 9. It is shown terminal voltage, rotor speed and rotor angle of the power system, respectively. The results are compared with those of the linear IEEE type 1 AVR+PSS and speed regulator. It is seen how dynamics of the terminal voltage and rotor speed exhibit large overshoots during post-fault state with the standard controller than with the nonlinear controller. It is evident that the proposed combined observer-controller can quickly and accurately converge to the desired terminal voltage and rotor speed for different operating points.

Robustness of the proposed observer-based controller is evaluated with respect to the variation of system parameters and error model. The values of the transmission line (L_e, R_e) and the inertia constant H increased by $+20$ and $-20\,\%$ from their original values, respectively. In addition to the abrupt and permanent variation of the power system parameters a three-phase short-circuit is simulated at the terminal of the generator. Figure 10 shows the performances of the terminal voltage and rotor speed of the combined observer-controller. It can be seen that the designed control scheme is not sensitive to the uncertainties of parameters and ensures the global stability of the system with good performances in transient and steady states.

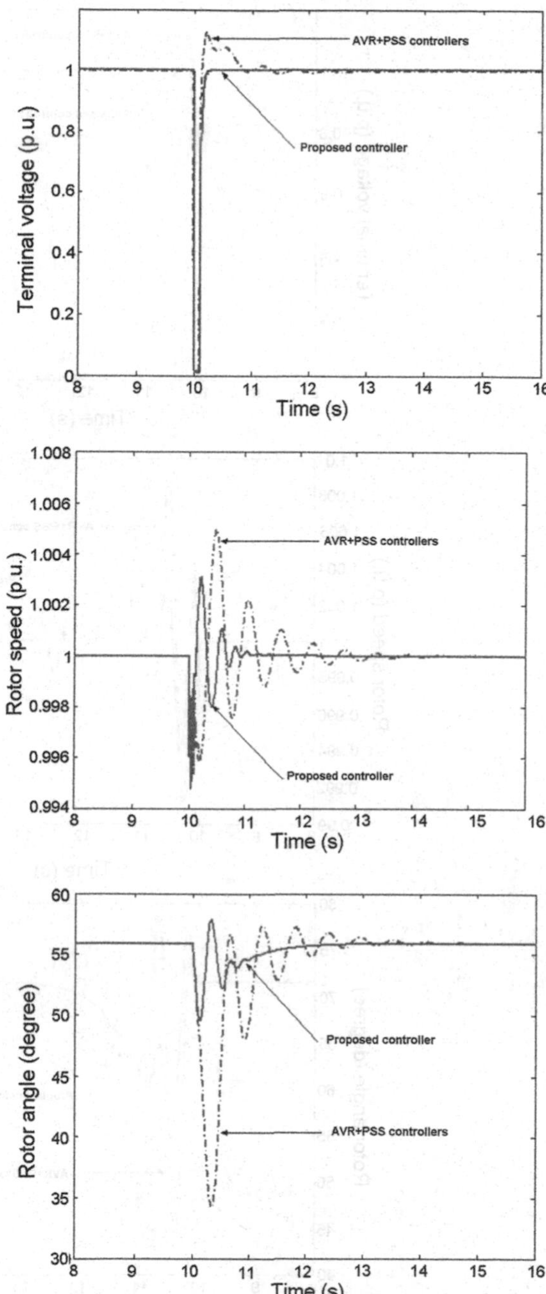

Fig. 8 Performance results of the proposed controller under large sudden fault for $P_m = 0.6\,p.u$

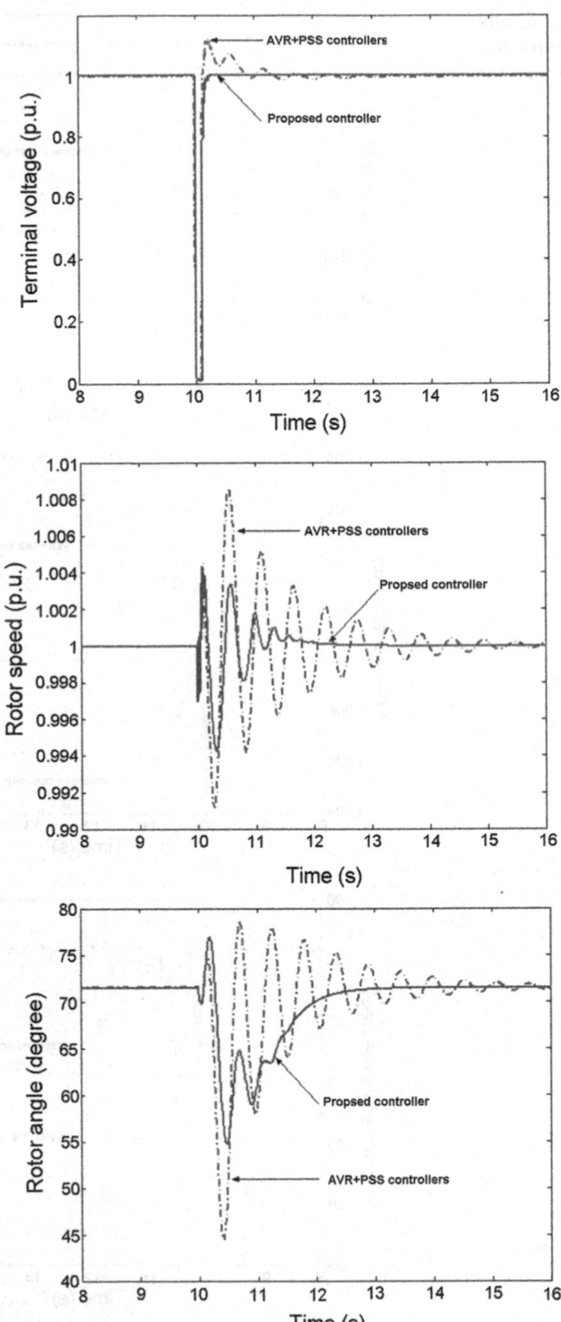

Fig. 9 Performance result of the proposed controller under large sudden fault for $P_m = 0.9$ p.u

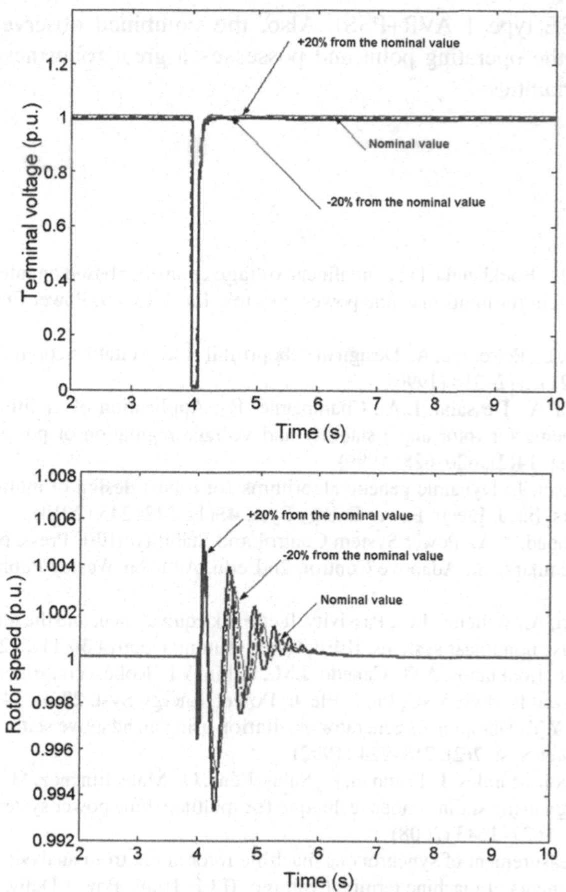

Fig. 10 Dynamic tracking performance control scheme under parameter perturbations

6 Conclusion

A nonlinear observer-controller has been developed and applied to the single machine infinite-bus power system. The synchronous generator is based on the complete 7th order model. The aim is to achieve both transient stability improvement and good post-fault performance of the generator terminal voltage and frequency.

The sliding mode technique was adopted to construct a nonlinear observer of damper currents winding. Then nonlinear control laws of terminal voltage and rotor speed has been provided. Global and exponential stability of both the control laws and the observer has been proven by applying Lyapunov stability theory.

Test results show the effectiveness of the proposed control strategy in improving transient stability of system under large disturbances in comparison with conventional controllers (IEEE type 1 AVR+PSS). Also, the combined observer-controller is independent of the operating point and possesses a great robustness to deal with parameter uncertainties.

References

Abbadi, A., Nezli, L., Boukhetala, D.: A nonlinear voltage controller based on interval type 2 fuzzy logic control system for multi-machine power systems. Int. J. Electr. Power Energy Syst. **45**(1), 456–467 (2013)

Ahmed, S.S., Chen, L., Petroianu, A.: Design of suboptimal H∞ excitation controllers. IEEE Trans. Power Syst. **11**(21), 312–318 (1996)

Akhrif, O., Okou, F.A., Dessaint, L.A., Champagne, R.: Application of mulitivariable feedback linearization scheme for rotor angle stability and voltage regulation of power systems. IEEE Trans. Power Syst. **14**(2), 620–628 (1999)

Alkhatib, H., Duveau, J.: Dynamic genetic algorithms for robust design of multi-machine power system stabilizers. Int. J. Electr. Power Energy Syst. **45**(1), 242–245 (2013)

Anderson, P.M., Fouad, A.A.: Power System Control and Stability. IEEE Press, New York (1994)

Astrom K.J., Wittenmark, B.: Adaptive Control, 2nd edn. Addison-Wesley Publishing Company, Chicago (1995)

Byrnes, C.I., Isidori, A., Willems, J.C.: Passivity, feedback equivalence, and the global stabilization of minimum phase non linear systems. IEEE Trans. Autom. Control **36**(11), 1228–1240 (1991)

Cabrera-Vazquez, J., Loukianov, A.G., Canedo, J.M., Utkin, V.I.: Robust controller for synchronous generator with local load via VSC. Int. J. Electr. Power. Energy Syst. **29**(4), 348–359 (2007)

Cheng, C.H., Hsu, Y.Y.: Damping of generator oscillation using an adaptive static var compensator. IEEE Trans. Power Syst. **7**(2), 718–724 (1992)

Colbia-Vega, de Léon-Morales, J., Fridman, L., Salas-Péna, O., Mata-Jiménez, M.T.: Robust excitation control design using sliding-mode technique for multimachine power systems. Electr. Power Syst. Res. **78**(9), 1627–1643 (2008)

De Mello, F.P.: Measurement of synchronous machine rotor angle from analysis of zero sequence harmonic components of machine terminal voltage. IEEE Trans. Power Deliv. **9**(4), 1770–1777 (1994)

Galaz, M., Ortega, R., Bazanella, A.S., Stankovic, A.M.: An energy-shaping approach to the design of excitation control of synchronous generators. Automatica **39**(1), 111–119 (2003)

Gao, L., Chen, L., Fan, Y., Ma, H.: A nonlinear control design for power systems. Automatica **28**(5), 975–979 (1992)

Ghandakly, A., Dai, J.: An adaptive synchronous generator stabilizer design by generalized multivariable pole shifting (GMPS) technique. IEEE Trans. Power Syst. **7**(3), 436–446 (2000)

Ghandakly, Adel,A., Farhoud, A. M.: A parametrically optimized self tuning regulator for power system stabilizers. IEEE Trans. Power Syst. **7**(3), 1245–1250 (1992)

Ghazizadeh, M.S., Hughes, F.M.: A generator transfer function regulator for improved excitation control. IEEE Trans. Power Syst. **13**(2), 437–441 (1998)

Guo, Y., Hill, D.J., Wang, Y.: Global transient stabilility and voltage regulation for power systems. IEEE Trans. Power Syst. **16**(4), 678–688 (2001)

Hill, D.J., Wang, Y.: Nonlinear decentralized control of large scale power systems. Automatica **36**(9), 1275–1289 (2000)

Huerta, H., Alexander, G., Loukianov, Cañedo, J. M.: Decentralized sliding mode block control of multimachine power systems. Int. J. Electr. Power Energy Syst. **32**(1), 1–11 (2010)

Isidori, A.: Nonlinear Control Systems, 3rd edn. Springer, New York (1995)

Jain, S., Khorrami, F., Fardanesh, B.: Adaptive nonlinear excitation control of power system with unknown interconnections. IEEE Trans. Control Syst. Technol. **2**(4), 436–446 (1994)

Jiao, X., Sun, Y., Shen, T.: Adaptive controller design for a synchronous generator with unknown perturbation in mechanical power. Int. J. Control Autom. Syst. **3**(2), 308–314 (2005)

Jiawei, Y., Zhu, C., Chengxiong, M., Dan, W., Jiming, L., Jianbo, S., Miao, L., Dah, L., Xiaoping, L.: Analysis and assessment of VSC excitation system for power system stability enhancement. Int. J. Electr. Power Energy Syst. **7**(5), 350–357 (2014)

Karimi, A., Feliachi, A.: Decentralized adaptive backstepping of electric power systems. Electr. Power Syst. Res. **78**(3), 484–493 (2008)

Khorrami, F., Jain, S., Fardanesh, B.: Adaptive nonlinear excitation control of power system with unknown interconnections. IEEE Trans. Control Syst. Technol. **2**(4), 436–446 (1994)

King, C.A., Chapman, J.W., Ilic, M.D.: Feedback linearizing excitation control on a full-scale power system model. IEEE Trans. Power Syst. **9**(2), 1102–1109 (1994)

Kokotovic, P.V.: The joy of feedback: non-linear and adaptive. IEEE Control Syst. Mag. **12**(3), 7–17 (1992)

Krstić, M., Kanellakopoulos, I., Kokotović, P.: Nonlinear and adaptive control design. Wiley Interscience Publication, New York (1995)

Kundur, G.P.: Power System Stability and Control. McGraw- Hill, New York (1994)

Loukianov, A.G., Canedo, J.M., Utkin, V.I., Cabrera-Vazquez, J.: Discontinuous controller for power systems: sliding mode block control approach. IEEE Trans. Ind. Electron. **51**(2), 340–353 (2004)

Mohagheghi, S., Valle, Y., Venayagamoorthy, G.K., Harley, R.G.: A proportional-integrator type adaptive critic design-based neuro-controller for a static compensator in a multimachine power system. IEEE Trans. Ind. Electron. **54**(1), 86–96 (2007)

Morales, J.D., Busawon, K., Acosta-Villarreal, G., Acha- daza, S.: Nonliear control for small synchronous generator. Int. J. Electr. Power Energy Syst. **23**(1), 1–11 (2001)

Mrad, F., Karaki, S., Copti, B.: An adaptive fuzzy-synchronous machine stabilizer. IEEE Trans. Syst. Man Cybern Part C **30**(1), 131–137 (2000)

Narendra, K.S., Balakrishnan, J.: Adaptive control using multiple models. IEEE Trans. Autom. Control **42**(2), 171–187 (1997)

Nickllasson, P.J., Ortega, R., Espinosa Perez, G.: Passivity-based control of a class of Blondel-Park transformable electric machines. IEEE Trans. Autom. Control **42**(5), 629–649 (1997)

Ohtsuk, K., Taniguchi, T., Sato, T.: A H∞ optimal theory based generator control system. IEEE Trans. Power Syst. **7**(1), 108–113 (1992)

Ortega, R., Lorı'a, A., Nicklasson, P., Sira-Ramıre, H.: Passivity-Based Control of Euler-Lagrange Systems. Springer, London (1998)

Ouassaid, M., Nejmi, A., Cherkaoui, M., Maaroufi, M.: A Nonlinear backstepping controller for power systems terminal voltage and rotor speed controls. Int. Rev. Autom. Control. **3**(1), 355–363 (2008)

Ouassaid, M., Maaroufi, M., Cherkaoui, M.: Decentralized nonlinear adaptive control and stability analysis of multimachine power system. Int. Rev. Electr. Eng. **5**(6), 2754–2763 (2010)

Ouassaid, M., Maaroufi, M., Cherkaoui, M.: Observer based nonlinear control of power system using sliding mode control strategy. Electr. Power Syst. Res. **84**(1), 153–143 (2012)

Park, J.W., Harley, R.G., Venayagamoorthy, G.K.: Adaptive-critic-based optimal neurocontrol for synchronous generators in a power system using MLP/RBF neural networks. IEEE Trans. Ind. Appl. **39**(5), 1529–1540 (2003)

Segal, R., Kothari, M.L., Madnani, S.: Radial basis function (RBF) network adaptive power system stabilizer. IEEE Trans. Power Syst. **15**(2), 722–727 (2000)

Shamsollahi, P., Malik, O.P.: An adaptive power system stabilizer using on-line trained neural network. IEEE Trans. Energy Convers. **12**(4), 382–389 (1997)

Shen, T., Mei, S., Lu, Q., Hu, W., Tamura, K.: Adaptive nonlinear excitation control with L2 disturbance attenuation for power systems. Automatica **39**(1), 81–89 (2003)

Slotine, J.J.E., Li, W.: Applied Nonlinear Control. Prentice-Hall, Englewoods Cliffs (1991)

Tan, Y., Wang, Y.: Augmentation of transient stability using a supperconduction coil and adaptive nonlinear control. IEEE Trans. Power Syst. **13**(2), 361–366 (1998)

Utkin, V.I.: Variable structure systems with sliding modes. IEEE Trans. Autom. Control **22**(2), 212–222 (1977)

Utkin, V.I., Guldner, J., Shi, J.: Sliding Mode Control in Electromechanical Systems. Taylor and Francis, London (1999)

Venayagamoorthy, G.K., Harley, R.G., Wunsch, D.C.: Dual heuristic programming excitation neurocontrol for generators in a multimachine power system. IEEE Trans. Ind. Appl. **39**(2), 382–394 (2003)

Wang, S.K.: A novel objectif function and algorithm for optimal PSS parameter design in a multimachine power system. IEEE Trans. Power Syst. **28**(1), 522–531 (2013)

Wang, Y., Cheng, D., Li, C., Ge, Y.: Dissipative Hamiltonian realization and energybased L2-disturbance attenuation control of multimachine power systems. IEEE Trans. Autom. Control **48**(8), 1428–1433 (2003)

Wu, B., Malik, O.P.: Multivariable adaptive control of synchronous machines in a multimachine power system. IEEE Trans. Power Syst. **21**(2), 1772–1787 (2006)

Xi, Z., Cheng, D., Lu, Q., Mei, S.: Nonlinear decentralized controller design for multimachine power systems using Hamiltonian function method. Automatica **38**(2), 527–534 (2002)

Zhao, Q., Jiang, J.: Robust controller design for generator excitation systems. IEEE Trans. Energy Convers. **28**(2), 201–207 (1995)

Switching Function Optimization of Sliding Mode Control to a Photovoltaic Pumping System

Asma Chihi, Adel Chbeb and Anis Sellami

Abstract The research deals with the performances of an asynchronous motor coupled to a pump in terms of optimal photovoltaic transfer, using the concept of variable structure systems by sliding mode. The main advantage is to implement a robust sliding mode control from a nonlinear system. The contribution of this work is modeling a new switching surface. The control law is based on adding an integral term for the considered surface in order to improve the performances of the system. Moreover, a sliding mode control technique associated with a boost converter is used to extract the Maximum Power Point Tracking (MPPT). In the first part of this chapter, a general modeling of the different elements of the photovoltaic pumping system is presented. In the second part, a methodology of synthetizing sliding mode control is developed with the choice of a novel switching function. The proposed control law acts on the duty cycle applied to a boost converter in order to transfer a maximum power delivered by the photovoltaic generator to the induction motor. Finally, the validation of the results is carried out with a comparative study to show the efficiency of the proposed control.

A. Chihi (✉) · A. Chbeb · A. Sellami
Research Unit on Control, Monitoring and Safety Systems (C3S), University of Tunis, 5 Avenue Taha Hussein, BP 56, 1008 Tunis, Tunisie
e-mail: asma.chihi@live.fr

A. Chihi · A. Chbeb · A. Sellami
High School of Sciences and Engineering of Tunis (ESSTT), University of Tunis, 5 Avenue Taha Hussein, BP 56, 1008 Tunis, Tunisie
e-mail: adel.chbeb@enib.rnu.tn

A. Chbeb
e-mail: adel.chbeb@enib.rnu.tn

A. Sellami
e-mail: Anis.Sellami@esstt.rnu.tn

A.T. Azar and Q. Zhu (eds.), *Advances and Applications in Sliding Mode Control systems*, Studies in Computational Intelligence 576, DOI 10.1007/978-3-319-11173-5_17

1 Introduction

The development and exploitation of clean sources of renewable energy becomes the subject of several studies. Many forms of renewable energy are operated in the literature such as: solar energy, wind, geothermal and biomass. A large part of energy from the sun is transmitted by infrared radiation. This energy is exploiting into electricity. That defined the photovoltaic effect.

The photovoltaic system is the most efficient source and well accepts renewable energy sources because of their suitability in distributed generation, transportation and satellite systems. Therefore, the major drawback of the PV system is the nonlinearity between the output voltage and current particularly shorted conditions. During this latter, the curve power voltage (P-V) has multiple peaks and admits a unique Maximum Power Point (MPP), which depends on irradiance and temperature conditions. When these items are changed, the operating point will change. Such, to overcome this problem, several researchers have investigated different methods for extracting the maximum power such as: Perturb & Observer, incremental conductance and Hill Climbing. So, a comparison of these techniques can result important information for the design of these types of systems.

The association in series or parallel connection of several photovoltaic cells helps to adapt the production of this energy at the demand of user. These associations provide a photovoltaic generator having a specific characteristic current – voltage (I-V), admits nonlinear models and represents a Maximum Power Point (MPP). This technique depends mainly on the irradiance level, cell temperature and the aging of the ensemble. The intersection of the electrical characteristic of the photovoltaic generator and the load provides the operating point of the system.

Recently, the use of the induction motor in many industrial applications has increases. This becomes that the induction motor has relatively low cost, reliable and rugged. But, controlling the induction motor is not a trivial task because it admits a nonlinear model and its physical parameters are most imprecisely known. Therefore, this problem opens the door of several studies in order to provide a robust control against parametric variations and uncertainties. Moreover, the indirect field oriented control by sliding mode presents satisfactory performances in presence of these internal disturbances. Many approaches consist to exploit the nonlinear control which admits robustness proprieties, such as: Sliding Mode Control (SMC), and Fuzzy Logic Control (FLC). The SMC theory is a best method which accomplished in practice.

Nowadays, the increasing need to use a photovoltaic system is more and more important in several applications such as photovoltaic system, water desalination. These items require a control law which has an important role for realization of such systems (Dal 2005; Moallem et al. 2001; Sabanovic and Izosimov 1993; Rao et al. 2008).

Therefore, in the practical control system, there is a difference between the mathematical and the real model. This difference comes from external disturbances and parametric variations. Therefore, it must be designed to maintain the desired performance in the closed loop with presence of these disturbances. The concept of sliding mode is one of the best solutions for this problem (Soltanpour and Fateh 2009). The main advantage of this kind of control is the robustness through parametric variations and convergence in a finite time. In this work, we applied a sliding mode control to a photovoltaic pumping system. So, two controls are developed; the first one is about controlling the boost converter via the Maximum Power Point Tracking (MPPT) with acting on the duty cycle, and the second one about controlling the induction motor by applying an indirect field oriented control based on sliding mode in order to obtain a constant speed and torque, respectively flow and pressure.

This chapter is organized as follows: Sect. 2 developed the problem formulation given in this type of system which required to a related work for putting the considered work in its context, Sect. 3 presents an overview of the photovoltaic pumping system which we talk about the different elements of the proposed system. The design of sliding mode control is describes in Sect. 4. The simulation results are interpreted and discussed in Sect. 5 to validate the best control method. Different techniques are compared and explained in Sect. 6. Finally, Sect. 7 includes the concluding remarks.

2 Problem Formulation

The most important advantage of the photovoltaic system is the production of electricity without harmful effects on the environment during all the exploitation period. So, the photovoltaic system is attached to natural parameter like solar irradiance and temperature which are varying randomly each instant.

In many studies, based approaches such as: (P&O algorithm, Inc. Cond. algorithm) are used to extract the Maximum Power Point (MPP) of the photovoltaic system (Aureliano et al. 2013). In order to find the maximum power to the load, an algorithm based on sliding mode control proposed by Bianconi et al. (2013), (Montoya et al., 2013), Mamarelis et al. (2014), Haroun et al. (2014) are developed. This technique has robustness against parametric variations and uncertainties.

Also, several researcher used the sliding mode to control the induction motor for different applications such as: pumping system (Mapurunga et al. 2014; Giannoursos and Manias 2014), irrigation (Melton et al. 2012), grid connected system (Yang et al. 2013; Mohan et al. 2013).

Moreover, our objective is to demonstrate that the sliding mode control applied to a moto-pump, admit a switching function that is presented with the difference between the selected variable and its reference. This control provides a satisfactory results that are reported by Haroun et al. (2014) as well as Mamarelis et al. 2014. But, each control must give some criteria to discuss its performances like robustness and accuracy. The items presented a dilemma. There is a robust system and also an accuracy system. The combination of both provides the desired solution. The novel technique is proposed using a modified sliding surface that is to add an integral error into the considered switching function, in order to achieve the solution for this dilemma.

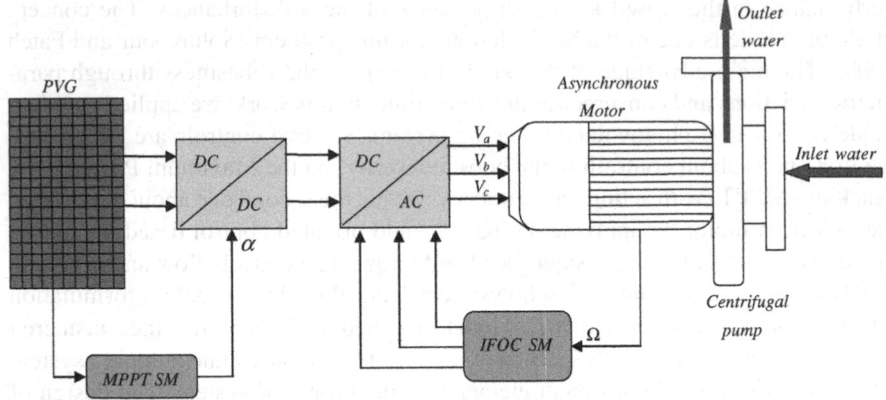

Fig. 1 Synoptic diagram of photovoltaic pumping system

3 Overview of the Photovoltaic Pumping System

We consider a nonlinear system shown in Fig. 1. It composed, like any system, of power part and control part. The power side system is the association, in a series; of the photovoltaic (PV) panel with a three phase asynchronous Moto-pump, through a boost converter and a three level AC inverter.

The regulation strategy is based on Sliding Mode Control (SMC). The PV is controlled via Maximum Power Point Tracking (MPPT) to extract the Maximum Power Point (MPP) with acting on the duty cycle α.The Moto-pump is controlled with Indirect Field Oriented Control (IFOC) by Sliding Mode (SM) techniques in order to extract the pressure and flow that correspond respectively to the torque and speed.

The general proposed scheme is as shown in Fig. 1.

3.1 Induction Motor—Pump Group

The pumps are an essential element in industrial and agricultural applications, such as oil sector, irrigation, water treatment. They are used for fluid transportation whether hot or cold, clean or dirty.

Various types of pumps are available in the market, such as volumetric pumps and centrifugal pumps, of surface or submerged.

The principal parameters characterized the pump are:

- The flow (Q), which presented by the following equation:

$$Q = \frac{q}{\rho} \tag{1}$$

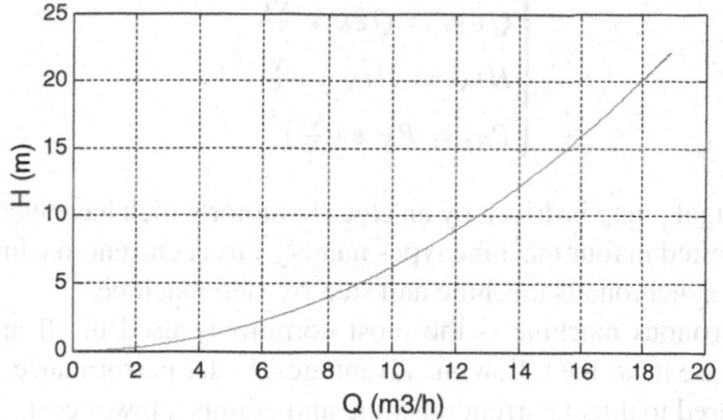

Fig. 2 Curves of pressure as a function of flow for different speed

where:
$\rho =$ the volumetric mass.

- The Head (H) can be written by the representative equation:

$$H = \frac{P(bar) * 10.2}{d} \text{ with } d = \frac{\rho}{1{,}000} \tag{2}$$

where
$d =$ the volumetric density.

- The efficiency (η), it provides by the following equation:

$$\eta = \frac{usefulpower}{powerconsumption} = \frac{QH\rho g}{P} \tag{3}$$

where
$g =$ gravity.

This work will focus on the use of a centrifugal pump which consists of a turbine disposed in a body system, which receives the fluid horizontally to drive it back, in the direction perpendicular to the inlet water.

The centrifugal pump have the following advantages: A good performance compared to the volumetric pump, easy maintenance due to its simple construction and small bulk.

A pump is characterized by a pressure corresponding to a flow which is bound by Fig. 2.

Similitude equations (4) can help in determining the pressure and flow in a different operating point.

$$\begin{cases} Qv_{N\prime} = Qv_N * \frac{N\prime}{N} \\ Ht_{N\prime} = Ht_N * \left(\frac{N\prime}{N}\right)^2 \\ P_{N\prime} = P_N * \left(\frac{N\prime}{N}\right)^3 \end{cases} \quad (4)$$

The centrifugal pump is driven by an electric motor which has diversified choice. The latter is limited in four machine types namely direct current machine, asynchronous machine, synchronous machine and step by step machine.

The asynchronous machine is the most commonly used in all applications of pumping, because it has the following advantages: better performance, simple maintenance compared to direct current machine and admits a lower cost.

The model of the asynchronous machine is shown by differential equations admitting a constant coefficient varying versus time.

To manipulate this machine, it is supposed to make a transformation of these equations in (a-b-c)–(d-q) frame. Park transformation is used in order to facilitate the manipulation of this machine.

3.2 The Inverters

An inverter is a Direct Current (DC) to Alternative Current (AC) converter. It is possible to impose across an alternative three phase load a voltage via a logic control.

The modulation technique is based on sinusoidal Pulse Width Modulation (PWM) with carriers.

The main principle of this technique is comparing a sinusoidal signal with triangular signal having the same magnitude and frequency.

The switches of each arm are controlled in a complementary way in order to prevent a short circuit in the input voltage E. The topology schema of the inverter is presented by Fig. 3.

In the studied system, the full-wave control is used; the operation of the switches is as follows:

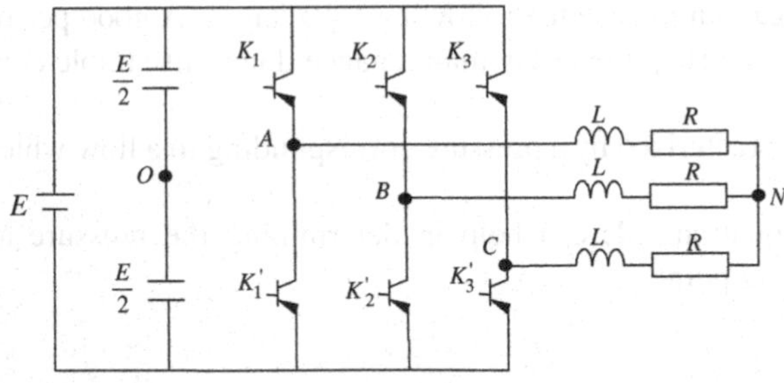

Fig. 3 Inverter DC–AC

- *K1* led for $t = 0$; $K1'$ led for $t = \pi$,
- *K2* led for $t = \frac{2\pi}{3}$; $K2'$ led for $t = \frac{2\pi}{3} + \pi$,
- *K3* led $for \quad t = \frac{4\pi}{3}$; $K3'led$ for $t = \frac{4\pi}{3} + \pi$.

The Three output phase voltages can be written as shown in Eq. (5).

$$\begin{cases} V_{AN} = V_{AO} - V_{NO} \\ V_{BN} = V_{BO} - V_{NO} \\ V_{CN} = V_{CO} - V_{NO} \end{cases} \tag{5}$$

where:

V_{AN}, V_{BN} and V_{CN} are respectively the phase voltages in the first, second and third arm.

The three line to line output voltages can be expressed by:

$$\begin{cases} U_{AB} = U_{AO} - U_{BO} \\ U_{BC} = U_{BO} - U_{CO} \\ U_{AC} = U_{AO} - U_{CO} \end{cases} \tag{6}$$

The three output phase voltages V_{AN}, V_{BN} and V_{CN} constitute a balanced three phase system.

The representative equation is expressed as follows:

$$V_{AN} + V_{BN} + V_{CN} = 0 \tag{7}$$

Also:

$$V_{AO} + V_{BO} + V_{CO} - 3V_{NO} = 0 \tag{8}$$

The following equation relative to the output phase voltage can be obtained:

$$V_{AN} = V_{AO} - \frac{1}{3}(V_{AO} + V_{BO} + V_{CO}) \tag{9}$$

Thus, the three phase voltage is effectively relaxed to:

$$\begin{cases} V_{AN} = \frac{2}{3}V_{AO} - \frac{1}{3}V_{BO} - \frac{1}{3}V_{CO} \\ V_{BN} = -\frac{1}{3}V_{AO} + \frac{2}{3}V_{BO} - \frac{1}{3}V_{CO} \\ V_{CN} = -\frac{1}{3}V_{AO} - \frac{1}{3}V_{BO} + \frac{2}{3}V_{CO} \end{cases} \tag{10}$$

The state of switches assumed perfect can be defined by three Boolean quantities $Ki = (1, 2, 3)$.

$Ki = 1$, where the switch in the top is closed and to bottom is open.

$Ki = 0$, where the switch in the top is open and to bottom is closed.

Under these conditions, we can write voltages V_{AO}, V_{BO} and V_{CO} according to control signals *Ki*, we get:

$$
\begin{aligned}
V_{AO} &= K_1 E - \tfrac{E}{2} \\
V_{BO} &= K_2 E - \tfrac{E}{2} \\
V_{CO} &= K_3 E - \tfrac{E}{2}
\end{aligned}
\tag{11}
$$

So, we can establish the instant equation for the output phase voltage:

$$
\begin{pmatrix} V_{AN} \\ V_{BN} \\ V_{CN} \end{pmatrix} = \frac{E}{3} \begin{pmatrix} 2 & -1 & -1 \\ -1 & 2 & -1 \\ -1 & -1 & 2 \end{pmatrix} \begin{pmatrix} K_1 \\ K_2 \\ K_3 \end{pmatrix}
\tag{12}
$$

3.3 Association Inverter—Induction Motor

The purpose of associating an inverter with induction motor is the control of the torque, speed, pressure and flow respectively.

Among the techniques used to control, field oriented vector control is presented in Fig. 4. This method is based on modeling the machine in the (d-q) stationary reference frame. That is to say $\phi_{rd} = \phi_r$ and $\phi_{rq} = 0$.

The mathematical model chosen to modeling the induction motor is relative to rotor flux—stator current. The state space of the induction motor is defined by relation (13).

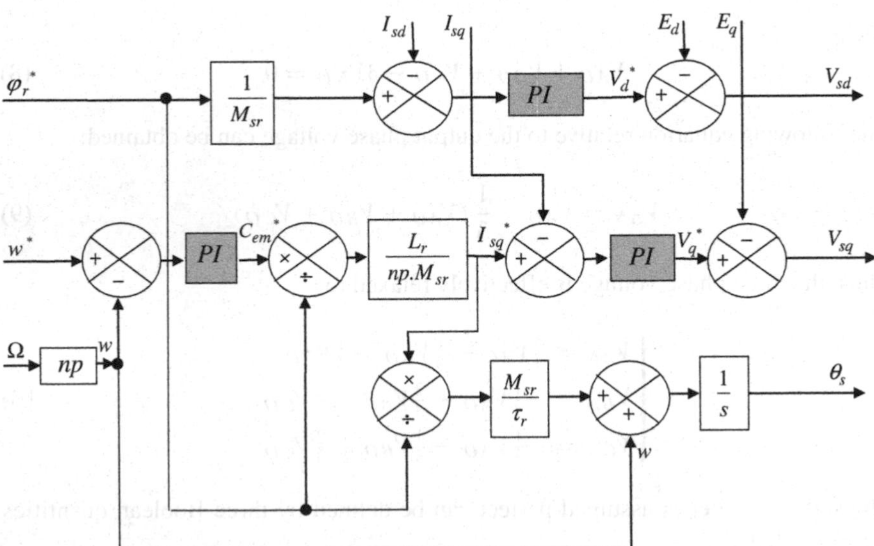

Fig. 4 Representative schema of indirect field oriented control

$$\begin{cases} \dfrac{di_{sd}}{dt} = -\left(\dfrac{1}{\sigma\tau_s} + \dfrac{1-\sigma}{\sigma\tau_r}\right) i_{sd} + \dfrac{1-\sigma}{\sigma M_{sr}\tau_r}\phi_{rd} + \dfrac{1-\sigma}{\sigma M_{sr}}\phi_{rq} w + \dfrac{1}{\sigma L_s} V_{sd} \\ \dfrac{di_{sq}}{dt} = -w_{dq} i_{sd} - \left(\dfrac{1}{\sigma\tau_s} + \dfrac{1-\sigma}{\sigma\tau_r}\right) i_{sq} - \dfrac{1-\sigma}{\sigma M_{sr}} w\phi_{rd} + \dfrac{1-\sigma}{\sigma M_{sr}\tau_r}\phi_{rq} + \dfrac{1}{\sigma L_s} V_{sq} \\ \dfrac{d\phi_{rd}}{dt} = \dfrac{M_{sr}}{\tau_r} i_{sd} - \dfrac{1}{\tau_r}\phi_{rd} - w\phi_{rq} \\ \dfrac{d\phi_{rq}}{dt} = \dfrac{M_{sr}}{\tau_r} i_{qs} + w\phi_{dr} - \dfrac{1}{\tau_r}\phi_{qr} \\ \dfrac{dw}{dt} = \frac{3}{2}\dfrac{np^2}{j}\dfrac{M_{sr}}{L_r}(\phi_{rd} i_{sq} - \phi_{rq} i_{sd}) - \dfrac{np}{j}(C_r - C_f) \end{cases} \quad (13)$$

where:

Ids, Iqs : d-, q-axis stator current components,
ϕ_{dr}, ϕ_{qr} : d-, q-axis rotor flux components,
wsl : slip angular speed (*wdq-wr*),
wdq : synchronous angular speed,
R_r, R_s : rotor and stator resistances,
M_{sr} : cyclic mutual inductance stator-rotor,
L_r, L_s : rotor and stator self-inductions,
τ_s, τ_r : stator and rotor time constant,
s : leakage coefficient,
np : pole-pair number,
j : inertia,
C_f : friction torque,
C_r : load torque,
t : Continuous time.

3.4 Photovoltaic System

The functioning of photovoltaic cells is characterized by the photons which absorb solar radiation and convert it into electricity.

The photovoltaic cells are constituted by semiconductor material. The technologies developed to this day are monocrystalline silicon, polycrystalline and silicon thin layer. These are developed to produce PV cells whose performance and lifetime are different.

- *Characteristics of PV Cell*

The diagram of the PV cell is presented by a current source in parallel with a diode. A shunt (R_{sh}) and a series resistance (R_s) are added to the model.
R_s is the intrinsic series resistance; its value is very small.
R_{Sh} is the equivalent shunt resistance which has a very high value, Fig. 5.
We require Kirchhoff's laws, the expression of the current in the PV cell is written as it's shown in relation (14).

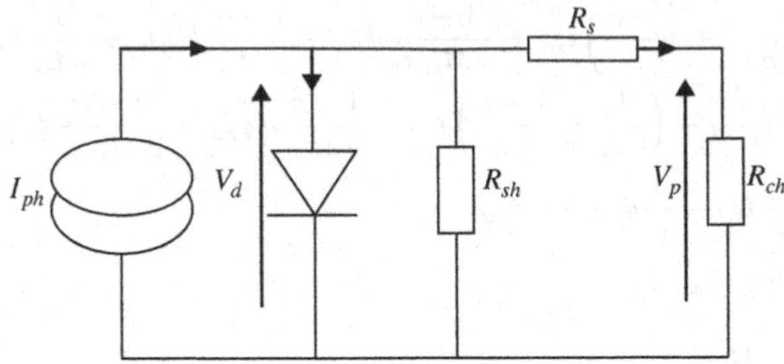

Fig. 5 PV cell schema

$$I_p = I_{ph} - I_{ss}\left(e^{\left(\frac{V_p + I_p R_s}{V_T}\right)} - 1\right) - \left(\frac{V_p + I_p R_s}{R_{sh}}\right) \tag{14}$$

where:

I_{ph} : Insolation current,
I_p : cell current,
I_{ss} : reverse saturation current,
V_p : cell voltage,
R_s : series resistance,
R_{sh} : parallel resistance,
V_T : Terminal voltage.

$$V_T = \frac{KT}{q} \tag{15}$$

where:

K : Bolzman constant,
T : temperature in Kelvin,
q : Charge of electron.
The power available at the terminals of "a cell" is very low. In this case, it's necessary to combine in series (N_s) or in parallel (N_p) cells for power modules compatible with the electrical equipment to be used.

Adding the $N_s$and N_p in Eq. (14), the expression of the current is defined by relation (16).

$$I_p = N_p\left[I_{ph} - I_s\left(e^{\left(\frac{1}{V_T}\left(\frac{V_p}{N_s} + \frac{R_s I_p}{N_p}\right)\right)} - 1\right)\right] - \frac{N_p V_p}{N_s R_{sh}} - \frac{R_s I_p}{R_{sh}} \tag{16}$$

The photovoltaic generator is a (series/parallel) combination of PV cells, which leads to increasing power and performance according to the user request. The series combination has the role of increasing voltage and parallel combination has the role of increasing the current as shown in Fig. 6.

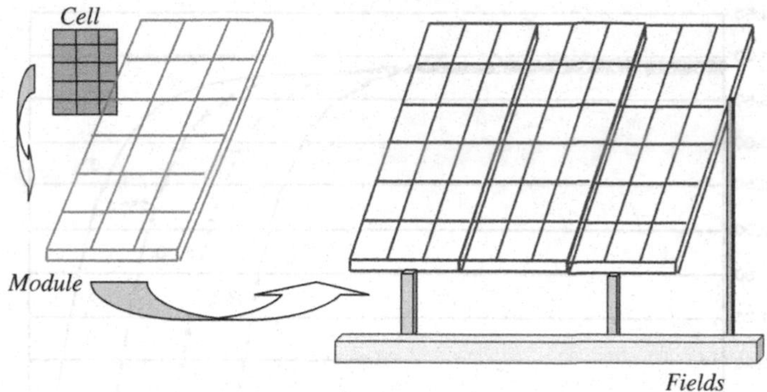

Fig. 6 PV Fields: assembling of cell and module

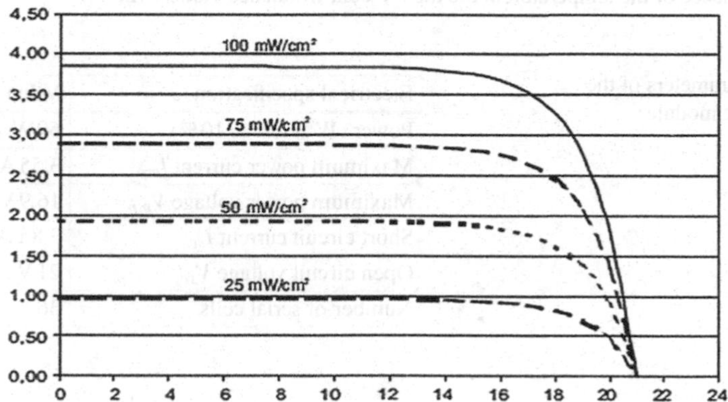

Fig. 7 Influence of the irradiance in the the PV cell, temperature 25 °C

The interconnection of photovoltaic cells defines a PV module and the interconnection of several modules led to a PV array.

The PV system can be used in two ways: Either with storage energy (electricity uses during the night and this makes it through batteries accumulation), or without storage, such as the PV pumping.

- *Effect of irradiance and tempera*ture

The temperature and the irradiance are two important parameters behavior of the PV cells, Fig. 7.

Figure 7 allows concluding that, for a fixed temperature equal to 25 °C (standard test condition), each change of illumination causes a change in the Maximum Power Point (MPP). And for fixing the solar irradiance in the 1k W/m^2 that is the value of standard test, we conclude also that the variation of temperature leads to change the (MPP), Fig. 8.

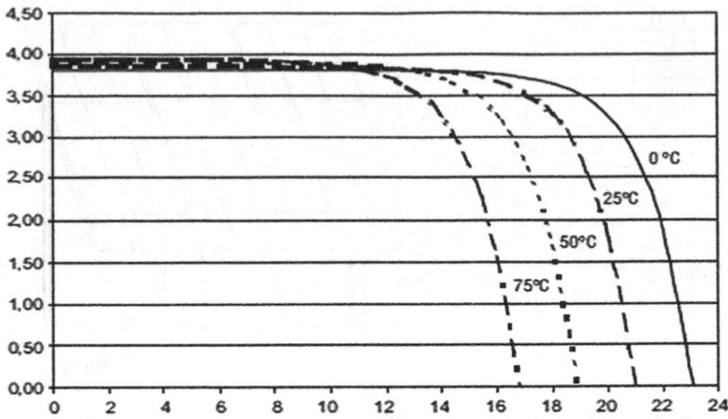

Fig. 8 Influence of the temperaturein the the PV cell, irradiance 1,000 W/m^2

Table 1 Parameters of the photovoltaic module

Electrical specification	Value
Power (W at test ± 10 %)	60 W
Maximum power current I_{mp}	3,55 A
Maximum power voltage V_{mp}	16,9 V
Short circuit current I_{sc}	3,85 A
Open circuit voltage V_{sc}	21 V
Number of serial cells	36

The model of the photovoltaic generator (PVG) used in this application is brand ATERA A-GO. The electrical parameters of the generator in the Standard Test Condition (STC) (1k W/m2 and 25 °C) are given by Table 1.

For the energy demands of the load, the power supplied with a single module is not enough. For this reason, we used a PV generator constant of 4 modules in series, so the maximum power supplied by the PVG is about 240 W.

3.5 DC–DC Converter

DC-DC converter is used to convert an unregulated DC voltage to a regulated DC output voltage.

The switching device used in this converter is usually IGBT, MOSFET. The switching loss increases with the switching frequency. So, the efficiency decreases.

The control voltage is obtained by comparing the representative output voltage with its reference value. The Pulse Width Modulation control signal is compared with a saw tooth voltage.

Generally, for this kind of system, there are three structures for switching controllers namely boost converter, buck converter and buck-boost converter (Nema et al. 2011).

The choice of the DC converter depends essentially on the characteristics of the load exactly for the motor and the pump employed.

At startup, the induction motor uses a high current, on the order of 6 to 8 times the nominal RMS current.

In the case of the pumping over sun, the system has a discontinuity and a variation of the illumination during the day. So, it is impossible to pump below a certain level of illumination. Thus, there is a loss of energy at the beginning and the end of the day with the risk of umbrage and interruption operation during the day.

In order to improve the performance of a photovoltaic pumping system, it is necessary to use the maximum power of the PV panels.

The optimization technique is the search of Maximum Power Point Tracking (MPPT). This technique is achieved by increasing a DC/DC converter between the PV panels and load. The principle is to vary the voltage in order to obtain the maximum power (Byung-Duk et al. 2008; Jong-Pil et al. 2010).

- *Boost Converter*

A boost converter produces a higher output voltage than the DC input voltage. The studied system admits an output voltage 200 V and input voltage 68 V. This verifies the use of the boost converter.

The circuit diagram of the boost converter is displayed in Fig. 9.

The output voltage is related to the input voltage by the following equation:

$$V_s = \frac{1}{1-\alpha} V_e \tag{17}$$

where:

α = duty cycle.

- *Inductor and Capacitor Design*

In The first part of the operating cycle [0; αT], the controlled switch is closed, hence we have:

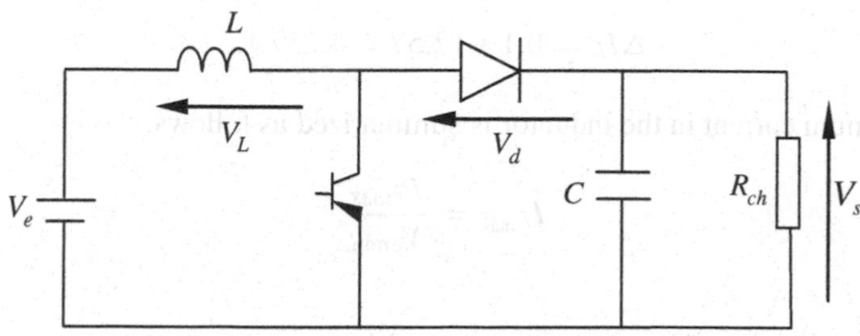

Fig. 9 Boost converter schema

$$V_e = V_s \tag{18}$$

where:

V_e : Input voltage of the boost converter.
V_s : output voltage of the boost converter.

The voltage across the inductor V_L can be written as:

$$V_L = L\frac{dI_L}{dt} = L\frac{\Delta I_L}{\Delta t} \tag{19}$$

where:

I_L : Input current in the inductor.
ΔI_L : Variation of the input current in the inductor.

The duty cycle can be expressed by:

$$\alpha = \frac{V_s - V_e}{V_s} \tag{20}$$

The interval which limits the duty cycle α is:

$$\alpha_{\max} = \frac{V_{s\max} - V_{e\min}}{V_{s\max}} = \frac{210 - 67.6}{210} = 0.678 \tag{21}$$

And

$$\alpha_{\min} = \frac{V_{s\min} - V_{e\max}}{V_{s\min}} = \frac{190 - 80}{190} = 0.578 \tag{22}$$

The maximum current in the inductor can be expressed as follows:

$$I_{L_{\max}} = \frac{P_{e\max}}{V_{e\min}} = \frac{850}{67.6} = 12.57A \tag{23}$$

We assume a 10 % ripple of the maximum current in the inductor, then:

$$\Delta I_L = 0.1 * 12.57 = 1.257A \tag{24}$$

The maximum current in the inductor is summarized as follows:

$$I_{L_{\max}} = \frac{P_{e\max}}{V_{e\min}} \tag{25}$$

where:

P_{emax} = the maximum power in the input of the boost converter.

The inductor value is given as:

$$L_{\max} = \frac{\alpha_{\max} V_{e\max}}{f \Delta I_L} = \frac{0.678 * 80}{16 * 10^3 * 1.257} = 2.7\,\text{mH} \quad (26)$$

with:
f = the switching frequency of the boost converter.

It is also assumed that the ripple of the capacitor voltage is equal to 5 % of the ripple output voltage:

$$\Delta V_s = 0.05\ V_{s\max} = 0.05 * 210 = 10.5\,\text{V} \quad (27)$$

The maximum current is expressed as follows:

$$I_{s\max} = \frac{P_{s\max}}{V_{s\min}} = \frac{850}{190} = 4.47\,\text{A} \quad (28)$$

The capacitor value is defined by:

$$C_{\max} = \frac{\alpha_{\max} I_{s\max}}{f \Delta V_s} = \frac{0.678 * 4.47}{16 * 10^3 * 10.5} = 18\,\mu\text{F} \quad (29)$$

where:
I_s = the output current of the boost converter.
ΔV_s = The ripple output voltage.

For sizing the input capacitor C_e which defined the continuous bus, we established the following equation:

$$C_e = \frac{\alpha_{\max}}{8 * L_{\max} * f \wedge 2 * 0.02 * \text{V}_{\text{pv}}} = \frac{0,678}{8 * 2.7 * 10^{-3} * (16 * 10^3)^2 * 0.02 * 80} = 0.1\,\mu F \quad (30)$$

3.6 Main Concept of the Sliding Mode Control

The sliding mode is a novel technique for modeling control. It is characterized by its robustness and it admits a convergence in a finite time.

The major problem of sliding mode is to design a control low in closed loop that drives asymptotically the mass to the origin. In other terms, the representative control is supposed to lead the state variable to zero.

The variable structure control is a system with variable structure according to the variation of the control. Moreover, such a system may have novel proprieties that do not exist in each structure (Sabanovic, 2011).

The main principle of variable structure control is to lead the state trajectory to a surface, called Sliding Surface, and to force them to stay in the vicinity of this surface.

The sliding mode control is a control law which is adapted to the variable structure systems. Indeed, to synthesize a control law by sliding mode, the first step is the choice of the sliding surface that converge the state trajectory of the system to the desired balanced point. In a second step, the establishment of the matching conditions of sliding mode which is connected to convergence of state trajectory of the balanced point. In the final step, we determine the control law, which represents the role allowing maintaining the attractiveness of the sliding mode.

- *Sliding Surface design*

The sliding surface represents the desired dynamic behavior. The state trajectory of the system should reach this surface. There aren't any specific criteria for choosing the sliding surface.

- *Matching condition*

The matching conditions of sliding mode are criteria that allow the dynamic of a studied system to convert to the sliding surface and stay there regardless of disturbances. These conditions can be defined by direct switching function or by the stability function of Lyapunov.

The Lyapunov function technique is chosen positive and decreasing to force the trajectory of the system to move towards the sliding surface. Therefore, the idea is to choose a scalar function *s(x)* which ensures the attraction of the variable x into a reference value. This is assured by a control function V, Eq. (31).

$$V(x) = \frac{1}{2}s^2(x) \tag{31}$$

With $\dot{V}(x)$ is given by Eq. (32).

$$\dot{V}(x) = s(x) \cdot \dot{s}(x) \tag{32}$$

- *Control law strategy*

The chosen surface is stable and converge the output to the desired output $y_d(t)$ which assures the convergence of the sliding mode. Then, we determine the control that will force the state system to reach the balanced point that reflects the existing condition s of sliding mode.

The control low U is composed by two components:

- U_{eq}: The equivalent control is used to maintain the state on the sliding surface $s(x) = 0$.
- U_{nl}: The non linear controller, it is the stabilizing control.

The global control U takes the final form (33).

$$U = U_{eq} + U_{nl} \tag{33}$$

3.7 Association PV—Boost Converter

The technique of extracting the Maximum Power Point Tracking (MPPT) is employed to a boost converter via duty cycle. Many methods have been used to obtain the MPPT such as P&O method and incremental conductance method. But, the output energy of the PV panels change frequently by the environmental change that is temperature (Ghazanfari and Farsangi 2012). For this reason, the employment of adequate control which is robust in the presence of parameter variation and disturbance, especially for a non-linear system, may improve the system performance, Fig. 10.

The principal schema of the boost converter is presented by the Fig. 11.

It is important to make a mathematical model of the association PV—boost converter. Therefore, Kirchhoff's laws are used, (34) and (35).

$$I_p = I_l + I_c \tag{34}$$

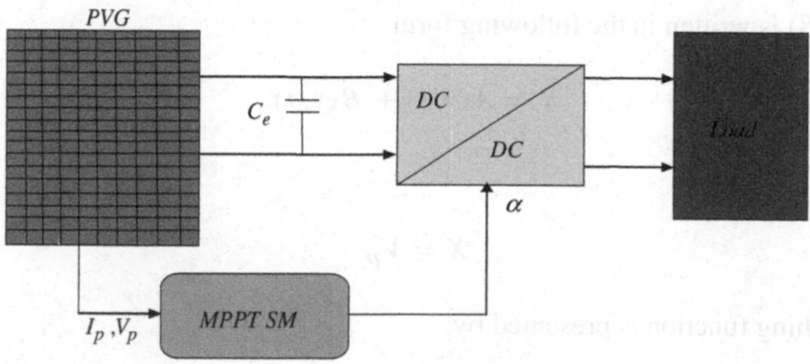

Fig. 10 Principal schema of MPPT control

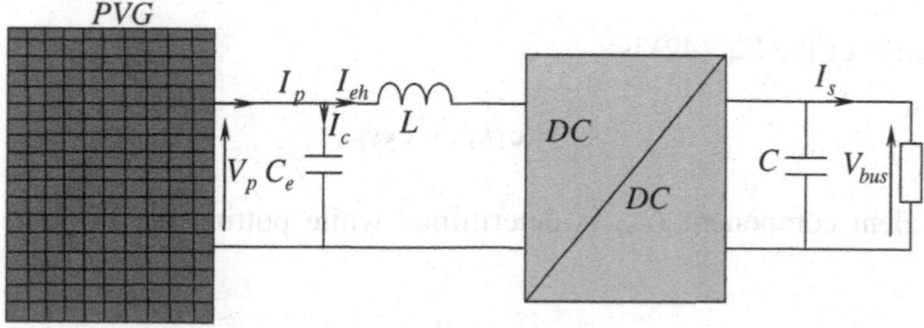

Fig. 11 Boost converter schema

So:

$$I_c = I_p - I_c \tag{35}$$

The input output current applied to the boost converter are given by the relation (36).

$$I_L = \frac{1}{1-\alpha} is \tag{36}$$

The value of the current across capacity C_e is expressed as follows:

$$I_c = C_e \frac{dVp}{dt} \tag{37}$$

Or, the expression of the output current of the PV panels is defined by:

$$I_p = I_{ph} - I_{ss}\left(e^{\left(\frac{V_P}{V_T}\right)} - 1\right) \tag{38}$$

From Eqs. (36), (37) and (38), the derivative voltage V_p is summarized as follows:

$$\frac{dV_p}{dt} = \frac{1}{C_e}(I_{ph} - I_{ss}\left(e^{\left(\frac{V_p}{V_T}\right)} - 1\right) - \frac{1}{C_e}\left(\frac{1}{1-\alpha}\right) is \tag{39}$$

So, Eq. (38) is written in the following form:

$$\dot{X} = A(x,t) + B(x,t) \tag{40}$$

where:

$$X = V_p \tag{41}$$

The switching function is presented by:

$$\mathrm{s}_v = \mathrm{c}_1\varepsilon_v + \mathrm{c}_5\int \varepsilon_v \mathrm{dt} \quad \text{where } \varepsilon_v = V_p - V_{pref} \tag{42}$$

The derivative of the Eq. (42) is:

$$\dot{\mathrm{s}}_v = \mathrm{c}_1\dot{\varepsilon}_v + \mathrm{c}_5\varepsilon_v \tag{43}$$

The equivalent component U_{eq} is determined while putting the Eq. (39) to zero, where:

$$u = \frac{1}{1-\alpha} \tag{44}$$

The nonlinear component is presented by:

$$U_{nl} = -k_v sign(s_v) \tag{45}$$

For minimizing the chattering given to the system, we replace the Signum function by:

$$U_{nl} = -k_v\,|s_v|^{\beta}\ \mathrm{sign}(s_v)\ \text{where}\quad 0 < \beta < 1 \tag{46}$$

The global control law can be written as:

$$u(t) = \frac{1}{is}\left(I_{ph} - I_{ss}\left(e^{\left(\frac{V_P}{V_T}\right)} - 1\right)\right) - k_v\,|s_v|^{\beta}\,sign(s_v) \tag{47}$$

4 Sliding Mode Control (SMC)

As mentioned in Paragraph 2.7, the concept of sliding mode control can be summarized in three steps: beginning by the choice of a sliding surface, reaching conditions and determination of control law (Utkin 1993; Lee et al. 1994; Fnaiech et al. 2006; Ahmed et al. 2010; Gao and Hung 1993 and Ellouze et al. 2010).

4.1 Sliding Surface Design

Generally, several researchers have used the sliding surfaces which are defined as (Chihi et al. 2012; Msaddek et al. 2013) :

$$\begin{cases} s_v = c_1\varepsilon_v \quad \text{with} \quad \varepsilon_v = V_p^* - V_p \\ s_w = c_2\varepsilon_w \quad \text{with} \quad \varepsilon_w = w^* - w \\ s_d = c_3\varepsilon_d \quad \text{with} \quad \varepsilon_d = I_{sd}^* - I_{sd} \\ s_q = c_4\varepsilon_q \quad \text{with} \quad \varepsilon_q = I_{sq}^* - I_{sq} \end{cases} \tag{48}$$

However, due to the complexity of the real system, we ameliorate the dynamic of the switching function by adding an integral term in order to obtain a faster response time and zero steady-state error. The new proposed sliding surfaces are presented as follows:

$$\begin{cases} s_v = c_1\varepsilon_v + c_5\int \varepsilon_v dt \\ s_w = c_2\varepsilon_w + c_6\int \varepsilon_w dt \\ s_d = c_3\varepsilon_d + c_7\int \varepsilon_d dt \\ s_q = c_4\varepsilon_q + c_8\int \varepsilon_q dt \end{cases} \tag{49}$$

where: $c_1, c_2, c_3, c_4, c_5, c_6, c_7, c_8$: proportional gains.

4.2 Determination of Control Design

The control law design of the photovoltaic system is constituted by four control loops relative to the four selected switching functions.

4.2.1 MPPT Control

The switching function is presented by Eq. (50):

$$s_v = c_1\varepsilon_v + c_5\int \varepsilon_v dt \tag{50}$$

The derivative of Eq. (50) is defined by:

$$\dot{s}_v = c_1\dot{\varepsilon}_v + c_5\varepsilon_v \tag{51}$$

We recall that the fundamental equation relative to the association of the photovoltaic is as follows:

$$\dot{V}_p = \frac{I_{ph}}{C_e} - \frac{I_{ss}}{C_e}(e^{\frac{V_p}{V_t}} - 1) + \frac{1}{C_e(1-\beta)}I_s \tag{52}$$

With

V_p: Output voltage of the PV cell,
I_{ph}: photo current of the PV cell,
I_{ss}: saturation current of the diode,
V_t: thermodynamic potential of PV cell,
β: Duty cycle.
The control law is determined by:

$$V_p^* = \frac{1}{I_s}(I_{ph} - I_{ss}(e^{\frac{V_p}{V_t}} - 1)) - k_v\,|s_v|^\alpha\ sign(s_v)\ with\ 0 < \alpha < 1 \tag{53}$$

4.2.2 Speed Controller

The representative switching function is expressed as follows:

$$s_\omega = c_2\varepsilon_\omega + c_6\int \varepsilon_\omega dt \tag{54}$$

The time derivative of Eq. (54) is given as:

$$\dot{s}_\omega = c_2\dot{\varepsilon}_\omega + c_6\varepsilon_\omega \tag{55}$$

The mechanical equation is defined by:

$$j\frac{dw}{dt} + f_r w = np(C_{em} - C_r) \tag{56}$$

The famous equation for the electromagnetic torque is given by:

$$C_{em} = \frac{3}{2}\frac{M_{sr}np}{L_r}(\phi_{rd}i_{sq} - \phi_{rq}i_{sd}) \tag{57}$$

where the control law is ${C_{em}}^*$:

$${C_{em}}^* = C_{em\_eq} + C_{em\_nl} \tag{58}$$

During the sliding mode, we have: $s_w = \dot{s}_w = 0$.
The global control functions as follows:

$${C_{em}}^* = \frac{J}{np}(\dot{w}^* + \frac{f}{J}w + \frac{c_6}{c_2}\varepsilon_w) - k_w\,|s_w|^{\alpha}\, sign(s_w) \text{ with } 0 < \alpha < 1 \tag{59}$$

where:

k_w : Proportional gain of the nonlinear control relative to the speed controller,
Sign (.) : Signum function.

4.2.3 Direct Stator Current Controller

The switching function relative to the direct stator current controller is defined as:

$$s_d = c_3\varepsilon_d + c_7\int \varepsilon_d \mathrm{dt} \tag{60}$$

The time derivative of the switching function in Eq. (60) is:

$$\dot{s}_d = c_3\dot{\varepsilon}_d + c_7\varepsilon_d \tag{61}$$

Let us use the time derivative of the direct stator current controller which is presented in the system Eq. (13).
The direct stator voltage is the difference between the direct voltage and the d-back electromotive force (EMF).

$$V_{sd} = V_d - E_d \tag{62}$$

where:

$$V_d = \sigma L_s(p + (\frac{1}{\sigma\tau_s} + \frac{1-\sigma}{\tau_r\sigma}))i_{sd} \tag{63}$$

And

$$E_d = \sigma L_s(w_{dq} i_{sq} + (\frac{1-\sigma}{\sigma M_{sr}\tau_r})\phi_r) \tag{64}$$

The control law $V_d^*$is given by:

$$V_d^* = V_{d\_eq} + V_{d\_nl} \tag{65}$$

From Eqs. (62)–(65), we get:

$$V_d^* = \sigma L_s(\dot{I}_{sd}^* + (\frac{1}{\sigma\tau_s} + \frac{(1-\sigma)}{\sigma\tau_r})i_{sd} + \frac{c_7}{c_3}\varepsilon_d) \\ - k_d \left|s_d\right|^{\alpha} sign(s_d) \text{ with } 0 < \alpha < 1 \tag{66}$$

where:
k_d: Proportional gain of the nonlinear control relative to the speed controller.

4.2.4 Quadratic Stator Current Controller

The representative equation defining the switching function relative to the quadratic stator current is given by:

$$s_q = c_4\varepsilon_q + c_8\int \varepsilon_q dt \tag{67}$$

We consider the time derivative of Eq. (67) is portrayed as:

$$\dot{s}_q = c_4\dot{\varepsilon}_q + c_8\varepsilon_q \tag{68}$$

Referring to the system Eq. (13), when we use the time derivative of the quadratic stator current, we obtain:

$$V_{sq} = V_q - E_q \tag{69}$$

where
E_q represents the q—back electromotive force (EMF).
The quadratic voltage can be determined by:

$$V_q = \sigma L_s(p + (\frac{1}{\sigma\tau_s} + \frac{1-\sigma}{\tau_r\sigma}))i_{sq} \tag{70}$$

The q-back EMF is defined by:

$$E_q = -\sigma L_s(w_{dq} i_{sd} + (\frac{1-\sigma}{\sigma M_{sr}})w_{sl}\phi_r) \tag{71}$$

The control law V_q^* is presents by:

$$V_q^* = V_{q\_eq} + V_{q\_nl} \tag{72}$$

During the sliding mode:

$$s_q = \dot{s}_q = 0 \tag{73}$$

The global control law is summarized as follows:

$$\begin{aligned} V_q{}^* = \sigma L_s(\dot{I}_{sq}^* + (\frac{1}{\sigma\tau_s} + \frac{(1-\sigma)}{\sigma\tau_r})i_{sq} + \frac{c_8}{c_4}\varepsilon_d) \\ - k_q\left|s_q\right|^\alpha sign(s_q) \;\; with \;\; 0 < \alpha < 1 \end{aligned} \tag{74}$$

where:
kq : Proportional gain of the nonlinear control relative to the speed controller.

4.3 Matching Condition

The existant condition of sliding mode advert that both s and $\dot{s}$ will tend to zero when t tend to infinity, Let us consider the Lyapunov function candidate presented by Eqs. (31) and (32).

4.3.1 MPPT Controller

The time derivative of the sliding surface related to the MPPT controller is:

$$\dot{s}_v = c_1(\dot{v_p}^* - \dot{v_p}) + c_5(v_p^* - v_p) \tag{75}$$

We note by:

$$\dot{s}_v = c_1\left[\frac{I_{ph}}{C_e} - \frac{I_{ss}}{C_e}(e^{\frac{x_1}{V_T}} - 1) - \frac{i_s}{C_e}u + c_5\varepsilon_v\right] \tag{76}$$

Also,

$$\dot{s}_v = c_1\left[\frac{I_{ph}}{C_e} - \frac{I_{ss}}{C_e}(e^{\frac{x_1}{V_T}} - 1) - \frac{i_s}{C_e}(u_{eq} + u_{nl}) + c_5\varepsilon_v\right] \tag{77}$$

To change the equivalent control by his expression, the time derivative of the surface is taken as:

$$\dot{s}_v = -c_1\frac{i_s}{C_e}K_v\left|s_v\right|^\alpha \tag{78}$$

We satisfy the Lyapunov conditions, we obtain:

$$s.\dot{s}_v = -c_1 \frac{i_s}{C_e} k_v \left| s_v \right|^{\alpha+1} \tag{79}$$

We request the inequality $s.\dot{s}_v < 0$, we get:

$$-c_1 \frac{i_s}{C_e} k_v \left| s_v \right|^{\alpha+1} < 0 \tag{80}$$

The inequality (80) means that in order to obtain a robust and stable control, the gains of the sliding mode controller can be chosen positive $k_v > 0$.

4.3.2 Speed Controller

The time derivative of the switching function is as follows:

$$\dot{s}_w = c_2 \dot{\varepsilon}_w + c_6 \varepsilon_w \tag{81}$$

From the mechanical Eq. (81), we present:

$$\dot{s} = c_2 \dot{w}^* - c_2 \left[\frac{j}{np} C_{em} - \frac{f}{j} w \right] + c_6 \varepsilon_w \tag{82}$$

We use the Lyapunov function, we obtain:

$$s_w.\dot{s}_w = s_w \left(-\frac{c_2 n p k_w}{j} \left| s_w \right|^{\alpha} sign(s_w)\right) < 0 \tag{83}$$

Thus, we get:

$$\left| s \right|^{\alpha+1} > 0 => k_w > \frac{j}{c_2 n p} \tag{84}$$

4.3.3 Direct Stator Current Controller

The representative time derivative of the switching surface s_d is:

$$\dot{s}_d = c_3 (\dot{I}_{sd}^* - \dot{I}_{sd}) + c_7 \varepsilon_d \tag{85}$$

We replace the equation which presented the direct stator current in the Eq. (13), we obtain:

$$\dot{s}_d = c_3 \dot{I}_{sd}^* - c_3 \begin{bmatrix} -(\frac{1}{\sigma \tau_s} + \frac{1-\sigma}{\sigma \tau_r}) I_{sd} + w_{dq} I_{sq} + \frac{1-\sigma}{\sigma M_{sr} \tau_r} \phi_r \\ + \frac{1}{\sigma L_s} (V_d - E_d) \end{bmatrix} + c_7 \varepsilon_d \tag{86}$$

We satisfy the Lyapunov condition, we have:

$$s_d.\dot{s}_d = -\frac{c_3 k_d}{\sigma L_s}|s_d|^{\alpha+1} < 0 \tag{87}$$

Therefore, we get:

$$|s_d|^{\alpha+1} > 0, k_d > \frac{\sigma L_s}{c_3} \tag{88}$$

4.3.4 Quadratic Stator Current Controller

The time derivative of s_q is as follows:

$$\dot{s}_q = c_4(\dot{I}_{sq}^* - \dot{I}_{sq}) + c_8 \varepsilon_q \tag{89}$$

We make the time derivative of quadratic stator current in the Eq. (13) to the Eq. (89), we obtain:

$$\dot{s}_q = c_4 \dot{I}_{sq}^* - c_4 \begin{bmatrix} -w_{dq} i_{sd} - (\frac{1}{\sigma \tau_s} + \frac{1-\sigma}{\sigma \tau_r}) i_{sq} - \frac{1-\sigma}{\sigma M_{sr}} w \phi_{rd} \\ + \frac{1-\sigma}{\sigma M_{sr} \tau_r} \phi_{rq} + \frac{1}{\sigma L_s}(V_q - E_q) \end{bmatrix} + c_8 \varepsilon_d \tag{90}$$

Applying the Lyapunov condition, we get:

$$s_q.\dot{s}_q = -\frac{c_4 k_q}{\sigma L_s}|s_q|^{\alpha+1} < 0 \tag{91}$$

The condition where the stability of the control strategy is presented, is expressed as:

$$|s_q|^{\alpha+1} > 0, k_q > \frac{\sigma L_s}{c_4} \tag{92}$$

5 Results and Discussion

In this section, we are going to examine the performance of the proposed control developed above.

The three phase induction motor machine under test is characterized by: 85/140 V, 3.5/6 A, $f = 50$ Hz, $R_s = 3.45\,\Omega$, $R_r = 2.95\,\Omega$, $L_s = 0.1442$ H, $L_r = 0.1442$ H, $M_{sr} = 0.1342$ H, $j = 0.01$ Kgm^2, np = 2. The coefficients k_v, k_w, k_d and k_q involved in the control law, are tuned to values: $k_v = 1000$, $k_w = 2000$, $k_d = 3000$ and $k_q = 3000$. These gains have been adjusted until the obtain of the validate results.

The result simulation of the proposed switching functions related to the maximum power point tracking controller, speed controller, direct stator current controller and

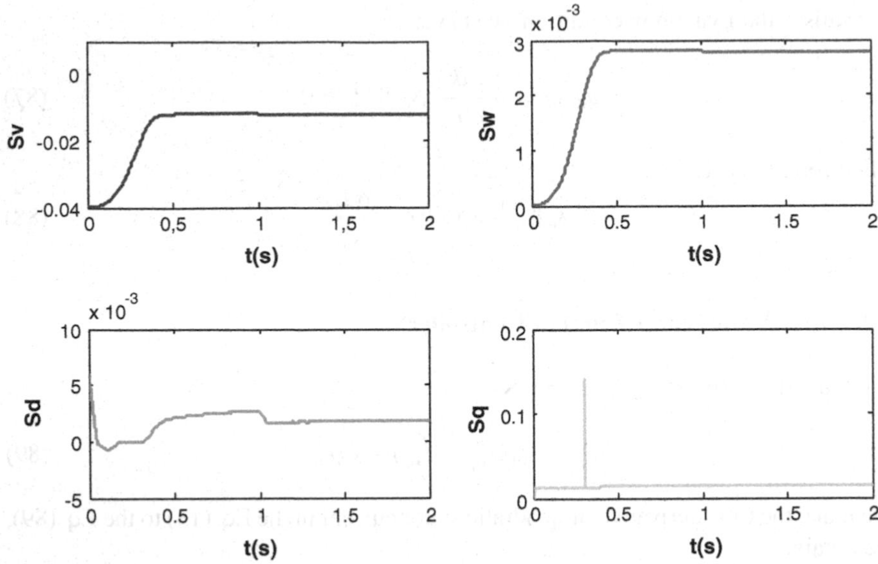

Fig. 12 Evolution of sliding surfaces

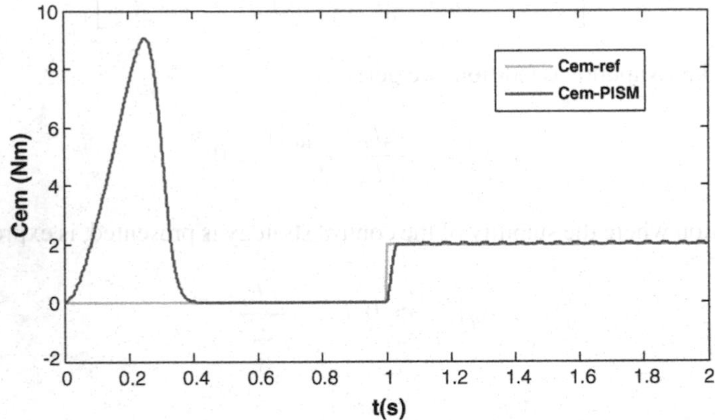

Fig. 13 Evolution of the torque

quadratic stator current controller are shown in Fig. 12. The representative characteristics of the sliding surfaces starts with a zero load torque and after that with 2 Nm from the instant 1s with a nominal speed equal to 157 rad/s and reference flux equal to 0.8 Wb. The sliding surfaces kept to zero, which verified the criteria of the sliding mode.

The electromagnetic torque admits a sinusoidal curve in the transitory regime and presents a peak startup 9.2 Nm in the instant 0.2 s. After that, it stabilized in

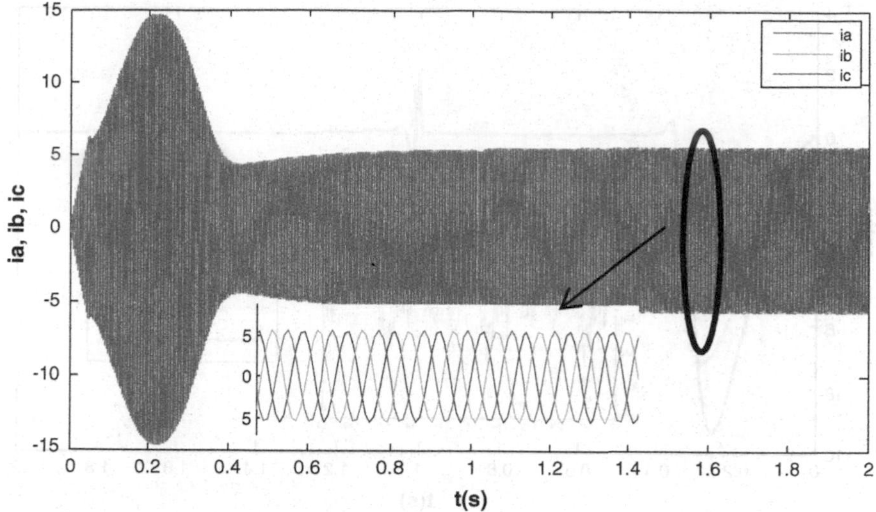

Fig. 14 Evolution of the three phase current i_a, i_b, i_c

zero. When we insert the load torque 2 Nm, the settling time is 0.23 and the torque stabilized with the magnitude of 10 Nm in the reference value.

The three phase current i_a, i_b and i_c admits a sinusoidal forms, it provides the same magnitude 5 A and when we insert the load torque the magnitude increases at 5.5 A, Fig. 14.

6 Comparative Study

The system model still relies on a number of approximations by neglecting some phenomena such as: the dynamic high frequency. Sometimes, we are forced to work on the dynamics of the system as it is. For this reason, it is important to know the reliability of some control and the degree of confidence that we can fix.

While sometimes some systems require robustness and other systems require precision. The dilemma between robustness and precision is still a research topic. The proposed control combines the advantages of both.

6.1 Accuracy

In this section, a comparative study between different regulators, such as conventional Proportional Integral (PI), Proportional Sliding Mode (PSM) and Proportional Integral Sliding Mode (PISM), is presented in order to test the system accuracy.

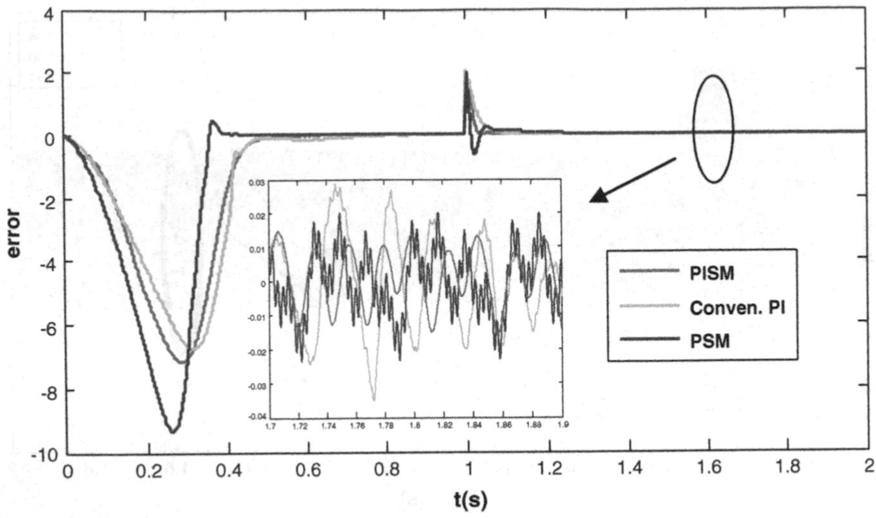

Fig. 15 Representative error of the torque with different methods

In this context, it is necessary to calculate the error applied to the torque. This error is the difference between the representative torque and its reference. We can conclude that the respective error with PSM is more elevated than the conventional PI and PISM, Fig. 15.

Table 2 compare our results with those obtained with different techniques, such as: conventional PI, PSM and PISM. The best results are provided with PISM. We interpreted that the addition of the integral action given to the switching function ameliorates the precision of the system.

6.2 Robustness

For any kind of control, it is necessary to test the robustness. In the case of induction motor, the robustness test is to vary some parameters of asynchronous machine (variation of resistances and inductances).

Generally, this variation is equal to $\pm 50\,\%$, relative to the rotor resistance and stator resistance, $+20\,\%$ relative to the rotor inductance and stator inductance and $+20\,\%$ in the case of the cyclic mutual inductance.

Table 2 Comparison results for precision test with different techniques

	Conventional PI	PSM	PISM
Steady state error	0.020	0.025	0.010
Settling time	0.450	0.350	0.210

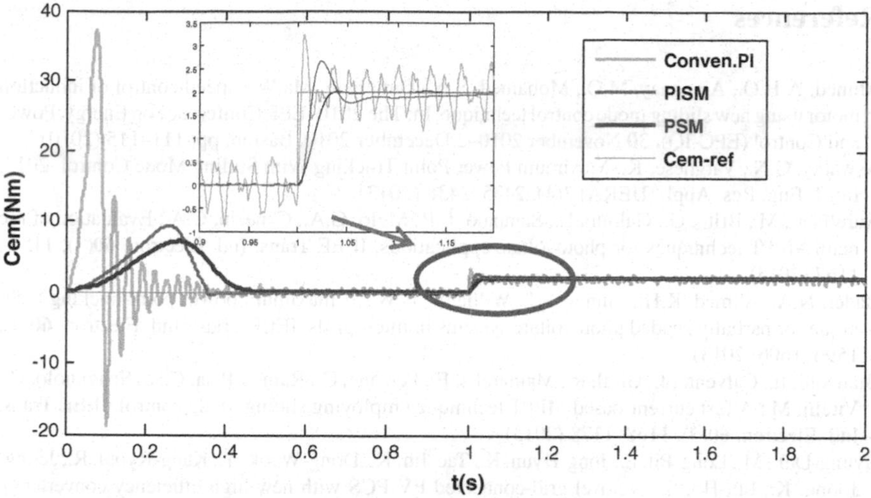

Fig. 16 Robustness test with +20 % of rotor inductance variation

Figure 16 presents the rotor inductance variation with +20 % its nominal value; we remark that the response becomes oscillatory; also the peak startup increases 36Nm. We concluded that the response with conventional PI is changed. However, with sliding mode has not changed. This verified the robustness provided by the proposed control.

7 Conclusion

The work presented in this chapter is about the photovoltaic pumping system which specifies the strategy of sliding mode control. This study addressed the problem raised by the design of the photovoltaic pumping system. This later requires a precision in the representative parameters such as flow and pressure respectively torque and speed, because of each instant it must know the desired flow and pressure to want pumping.

For that, it must design some control which approved these criteria to the considered system. The sliding mode control gives best performances. We are developed a novel control based for the choice of the switching function which is ameliorate by the adding of an integral error. It's for this objective the chapter is designed. In fact, we presented an overview of the pumping system which we detail the constitution of different elements of the system. Then, an indirect field oriented control based on sliding mode is presented in order to control the asynchronous motor. This structure synthetize a three control loops related to a speed controller, direct stator controller and quadratic stator controller. Yet, the robustness and accuracy are demonstrated by simulation results. Thus, the proposed method presents satisfactory results compared with the techniques reported in the literature.

References

Ahmed, A.H.O., Ajangnay, M.O., Mohamed, S.A., Dunnigan, Ma.W.: Speed control of induction motor using new sliding mode control technique. In: The 2010 IEEE Conference on Energy, Power and Control (EPC-IQ), 30 November 2010–2 December 2010, Basrah, pp. 111–115 (2010).

Aswathy, G.N., Varghese, K.: Maximum Power Point Tracking With Sliding Mode Control, 2013. Int. J. Eng. Res. Appl. (IJERA) **3**(4), 2435–2438 (2013)

Aureliano, M., Brito, G., Galotto, L., Sampaio, L.P., Melo, G.A., Canesin, C.A.: Evaluation of the main MPPT techniques for photovoltaic applications. IEEE Trans. Ind. Electron. **60**(3), 1156–1167 (2013)

Bader, N.A., Ahmed, K.H., Finney, S.J., Williams, B.W.: A maximum power point tracking technique for partially shaded photovoltaic systems in microgrids. IEEE Trans. Ind. Electron. **60**(4), 1596–1606 (2013)

Bianconi, E., Calvente, J., Giral, R., Mamarelis, E., Petrone, G., Ramos-Paja, C.A., Spagnuolo, G., Vitelli, M.: A fast current-based MPPT technique employing sliding mode control. IEEE Trans. Ind. Electron. **60**(3), 1168–1178 (2013)

Byung-Duk, M., Long-Pil, L., Jong-Hyun, K., Tae-Jin, K., Dong-Wook, Y., Kang-Ryoul, R., Jeong-Joong, K., Eui-Ho, S.: A novel grid-connected PV PCS with new high efficiency converter. J. Power Electron. **8**(4), 309–316 (2008)

Chihi, A., Sallami, A., Kalfa, A.: Sliding mode control of a photovoltaic pumping system. The 2012 IEEE - Mediterranean Elecrotechnical Conference (MELECON), 25–28 March 2012, Yasmine Hammamet, pp. 936–939. doi:10.1109/MELCON.2012.6196581

Dal, D.: Sensorless sliding mode direct torque control (DTC) of induction motor. In: The 2005 Proceeding of the IEEE International Symposium on Industrial Electronics (ISIE), 20–23 June 2005, pp. 911–916. doi:10.1109/ISIE.2005.1529045

Ellouze, M., Gamoudi, R., Mami, A.: Sliding mode control applied to a photovoltaic water pumping system. Int. J. Phys. Sci. **5**(4), 334–344 (2010)

Fnaiech, M.A., Betin, F., Fnaiech, F., Capolino, G.A.: Sliding mode control for dual three-phase induction motor drives. The 2006 IEEE International Symposium on Industrial Electronics (ISIE), 9–13 July 2006, Montreal, Que, pp. 2281–2285 (2006).doi:10.1109/ISIE.2006.295928

Gao, W., Hung, J.C.: Variable structure control for nonlinear systems: a new approach. IEEE Trans. Ind. Electron. **40**(1), 45–55 (1993)

Ghazanfari, J., Farsangi, M.M.: Maximum power point tracking using sliding mode control for photovoltaic array. Iranian J. Electric. Electron. Eng. **9**(3), 189–196 (2012)

Giannoursos, S.V., Manias, S.N.: A data-driven process controller for energy efficient variable-speed pump operation in central cooling water system of marine vessels. IEEE Trans. Ind. Electron. **1**, 99 (2014)

Haroun, R., El Aroudi, A., Cid-Pastor, A.: sliding mode control of output-parallel-connected two-stage boost converters for PV systems. In: The 2014 Multi-Conference on Systems, Signals and Devices (SSD), 11–14 Feb 2014, Barcelona, pp. 1–6 (2014). doi:10.1109/SSD.2014.6808810

Jong-Pil, L., Byung-Duk, M., Tae-Jin, K., Dong-Wook, Y., Ji-Yoon, Y.: Input-series-output-parallel connected DC/DC converter for a photovoltaic PCS with high efficiency under a wide load range. J. Power Electron. **10**(1), 9–13 (2010)

Juan, C.Y., Hugo, C.G., Leobardo, H.G., José, A.O.: Design and simulation by photovoltaic system with tapped topology. Int. J. Modern Eng. Res. (IJMER) **3**(2), 1238–1244 (2013)

Lee, D.C., Sul, S.K., Park, M.: High performance current regulator for a field-oriented controlled induction motor drive. IEEE Trans. Ind. Appl. **30**(5), 1247–1257 (1994)

Levron, Y., Shmilovitz, D.: Maximum power point tracking employing sliding mode control. IEEE Trans. Circuits Syst. **60**(3), 732–734 (2013)

Mamarelis, E., Petrone, G., Spagnuolo, G.: Design of a sliding-mode-controlled SEPIC for PV MPPT applications. IEEE Trans. Ind. Appl. **60**(7), 3387–3398 (2014)

Mapurunga, C.J.V., De Carvalho, F.G., Moreira, T.L.F., de Souza, R.L.A.: Implementation of a high-efficiency, high-lifetime, and low-cost converter for an autonomous photovoltaic water pumping system. Ieee Trans. Ind. Appl. **50**(1), 631–641 (2014)

Melton, F.S., Johnson, L.F., Lund, C.P., Pierce, L.L., Michaelis, A.R., Hiatt, S.H., Guzman, A., Adhikari, D., Purdy, A.J., Rosevelt, C., Votava, P., Trout, T.J., Temesgen, B., Frame, K., Sheffner, E.J., Nemani, R.R.: Satellite irrigation management support with the terrestrial observation and prediction system: a framework for integration of satellite and surface observations to support improvements in agricultural water resource management. IEEE J. Select. Topics Appl. Earth Obser. Remote Sens. **5**(6), 1709–1721 (2012)

Moacyr, A., Galotto, L., Luigi, G., Leonardo, P.S., Azevedo, M., Carlos, A.C.: Evaluation of the main MPPT techniques for photovoltaic applications (2013). IEEE Trans. Ind. Electron. **60**(3), 1156–1167 (2013)

Moallem, M., Mirzaeian, B., Mohammed, O.A., Lucas, C.: Multi-objective genetic-fuzzy optimal design of PI controller in the indirect field oriented control of an induction motor. IEEE Trans. Magn. **37**(5), 3608–3612 (2001)

Mohan, a., Mathew, D., Nair, V.M.: Grid connected PV inverter using adaptive totalsliding mode controller. In: The 2014 Conference on Control Communication and Computing (ICCC), 13–15 December 2013, thruvananthapuran, pp. 457–462 (2013). doi:10.1109/ICCC.2013.6731698

Montoya, D.G., Paja, C.A.R., Giral, R.: A new solution of maximum power point tracking based on sliding mode control. In: The 2013 Industrial Electronics Society (IECON), 10–13 Nov 2013, Vienna, pp. 8350–8355 (2013). doi:10.1109/IECON.2013.6700532

Msaddek, A., Gaaloul, A., M'sahli, F.:A novel higher order sliding mode control: application to an induction motor. In: The International conference on Control, Engineering and Information Technology (CEIT'13), 4–7 June 2013, Sousse, pp. 13–21 (2013)

Nema, S., Nema, R.K., Agnihotri, G.: Inverter topologies and control structure in photovoltaic applications : a review. J. renwable sustain. Energ. **3**(1) (2011)

Rao, S., Buss, M. Utkin, V.: Sliding mode based stator flux and speed observer for induction machines. In: The 2008 IEEE International Workshop on Variable Structure Systems, 8–10 June 2008, Antalya, pp. 95–99. doi:10.1109/VSS.2008.4570689

Sabanovic, A., Izosimov, D.B.: Application of sliding modes to induction motor control, (1981). IEEE Trans. Ind. Appl. **IA–17**(1), 41–49 (1993)

Sabanovic, A.: Variable structure systems with sliding modes in motion control - a survey. IEEE Trans. Ind. Inf. **7**(2), 212–223 (2011)

Soltanpour, M.R., Fateh, M.M.: Sliding mode robust control of robot manipulators in the task space by support of feedback linearization and backstepping control, (2009). World Appl. Sci. J. **6**(1), 70–76 (2009)

Soltanpour, M.R., Zolfaghari, B., Soltani, M., Khooban, M.: Fuzzy sliding mode control design for a class of nonlinear systems with structured and unstructured uncertainties. Int. J. Innov. Comput. Inf. Control **9**(7), 2713–2726 (2013)

Utkin, V.I.: Sliding mode control design principles and applications to electric drives. IEEE trans. Ind. Electron. **40**(1), 23–36 (1993)

Yang L., Xiang H., Yang, X., Xie, R., Liu, T.: A variable-band hysteresis modulated multi-resonant sliding-mode controller for three-phasegrid-connected VSI with an LCL-filter. In: The 2013 ECCE Asia Downunder (ECCE Asia), 3–6 June 2013, Melbourne, VIC, pp. 670–674 (2013).doi:10.1109/ECCE-Asia.6579172

Zhang, J., Shi, P., Xia, Y.: Robust adaptive sliding-mode control for fuzzy systems with mismatched uncertainties. IEEE Trans. Fuzzy Syst. **18**(4), 700–711 (2010)

Zhang, Z., Zhao, Y., Qiao, W., Qu, L.: A space-vector modulated sensorless direct-torque control for direct-drive pmsg wind turbines. IEEE Trans. Ind. Electron. **90**, 1–11 (2014)

Contribution to Study Performance of the Induction Motor by Sliding Mode Control and Field Oriented Control

Oukaci Assia, Toufouti Riad and Dib Djalel

Abstract The induction motor squirrel cage that is deemed by its strength, high torque mass, robustness, and its relatively low cost ... etc., meanwhile, it benefited from the support of industry since its invention (invention by Tesla the late nineteenth century). Unfortunately, these advantages are accompanied by a high complexity of the physical interactions between the stator and the rotor. Therefore, dynamic control requires complex control algorithms in contrast to its structural simplicity. In recent decades, many techniques of control of the induction machine, such as technical oriented control or Field Oriented control, have emerged and are currently used to enjoy the benefits of the asynchronous machine for applications where variable speed is essential. The high operating control of the induction machine began with the invention of the oriented vector control in the late 60s flux. Before that time control of the induction machine was limited to scalar commands. This operating control does not provide a decoupling between the flux and torque. To illustrate this, the torque of a cage induction motor has to be increased by increasing the slip, the flux is affected by a decrease; therefore the torque control is dependent of the stream, for this the inherent coupling between these two variables makes conventional techniques less efficient. To solve these problems this paper seeks to analyze dynamical performances and sensitivity to induction motor parameter changes, two techniques are applied Sliding Mode Control and Field Oriented Control. For this, this design on the basis of some simulations results is illustrated with different functions in order to illustrate its efficiency and make comparison between the two techniques; Numerical simulations are presented to validate the proposed methods. The objective of this paper is to

O. Assia
Department of Electrical Engineering, Universty Mentouri of Constantine,
City Rebahi Nouar, 29 street Bouthlidja Tayeb, 41000 Constantine, Algeria
e-mail: assia_oukaci@yahoo.fr

T. Riad
Department of Electrical Engineering, Universty Mouhamed Chrif
Msaadia of Souk Ahras, 41000 Souk Ahras, Algeria
e-mail: toufoutidz@yahoo.fr

D. Djalel (✉)
Laboratory LABGET, Department of Electrical Engineering, Universty of Tebessa,
12000 Tebessa, Algeria
e-mail: dibdjalel@gmail.com

A.T. Azar and Q. Zhu (eds.), *Advances and Applications in Sliding Mode Control systems*,
Studies in Computational Intelligence 576, DOI 10.1007/978-3-319-11173-5_18

guarantee the desired performance of the induction motor, robust to the parameters variations, disturbances, and reach the speed of rotation at the speed desired in a minimum response time.

Nomenculture

r, s : Subscripts stand for rotor and stator;
R_r, R_s : Rotor and stator resistances;
L_r, L_s, L_m : Rotor, stator and mutual inductances;
C_{em} : Electromagnetic torque;
C_r : Load torque;
σ : Total leakage coefficient;
J : Moment of inertia;
v, i : Voltage and current;
ϕ : Flux linkage;
ω_r : Electrical angular rotor speed;
ω_s : Synchronously rotating angular speed;
p : Number of poles pair.

1 Introduction

Currently, induction motors (IM) are widely used in many industrial applications, including transportation, conveyor systems, actuators, material han-dling, pumping of liquid metal, and others, with satisfactory performance (Boucheta et al. 2012). The electromechanical systems of this motor are suitable for a large spectrum of industrial applications (Isidori 1995). However, induction motors are multivariable nonlinear and strongly coupled time-varying systems, mainly, in variable speed applications. However, its dynamic control requires complex control algorithms, facing its structural simplicity, since there is a complex coupling between the input variables, output variables and the internal variables of the machine (Leonhard 1994; Meziane et al. 2008; Maher 2012).

Induction motors are suitable electromechanical systems for a large spectrum of industrial applications. However, induction motors are multivariable nonlinear and strongly coupled time-varying systems, mainly, in variable speed applications (Leonhard 1994; Hautier and Caron 1995). However, its dynamic control requires complex control algorithms, facing its structural simplicity, since there is a complex coupling between the input variables, output variables and the internal variables of the machine (Isidori 1995; Meziane et al. 2008; Maher 2012).

So with the invention of power electronics, and advances in computing, has made a radical revolution. Its goal is to develop control strategies for induction motors. The design of suitable control algorithms for those motors (IM) has been widely investigated for more than two decades. Since the beginning of field oriented control (FOC) of AC drives works like a separately excited DC motor and it was proposed by Blaschke (Direct FOC) and Hasse (Indirect FOC) in early 1970s, (Ramesh et al. 2013), seen as a viable replacement of the traditional DC drives, several techniques from linear control theory have been used in the different control loops of the FOC scheme, such as Proportional Integral (PI) regulators, and exact feedback linearization (Ouhrouche and Volat 2000; Duarte-Mermoud and Travieso-Torres 2012). Due to their linear characteristics, these techniques do not guarantee suitable machine operation for the whole operation range, and do not consider the parameter variations of the motor-load set. The field oriented control technique has a major disadvantage, such as; requirement of co-ordinate transformations, current controllers, sensitive to parameter variations. This drawbacks of FOC schemes are minimized with the new control strategy i.e., DTFC scheme, which is introduced by Isao Takahashi and Toshihiko Noguchi, in the mid 1980s, (Ramesh et al. 2013).

Direct torque and flux control of an IM is requires the rotor shaft angular position information. The rotor shaft position can be measured through either speed sensors (i.e., speed encoder) or from an estimator/observer using current and voltage signals and information of the IM parameters. The use of speed encoder is associated with some drawbacks, such as, requirement of shaft extension, reduction of mechanical robustness of the motor drive, reduces the drive reliability and not suitable for hostile environments, and also costlier. These drawbacks have made speed sensorless direct torque and flux controlled IM very attractive over the conventional direct torque and flux control (DTFC) drive. There are some applications for sensorless drive, where there is no sufficient space to put the speed sensor or the nature of the environment does not allow the use of any additional rotor speed sensors, (Ramesh et al. 2013).

Over the past years, several schemes have proposed for rotor speed estimation in the sensorless vector controlled IMs, (Caruana et al. 2006; Kojabadi 2005). They are: (i) signal injection based method (Caruana et al. 2006), (ii) state observer based method (Rojas et al. 2004), and (iii) model based method (Kojabadi 2005). The signal injection method is suffers from computational complexity and requirement of external hardware for signal injection, (Ramesh et al. 2013). Among these methods, the sliding mode technique is one of the nonlinear control techniques has also been proposed to solve the problems mentioned above (Rao et al. 2009; Saiad 2012).

Much research has been done in recent years to apply various approaches to attenuate the effect of uncertainties. On the basic aspect, the conventional proportional-integral-derivative (PID) controllers are widely used in industry due to their simple control structure, ease of design and low cost. (El-Sousy 2013; Holmes et al. 2012). However, the PID controller cannot provide perfect control performance if the controlled system is highly nonlinear and uncertain as in the case of IM. In addition, an objection to the real-time use of such control scheme is the lack of knowledge of uncertainties (El-Sousy 2013).

Due to the existence of nonlinearities, uncertainties, and disturbances, conventional linear control methods, including the PID control, cannot guarantee a sufficiently high performance for the IM servo drive system. To deal with these uncertainties, and to enhance the control performance, in recent years, many nonlinear control methods have been developed for the IM drive system, such as variable structure control (Liaw et al. 2001), adaptive and robust control (Xia et al. 2000; Ravi Teja and Chakraborty 2012), sliding mode control (SMC) (Li et al. 2005; Comanescu et al. 2008), higher-order sliding-mode control (Rashed et al. 2005; Traoré et al. 2008), fuzzy control, neural network control, wavelet neural net work control (Lin and Hsu 2002; Castillo-Toledo et al. 2008), hybrid control (Wai 2001; Wai et al. 2002), optimal control (Bottura et al. 2000; Attaianese and Timasso 2001), intelligent SMC (Wai et al. 2002; Zhu et al. 2011), supervisory control using genetic algorithm (Wai 2003; Su and Kung 2005) and so on. These approaches improve the control performance of the IM drive from different aspects. The sliding-mode control (SMC) is one of the effective nonlinear robust control approaches since it provides system dynamics with an invariance property to uncertainties once the system dynamics are controlled in the sliding mode (El-Sousy 2013).

Sliding-mode control has received much attention in the control of IM drives. It is well known that the major advantage of sliding mode control (SMC) systems is its insensitivity to parameter variations and external disturbance once the system trajectory reaches and stays on the sliding surface (Slotine and Li 1991; Veselic et al. 2010).

The robustness of the SMC is guaranteed usually by using a large switching control gain. This switching strategy often in the hitting control law (Slotine and Li 1991; Rao et al. 2009; Astrom and Wittenmark 1995; Corradini et al. 2012; El-Sousy 2013).

The sliding mode control (SMC) is a nonlinear control and based on the switching functions of state variables, used to create a variety or hyper sliding surface, whose purpose is to force the system dynamism to correspond with the defined by the equation of the hyper-surface. When the state is maintained on the hyper surface, the system is in sliding regime. Its dynamic is so insensitive to external disturbances and parametric conditions as sliding regime are carried out. In the synthesis of the control law by way of sliding, the sliding surface is defined as an independent and stable linear system. However, the dynamics imposed by such a system is slower than that imposed by a nonlinear system, hence the importance of using the latter type of systems to synthesize the sliding surface in some applications (Hautier and Caron 1995; Araujo and Freitas 2000; Aurora and Ferrara 2007; Bounar et al. 2012; Saiad 2012; Talhaoui et al. 2013).

1.1 Chapter Objectives

The main objective of this work is to improve the performance of the electric machine converter association. Indeed, a new technique for robust control by sliding mode is presented. The application example is given in this work is that the induction motor.

The bulk of this work thereafter is to arrive at clear a comparative study between conventional vector control orientation of the rotor flux and the new technique of sliding mode control. This leads to show fields and limits of use of each control, while by highlighting all the features that differentiate them both. For this, the proposed controls are applied to achieve a speed- and flux-tracking in a minimum response time objective under parameter uncertainties and disturbance of load thrust force

1.2 Chapter's Structure

The reminder of the current chapter is organized as follows: The first part focuses on the development of a mathematical model for an induction motor, the second part focuses on the development of the mathematical model for the field oriented control technique to induction motor, the third part focuses on the development of the sliding mode control to the induction motor, the last part of this chapter focuses on the performance analysis of theoretical results for both techniques to make a comparison between them. They have been validated by numerical simulations in Matlab/Simulink environment.

2 Nonlinear Induction Motor Model

Induction motor as various electric machines constitutes a theoretically interesting and practically important class of nonlinear dynamic systems. Induction motor is known as a complex nonlinear system in which time-varying parameters entail additional difficulty for induction motor system control, conditions monitoring and faults diagnosis (Leonhard 1994; Isidori 1995). Based on the fact that the nonlinear model of the induction motor system can be significantly simplified, if only one applies the d-q Park transformation (Appendix 1) (Jimoh et al. 2012), different structures of the nonlinear model are investigated. The choice of a model depends on measurement possibilities, selected state variables of the machine and the problem at hand. In this paper, the considered induction motor model has stator current, rotor flux and rotor angular velocity as selected state variables. The control inputs are the stator voltage and load torque. The available stator current measurements are the induction motor system outputs. The nonlinear state space model of the induction motor is expressed as the following (Hautier and Caron 1995; Meziane et al. 2008; Mira and Duarte-Mermoud 2009):

$$\begin{cases} \dot{x} = f(x) + g(x).v \\ y = h(x) \end{cases} \tag{1}$$

With:

$$v = \begin{bmatrix} V_{s\alpha} \ V_{s\beta} \end{bmatrix}^T = \begin{bmatrix} U_1 \ U_2 \end{bmatrix}^T$$
$$x = \begin{bmatrix} i_{s\alpha} \ i_{s\beta} \ \phi_{r\alpha} \ \phi_{r\beta} \Omega_r \end{bmatrix}^T = \begin{bmatrix} x_1 \ x_2 \ x_3 \ x_4 x_5 \end{bmatrix}^T$$

Such as :
x : State vector.
v : Vector control.
y : Output selected.
$h(x)$: An analytic function.

$$f(x) = \begin{bmatrix} a_{11} \cdot x_1 + a_{13} \cdot x_3 + a_{14} \cdot x_4.x_5 \\ a_{11} \cdot x_2 - a_{14} \cdot x_3.x_5 + a_{13} \cdot x_4 \\ a_{31} \cdot x_1 + a_{33} \cdot x_3 + a_{34} \cdot x_4.x_5 \\ a_{31} \cdot x_2 - a_{34} \cdot x_3.x_5 + a_{33} \cdot x_4 \\ \mu \cdot (x_2 \cdot x_3 - x_1 \cdot x_4) - \frac{C_r}{J} \end{bmatrix}; g(x) = \begin{bmatrix} b_{11} & 0 \\ 0 & b_{11} \\ 0 & 0 \\ 0 & 0 \\ 0 & 0 \end{bmatrix}$$

Such as:

$$a_{11} = -\left[\frac{1}{\sigma \cdot T_s} + \frac{1}{T_r} \cdot \left(\frac{1-\sigma}{\sigma}\right)\right], \quad a_{13} = \frac{1-\sigma}{\sigma} \cdot \frac{1}{M \cdot T_r}, \quad a_{14} = \frac{1}{M} \cdot \frac{1-\sigma}{\sigma} \cdot p,$$

$$a_{31} = \frac{M}{T_r}, \quad a_{33} = -\frac{1}{T_r}, \quad a_{34} = -p, \quad b_{11} = \frac{1}{L_s.\sigma}, \quad \sigma = 1 - \frac{M^2}{L_s.L_r}, \quad T_s = \frac{L_s}{R_s},$$

$$T_r = \frac{L_r}{R_r}, \quad \mu = \frac{p \cdot M}{J \cdot L_r}.$$

3 Principle of Field Oriented Control FOC

The basic idea of this method of control is to bring the behavior of the asynchronous machine similar to that of a DC machine with separate excitation or decoupling is natural. This method is based on transforming the electric machine variables to a repository that rotates with the rotor flux vector oriented (Blaschke 1977; Hautier and Caron 1995; Rong-Jong et al. 2005; Meziane et al. 2008; Bouchhida et al. 2012). Therefore, it can control the flux of the machine with the component I_{sd} of the stator current which is the equivalent of the inductor DC current machine. While, the component I_{sq} to control the armature current corresponding to the DC machine, electromagnetic torque (Mira and Duarte-Mermoud 2009). This is shown in Fig. 1

The relationship of the electromagnetic torque of the DC machine is given by, (Hautier and Caron 1995):

$$C_e = \kappa \cdot \phi \cdot i_a = K \cdot i_f \cdot i_a \tag{2}$$

With:
ϕ : Flux imposed by the excitation current i_f and i_a : Inductive current.

The inductive current i_a is the magnitude of the torque generator and the excitation current i_f is the magnitude of the flux generator. Thus, in a DC machine everything happens as if the control variables i_f and i_a are orthogonal. This means that the current

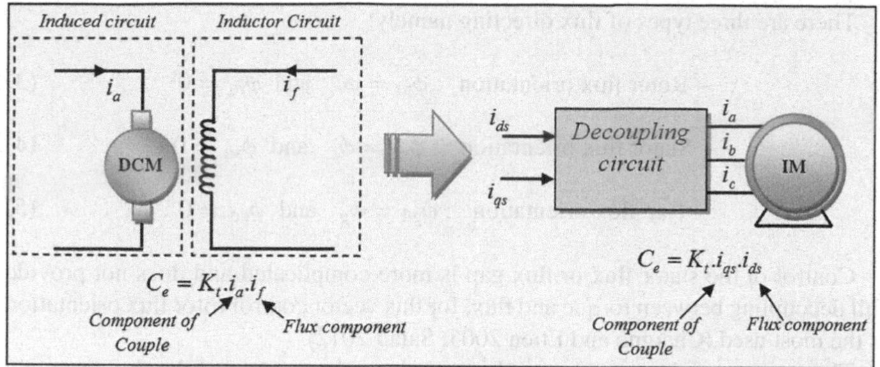

Fig. 1 Analogy induction machines with DC motor, (Hautier and Caron 1995; Meziane et al. 2008)

flux controlled by the i_f, and the current torque by the i_a. It is said that the armature and inductor are naturally decoupled.

Thus dissociates the stator current of the asynchronous machine in two components I_{sd} and I_{sq} in quadrature such that the current I_{sd} is oriented along the axis of the flux guide. A constant rotor flux, the couple then depend only aware I_{sq} (Hautier and Caron 1995; Chaigne and Etien 2005; Meziane et al. 2008).

3.1 Field Oriented Control Structure

The choice of the reference model of the induction machine from the template in the Park transformation is such that the axis (*d*) coincides with the desired direction of the flux (the rotor flux, the stator flux or air gap flux) as shown in the following Fig. 2, (Araujo and Freitas 2000; Mira and Duarte-Mermoud 2009; El-Sousy and Salem 2004):

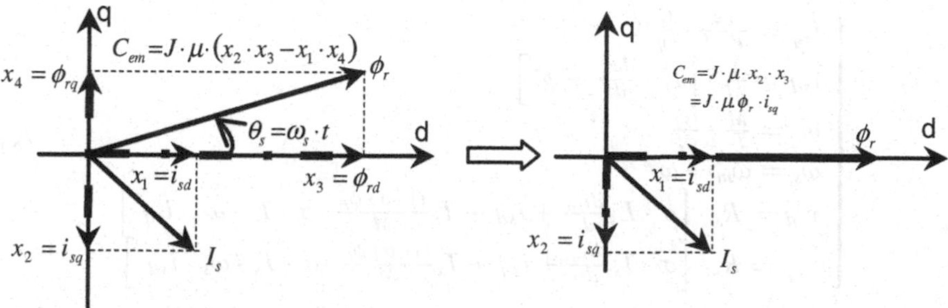

Fig. 2 Angular relations of current vectors, (Araujo and Freitas 2000)

There are three types of flux directing namely:

$$\text{– Rotor flux orientation} \ : \phi_{\text{rd}} = \phi_r \quad \text{and} \quad \phi_{\text{rq}} = 0 \tag{3}$$

$$\text{– Stator flux orientation} \ : \phi_{\text{sd}} = \phi_s \quad \text{and} \quad \phi_{\text{sq}} = 0 \tag{4}$$

$$\text{– Gap flux orientation} \ : \phi_{\text{gd}} = \phi_g \quad \text{and} \quad \phi_{\text{gq}} = 0 \tag{5}$$

Control of the stator flux or flux gap is more complicated and does not provide full decoupling between torque and flux, for this vector control rotor flux orientation is the most used (Chaigne and Etien 2005; Saiad 2012).

The strategy of the vector control is to independently control the flux term and the current term to impose a couple. Keeping the control variables as (V_{sd}, V_{sq}) and state variables such as stator currents (i_{sd}, i_{sq}), the flux ϕ_r and the mechanical speed. When an electric motor drives a mechanical load it is essential to properly control the dynamics of it, to master the instantaneous torque of it. The thrust of the vector control is to have the asynchronous machine for a couple proportional to flux engine and a current like the DC machine. So, let's take the expression of the electromagnetic torque of the induction machine (Rao et al. 2009):

$$C_{em} = J \cdot \mu \cdot (x_2 \cdot x_3 - x_1 \cdot x_4) \tag{6}$$

In the reference dq which are projected the rotor flux and the stator current running at the speed of the rotating field, either in: $\theta_s = \omega_s \cdot t$

In order to have expression analogous to that electromagnetic torque of a DC motor the axis will be directed of the rotor flux, air gap torque of the expression becomes:

$$C_{em} = J \cdot \mu \cdot x_2 \cdot x_3 \tag{7}$$

By imposing the condition (3) to state the model of IM (1) supplied with voltage equations at the following reduced system is realized, (Boukettaya et al. 2008; Meziane et al. 2008; Bouchhida et al. 2012):

$$\begin{cases} i_{sq} = \frac{L_r}{P \cdot M} \cdot \frac{C_{em}^*}{\phi_r^*} \\ i_{sd} = \frac{1}{M} \cdot \left[T_r \cdot \frac{d\phi_r^*}{dt} + \phi_r^* \right] \\ \omega_r = \frac{M}{T_r} \cdot \frac{i_{sq}}{\phi_r^*} \\ \omega_s = \omega_m + \omega_r \\ v_{sd} = R_s \cdot \left[\sigma \cdot L_s \frac{di_{sd}}{dt} + i_{sd} + T_s \frac{(1-\sigma) \cdot \phi_r^*}{M} \text{-} \sigma \cdot T_s \cdot \omega_s \cdot i_{sq} \right] \\ v_{sq} = R_s \cdot \left[\sigma \cdot T_s \frac{di_{sq}}{dt} + i_{sq} + T_s \frac{(1-\sigma) \cdot \phi_r^*}{M} \text{-} \sigma \cdot T_s \cdot \omega_s \cdot i_{sd} \right] \end{cases} \tag{8}$$

The diagram of vector control with a flux model is given in Fig. 3:

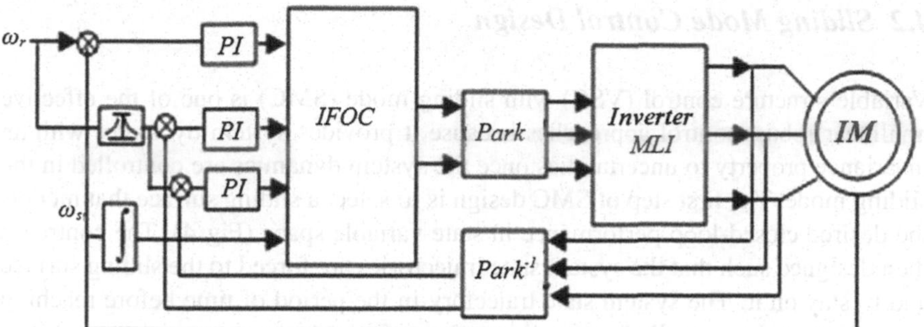

Fig. 3 Block diagram of indirect field oriented control structure (Meziane et al. 2008; Bouchhida et al. 2012)

4 General Concept of Sliding Mode Control

4.1 Condition Existence of Sliding Mode Control

The sliding mode exists when the switching takes place continuously between U_{max} and U_{min} . This is illustrated in Fig. 4 for the case of a control system of the second order with two state variables x_1 and x_2, (Wai 2003)

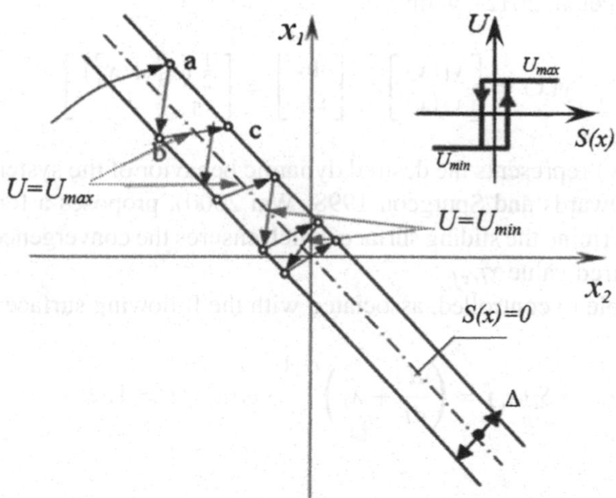

Fig. 4 Path of steady state sliding mode

4.2 Sliding Mode Control Design

Variable structure control (VSC) with sliding mode (SMC) is one of the effective nonlinear robust control approaches because it provides system dynamics with an invariance property to uncertainties once the system dynamics are controlled in the sliding mode. The first step of SMC design is to select a sliding surface that models the desired closed-loop performance in state variable space (Fig. 4). The control is then designed such that the system state trajectories are forced to the sliding surface and to stay on it. The system state trajectory in the period of time before reaching the sliding surface is called the reaching phase. Once the system trajectory reaches the sliding surface, it stays on it and slides along it toward the origin. The system trajectory sliding along the sliding surface toward the origin is the sliding mode. The insensitivity of the controlled system to uncertainties exists in the sliding mode, but not during the reaching phase. Thus, the system dynamic in the reaching phase continues to be influenced by uncertainties (Utkin 1993; Wai 2000; Ghanes and Zheng 2009; Boucheta et al. 2012).

4.2.1 The Choice of the Surface

The choice of the sliding surface for the necessary number and shape, depending on the application and purpose. In general, for a system defined by the state Eq. (1), choose "m" sliding surfaces for a vector of dimension "m", (Dwards and Spurgeon 1998; Boucheta et al. 2012), with:

$$y(x) = \begin{bmatrix} y_1(x) \\ y_2(x) \end{bmatrix} = \begin{bmatrix} \Phi_r \\ \Omega_r \end{bmatrix} = \begin{bmatrix} \frac{1}{2}\left(x_3^2 + x_4^2\right) \\ x_5 \end{bmatrix} \tag{9}$$

The surface $S(x)$ represents the desired dynamic behavior of the system. J. J Slotine (Utkin 1993; Dwards and Spurgeon 1998; Wai 2000), proposes a form of general equation to determine the sliding surface which ensures the convergence of a variable towards its desired value x_{iref} .

If x_i a variable to controlled, associated with the following surface:

$$S_i(x_i) = \left(\frac{d}{dt} + \lambda_i\right)^{r-1} \cdot e_i(x_i)/i = 1, 2. \tag{10}$$

With:

λ_i : is a positive constant.

r : is the relative degree (Appendix 2).

And that for:

$$\begin{aligned} r = 1 \quad &\Rightarrow \quad S(x) = e(x) \\ r = 2 \quad &\Rightarrow \quad S(x) = \lambda_e(x) + \dot{e}(x) \\ r = 3 \quad &\Rightarrow \quad S(x) = \lambda^2 e(x) + 2\lambda_{\dot{e}}(x) + \ddot{e}(x) \end{aligned} \tag{11}$$

The difference between the controlled variable and its reference is:

$$e_i(x) = x_i - x_{iref} \tag{12}$$

The purpose of this paper is to determine a control law to force the system states, i.e., the rotor flux and the electromagnetic torque to follow the sliding surface as: $S = [S_1 \quad S_2]^T$ (Ghanes and Zheng 2009).

4.2.2 Area Calculation

After the calculation of the relative degree (given in [Appendix 2]). The sliding surfaces of the Eq. (12), can be determined as follows:

$$\begin{cases} S_1 = \lambda_1 \cdot e_1 + \dot{e}_1 \\ S_2 = \lambda_2 \cdot e_2 + \dot{e}_2 \end{cases} \tag{13}$$

With :

$$\begin{cases} e_1 = \Phi_r - \Phi_{iref} \\ e_2 = \Omega_r - \Omega_{iref} \end{cases} \tag{14}$$

Are successively error flux (e_1) and error rate (e_2).

When substituting (1) and (14) into (13) the following result is as follows:

$$\begin{cases} S_1 = \lambda_1 \cdot (\Phi_r - \Phi_{iref}) + a_{31} \cdot (x_1 \cdot x_3 + x_2 \cdot x_4) + 2 \cdot a_{33} \cdot \Phi_r - \dot{\Phi}_{iref} \\ S_2 = \lambda_2 \cdot (\Omega_r - \Omega_{iref}) + \mu \cdot (x_2 \cdot x_3 - x_1 \cdot x_4) - \frac{c_r}{J} - \frac{f}{J} . \Omega_r - \dot{\Omega}_{iref} \end{cases} \tag{15}$$

However, to continue Φ_{iref} and Ω_{iref}, it suffices to make the sliding surface attractive and invariant.

4.2.3 Equivalent Command for the Invariance

Once the sliding surface is chosen, it remains to determine the control necessary to attract the controlled variable to the surface and then to his balance point, the relation will be as follows:

$$u = u_{eq} + u_n \tag{16}$$

where u_{eq} is called the equivalent control, which dictates the motion of the state trajectory along the sliding surface, and u_n is a term introduced to satisfy the condition following convergence $\dot{S}(x).S(x) < 0$, and it determines the dynamic behavior of the system during the convergence mode. So this command guarantees the attractiveness of the variable to be controlled to the sliding surface, (Utkin 1993; Dwards and Spurgeon 1998; Wai 2000).

The necessary condition for the system states follow the path defined by the sliding surfaces is:

$$\dot{S} = 0 \tag{17}$$

The equivalent command is the commands ensure the condition (17). Then the derivation of Eq. (15) gives:

$$\begin{cases} \dot{S}_1 = 2 \cdot (\lambda \cdot a_{33} + 2 \cdot a_{33}^2 + a_{31} \cdot a_{13}) \cdot \Phi_r + (\lambda \cdot a_{31} + 3 \cdot a_{33} \cdot a_{31} + a_{11} \cdot a_{31}) \cdot f_1 \\ \quad -(a_{31} \cdot a_{34}) \cdot f_2 + a_{31}^2 \cdot f_3 - \lambda \cdot \dot{\Phi}_{rref} - \ddot{\Phi}_{rref} + b_{11} \cdot a_{31} \cdot (x_3 \cdot U_1 \ + x_4 \cdot U_2\ ) \\ \dot{S}_2 = \mu \cdot \left(\lambda_1 + a_{33}^{+} a_{11}\right) \ \cdot f_2 - \lambda_2 . \frac{c_r}{J} + \cdot \mu \cdot a_{34} . x_5 \cdot f_1 + - (a_{34} \cdot a_{14} \cdot \mu) \cdot \Phi_r \\ \quad -\lambda_2 . \dot{\Omega}_{rref} - \ddot{\Omega}_{rref} + b_{11} \cdot \mu \cdot (x_3 \cdot U_1 \ - x_4 \cdot U_2\ ) \end{cases} \tag{18}$$

Such as:

$$\begin{cases} f_1 = x_1 \cdot x_3 + x_2 \cdot x_4 \\ f_2 = x_2 \cdot x_3 - x_1 \cdot x_4 \\ f_3 = x_1^2 + x_2^2 \end{cases} \tag{19}$$

The ideal diet is almost never possible. Therefore, the second term of the command must be used to restore the system state to the surface whenever it deviates. Thus, it should be taken as follows:

$$u_n = M_i \cdot sign\,(S_i(x)) \tag{20}$$

M_i is a constant, representing the maximum controller output required to overcome parameter uncertainties and disturbances; and $S_i\ (x)$ is called the switching function because the control action switches its sign on the two sides of the switching surface $S = 0$. A second-order system S is defined in Eq. (13), (Utkin 1993; Dwards and Spurgeon 1998; Wai 2000; Boucheta et al. 2012).

$S(x)$ Slip function is selected such that it is a solution of the following differential equation:

$$\dot{S}_i\ (x) = -M_i \cdot sign\ (S_i\ (x)) \ \ / i = 1, 2. \tag{21}$$

Then equation (17) can be written:

$$\begin{cases} -M_1 \cdot sign\,(S_1\,(x)) = 2 \cdot (\lambda \cdot a_{33} + 2 \cdot a_{33}^2 + a_{31} \cdot a_{13}) \cdot \Phi_r + (\lambda \cdot a_{31} + 3 \cdot a_{33} \cdot a_{31} + a_{11} \cdot a_{31}) \cdot f_1 \\ \qquad -(a_{31} \cdot a_{34}) \cdot f_2 + a_{31}^2 \cdot f_3 - \lambda \cdot \dot{\Phi}_{rref} - \ddot{\Phi}_{rref} + b_{11} \cdot a_{31} \cdot (x_3 \cdot U_1 \ + x_4 \cdot U_2\ ) \\ -M_2 \cdot sign\,(S_2\,(x)) = \mu \cdot (\lambda_1 + a_{33} + a_{11}) \ \cdot f_2 - \lambda_2 . \frac{c_r}{J} + \cdot \mu \cdot a_{34} . x_5 \cdot f_1 + - (a_{34} \cdot a_{14} \cdot \mu) \cdot \Phi_r \\ \qquad -\lambda_2 . \dot{\Omega}_{rref} - \ddot{\Omega}_{rref} + b_{11} \cdot \mu \cdot (x_3 \cdot U_1 \ - x_4 \cdot U_2\ ) \end{cases} \tag{22}$$

According to the Eqs. (19), (20) and (22) the equivalent command (16) for this invariance cans determined as (Araujo and Freitas 2000; Boucheta et al. 2012):

$$u = \begin{bmatrix} U_1 \\ U_2 \end{bmatrix} = G^{-1} \cdot \begin{bmatrix} X \\ Y \end{bmatrix} \tag{23}$$

With:

$$\begin{cases} X = 2\cdot(\lambda\cdot\left(\frac{a_{33}}{a_{31}}\right) + 2\cdot\left(\frac{a_{33}^2}{a_{31}}\right) + a_{13})\cdot\Phi_r + (\lambda + 3\cdot a_{33} + a_{11})\cdot f_1 - (a_{34})\cdot f_2 \\ \quad + a_{31}\cdot f_3 - \left(\frac{\lambda}{a_{31}}\right)\cdot\dot{\Phi}_{rref} - \left(\frac{1}{a_{31}}\right)\ddot{\Phi}_{rref} + \left(\frac{M_1}{a_{31}}\right)\cdot sign\,(S_1\,(x)) \\ Y = (\lambda_1 + a_{33} + a_{11})\ \cdot f_2 - \lambda_2.\frac{C_r}{J.\mu} + a_{34}.x_5\cdot f_1 + -\,(a_{34}\cdot a_{14})\cdot\Phi_r \\ \quad -\left(\frac{\lambda_2.}{\mu}\right).\dot{\Omega}_{rref} - \left(\frac{1}{\mu}\right)\ddot{\Omega}_{rref} + \left(\frac{M_2}{\mu}\right)\cdot sign\,(S_2\,(x)) \end{cases} \tag{24}$$

And

$$G = \begin{bmatrix} -b_{11}\cdot x_3 & -b_{11}\cdot x_4 \\ b_{11}\cdot x_4 & -b_{11}\cdot x_3 \end{bmatrix} \tag{25}$$

In taking into consideration the condition transversal matrix (25), then:

$$\det G \neq 0 \tag{26}$$

Therefore:

$$b_{11}^2\cdot\left(x_3^2 + x_4^2\right) \neq 0$$

With:

$$\begin{cases} x_3 = \phi_{r\alpha} \\ x_4 = \phi_{r\beta} \end{cases} \tag{27}$$

The determinant is non-zero, therefore, the matrix G is invertible, unless x = 0 and/or stopping the motor; the current is zero so the flux is zero, there must be provided the initial conditions of flux at startup.

For the switching law to intervene in the law of overall control, choose and sufficiently large, convergence criteria (Aurora and Ferrara 2007; Boucheta et al. 2012).

$$\begin{cases} M_1 > \left| \begin{array}{c} 2\cdot\left(\lambda\cdot\left(\frac{a_{33}}{a_{31}}\right) + 2\cdot\left(\frac{a_{33}^2}{a_{31}}\right) + a_{13}\right)\cdot\Phi_r \\ + (\lambda + 3\cdot a_{33} + a_{11})\cdot f_1 \\ -(a_{34})\cdot f_2 + a_{31}^{\cdot} f_3 - \left(\frac{\lambda}{a_{31}}\right)\cdot\dot{\Phi}_{rref} \\ -\left(\frac{1}{a_{31}}\right)\ddot{\Phi}_{rref} \end{array} \right| \\ M_2 > \left| \begin{array}{l} (\lambda_1 + a_{33} + a_{11})\ \cdot f_2 - \lambda_2.\frac{C_r}{J.\mu} \\ + a_{34}.x_5\cdot f_1 - (a_{34}\cdot a_{14})\cdot\Phi_r \\ -\left(\frac{\lambda_2.}{\mu}\right).\dot{\Omega}_{rref} - \left(\frac{1}{\mu}\right)\ddot{\Omega}_{rref} \end{array} \right| \end{cases} \tag{28}$$

The diagram of sliding mode control is given in Fig. 5 (Saiad 2012):

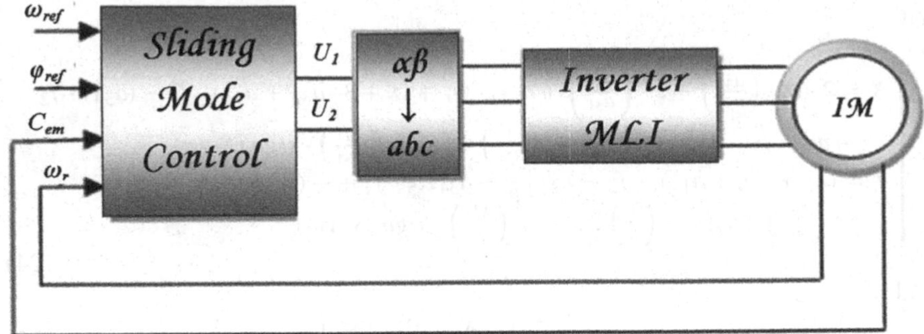

Fig. 5 Block diagram of the sliding mode control

5 Simulations Results

The simulation analysis of the mathematical model of the induction machine, were carried out in MATLAB/SIMULINK to demonstrate the effectiveness of the proposed control scheme for speed control of the IM described above, and allows seeing the performance comparison of control: Sliding Mode Control (SMC) and field oriented control (FOC), with test of two mode of operation, speed variation and other inversion of rotation with variable load torque C_r and rotor resistance R_r.

5.1 Simulation Results Without Inversion Speed

This simulation is realized with the reference speed given in Table 1 as follow:

5.1.1 Analysis and Discussions of the Results

The results simulations are given in the following Figs. 6, 7, 8, 9, 10, 11 12:

The proposed controller has been tested also with detuned rotor resistance. The rotor resistance is considered the most effective parameter in the indirect vector control as the slip calculator depends mainly on it. In classical indirect vector control, the variation of this parameter will adversely affect the motor performance. In this test, the rotor resistance is assumed to increase as follows: at 0.8 s will rise to 50 %

Table 1 The reference speed at the first simulation

Times t (s)	0 → 1.5	1.5 → 2.75	2.75 → 4
Reference speed Ω_{ref} (rad/s)	157	170	100

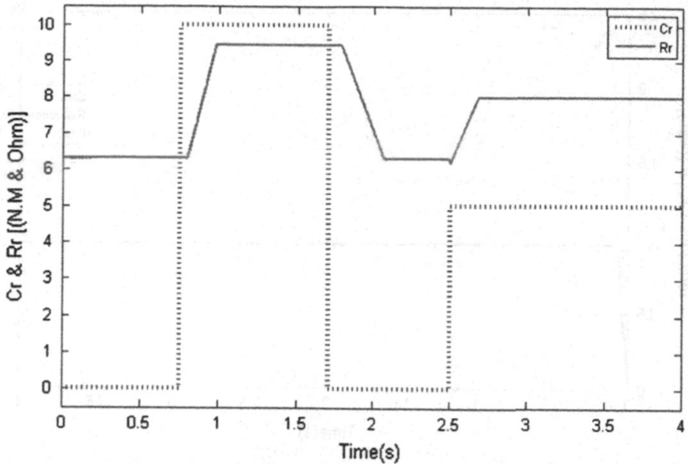

Fig. 6 Load and rotor resistance variations

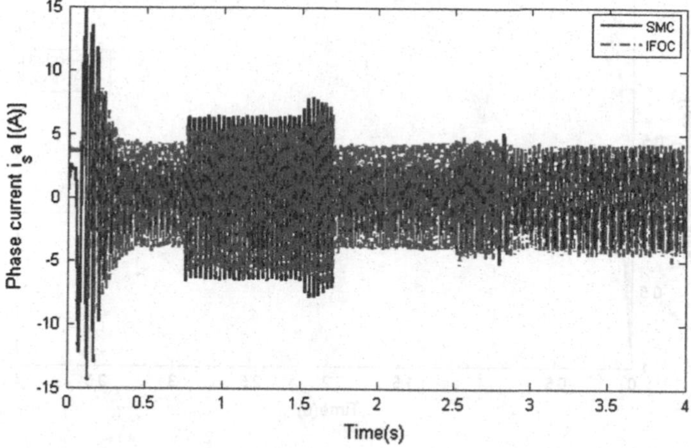

Fig. 7 Current of one phase

R_r , and at 2.1 s will increase to 30 % R_r (Fig. 6), in the machine model and keeping nominal value in the slip calculator.

In this test too, the load torque is assumed to change from 0 to 10 Nm. at 0.75 s and stepped again to no load at 1.75 s, and change again from 0 to 5 Nm. at 2.5 s as shown in Fig. 6.

The Fig. 7 reports an enlarged view of the phase stator current during high speed operation. It is seen that, sinusoidal current waveform is obtained with less distortion.

The Fig. 8 indicates that the control Sliding Mode Control (SMC) provides a successful prosecution at its rotor flux reference (unlike the control Indirect Field Oriented Control (IFOC)). Thus, the d-axis rotor flux linkage is kept constant at

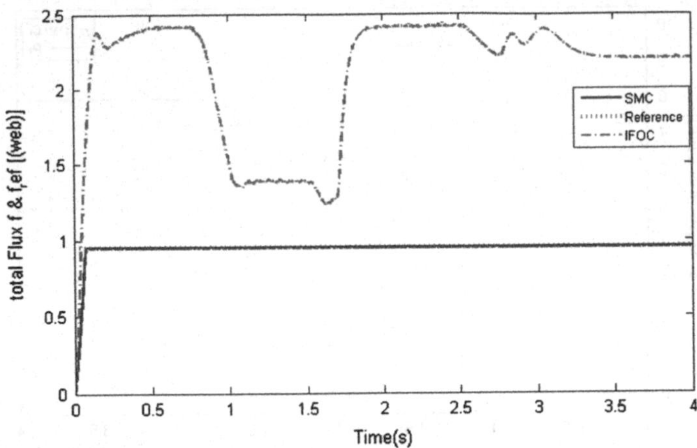

Fig. 8 Rotor flux tracking performance

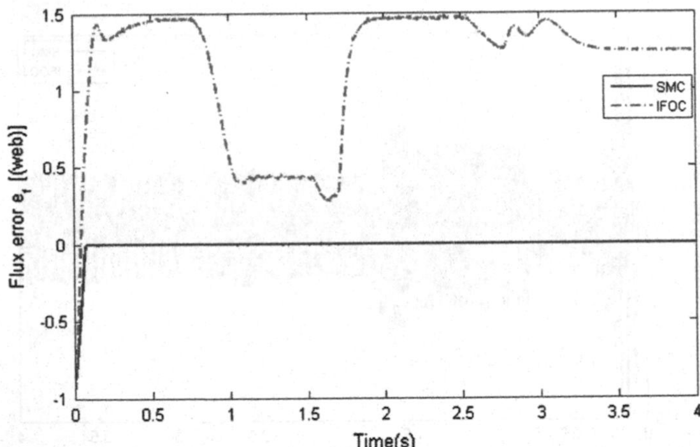

Fig. 9 Rotor flux error

the rated value, while the q-axis flux is kept zero in all the simulation period; only, small notches in the d-axis flux have been noticed at the instants of load disturbance application, (in the static regime the error is zero as shown in Fig. 9). In other words, the decoupling condition between the speed and rotor flux has been realized.

Figure 10 shows the results of simulation speed with the two types of controls (IFOC and SMC), when the machine is operated at different speeds as it's given in Table 1. It can be seen that the speed follows its reference reasonably well, but the SMC controller has a response time better than IFOC (0.35s), this time can be explained by the speed of this technique. As shown in Fig. 10 Zoom 1.

By applying a resisting torque which produces the heating of the machine, who led to varied the rotor see Fig. 6. In these circumstances, finds that the speed perfectly

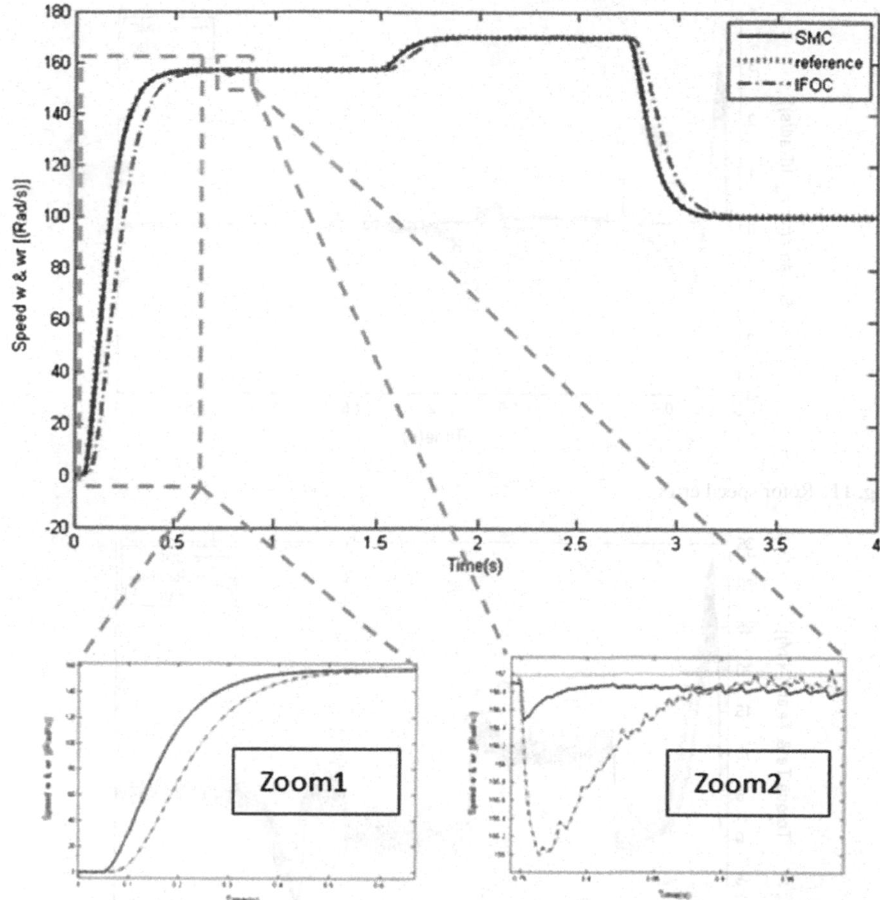

Fig. 10 Rotor speed tracking performance

follows its reference in the two types of control that is clearly demonstrated by the speed error is zero see Fig. 11, but here too the SMC has a time response is perfect and the maximum overshoot is about 2 % compared to IFOC as shown in Fig. 10 Zoom 2, good tracking performance has been achieved with the proposed sliding controller in spite of the mismatched rotor resistance.

It has been indicated in the Fig. 12 that excellent tracking performance has been achieved in spite of the load torque disturbance. Only, small notches have been noticed at the instants of load disturbance application, peaks 30 % of the load torque for SMC and 20 % for IFOC.

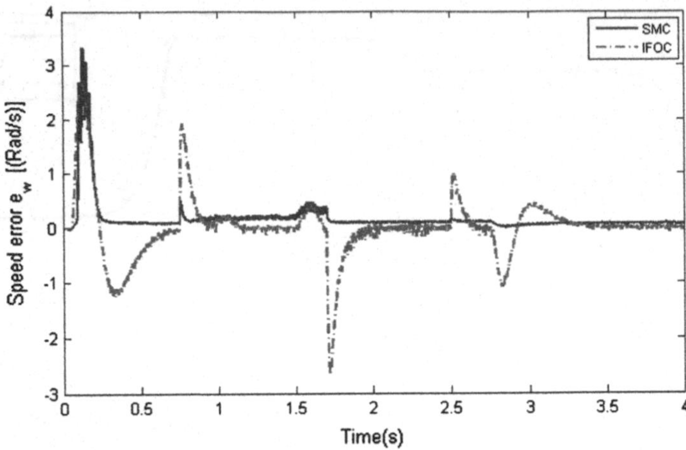

Fig. 11 Rotor speed error

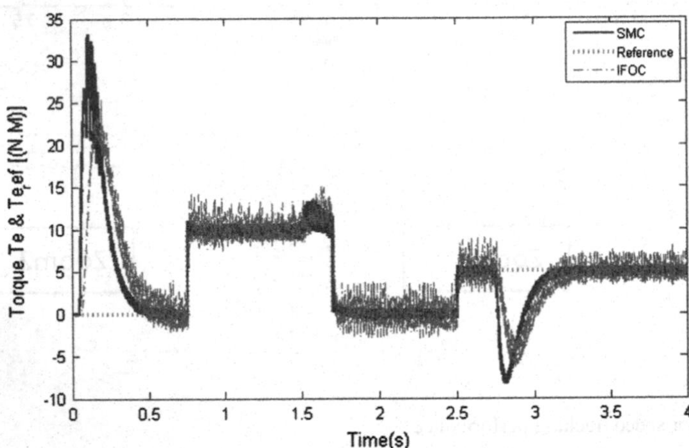

Fig. 12 Electromechanical torque to the load variation

5.2 Simulation Results with Inversion Speed

This simulation is realized with the reference speed given in Table 2 as follow:

5.2.1 Analysis and Discussions of the Results

The results simulations are given as following Figs. 13, 14, 15, 16, 17, 18 19:

The proposed controller has been tested also with detuned rotor resistance. The rotor resistance is considered the most effective parameter in the indirect vector

Table 2 The reference speed variation at the second simulation

Times t (s)	0 → 1.5	1.5 → 3	3 → 4
Reference speed Ω_{ref} (rad/s)	157	−157	157

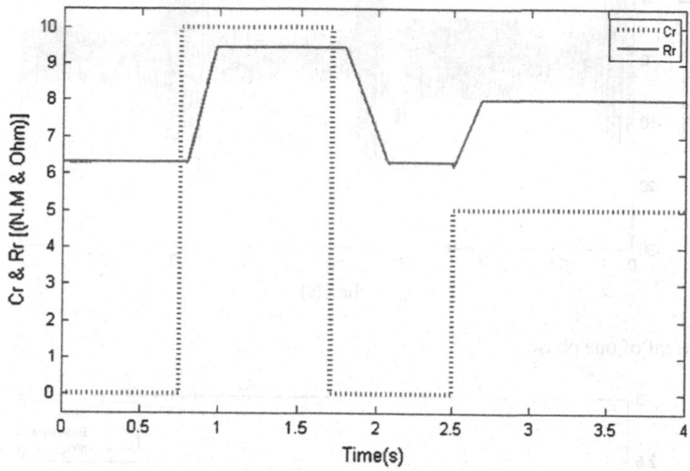

Fig. 13 Load and rotor resistance variations

control as the slip calculator depends mainly on it. In classical indirect vector control, the variation of this parameter will adversely affect the motor performance. In this test, the rotor resistance is assumed to increase as follows: at 0.8 s will rise to 50 % R_r , and at 2.1 s will increase to 30 % R_r (Fig. 13), in the machine model and keeping nominal value in the slip calculator.

In this test too, the load torque is assumed to change from 0 to 10 Nm. at 0.75 s and stepped again to no load at 1.75 s, and change again from 0 to 5 Nm. at 2.5 s as shown in Fig. 13.

The Fig. 14 reports an enlarged view of the phase stator current during high speed operation. It is seen that, sinusoidal current waveform is obtained with less distortion for both techniques, with peaks in the transitional regime and at the inversion of speed.

The Fig. 15 indicates that the control Sliding Mode Control (SMC) provides a successful prosecution at its rotor flux reference (unlike the control Indirect Field Oriented Control (IFOC)). Thus, the d-axis rotor flux linkage is kept constant at the rated value, while the q-axis flux is kept zero in all the simulation period; only, small notches in the d-axis flux have been noticed at the instants of load disturbance application and moment of inversion the direction of rotation, (in the static regime the error is zero as shown in Fig. 16). In other words, the decoupling condition between the speed and rotor flux has been realized.

Figure 17 shows the results of simulation speed with the two types of controls (IFOC and SMC), when the machine is operated at different speeds as it's given in Table 2. It can be seen that the speed follows its reference reasonably well, but

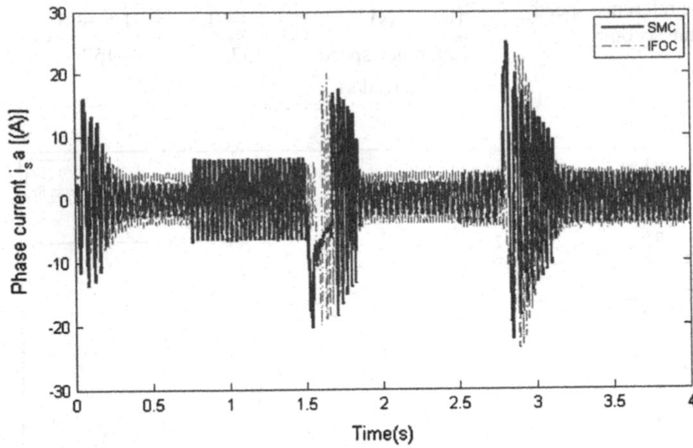

Fig. 14 Current of one phase

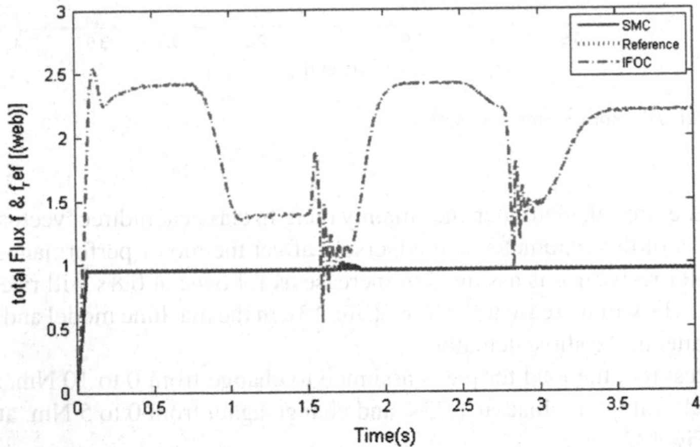

Fig. 15 Rotor flux tracking performance

the SMC controller has a response time better than IFOC (0.35s), this time can be explained by the speed of this technique. As shown in Fig. 17 Zoom 1.

By applying a resisting torque which produces the heating of the machine, who led to varied the rotor see Fig. 13. In those circumstances, finds that the speed perfectly follows its reference in the two types of control that is clearly demonstrated by the speed error is zero see Fig. 18, but here too the SMC has a time response is perfect and the maximum overshoot is about 2 % compared to IFOC as shown in Fig. 17 Zoom 2, good tracking performance has been achieved with the proposed sliding controller in spite of the mismatched rotor resistance and the reversing the direction of rotation.

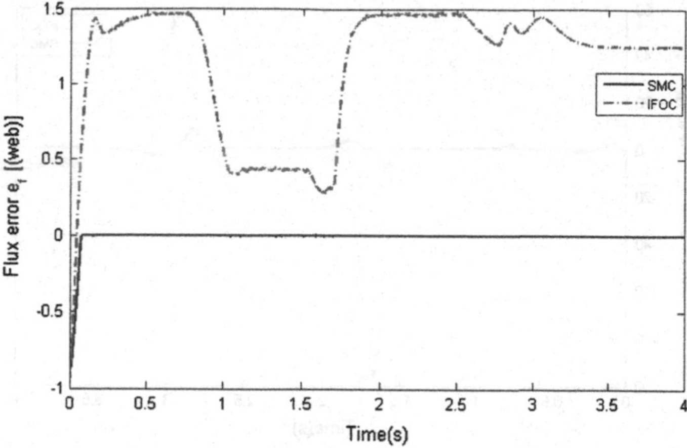

Fig. 16 Rotor flux error

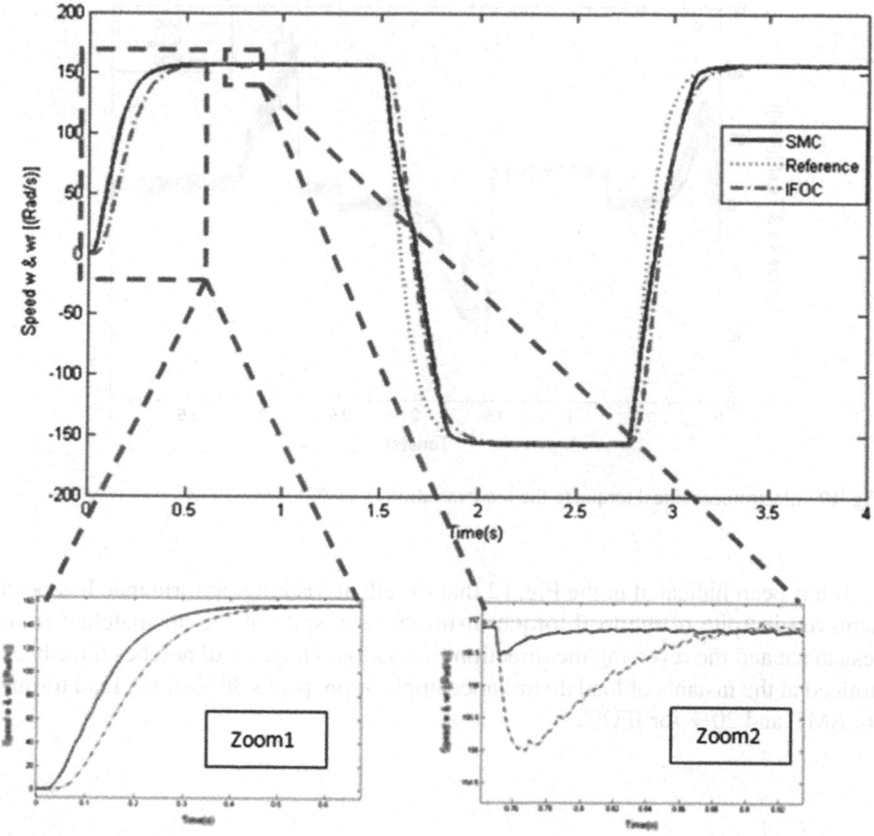

Fig. 17 Rotor speed tracking performance

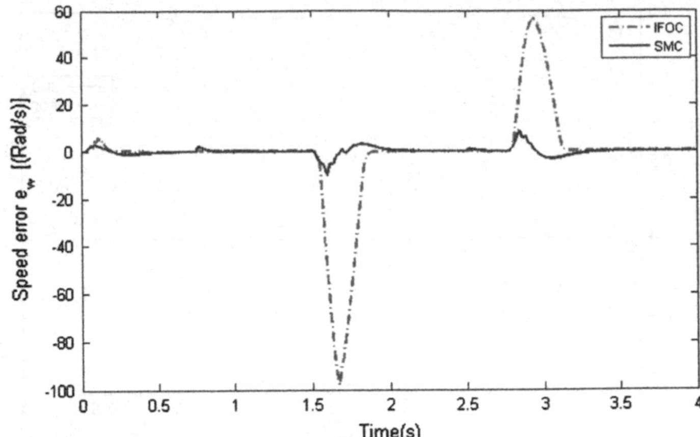

Fig. 18 Rotor speed error

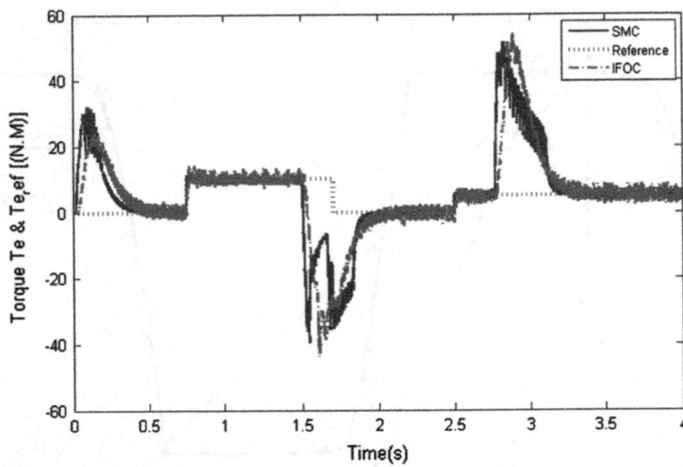

Fig. 19 Electromechanical torque to the load variation

It has been indicated in the Fig. 12 that excellent tracking performance has been achieved in spite of the load torque disturbance in spite of the mismatched rotor resistance and the reversing the direction of rotation. Only, small notches have been noticed at the instants of load disturbance application, peaks 30 % of the load torque for SMC and 20 % for IFOC.

6 Conclusion

Industrial systems are often significantly nonlinear behavior. The linearization around an operating point is often inadequate for the needs of the command, therefore it is important to develop control methods for nonlinear systems.

Controlling an IM can be done using several techniques, each of which offers dynamic and static performances with well-defined limits applications. The problem arises in the choice of a particular method. The use of a method or the other is normally based on the constraints of the specifications, which are sometimes added new requirements of energy saving and material economy that should be taken into account.

It is with this understanding that this work has been made. Indeed, the main objective of this chapter is the development of a new robust control by sliding mode. This type of control has been sufficiently discussed compared to the vector control of rotor flux orientation.

First vector control by indirect rotor flux orientation gave lower performance. Indeed, the strength of the test IFOC which has opposite variation of the rotor resistance shows that control loses its linearity property and affects more decoupling between rotor flux and torque. To improve the performance of this command and achieve better results, the online identification of parameters of the machine is essential.

In this sense, our contribution is to propose a methodology of robust control systems related to variable structures whose purpose is to overcome the disadvantages of conventional controls, as the SMC is by nature a non-linear control and their control law is changed in a discontinuous manner, in this case, the sliding mode control. This control is characterized by its robustness against external and internal disturbances. The sliding surface is determined depending on the desired performance. While, the control law is chosen in order to ensure the convergence conditions and sliding i.e., attractiveness and the invariance of switching surfaces.

Finally, the study of the sliding mode control of the induction machine consists in defining a sliding surface on which the system converges. The corresponding switching function allows the system to always tend towards the sliding surface. The technique of sliding mode control used for control of the induction motor has led to good performance, in many cases obtained in a better quality of adjustment relative to the vector control, it offers some advantages: first, robustness with respect to variations of system parameters, second, a high-performance dynamic "acceptable response time and error stationary practically zero", and finally a simple implementation of the law switching.

In addition, the objective of this work is achieved, because there is a perfect prosecution of the rotor flux, for its reference which makes for a good decoupling between the rotor flux and the electromagnetic torque, good trajectory tracking the desired output speed with a minimum response time, the robustness to variations in parameters, which has been shown by simulation results. The performance of this technique depends on a suitable choice of the coefficients of the sliding surface and the speed of the response depends on the maximum torque that can give the machine.

Perspectives From this, labor perspective can be considered, it would be interesting to:

- Propose a synthesis of an observer flux sliding mode. Because in the design of the sliding mode control, it is assumed that all states were measured. Since only the current measurements are available, then the estimated for the rotor flux of an application in real time is necessary.
- In an effort to reduce cost, speed sensor can be replaced by a flux observer and speed.
- And finally, it would be interesting to use this machine (induction machine IM) and this type of control (Sliding Mode Control SMC) for the adopted to renewable energies (wind).

Appendix: A.1 Arbitrary Reference Frame Theory

Arbitrary reference frame theory is mainly used in the dynamic analysis of electrical machines. Because of the highly coupled nature of the machine, especially the inductances within the winding make it rather impossible to perform dynamic simulations and analysis on electrical machines.

Arbitrary reference frame theory was discovered by Blondel, Dreyfus, Doherty and Nickle as mentioned in the classical paper (Park 1929). This newly found theory was generalized by Park on synchronous machines and this method was later extended by Stanley to the application of dynamic analysis of induction machines (Stanley 1938).

By using this method a poly-phase machine is transformed to a two-phase machine with their magnetic axis in quadrature as illustrated in Fig. 20. This method is also commonly referred to as the dq method in balanced systems and to the dq0 method in unbalanced systems with the '0' relating to the zero sequence or homopolar components in the Fortes cue Transformation (Jimoh et al. 2012).

This transformation eliminates mutual magnetic coupling between the phases and therefore makes the magnetic flux linkage of one winding independent of the current of another winding.

The transformation is done by applying a transformation matrix, Eq. (29) while the inverse transformation matrix, Eq. (30) will be transformed back to the natural reference frame. Eqs. (29) and (30) applly to a three phase system but can be modified to accommodate a system with any number of phases which might be useful in the case of the machine having an auxiliary winding as proposed in this work (Jimoh et al. 2012).

$$[P] = \sqrt{\frac{2}{3}} \begin{bmatrix} \frac{1}{\sqrt{2}} & \frac{1}{\sqrt{2}} & \frac{1}{\sqrt{2}} \\ \cos\theta & \cos\left(\theta - \frac{2\pi}{3}\right) & \cos\left(\theta + \frac{2\pi}{3}\right) \\ -\sin\theta & -\sin\left(\theta - \frac{2\pi}{3}\right) & -\sin\left(\theta + \frac{2\pi}{3}\right) \end{bmatrix} \tag{29}$$

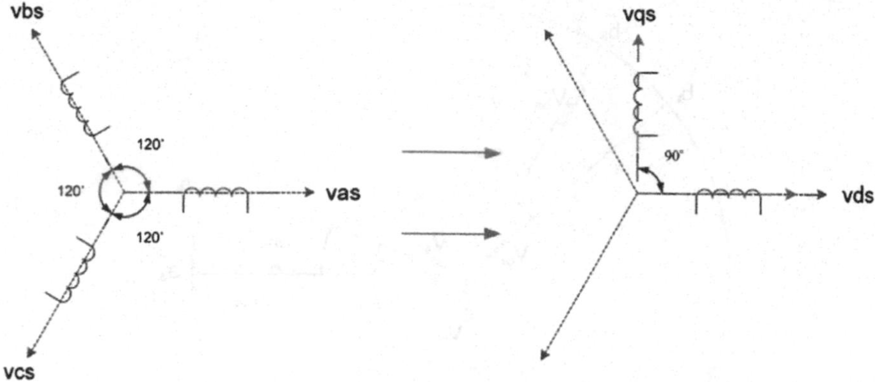

Fig. 20 Park's transform

$$[P]^{-1} = \sqrt{\frac{2}{3}} \begin{bmatrix} \frac{1}{\sqrt{2}} & \cos\theta & -\sin\theta \\ \frac{1}{\sqrt{2}} & \cos\left(\theta - \frac{2\pi}{3}\right) & -\sin\left(\theta - \frac{2\pi}{3}\right) \\ \frac{1}{\sqrt{2}} & \cos\left(\theta + \frac{2\pi}{3}\right) & -\sin\left(\theta + \frac{2\pi}{3}\right) \end{bmatrix} \quad (30)$$

A.2 Modeling of Three-Phase Induction Motor

The winding arrangement of a symmetrical induction machine is shown in Fig. 21. The stator windings are identical and sinusoidally distributed, displaced 120° apart, with *Ns* equivalent turns and resistance *Rs* per winding, per phase. Similarly the rotor windings are also considered as three identical sinusoidally distributed windings, displaced 120° apart, with *Nr* equivalent turns and resistance of *Rr* per, winding per phase (Jimoh et al. 2012).

In developing the equations which describe the behaviour of the induction machine the following assumptions are made (Jimoh et al. 2012):

1. The airgap is uniform.
2. Eddy currents, friction and windage losses and saturation are neglected.
3. The windings are distributed sinusoidally around the air gap.
4. The windings are identical

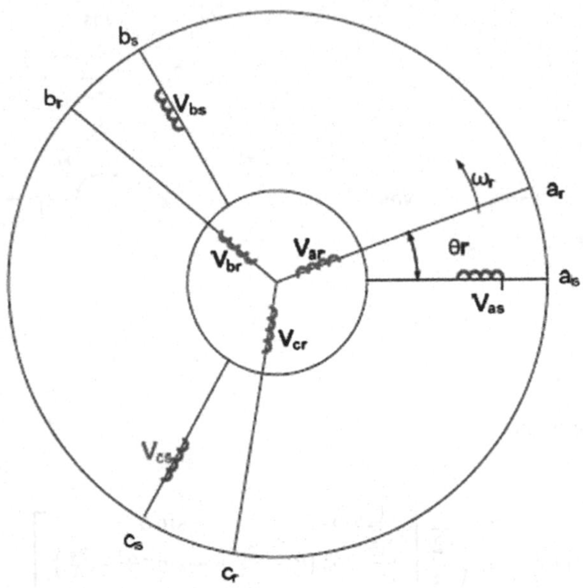

Fig. 21 Three-phase winding Arrangement

Appendix: B Calculates of the Relative degree

B.1 Derivation of Lie

Is $h : \Re^n \rightarrow \Re$ a scalar function and $f = \left[f_1 \; f_2 \ldots f_n \right]^T : \Re^n \rightarrow \Re^n$ a vector field with n is the order of the system (Pietrzak-David et al. 2000; Meziane et al. 2008).

We use the notation $L_f \cdot h\,(x) : \Re^n \rightarrow \Re$ to denote the given scalar function:

$$L_f \cdot h\,(x) = \frac{\partial h}{\partial x} \cdot f\,(x) = \left[\frac{\partial h}{\partial x_1}, \ldots, \frac{\partial h}{\partial x_n} \right] \cdot \begin{bmatrix} f_1\,(x) \\ \vdots \\ f_n\,(x) \end{bmatrix} = \sum_{i=1}^{n} \frac{\partial h}{\partial x_i} \cdot f_i\,(x) \quad (31)$$

where: $x = [x_1, x_2, \ldots, x_n]^T$

And $L_f \cdot h\,(x)$ is called the Lie derivative in the direction of the vector field f.

Similarly, it may be noted, for $k = 0, 1, 2, 3 \ldots$:

$$\begin{cases} L_f^k \cdot h\,(x) = \frac{\partial^k h}{\partial x^k} \cdot f\,(x) = \frac{\partial}{\partial x} \left(L_f^{k-1} \cdot h(x) \right) \cdot f(x) \\ L_f^0 \cdot h\,(x) = h(x) \end{cases} \quad (32)$$

B.2 Relative Degree

Consider the following nonlinear system:

$$\begin{cases} \overset{\bullet}{x} = f(x) + g(x)u \\ y = h(x) \end{cases} \quad (33)$$

Now consider the output $y = h\,(x)^{\in}\,\Re$. It is said that $y = h\,(x)$ has a degree relative r with respect to the input scalar u where:

$$\begin{cases} L_g L_f^k \cdot h\,(x) = 0\ ,\ 0 \leq k \leq r-1 \\ L_g L_f^{r-1} \cdot h\,(x) \neq 0 \end{cases} \quad (34)$$

Lie derivative of the scalar function $h\,(x)$ taken along first and then along the second vector g is defined by:

$$L_g L_f h\,(x) = \frac{\partial \left(L_f h\right)}{\partial x} \cdot g\,(x) \quad (35)$$

Notes

- The concept of relative degree r is very important during the linearization because it lets us know if our system is linearizable completely or partially.
- It should be noted that the relative degree r is the number of times to derive the output y for the u control appears, it is verified as follows:

$$\dot{y} = \frac{\partial h}{\partial x}\dot{x} = \frac{\partial h}{\partial x} \cdot f\,(x) + \frac{\partial h}{\partial x} \cdot g(x) \cdot u \quad (36)$$

$$\dot{y} = L_f \cdot h\,(x) + L_g \cdot h\,(x) \cdot u \quad (37)$$

And since $L_g \cdot h\,(x) = 0$, then:

$$\dot{y} = L_f \cdot h\,(x) \quad (38)$$

Likewise we find:

$$\begin{aligned} \ddot{y} &= L_f^2 \cdot h\,(x) \\ &\vdots \\ y^{(r-1)} &= L_f^{(r-1)} \cdot h\,(x) \\ y^r &= \tfrac{\partial}{\partial x}\left(L_f^{(r-1)} \cdot h\,(x)\right)\overset{\bullet}{x} \\ &= L_f^r \cdot h\,(x) + L_g \cdot L_f^{r-1} \cdot h\,(x) \cdot u \end{aligned} \quad (39)$$

Calculate the Relative Degree of the Induction Motor System The relative degree r_i of system corresponding to outputs y_i of the induction machine to the relationship (33) is given as follows:

(a) *Relative degree of module rotor flux*

$$h_1(x) = (x_3^2 + x_4^2) \tag{40}$$

From Eq. (36), derived the Eq. (40) found:

$$\dot{h}_1(x) = a_{31}.(x_1.x_3 + x_2x_4) + a_{33}.(x_3^2 + x_4^2) \tag{41}$$

The first derivative of h_1 *(x)* does not involve the input *v*. then it must derive a second time, as find in Eq. (39).

$$\ddot{h}_1 = (x_1x_3 + x_2x_4).(a_{11}a_{33} + 3a_{33}a_{31}) - (x_2x_3 - x_1x_4).(a_{31}a_{34}) + (x_3^2 + x_4^2). \\ (2.a_{33}^2 + a_{13}a_{31}) + a_{31}b_{11}x_3U_1 + a_{31}b_{11}x_4U_2 \tag{42}$$

The U_1, U_2 commands appear after the second derivative; therefore, the relative degree with respect to $h_1(x)$ is $r_1 = 2$.

(b) *Relative degree of the rotation speed.*

$$h_2(x) = \Omega_r \tag{43}$$

From Eq. (36), derived the Eq. (43) found:

$$\dot{h}_2(x) = \mu.(x_2.x_3 - x_1.x_4) - \frac{c_r}{J} \tag{44}$$

Again the first derivative of h_2 *(x)* does not involve the input *v*. then it must derive a second time, as find in Eq. (39).

$$\ddot{h}_2 = \dot{\Omega} = J \cdot \mu \cdot a_{34} \cdot (x_1 \cdot x_3 + x_2 \cdot x_4) - J \cdot \mu \cdot a_{14} \cdot (x_3^2 + x_4^2) + +J \cdot \mu \cdot b_{11} \\ (x_3 \cdot U_2 - x_4 \cdot U_1) + J \cdot \mu \cdot (a_{11} + a_{33}) \cdot (x_2 \cdot x_3 - x_1 \cdot x_4) \tag{45}$$

The U_1, U_2 commands appear at the end of the second derivative, so the relative degree with respect to $h_2\ (x)$ is $r_2 = 2$.

References

Araujo, R.E., Freitas, D.: Non-linear control of an induction motor: sliding mode theory leads to robust and simple solution. Int. J. Adapt. Control Signal Process. **14**(2), 331–353 (2000). ISSN: 0890–6327

Aurora, C., Ferrara, A.: A sliding mode observer for sensorless induction motor speed regulation. Int. J. Syst. Sci. **38**(11), 913–929 (2007). doi:10.1080/00207720701620043
Astrom, K.J., Wittenmark, B.: Adaptive Control. Addison-Wesley, New York (1995)
Attaianese, C.,Timasso, G. (2001) Control of induction motor. In: Proceedings - Electric Power Applications, vol. 148 (3), pp. 272–278, May 2001
Blaschke, F.: The principle of field orientation control as applied to the new transvector closed loop control system for rotating machines. Siemens Rev. **39**(5), 217–220 (1977)
Bottura, C.P., Neto, M.F.S., Filho, S.A.A.: Robust speed control of an induction motor: An control theory approach with field orientation and -analysis. IEEE Trans. Power Elect. **15**(5), 908–9152 (2000)
Boucheta, A., Bousserhane, I.K., Hazzab, A., Sicard, P., Fellah, M.K.: Speed control of linear induction motor using sliding modecontroller considering the end effects. J. Electr. Eng. Technol. **7**(1), 34–45 (2012)
Bouchhida, O., Boucherit, M.S., & Cherifi, A. (2012). Minimizing Torque-Ripple in Inverter-Fed Induction Motor Using Harmonic Elimination PWM Technique. First published Induction Motors - Modeling and Control, Edited by Rui Esteves Araújo, Croatia, pp. 465–486,doi:10.5772/37883.
Boukettaya, G., Andoulsi, R., Ouali, V. (2008) Commande vectorielle avec observateur de vitesse d'une pompe asynchrone couplée à un générateur photovoltaïque. In: Revue des Energies Renouvelables CICME'08 Sousse, pp. 75–85
Bounar, N., Boulkroune, A., Boudjema, F., (2012) Fuzzy slinding mode control of double-fed induction machine. In: International conference on Information Processing and Electrical Engineering, ICIPEE'12, pp. 31–36, ISBN : 978-9931-9068-0-9
Caruana, C., Asher, G.M., Sumner, M.: Performance of high frequency signal injection techniques for zero-low-frequency vector control induction machines under sensorless conditions. IEEE Trans. Ind. Electr. **53**, 225–238 (2006)
Castillo-Toledo, B., Di Gennaro, S., Loukianov, A.G., Rivera, J.: Hybrid control of induction motors via sampled closed representations. IEEE Trans. Ind. Electr. **55**(10), 3756–3771 (2008)
Chaigne, C., Etien, E., Cauët, S., Rambaul, L.: Commande Vectorielle Sans Capteur des Machines. Asynchrones edition. Hermes Science Publisching, London (2005)
Comanescu, M., Xu, L., Batzel, T.D.: Decoupled current control of sensorless induction-motor drives by integral sliding mode. IEEE Trans. Ind. Electr. **55**(11), 3836–3845 (2008)
Corradini, M. L., Ippoliti, G., Longhi, S., & Orlando, G. (2012). A quasisliding mode approach for robust control and speed estimation of PM synchronous motors. Feb. 2012, IEEE Trans. Ind. Electron., 59 (2), pp. 1096–1104.
Duarte-Mermoud, M. A., Travieso-Torres, J. C. (2012) Advanced Control Techniques for Induction Motors. First published Induction Motors - Modeling and Control, Edited by Rui Esteves Araújo, Croatia, pp. 295–325. ISBN: 978-953-51-0843-6
Dwards, C.E., Spurgeon, S.K.: Sliding Mode Control: Theory and Application. Taylor & Francis, London (1998)
El-Sousy, F.F.M.: Adaptive Dynamic Sliding-Mode Control System Using Recurrent RBFN for High-Performance Induction Motor Servo Drive. November 2013. IEEE Transactions on Industrial Informatics **9**(4), (2013). 1551-3203
El-Sousy, F.F.M., Salem, M.M.: Simple neuro-controllers for field oriented induction motor servo drive system. J. Power Electr. **4**(1), 28–38 (2004)
Ghanes, M., Zheng, G.: On sensorless induction motor drives: Sliding-mode observer and output feedback controller, industrial electronics. IEEE Trans. Ind. Electron **56**(9), 3404–3413 (2009)
Hautier, J. P., Caron, J. P. (1995). Modélisation et commande de la machine asynchrone. Editions Technip.
Holmes, D. G., McGrath, B. P., & Parker, S. G. (2012). Current regulation strategies for vector-controlled induction motor drives. Oct. 2012, IEEE Trans. Ind. Electron, 59 (10), pp. 3680–3689.
Isidori, A.: Nonlinear Control Systems, 3rd edn. Springer, New York (1995). ISBN-10: 3540199160, ISBN-13: 978–3540199168

Jimoh, Adisa A., Pierre-Jac Venter, P.J., and Edward K. Appiah. (2012) Modelling and Analysis of Squirrel Cage Induction Motor with Leading Reactive Power Injection. First published Induction Motors - Modelling and Control, Edited by Rui Esteves Araújo, Printed in Croatia, pp. 99–126. ISBN: 978-953-51-0843-6

Kojabadi, H.M.: Simulation and experimental studies of model reference adaptive system for sensorless induction motor drive. Simulat. Model. Practice Theory **13**, 451–464 (2005)

Leonhard, W.: Control of machines with the help of microelectronics. In: Third IFAC Symposium on Control in Power Electronics and Electrical Drives, Lausanne, pp. 35–58 (1994)

Li, J., Xu, L., Zhang, Z.: An adaptive sliding-mode observer for induction motor sensorless speed control. IEEE Trans. Ind. Appl. **41**(4), 1039–1046 (2005)

Liaw, C.M., Lin, Y.M., Chao, K.H.: A VSS speed controller with model reference response for induction motor drive. IEEE Trans. Ind. Electr. **48**(6), 1136–1147 (2001)

Lin, C.M., Hsu, C.F.: Neural-network-based adaptive control for induction servomotor drive system. IEEE Trans. Ind. Electr. **4**(91), 115–123 (2002)

Maher, R., Emar, A., Awad, M.: Indirect Field Oriented Control of an Induction Motor Sensing DC-link Current with PI Controller. Int. J. Control Sci. Eng. **2**(3), 19–25 (2012). doi:10.5923/j.control.20120203.01

Meziane, S., Toufouti, R., Benalla, H.: Generalized nonlinear predictive control of induction motor. Int. Rev. Autom. Control **1**(2), 65–71 (2008)

Mira, F.J., Duarte-Mermoud, M.A. (2009) Speed control of an asynchronous motor using a field oriented control scheme together with a fractional order PI controller. Ann. Chilean Inst. Eng. Spania. **121**(1), pp. 1–13, ISSN: 0716–3290

Ouhrouche, M. A., Volat, C., (2000). Simulation of a direct field-oriented controller for an induction motor using matlab/simulink software package. In: Proceeding of the IASTED International Conference Modelling and Simulation, Pennsylvania, USA (May 15–17, 2000)

Park, R.H.: Two-reaction theory of synchronous machines generalized method of analysis-part I. IEEE Trans. Ind. Electr. **48**(6), 716–727 (1929)

Pietrzak-David, M., De Fornel, B., Purwoadi, M.A., (2000). Nonlinear control for sensorless induction motor drives. In: 9th International Conference on Power Electronic and Motion- EPE PEMC, Kosic

Ramesh, T., Panda, A.K., Kumar, S.S. (2013) Sliding-mode and fuzzy logic control based MRAS speed estimators for sensorless direct torque and flux control of an induction motor drive. In: 2013 Annual IEEE India Conference (INDICON). 978.1-4799-2275-8/13/31.00

Rao, S., Buss, M., Utkin, V.: Simultaneous state and parameter estimation in induction motors using first- and second-order sliding modes. IEEE Trans. Ind. Electron. **56**(9), 3369–3376 (2009)

Rashed, M., Goh, K. B., Dunnigan, M. W., Mac Connell, P. F. A., Stronach, A. F.& Williams, B.W. (2005). Sensorless second-order sliding-mode speed control of a voltage-fed induction-motor drive using nonlinear state feedback. Sep. 2005, in IEE Proc.–Electr. Power Appl., 152 (5), pp. 1127–1136.

Ravi Teja, A.V., Chakraborty, C., Maiti, S., Hori, Y.: A new model reference adaptive controller for four quadrant vector controlled induction motor drives. IEEE Trans. Ind. Electron. **59**(10), 3757–3767 (2012)

Rojas, S.I., Moreno, J., Espinosa-Perez, G.: Global observability analysis of sensorless induction motors. Automatica **40**, 1079–1085 (2004)

Rong-Jong, W., Jeng-Dao, L., Kuo-Min, L.: Robust decoupled control of direct field oriented induction motor drive industrial electronics. IEEE Trans. Ind. Electron. **52**(3), 837–854 (2005)

Saiad, A. (2012) Sliding mode controller design of an induction motor. In: International conference on information processing and electrical engineering. ICIPEE'12, pp. 396–399. ISBN : 978-9931-9068-0-9

Slotine, J.J.E., Li, W.: Applied Nonlinear Control. Prentice-Hall, Englewood Cliffs (1991)

Stanley, H.C.: An Analysis of the Induction Machine. American Institute of Electrical Engineers, Transactions on Industrial Electronics **57**(12), 751–757 (1938)

Su, K.-H., Kung, C.-C. (2005) Supervisory enhanced genetic algorithm controller design and its application to decoupling induction motor drive. In: Proceedings of IEE–Electrical Power Applications, vol. 152 (4), pp. 1015–1026, Jul. 2005

Talhaoui, H., Menacer, A., Kechida, R. (2013) Rotor resistance estimation using ekf for the rotor fault diagnosis in sliding mode control induction motor. In: Proceedings of the 3rd International Conference on Systems and Control, Algiers, Algeria, October 29–31, 2013. 978-1-4799-0275-0/13/31.00

Traoré, D., Plestan, F., Glumineau, A., de Leon, J.: Sensorless induction motor: High-order sliding-mode controller and adaptive interconnected observer. IEEE Trans. Ind. Electron. **55**(11), 3818–3827 (2008)

Utkin, V.I.: Sliding mode control design principles and applications to electric drive. IEEE Trans. Ind. Electron. **40**(1), 23–36 (1993)

Veselic, B., Perunicic-Drazenovic, B., Milosavljevic, C.: Improved discrete-time sliding-mode position control using Euler velocity estimation. IEEE Trans. Ind. Electron. **57**(11), 3840–3847 (2010)

Wai, R.J.: Adaptive sliding-mode control for induction servomotor drives. IEEE Proc Electr. Power Appl. **147**, 553–562 (2000)

Wai, R.J.: Hybrid control for speed sensorless induction motor drive. IEEE Trans. Fuzzy Syst. **9**(1), 116–138 (2001)

Wai, R.J.: Supervisory genetic evolution control for indirect field-oriented induction motor drive. Proc. IEE–Electr. Power Appl. **150**(2), 215–226 (2003)

Wai, R.J., Lin, C.M., Hsu, C.F.: Hybrid control for induction servomotor drive. Proc. IEE–Control Theor. Appl. **149**(6), 555–562 (2002)

Wai, R.J., Duan, R.Y., Chang, H.H.: Wavelet neural network control for induction motor drive using sliding-mode design technique. IEEE Trans. Ind. Electron. **50**(4), 733–748 (2003)

Xia, Y., Yu, X., Oghanna, W.: Adaptive robust fast control for induction motors. IEEE Trans. Ind. Electron. **47**(4), 854–862 (2000)

Zhu, Z., Xia, Y., Fu, M.: Adaptive sliding mode control for attitude stabilization with actuator saturation. IEEE Trans. Ind. Electron. **58**(10), 4898–4907 (2011)

Su, K.-H., Kung, C.-C. (2005) Supervisory enhanced genetic algorithm controller design and its application to decoupling induction motor drive. In: Proceedings of IEE-Electrical Power Applications, vol. 152(4), pp. 1015–1026, Jul. 2005

Ibtissem, B., Mendaci, A., Kechida, R. (2013) Rotor resistance estimation using EKF for the rotor fault diagnosis in sliding mode control induction motor. In: Proceedings of the 3rd International Conference on Systems and Control, Algiers, Algeria, October 29–31, 2013, 978-1-4799-0275-0/13/$31.00

Traoré, D., Leon, J. de, Glumineau, A., Loron, L.: Sensorless induction motor: High-order sliding mode controller and adaptive interconnected observer. IEEE Trans. Ind. Electron. 55(11), 3818–3827 (2008)

Utkin, V.I.: Sliding mode control design principles and applications to electric drives. IEEE Trans. Ind. Electron. 40(1), 23–36 (1993)

Veselic, B., Perunicic-Drazenovic, B., Milosavljevic, C.: Improved discrete-time sliding-mode position control using Euler velocity estimation. IEEE Trans. Ind. Electron. 57(11), 3840–3847 (2010)

Wai, R.J.: Adaptive sliding-mode control for induction servomotor drives. IEE Proc. Electr. Power Appl. 147(6), 553–562 (2000)

Wai, R.J.: [illegible] control [illegible] induction motor drive. IEEE Trans. Fuzzy Syst. 9(1), [illegible] (2001)

Wai, R.J.: Supervisory genetic evolution control for indirect field-oriented induction motor drive. Proc. IEE Electr. Power Appl. 150(2), 215–226 (2003)

Wai, R.J., Lin, C.M., Hsu, C.F.: Hybrid control for induction servomotor drive. Proc. IEE Control Theory Appl. 149(6), 555–562 (2002)

Wai, R.J., Duan, R.Y., Chang, H.H.: Wavelet neural network control for induction motor drive using sliding-mode design technique. IEEE Trans. Ind. Electron. 50(4), 733–748 (2003)

Xue, Y., Yu, X., Gaudiano, N.: Adaptive robust fast control for induction motors. IEEE Trans. Ind. Electron. 47(4), 854–862 (2000)

Zhu, Z., Xia, Y., Fu, M.: Adaptive sliding mode control for attitude stabilization with actuator saturation. IEEE Trans. Ind. Electron. 58(10), 4898–4907 (2011)

Anti-synchronization of Identical Chaotic Systems Using Sliding Mode Control and an Application to Vaidyanathan–Madhavan Chaotic Systems

Sundarapandian Vaidyanathan and Ahmad Taher Azar

Abstract Anti-synchronization is an important type of synchronization of a pair of chaotic systems called the master and slave systems. The anti-synchronization characterizes the asymptotic vanishing of the sum of the states of the master and slave systems. In other words, anti-synchronization of master and slave system is said to occur when the states of the synchronized systems have the same absolute values but opposite signs. Anti-synchronization has applications in science and engineering. This work derives a general result for the anti-synchronization of identical chaotic systems using sliding mode control. The main result has been proved using Lyapunov stability theory. Sliding mode control (SMC) is well-known as a robust approach and useful for controller design in systems with parameter uncertainties. Next, as an application of the main result, anti-synchronizing controller has been designed for Vaidyanathan–Madhavan chaotic systems (2013). The Lyapunov exponents of the Vaidyanathan–Madhavan chaotic system are found as $L_1 = 3.2226$, $L_2 = 0$ and $L_3 = -30.3406$ and the Lyapunov dimension of the novel chaotic system is found as $D_L = 2.1095$. The maximal Lyapunov exponent of the Vaidyanathan–Madhavan chaotic system is $L_1 = 3.2226$. As an application of the general result derived in this work, a sliding mode controller is derived for the anti-synchronization of the identical Vaidyanathan–Madhavan chaotic systems. MATLAB simulations have been provided to illustrate the qualitative properties of the novel 3-D chaotic system and the anti-synchronizer results for the identical novel 3-D chaotic systems.

S. Vaidyanathan (✉)
Research and Development Centre, Vel Tech University,
Avadi, Chennai 600062, Tamil Nadu, India
e-mail: sundarvtu@gmail.com

A.T. Azar
Faculty of Computers and Information, Benha University, Banha, Egypt
e-mail: ahmad_t_azar@ieee.org

A.T. Azar and Q. Zhu (eds.), *Advances and Applications in Sliding Mode Control systems*,
Studies in Computational Intelligence 576, DOI 10.1007/978-3-319-11173-5_19

1 Introduction

Chaos is an interesting phenomenon of nonlinear dynamical systems. Chaotic systems are nonlinear dynamical systems which are sensitive to initial conditions, topologically mixing and with dense periodic orbits. Sensitivity to initial conditions of chaotic systems is popularly known as the *butterfly effect*. Small changes in an initial state will make a very large difference in the behavior of the system at future states. Chaotic behaviour was suspected well over hundred years ago in the study of three bodies problem, but it was established only a few decades ago in the study of 3-D weather models (Lorenz 1963).

The Lyapunov exponent is a measure of the divergence of phase points that are initially very close and can be used to quantify chaotic systems. It is common to refer to the largest Lyapunov exponent as the maximal Lyapunov exponent (MLE). A positive maximal Lyapunov exponent and phase space compactness are usually taken as defining conditions for a chaotic system.

Since the discovery of Lorenz system in 1963, there is a great deal of interest in the chaos literature in finding new chaotic systems. Some well-known paradigms of 3-D chaotic systems in the literature are (Arneodo et al. 1981; Cai and Tan 2007; Chen and Ueta 1999; Chen and Lee 2004; Li 2008; Liu et al. 2004; Lü and Chen 2002; Rössler 1976; Sprott 1994; Sundarapandian and Pehlivan 2012; Tigan and Opris 2008; Vaidyanathan 2013a, b, 2014; Zhou et al. 2008; Zhu et al. 2010).

Chaotic systems have several important applications in science and engineering such as oscillators (Kengne et al. 2012; Sharma et al. 2012), lasers (Li et al. 2014; Yuan et al. 2014), chemical reactions (Gaspard 1999; Petrov et al. 1993), cryptosystems (Rhouma and Belghith 2011; Usama et al. 2010), secure communications (Feki 2003; Murali and Lakshmanan 1998; Zaher and Abu-Rezq 2011), biology (Das et al. 2014; Kyriazis 1991), ecology (Gibson and Wilson 2013; Suérez 1999), robotics (Mondal and Mahanta 2014; Nehmzow and Walker 2005; Volos et al. 2013), cardiology (Qu 2011; Witte and Witte 1991), neural networks (Huang et al. 2012; Kaslik and Sivasundaram 2012; Lian and Chen 2011), finance (Guégan 2009; Sprott 2004), etc.

Synchronization of chaotic systems is a phenomenon that occurs when two or more chaotic systems are coupled or when a chaotic system drives another chaotic system. Because of the butterfly effect which causes exponential divergence of the trajectories of two identical chaotic systems started with nearly the same initial conditions, the synchronization of chaotic systems is a challenging research problem in the chaos literature.

Major works on synchronization of chaotic systems deal with the complete synchronization (CS) which has the goal of using the output of the master system to control the slave system so that the output of the slave system tracks the output of the master system asymptotically. Thus, if $x(t)$ and $y(t)$ denote the states of the master and slave systems, then the design goal of complete synchronization (CS) problem is to satisfy the condition

$$\lim_{t\to\infty} \|x(t) - y(t)\| = 0, \quad \forall x(0), y(0) \in \mathbb{R}^n \tag{1}$$

Anti-synchronization (AS) is an important type of synchronization of a pair of chaotic systems called the master and slave systems. The anti-synchronization characterizes the asymptotic vanishing of the sum of the states of the master and slave systems. In other words, anti-synchronization of master and slave system is said to occur when the states of the synchronized systems have the same absolute values but opposite signs. Thus, if $x(t)$ and $y(t)$ denote the states of the master and slave systems, then the design goal of anti-synchronization problem (AS) is to satisfy the condition

$$\lim_{t\to\infty} \|x(t) + y(t)\| = 0, \quad \forall x(0), y(0) \in \mathbb{R}^n \tag{2}$$

Pecora and Carroll pioneered the research on synchronization of chaotic systems with their seminal papers in 1990s (Carroll and Pecora 1991; Pecora and Carroll 1990). The active control method (Liu et al. 2007; Rafikov and Balthazar 2007; Sundarapandian 2010; Ucar et al. 2007; Vaidyanathan 2012c; Wang and Liu 2006) is commonly used when the system parameters are available for measurement and the adaptive control method (Wu et al. 2008; Huang 2008; Lin 2008; Sarasu and Sundarapandian 2012a, b, c) is commonly used when some or all the system parameters are not available for measurement and estimates for unknown parameters of the systems.

Other popular methods for chaos synchronization are the sampled-data feedback method (Gan and Liang 2012; Li et al. 2011; Xiao et al. 2014; Zhang and Zhou 2012), time-delay feedback method (Chen et al. 2014; Jiang et al. 2004; Shahverdiev et al. 2009; Shahverdiev and Shore 2009), backstepping method (Njah et al. 2010; Tu et al. 2014; Vaidyanathan 2012a; Zhang et al. 2004), etc.

Complete synchronization (Rasappan and Vaidyanathan 2012; Suresh and Sundarapandian 2013; Vaidyanathan and Rajagopal 2011) is characterized by the equality of state variables evolving in time, while anti-synchronization (Vaidyanathan 2011, 2012b; Vaidyanathan and Sampath 2012) is characterized by the disappearance of the sum of relevant state variables evolving in time.

This research work is organized as follows. Section 2 gives a brief introduction about sliding mode control. Section 3 discusses the problem statement for the anti-synchronization of two identical chaotic systems and our design methodology. Section 4 contains the main result of this work, namely, sliding controller design for the global anti-synchronization of identical chaotic systems. Our sliding mode control law is designed by considering constant-plus-proportional sliding law. The main result for the global anti-synchronization of chaotic systems is established using Lyapunov stability theory.

Section 5 introduces the Vaidyanathan–Madhavan chaotic system (Vaidyanathan and Madhavan 2013), which is a seven-term novel 3-D chaotic system with three quadratic nonlinearities. Section details the qualitative properties of the Vaidyanathan–Madhavan 3-D chaotic system. The Lyapunov exponents of the

Vaidyanathan–Madhavan chaotic system are found as $L_1 = 3.2226$, $L_2 = 0$ and $L_3 = -30.3406$ and the Lyapunov dimension of the novel chaotic system is found as $D_L = 2.1095$. The maximal Lyapunov exponent of the Vaidyanathan–Madhavan chaotic system is $L_1 = 3.2226$.

In Sect. 7, we describe the sliding mode controller design for the global anti-synchronization of identical Vaidyanathan–Madhavan chaotic systems. MATLAB simulations are shown to validate and illustrate the sliding mode controller design for the anti-synchronization of the Vaidyanathan–Madhavan chaotic systems. Section 8 contains a summary of the main results derived in this research work.

2 Sliding Mode Control and Chaos Anti-synchronization

In control theory, the sliding mode control approach is recognized as an efficient tool for designing robust controllers for linear or nonlinear control systems operating under uncertainty conditions (Perruquetti and Barbot 2002; Utkin 1992).

The started steps of sliding mode control theory originated in the early 1950 s and this was initiated by S.V. Emel'yanov as *Variable Structure Control* (Itkis 1976; Utkin 1978; Zinober 1993). Variable structure control (VSC) is a form of discontinuous nonlinear control and this method alters the dynamics of a nonlinear system by application of a high-frequency switching control.

Sliding mode control method has a major advantage of low sensitivity to parameter variations in the plant and disturbances affecting the plant, which eliminates the necessity of exact modeling of the plant.

In the sliding mode control theory, the control dynamics has two sequential modes, viz. (i) the *reaching mode*, and (ii) the *sliding mode*. Basically, a sliding mode controller (SMC) design consists of two parts: hyperplane (or sliding surface) design and controller design.

A hyperplane is first designed via the pole-placement approach in the modern control theory and a controller is then designed based on the sliding condition. The stability of the overall control system is ensured by the sliding condition and by a stable hyperplane. Sliding mode control theory has been used to deal with many research problems of control literature (Bidarvatan et al. 2014; Feng et al. 2014; Hamayun et al. 2013; Lu et al. 2014; Ouyang et al. 2014; Zhang et al. 2014).

3 Problem Statement

This section gives a problem statement of global anti-synchronization of a pair of identical chaotic systems called the *master* and *slave* systems.

The *master* system is taken as the chaotic system

$$\dot{x} = Ax + f(x), \tag{3}$$

where $x \in \mathbb{R}^n$ is the state of the system, A is the $n \times n$ matrix of system parameters and f is a vector field that contains the nonlinear parts of the system and satisfies $f(0) = 0$.

The *slave* system is taken as the controlled chaotic system

$$\dot{y} = Ay + f(y) + u, \tag{4}$$

where $y \in \mathbb{R}^n$ is the state of the system, and u is the controller to be determined.

The *anti-synchronization error* between the master and slave systems is defined by

$$e = y + x \tag{5}$$

Differentiating (5) and simplifying, the error dynamics is obtained as

$$\dot{e} = Ae + \eta(x, y) + u \tag{6}$$

where

$$\eta(x, y) = f(x) + f(y) \tag{7}$$

The design problem is to determine a feedback control u so that the anti-synchronization error dynamics (6) is globally asymptotically stable at the origin for all initial conditions $e(0) \in \mathbb{R}^n$.

For the SMC design for the global anti-synchronization of the systems (3) and (4), the control u is taken as

$$u(t) = -\eta(x, y) + Bv(t), \tag{8}$$

where B is an $(n \times 1)$ column vector chosen such that (A, B) is controllable.

Upon substituting (8) into (6), the closed-loop error system is obtained as

$$\dot{e}(t) = Ae(t) + Bv(t), \tag{9}$$

which is a linear time-invariant control system with a single input v.

Hence, by the use of the nonlinear control law (8), original problem of global anti-synchronization of identical chaotic systems (3) and (4) has been converted into an equivalent problem of globally stabilizing the error dynamics (9).

4 Sliding Controller Design for Global Anti-synchronization

This section derives the main result, *viz.* sliding controller design for the global anti-synchronization of the identical chaotic systems (3) and (4). After applying the control (8) with (A, B) a controllable pair, it is supposed that the nonlinear error dynamics (6) has been simplified as the linear error dynamics (9).

In the sliding controller design, the sliding variable is first defined as

$$s(e) = Ce = c_1 e_1 + c_2 e_2 + \cdots + c_n e_n, \tag{10}$$

where C is an $(1 \times n)$ row vector to be determined.

The sliding manifold S is defined as the hyperplane

$$S = \{e \in \mathbb{R}^n \ : \ s(e) = Ce = 0\} \tag{11}$$

If a sliding motion occurs on S, then the sliding mode conditions must be satisfied, which are given by

$$s \equiv 0 \ \text{ and } \ \dot{s} = CAe + CBv = 0 \tag{12}$$

It is assumed that the row vector C is chosen so that $CB \neq 0$.

The sliding motion is affected by the so-called equivalent control given by

$$v_{\text{eq}}(t) = -(CB)^{-1}CAe(t) \tag{13}$$

As a consequence, the equivalent dynamics in the sliding phase is defined by

$$\dot{e} = \left[I - B(CB)^{-1}C\right]Ae = Ee, \tag{14}$$

where

$$E = \left[I - B(CB)^{-1}C\right]A \tag{15}$$

It can be easily verified that E is independent of the control and has at most $(n-1)$ nonzero eigenvalues, depending on the chosen switching surface, while the associated eigenvectors belong to $\ker(C)$.

Since (A, B) is controllable, the matrices B and C can be chosen so that E has any desired $(n-1)$ stable eigenvalues.

Thus, the dynamics in the sliding mode is globally asymptotically stable.

Finally, for the sliding mode controller (SMC) design, the constant plus proportional rate reaching law is used, which is given by

$$\dot{s} = -\beta\,\text{sgn}(s) - \alpha s \tag{16}$$

where $\text{sgn}(\cdot)$ denotes the sign function and the gains $\alpha > 0$, $\beta > 0$ are found so that the sliding condition is satisfied and the sliding motion will occur.

From the Eqs. (12) and (16), sliding control v is found as

$$CAe + CBv = -\beta \operatorname{sgn}(s) - \alpha s \tag{17}$$

Since $s = Ce$, the Eq. (17) can be simplified to get

$$v = -(CB)^{-1}\left[C(\alpha I + A)e + \beta \operatorname{sgn}(s)\right] \tag{18}$$

Next, the main result of this section is established as follows.

Theorem 1 *A sliding mode control law that achieves global anti-synchronization between the identical chaotic systems* (3) *and* (4) *for all initial conditions* $x(0)$, $y(0)$ *in* $\mathbb{R}^n$ *is given by the equation*

$$u(t) = -\eta(x(t), y(t)) + Bv(t), \tag{19}$$

where v *is defined by* (18), B *is an* $(n \times 1)$ *vector such that* (A, B) *is controllable,* C *is an* $(1 \times n)$ *vector such that* $CB \neq 0$ *and that the matrix* E *defined by Eq.* (15) *has* $(n - 1)$ *stable eigenvalues.*

Proof The proof is carried out using Lyapunov stability theory (Khalil 2001).

Substituting the sliding control law (19) into the error dynamics (6) leads to

$$\dot{e} = Ae + Bv \tag{20}$$

Substituting for v from (18) into (20), the error dynamics is obtained as

$$\dot{e} = Ae - B(CB)^{-1}\left[C(\alpha I + A)e + \beta \operatorname{sgn}(s)\right] \tag{21}$$

The global asymptotic stability of the error system (21) is proved by taking the candidate Lyapunov function

$$V(e) = \frac{1}{2}\, s^2(e), \tag{22}$$

which is a non-negative definite function on $\mathbb{R}^n$.

It is noted that

$$V(e) = 0 \iff s(e) = 0 \tag{23}$$

The sliding mode motion is characterized by the equations

$$s(e) = 0 \text{ and } \dot{s}(e) = 0 \tag{24}$$

By the choice of E, the dynamics in the sliding mode given by (14) is globally asymptotically stable.

When $s(e) \neq 0$, $V(e) > 0$.

Also, when $s(e) \neq 0$, differentiating V along the error dynamics (21) or the equivalent dynamics (16), the following dynamics is obtained:

$$\dot{V} = s\dot{s} = -\beta s \text{ sgn}(s) - \alpha s^2 < 0 \tag{25}$$

Hence, by Lyapunov stability theory (Khalil 2001), it is concluded that the error dynamics (21) is globally asymptotically stable for all initial conditions $e(0) \in \mathbb{R}^n$.

This completes the proof. □

5 Vaidyanathan–Madhavan 3-D Chaotic System

This section describes the equations and phase portraits of Vaidyanathan–Madhavan 3-D chaotic system (Vaidyanathan and Madhavan 2013).

The Vaidyanathan–Madhavan chaotic system is a described by the 3-D dynamics

$$\begin{aligned} \dot{x}_1 &= a(x_2 - x_1) + x_2 x_3, \\ \dot{x}_2 &= b x_1 + c x_1 x_3, \\ \dot{x}_3 &= -d x_3 - x_1 x_2 - x_1^2, \end{aligned} \tag{26}$$

where x_1, x_2, x_3 are the states and a, b, c, d are constant, positive, parameters.

The system (26) is a seven-term polynomial chaotic system with three quadratic nonlinearities.

The system (26) depicts a strange chaotic attractor when the constant parameter values are taken as

$$a = 22, \quad b = 400, \quad c = 50, \quad d = 0.5 \tag{27}$$

For simulations, the initial values of the Vaidyanathan–Madhavan chaotic system (26) are taken as

$$x_1(0) = 0.6, \quad x_2(0) = 1.8, \quad x_3(0) = 1.2 \tag{28}$$

The novel 3-D chaotic system (26) exhibits a 2-scroll chaotic attractor. Figure 1 describes the 2-scroll chaotic attractor of the Vaidyanathan–Madhavan chaotic system (26) in 3-D view.

Figure 2 describes the 2-D projection of the strange chaotic attractor of the Vaidyanathan–Madhavan chaotic system (26) in (x_1, x_2)-plane. In the projection on the (x_1, x_2)-plane, a 2-scroll chaotic attractor is clearly seen.

Figure 3 describes the 2-D projection of the strange chaotic attractor of the Vaidyanathan–Madhavan chaotic system (26) in (x_2, x_3)-plane. In the projection on the (x_2, x_3)-plane, a 2-scroll chaotic attractor is clearly seen.

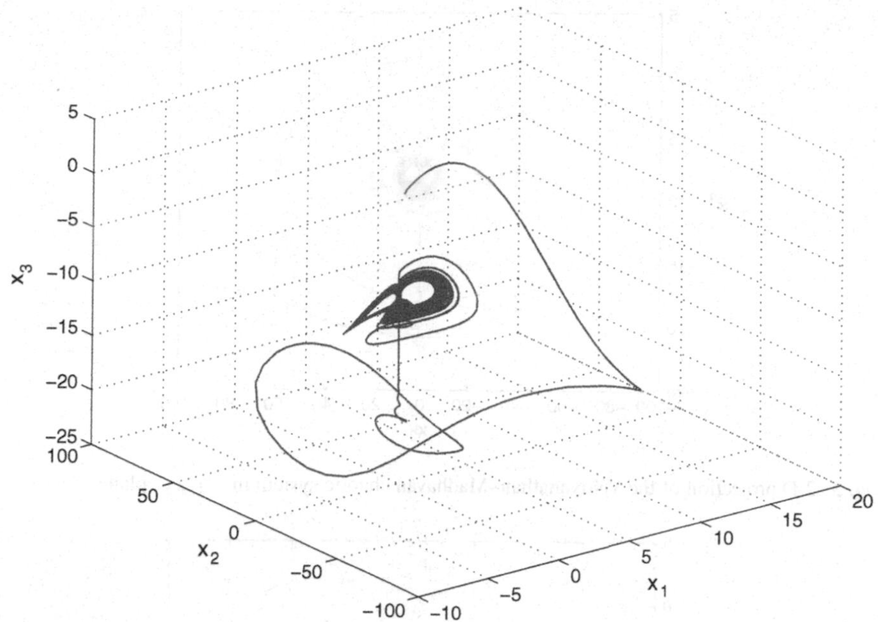

Fig. 1 Strange attractor of the Vaidyanathan–Madhavan chaotic system in $\mathbf{R}^3$

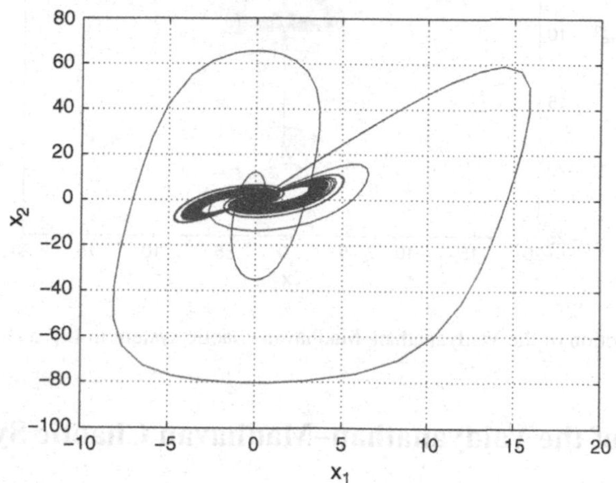

Fig. 2 2-D projection of the Vaidyanathan–Madhavan chaotic system in (x_1, x_2)-plane

Figure 4 describes the 2-D projection of the strange chaotic attractor of the Vaidyanathan–Madhavan chaotic system (26) in (x_1, x_3)-plane. In the projection on the (x_1, x_3)-plane, a 2-scroll chaotic attractor is clearly seen.

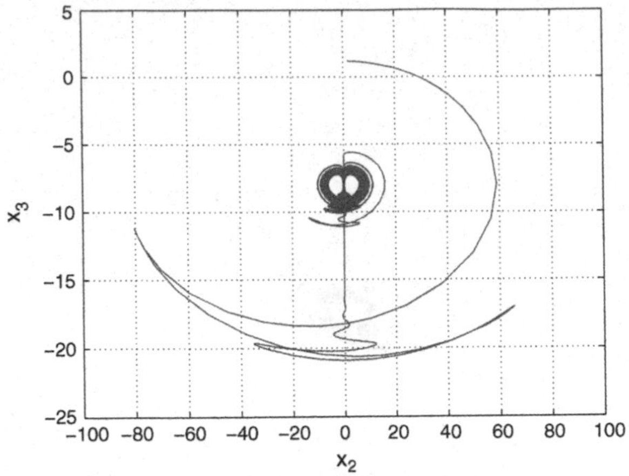

Fig. 3 2-D projection of the Vaidyanathan–Madhavan chaotic system in (x_2, x_3)-plane

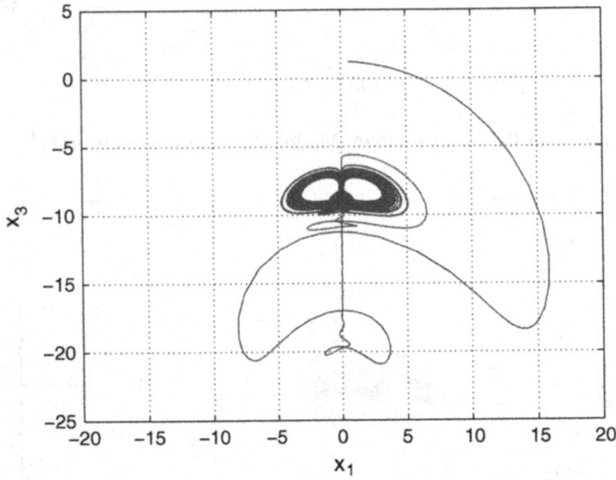

Fig. 4 2-D projection of the Vaidyanathan–Madhavan chaotic system in (x_1, x_3)-plane

6 Analysis of the Vaidyanathan–Madhavan Chaotic System

This section gives the qualitative properties of the Vaidyanathan–Madhavan 3-D chaotic system (2013).

6.1 Symmetry and Invariance

The Vaidyanathan system (26) is invariant under the coordinates transformation

$$(x_1, x_2, x_3) \to (-x_1, -x_2, x_3). \tag{29}$$

The transformation (29) persists for all values of the system parameters. Thus, the Vaidyanathan system (26) has rotation symmetry about the x_3-axis.

Hence, it follows that any non-trivial trajectory of the system (26) must have a twin trajectory.

It is easy to check that the x_3-axis is invariant for the flow of the Vaidyanathan system (26). Hence, all orbits of the system (26) starting from the x_3 axis stay in the x_3 axis for all values of time.

6.2 Equilibria

The equilibrium points of the Vaidyanathan–Madhavan system (26) are obtained by solving the nonlinear equations

$$\begin{aligned} f_1(x) &= a(x_2 - x_1) + x_2 x_3 = 0 \\ f_2(x) &= bx_1 + cx_1 x_3 \quad\;\; = 0 \\ f_3(x) &= -dx_3 - x_1 x_2 - x_1^2 = 0 \end{aligned} \tag{30}$$

We take the parameter values as in the chaotic case, viz.

$$a = 22, \quad b = 400, \quad c = 50, \quad d = 0.5 \tag{31}$$

Solving the nonlinear system of Eqs. (30) with the parameter values (31), we obtain three equilibrium points of the Vaidyanathan–Madhavan system (26) as

$$E_0 = \begin{bmatrix} 0 \\ 0 \\ 0 \end{bmatrix}, \quad E_1 = \begin{bmatrix} 1.2472 \\ 1.9599 \\ -8.0000 \end{bmatrix} \text{ and } E_2 = \begin{bmatrix} -1.2472 \\ -1.9599 \\ -8.0000 \end{bmatrix} \tag{32}$$

The Jacobian matrix of the Vaidyanathan system (26) at $(x_1^\star, x_2^\star, x_3^\star)$ is obtained as

$$J(x^\star) = \begin{bmatrix} -22 & 22 + x_3^\star & x_2^\star \\ 400 + 50x_3^\star & 0 & 50x_1^\star \\ -x_2^\star - 2x_1^\star & -x_1^\star & -0.5 \end{bmatrix} \tag{33}$$

The Jacobian matrix at E_0 is obtained as

$$J_0 = J(E_0) = \begin{bmatrix} -22 & 22 & 0 \\ 400 & 0 & 0 \\ 0 & 0 & -0.5 \end{bmatrix} \tag{34}$$

The matrix J_0 has the eigenvalues

$$\lambda_1 = -0.5, \quad \lambda_2 = -105.451, \quad \lambda_3 = 83.451 \tag{35}$$

This shows that the equilibrium point E_0 is a saddle-point, which is unstable.

The Jacobian matrix at E_1 is obtained as

$$J_1 = J(E_1) = \begin{bmatrix} -22 & 14 & 1.9599 \\ 0 & 0 & 62.36 \\ -4.4543 & -1.2472 & -0.5 \end{bmatrix} \tag{36}$$

The matrix J_1 has the eigenvalues

$$\lambda_1 = -26.7022, \quad \lambda_{2,3} = 2.1011 \pm 14.3283i \tag{37}$$

This shows that the equilibrium point E_1 is a saddle-focus, which is unstable.

The Jacobian matrix at E_2 is obtained as

$$J_2 = J(E_2) = \begin{bmatrix} -22 & 14 & -1.9599 \\ 0 & 0 & -62.36 \\ 4.4543 & 1.2472 & -0.5 \end{bmatrix} \tag{38}$$

The matrix J_2 has the eigenvalues

$$\lambda_1 = -26.7022, \quad \lambda_{2,3} = 2.1011 \pm 14.3283i \tag{39}$$

This shows that the equilibrium point E_2 is a saddle-focus, which is unstable.

Hence, E_0, E_1, E_2 are all unstable equilibrium points of the Vaidyanathan–Madhavan chaotic system (26), where E_0 is a saddle point and E_1, E_2 are saddle-focus points.

6.3 Lyapunov Exponents and Lyapunov Dimension

We take the initial values of the Vaidyanathan–Madhavan system as in (28) and the parameter values of the Vaidyanathan–Madhavan system as (27).

Then the Lyapunov exponents of the Vaidyanathan system (26) are numerically obtained as

$$L_1 = 3.3226, \quad L_2 = 0, \quad L_3 = -30.3406 \tag{40}$$

Thus, the maximal Lyapunov exponent of the Vaidyanathan–Madhavan system (26) is $L_1 = 3.3226$.

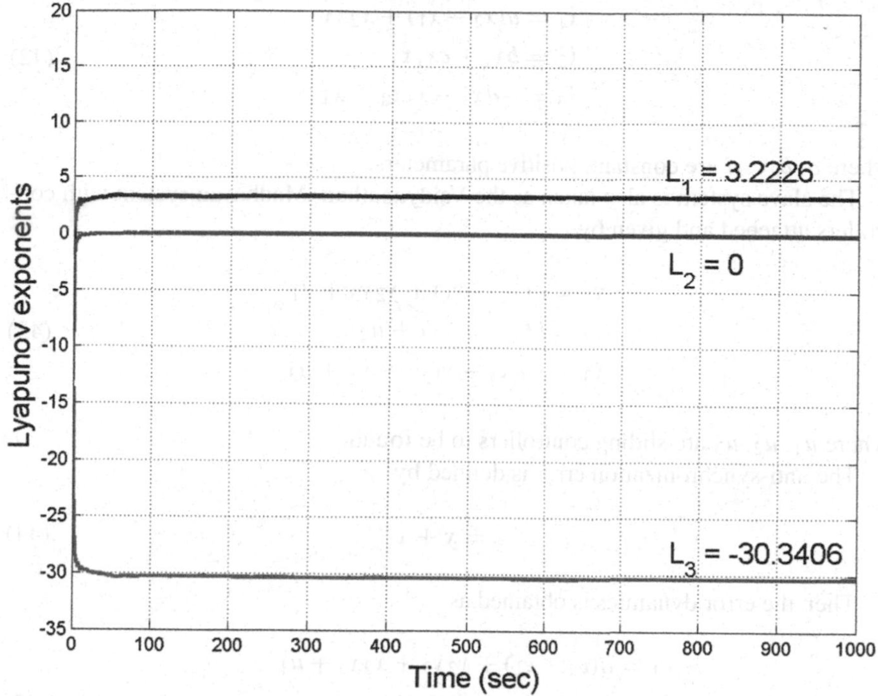

Fig. 5 Dynamics of the lyapunov exponents of the Vaidyanathan–Madhavan system

Since $L_1 + L_2 + L_3 = -27.018 < 0$, the system (26) is dissipative.
Also, the Lyapunov dimension of the system (26) is obtained as

$$D_L = 2 + \frac{L_1 + L_2}{|L_3|} = 2.1095 \tag{41}$$

Figure 5 depicts the dynamics of the Lyapunov exponents of the Vaidyanathan–Madhavan system (26).

7 Anti-synchronization of Vaidyanathan–Madhavan Chaotic Systems via SMC

This section details the construction of an anti-synchronizer for identical Vaidyanathan–Madhavan chaotic systems via sliding mode control method.

The master system is taken as the Vaidyanathan–Madhavan system given by

$$
\begin{aligned}
\dot{x}_1 &= a(x_2 - x_1) + x_2 x_3 \\
\dot{x}_2 &= b x_1 + c x_1 x_3 \\
\dot{x}_3 &= -d x_3 - x_1 x_2 - x_1^2
\end{aligned}
\tag{42}
$$

where a, b, c, d are constant, positive parameters.

The slave system is also taken as the Vaidyanathan–Madhavan system with controllers attached and given by

$$
\begin{aligned}
\dot{y}_1 &= a(y_2 - y_1) + y_2 y_3 + u_1 \\
\dot{y}_2 &= b y_1 + c y_1 y_3 + u_2 \\
\dot{y}_3 &= -d y_3 - y_1 y_2 - y_1^2 + u_3
\end{aligned}
\tag{43}
$$

where u_1, u_2, u_3 are sliding controllers to be found.

The anti-synchronization error is defined by

$$
e = y + x \tag{44}
$$

Then the error dynamics is obtained as

$$
\begin{aligned}
\dot{e}_1 &= a(e_2 - e_1) + y_2 y_3 + x_2 x_3 + u_1 \\
\dot{e}_2 &= b e_1 + c(y_1 y_3 + x_1 x_3) + u_2 \\
\dot{e}_3 &= -d e_3 - y_1 y_2 - x_1 x_2 - y_1^2 - x_1^2 + u_3
\end{aligned}
\tag{45}
$$

The error dynamics (45) can be expressed in matrix form as

$$
\dot{e} = Ae + \eta(x, y) + u \tag{46}
$$

where

$$
A = \begin{bmatrix} -a & a & 0 \\ b & 0 & 0 \\ 0 & 0 & -d \end{bmatrix}, \quad \eta(x, y) = \begin{bmatrix} y_2 y_3 + x_2 x_3 \\ c(y_1 y_3 + x_1 x_3) \\ -y_1 y_2 - x_1 x_2 - y_1^2 - x_1^2 \end{bmatrix}, \quad u = \begin{bmatrix} u_1 \\ u_2 \\ u_3 \end{bmatrix} \tag{47}
$$

The parameter values of a, b, c, d are taken as in the chaotic case, i.e.

$$
a = 22, \quad b = 400, \quad c = 50, \quad d = 0.5 \tag{48}
$$

First, the control u is set as

$$
u = -\eta(x, y) + Bv, \tag{49}
$$

where B is chosen such that (A, B) is controllable.

A simple choice for B is

$$B = \begin{bmatrix} 1 \\ 1 \\ 1 \end{bmatrix} \tag{50}$$

The sliding variable is picked as

$$s = Ce = \begin{bmatrix} 1 & 2 & -2 \end{bmatrix} e = e_1 + 2e_2 - 2e_3 \tag{51}$$

The choice of the sliding variable indicated by (51) renders the sliding mode dynamics globally asymptotically stable.

Next, we choose the SMC gains as

$$\alpha = 6 \quad \text{and} \quad \beta = 0.2 \tag{52}$$

Using the formula (18), the control v is obtained as

$$v(t) = -784e_1 - 34e_2 + 11e_3 - 0.2\,\text{sgn}(s) \tag{53}$$

As a consequence of Theorem 1 (Sect. 4), the following result is obtained.

Theorem 2 *The control law defined by* (49)*, where* v *is defined by* (53)*, renders the Vaidyanathan systems* (42) *and* (43) *globally and asymptotically anti-synchronized for all values of the initial states* $x(0), y(0) \in \mathbb{R}^3$.

For numerical simulations, the classical fourth-order Runge-Kutta method with step-size $h = 10^{-8}$ is used in the MATLAB software.

The parameter values are taken as in the chaotic case of the Vaidyanathan systems (42) and (43), i.e.

$$a = 22, \quad b = 400, \quad c = 50, \quad d = 0.5$$

The sliding mode gains are taken as $\alpha = 6$ and $\beta = 0.2$.

The initial values of the master system (42) are taken as

$$x_1(0) = 5.2, \quad x_2(0) = 2.7, \quad x_3(0) = -3.2$$

The initial values of the slave system (43) are taken as

$$y_1(0) = 3.4, \quad y_2(0) = 3.1, \quad y_3(0) = -8.4$$

Figures. 6, 7 and 8 show the anti-synchronization of the Vaidyanathan systems (42) and (43). Figure 9 shows the time-history of the anti-synchronization errors e_1, e_2 and e_3.

In Fig. 6, it is seen that the odd states $x_1(t)$ and $y_1(t)$ are anti-synchronized in 1 s.

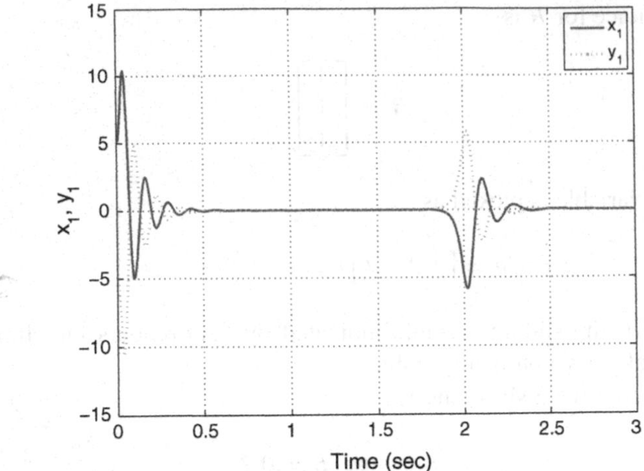

Fig. 6 Anti-synchronization of the states x_1 and y_1

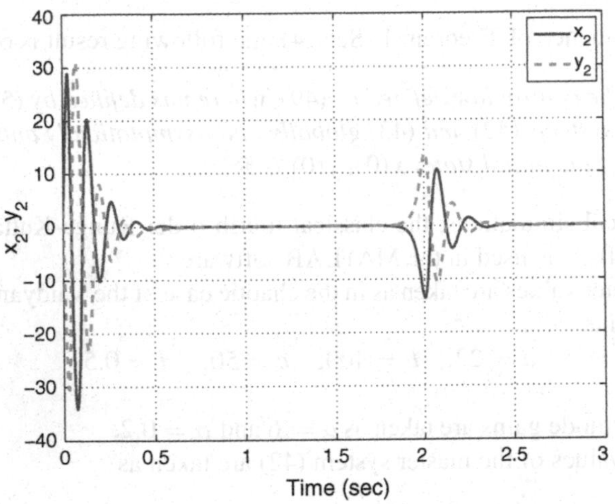

Fig. 7 Anti-synchronization of the states x_2 and y_2

In Fig. 7, it is seen that the even states $x_2(t)$ and $y_2(t)$ are anti-synchronized in 1 s.

In Fig. 8, it is seen that the odd states $x_3(t)$ and $y_3(t)$ are anti-synchronized in 1 s.

Figure 9 shows the time-history of the anti-synchronization errors e_1, e_2 and e_3. It is seen that the anti-synchronization errors converge to zero in 1 s.

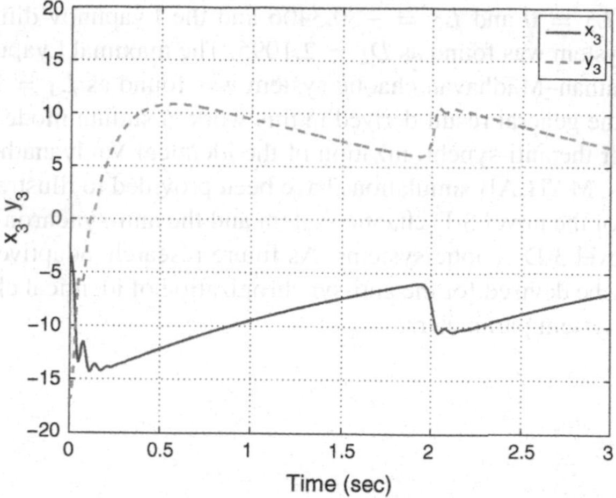

Fig. 8 Anti-synchronization of the states x_3 and y_3

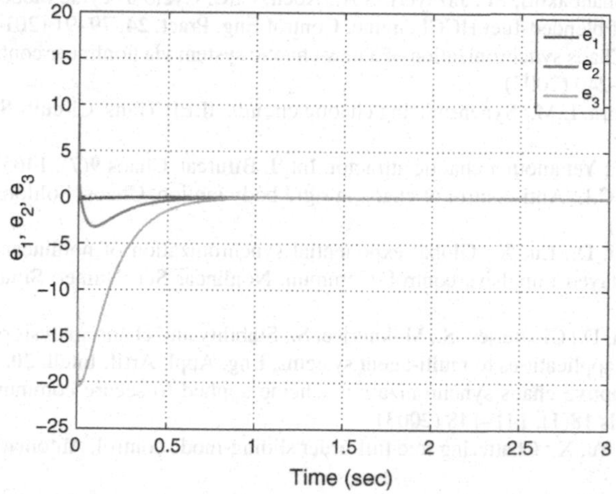

Fig. 9 Time-history of the anti-synchronization errors e_1, e_2, e_3

8 Conclusions

A general result has been derived in this work for the anti-synchronization of identical chaotic systems using sliding mode control. The main result has been proved using Lyapunov stability theory. Sliding mode control (SMC) is well-known as a robust approach and useful for controller design in systems with parameter uncertainties. Next, as an application of the main result, anti-synchronizing controller has been designed for Vaidyanathan–Madhavan chaotic systems (2013). The Lyapunov exponents of the Vaidyanathan–Madhavan chaotic system were found as

$L_1 = 3.2226$, $L_2 = 0$ and $L_3 = -30.3406$ and the Lyapunov dimension of the novel chaotic system was found as $D_L = 2.1095$. The maximal Lyapunov exponent of the Vaidyanathan–Madhavan chaotic system was found as $L_1 = 3.2226$. As an application of the general result derived in this work, a sliding mode controller has been derived for the anti-synchronization of the identical Vaidyanathan–Madhavan chaotic systems. MATLAB simulations have been provided to illustrate the qualitative properties of the novel 3-D chaotic system and the anti-synchronizer results for the identical novel 3-D chaotic systems. As future research, adaptive sliding mode controllers may be devised for the anti-synchronization of identical chaotic systems with unknown system parameters.

References

Arneodo, A., Coullet, P., Tresser, C.: Possible new strange attractors with spiral structure. Common. Math. Phys. **79**(4), 573–576 (1981)

Bidarvatan, M., Shahbakhti, M., Jazayeri, S.A., Koch, C.R.: Cycle-to-cycle modeling and sliding mode control of blended-fuel HCCI engine. Control Eng. Pract. **24**, 79–91 (2014)

Cai, G., Tan, Z.: Chaos synchronization of a new chaotic system via nonlinear control. J. Uncertain Syst. **1**(3), 235–240 (2007)

Carroll, T.L., Pecora, L.M.: Synchronizing chaotic circuits. IEEE Trans. Circuits Syst. **38**(4), 453–456 (1991)

Chen, G., Ueta, T.: Yet another chaotic attractor. Int. J. Bifurcat. Chaos **9**(7), 1465–1466 (1999)

Chen, H.K., Lee, C.I.: Anti-control of chaos in rigid body motion. Chaos, Solitons Fractals **21**(4), 957–965 (2004)

Chen, W.-H., Wei, D., Lu, X.: Global exponential synchronization of nonlinear time-delay lure systems via delayed impulsive control. Commun. Nonlinear Sci. Numer. Simul. **19**(9), 3298–3312 (2014)

Das, S., Goswami, D., Chatterjee, S., Mukherjee, S.: Stability and chaos analysis of a novel swarm dynamics with applications to multi-agent systems. Eng. Appl. Artif. Intell. **30**, 189–198 (2014)

Feki, M.: An adaptive chaos synchronization scheme applied to secure communication. Chaos, Solitons Fractals **18**(1), 141–148 (2003)

Feng, Y., Han, F., Yu, X.: Chattering free full-order sliding-mode control. Automatica **50**(4), 1310–1314 (2014)

Gan, Q., Liang, Y.: Synchronization of chaotic neural networks with time delay in the leakage term and parametric uncertainties based on sampled-data control. J. Franklin Inst. **349**(6), 1955–1971 (2012)

Gaspard, P.: Microscopic chaos and chemical reactions. Physica A: Stat. Mech. Appl. **263**(1–4), 315–328 (1999)

Gibson, W.T., Wilson, W.G.: Individual-based chaos: Extensions of the discrete logistic model. J. Theor. Biol. **339**, 84–92 (2013)

Guégan, D.: Chaos in economics and finance. Annu. Rev. Control **33**(1), 89–93 (2009)

Hamayun, M.T., Edwards, C., Alwi, H.: A fault tolerant control allocation scheme with output integral sliding modes. Automatica **49**(6), 1830–1837 (2013)

Huang, J.: Adaptive synchronization between different hyperchaotic systems with fully uncertain parameters. Phys. Lett. A **372**(27–28), 4799–4804 (2008)

Huang, X., Zhao, Z., Wang, Z., Li, Y.: Chaos and hyperchaos in fractional-order cellular neural networks. Neurocomputing **94**, 13–21 (2012)

Itkis, U.: Control Systems of Variable Structure. Wiley, New York (1976)

Jiang, G.-P., Zheng, W.X., Chen, G.: Global chaos synchronization with channel time-delay. Chaos, Solitons Fractals **20**(2), 267–275 (2004)

Kaslik, E., Sivasundaram, S.: Nonlinear dynamics and chaos in fractional-order neural networks. Neural Networks **32**, 245–256 (2012)

Kengne, J., Chedjou, J.C., Kenne, G., Kyamakya, K.: Dynamical properties and chaos synchronization of improved Colpitts oscillators. Commun. Nonlinear Sci. Numer. Simul. **17**(7), 2914–2923 (2012)

Khalil, H.K.: Nonlinear Systems. Prentice Hall, Upper Saddle River (2001)

Kyriazis, M.: Applications of chaos theory to the molecular biology of aging. Exp. Gerontol. **26**(6), 569–572 (1991)

Li, D.: A three-scroll chaotic attractor. Phys. Lett. A **372**(4), 387–393 (2008)

Li, N., Pan, W., Yan, L., Luo, B., Zou, X.: Enhanced chaos synchronization and communication in cascade-coupled semiconductor ring lasers. Commun. Nonlinear Sci. Numer. Simul. **19**(6), 1874–1883 (2014)

Li, N., Zhang, Y., Nie, Z.: Synchronization for general complex dynamical networks with sampled-data. Neurocomputing **74**(5), 805–811 (2011)

Lian, S., Chen, X.: Traceable content protection based on chaos and neural networks. Appl. Soft Comput. **11**(7), 4293–4301 (2011)

Lin, W.: Adaptive chaos control and synchronization in only locally lipschitz systems. Phys. Lett. A **372**(18), 3195–3200 (2008)

Liu, C., Liu, T., Liu, L., Liu, K.: A new chaotic attractor. Chaos, Solitions Fractals **22**(5), 1031–1038 (2004)

Liu, L., Zhang, C., Guo, Z.A.: Synchronization between two different chaotic systems with nonlinear feedback control. Chin. Phys. **16**(6), 1603–1607 (2007)

Lorenz, E.N.: Deterministic periodic flow. J. Atmos. Sci. **20**(2), 130–141 (1963)

Lü, J., Chen, G.: A new chaotic attractor coined. Int. J. Bifurcat. Chaos **12**(3), 659–661 (2002)

Lu, W., Li, C., Xu, C.: Sliding mode control of a shunt hybrid active power filter based on the inverse system method. Int. J. Electr. Power Energy Syst. **57**, 39–48 (2014)

Mondal, S., Mahanta, C.: Adaptive second order terminal sliding mode controller for robotic manipulators. J. Franklin Inst. **351**(4), 2356–2377 (2014)

Murali, K., Lakshmanan, M.: Secure communication using a compound signal from generalized chaotic systems. Phys. Lett. A **241**(6), 303–310 (1998)

Nehmzow, U., Walker, K.: Quantitative description of robotenvironment interaction using chaos theory. Robotics Auton. Syst. **53**(3–4), 177–193 (2005)

Njah, A.N., Ojo, K.S., Adebayo, G.A., Obawole, A.O.: Generalized control and synchronization of chaos in RCL-shunted Josephson junction using backstepping design. Physica C: Supercond. **470**(13–14), 558–564 (2010)

Ouyang, P.R., Acob, J., Pano, V.: PD with sliding mode control for trajectory tracking of robotic system. Robotics Comput. Integr. Manuf. **30**(2), 189–200 (2014)

Pecora, L.M., Carroll, T.L.: Synchronization in chaotic systems. Phys. Rev. Lett. **64**(8), 821–824 (1990)

Perruquetti, W., Barbot, J.P.: Sliding Mode Control in Engineering. Marcel Dekker, New York (2002)

Petrov, V., Gaspar, V., Masere, J., Showalter, K.: Controlling chaos in belousov-zhabotinsky reaction. Nature **361**, 240–243 (1993)

Qu, Z.: Chaos in the genesis and maintenance of cardiac arrhythmias. Prog. Biophys. Mol. Biol. **105**(3), 247–257 (2011)

Rafikov, M., Balthazar, J.M.: On control and synchronization in chaotic and hyperchaotic systems via linear feedback control. Commun. Nonlinear Sci. Numer. Simul. **13**(7), 1246–1255 (2007)

Rasappan, S., Vaidyanathan, S.: Global chaos synchronization of WINDMI and Coullet chaotic systems by backstepping control. Far East J. Math. Sci. **67**(2), 265–287 (2012)

Rhouma, R., Belghith, S.: Cryptoanalysis of a chaos based cryptosystem on DSP. Commun. Nonlinear Sci. Numer. Simul. **16**(2), 876–884 (2011)

Rössler, O.E.: An equation for continuous chaos. Phys. Lett. **57A**(5), 397–398 (1976)
Sarasu, P., Sundarapandian, V.: Adaptive controller design for the generalized projective synchronization of 4-scroll systems. Int. J. Syst. Signal Control Eng. Appl. **5**(2), 21–30 (2012a)
Sarasu, P., Sundarapandian, V.: Generalized projective synchronization of two-scroll systems via adaptive control. Int. J. Soft Comput. **7**(4), 146–156 (2012b)
Sarasu, P., Sundarapandian, V.: Generalized projective synchronization of two-scroll systems via adaptive control. Eur. J. Sci. Res. **72**(4), 504–522 (2012c)
Shahverdiev, E.M., Bayramov, P.A., Shore, K.A.: Cascaded and adaptive chaos synchronization in multiple time-delay laser systems. Chaos, Solitons Fractals **42**(1), 180–186 (2009)
Shahverdiev, E.M., Shore, K.A.: Impact of modulated multiple optical feedback time delays on laser diode chaos synchronization. Opt. Commun. **282**(17), 2572–3568 (2009)
Sharma, A., Patidar, V., Purohit, G., Sud, K.K.: Effects on the bifurcation and chaos in forced duffing oscillator due to nonlinear damping. Commun. Nonlinear Sci. Numer. Simul. **17**(6), 2254–2269 (2012)
Sprott, J.C.: Some simple chaotic flows. Phys. Rev. E **50**(2), 647–650 (1994)
Sprott, J.C.: Competition with evolution in ecology and finance. Phys. Lett. A **325**(5–6), 329–333 (2004)
Suérez, I.: Mastering chaos in ecology. Ecol. Model. **117**(2–3), 305–314 (1999)
Sundarapandian, V.: Output regulation of the Lorenz attractor. Far East J. Math. Sci. **42**(2), 289–299 (2010)
Sundarapandian, V., Pehlivan, I.: Analysis, control, synchronization, and circuit design of a novel chaotic system. Math. Comput. Model. **55**(7–8), 1904–1915 (2012)
Suresh, R., Sundarapandian, V.: Global chaos synchronizatoin of a family of n-scroll hyperchaotic chua circuits using backstepping control with recursive feedback. Far East J. Math. Sci. **73**(1), 73–95 (2013)
Tigan, G., Opris, D.: Analysis of a 3D chaotic system. Chaos, Solitons Fractals **36**, 1315–1319 (2008)
Tu, J., He, H., Xiong, P.: Adaptive backstepping synchronization between chaotic systems with unknown lipschitz constant. Appl. Math. Comput. **236**, 10–18 (2014)
Ucar, A., Lonngren, K.E., Bai, E.W.: Chaos synchronization in RCL-shunted josephson junction via active control. Chaos, Solitons Fractals **31**(1), 105–111 (2007)
Usama, M., Khan, M.K., Alghatbar, K., Lee, C.: Chaos-based secure satellite imagery cryptosystem. Comput. Math. Appl. **60**(2), 326–337 (2010)
Utkin, V.: Sliding Modes and Their Applications in Variable Structure Systems. MIR Publishers, Moscow (1978)
Utkin, V.I.: Sliding Modes in Control and Optimization. Springer, New York (1992)
Vaidyanathan, S.: Anti-synchronization of Newton-Leipnik and Chen-Lee chaotic systems by active control. Int. J. Control Theory Appl. **4**(2), 131–141 (2011)
Vaidyanathan, S.: Adaptive backstepping controller and synchronizer design for arneodo chaotic system with unknown parameters. Int. J. Comput. Sci. Inf. Technol. **4**(6), 145–159 (2012a)
Vaidyanathan, S.: Anti-synchronization of sprott-L and sprott-M chaotic systems via adaptive control. Int. J. Control Theory Appl. **5**(1), 41–59 (2012b)
Vaidyanathan, S.: Output regulation of the liu chaotic system. Appl. Mech. Mater. **110–116**, 3982–3989 (2012c)
Vaidyanathan, S.: A new six-term 3-D chaotic system with an exponential nonlinearity. Far East J. Math. Sci. **79**(1), 135–143 (2013a)
Vaidyanathan, S.: Analysis and adaptive synchronization of two novel chaotic systems with hyperbolic sinusoidal and cosinusoidal nonlinearity and unknown parameters. J. Eng. Sci. Technol. Rev. **6**(4), 53–65 (2013b)
Vaidyanathan, S.: A new eight-term 3-D polynomial chaotic system with three quadratic nonlinearities. Far East J. Math. Sci. **84**(2), 219–226 (2014)
Vaidyanathan, S., Madhavan, K.: Analysis, adaptive control and synchronization of a seven-term novel 3-D chaotic system. Int. J. Control Theory Appl. **6**(2), 121–137 (2013)

Vaidyanathan, S., Rajagopal, K.: Global chaos synchronization of four-scroll chaotic systems by active nonlinear control. Int. J. Control Theory Appl. **4**(1), 73–83 (2011)
Vaidyanathan, S., Sampath, S.: Anti-synchronization of four-scroll chaotic systems via sliding mode control. Int. J. Autom. Comput. **9**(3), 274–279 (2012)
Volos, C.K., Kyprianidis, I.M., Stouboulos, I.N.: Experimental investigation on coverage performance of a chaotic autonomous mobile robot. Robotics Auton. Syst. **61**(12), 1314–1322 (2013)
Wang, F., Liu, C.: A new criterion for chaos and hyperchaos synchronization using linear feedback control. Phys. Lett. A **360**(2), 274–278 (2006)
Witte, C.L., Witte, M.H.: Chaos and predicting varix hemorrhage. Med. Hypotheses **36**(4), 312–317 (1991)
Wu, X., Guan, Z.-H., Wu, Z.: Adaptive synchronization between two different hyperchaotic systems. Nonlinear Anal.: Theory, Methods Appl. **68**(5), 1346–1351 (2008)
Xiao, X., Zhou, L., Zhang, Z.: Synchronization of chaotic Lure systems with quantized sampled-data controller. Commun. Nonlinear Sci. Numer. Simul. **19**(6), 2039–2047 (2014)
Yuan, G., Zhang, X., Wang, Z.: Generation and synchronization of feedback-induced chaos in semiconductor ring lasers by injection-locking. Optik - Int. J. Light Electron Optics **125**(8), 1950–1953 (2014)
Zaher, A.A., Abu-Rezq, A.: On the design of chaos-based secure communication systems. Commun. Nonlinear Sci. Numer. Simul. **16**(9), 3721–3727 (2011)
Zhang, H., Zhou, J.: Synchronization of sampled-data coupled harmonic oscillators with control inputs missing. Syst. Control Lett. **61**(12), 1277–1285 (2012)
Zhang, J., Li, C., Zhang, H., Yu, J.: Chaos synchronization using single variable feedback based on backstepping method. Chaos, Solitons Fractals **21**(5), 1183–1193 (2004)
Zhang, X., Liu, X., Zhu, Q.: Adaptive chatter free sliding mode control for a class of uncertain chaotic systems. Appl. Math. Comput. **232**, 431–435 (2014)
Zhou, W., Xu, Y., Lu, H., Pan, L.: On dynamics analysis of a new chaotic attractor. Phys. Lett. A **372**(36), 5773–5777 (2008)
Zhu, C., Liu, Y., Guo, Y.: Theoretic and numerical study of a new chaotic system. Intell. Inf. Manage. **2**, 104–109 (2010)
Zinober, A.S.: Variable Structure and Lyapunov Control. Springer, New York (1993)

Hybrid Synchronization of Identical Chaotic Systems Using Sliding Mode Control and an Application to Vaidyanathan Chaotic Systems

Sundarapandian Vaidyanathan and Ahmad Taher Azar

Abstract Hybrid phase synchronization is a new type of synchronization of a pair of chaotic systems called the master and slave systems. In hybrid phase synchronization, the odd numbered states of the master and slave systems are completely synchronized (CS), while their even numbered states are anti-synchronized (AS). The hybrid phase synchronization has applications in secure communications and cryptosystems. This work derives a new result for the hybrid phase synchronization of identical chaotic systems using sliding mode control. The main result has been proved using Lyapunov stability theory. Sliding mode control (SMC) is well-known as a robust approach and useful for controller design in systems with parameter uncertainties. As an application of this general result, a sliding mode controller is derived for the hybrid phase synchronization of the identical 3-D Vaidyanathan chaotic systems (2014). MATLAB simulations have been provided to illustrate the Vaidyanathan system and the hybrid synchronizer results for the identical Vaidyanathan systems.

1 Introduction

Chaotic behaviour is an important feature, which is observed in some nonlinear dynamical systems. Chaotic behaviour was suspected well over hundred years ago in the study of three bodies problem, but it was established only a few decades ago in the study of 3-D weather models (Lorenz 1963).

A chaotic system is usually characterized by its extreme sensitivity of behavior to initial conditions. Small changes in an initial state will make a very large difference in the behavior of the system at future states.

S. Vaidyanathan (✉)
Research and Development Centre, Vel Tech University, Avadi,
Chennai 600062, Tamil Nadu, India
e-mail: sundarvtu@gmail.com

A.T. Azar
Faculty of Computers and Information, Benha University, Benha, Egypt
e-mail: ahmad_t_azar@ieee.org

A.T. Azar and Q. Zhu (eds.), *Advances and Applications in Sliding Mode Control systems*,
Studies in Computational Intelligence 576, DOI 10.1007/978-3-319-11173-5_20

The Lyapunov exponent is a measure of the divergence of phase points that are initially very close and can be used to quantify chaotic systems. It is common to refer to the largest Lyapunov exponent as the maximal Lyapunov exponent (MLE). A positive maximal Lyapunov exponent and phase space compactness are usually taken as defining conditions for a chaotic system.

In 1963, Lorenz found out that a very small difference in the initial conditions of his 3-D deterministic weather model led to large changes in the phase space (Lorenz 1963). This was followed by the discoveries of many well-known paradigms of 3-D chaotic systems in the literature (Rössler 1976; Arneodo et al. 1981; Sprott 1994; Chen and Ueta 1999; Lü and Chen 2002; Liu et al. 2004; Cai and Tan 2007; Chen and Lee 2004; Tigan and Opris 2008; Zhou et al. 2008; Sundarapandian and Pehlivan 2012; Vaidyanathan 2013a, b, 2014).

Chaotic systems have several applications in science and engineering. Some important applications can be mentioned as cryptosystems (Usama et al. 2010; Rhouma and Belghith 2011), secure communications (Murali and Lakshmanan 1998; Feki 2003; Zaher and Abu-Rezq 2011), chemical reactions (Petrov et al. 1993; Gaspard 1999), oscillators (Kengne et al. 2012; Sharma et al. 2012), lasers (Yuan et al. 2014; Li et al. 2014), biology (Das et al. 2014; Kyriazis 1991), ecology (Suérez 1999; Gibson and Wilson 2013), robotics (Nehmzow and Walker 2005; Volos et al. 2013; Mondal and Mahanta 2014), cardiology (Qu 2011; Witte and Witte 1991), neural networks (Kaslik and Sivasundaram 2012; Huang et al. 2012; Lian and Chen 2011), finance (Sprott 2004; Guégan 2009), etc.

Synchronization of chaotic systems is a phenomenon that occurs when two or more chaotic systems are coupled or when a chaotic system drives another chaotic system. Because of the butterfly effect which causes exponential divergence of the trajectories of two identical chaotic systems started with nearly the same initial conditions, the synchronization of chaotic systems is a challenging research problem in the chaos literature.

The master-slave or drive-response formalism is used in most of the chaos synchronization approaches. If a particular chaotic system is called the master or drive system and another chaotic system is called the slave or response system, then the goal of chaos synchronization is to use the output of the master system to control the slave system so that the output of the slave system tracks the output of the master system asymptotically.

Pecora and Carroll pioneered the research on synchronization of chaotic systems with their seminal papers in 1990s (Pecora and Carroll 1990; Carroll and Pecora 1991). The active control method (Ucar et al. 2007; Liu et al. 2007; Sundarapandian 2010; Vaidyanathan 2012c; Wang and Liu 2006; Rafikov and Balthazar 2007) is commonly used when the system parameters are available for measurement and the adaptive control method (Wu et al. 2008; Huang 2008; Lin 2008; Sarasu and Sundarapandian 2012a, b, c) is commonly used when some or all the system parameters are not available for measurement and estimates for unknown parameters of the systems.

Other popular methods for chaos synchronization are the sampled-data feedback method (Xiao et al. 2014; Zhang and Zhou 2012; Li et al. 2011; Gan and Liang

2012), time-delay feedback method (Shahverdiev and Shore 2009; Jiang et al. 2004; Chen et al. 2014; Shahverdiev et al. 2009), backstepping method (Njah et al. 2010; Tu et al. 2014; Zhang et al. 2004; Vaidyanathan 2012a), etc.

Complete synchronization (Vaidyanathan and Rajagopal 2011a; Rasappan and Vaidyanathan 2012a; Suresh and Sundarapandian 2013) is characterized by the equality of state variables evolving in time, while anti-synchronization (Vaidyanathan 2011; Vaidyanathan and Sampath 2012; Vaidyanathan 2012b) is characterized by the disappearance of the sum of relevant state variables evolving in time.

In hybrid synchronization of the master and slave systems, the odd numbered states of the two systems are completely synchronized while the even numbered states are anti-synchronized so that the complete synchronization (CS) and anti-synchronization (AS) co-exist in the synchronization process. Thus, the hybrid synchronization (Vaidyanathan and Rajagopal 2011b; Sundarapandian and Karthikeyan 2012; Karthikeyan and Sundarapandian 2014; Rasappan and Vaidyanathan 2012b) is an important type of synchronization of chaotic systems, which has applications in secure communication devices.

This research work is organized as follows. Section 2 gives a basic introduction into sliding mode control and chaos synchronization. Section 3 discusses the problem statement for the synchronization of two identical chaotic systems and our design methodology. Section 4 contains the main result of this work, namely, sliding controller design for the global chaos synchronization of identical chaotic systems. Section 5 summarizes the qualitative properties of the Vaidyanathan chaotic system (Vaidyanathan 2014). In Sect. 6, we describe the sliding mode controller design for the global chaos synchronization of identical Vaidyanathan systems. MATLAB simulations are shown to validate and illustrate the sliding mode controller design for the synchronization of the Vaidyanathan systems. Section 7 contains a summary of the main results derived in this research work.

2 Sliding Mode Control and Chaos Synchronization

In control theory, the sliding mode control approach is recognized as an efficient tool for designing robust controllers for linear or nonlinear control systems operating under uncertainty conditions (Perruquetti and Barbot 2002; Utkin 1992).

Sliding mode control method has a major advantage of low sensitivity to parameter variations in the plant and disturbances affecting the plant, which eliminates the necessity of exact modeling of the plant.

In the sliding mode control theory, the control dynamics has two sequential modes, viz. (i) the *reaching mode*, and (ii) the *sliding mode*. Basically, a sliding mode controller (SMC) design consists of two parts: hyperplane (or sliding surface) design and controller design.

A hyperplane is first designed via the pole-placement approach in the modern control theory and a controller is then designed based on the sliding condition. The stability of the overall control system is ensured by the sliding condition and by a stable hyperplane. Sliding mode control theory has been used to deal with many research problems of control literature (Feng et al. 2014; Ouyang et al. 2014; Bidarvatan et al. 2014; Lu et al. 2014; Zhang et al. 2014; Hamayun et al. 2013).

3 Problem Statement

This section gives a problem statement of global hybrid-phase synchronization of a pair of identical chaotic systems called the *master* and *slave* systems.

The *master* system is taken as the chaotic system

$$\dot{x} = Ax + f(x), \tag{1}$$

where $x \in \mathbb{R}^n$ is the state of the system, A is the $n \times n$ matrix of system parameters and f is a vector field that contains the nonlinear parts of the system and satisfies $f(0) = 0$.

The *slave* system is taken as the controlled chaotic system

$$\dot{y} = Ay + f(y) + u, \tag{2}$$

where $y \in \mathbb{R}^n$ is the state of the system, and u is the controller to be determined.

The *hybrid synchronization error* is defined by

$$e_i = \begin{cases} y_i - x_i & \text{if } i \text{ is odd} \\ y_i + x_i & \text{if } i \text{ is even} \end{cases} \tag{3}$$

Differentiating (3) and simplifying, the error dynamics is obtained as

$$\dot{e} = Ae + \eta(x, y) + u \tag{4}$$

The design problem is to determine a feedback control u so that the error dynamics (4) is globally asymptotically stable at the origin for all initial conditions $e(0) \in \mathbb{R}^n$.

For the SMC design for the hybrid phase synchronization of the systems (1) and (2), the control u is taken as

$$u(t) = -\eta(x, y) + Bv(t), \tag{5}$$

where B is an $(n \times 1)$ column vector chosen such that (A, B) is controllable.

Upon substituting (5) into (4), the closed-loop error system is obtained as

$$\dot{e}(t) = Ae(t) + Bv(t), \tag{6}$$

which is a linear time-invariant control system with a single input v.

Hence, by the use of the nonlinear control law (5), original problem of hybrid phase synchronization of identical chaotic systems (1) and (2) has been converted into an equivalent problem of globally stabilizing the error dynamics (6).

4 Sliding Controller Design for Hybrid Phase Synchronization

This section derives the main result, *viz.* sliding controller design for the hybrid phase synchronization of the identical chaotic systems (1) and (2). After applying the control (5) with (A, B) a controllable pair, it is supposed that the nonlinear error dynamics (4) has been simplified as the linear error dynamics (6).

In the sliding controller design, the sliding variable is first defined as

$$s(e) = Ce = c_1 e_1 + c_2 e_2 + \cdots + c_n e_n, \tag{7}$$

where C is an $(1 \times n)$ row vector to be determined.

The sliding manifold S is defined as the hyperplane

$$S = \{e \in \mathbb{R}^n : \quad s(e) = Ce = 0\} \tag{8}$$

If a sliding motion occurs on S, then the sliding mode conditions must be satisfied, which are given by

$$s \equiv 0 \text{ and } \dot{s} = CAe + CBv = 0 \tag{9}$$

It is assumed that the row vector C is chosen so that $CB \neq 0$.

The sliding motion is affected by the so-called equivalent control given by

$$v_{\text{eq}}(t) = -(CB)^{-1} CAe(t) \tag{10}$$

As a consequence, the equivalent dynamics in the sliding phase is defined by

$$\dot{e} = \left[I - B(CB)^{-1}C\right] Ae = Ee, \tag{11}$$

where

$$E = \left[I - B(CB)^{-1}C\right] A \tag{12}$$

It can be easily verified that E is independent of the control and has at most $(n - 1)$ nonzero eigenvalues, depending on the chosen switching surface, while the associated eigenvectors belong to $\ker(C)$.

Since (A, B) is controllable, the matrices B and C can be chosen so that E has any desired $(n - 1)$ stable eigenvalues.

Thus, the dynamics in the sliding mode is globally asymptotically stable.

Finally, for the sliding mode controller (SMC) design, the constant plus proportional rate reaching law is used, which is given by

$$\dot{s} = -\beta\,\text{sgn}(s) - \alpha s \tag{13}$$

where sgn(·) denotes the sign function and the gains $\alpha > 0, \beta > 0$ are found so that the sliding condition is satisfied and the sliding motion will occur.

From the equations (9) and (13), sliding control v is found as

$$CAe + CBv = -\beta\,\text{sgn}(s) - \alpha s \tag{14}$$

Since $s = Ce$, the equation (14) can be simplified to get

$$v = -(CB)^{-1}\left[C(\alpha I + A)e + \beta\,\text{sgn}(s)\right] \tag{15}$$

Next, the main result of this section is established as follows.

Theorem 1 *A sliding mode control law that achieves hybrid phase synchronization between the identical chaotic systems* (1) *and* (2) *for all initial conditions* $x(0), y(0) \in \mathbb{R}^n$ *is given by the equation*

$$u(t) = -\eta(x(t), y(t)) + Bv(t), \tag{16}$$

where v *is defined by* (15)*,* B *is an* $(n \times 1)$ *vector such that* (A, B) *is controllable,* C *is an* $(1 \times n)$ *vector such that* $CB \neq 0$ *and that the matrix* E *defined by Eq.* (12) *has* $(n-1)$ *stable eigenvalues.*

Proof The proof is carried out using Lyapunov stability theory (Khalil 2001).

Substituting the sliding control law (16) into the error dynamics (4) leads to

$$\dot{e} = Ae + Bv \tag{17}$$

Substituting for v from (15) into (17), the error dynamics is obtained as

$$\dot{e} = Ae - B(CB)^{-1}\left[C(\alpha I + A)e + \beta\,\text{sgn}(s)\right] \tag{18}$$

The global asymptotic stability of the error system (18) is proved by taking the candidate Lyapunov function

$$V(e) = \frac{1}{2}\,s^2(e), \tag{19}$$

which is a non-negative definite function on $\mathbb{R}^n$.

It is noted that

$$V(e) = 0 \iff s(e) = 0 \tag{20}$$

The sliding mode motion is characterized by the equations

$$s(e) = 0 \text{ and } \dot{s}(e) = 0 \tag{21}$$

By the choice of E, the dynamics in the sliding mode given by (11) is globally asymptotically stable.

When $s(e) \neq 0$, $V(e) > 0$.

Also, when $s(e) \neq 0$, differentiating V along the error dynamics (18) or the equivalent dynamics (13), the following dynamics is obtained:

$$\dot{V} = s\dot{s} = -\beta s \operatorname{sgn}(s) - \alpha s^2 < 0 \tag{22}$$

Hence, by Lyapunov stability theory (Khalil 2001), it is concluded that the error dynamics (18) is globally asymptotically stable for all initial conditions $e(0) \in \mathbb{R}^n$.

This completes the proof. □

5 Analysis of the Vaidyanathan Chaotic System

This section gives details and qualitative properties of the Vaidyanathan chaotic system (Vaidyanathan 2014), which is a novel eight-term 3-D polynomial system with three quadratic nonlinearities.

The Vaidyanathan 3-D chaotic system is a polynomial system described by

$$\begin{aligned} \dot{x}_1 &= a(x_2 - x_1) + x_2 x_3, \\ \dot{x}_2 &= b x_1 + c x_2 - x_1 x_3, \\ \dot{x}_3 &= -d x_3 + x_1^2, \end{aligned} \tag{23}$$

where x_1, x_2, x_3 are the states and a, b, c, d are constant, positive, parameters.

The Vaidyanathan system (23) depicts a strange chaotic attractor when the constant parameter values are taken as

$$a = 25, \quad b = 33, \quad c = 11, \quad d = 6. \tag{24}$$

For simulations, the initial values of the Vaidyanathan system (23) are taken as

$$x_1(0) = 1.5, \quad x_2(0) = 3.2, \quad x_3(0) = 2.7 \tag{25}$$

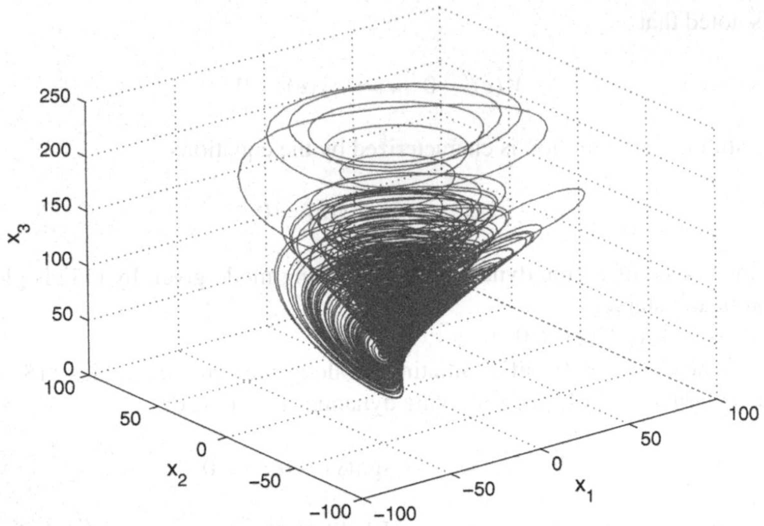

Fig. 1 Strange attractor of the Vaidyanathan system in $\mathbf{R}^3$

The Vaidyanathan 3-D chaotic system (23) exhibits a 3-scroll chaotic attractor. Figure 1 describes the 3-scroll chaotic attractor of the Vaidyanathan system (23) in 3-D view.

Figure 2 describes the 2-D projection of the strange chaotic attractor of the novel system (23) in (x_1, x_2)-plane. In the projection on the (x_1, x_2)-plane, a 3-scroll chaotic attractor is clearly seen.

Figure 3 describes the 2-D projection of the strange chaotic attractor of the novel system (23) in (x_2, x_3)-plane. In the projection on the (x_2, x_3)-plane, a 3-scroll chaotic attractor is clearly seen.

Figure 4 describes the 2-D projection of the strange chaotic attractor of the novel system (23) in (x_1, x_3)-plane. In the projection on the (x_2, x_3)-plane, a 3-scroll chaotic attractor is clearly seen.

5.1 Symmetry and Invariance

The Vaidyanathan system (23) is invariant under the coordinates transformation

$$(x_1, x_2, x_3) \to (-x_1, -x_2, x_3). \tag{26}$$

The transformation (26) persists for all values of the system parameters. Thus, the Vaidyanathan system (23) has rotation symmetry about the x_3-axis.

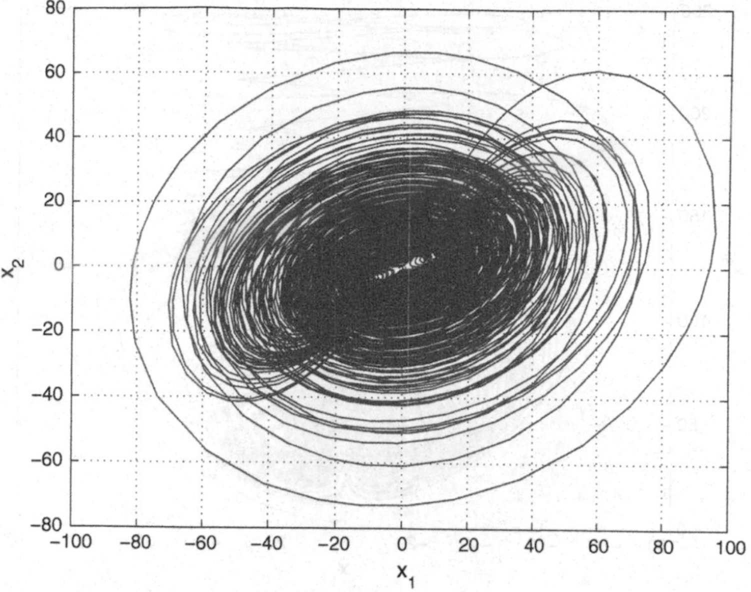

Fig. 2 2-D projection of the Vaidyanathan system in (x_1, x_2)-plane

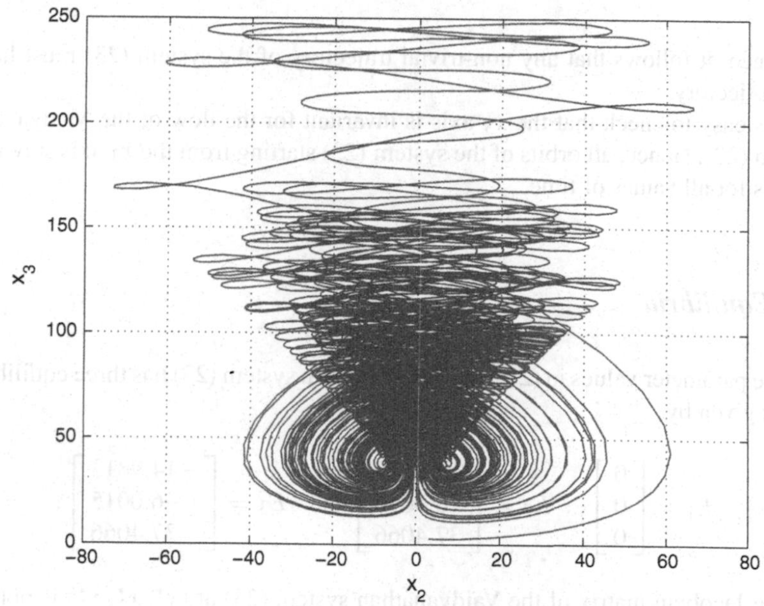

Fig. 3 2-D projection of the Vaidyanathan system in (x_2, x_3)-plane

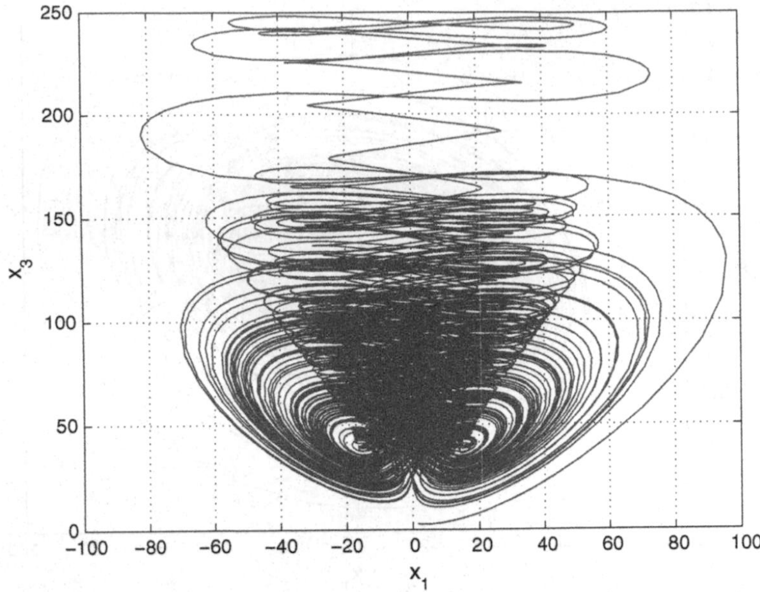

Fig. 4 2-D projection of the Vaidyanathan system in (x_1, x_3)-plane

Hence, it follows that any non-trivial trajectory of the system (23) must have a twin trajectory.

It is easy to check that the x_3-axis is invariant for the flow of the Vaidyanathan system (23). Hence, all orbits of the system (23) starting from the x_3 axis stay in the x_3 axis for all values of time.

5.2 Equilibria

For the parameter values in (24), the Vaidyanathan system (23) has three equilibrium points given by

$$E_1 = \begin{bmatrix} 0 \\ 0 \\ 0 \end{bmatrix}, \quad E_2 = \begin{bmatrix} 14.9813 \\ 6.0015 \\ 37.4066 \end{bmatrix} \text{ and } E_3 = \begin{bmatrix} -14.9813 \\ -6.0015 \\ 37.4066 \end{bmatrix}. \tag{27}$$

The Jacobian matrix of the Vaidyanathan system (23) at $(x_1^\star, x_2^\star, x_3^\star)$ is obtained as

$$J(x^\star) = \begin{bmatrix} -25 & 25 + x_3^\star & x_2^\star \\ 33 - x_3^\star & 11 & -x_1^\star \\ 2x_1^\star & 0 & -6 \end{bmatrix} \tag{28}$$

The Jacobian matrix at E_1 is obtained as

$$J_1 = J(E_1) = \begin{bmatrix} -25 & 25 & 0 \\ 33 & 11 & 0 \\ 0 & 0 & -6 \end{bmatrix} \tag{29}$$

The matrix J_1 has the eigenvalues

$$\lambda_1 = -40.8969, \quad \lambda_2 = -6, \quad \lambda_3 = 26.8969 \tag{30}$$

This shows that the equilibrium point E_1 is a saddle-point.
The Jacobian matrix at E_2 is obtained as

$$J_2 = J(E_2) = \begin{bmatrix} -25 & 62.4066 & 6.0015 \\ -4.4066 & 11 & -14.9813 \\ 29.9626 & 0 & -6 \end{bmatrix} \tag{31}$$

The matrix J_2 has the eigenvalues

$$\lambda_1 = -40.5768, \quad \lambda_{2,3} = 10.2884 \pm 25.1648i \tag{32}$$

This shows that the equilibrium point E_2 is a saddle-focus.
The Jacobian matrix at E_3 is obtained as

$$J_3 = J(E_3) = \begin{bmatrix} -25 & 62.4066 & -6.0015 \\ -4.4066 & 11 & 14.9813 \\ -29.9626 & 0 & -6 \end{bmatrix} \tag{33}$$

The matrix J_3 has the eigenvalues

$$\lambda_1 = -40.5768, \quad \lambda_{2,3} = 10.2884 \pm 25.1648i \tag{34}$$

This shows that the equilibrium point E_3 is a saddle-focus.
Hence, all the three equilibria of the Vaidyanathan system (23) are unstable.

5.3 Lyapunov Exponents and Lyapunov Dimension

For the parameter values as given by Eq. (24) and the initial state as given by Eq. (25), the Lyapunov exponents of the Vaidyanathan system (23) are numerically obtained as

$$L_1 = 6.5023, \quad L_2 = 0, \quad L_3 = -26.4352 \tag{35}$$

Thus, the maximal Lyapunov exponent of the Vaidyanathan system (23) is $L_1 = 6.5023$.

Since $L_1 + L_2 + L_3 = -19.9329 < 0$, the system (23) is dissipative.

Also, the Lyapunov dimension of the Vaidyanathan system (23) is obtained as

$$D_L = 2 + \frac{L_1 + L_2}{|L_3|} = 2.2467 \tag{36}$$

which is fractional.

Figure 5 depicts the dynamics of the Lyapunov exponents of the Vaidyanathan system (23).

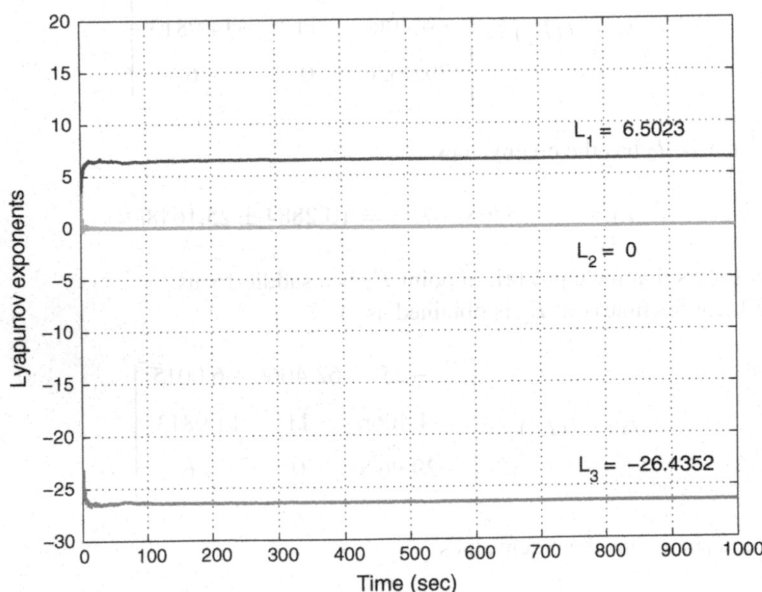

Fig. 5 Dynamics of the Lyapunov exponents of the Vaidyanathan system

6 SMC Design of Synchronization of Vaidyanathan Chaotic Systems

This section details the construction of a hybrid synchronizer for identical Vaidyanathan chaotic systems via sliding mode control method.

The master system is taken as the Vaidyanathan system given by

$$\begin{aligned} \dot{x}_1 &= a(x_2 - x_1) + x_2 x_3 \\ \dot{x}_2 &= b x_1 + c x_2 - x_1 x_3 \\ \dot{x}_3 &= -d x_3 + x_1^2 \end{aligned} \tag{37}$$

where a, b, c, d are constant, positive parameters.

The slave system is also taken as the Vaidyanathan system with controllers attached and given by

$$\begin{aligned} \dot{y}_1 &= a(y_2 - y_1) + y_2 y_3 + u_1 \\ \dot{y}_2 &= b y_1 + c y_2 - y_1 y_3 + u_2 \\ \dot{y}_3 &= -d y_3 + y_1^2 + u_3 \end{aligned} \tag{38}$$

where u_1, u_2, u_3 are sliding controllers to be found.

The hybrid phase synchronization error is defined by

$$\begin{aligned} e_1 &= y_1 - x_1 \\ e_2 &= y_2 + x_2 \\ e_3 &= y_3 - x_3 \end{aligned} \tag{39}$$

Then the error dynamics is obtained as

$$\begin{aligned} \dot{e}_1 &= a(e_2 - e_1) - 2a x_2 + y_2 y_3 - x_2 x_3 + u_1 \\ \dot{e}_2 &= b e_1 + c e_2 + 2b x_1 - y_1 y_3 - x_1 x_3 + u_2 \\ \dot{e}_3 &= -d e_3 + y_1^2 - x_1^2 + u_3 \end{aligned} \tag{40}$$

The error dynamics (40) can be expressed in matrix form as

$$\dot{e} = Ae + \eta(x, y) + u \tag{41}$$

where

$$A = \begin{bmatrix} -a & a & 0 \\ b & c & 0 \\ 0 & 0 & -d \end{bmatrix}, \quad \eta(x, y) = \begin{bmatrix} -2a x_2 + y_2 y_3 - x_2 x_3 \\ 2b x_1 - y_1 y_3 + x_1 x_3 \\ y_1^2 - x_1^2 \end{bmatrix}, \quad u = \begin{bmatrix} u_1 \\ u_2 \\ u_3 \end{bmatrix} \tag{42}$$

A hybrid synchronizing sliding controller can be designed by the procedure outlined in Sect. 5.

The parameter values of a, b, c, d are taken as in the chaotic case, i.e.

$$a = 25, \quad b = 33, \quad c = 11, \quad d = 6. \tag{43}$$

First, the control u is set as

$$u = -\eta(x, y) + Bv, \tag{44}$$

where B is chosen such that (A, B) is controllable.

A simple choice for B is

$$B = \begin{bmatrix} 1 \\ 1 \\ 1 \end{bmatrix} \tag{45}$$

The sliding variable is picked as

$$s = Ce = \begin{bmatrix} 1 & 2 & -1 \end{bmatrix} e = e_1 + 2e_2 - e_3 \tag{46}$$

Then the matrix E defined by (12) has the eigenvalues

$$\lambda_1 = -47, \quad \lambda_2 = -20, \quad \lambda_3 = 0 \tag{47}$$

The choice of the sliding variable indicated by (46) renders the sliding mode dynamics globally asymptotically stable.

Next, we choose the SMC gains as

$$\alpha = 6, \quad \beta = 0.2 \tag{48}$$

Using the formula (15), the control v is obtained as

$$v(t) = -23.5e_1 - 29.5e_2 - 0.1 \,\mathrm{sgn}(s) \tag{49}$$

As a consequence of Theorem 1 (Sect. 4), the following result is obtained.

Theorem 2 *The control law defined by* (44), *where v is defined by* (49), *renders the Vaidyanathan systems* (37) *and* (38) *globally and asymptotically hybrid phase synchronized for all values of the initial states $x(0), y(0) \in \mathbb{R}^3$.* □

For numerical simulations, the classical fourth-order Runge-Kutta method with step-size $h = 10^{-8}$ is used in the MATLAB software.

The parameter values are taken as in the chaotic case of the Vaidyanathan systems (37) and (38), i.e.

$$a = 25, \quad b = 33, \quad c = 11, \quad d = 6$$

The sliding mode gains are taken as

$$\alpha = 6 \text{ and } \beta = 0.2$$

The initial values of the master system (37) are taken as

$$x_1(0) = 2.5, \quad x_2(0) = -3.7, \quad x_3(0) = -3.2$$

The initial values of the slave system (38) are taken as

$$y_1(0) = 4.3, \quad y_2(0) = -1.6, \quad y_3(0) = 2.4$$

Figures 6, 7 and 8 show the hybrid synchronization of the Vaidyanathan systems (37) and (38).

In Fig. 6, it is seen that the odd states $x_1(t)$ and $y_1(t)$ are completely synchronized in 0.5 s.

In Fig. 7, it is seen that the even states $x_2(t)$ and $y_2(t)$ are anti-synchronized in 0.5 s.

In Fig. 8, it is seen that the odd states $x_3(t)$ and $y_3(t)$ are completely synchronized in 0.5 s.

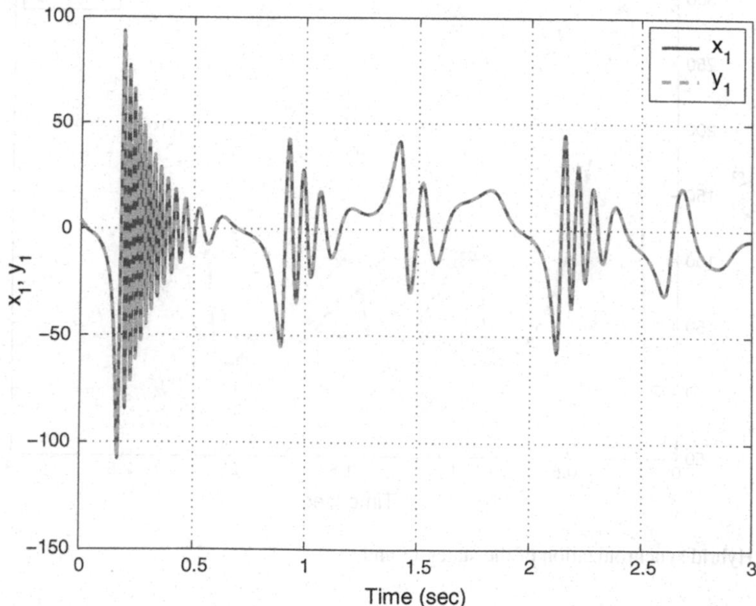

Fig. 6 Hybrid synchronization of the states x_1 and y_1

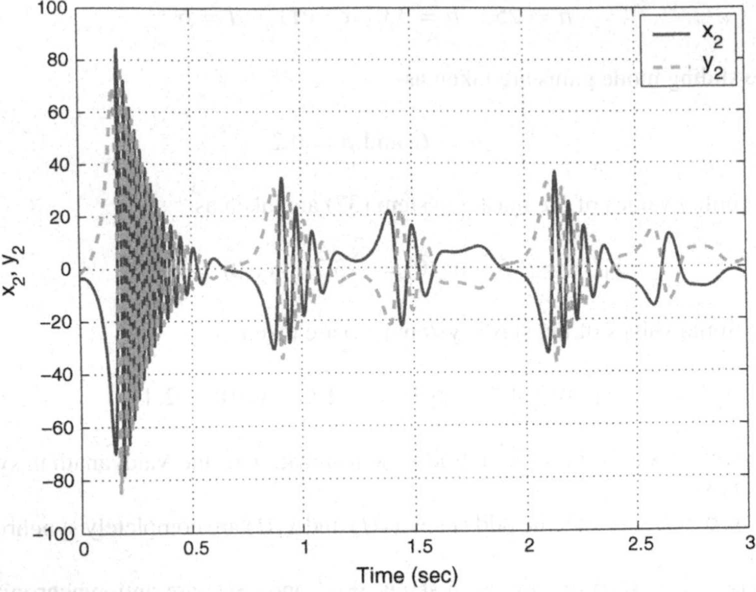

Fig. 7 Hybrid synchronization of the states x_2 and y_2

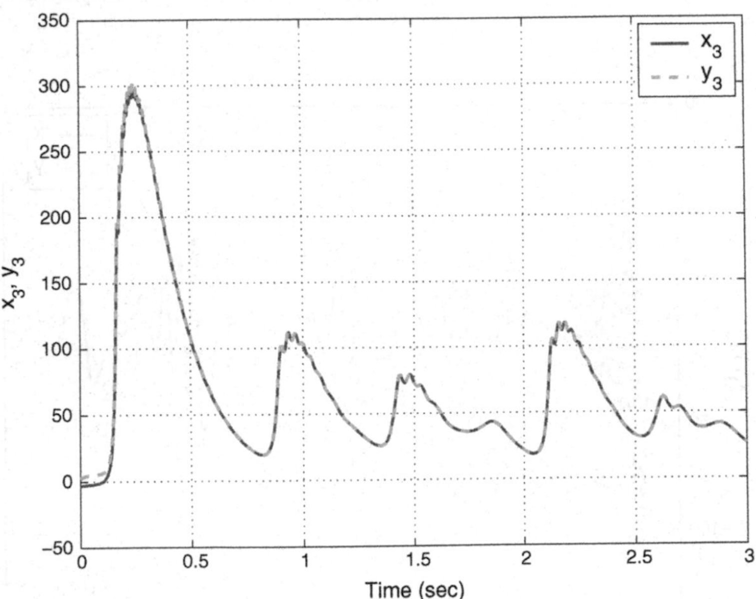

Fig. 8 Hybrid synchronization of the states x_3 and y_3

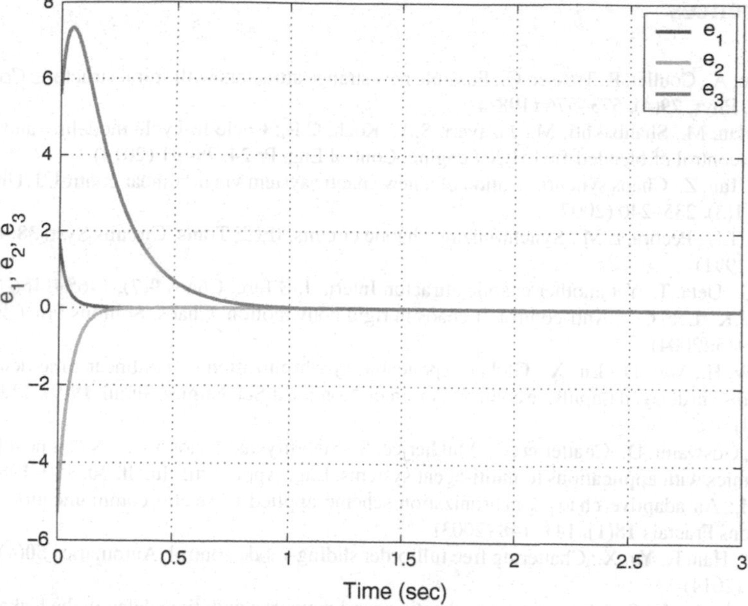

Fig. 9 Time-history of the hybrid synchronization errors e_1, e_2, e_3

Figure 9 shows the time-history of the hybrid synchronization errors e_1, e_2 and e_3. It is seen that the hybrid synchronization errors converge to zero in 0.5 s. Thus, the sliding controller for hybrid synchronization of identical Vaidyanathan systems yields very fast convergence.

7 Conclusions

Hybrid phase synchronization is a new type of synchronization of a pair of chaotic systems called the master and slave systems, where the odd states are completely synchronized and the even states anti-synchronized. In this research work, a general theorem has been developed for the hybrid phase synchronization of identical chaotic systems via sliding mode controller. The main result was proved using Lyapunov stability theory. As an application of our general result, a sliding mode controller has been designed for the hybrid phase synchronization of identical Vaidyanathan chaotic systems (2014). MATLAB simulations were shown to illustrate the qualitative properties of the Vaidyanathan system and the hybrid synchronizer results for the identical Vaidyanathan systems. As future research, adaptive sliding mode controllers may be devised for the hybrid chaos synchronization of identical chaotic systems with unknown system parameters.

References

Arneodo, A., Coullet, P., Tresser, C.: Possible new strange attractors with spiral structure. Common. Math. Phys. **79**(4), 573–576 (1981)

Bidarvatan, M., Shahbakhti, M., Jazayeri, S.A., Koch, C.R.: Cycle-to-cycle modeling and sliding mode control of blended-fuel HCCI engine. Control Eng. Pr **24**, 79–91 (2014)

Cai, G., Tan, Z.: Chaos synchronization of a new chaotic system via nonlinear control. J. Uncertain Syst. **1**(3), 235–240 (2007)

Carroll, T.L., Pecora, L.M.: Synchronizing chaotic circuits. IEEE Trans. Circuits Syst. **38**(4), 453–456 (1991)

Chen, G., Ueta, T.: Yet another chaotic attractor. Intern. J. Bifurc. Chaos **9**(7), 1465–1466 (1999)

Chen, H.K., Lee, C.I.: Anti-control of chaos in rigid body motion. Chaos, Solitons Fractals **21**(4), 957–965 (2004)

Chen, W.-H., Wei, D., Lu, X.: Global exponential synchronization of nonlinear time-delay Lure systems via delayed impulsive control. Commun. Nonlinear Sci. Numer. Simul. **19**(9), 3298–3312 (2014)

Das, S., Goswami, D., Chatterjee, S., Mukherjee, S.: Stability and chaos analysis of a novel swarm dynamics with applications to multi-agent systems. Eng. Appl. Artif. Intell. **30**, 189–198 (2014)

Feki, M.: An adaptive chaos synchronization scheme applied to secure communication. Chaos, Solitons Fractals **18**(1), 141–148 (2003)

Feng, Y., Han, F., Yu, X.: Chattering free full-order sliding-mode control. Automatica **50**(4), 1310–1314 (2014)

Gan, Q., Liang, Y.: Synchronization of chaotic neural networks with time delay in the leakage term and parametric uncertainties based on sampled-data control. J. Frankl. Inst. **349**(6), 1955–1971 (2012)

Gaspard, P.: Microscopic chaos and chemical reactions. Physica A: Stat. Mech. Appl **263**(1–4), 315–328 (1999)

Gibson, W.T., Wilson, W.G.: Individual-based chaos: Extensions of the discrete logistic model. J. Theor. Biol. **339**, 84–92 (2013)

Guégan, D.: Chaos in economics and finance. Ann. Rev. Control **33**(1), 89–93 (2009)

Hamayun, M.T., Edwards, C., Alwi, H.: A fault tolerant control allocation scheme with output integral sliding modes. Automatica **49**(6), 1830–1837 (2013)

Huang, J.: Adaptive synchronization between different hyperchaotic systems with fully uncertain parameters. Phys. Lett. A **372**(27–28), 4799–4804 (2008)

Huang, X., Zhao, Z., Wang, Z., Li, Y.: Chaos and hyperchaos in fractional-order cellular neural networks. Neurocomputing **94**, 13–21 (2012)

Jiang, G.-P., Zheng, W.X., Chen, G.: Global chaos synchronization with channel time-delay. Chaos, Solitons Fractals **20**(2), 267–275 (2004)

Karthikeyan, R., Sundarapandian, V.: Hybrid chaos synchronization of four-scroll systems via active control. J. Elect. Eng. **65**(2), 97–103 (2014)

Kaslik, E., Sivasundaram, S.: Nonlinear dynamics and chaos in fractional-order neural networks. Neural Netw **32**, 245–256 (2012)

Kengne, J., Chedjou, J.C., Kenne, G., Kyamakya, K.: Dynamical properties and chaos synchronization of improved Colpitts oscillators. Commun. Nonlinear Sci. Numer Simul. **17**(7), 2914–2923 (2012)

Khalil, H.K.: Nonlinear Systems. Prentice Hall, Upper Saddle River (2001)

Kyriazis, M.: Applications of chaos theory to the molecular biology of aging. Exp. Gerontol. **26**(6), 569–572 (1991)

Li, N., Pan, W., Yan, L., Luo, B., Zou, X.: Enhanced chaos synchronization and communication in cascade-coupled semiconductor ring lasers. Commun. Nonlinear Sci. Numer. Simul. **19**(6), 1874–1883 (2014)

Li, N., Zhang, Y., Nie, Z.: Synchronization for general complex dynamical networks with sampled-data. Neurocomputing **74**(5), 805–811 (2011)

Lian, S., Chen, X.: Traceable content protection based on chaos and neural networks. Appl. Soft Comput. **11**(7), 4293–4301 (2011)
Lin, W.: Adaptive chaos control and synchronization in only locally Lipschitz systems. Phys. Lett. A **372**(18), 3195–3200 (2008)
Liu, C., Liu, T., Liu, L., Liu, K.: A new chaotic attractor. Chaos, Solitions Fractals **22**(5), 1031–1038 (2004)
Liu, L., Zhang, C., Guo, Z.A.: Synchronization between two different chaotic systems with nonlinear feedback control. Chin. Phys. **16**(6), 1603–1607 (2007)
Lorenz, E.N.: Deterministic periodic flow. J. Atmos. Sci. **20**(2), 130–141 (1963)
Lü, J., Chen, G.: A new chaotic attractor coined. Intern. J. Bifurc. Chaos **12**(3), 659–661 (2002)
Lu, W., Li, C., Xu, C.: Sliding mode control of a shunt hybrid active power filter based on the inverse system method. Intern. J. Elect. Power Energy Syst. **57**, 39–48 (2014)
Mondal, S., Mahanta, C.: Adaptive second order terminal sliding mode controller for robotic manipulators. J. Franklin Inst. **351**(4), 2356–2377 (2014)
Murali, K., Lakshmanan, M.: Secure communication using a compound signal from generalized chaotic systems. Phys. Lett. A **241**(6), 303–310 (1998)
Nehmzow, U., Walker, K.: Quantitative description of robotenvironment interaction using chaos theory. Robot. Auton. Syst. **53**(3–4), 177–193 (2005)
Njah, A.N., Ojo, K.S., Adebayo, G.A., Obawole, A.O.: Generalized control and synchronization of chaos in RCL-shunted Josephson junction using backstepping design. Physica C **470**(13–14), 558–564 (2010)
Ouyang, P.R., Acob, J., Pano, V.: PD with sliding mode control for trajectory tracking of robotic system. Robot. Comput. Integr. Manuf. **30**(2), 189–200 (2014)
Pecora, L.M., Carroll, T.L.: Synchronization in chaotic systems. Phys. Rev. Lett. **64**(8), 821–824 (1990)
Perruquetti, W., Barbot, J.P.: Sliding Mode Control in Engineering. Marcel Dekker, New York (2002)
Petrov, V., Gaspar, V., Masere, J., Showalter, K.: Controlling chaos in Belousov-Zhabotinsky reaction. Nature **361**, 240–243 (1993)
Qu, Z.: Chaos in the genesis and maintenance of cardiac arrhythmias. Prog. Biophys. Mol. Biol. **105**(3), 247–257 (2011)
Rafikov, M., Balthazar, J.M.: On control and synchronization in chaotic and hyperchaotic systems via linear feedback control. Commun. Nonlinear Sci. Numer. Simul. **13**(7), 1246–1255 (2007)
Rasappan, S., Vaidyanathan, S.: Global chaos synchronization of WINDMI and Coullet chaotic systems by backstepping control. Far East J. Math. Sci. **67**(2), 265–287 (2012a)
Rasappan, S., Vaidyanathan, S.: Hybrid synchronization of n-scroll Chua and Lur'e chaotic systems via backstepping control with novel feedback. Arch. Control Sci. **22**(3), 343–365 (2012b)
Rhouma, R., Belghith, S.: Cryptoanalysis of a chaos based cryptosystem on DSP. Commun. Nonlinear Sci. Numer. Simul. **16**(2), 876–884 (2011)
Rössler, O.E.: An equation for continuous chaos. Phys. Lett. **57A**(5), 397–398 (1976)
Sarasu, P., Sundarapandian, V.: Adaptive controller design for the generalized projective synchronization of 4-scroll systems. Intern. J. Syst. Signal Control Eng. Appl. **5**(2), 21–30 (2012a)
Sarasu, P., Sundarapandian, V.: Generalized projective synchronization of two-scroll systems via adaptive control. Int. J. Soft Comput. **7**(4), 146–156 (2012b)
Sarasu, P., Sundarapandian, V.: Generalized projective synchronization of two-scroll systems via adaptive control. Eur. J. Sci. Res. **72**(4), 504–522 (2012c)
Shahverdiev, E.M., Bayramov, P.A., Shore, K.A.: Cascaded and adaptive chaos synchronization in multiple time-delay laser systems. Chaos, Solitons Fractals **42**(1), 180–186 (2009)
Shahverdiev, E.M., Shore, K.A.: Impact of modulated multiple optical feedback time delays on laser diode chaos synchronization. Optics Commun. **282**(17), 3568–3572 (2009)
Sharma, A., Patidar, V., Purohit, G., Sud, K.K.: Effects on the bifurcation and chaos in forced Duffing oscillator due to nonlinear damping. Commun. Nonlinear Sci. Numer. Simul. **17**(6), 2254–2269 (2012)

Sprott, J.C.: Some simple chaotic flows. Phys. Rev. E **50**(2), 647–650 (1994)
Sprott, J.C.: Competition with evolution in ecology and finance. Phys. Lett. A **325**(5–6), 329–333 (2004)
Suérez, I.: Mastering chaos in ecology. Ecol. Model. **117**(2–3), 305–314 (1999)
Sundarapandian, V.: Output regulation of the Lorenz attractor. Far East J. Math. Sci. **42**(2), 289–299 (2010)
Sundarapandian, V., Karthikeyan, R.: Hybrid synchronization of hyperchaotic Lorenz and hyperchaotic Chen systems via active control. J. Eng. Appl. Sci. **7**(3), 254–264 (2012)
Sundarapandian, V., Pehlivan, I.: Analysis, control, synchronization, and circuit design of a novel chaotic system. Math. Comput. Model. **55**(7–8), 1904–1915 (2012)
Suresh, R., Sundarapandian, V.: Global chaos synchronizatoin of a family of n-scroll hyperchaotic Chua circuits using backstepping control with recursive feedback. Far East J. Math. Sci. **73**(1), 73–95 (2013)
Tigan, G., Opris, D.: Analysis of a 3D chaotic system. Chaos, Solitons Fractals **36**, 1315–1319 (2008)
Tu, J., He, H., Xiong, P.: Adaptive backstepping synchronization between chaotic systems with unknown Lipschitz constant. Appl. Math. Comput. **236**, 10–18 (2014)
Ucar, A., Lonngren, K.E., Bai, E.W.: Chaos synchronization in RCL-shunted Josephson junction via active control. Chaos, Solitons Fractals **31**(1), 105–111 (2007)
Usama, M., Khan, M.K., Alghatbar, K., Lee, C.: Chaos-based secure satellite imagery cryptosystem. Comput. Math. Appl. **60**(2), 326–337 (2010)
Utkin, V.I.: Sliding Modes in Control and Optimization. Springer, New York (1992)
Vaidyanathan, S.: Anti-synchronization of Newton-Leipnik and Chen-Lee chaotic systems by active control. Intern. J. Control Theory Appl. **4**(2), 131–141 (2011)
Vaidyanathan, S.: Adaptive backstepping controller and synchronizer design for Arneodo chaotic system with unknown parameters. Intern. J. Comput. Sci. Inform. Technol. **4**(6), 145–159 (2012a)
Vaidyanathan, S.: Anti-synchronization of Sprott-L and Sprott-M chaotic systems via adaptive control. Intern. J. Control Theory Appl. **5**(1), 41–59 (2012b)
Vaidyanathan, S.: Output regulation of the Liu chaotic system. Appl. Mech. Mater. **110–116**, 3982–3989 (2012c)
Vaidyanathan, S.: A new six-term 3-D chaotic system with an exponential nonlinearity. Far East J. Math. Sci. **79**(1), 135–143 (2013a)
Vaidyanathan, S.: Analysis and adaptive synchronization of two novel chaotic systems with hyperbolic sinusoidal and cosinusoidal nonlinearity and unknown parameters. J. Eng. Sci. Technol. Rev. **6**(4), 53–65 (2013b)
Vaidyanathan, S.: A new eight-term 3-D polynomial chaotic system with three quadratic nonlinearities. Far East J. Math. Sci. **84**(2), 219–226 (2014)
Vaidyanathan, S., Rajagopal, K.: Global chaos synchronization of four-scroll chaotic systems by active nonlinear control. Intern. J. Control Theory Appl. **4**(1), 73–83 (2011a)
Vaidyanathan, S., Rajagopal, K.: Hybrid synchronization of hyperchaotic Wang-Chen and hyperchaotic Lorenz systems by active nonlinear control. Intern. J. Syst. Signal Control Eng. Appl. **4**(3), 55–61 (2011b)
Vaidyanathan, S., Sampath, S.: Anti-synchronization of four-scroll chaotic systems via sliding mode control. Int. J. Autom. Comput. **9**(3), 274–279 (2012)
Volos, C.K., Kyprianidis, I.M., Stouboulos, I.N.: Experimental investigation on coverage performance of a chaotic autonomous mobile robot. Robot. Autonom. Syst. **61**(12), 1314–1322 (2013)
Wang, F., Liu, C.: A new criterion for chaos and hyperchaos synchronization using linear feedback control. Phys. Lett. A **360**(2), 274–278 (2006)
Witte, C.L., Witte, M.H.: Chaos and predicting varix hemorrhage. Med. Hypotheses **36**(4), 312–317 (1991)
Wu, X., Guan, Z.-H., Wu, Z.: Adaptive synchronization between two different hyperchaotic systems. Nonlinear Analysis: Theory, Methods Appl. **68**(5), 1346–1351 (2008)

Xiao, X., Zhou, L., Zhang, Z.: Synchronization of chaotic Lure systems with quantized sampled-data controller. Commun. Nonlinear Sci. Numer. Simul. **19**(6), 2039–2047 (2014)

Yuan, G., Zhang, X., Wang, Z.: Generation and synchronization of feedback-induced chaos in semiconductor ring lasers by injection-locking. Optik Intern. J. Light Electron Optics **125**(8), 1950–1953 (2014)

Zaher, A.A., Abu-Rezq, A.: On the design of chaos-based secure communication systems. Commun. Nonlinear Syst. Numer. Simul. **16**(9), 3721–3727 (2011)

Zhang, H., Zhou, J.: Synchronization of sampled-data coupled harmonic oscillators with control inputs missing. Syst. Control Lett. **61**(12), 1277–1285 (2012)

Zhang, J., Li, C., Zhang, H., Yu, J.: Chaos synchronization using single variable feedback based on backstepping method. Chaos, Solitons Fractals **21**(5), 1183–1193 (2004)

Zhang, X., Liu, X., Zhu, Q.: Adaptive chatter free sliding mode control for a class of uncertain chaotic systems. Appl. Math. Comput. **232**, 431–435 (2014)

Zhou, W., Xu, Y., Lu, H., Pan, L.: On dynamics analysis of a new chaotic attractor. Phys. Lett. A **372**(36), 5773–5777 (2008)

Global Chaos Control of a Novel Nine-Term Chaotic System via Sliding Mode Control

Sundarapandian Vaidyanathan, Christos K. Volos and Viet-Thanh Pham

Abstract Chaotic systems are nonlinear dynamical systems which are very sensitive to even small changes in the initial conditions. The control of chaotic systems is to design state feedback control laws that stabilize the chaotic systems around the unstable equilibrium points. This work derives a general result for the global chaos control of novel chaotic systems using sliding mode control. The main result has been proved using Lyapunov stability theory. Sliding mode control (SMC) is well-known as a robust approach and useful for controller design in systems with parameter uncertainties. Next, a novel nine-term 3-D chaotic system has been proposed in this paper and its properties have been detailed. The Lyapunov exponents of the novel chaotic system are found as $L_1 = 6.8548$, $L_2 = 0$ and $L_3 = -32.8779$ and the Lyapunov dimension of the novel chaotic system is found as $D_L = 2.2085$. The maximal Lyapunov exponent of the novel chaotic system is $L_1 = 6.8548$. As an application of the general result derived in this work, a sliding mode controller is derived for the global chaos control of the identical novel chaotic systems. MATLAB simulations have been provided to illustrate the qualitative properties of the novel 3-D chaotic system and the sliding controller results for the stabilizing control developed for the novel 3-D chaotic system.

S. Vaidyanathan (✉)
Research and Development Centre, Vel Tech University, Avadi, Chennai
600062, Tamil Nadu, India
e-mail: sundarvtu@gmail.com

C.K. Volos
Faculty of Sciences, School of Physics, Aristotle University of Thessaloniki,
14451 Thessaloniki, Greece
e-mail: chvolos@gmail.com

V.-T. Pham
School of Electronics and Telecommunications, Hanoi University of Science and Technology,
01 Dai Co Viet, Hanoi, Vietnam
e-mail: pvt3010@gmail.com

A.T. Azar and Q. Zhu (eds.), *Advances and Applications in Sliding Mode Control systems*,
Studies in Computational Intelligence 576, DOI 10.1007/978-3-319-11173-5_21

1 Introduction

Chaotic systems are nonlinear dynamical systems which are very sensitive to even small changes in the initial conditions. Sensitivity to initial conditions of chaotic systems is popularly known as the *butterfly effect*. Small changes in an initial state will make a very large difference in the behavior of the chaotic system at future states.

Chaotic systems are nonlinear dynamical systems which are sensitive to initial conditions, topologically mixing and with dense periodic orbits.

Chaotic behaviour was suspected well over 100 years ago in the study of three bodies problem, but it was established only a few decades ago in the study of 3-D weather models (Lorenz 1963).

The Lyapunov exponent is a measure of the divergence of phase points that are initially very close and can be used to quantify chaotic systems. It is common to refer to the largest Lyapunov exponent as the maximal Lyapunov exponent (MLE). A positive maximal Lyapunov exponent and phase space compactness are usually taken as defining conditions for a chaotic system. A chaotic system is also defined as a nonlinear dynamical system having at least one positive Lyapunov exponent.

Since the discovery of Lorenz system in 1963, there is a great deal of interest in the chaos literature in finding new chaotic systems. Some well-known paradigms of 3-D chaotic systems in the literature are Rössler (1976), Arneodo et al. (1981), Sprott (1994), Chen and Ueta (1999), Lü and Chen (2002), Liu et al. (2004), Cai and Tan (2007), Chen and Lee (2004), Tigan and Opris (2008), Zhou et al. (2008), Sundarapandian and Pehlivan (2012), Vaidyanathan (2013a, b, 2014), Zhu et al. (2010), Li (2008).

This paper introduces a novel nine-term 3-D chaotic system having four nonlinearities. The Lyapunov exponents of the novel 3-D chaotic system are found as $L_1 = 6.8548$, $L_2 = 0$ and $L_3 = -32.8779$ and the Lyapunov dimension of the novel chaotic system is found as $D_L = 2.2085$. The maximal Lyapunov exponent of the novel chaotic system is $L_1 = 6.8548$.

Chaotic systems have several important applications in science and engineering such as oscillators (Kengne et al. 2012; Sharma et al. 2012), lasers (Yuan et al. 2014; Li et al. 2014), chemical reactions (Petrov et al. 1993; Gaspard 1999), cryptosystems (Usama et al. 2010; Rhouma and Belghith 2011), secure communications (Murali and Lakshmanan 1998; Feki 2003; Zaher and Abu-Rezq 2011), biology (Das et al. 2014; Kyriazis 1991), ecology (Suérez 1999; Gibson and Wilson 2013), robotics (Nehmzow and Walker 2005; Volos et al. 2013; Mondal and Mahanta 2014), cardiology (Qu 2011; Witte and Witte 1991), neural networks (Kaslik and Sivasundaram 2012; Huang et al. 2012; Lian and Chen 2011), finance (Sprott 2004; Guégan 2009), etc.

The control of chaotic systems is to design state feedback control laws that stabilize the chaotic systems around the unstable equilibrium pints. Chaos and control of chaotic dynamical systems that have both received great attention in the last few decades (Alekseev and Loskutov 1987; Lima and Pettini 1990; Weeks and Burgess

1997; Lima and Pettini 1998; Basios et al. 1998; Mirus and Sprott 1999; Ge et al. 2000; Sun and Cao 2008; Sundarapandian and Pehlivan 2012).

Synchronization of chaotic systems is a phenomenon that occurs when two or more chaotic systems are coupled or when a chaotic system drives another chaotic system. Because of the butterfly effect which causes exponential divergence of the trajectories of two identical chaotic systems started with nearly the same initial conditions, the synchronization of chaotic systems is a challenging research problem in the chaos literature.

Pecora and Carroll pioneered the research on synchronization of chaotic systems with their seminal papers in 1990s (Pecora and Carroll 1990; Carroll and Pecora 1991). The active control method (Ucar et al. 2007; Liu et al. 2007; Sundarapandian 2010; Vaidyanathan 2012c; Wang and Liu 2006; Rafikov and Balthazar 2007) is commonly used when the system parameters are available for measurement and the adaptive control method (Wu et al. 2008; Huang 2008; Lin 2008; Sarasu and Sundarapandian 2012a, b, c) is commonly used when some or all the system parameters are not available for measurement and estimates for unknown parameters of the systems.

Other popular methods for chaos synchronization are the sampled-data feedback method (Xiao et al. 2014; Zhang and Zhou 2012; Li et al. 2011; Gan and Liang 2012), time-delay feedback method (Shahverdiev and Shore 2009; Jiang et al. 2004; Chen et al. 2014; Shahverdiev et al. 2009), backstepping method (Njah et al. 2010; Tu et al. 2014; Zhang et al. 2004; Vaidyanathan 2012a), etc.

Complete synchronization (Vaidyanathan and Rajagopal 2011; Rasappan and Vaidyanathan 2012; Suresh and Sundarapandian 2013) is characterized by the equality of state variables evolving in time, while anti-synchronization (Vaidyanathan 2011; Vaidyanathan and Sampath 2012; Vaidyanathan 2012b) is characterized by the disappearance of the sum of relevant state variables evolving in time.

Both control and synchronization of chaotic systems are important research problems. This work deals with the global chaos control of the chaotic systems. Explicitly, sliding mode control theory has been used for the derivation of state feedback-based sliding controllers for the global stabilization of the chaotic systems about the unstable equilibrium points.

This research work is organized as follows. Section 2 gives an introduction to sliding mode control for global chaos control of nonlinear control systems. Section 3 contains the main result of this work, namely, sliding controller design for the global chaos control of chaotic systems. Section 4 introduces the novel nine-term 3-D chaotic system with four quadratic nonlinearities. Section 5 details the qualitative properties of the novel 3-D chaotic system. The Lyapunov exponents of the novel chaotic system are found as $L_1 = 6.8548$, $L_2 = 0$ and $L_3 = -32.8779$ and the Lyapunov dimension of the novel chaotic system is found as $D_L = 2.2085$. The maximal Lyapunov exponent of the novel chaotic system is $L_1 = 6.8548$. In Sect. 6, we describe the sliding mode controller design for the global chaos control of the novel nine-term 3-D chaotic system. Section 7 contains a summary of the main results derived in this research work.

2 Sliding Mode Controller for Global Chaos Control

In control theory, the sliding mode control approach is recognized as an efficient tool for designing robust controllers for linear or nonlinear control systems operating under uncertainty conditions (Perruquetti and Barbot 2002; Utkin 1992).

The basic steps of sliding mode control theory originated in the early 1950s and this was initiated by S.V. Emel'yanov as *Variable Structure Control* (Itkis 1976; Utkin 1978; Zinober 1993). Variable structure control (VSC) is a form of discontinuous nonlinear control and this method alters the dynamics of a nonlinear system by application of a high-frequency switching control. The state-feedback control law is not a continuous function of time and it switches from one smooth condition to another. So the structure of the variable structure control law varies based on the position of the state trajectory.

Sliding mode control method has a major advantage of low sensitivity to parameter variations in the plant and disturbances affecting the plant, which eliminates the necessity of exact modeling of the plant.

In the sliding mode control theory, the control dynamics has two sequential modes, viz. (i) the *reaching mode*, and (ii) the *sliding mode*.

Basically, a sliding mode controller (SMC) design consists of two parts: hyperplane (or sliding surface) design and controller design.

A hyperplane is first designed via the pole-placement approach in the modern control theory and a controller is then designed based on the sliding condition. The stability of the overall control system is ensured by the sliding condition and by a stable hyperplane.

Sliding mode control theory has been used to deal with many research problems of control literature (Feng et al. 2014; Ouyang et al. 2014; Bidarvatan et al. 2014; Lu et al. 2014; Zhang et al. 2014; Hamayun et al. 2013).

3 Sliding Mode Controller Design for Global Chaos Control

In this work, we consider a controlled chaotic system given by the dynamics

$$\dot{x} = Ax + f(x) + u, \tag{1}$$

where $x \in \mathbb{R}^n$ is the state of the system, A is the $n \times n$ matrix of system parameters and f is a vector field that contains the nonlinear parts of the system and satisfies $f(0) = 0$. Also, u is the sliding mode controller to be designed.

The design problem is to determine a feedback control u so that the plant dynamics (1) is globally asymptotically stable at the origin for all initial conditions $x(0) \in \mathbb{R}^n$.

For the SMC design for the global chaos control of the system (1), the control u is taken as

$$u(t) = -f(x) + Bv(t) \tag{2}$$

In Eq. (2), B is an $(n \times 1)$ column vector chosen such that (A, B) is controllable. Upon substituting (2) into (1), the closed-loop plant dynamics is obtained as

$$\dot{x} = Ax + Bv, \tag{3}$$

which is a linear time-invariant control system with a single input v.

Hence, by the use of the nonlinear control law (2), original problem of global chaos control of the chaotic system (1) has been converted into an equivalent problem of globally stabilizing the linear dynamics (3).

In the sliding controller design, the sliding variable is first defined as

$$s(x) = Cx = c_1x_1 + c_2x_2 + \cdots + c_nx_n, \tag{4}$$

where C is an $(1 \times n)$ row vector to be determined.

The sliding manifold S is defined as the hyperplane

$$S = \{x \in \mathbb{R}^n : s(x) = Cx = 0\} \tag{5}$$

If a sliding motion occurs on S, then the sliding mode conditions must be satisfied, which are given by

$$s \equiv 0 \text{ and } \dot{s} = CAx + CBv = 0 \tag{6}$$

It is assumed that the row vector C is chosen so that $CB \neq 0$.

The sliding motion is affected by the so-called equivalent control given by

$$v_{\text{eq}}(t) = -(CB)^{-1}CAx(t) \tag{7}$$

As a consequence, the equivalent dynamics in the sliding phase is defined by

$$\dot{x} = \left[I - B(CB)^{-1}C\right]Ax = Ex, \tag{8}$$

where

$$E = \left[I - B(CB)^{-1}C\right]A \tag{9}$$

It can be easily verified that E is independent of the control and has at most $(n-1)$ nonzero eigenvalues, depending on the chosen switching surface, while the associated eigenvectors belong to $\ker(C)$.

Since (A, B) is controllable, the matrices B and C can be chosen so that E has any desired $(n-1)$ stable eigenvalues.

Thus, the dynamics in the sliding mode is globally asymptotically stable.

Finally, for the sliding mode controller (SMC) design, the constant plus proportional rate reaching law is used, which is given by

$$\dot{s} = -\beta \operatorname{sgn}(s) - \alpha s \tag{10}$$

where $\operatorname{sgn}(\cdot)$ denotes the sign function and the gains $\alpha > 0$, $\beta > 0$ are found so that the sliding condition is satisfied and the sliding motion will occur.

From the Eqs. (6) and (10), sliding control v is found as

$$CAx + CBv = -\beta \operatorname{sgn}(s) - \alpha s \tag{11}$$

Since $s = Cx$, the Eq. (11) can be simplified to get

$$v = -(CB)^{-1}\left[C(\alpha I + A)x + \beta \operatorname{sgn}(s)\right] \tag{12}$$

Next, the main result of this section is established as follows.

Theorem 1 *A sliding mode control law that achieves global chaos control for the chaotic system* (1) *for all initial conditions* $x(0)$ *in* $\mathbb{R}^n$ *is given by the equation*

$$u(t) = -f(x(t)) + Bv(t), \tag{13}$$

where v *is defined by* (12), B *is an* $(n \times 1)$ *vector such that* (A, B) *is controllable,* C *is an* $(1 \times n)$ *vector such that* $CB \neq 0$ *and that the matrix* E *defined by* Eq. (9) *has* $(n - 1)$ *stable eigenvalues.*

Proof The proof is carried out using Lyapunov stability theory (Khalil 2001).

Substituting the sliding control law (13) into the plant dynamics (1) leads to the closed-loop dynamics

$$\dot{x} = Ax + Bv \tag{14}$$

Substituting for v from (12) into (14), the closed-loop plant dynamics is obtained as

$$\dot{x} = Ax - B(CB)^{-1}\left[C(\alpha I + A)x + \beta \operatorname{sgn}(s)\right] \tag{15}$$

The global asymptotic stability of the error system (15) is proved by taking the candidate Lyapunov function

$$V(x) = \frac{1}{2}\, s^2(x), \tag{16}$$

which is a non-negative definite function on $\mathbb{R}^n$.

It is noted that

$$V(x) = 0 \iff s(x) = 0 \tag{17}$$

The sliding mode motion is characterized by the equations

$$s(x) = 0 \text{ and } \dot{x}(e) = 0 \tag{18}$$

By the choice of E, the dynamics in the sliding mode given by (8) is globally asymptotically stable.

When $s(x) \neq 0$, $V(x) > 0$.

Also, when $s(x) \neq 0$, differentiating V along the plant dynamics (15) or the equivalent dynamics (10), the following dynamics is obtained:

$$\dot{V} = s\dot{s} = -\beta s \,\operatorname{sgn}(s) - \alpha s^2 < 0 \tag{19}$$

Hence, by Lyapunov stability theory (Khalil 2001), it is concluded that the closed-loop plant dynamics (15) is globally asymptotically stable for all initial conditions $x(0) \in \mathbb{R}^n$.

This completes the proof. □

4 A Novel 3-D Chaotic System

The nine-term novel chaotic system is a described by the 3-D dynamics

$$\begin{aligned} \dot{x}_1 &= a(x_2 - x_1) - x_2 x_3, \\ \dot{x}_2 &= b x_1 - x_2 - x_1 x_3, \\ \dot{x}_3 &= -c x_3 + d(x_2^2 - x_1^2), \end{aligned} \tag{20}$$

where x_1, x_2, x_3 are the states and a, b, c, d are constant, positive, parameters.

The system (20) is a nine-term polynomial chaotic system with four quadratic nonlinearities.

The system (20) depicts a strange chaotic attractor when the constant parameter values are taken as

$$a = 22, \;\; b = 600, \;\; c = 3, \;\; d = 11 \tag{21}$$

For simulations, the initial values of the novel 3-D chaotic system (20) are taken as

$$x_1(0) = 1.6, \;\; x_2(0) = 0.8, \;\; x_3(0) = 1.2 \tag{22}$$

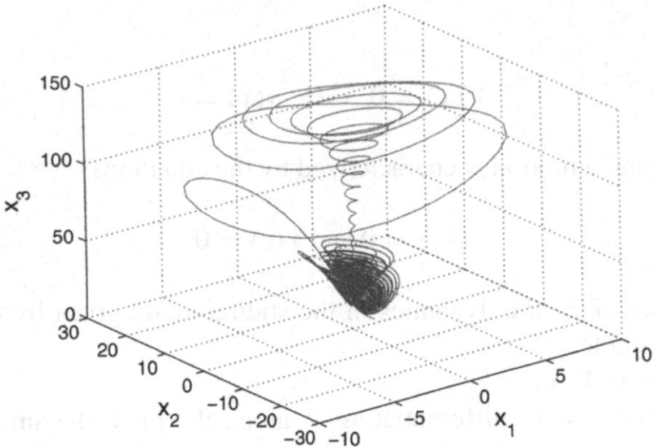

Fig. 1 Strange attractor of the novel chaotic system in $\mathbf{R}^3$

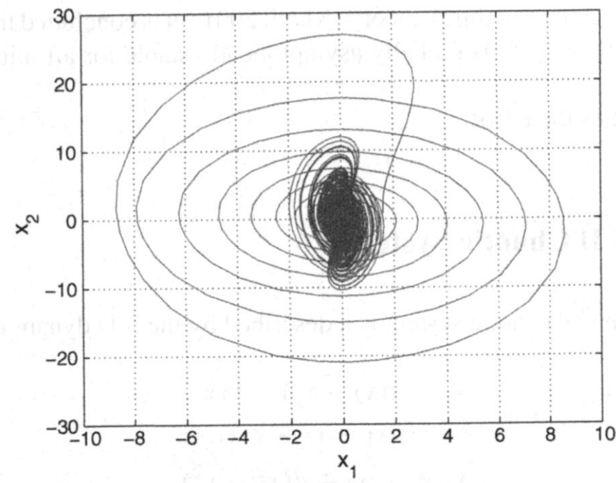

Fig. 2 2-D projection of the novel chaotic system in (x_1, x_2)-plane

The novel 3-D chaotic system (20) exhibits a three-scroll chaotic attractor. Figure 1 describes the three-scroll chaotic attractor of the novel chaotic system (20) in 3-D view.

Figure 2 describes the 2-D projection of the strange chaotic attractor of the novel system (20) in (x_1, x_2)-plane. In the projection on the (x_1, x_2)-plane, a three-scroll chaotic attractor is clearly seen.

Figure 3 describes the 2-D projection of the strange chaotic attractor of the novel system (20) in (x_2, x_3)-plane. In the projection on the (x_2, x_3)-plane, a three-scroll chaotic attractor is clearly seen.

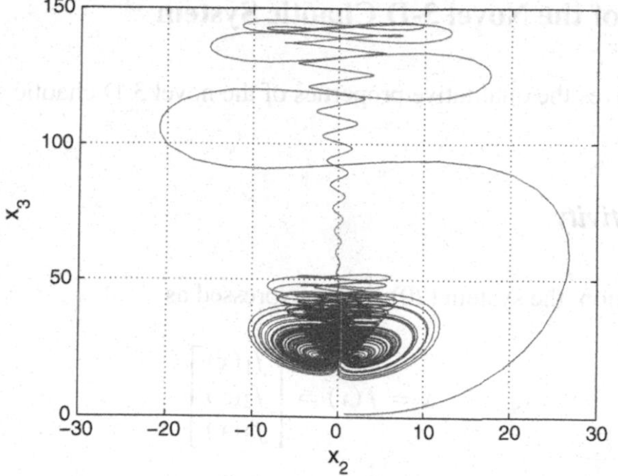

Fig. 3 2-D projection of the novel chaotic system in (x_2, x_3)-plane

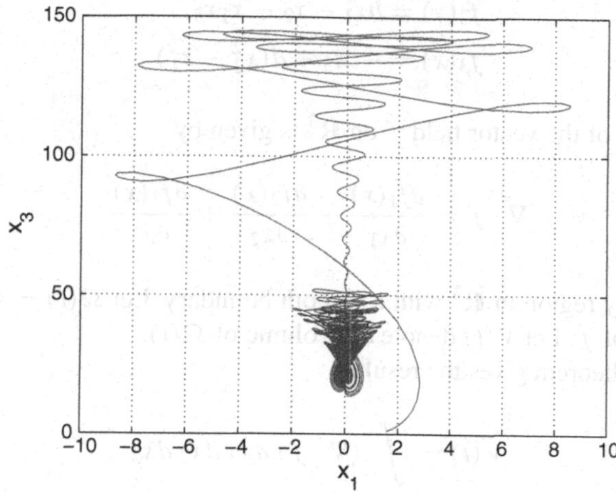

Fig. 4 2-D projection of the novel chaotic system in (x_1, x_3)-plane

Figure 4 describes the 2-D projection of the strange chaotic attractor of the novel system (20) in (x_1, x_3)-plane. In the projection on the (x_2, x_3)-plane, a three-scroll chaotic attractor is clearly seen.

Combining, Figs. 1, 2, 3, and 4 represent a strongly chaotic system given by (20).

5 Analysis of the Novel 3-D Chaotic System

This section gives the qualitative properties of the novel 3-D chaotic system.

5.1 Dissipativity

In vector notation, the system (20) can be expressed as

$$\dot{x} = f(x) = \begin{bmatrix} f_1(x) \\ f_2(x) \\ f_3(x) \end{bmatrix} \tag{23}$$

where

$$\begin{aligned} f_1(x) &= a(x_2 - x_1) - x_2 x_3 \\ f_2(x) &= b x_1 - x_2 - x_1 x_3 \\ f_3(x) &= -c x_3 + d(x_2^2 - x_1^2) \end{aligned} \tag{24}$$

The divergence of the vector field f on $\mathbb{R}^3$ is given by

$$\nabla \cdot f = \frac{\partial f_1(x)}{\partial x_1} + \frac{\partial f_2(x)}{\partial x_2} + \frac{\partial f_3(x)}{\partial x_3} \tag{25}$$

Let Ω be any region in $\mathbb{R}^3$ with a smooth boundary. Let $\Omega(t) = \Phi_t(\Omega)$, where Ω_t is the flow of f. Let $V(t)$ denote the volume of $\Omega(t)$.

Liouville's theorem gives the result

$$\dot{V}(t) = \int\limits_{\Omega(t)} (\nabla \cdot f)\, dx_1\, dx_2\, dx_3 \tag{26}$$

The divergence of the flow of the novel system (20) is determined as

$$\nabla \cdot f = \frac{\partial f_1(x)}{\partial x_1} + \frac{\partial f_2(x)}{\partial x_2} + \frac{\partial f_3(x)}{\partial x_3} = -a - 1 - c = -\mu < 0 \tag{27}$$

where

$$\mu = a + 1 + c > 0 \tag{28}$$

as a, b, c, d are positive parameters.

Substituting the value of $(\nabla \cdot f)$ in (26), it follows that

$$\dot{V}(t) = \int\limits_{\Omega(t)} (-\mu)\, dx_1\, dx_2\, dx_3 = -\mu V(t) \tag{29}$$

Integration of the linear differential Eq. (29) yields

$$V(t) = \exp(-\mu t) V(0) \tag{30}$$

Since $\mu > 0$, it follows from Eq. (30) that $V(t) \to 0$ exponentially as $t \to \infty$. Thus, the novel 3-D chaotic system (20) is dissipative. Thus, the system limit sets are ultimately confined into a specific limit set of zero volume, and the asymptotic motion of the novel chaotic system (30) settles onto a strange attractor of the system.

5.2 Symmetry and Invariance

The novel chaotic system (20) is invariant under the coordinates transformation

$$(x_1, x_2, x_3) \to (-x_1, -x_2, x_3). \tag{31}$$

The transformation (31) persists for all values of the system parameters. Thus, the novel system (20) has rotation symmetry about the x_3-axis. As a consequence, any non-trivial trajectory of the system (20) must have a twin trajectory.

It is easy to check that the x_3-axis is invariant for the flow of the novel chaotic system (20). Hence, all orbits of the system (20) starting from the x_3 axis stay in the x_3 axis for all values of time.

5.3 Equilibria

The equilibrium points of the novel 3-D chaotic system (20) are obtained by solving the nonlinear equations

$$\begin{aligned} f_1(x) &= a(x_2 - x_1) - x_2 x_3 = 0 \\ f_2(x) &= b x_1 - x_2 - x_1 x_3 = 0 \\ f_3(x) &= -c x_3 + d(x_2^2 - x_1^2) = 0 \end{aligned} \tag{32}$$

We take the parameter values as in the chaotic case, viz.

$$a = 22, \quad b = 600, \quad c = 3, \quad d = 11 \tag{33}$$

Solving the nonlinear system of equations (32) with the parameter values (33), the equilibrium points of the novel chaotic system (20) are obtained as

$$E_0 = \begin{bmatrix} 0 \\ 0 \\ 0 \end{bmatrix}, \quad E_1 = \begin{bmatrix} 0.0042 \\ 2.4474 \\ 21.9619 \end{bmatrix} \text{ and } E_2 = \begin{bmatrix} -0.0042 \\ -2.4474 \\ 21.9619 \end{bmatrix}. \tag{34}$$

The Jacobian matrix of the novel chaotic system (20) at $(x_1^\star, x_2^\star, x_3^\star)$ is obtained as

$$J(x^\star) = \begin{bmatrix} -22 & 22 - x_3^\star & -x_2^\star \\ 600 - x_3^\star & -1 & -x_1^\star \\ -22x_1^\star & 22x_2^\star & -3 \end{bmatrix} \tag{35}$$

The Jacobian matrix at E_0 is obtained as

$$J_0 = J(E_0) = \begin{bmatrix} -22 & 22 & 0 \\ 600 & -1 & 0 \\ 0 & 0 & -3 \end{bmatrix} \tag{36}$$

The matrix J_0 has the eigenvalues

$$\lambda_1 = -126.8701, \quad \lambda_2 = -3, \quad \lambda_3 = 103.8701 \tag{37}$$

This shows that the equilibrium point E_0 is a saddle-point, which is unstable. The Jacobian matrix at E_1 is obtained as

$$J_1 = J(E_1) = \begin{bmatrix} -22.0000 & 0.0381 & -2.4474 \\ 578.0381 & -1.0000 & -0.0042 \\ -0.0924 & 53.8428 & -3.0000 \end{bmatrix} \tag{38}$$

The matrix J_1 has the eigenvalues

$$\lambda_1 = -52.4131, \quad \lambda_{2,3} = 13.2065 \pm 35.7625i \tag{39}$$

This shows that the equilibrium point E_1 is a saddle-focus, which is unstable. The Jacobian matrix at E_2 is obtained as

$$J_2 = J(E_2) = \begin{bmatrix} -22.0000 & 0.0381 & 2.4474 \\ 578.0381 & -1.0000 & 0.0042 \\ 0.0924 & -53.8428 & -3.0000 \end{bmatrix} \tag{40}$$

The matrix J_2 has the eigenvalues

$$\lambda_1 = -52.4131, \quad \lambda_{2,3} = 13.2065 \pm 35.7625i \tag{41}$$

This shows that the equilibrium point E_2 is a saddle-focus, which is unstable.

Hence, E_0, E_1, E_2 are all unstable equilibrium points of the novel 3-D chaotic system (20), where E_0 is a saddle point and E_1, E_2 are saddle-focus points.

5.4 Lyapunov Exponents and Lyapunov Dimension

We take the initial values of the novel chaotic system (20) as in (22) and the parameter values of the novel chaotic system (20) as (21).

Then the Lyapunov exponents of the novel chaotic system (20) are numerically obtained as

$$L_1 = 6.8548, \quad L_2 = 0, \quad L_3 = -32.8779 \tag{42}$$

Thus, the maximal Lyapunov exponent of the novel chaotic system (20) is $L_1 = 6.8548$.

Since $L_1 + L_2 + L_3 = -26.0231 < 0$, the system (20) is dissipative.

Also, the Lyapunov dimension of the system (20) is obtained as

$$D_L = 2 + \frac{L_1 + L_2}{|L_3|} = 2.2085 \tag{43}$$

Figure 5 depicts the dynamics of the Lyapunov exponents of the novel chaotic system (20).

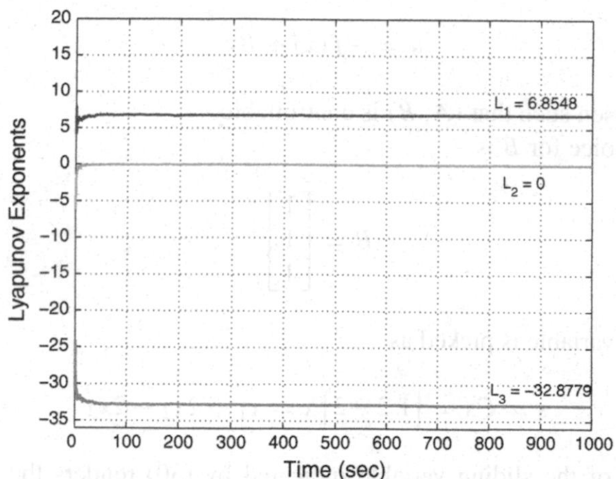

Fig. 5 Dynamics of the Lyapunov exponents of the novel chaotic system

6 Global Chaos Control of the Novel Chaotic System via SMC

This section details the construction of a sliding mode controller for the global chaos control of the novel 3-D chaotic system (20).

Thus, the controlled novel chaotic system is taken as

$$\begin{aligned} \dot{x}_1 &= a(x_2 - x_1) - x_2 x_3 + u_1, \\ \dot{x}_2 &= b x_1 - x_2 - x_1 x_3 + u_2, \\ \dot{x}_3 &= -c x_3 + d(x_2^2 - x_1^2) + u_3, \end{aligned} \tag{44}$$

where a, b, c, d are constant, positive parameters and u_1, u_2, u_3 are the sliding controllers to be designed.

The plant dynamics (44) can be expressed in matrix form as

$$\dot{x} = Ax + f(x) + u \tag{45}$$

where

$$A = \begin{bmatrix} -a & a & 0 \\ b & -1 & 0 \\ 0 & 0 & -c \end{bmatrix}, \quad f(x) = \begin{bmatrix} -x_2 x_3 \\ -x_1 x_3 \\ d(x_2^2 - x_1^2) \end{bmatrix}, \quad u = \begin{bmatrix} u_1 \\ u_2 \\ u_3 \end{bmatrix} \tag{46}$$

The parameter values of a, b, c, d are taken as in the chaotic case, i.e.

$$a = 22, \quad b = 600, \quad c = 3, \quad d = 11 \tag{47}$$

First, the control u is set as

$$u = -f(x) + Bv, \tag{48}$$

where B is chosen such that (A, B) is controllable.

A simple choice for B is

$$B = \begin{bmatrix} 1 \\ 1 \\ 1 \end{bmatrix} \tag{49}$$

The sliding variable is picked as

$$s = Cx = \begin{bmatrix} 1 & 2 & -2 \end{bmatrix} x = x_1 + 2x_2 - 2x_3 \tag{50}$$

The choice of the sliding variable indicated by (50) renders the sliding mode dynamics globally asymptotically stable.

Next, we choose the SMC gains as

$$\alpha = 5 \text{ and } \beta = 0.2 \tag{51}$$

Using the formula (12), the control v is obtained as

$$v(t) = -1183x_1 - 30x_2 + 4x_3 - 0.2\text{sgn}(s) \tag{52}$$

As a consequence of Theorem 1 (Sect. 3), the following result is obtained.

Theorem 2 *The control law defined by* (48), *where v is defined by* (52), *renders the novel 3-D chaotic system* (44) *globally and asymptotically stable for all values of the initial states* $x(0) \in \mathbb{R}^3$. □

For numerical simulations, the classical fourth-order Runge–Kutta method with step-size $h = 10^{-8}$ is used in the MATLAB software.

The parameter values are taken as in the chaotic case for the novel chaotic system (44), i.e.

$$a = 22, \quad b = 600, \quad c = 3, \quad d = 11$$

The sliding mode gains are taken as $\alpha = 5$ and $\beta = 0.2$.

The initial values of the master system (44) are taken as

$$x_1(0) = 5.4 \quad x_2(0) = 8.5, \quad x_3(0) = -4.7$$

Figure 6 shows the global chaos control of the novel chaotic system (44). It is clear from Fig. 6 that the controlled states $x_1(t)$, $x_2(t)$ and $x_3(t)$ converge to zero in 1.5 s.

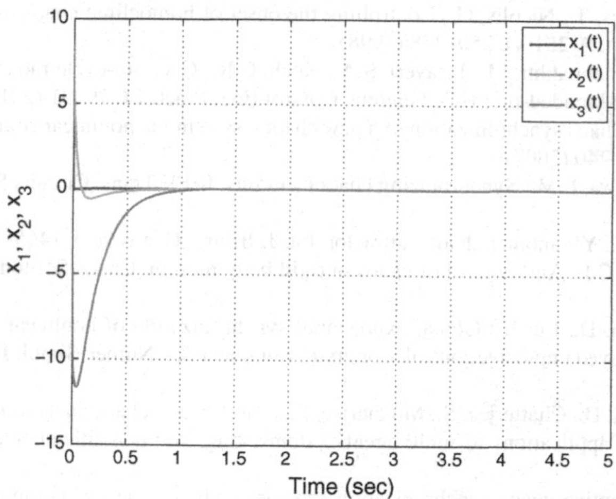

Fig. 6 Time-history of the controlled states $x_1(t)$, $x_2(t)$, $x_3(t)$

7 Conclusions

The control of chaotic systems is to design state feedback control laws that stabilize the chaotic systems around the unstable equilibrium points. In this work, a general result has been derived for the global chaos control of chaotic systems using sliding mode control. The main result has been proved using Lyapunov stability theory. Sliding mode control (SMC) is well-known as a robust approach and useful for controller design in systems with parameter uncertainties. Next, as an application of the main result, a global chaos controller has been derived for the nine-term polynomial novel 3-D chaotic system proposed in this work. The Lyapunov exponents of the novel chaotic system were found as $L_1 = 6.8548$, $L_2 = 0$ and $L_3 = -32.8779$ and the Lyapunov dimension of the novel chaotic system was found as $D_L = 2.2085$. The maximal Lyapunov exponent of the novel chaotic system was found as $L_1 = 6.8548$. As an application of the general result derived in this work, a sliding mode controller has been derived for the global chaos control of the novel 3-D chaotic system. MATLAB simulations have been provided to depict the basic qualitative properties of the novel 3-D chaotic system and the global controller results for the novel chaotic system. As future research, adaptive sliding mode controllers may be devised for the global chaos control of chaotic systems with unknown system parameters.

References

Alekseev, V.V., Loskutov, A.Y.: Control of a system with a strange attractor through periodic parametric perturbations. Sov. Phys. Dokl. **32**, 270–271 (1987)

Arneodo, A., Coullet, P., Tresser, C.: Possible new strange attractors with spiral structure. Common. Math. Phys. **79**(4), 573–576 (1981)

Basios, V., Bountis, T., Nicolis, G.: Controlling the onset of homoclinic chaos due to parametric noise. Phys. Lett. A **251**(4), 250–258 (1998)

Bidarvatan, M., Shahbakhti, M., Jazayeri, S.A., Koch, C.R.: Cycle-to-cycle modeling and sliding mode control of blended-fuel HCCI engine. Control Eng. Pract. **24**, 79–91 (2014)

Cai, G., Tan, Z.: Chaos synchronization of a new chaotic system via nonlinear control. J. Uncertain Syst. **1**(3), 235–240 (2007)

Carroll, T.L., Pecora, L.M.: Synchronizing chaotic circuits. IEEE Trans. Circuits Syst. **38**(4), 453–456 (1991)

Chen, G., Ueta, T.: Yet another chaotic attractor. Int. J. Bifurc. Chaos **9**(7), 1465–1466 (1999)

Chen, H.K., Lee, C.I.: Anti-control of chaos in rigid body motion. Chaos Solitons Fractals **21**(4), 957–965 (2004)

Chen, W.-H., Wei, D., Lu, X.: Global exponential synchronization of nonlinear time-delay Lure systems via delayed impulsive control. Commun. Nonlinear Sci. Numer. Simul. **19**(9), 3298–3312 (2014)

Das, S., Goswami, D., Chatterjee, S., Mukherjee, S.: Stability and chaos analysis of a novel swarm dynamics with applications to multi-agent systems. Eng. Appl. Artificial Intell. **30**, 189–198 (2014)

Feki, M.: An adaptive chaos synchronization scheme applied to secure communication. Chaos Solitons Fractals **18**(1), 141–148 (2003)

Feng, Y., Han, F., Yu, X.: Chattering free full-order sliding-mode control. Automatica **50**(4), 1310–1314 (2014)
Gan, Q., Liang, Y.: Synchronization of chaotic neural networks with time delay in the leakage term and parametric uncertainties based on sampled-data control. J. Frankl. Inst. **349**(6), 1955–1971 (2012)
Gaspard, P.: Microscopic chaos and chemical reactions. Phys. A: Stat. Mech. Its Appl. **263**(1–4), 315–328 (1999)
Ge, S.S., Wang, C., Lee, T.H.: Adaptive backstepping control of a class of chaotic systems. Int. J. Bifurc. Chaos **10**(5), 1149–1156 (2000)
Gibson, W.T., Wilson, W.G.: Individual-based chaos: Extensions of the discrete logistic model. J. Theor. Biol. **339**, 84–92 (2013)
Guégan, D.: Chaos in economics and finance. Ann. Rev. Control **33**(1), 89–93 (2009)
Hamayun, M.T., Edwards, C., Alwi, H.: A fault tolerant control allocation scheme with output integral sliding modes. Automatica **49**(6), 1830–1837 (2013)
Huang, J.: Adaptive synchronization between different hyperchaotic systems with fully uncertain parameters. Phys. Lett. A **372**(27–28), 4799–4804 (2008)
Huang, X., Zhao, Z., Wang, Z., Li, Y.: Chaos and hyperchaos in fractional-order cellular neural networks. Neurocomputing **94**, 13–21 (2012)
Itkis, U.: Control Systems of Variable Structure. Wiley, New York (1976)
Jiang, G.-P., Zheng, W.X., Chen, G.: Global chaos synchronization with channel time-delay. Chaos Solitons Fractals **20**(2), 267–275 (2004)
Kaslik, E., Sivasundaram, S.: Nonlinear dynamics and chaos in fractional-order neural networks. Neural Netw. **32**, 245–256 (2012)
Kengne, J., Chedjou, J.C., Kenne, G., Kyamakya, K.: Dynamical properties and chaos synchronization of improved Colpitts oscillators. Commun. Nonlinear Sci. Numer. Simul. **17**(7), 2914–2923 (2012)
Khalil H. K.: Nonlinear Systems. Prentice Hall, Englewood Cliffs (2001)
Kyriazis, M.: Applications of chaos theory to the molecular biology of aging. Experimental Gerontol. **26**(6), 569–572 (1991)
Li, D.: A three-scroll chaotic attractor. Phys. Lett. A **372**(4), 387–393 (2008)
Li, N., Pan, W., Yan, L., Luo, B., Zou, X.: Enhanced chaos synchronization and communication in cascade-coupled semiconductor ring lasers. Commun. Nonlinear Sci. Numer. Simul. **19**(6), 1874–1883 (2014)
Li, N., Zhang, Y., Nie, Z.: Synchronization for general complex dynamical networks with sampled-data. Neurocomputing **74**(5), 805–811 (2011)
Lian, S., Chen, X.: Traceable content protection based on chaos and neural networks. Appl. Soft Comput. **11**(7), 4293–4301 (2011)
Lima, R., Pettini, M.: Suppression of chaos by resonant parameteric perturbations. Phys. Rev. A **41**, 726–733 (1990)
Lima, R., Pettini, M.: Parametric resonant control of chaos. Int. J. Bifurc. Chaos **8**(8), 1675–1684 (1998)
Lin, W.: Adaptive chaos control and synchronization in only locally Lipschitz systems. Phys. Lett. A **372**(18), 3195–3200 (2008)
Liu, C., Liu, T., Liu, L., Liu, K.: A new chaotic attractor. Chaos Solitions Fractals **22**(5), 1031–1038 (2004)
Liu, L., Zhang, C., Guo, Z.A.: Synchronization between two different chaotic systems with nonlinear feedback control. Chin. Phys. **16**(6), 1603–1607 (2007)
Lorenz, E.N.: Deterministic periodic flow. J. Atmos. Sci. **20**(2), 130–141 (1963)
Lü, J., Chen, G.: A new chaotic attractor coined. Int. J. Bifurc. Chaos **12**(3), 659–661 (2002)
Lu, W., Li, C., Xu, C.: Sliding mode control of a shunt hybrid active power filter based on the inverse system method. Int. J. Electr. Power Energ. Syst. **57**, 39–48 (2014)
Mirus, K.A., Sprott, J.C.: Controlling chaos in a high dimensional system with periodic parametric perturbations. Phys. Lett. A **254**(5), 275–278 (1999)

Mondal, S., Mahanta, C.: Adaptive second order terminal sliding mode controller for robotic manipulators. J. Frankl. Inst. **351**(4), 2356–2377 (2014)

Murali, K., Lakshmanan, M.: Secure communication using a compound signal from generalized chaotic systems. Phys. Lett. A **241**(6), 303–310 (1998)

Nehmzow, U., Walker, K.: Quantitative description of robotenvironment interaction using chaos theory. Robot. Auton. Syst. **53**(3–4), 177–193 (2005)

Njah, A.N., Ojo, K.S., Adebayo, G.A., Obawole, A.O.: Generalized control and synchronization of chaos in RCL-shunted Josephson junction using backstepping design. Phys. C: Supercond. **470**(13–14), 558–564 (2010)

Ouyang, P.R., Acob, J., Pano, V.: PD with sliding mode control for trajectory tracking of robotic system. Robot. Comput. Integr. Manuf. **30**(2), 189–200 (2014)

Pecora, L.M., Carroll, T.L.: Synchronization in chaotic systems. Phys. Rev. Lett. **64**(8), 821–824 (1990)

Perruquetti, W., Barbot, J. P.: Sliding Mode Control in Engineering. Marcel Dekker, New York (2002)

Petrov, V., Gaspar, V., Masere, J., Showalter, K.: Controlling chaos in Belousov–Zhabotinsky reaction. Nature **361**, 240–243 (1993)

Qu, Z.: Chaos in the genesis and maintenance of cardiac arrhythmias. Prog. Biophys. Mol. Biol. **105**(3), 247–257 (2011)

Rafikov, M., Balthazar, J.M.: On control and synchronization in chaotic and hyperchaotic systems via linear feedback control. Commun. Nonlinear Sci. Numer. Simul. **13**(7), 1246–1255 (2007)

Rasappan, S., Vaidyanathan, S.: Global chaos synchronization of WINDMI and Coullet chaotic systems by backstepping control. Far East J. Math. Sci. **67**(2), 265–287 (2012)

Rhouma, R., Belghith, S.: Cryptoanalysis of a chaos based cryptosystem on DSP. Commu. Nonlinear Sci. Numer. Simul. **16**(2), 876–884 (2011)

Rössler, O.E.: An equation for continuous chaos. Phys. Lett. **57A**(5), 397–398 (1976)

Sarasu, P., Sundarapandian, V.: Adaptive controller design for the generalized projective synchronization of 4-scroll systems. Int. J. Syst. Signal Control Eng. Appl. **5**(2), 21–30 (2012a)

Sarasu, P., Sundarapandian, V.: Generalized projective synchronization of two-scroll systems via adaptive control. Int. J. Soft Comput. **7**(4), 146–156 (2012b)

Sarasu, P., Sundarapandian, V.: Generalized projective synchronization of two-scroll systems via adaptive control. Eur. J. Sci. Res. **72**(4), 504–522 (2012c)

Shahverdiev, E.M., Bayramov, P.A., Shore, K.A.: Cascaded and adaptive chaos synchronization in multiple time-delay laser systems. Chaos Solitons Fractals **42**(1), 180–186 (2009)

Shahverdiev, E.M., Shore, K.A.: Impact of modulated multiple optical feedback time delays on laser diode chaos synchronization. Optics Commun. **282**(17), 3568–3572 (2009)

Sharma, A., Patidar, V., Purohit, G., Sud, K.K.: Effects on the bifurcation and chaos in forced Duffing oscillator due to nonlinear damping. Commun. Nonlinear Sci. Numer. Simul. **17**(6), 2254–2269 (2012)

Sprott, J.C.: Some simple chaotic flows. Phys. Rev. E **50**(2), 647–650 (1994)

Sprott, J.C.: Competition with evolution in ecology and finance. Phys. Lett. A **325**(5–6), 329–333 (2004)

Suérez, I.: Mastering chaos in ecology. Ecol. Model. **117**(2–3), 305–314 (1999)

Sun, H., Cao, H.: Chaos control and synchronization of a modified chaotic system. Chaos Solitons Fractals **37**(5), 1442–1455 (2008)

Sundarapandian, V.: Output regulation of the Lorenz attractor. Far East J. Math. Sci. **42**(2), 289–299 (2010)

Sundarapandian, V., Pehlivan, I.: Analysis, control, synchronization, and circuit design of a novel chaotic system. Math. Comput. Model. **55**(7–8), 1904–1915 (2012)

Suresh, R., Sundarapandian, V.: Global chaos synchronizatoin of a family of n-scroll hyperchaotic Chua circuits using backstepping control with recursive feedback. Far East J. Math. Sci. **73**(1), 73–95 (2013)

Tigan, G., Opris, D.: Analysis of a 3D chaotic system. Chaos Solitons Fractals **36**(5), 1315–1319 (2008)

Tu, J., He, H., Xiong, P.: Adaptive backstepping synchronization between chaotic systems with unknown Lipschitz constant. Appl. Math. Comput. **236**, 10–18 (2014)

Ucar, A., Lonngren, K.E., Bai, E.W.: Chaos synchronization in RCL-shunted Josephson junction via active control. Chaos Solitons Fractals **31**(1), 105–111 (2007)

Usama, M., Khan, M.K., Alghatbar, K., Lee, C.: Chaos-based secure satellite imagery cryptosystem. Comput. Math. Appl. **60**(2), 326–337 (2010)

Utkin, V.: Sliding Modes and Their Applications in Variable Structure Systems. MIR Publishers, Moscow (1978)

Utkin, V. I.: Sliding Modes in Control and Optimization. Springer, New York (1992)

Vaidyanathan, S.: Anti-synchronization of Newton-Leipnik and Chen-Lee chaotic systems by active control. Int. J. Control Theory Appl. **4**(2), 131–141 (2011)

Vaidyanathan, S.: Adaptive backstepping controller and synchronizer design for Arneodo chaotic system with unknown parameters. Int. J. Comput. Sci. Inf. Technol. **4**(6), 145–159 (2012a)

Vaidyanathan, S.: Anti-synchronization of Sprott-L and Sprott-M chaotic systems via adaptive control. Int. J. Control Theory Appl. **5**(1), 41–59 (2012b)

Vaidyanathan, S.: Output regulation of the Liu chaotic system. Appl. Mech. Mater. **110–116**, 3982–3989 (2012c)

Vaidyanathan, S.: A new six-term 3-D chaotic system with an exponential nonlinearity. Far East J. Math. Sci. **79**(1), 135–143 (2013a)

Vaidyanathan, S.: Analysis and adaptive synchronization of two novel chaotic systems with hyperbolic sinusoidal and cosinusoidal nonlinearity and unknown parameters. J. Eng. Sci. Technol. Rev. **6**(4), 53–65 (2013b)

Vaidyanathan, S.: A new eight-term 3-D polynomial chaotic system with three quadratic nonlinearities. Far East J. Math. Sci. **84**(2), 219–226 (2014)

Vaidyanathan, S., Rajagopal, K.: Global chaos synchronization of four-scroll chaotic systems by active nonlinear control. Int. J. Control Theory Appl. **4**(1), 73–83 (2011)

Vaidyanathan, S., Sampath, S.: Anti-synchronization of four-scroll chaotic systems via sliding mode control. Int. J. Autom. Comput. **9**(3), 274–279 (2012)

Volos, C.K., Kyprianidis, I.M., Stouboulos, I.N.: Experimental investigation on coverage performance of a chaotic autonomous mobile robot. Robot. Auton. Syst. **61**(12), 1314–1322 (2013)

Wang, F., Liu, C.: A new criterion for chaos and hyperchaos synchronization using linear feedback control. Phys. Lett. A **360**(2), 274–278 (2006)

Weeks, E.R., Burgess, J.M.: Evolving artificial neural networks to control chaotic systems. Phys. Rev. E **56**(2), 1531–1540 (1997)

Witte, C.L., Witte, M.H.: Chaos and predicting varix hemorrhage. Med. Hypotheses **36**(4), 312–317 (1991)

Wu, X., Guan, Z.-H., Wu, Z.: Adaptive synchronization between two different hyperchaotic systems. Nonlinear Anal. Theory Methods Appl. **68**(5), 1346–1351 (2008)

Xiao, X., Zhou, L., Zhang, Z.: Synchronization of chaotic Lure systems with quantized sampled-data controller. Commun. Nonlinear Sci. Numer. Simul. **19**(6), 2039–2047 (2014)

Yuan, G., Zhang, X., Wang, Z.: Generation and synchronization of feedback-induced chaos in semiconductor ring lasers by injection-locking. Optik Int. J. Light Electron Opt. **125**(8), 1950–1953 (2014)

Zaher, A.A., Abu-Rezq, A.: On the design of chaos-based secure communication systems. Commun. Nonlinear Syst. Numer. Simul. **16**(9), 3721–3727 (2011)

Zhang, H., Zhou, J.: Synchronization of sampled-data coupled harmonic oscillators with control inputs missing. Syst. Control Lett. **61**(12), 1277–1285 (2012)

Zhang, J., Li, C., Zhang, H., Yu, J.: Chaos synchronization using single variable feedback based on backstepping method. Chaos Solitons Fractals **21**(5), 1183–1193 (2004)

Zhang, X., Liu, X., Zhu, Q.: Adaptive chatter free sliding mode control for a class of uncertain chaotic systems. Appl. Math. Comput. **232**, 431–435 (2014)

Zhou, W., Xu, Y., Lu, H., Pan, L.: On dynamics analysis of a new chaotic attractor. Phys. Lett. A **372**(36), 5773–5777 (2008)
Zhu, C., Liu, Y., Guo, Y.: Theoretic and numerical study of a new chaotic system. Intell. Inf. Manag. **2**, 104–109 (2010)
Zinober, A. S.: Variable Structure and Lyapunov Control. Springer, New York (1993)

Printed in the United States
By Bookmasters